Residual Stresses 2016
ICRS-10

Edited by

Thomas M. Holden
Ondrej Muránsky
Lyndon Edwards

This book presents the proceedings of the Tenth International Conference on Residual Stresses which was devoted to the prediction/modelling, evaluation, control, and application of residual stresses over a wide range of applications. New developments on stress-measurement techniques, on the modelling and prediction of residual stresses and on progress made in the fundamental understanding of the nature of residual stresses are highlighted. The proceedings offer an overview of the current understanding of how residual stresses affect the properties of materials, components and structures.

Residual Stresses 2016
ICRS-10

10th International Conference on Residual Stresses (ICRS10), Sydney, Australia, 3-7 July, 2016.

Thomas M. Holden, Ondrej Muránsky, Lyndon Edwards

Institute of Materials Engineering Australian Nuclear Science & Technology Organization, Locked Bag 2001, Kirrawee DC, Sydney NSW 2232, Australia

Peer review statement

All papers published in this volume of "Materials Research Proceedings" have been peer reviewed. The process of peer review was initiated and overseen by the above proceedings editors. All reviews were conducted by expert referees in accordance to Materials Research Forum LLC high standards.

Published under License by **Materials Research Forum LLC**
Millersville, PA 17551, USA

Published as part of the proceedings series
Materials Research Proceedings
Volume 2 (2017)

ISSN 2474-3941 (Print)
ISSN 2474-395X (Online)

ISBN 978-1-94529116-6 (Print)
ISBN 978-1-94529117-3 (eBook)

Distributed worldwide by

Materials Research Forum LLC
105 Springdale Lane
Millersville, PA 17551
USA
http://www.mrforum.com

Manufactured in the United State of America
10 9 8 7 6 5 4 3 2 1

Table of Contents

Preface

ICRS-10, the 10th International Conference on Residual Stresses, was held at the Novotel Sydney Brighton Beach Hotel at Brighton Le Sands, Sydney between the 3rd and the 7th of July 2016.

ICRS-10 was the latest of a highly successful series which started in Garmisch-Partenkirchen (Germany) thirty years ago and continued in Nancy (France, 1988), Tokushima (Japan, 1991), Baltimore (USA, 1994), Linkoping (Sweden, 1997), Oxford (UK, 2000), Xi'an (China, 2004) and Denver (USA, 2008) and, again in Garmisch-Partenkirchen (Germany, 2012).

The conference continues to be the key forum for scientists and engineers interested in the prediction, evaluation, control, and application of residual stresses. ICRS-10 featured 149 oral and 30 poster presentations given to 206 predominantly international attendees from 31 countries.

Publication in these proceedings was voluntary and after peer review a total of 102 papers are included in these proceedings. This could not have been achieved without the very significant efforts of both reviewers and authors. Without this work these proceedings could not exist and I would like to thanks them on behalf of the community.

I hope that all those who attended would share my opinion that ICRS-10 was a vibrant, exciting and productive conference and that these proceedings give some sense of the success of ICRS-10 to the reader.

Finally, I would like to thank the members of both the ICRS Scientific Board the ICRS-10 Organizing Committee without whose tireless work ICRS-10 would not have succeeded

Lyndon Edwards
Chair, ICRS-10

December, 2016
Sydney, Australia.

Committees

Local Scientific Committee:
L. Edwards
O. Muránsky
C.J. Hamelin
A. Paradowska

International Scientific Committee:
Y. Akiniwa (Japan)
K. Akita (Japan)
S. Denis (France)
L. Edwards (Australia)
C. Hubbard (USA)
Z. Kowalewski (Poland)
J.-L. Lebrun (France)
E. Mittemeijer (Germany)
C. Noyan (USA)
R. Lin Peng (Sweden)
W. Reimers (Germany)
P. Scardi (Italy)
B. Scholtes (Germany)
O. Thomas (France)
P. Withers (United Kingdom)
K. Xu (China)

Sponsors

ICRS-10 Partner
PROTO Manufacturing Ltd.
2175 Solar Crescent Oldcastle,
Ontario, N0R 1L0, Canada

ICRS-10 Sponsors
Rigaku Corporation
3-9-12 Matsubara-Unit 8, Akishima-Shi
Tokyo 196-8666, Japan

VEQTER Limited
Unicorn Business Park, Whitby Road
Bristol, BS4 4EX, UK

Stresstech Oy
Tikkutehtaantie
1 40800 Vaajakoski, Finland

ANSTO
New Illawarra Road
Sydney NSW 2234 Australia

Residual Stresses 2016: ICRS-10 Materials Research Forum LLC
Materials Research Proceedings **2** (2016) 1-6 doi: http://dx.doi.org/10.21741/9781945291173-1

Numerical Investigation of Residual Stresses in Chain-die Formed AHSS U-Channels

Yong Sun[1, a], Zhen Qian[1,b], Vladimir Luzin[1,c], William J.T Daniel[1,d], Mingxing Zhang[1,e] and Shichao Ding[1,f *]

[1]School of mechanical and Mining Engineering, The University of Queensland, QLD, Australia

[2]Australian Nuclear Science and Technology Organization, Lucas Heights, NSW, Australia

[a]y.sun@uq.edu.au, [b]z.qian@uq.edu.au, [c]Vladimir.Luzin@ansto.gov.au, [d]billd@uq.edu.au, [e]mingxing.zhang@uq.edu.au, [f]s.ding@uq.edu.au

Keywords: Residual Stresses, AHSS, U-Channel, Chain-Die Forming, Finite Element Analysis

Abstract. Advanced high-strength steel is increasingly being used in automotive structural components due to its excellent strength-to-weight ratios. However, the variations of residual stresses magnitude in AHSS products are usually very complex and unpredictable due to the fabrication process and the material's high strength. Consequently, unbalanced residual stresses are responsible for a series of product defects. Chain-die forming is a novel AHSS fabrication method which has the characteristics of preserving the material's ductility maximally and also reducing the residual stresses in the product. In this study, the finite element method is employed to investigate the equivalent residual stresses (Von Mises) in Chain-die formed AHSS U-channels. Finite element simulation of roll forming of the same type as AHSS U-channel forming is performed to make a comparison of the residual stresses distribution characterizations in AHSS U-channels which are fabricated by roll forming and by Chain-die forming. The results indicate that the residual stresses in Chain-die formed U-channels stay at a very low level and are almost negligible. In the meanwhile, due to the bending, reverse bending and other unpredictable redundant deformation types in the roll forming process, the residual stresses are more significant than those of Chain-die formed AHSS channels. The comparison of the longitudinal strain developments of flange edges of roll formed and Chain-die formed U-channels is given to explain the differences between the residual stress distributions in the roll formed and Chain-die formed U-channels. This paper gives a comprehensive understanding of the characteristics of the residual stress distribution in Chain-die formed AHSS U-channels. It provides a clear evidence to illustrate the superiority of Chain-die forming in reducing the residual stresses in AHSS products.

Introduction

Advanced High-Strength Steel (AHSS), especially dual phase steel, is more and more frequently being employed by automakers for structural parts of a motor vehicle due to its advantages for weight reduction and safety improvement. AHSS has the required high strength but the elongation is not large enough. The characteristics bring more challenges in fabricating AHSS products.

Roll forming is a highly efficient fabricating method for mass production of long and straight metal products. It can be understood as a continuous bending operation which the metal strips are progressively formed through consecutive sets of rolls into various profiles [1]. Although roll forming has been developed over a century, there are still unsolved issues in predictability and control of the redundant plastic deformation occurring during a roll forming process due to relaxation between roll stands [2]. Consequently, this becomes a bottleneck which greatly restricts the development and application of roll forming. Specifically, the non-uniform permanent plastic deformation results in the residual stresses in a roll formed product. The distributions of residual

Materials Research Proceedings **2** (2016) 1-6 doi: http://dx.doi.org/10.21741/9781945291173-1

stresses are responsible for a series of typical roll formed product defects and/or distortions from the desired product shapes, such as curvature, end flare, edge wave, torsion, springback and so on [3]. It should be also noted that the existence of longitudinal residual stresses would adversely affect structural stability, fatigue performance, stiffness and other properties [4].

Chain-die forming, as a novel sheet metal forming method, was therefore proposed by Ding et al. [5] to break the bottleneck of fabricating AHSS products. It is considered to have a high potential to be a more economical and energy-saving method compared to conventional roll forming, due to its technical characteristics of both bending and stamping. Ding et al. [5] performed a series of experiments of Chain-die forming AHSS U-channels with pre-made holes. They proved that the technology has the advantage of conserving the material's ductility. This shows that Chain-die forming can be used to fabricate AHSS products, even though some of them have poor ductility, as there is almost no redundant plastic deformation and low residual stresses in the non-deformed areas.

As Ding et al. [5] introduced, the principle of Chain-die forming is to prolong the effective forming distance by extending the virtual roll radii. Specifically, the technology introduces a series of discontinuous forming tools moving on a track board to implement the virtual large rolls, as seen in Fig. 1 (a). The increase of deformation length shown as in Fig. 1 (b) results in the reduction of the peak longitudinal strain and the residual longitudinal strain. That is, the large roll radii can significantly reduce or even eliminate the redundant plastic deformation occurring in roll forming. Consequently, the corresponding residual stresses are significant reduced. The typical defects of roll formed products can be therefore theoretically removed, thus improving the quality of products.

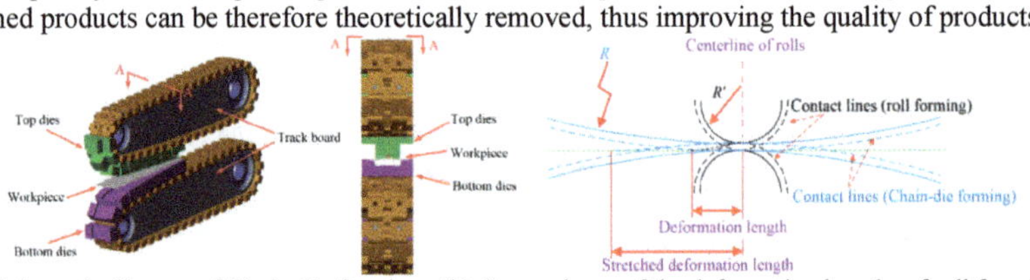

(a) Schematic diagram of Chain-die forming (b) Comparisons of the deformation lengths of roll forming and Chain-die forming

Fig 1.The principle of Chain-die forming. (Note: R are the radii of the top and bottom rolls of Chain-die forming and R′ is the radius of a roll of conventional roll forming)

The structure of the paper is as follows. A brief background of residual stresses in roll forming and the principles of Chain-die forming are introduced in this section firstly, followed by the clarifications of finite element (FE) modeling of roll forming and Chain-die forming AHSS U-channels. Then the equivalent residual stresses (Von-Mises) in the non-deformation areas of roll formed and Chain-die formed AHSS U-channels are investigated. The comparison of the longitudinal strain developments of the flange edges of roll formed and Chain-die formed AHSS U-channels are then studied and discussed. Finally, some important concluding remarks are summarized.

FE modeling

The FE modelling of roll forming and Chain-die forming the U-channels with same product parameters are introduced in detail in this section. Then the material properties of the working piece adopted in this study are illustrated followed by the clarifications of the contact properties employed between the forming tools and the workpiece.

Simulation set-up.

The simulation verifications were carried out in ABAQUS/Standard. Specific python scripts were imported to ABAQUS/CAE to complete the FE modelling processes. The implicit solution algorithm is employed to improve the accuracy of results. The forming tools are defined as the rigid shell bodies and the workpiece is defined as a deformable body. Taking advantages of symmetry to improve the computing efficiency, only one half of the workpiece and rolls have been modelled.

In FE modelling of the roll forming process, all the base radii of the forming tools are 0.30m. The center points of the rolls were assigned as the reference rotation points of the corresponding forming tools respectively. The angular velocities,ω, of the rotations of the forming tools were set as the 0.5rad/s, and in meanwhile the horizontal forward velocity of the working piece ,v, was set as 15mm/s ($v = \omega r$). As the main plastic deformation occurs in the bend corners, the initial flat blank had been meshed using different mesh sizes. A very fine mesh has been adopted in the bend corners and flange edge portions and a coarse mesh had been applied in the web of the blank. The workpiece was meshed with 97500 eight node linear brick elements with reduced integration and hourglass control (C3D8R). There were 5 mesh layers (elements) along the thickness direction to guarantee a good convergence. To avoid penetration into the sheet, the mesh type of the forming tools was the four node 3D bilinear rigid quadrilateral element (R3D4).

In FE modelling of Chain-die forming process, the forming tools were part of the rolls with 35 meter radii. A workpiece identical to the one applied in FE modelling of roll forming process was adopted in FE modelling of Chain-die forming process. The same meshing strategies and similar boundaries conditions were applied in the FE model of Chain-die forming process. The details of the FE models of roll forming and Chain-die forming processes and the mesh of the workpiece are shown in Fig.2 (a) and (b) respectively.

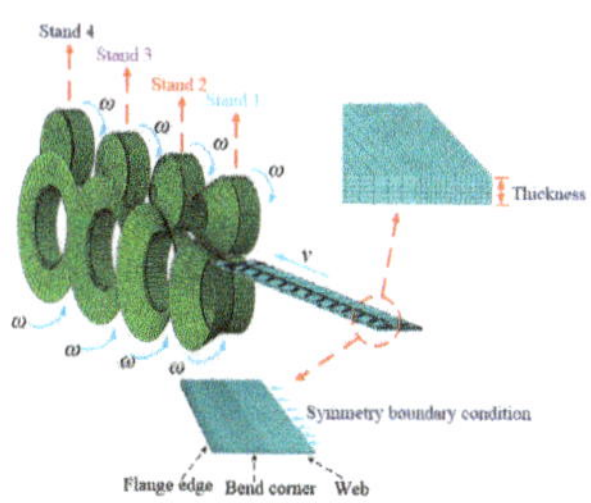

(a) FE model of Roll forming

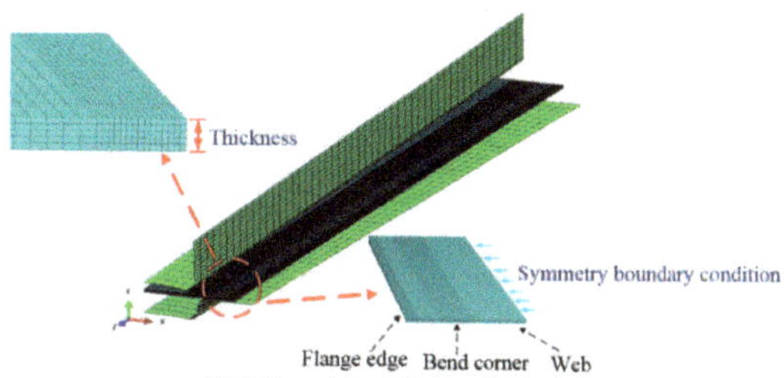

(b) FE model of Chain-die forming

Fig 2.FE models of roll forming and Chain-die forming

Material properties of the workpiece.

Material used in the FE models is an elastic-plastic strain hardening material. The true plastic stress-strain curve of the sheet metal can be calculated and interpolated using Swift's isotropic strain hardening law, as shown in Eq.1. As isotropic hardening is applied in all simulations, Von Mises yield function is sufficient to describe the yield criteria of the material.

$$\sigma_e = K\,(\varepsilon_e + \varepsilon_0)^n \qquad \text{(Eq.1)}$$

where σ_e is the equivalent stress, ε_e is the equivalent strain, ε_0 is the initial strain, K is the strength coefficient and n is the strain hardening exponent. Those values are given in Table 3.

Contact and friction properties between forming tools and workpiece.

The types of interaction are set as 'surface-to-surface' (implicit) contacts between the deformable surfaces (sheet metal) and the rigid surfaces (forming tools). The penalty method and the Coulomb friction model are applied to the contact constraints. The friction coefficient of the FE models is assumed as 0.1. The details of the parameters of the FE models have been summarized in Table 1.

Materials Research Proceedings **2** (2016) 1-6 doi: http://dx.doi.org/10.21741/9781945291173-1

Table 1. Summary FEM simulation parameters

Simulation Parameters	Chain-die forming	Roll forming
Product width - f (mm)	80	80
Roll radius (mm)	35000	30
Blank thickness - t (mm)	1.2	1.2
Blank length - L (mm)	300	300
Tensile strength of blank - TS (MPa)	1180MPa	1180MPa
Algorithm method	Implicit	Implicit
Mesh number (blank)	13050	13050
Span space (mm)	---	200
Yield strength – σ_s (MPa)	845	845
Strength coefficient – K (MPa)	1864	1864
Young's modulus – E (GPa)	206	206
Poisson's ratio – υ	0.3	0.3
Initial strain – ε_0	0.000367	0.000367
Strain hardening exponent – n	0.11	0.11
Elastic Strain	0.45%	0.45%

In the FE model of roll forming and Chain-die forming processes, the whole simulations include the loading and unloading processes. Specifically, the workpiece was firstly pulled into the former followed by being gradually formed to a U channel by the friction forces with the rotations of the forming tools, the same as the real forming process. The accumulated stress-strain of the workpiece starts to be released by springback while the forming tools move away from the workpiece. The workpiece with the last stress-strain state in the previous model is finally imported into a new simulation model with an implicit solution algorithm. The stress-strain of the workpiece is further released by springback until it reaches a steady state in the new FE model.

Results and discussions

The equivalent residual stresses (Von-Mises) in AHSS U channels induced by roll forming and Chain-die forming are presented and discussed in details in this section. The comparison of the longitudinal strain developments of flange edges of roll formed and Chain-die formed U channels is also clarified to explain the phenomenon.

Equivalent residual stresses comparison.

The residual stresses in AHSS U-channels induced by roll forming and Chain-die forming are presented in Fig.3 (a) and Fig.3 (b) respectively. As shown in Fig.3 (b), the residual stresses in non-deformation areas of Chain-die formed U-channels are negligible while there are significant and uneven residual stresses distribution in non-deformation areas of roll formed AHSS channels.

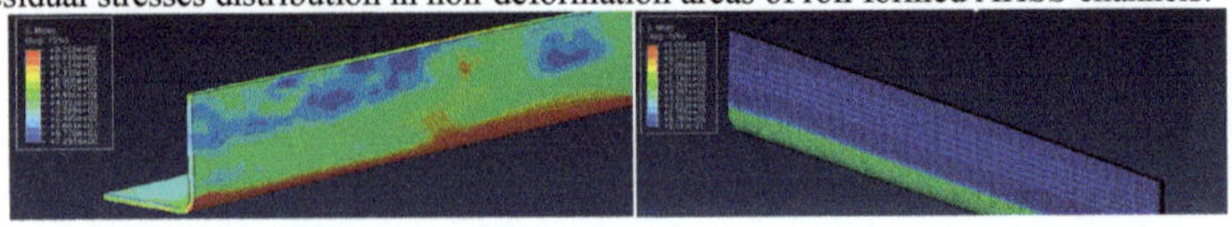

(a) Roll formed U-channels (b) Chain-die formed U-channels

Fig 3. Equivalent residual stresses comparison

As the Chain-die forming process can be considered as a gradually stamping process, there are nearly no redundant deformations in Chain-die forming process. Consequently, the low residual stresses in Chain-die formed U-channels are accumulated and the product defects induced in roll forming process can therefore be avoided, as seen in Fig.4.

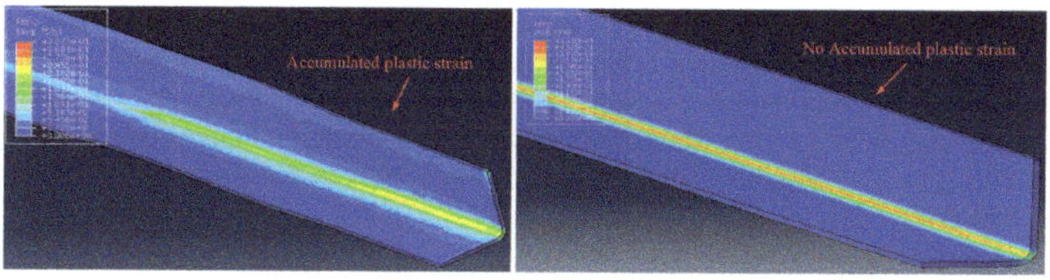

(a) Roll formed U channels (b) Chain-die formed U channels

Fig 4.Equivalent residual plastic strain comparison (PEEQ)

Longitudinal residual strains comparison.

As the longitudinal strain development indicates the severity of the relevant forming process, it can be used to explain the characteristic differences of the residual stress distributions in two types of AHSS U-channels. The comparison of the longitudinal strain developments of flange edges of Chain-die formed and roll formed the AHSS U-channels is shown in Fig.5. It is found that the longitudinal strain developed during Chain-die forming is much smoother, the peak strain and residual strain are smaller than in roll forming, even just in one forming pass rather than 4 passes of roll forming. The unavoidable redundant plastic deformation in a roll forming process can also be avoided in Chain-die forming.

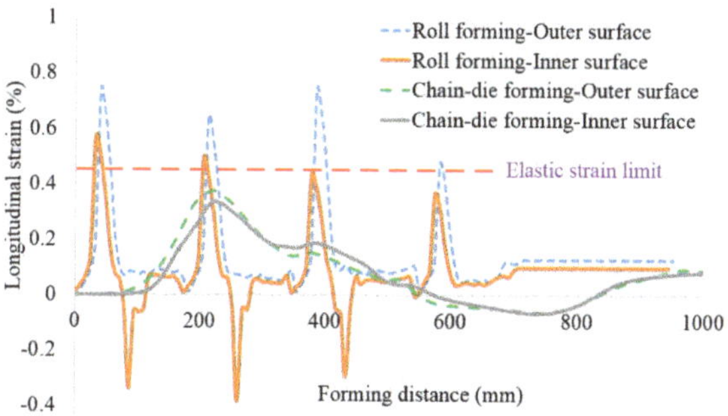

Fig 5.The longitudinal strain development between roll formed and Chain-die formed U-channels

As shown as in Fig.5, the peak longitudinal strain in the roll forming process is nearly 2 times larger than that in the Chain-die forming process. It should be noted that the bending on the flange phenomenon in Chain-die forming is much lighter and gentler than in roll forming, the corresponding bending strain is much smaller than in roll forming. That is, the surface longitudinal strain superposing the membrane strain and bending strain on the flange during Chain-die forming is well below the material's elastic limit but not in roll forming. It is therefore hard to predict and control the quality of a roll formed product due to its complicated longitudinal strain development. On the contrary, it is possible to make a product with zero residual stress in the non-deformed area by Chain-die forming and give a better quality product than in roll forming.

Concluding Remarks

The residual stresses in Chain-die formed AHHS U-channels are investigated numerically in this paper. Finite element analysis is employed to simulate the roll forming and Chain-die forming processes. The equivalent residual stresses of the roll formed and Chain-die formed products are discussed. Experimental investigation of the residual stresses in Chain-die formed U-channels is very necessary to verify the simulations. Some conclusions are reached by this study:

- Using Chain-die forming can achieve much more bending without product defects than roll forming in a single pass. Hence Chain-die forming can maximally reserve the material's ductility;
- Unlike the high residual stresses distribution in the non-deformation areas in roll formed U-channels, the low redundant strains accumulated in Chain-die formed forming makes the residual stresses in U-channels well below the corresponding material's yield point;
- It is hard to predict the residual stresses in roll formed product due to its complicated fabrication processes, including bending, reverse bending and other unnecessary deformation types. On the contrary, it is possible to make a product with zero residual stress in the non-deformed area by Chain-die forming and that gives a better quality product than in roll forming.

The determination of residual stresses in Chain-die formed U-channels by the neutron diffraction method is currently in progress and the experiments are supported by Bragg Institute beam time proposal 4865. The experimental verification of the numerical simulations will be presented in a future paper.

Acknowledgement

The authors would like to thank for the financial support from the Baosteel-Australia Joint Research and Development Centre (BAJC), Ningbo SaiRolf Metal Forming Co., Ltd., Bragg Institute beam time proposal 4865 and China Scholarship Council. Colleagues who contributed their efforts and support to this paper are also appreciated. Finally, the authors would like to give special thanks to Baosteel Co.,Ltd. for providing the material properties of the experimental samples.

References

[1] Paralikas, J., Salonitis, K., Chryssolouris, G., 2009. Investigation of the effects of main roll-forming process parameters on quality for a V-section profile from AHSS. The International Journal of Advanced Manufacturing Technology 2009, 44(3-4), 223-237. *http://dx.doi.org/10.1007/s00170-008-1822-9*

[2] Panton, S. M.; Zhu, S. D.; Duncan, J. L. Fundamental deformation types and sectional properties in roll forming. International journal of mechanical sciences 1994, 36(8), 725-735. *http://dx.doi.org/10.1016/0020-7403(94)90088-4*

[3] Semiatin, S.L. ASM Handbook; American Society for Metals: Materials Park, United States, 1988; 2110 pp.

[4] Hou, G., Jing, X., Yan, D., & Shen, K. Prediction of longitudinal residual stress distribution in cold-formed square hollow sections. In Electric Technology and Civil Engineering (ICETCE), 2011 International Conference on (pp. 557-562). IEEE.

[5] Zhang, Y. Ding, S., 2012. A comparison study of Chain-die Forming and roll forming by forming a top hat section. Proceedings of the 14th International Conference on Metal Forming, Wiley, London, England, 703-706.

Residual Stresses 2016: ICRS-10 Materials Research Forum LLC
Materials Research Proceedings **2** (2016) 7-12 doi: http://dx.doi.org/10.21741/9781945291173-2

Quasi-static Process Modelling of Deep Cold Rolling on Ti-6Al-4V

A. Lim[1,2,a*], S. Castagne[1], C.C. Wong[2]

[1]School of Mechanical and Aerospace Engineering, Nanyang Technological University, Singapore

[2]Advanced Technology Centre, Rolls-Royce Singapore Pte Ltd, Singapore

[a]lims0200@e.ntu.edu.sg

Keywords: Deep Cold Rolling, Residual Stress Modelling Methods, Residual Stresses in Manufacturing Process, Simulation, Abaqus

Abstract. This paper proposes a 3D Finite Element (FE) model for the deep cold rolling process, modelled using the commercially available Abaqus/Standard FE software, and used the afore mentioned model to study the effects of pressure and tool diameter on the residual stress profile predicted. The model was validated with experimental data. The predicted results showed about 20% deviation for the measured data in the near surface region, but showed good correlation at the compressive-tensile cross-over point. Reasons for this deviation are explored and analysed with several recommendations made to improve future FE models.

Introduction

Mechanical surface treatment (MST) methods physically alter the surface and near sub-surface regions of a component, creating a region of compressive residual stress. This compressive residual stress improves fatigue life and foreign object damage tolerance. Deep cold rolling (DCR) is a mechanical surface treatment and its main objective is to induce deep compressive residual stresses in the surface and sub-surface layers (up to 1 mm) of the component [1].

While the high compressive stresses generated by DCR in the near surface layers can extend fatigue life, it is crucial that the location and magnitude of the corresponding balancing tensile residual stresses are carefully managed and understood, as they can negate the benefits of the near surface compressive residual stress and adversely affect the fatigue life of the component [2]. However, experimental determination of residual stress distributions by the hole drilling method and the X-ray diffraction method is time consuming, expensive and limited to discrete measurement points. Hence, FE simulation is a crucial tool in the process optimization of the DCR method as it allows the user to analyse and predict the residual stress profile of the component.

DCR creates different directional stress profiles and is reliant on both plastic deformation at the surface and Hertzian stress at the sub-surface region to generate the resultant residual stress profiles. Previous quasi-static finite element studies of the DCR process focused on the mechanics of a single pass [3]. While useful for understanding the mechanics of the DCR process, modelling of a single pass is insufficient to study the broader residual stress distributions.

The objective of this study is to propose and validate a FE model to simulate the DCR process. This model will enable a deeper understanding of the broader residual stress distributions caused by the DCR process.

Methodology

Material Model. The material model attributed to the model is isotropic elastic-plastic with strain hardening. A Young's modulus of 113.8 GPa and a Poisson ratio of 0.342 are prescribed for the elastic portion of the material model. The strain hardening portion of the material model was experimentally determined via uniaxial compression of identical Ti-6Al-4V material in cylinder geometry measuring 6.75 mm height by 4.5 mm diameter. The machine stiffness is taken into account using plate to plate compression. Figure 1 shows the determined yield stress against plastic strain.

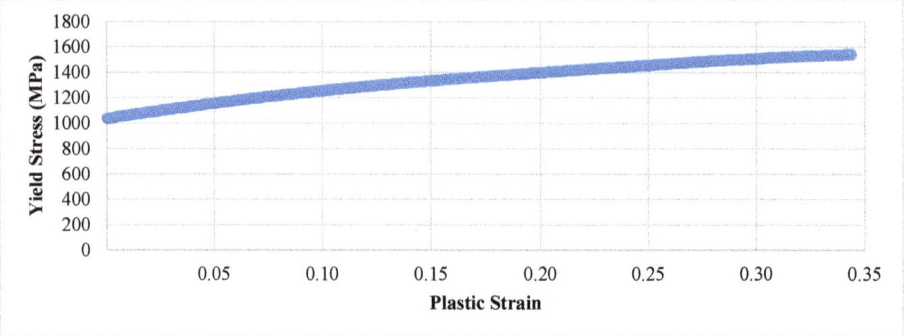

Figure 1: Yield stress against plastic strain for Ti-6Al-4V

Finite Element Model. The FE DCR model proposed is a 3-dimensional rectangular test coupon measuring 35 mm x 30 mm x 10 mm. commercially available FE code Abaqus/Standard V6.14 was used to model the test coupon. The 3D deformable element type used was the C3D8R element, an 8-node linear brick, using 56916 elements to model the test coupon. Uniform elements were used in the DCR zone to increase the resolution of the treated area, as shown in Figure 2a. Sensitivity analysis on the thickness of the test coupon was carried out to ensure that the thickness did not affect the residual stress profile generated.

To reduce computational time, a treated area of 15 mm lengthwise (LW – parallel to direction of rolling) by 10 mm crosswise (CW – perpendicular to direction of rolling) was prescribed. The deep cold rolled zone allowed enough area for a strain gage rosette to be applied onto the experimental test block for experimental validation.

The DCR tool is made of tungsten carbide and has a Vickers hardness of approximately 2000 HV. Since this is several orders of magnitude harder than the test coupon, the tool was modelled as an analytical rigid surface. The centre node of the modelled tool was used as the main reference and control point. The equivalent concentrated force can be calculated using the following formula:

$$F = \left[\frac{\pi}{4} * d_{piston}^2 * P\cos(\theta)\right] \quad (1)$$

where, d_{piston} is the piston diameter, P is the applied pressure and θ is the angle of the tool to the normal.

The movement of the ball is controlled via translating the centre node of the ball, in the length and breadth direction. The DCR tool has an inner assembly containing the burnishing ball which is free floating (as shown in Figure 2b). The force prescribed determines the vertical position of the ball, relative to the component surface at equilibrium. Therefore, the displacement of the ball in the Z direction (normal to surface of the component) was left free. This is to mimic the operating mechanics of the DCR tool where the inner tool assembly is free to move in an up and down motion within the outer tool housing up to the limits of the stroke length. This allows the DCR tool to adapt to minor variations in geometry.

Materials Research Proceedings **2** (2016) 7-12 doi: http://dx.doi.org/10.21741/9781945291173-2

The rotational degrees of freedom of the ball were also left free, ensuring that the ball is able to roll across the surface. The penalty formulation with a friction coefficient of 0.2 was used to model the surface to surface contact between the DCR ball and the work piece [4]. This also ensures that the ball rolls rather than slides across the component surface.

The boundary conditions were imposed such that the bottom surface of the test coupon is pinned. Since the test coupon is of sufficient thickness, it was assumed that the spring back of the coupon after treatment is negligible. The dimensions of the test coupon, the mesh, elements used and the boundary conditions are illustrated in Figure 2.

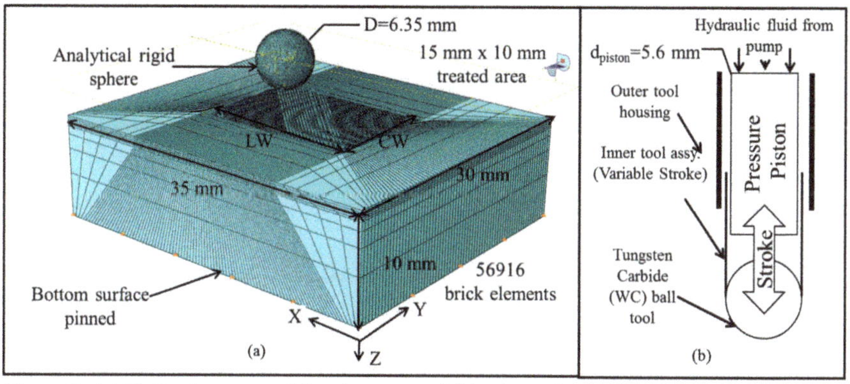

Figure 2: (a) Finite element model and set-up and (b) schematic of DCR tool

Experimental Setup and DCR Tool. Experiments were carried out to validate the proposed FE model. A flat rectangular test coupon of Ti-6Al-4V, measuring 35 mm x 30 mm x 20 mm was mounted on a work piece holding fixture and deep cold rolled. The DCR tool (6.35 mm ball diameter), supplied by Ecoroll AG®, at 20.0 MPa and 38.0 MPa of pressure was used to perform the treatment. The test coupon was orientated such that the rolling direction of the DCR process is identical to the rolling direction that was used in the manufacturing of the test coupons. Figure 3 shows the DCR process experimental set up.

The residual stress profile of the test coupon was determined at the centre of the treated area using the central hole drilling method (CHD). The strain relaxation via the incremental drill method was measured and the residual stresses were back calculated as described in ASTM 837. A final hole depth of 1.4mm was drilled in order to determine the stresses up to a depth of 1024μm.

Results and Discussion

FE Model Validation. The finite element model was validated using experimental CHD data. In this study, the experimental and finite element data was normalized against the absolute maximum residual stress determined experimentally in the CW direction. The normalized graph is presented in Figure 4.

Materials Research Proceedings **2** (2016) 7-12 doi: http://dx.doi.org/10.21741/9781945291173-2

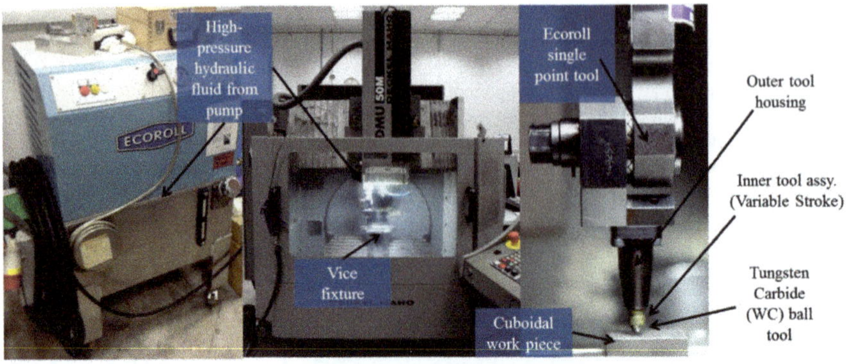

Figure 3: Experimental DCR set-up on a 3-axis CNC mill

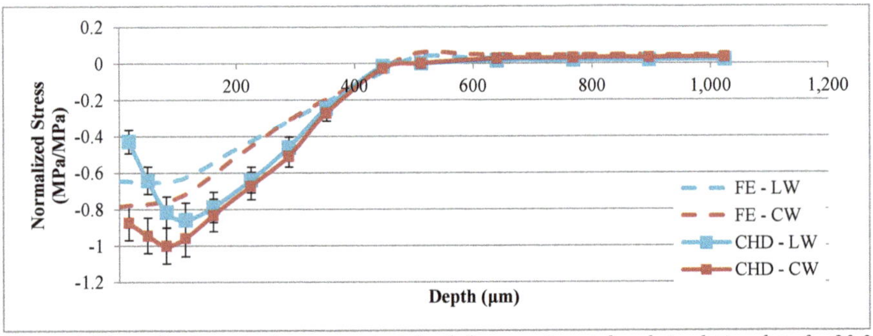

Figure 4: Normalized residual stress against depth for experimental and simulation data for 20.0 MPa

From Figure 4, it can be seen that there is some divergence of the residual stress in the near surface region. The FE model over-predicted the surface stress in the LW direction (parallel to rolling) at -0.43 MPa/MPa measured against -0.63 MPa/MPa predicted, and under predicted the surface stress in the CW direction (perpendicular to rolling) at -0.87 MPa/MPa measured against -0.78 MPa/MPa predicted. Figure 5 shows the residual stress profiles when the pressure is increased to 38.0 MPa. There is a similar deviation between the measured and predicted residual stress profiles for both cases.

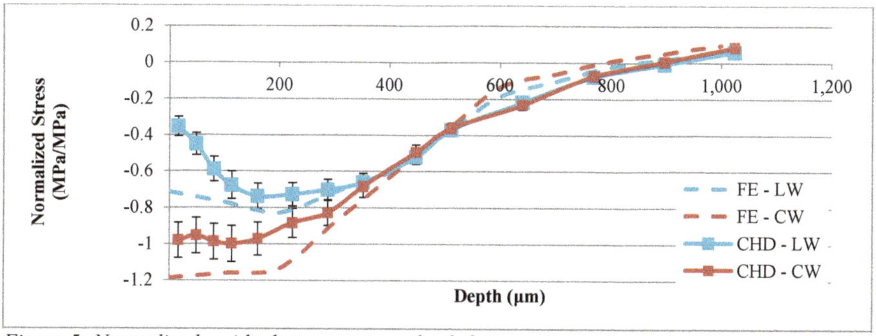

Figure 5: Normalized residual stress against depth for experimental and simulation data for 38.0 MPa

Deeper into the material, both FE models are able to predict the general trend of the residual stress profile and the Compressive-Tensile Cross-Over point (CTCO). The CTCO point is of particular importance as the location of the balancing tensile stress needs to be able to be understood and predicted.

Regarding the deviation between the FE and CHD data, the material model used assumed that the material was isotropic. Further compression tests taking into account the rolling direction of the manufacturing process should be carried out to determine if this simplifying assumption holds valid. A more detailed model, such as the Johnson-Cook model, could be considered for future FE simulations.

In this study, the implicit FE code is used and it does not take time into consideration when performing iterations. The time step used in the code relates to the relative position of the components and forces applied rather than actual time as a parameter. The implicit code docs not take strain rate into consideration

Another potential cause of the deviation could be the measurement method used to determine the experimental residual stress profiles. While CHD is an affordable and quick way of determining residual stress depth profiles, it has limited resolution in the near surface region (up to about 200 μm). This can be seen in the error bars shown in Figure 4 for the experimental graphs. At the near surface region, CHD has an error of about 10 – 15% σ_{max}. X-ray diffraction (XRD) or Focused Ion Beam (FIB) methods could potentially be used to attain more accurate data in the near surface region.

A final source of error that may have resulted in the deviation is the concentrated force applied based on equation (1). The equation calculates the theoretical force exerted by the DCR ball element on the component based on the pressure applied and the projected area of the pressure piston. A dynamometer could be placed under the component fixture to determine the actual force applied on the test coupon. This measured force can then be used to further improve the FE model.

The motivation behind process modelling the DCR process can be seen in the stress plots of the component shown in Figure 6. There is a zone of tensile residual stress that was formed near the boundary region at 1.0mm in depth for the 38.0 MPa case that was not present for the 20.0 MPa case. Knowledge of this zone of tensile stress will assist the process engineer in designing of the DCR zone and ensure that the component does not prematurely fail due to the tensile stress.

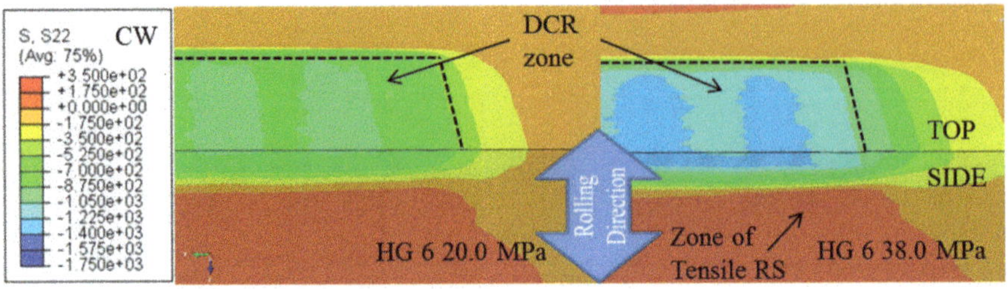

Figure 6: FE stress plot for HG 6, 20.0 MPa and 38.0 MPa, cut at x = 17.5 cm (centre of treated zone)

Conclusion

This paper proposed a 3D FE model to simulate the DCR process. There is a maximum deviation of 0.2 MPa/MPa at the surface region between the predicted FE results and the CHD measured data. There is good agreement between the FE and CHD data beyond 400 µm. Some causes of the deviation between the FE predicted values and the experimental data have been proposed and will need to be verified. Different material models will be studied in order to better understand the effects of the material models on the accuracy of the FE simulation.

Acknowledgement

This study was conducted using funding from the Singapore Economic Development Board (EDB) and Rolls-Royce Singapore Pte. Ltd. Joint Industrial Post-graduate Programme (IPP) grant. Finite element modelling for this study was conducted in Computer Aided Engineering Laboratory 1 of School of Mechanical and Aerospace Engineering, Nanyang Technological University.

References

[1] Altenberger, I. Deep rolling—the past, the present and the future. in Proceedings of 9th International Conference on Shot Peening, Sept. 2005.

[2] Webster, G. and A. Ezeilo, Residual stress distributions and their influence on fatigue lifetimes. International Journal of Fatigue, 2001. **23**: p. 375-383. *http://dx.doi.org/10.1016/S0142-1123(01)00133-5*

[3] Balland, P., et al., Mechanics of the burnishing process. Precision Engineering, 2013. **37**(1): p. 129-134. *http://dx.doi.org/10.1016/j.precisioneng.2012.07.008*

[4] Lim, A., S. Castagne, and C.C. Wong, Effect of Friction Coefficient on Finite Element Modeling of the Deep Cold Rolling Process, in Shot Peening: 12th International Conference, ICSP 12, Goslar, Germany, September 2014, Proceedings, L. Wagner, Editor. 2014: Clausthal-Zellerfeld, Germany. p. 376-380.

Residual Stresses 2016: ICRS-10 Materials Research Forum LLC
Materials Research Proceedings **2** (2016) 13-18 doi: http://dx.doi.org/10.21741/9781945291173-3

Effect of Cooling and Shot Peening on Residual Stresses and Fatigue Performance of Milled Inconel 718

Zhe Chen[1, a *], Ru Lin Peng[1, b], Johan Moverare[1, c], Olle Widman[2, d], David Gustafsson[3, e], Sten Johansson[1, f]

[1]Division of Engineering Materials, Linköping University, 58183 Linköping, Sweden

[2]Ytstruktur Arboga AB, 73222 Arboga, Sweden

[3]Siemens Industrial Turbomachinery AB, 61283 Finspång, Sweden

[a]zhe.chen@liu.se, [b]ru.peng@liu.se, [c]johan.moverare@liu.se, [d]olle.widman@ytstrukturarboga.se, [e]david.gustafsson@siemens.com, [f]sten.johansson@liu.se

Keywords: Residual Stresses, Surface Integrity, Milling, Shot Peening, Superalloys, Fatigue

Abstract. The present study highlights the effect of cooling and post-machining surface treatment of shot peening on the residual stresses and corresponding fatigue life of milled superalloy Inconel 718. It was found that tensile residual stresses were created on the milled surface, regardless of the use of coolant, however, the wet milling operation led to a lower surface tension and a reduced thickness of the tensile layer. The shot peening performed on the dry-milled specimens completely annihilated the surface tensile residual tresses and introduced a high level of surface compression. A comparable fatigue life for the wet-milled specimens was obtained as compared with the specimens prepared by dry milling. This is very likely attributed to that the milling-induced surface damage with respect to cracked non-metallic inclusions is the predominant cause of the fatigue failure. The presence of the compressive layer induced by shot peening resulted in a significant increase of the fatigue life and strength, while the extent to which the lifetime was prolonged was decreased as the applied load was increased.

Introduction

Fatigue is one of the main causes of failure to various structures in turbine engines. The fatigue life of a component strongly depends on its surface condition produced by machining since in most cases fatigue crack initiation starts on free surfaces. Residual stress is one of the most relevant practical parameters to assess the surface quality of a machined surface; it is superimposed on the applied cyclic loads, altering the driving force for crack initiation and propagation during fatigue. Generally, tensile residual stresses are perceived to be detrimental to the fatigue performance, whereas compressive residual stresses have a beneficial effect. The formation of residual stresses in machining processes is essentially dominated by the plastic deformation in subsurface of the workpiece material together with the thermal impact at surface [1]. The thermally-induced residual stresses are usually in tension, thus sufficient cooling could effectively reduce the surface tension on a machined surface by lowering the cutting temperature, or even introduce compressive residual stresses [2].

On the other hand, mechanical surface treatments, such as shot peening, are nowadays widely used on machined components by which compressive residual stresses are induced as it produces a work-hardened layer and misfit strains between the bulk and surface material. An enhanced fatigue life and strength by shot peening have been found for a variety of engineering materials [3-5].

Inconel 718 is a polycrystalline nickel-based superalloy and has wide applications in aerospace and power generation industries because of its superior mechanical properties and good resistance to oxidation/corrosion environments. A great number of studies have been conducted to improve the surface integrity of machined Inconel 718 by approaches with process optimization or post-machining surface treatments (like shot peening) [6]. However, further investigations on the effect of changes in

Materials Research Proceedings **2** (2016) 13-18 doi: http://dx.doi.org/10.21741/9781945291173-3

surface integrity on the fatigue properties of the components are somewhat limited although it has a great practical importance for the assessment of the component life and also the knowledge obtained can be used backwards to guide the surface integrity modification. The purpose of the current study is to characterize the residual stresses generated on milled Inconel 718 as influenced by the use of coolant and a subsequent shot-peening treatment. Meanwhile, the fatigue performance of the specimens corresponding to the different surface conditions has also been studied in a four-point bending mode.

Experimental work

The material used in this study was taken from a disc forging of Inconel 718 with chemical composition given in Table 1. The forging was solution annealed at 970 °C followed by air cooling to room temperature, and then a two-stage ageing was performed first at 720 °C for 8 h, further at 620 °C for another 8 h, and finally air cooled to room temperature.

Table 1. Chemical composition [wt%] of the Inconel 718 disc forging.

	Fe	Ni	Cr	Mo	Nb	Ti	Al	C
Min. (%)	Bal.	50	17	2.8	4.75	0.65	0.2	
Max. (%)		55	21	3.3	5.5	1.15	0.8	0.08

Fatigue test bars with a dimension of $10 \times 10 \times 80$ mm^3 were pre-manufactured from the heat-treated forging by wire electric discharge machining. The surface to be loaded in tension during fatigue was then machined by face milling using a 20 mm diameter cutter with two uncoated cemented carbide inserts. The cutting speed was fixed at 30 m/min (corresponding spindle speed was 382 rpm) and the depth of cut was 0.5 mm. The feed direction was along the longitudinal direction of the bar with a feed rate of 76 mm/min. Chamfers on the tensile side were introduced in order to avoid corner crack initiation. Three groups of specimens were prepared; the specimens of the first two groups were dry milled and milled under coolant respectively, while for the last group, the surface that has been machined by dry milling was subsequently shot peened using spherical S170 H cast steel shots with 150 to 200 % surface coverage, while the shot-peening intensity was varied from 0.2 to 0.3 mmA.

The microstructure beneath the dry-milled, wet-milled and shot-peened surface was characterized on polished cross-sections prior to fatigue testing using a scanning electron microscope (SEM) together with electron channeling contrast imaging (ECCI). In addition, the In-depth residual stresses created by milling and shot peening were measured by using X-ray diffraction, combined with layer removal by electrolytical polishing. Cr-Kα radiation was chosen, giving a diffraction peak at $2\theta \sim 128°$ for the {220} family of lattice planes of the nickel-based matrix. Peaks were measured at nine ψ-angles between $\psi = \pm 55°$, and residual stresses were calculated based on the "$\sin^2 \psi$" method [7] with an X-ray elastic constant of 4.65×10^{-6} MPa^{-1}. Deviation in the measured residual stresses due to the layer removal were corrected in the case of a flat plate.

All fatigue tests were conducted at room temperature under load control using a sinusoidal waveform at a load ratio of 0.1 and a frequency of 20 Hz. The distance between the two loading and two supporting rollers was 12 mm and 60 mm, respectively. For each group of the three, four specimens were tested at different peak loads in the range of 8 kN to 16 kN. The corresponding peak stress at the surface, calculated assuming pure elastic loading, were approximately 600 MPa to 1200 MPa. The yield strength of the Inconel 718 forging at room temperature, on the other hand, is slightly above 1000 MPa. All specimens were fatigued until rupture and the specimen deflection at the maximum/minimum load versus the number of cycles was recorded. A line was fitted to the initial linear part of the deflection range-number of cycles curve and extrapolated to the larger cycle region. The number of cycles corresponding to 1% increase of the deflection range from the fitted line was then defined as the fatigue life in the present study. Accordingly, the lifetime of the specimens is largely dominated by the fatigue cycles spent on crack initiation. The failed specimens were examined under SEM in order to identify the preferential sites where fatigue cracks may initiate.

Materials Research Proceedings **2** (2016) 13-18 doi: http://dx.doi.org/10.21741/9781945291173-3

Results and discussions

Fig. 1(a) and (b) shows the in-depth residual stresses induced by dry milling and wet milling. Stress components in two in-plane directions, i.e. transverse direction (TD) and longitudinal direction (LD) (corresponding to the cutting direction and feed direction), were measured. In general, tensile residual stresses were created on the milled surface, regardless of the application of coolant, but it is clear that the wet milling operation led to a lower surface tension and a reduced thickness of the tensile layer. As the depth increases, the residual stresses gradually shift to compression until stabilizing at ~ 0 MPa.

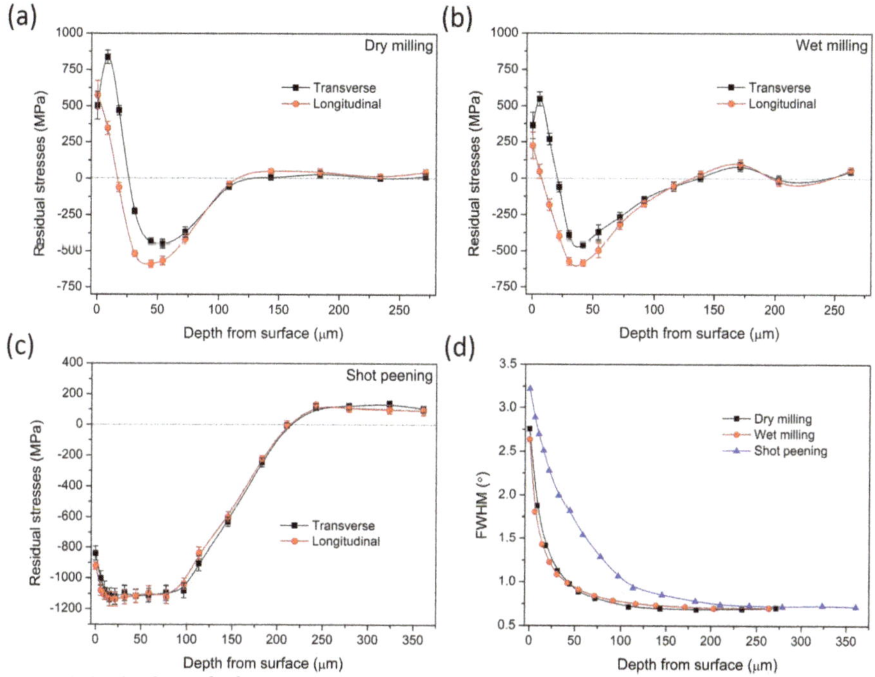

Figure 1. In-depth residual stresses generated by (a) dry milling, (b) wet milling, and (c) shot peening; (d) A comparison of the full width at half maximum intensity (FWHM) obtained from the measured diffraction peaks.

The post-milling surface treatment by shot peening annihilated the high tension on the dry-milled surface and introduced a surface plateau, extending to a depth of 100 μm, with great compressive residual stresses in both TD and LD, see Fig. 1(c). The high level of surface compression was created as a consequence of the mechanically-induced plastic deformation during shot peening. This can be seen from the dramatically increased broadening of the diffraction peaks, i.e. full width at half maximum intensity (FWHM), measured in the shot-peened surface layer, see Fig. 1(d). The formation of the tensile residual stresses on the milled surface is most likely to be of thermal origin associated with the great heat generation during machining [1]. From Fig. 1(d), one can see that the wet-milled surface underwent less plastic deformation compared with that in the case of dry milling. A very likely explanation is that the coolant could contribute to lowering the friction and dissipating the generated heat, leading to a relatively low cutting temperature. As an effect of the reduced cutting temperature, the thermally-induced residual stresses became less in tension on the surface produced by wet milling. The reduced thermal impact during wet milling can be further supported by the microstructural characterization beneath the milled surfaces. Instead of a continuous thick white layer, approximately 4 to 5 µm, as observed on the dry-milled surface (Fig. 2(a)), the surface white layer appeared discontinuously with a thickness less than 1 µm on the wet-milled surface (Fig. 2(b)). As suggested by Bushlya et al. [8], the development of white layers takes place in machining of Inconel 718 normally when the cutting temperature is increased, e.g. at high cutting speeds, cutting with worn tools or in dry machining operations. The shot-peened surface compared with the dry-milled surface showed significantly increased plastic deformation in microstructure, see Fig. 2(c), which is consistent with the greatly higher FWHM measured in the shot-peened layer.

Figure 2. Electron channeling contrast imaging (ECCI) micrographs showing the microstructure beneath (a) the dry-milled surface, (b) the wet-milled surface, and (c) the shot-peened surface. In (a) and (b), dash lines are drawn to compare the thickness of the superficial white layer formed in dry and wet milling.

A comparison of the fatigue performance as influenced by the use of coolant and post-machining surface treatment of shot peening is presented in Fig. 3. Although relatively low surface residual stresses were obtained by wet milling, the fatigue life was observed to be comparable with that for the specimens prepared by dry milling. However, in the high-cycle regime with a lower applied load, it showed a slight increase in fatigue resistance for the wet-milled condition which is very likely due to a stronger effect of residual stresses. The shot peening, on the other hand, led to a great increase of the fatigue life, particularly in the high-cycle regime; the enhancement could be up to roughly two orders of magnitude compared with the lifetime of the dry-milled specimens.

Fracture surface examinations can offer insights into the fatigue failure mechanism of the specimens with different surface conditions. Shown in Fig. 4 is an example of the typical fracture appearance observed on the fatigued specimens with either a dry-milled or wet-milled surface. It can be clearly seen that multiple cracks were initiated at the milled surface during fatigue loading and the coalescence of the cracks led to a macroscopic fluctuant fracture surface. Close examinations further revealed that the initiation of fatigue cracks took place primarily associated with the cracking of surface non-metallic inclusions (Nb-rich carbides and/or Ti-rich nitrides). Previous studies by the authors [9] have shown that the giant plastic work during machining of Inconel 718 could cause cracking of non-metallic inclusions on the machined surface. In this study, substantial cracked carbides as well as a few cracked nitrides (due to the much lower amount of nitrides in the alloy) were also observed after the milling operations. These pre-existing surface defects provide multiple sites where fatigue cracks preferably initiate, or could even start to grow without an incubation of nucleation. Based on such predominance of the failure mechanism, the comparable fatigue life of the specimens prepared by dry and wet milling is very likely attributed to the similar damage on the milled surfaces with respect to the non-metallic inclusion cracking. The effect of residual stresses in this case appears to be less significant.

In the case of the shot-peened specimens, the surface compression was deep and strong enough to shift the crack initiation sites to subsurface regions corresponding to the depth of the compressive layer, see Fig. 5. The surface microstructure of the shot-peened specimens still consists of a large amount of cracked inclusions, however, the development of fatigue cracks from these flaws was retarded due to the

Materials Research Proceedings **2** (2016) 13-18 doi: http://dx.doi.org/10.21741/9781945291173-3

presence of the great compressive residual stresses. As a result, an enhanced fatigue life and strength were obtained, as shown in Fig. 3. The beneficial effect of compressive residual stresses in terms of retarding surface cracking during fatigue and increasing the resistance of the component to fatigue failures is well consistent with the previous findings in shot peening of other metallic materials [3,4]. The extent to which the lifetime was prolonged was decreased as the applied load was increased due to the residual stress relaxation in low cycle fatigue resulting from significant cycling strains [10].

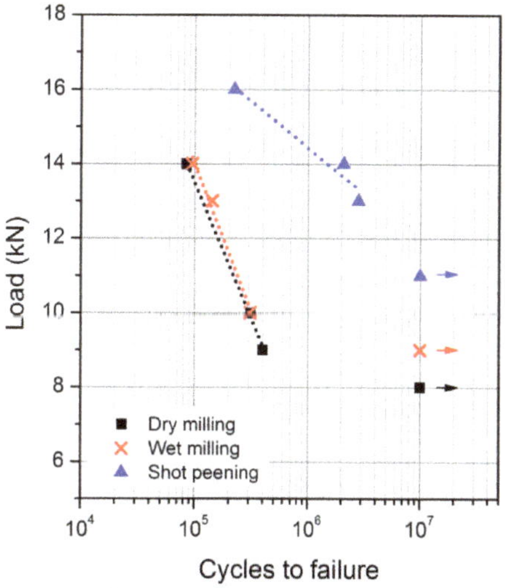

Figure 3. A comparison of the obtained fatigue life as influenced by the use of coolant as well as the post-machining surface treatment of shot peening.

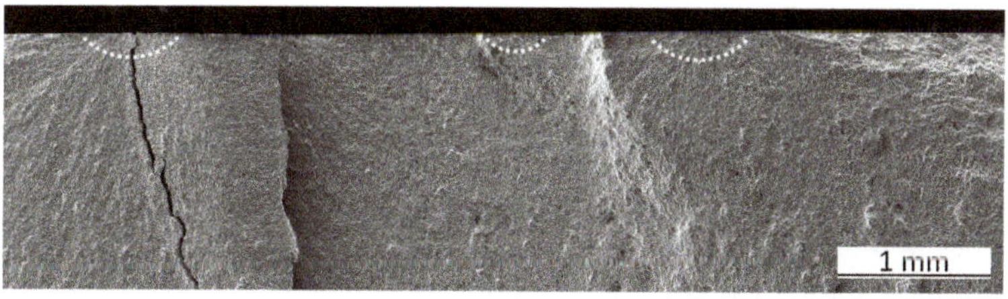

Figure 4. Typical fracture appearance after the fatigue of the milled specimens regardless of the use of coolant where multiple crack initiation sites at the milled surface were observed; some of them are located by dash lines.

Conclusions

The present work investigated the residual stresses generated on milled Inconel 718 as influenced by the use of coolant in machining or by post-machining surface treatment with shot peening. The corresponding fatigue performance of the specimens was also investigated. The results showed that the wet milling led to reduced tensile residual stresses on the machined surface compared with that in the case of dry milling. However, a comparable fatigue life was obtained for the specimens milled with or without the use of the coolant. This is very likely due to that for both conditions the milling-induced surface damage with respect to cracked non-metallic inclusions dominated the crack initiation during fatigue. The shot-peening treatment annihilated the surface tension induced by milling and introduced high compressive residual stresses. The presence of the compressive layer retards surface cracking from the pre-existing cracked carbides and/or nitrides and shifts the crack initiation sites to sub-surface regions, leading to a significant increase of the fatigue life and strength for the shot-peened specimens. The extent to which the lifetime was prolonged was reduced as the applied load was increased.

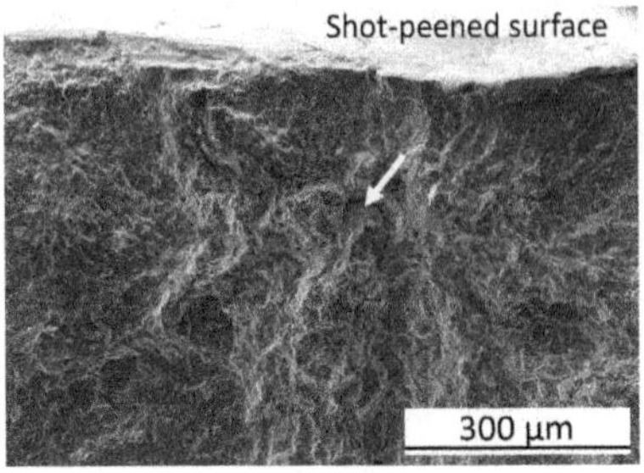

Figure 5. Fatigue fracture surface of a shot-peened specimen where it shows a transition of the crack initiation site from surface to subsurface regions (pointed by the arrow) compared with the observation in Fig. 4.

Acknowledgements

Faculty grant SFO-MAT-LiU#2009-00971 at Linköping University, Ytstruktur Arboga AB and Siemens Industrial Turbomachinery AB are acknowledged for their valuable supports in this study.

References

[1] K. Jacobus, R. DeVor, S. Kapoor, Machining-induced residual stress: experimentation and modeling, J. Manuf. Sci. Eng. 122 (2000) 20-31.
http://dx.doi.org/10.1115/1.538906

[2] F. Pusavec, H. Hamdi, J. Kopac, I. Jawahir, Surface integrity in cryogenic machining of nickel based alloy-Inconel 718, J. Mater. Process. Technol. 211 (2011) 773-783.
http://dx.doi.org/10.1016/j.jmatprotec.2010.12.013

[3] J. Lindemann, C. Buque, F. Appel, Effect of shot peening on fatigue performance of a lamellar titanium aluminide alloy, Acta Mater. 54 (2006) 1155-1164.
http://dx.doi.org/10.1016/j.actamat.2005.10.043

[4] L. Wagner, Mechanical surface treatments on titanium, aluminum and magnesium alloys, Mater. Sci. Eng. A 263 (1999) 210-216.
http://dx.doi.org/10.1016/S0921-5093(98)01168-X

[5] Y. Choi, A study on the effects of machining-induced residual stress on rolling contact fatigue, Int. J. Fatigue 31 (2009) 1517-1523.
http://dx.doi.org/10.1016/j.ijfatigue.2009.05.001

[6] A. Thakur, S. Gangopadhyay, State-of-the-art in surface integrity in machining of nickel-based super alloys, Int. J. Mach. Tools Manuf. 100 (2016) 25-54.
http://dx.doi.org/10.1016/j.ijmachtools.2015.10.001

[7] I.C. Noyan, J.B. Cohen, Residual stress-measurement by diffraction and interpretation, (1986).

[8] V. Bushlya, J. Zhou, F. Lenrick, P. Avdovic, J. Ståhl, Characterization of white layer generated when turning aged Inconel 718, Procedia Eng. 19 (2011) 60-66.
http://dx.doi.org/10.1016/j.proeng.2011.11.080

[9] Z. Chen, R.L. Peng, J. Moverare, P. Avdovic, J. Zhou, S. Johansson, Surface integrity and structural stability of broached Inconel 718 at high temperatures, Metall. Mater. Trans. A 47 (2016) 3664-3676.
http://dx.doi.org/10.1007/s11661-016-3515-6

[10] R. McClung, A literature survey on the stability and significance of residual stresses during fatigue, Fatigue Fract. Eng. Mater. Struct. 30 (2007) 173-205.
http://dx.doi.org/10.1111/j.1460-2695.2007.01102.x

Residual Stresses 2016: ICRS-10 | Materials Research Forum LLC
Materials Research Proceedings 2 (2016) 19-24 | doi: http://dx.doi.org/10.21741/9781945291173-4

Effect of Ultrasonic Peening on Residual Stresses at a T-Butt Weld Toe

Alan K. HELLIER[1,a *], B. Gangadhara PRUSTY[1,b], Garth M. PEARCE[1,c], Mark REID[2,d], Anna M. PARADOWSKA[1,2,e] and Peter SIMONS[3,f]

[1]School of Mechanical and Manufacturing Engineering, UNSW Australia, (The University of New South Wales), Sydney, NSW 2052, Australia

[2]Bragg Institute, Australian Nuclear Science and Technology Organisation (ANSTO), Lucas Heights, NSW 2234, Australia

[3]Applied Ultrasonics Australia, Suite 11/20 Narabang Way, Belrose, NSW 2085, Australia

[a]a.hellier@unsw.edu.au, [b]g.prusty@unsw.edu.au, [c]g.pearce@unsw.edu.au, [d]markr@ansto.gov.au, [e]anp@ansto.gov.au, [f]psimons@appliedultrasonics.com.au

Keywords: A350 Grade, Black Mild Steel Plate, T-Butt Welded Joint, Ultrasonic Peening, Residual Stresses, Fatigue Crack Growth, Parametric Equations

Abstract. The current paper presents the results of neutron diffraction measurements of the through-thickness residual stress field at the toe of a T-butt weld, made from 10mm thick A350 grade black mild steel plates, after successful ultrasonic peening. A single ultrasonic peening treatment was carried out at the weld toe. Residual stresses were measured using the KOWARI instrument at ANSTO. The neutron diffraction technique was chosen for this study because of its ability to measure three-dimensional residual stress deep within the component at high resolutions.
Although the nominal yield stress of the A350 grade plate is 350 MPa the actual yield stress is generally higher, in this case averaging out to about 400 MPa. Ultrasonic peening was highly effective, leading to a residual stress redistribution with a maximum compressive stress of about 250 MPa at the weld toe surface and a maximum tensile stress of 220 MPa, at a depth of almost 3mm into the base plate. The resulting compressive residual stresses at the weld toe surface will almost certainly increase substantially both the fatigue initiation and propagation lives, or may *prevent fatigue completely*. Since A350 steel is widely used in buildings, bridges and offshore structures, ultrasonic peening shows great promise as an in-situ peening method in order to improve weld fatigue performance.

Introduction

A350 grade black mild steel is one of the most widely used structural materials in the world, being commonly found in buildings, bridges and offshore structures. Welding is typically used to join two plates of structural steel and this often takes the form of a T-butt weld. In addition, other more complex geometrical joints are often simplified for stress analysis purposes by approximating them as two-dimensional T-butt plate models (e.g. skewed T-joints, tubular welded joints, pipe–plate joints, etc.). However, all such welds are potentially susceptible to fatigue crack initiation and slow but accelerating growth arising as a result of fluctuating service loads, often eventually resulting in fast fracture.

Ultrasonic peening, more properly known as ultrasonic impact treatment (UIT), is a recent development of the well-established shot peening process. It was originally invented in 1972 in the former USSR to improve the fatigue and corrosion performance of ship and submarine hulls. UIT is similar to conventional needle or hammer peening in many respects. An important difference is that rather than using a pneumatic tool, which causes the needles or a single hammer-like rod to impact the weld surface at a frequency of 25-100 Hz, with UIT, the weld is impacted by a small number of rods vibrating at a much higher frequency on the order of 18,000-27,000 Hz. This makes it a much quieter device, which vibrates at a lower intensity, so that the operator can use it for longer periods of time before tiring [1]. Ultrasonic peening is relatively cheap, can be applied in-situ and offers significant improvements in the lifespan of welded components when applied correctly. This occurs for three different reasons: removal of weld defects; reduction of stress concentration; and redistribution of tensile stresses and/or introduction of compressive stresses [2].

Experimental Methods

Welding. Dimensions were 800×160×10mm for the base plate and 100×160×10mm for the attachment plate of each T-butt. During fabrication two base plates were tack welded back-to-back prior to welding of the double-beveled attachment plate to each, in order to minimize distortion. Balanced full-penetration GMAW fillet welding was employed, with six passes on each side. All welds passed ultrasonic testing (UT) for internal flaws and magnetic particle inspection (MPI) for surface flaws. Fig. 1 shows the weld detail of a typical T-butt joint before buffing.

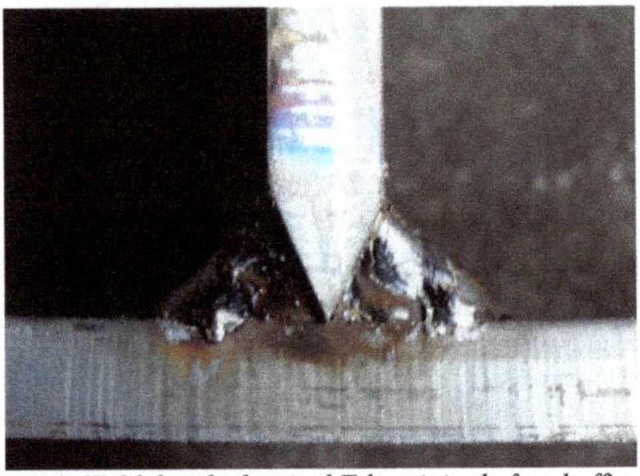

Fig. 1. Weld detail of typical T-butt joint before buffing.

Ultrasonic Peening. Ultrasonic peening treatments were carried out at the base plate and attachment plate weld toes on both sides of one of the T-butts using an Applied Ultrasonics Esonix UIT apparatus. The impact pins were oriented 45° from the treated weld area (perpendicular to the toe line) and the tool was continuously moved in an oscillating motion in a path parallel to the direction of the weld. In order to develop the desired groove, the tool was worked back and forth between a 30° and a 60° angle while maintain the oscillating motion described above. Groove formation was continuously observed during the application process. A properly formed, shiny groove at the weld toe was obtained; this is the main quality assurance inspection point for treatment verification [3].

Neutron Diffraction. It was attempted to measure through-thickness residual stresses from the weld toe into the base plate for both as-welded and ultrasonically peened specimens using the KOWARI instrument at the ANSTO Bragg Institute. The (non-destructive) neutron diffraction technique was chosen for this study because of its ability to measure three-dimensional residual stress deep within the component at high resolutions. For the neutron measurements a $0.5\times0.5\times1\text{mm}^3$ gauge volume was used for the longitudinal, transverse and normal components. The experimental measurements on the ultrasonically peened sample were successful but those on the as-welded sample were unfortunately shown to be incorrect, as the scan was made slightly inside the weld rather than exactly at the weld toe.

Materials Research Proceedings **2** (2016) 19-24 doi: http://dx.doi.org/10.21741/9781945291173-4

Parametric Equations. Brennan-Dover-Karé-Hellier (BDKH) parametric equations [4] are available for the stress intensity factor (SIF) geometric Y-factor at the deepest point of a semi-elliptical flaw at the toe of a T-butt weld, accurate for a wide range of geometric parameters under both tension (membrane) and pure bending loadings. These were derived from the results of eighty 8-noded shell finite element analyses in conjunction with the Niu-Glinka weight function [5]. From the models studied, the geometry validity limits for the equations developed are:

Weld angle:	$30° \leq \alpha \leq 60°$	
Crack aspect ratio:	$0 \leq a/c \leq 1.0$	
Crack depth:	$0.01 \leq a/T \leq 1.0$	(1)
Weld toe radius:	$0.01 \leq \rho/T \leq 0.066$	
Attachment width:	$0.3 \leq L/T \leq 4.0$	

Fig. 2 shows the geometry of the T-butt weldments studied including the crack geometry, with all the geometric parameters above defined.

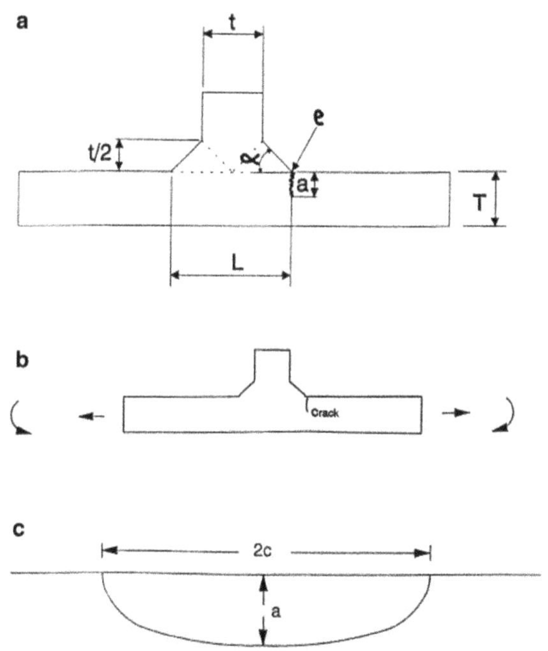

Fig. 2. (a) Local weld geometry studied (b) Geometry and loading used to derive stress intensity factors (c) Crack geometry (semi-elliptical crack).

Recently developed and unique Hellier-Brennan-Carr (HBC) T-butt weld toe surface stress concentration factor (SCF) [6] and stress distribution parametric equations [7] through the base plate thickness (i.e. the potential Mode I crack plane) are also available for similar ranges of geometric parameters and tensile (membrane) loading.

Crack Growth Equations. The Paris Law [8] is commonly used to predict the (Stage 2) fatigue propagation life for a component or structure containing a sharp initial crack. It does not take into account the influence of mean stress. Another equation which does incorporate the (second order) influence of mean stress on the propagation rate is the Forman Equation [9].

Computer Programs to Predict Fatigue Life. Two FORTRAN computer programs were written to predict the remaining fatigue life for a T-butt welded joint containing a semi-elliptical crack at the weld toe under tension (membrane) loading. The first of these programs uses the BDKH tension parametric equation in conjunction with the Paris Law, and is applicable to a stress-relieved joint. The second program uses the BDKH tension parametric equation together with the HBC tension parametric equation and the Forman Equation, and is applicable to both the as-welded and ultrasonically peened conditions where residual stresses are present.

Experimental Results

Residual Stresses. Although the nominal yield stress of the A350 grade plate is 350 MPa the actual yield stress is generally higher, in this case averaging out to about 400 MPa. Ultrasonic peening was highly effective, leading to a residual stress redistribution with a maximum compressive stress of about 250 MPa at the weld toe surface and a maximum tensile stress of 220 MPa, at a depth of almost 3mm into the base plate (refer to Fig. 3).

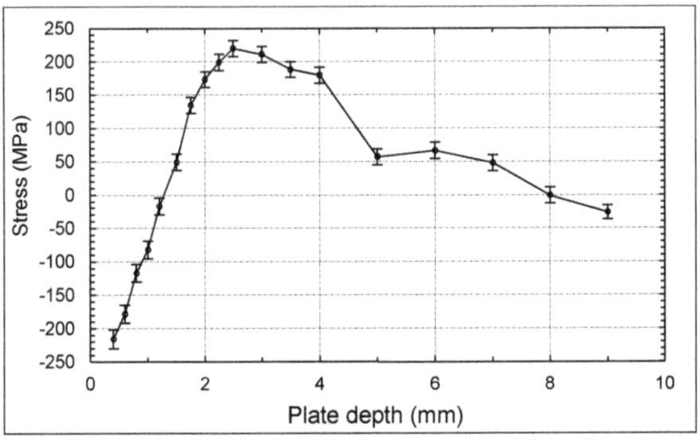

Fig. 3. T-butt weld toe residual stresses through the base plate after ultrasonic peening.

Since there are no measurements in the as-welded state, a typical as-welded residual stress distribution from the literature [10] was used instead, as shown in Fig. 4. The sample investigated using neutron diffraction with a 2×2×2mm^3 gauge volume was a T-plate fillet weld, joining two 25mm thick plates. The plate material was BS 7191 grade 355 EMZ structural steel, which is very similar to A350 grade steel, and represents a large group of steels commonly used in the nuclear and offshore industries. SMAW welding with standard filler metal was used. Both welds consisted of four weld passes in an alternating sequence between the two sides. For the purpose of this work, the residual stress distribution at the T-butt weld toe centre line was scaled horizontally to fit a 10mm thick base plate.

Geometric, Material and Loading Parameters. Table 1 contains values of the geometric parameters selected for the present analyses.

The fatigue threshold, ΔK_{Ith}, was taken as 3.2 MPa√m. The Paris Law crack growth coefficient and exponent were $C = 8.57 \times 10^{-12}$ (SI units) and $m = 3$, respectively. Applied membrane stress varied from 0 to 100 MPa. The number of crack increments employed in the modelling was 10,000.

Preliminary Results. Fig. 5 shows the numerical crack growth results superimposed for: (a) as-welded; (b) stress-relieved; and (c) ultrasonically peened conditions. The analyses conducted assumed that a fatigue crack grows through a static residual stress field. As expected, it is apparent that the ultrasonically peened T-butt exhibited a longer fatigue life than the stress-relieved specimen, which in turn had a longer fatigue life than the as-welded specimen. It should be noted that the ultrasonically peened specimen had an initial crack depth of 1mm; when it had the same 0.1mm initial crack depth as the other two specimens it exhibited *no fatigue crack growth at all.*

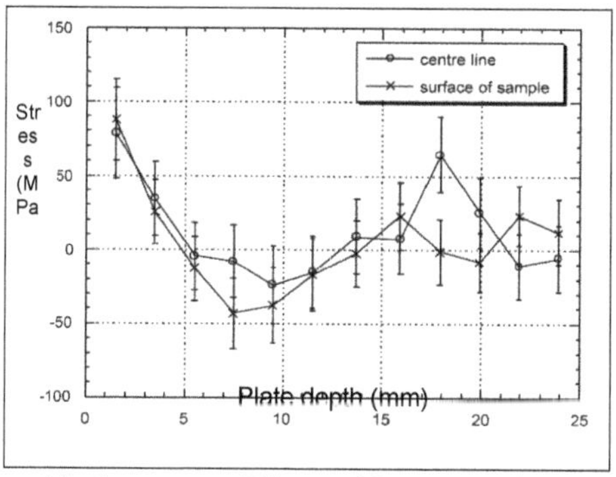

Fig. 4. Through-thickness residual stresses at T-butt weld toe centre line and surface of the sample as a function of base plate depth [10].

Table 1. Geometric parameters of T-butt joints.

Geometric Parameter	Value
Weld angle [α]	45°
Weld toe radius [ρ]	0.1mm
Plate thickness [T]	10mm
Welded attachment width [L]	30mm
Initial semi-elliptical crack depth [a_i]	0.1mm or 1mm
Initial semi-elliptical crack width [$2c_i$]	2mm or 20mm

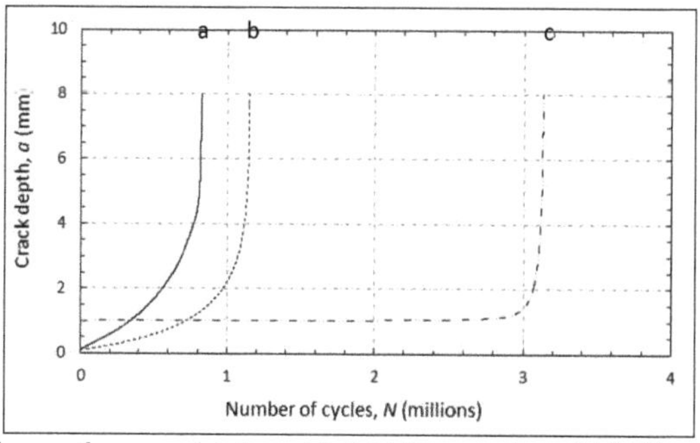

Fig. 5. Fatigue crack growth curves (a) As-welded (b) Stress-relieved (c) Ultrasonically peened specimens.

Ultrasonic peening was found to be very effective at modifying the residual stress distribution. As expected the as-welded residual stresses shortened the fatigue propagation life, whereas ultrasonic peening extended it or *completely prevented fatigue crack growth.*

References

[1] R.T. Yekta, S. Walbridge, Acceptance Criteria for Ultrasonic Impact Treatment (UIT), Ontario Ministry of Transportation, Provincial Highways Management Division, Highway Infrastructure Innovation Funding Program, Final Report No. HIIFP-110, Canada, February 2013.

[2] L. Lopez Martinez, Z. Barsoum, A. Paradowska, State-of-the-art: Fatigue life extension of offshore installations, in: Proc. 31st International Conference on Ocean, Offshore and Arctic Engineering (OMAE2012), Rio de Janeiro, Brazil, 1-6 July 2012, Paper 83044.
http://dx.doi.org/10.1115/omae2012-83044

[3] Esonix UIT, Application Guide, Post Weld Treatment for Fatigue Enhancement, Carbon Steel Welded Structures, Applied Ultrasonics, Revision 2.0, Revision Date: 28 June 2006.

[4] F.P. Brennan, W.D. Dover, R.F. Karé, A.K. Hellier, Parametric equations for T-butt weld toe stress intensity factors, Int. J. Fatigue 21 (1999) 1051-1062.
http://dx.doi.org/10.1016/S0142-1123(99)00090-0

[5] X. Niu, G. Glinka, Theoretical and experimental analyses of surface fatigue cracks in weldments, in: W.G. Reuter, J.H. Underwood, J.C. Newman (Eds.), Proc. Symposium on Surface-Crack Growth: Models, Experiments, and Structures, Sparks, NV, 25 April 1988, ASTM STP 1060, American Society for Testing and Materials, Philadelphia, PA, 1990, pp. 390-413.
http://dx.doi.org/10.1520/stp23445s

[6] F.P. Brennan, P. Peleties, A.K. Hellier, Predicting weld toe stress concentration factors for T and skewed T-joint plate connections, Int. J. Fatigue 22 (2000) 573-584.
http://dx.doi.org/10.1016/S0142-1123(00)00031-1

[7] A.K. Hellier, F.P. Brennan, D.G. Carr, Weld toe SCF and stress distribution parametric equations for tension (membrane) loading, in: Proc. Fatigue 2014, 11th International Fatigue Congress, MCG, Melbourne, 2-7 March 2014, Adv. Mater. Research 891-892 (2014) 1525-1530.

[8] P. Paris, F. Erdogan, A critical analysis of crack propagation laws, Trans. ASME, J. Basic Eng. 85 (1963) 528-534.
http://dx.doi.org/10.1115/1.3656900

[9] R.G. Forman, V.E. Kearney, R.M. Engle, Numerical analysis of crack propagation in cyclic-loaded structures, Trans. ASME, J. Basic Eng. 89 (1967) 459-463.
http://dx.doi.org/10.1115/1.3609637

[10] P.S. May, R.C. Wimpory, G.A. Webster, N.P. O'Dowd, Determination of the Residual Stress Distribution in a Welded T-Plate Joint, Experimental Report, REST Instrument, Experiment Number 425, The Studsvik Neutron Research Laboratory (NFL), University of Uppsala, Studsvik, Sweden, 2000.

Structural Engineering Studies on Reinforced Concrete Structure using Neutron Diffraction

Hiroshi Suzuki[1,a *], Koichi Kusunoki[2,b], Manabu Kanematsu[3,c], Tomohisa Mukai[4,d] and Stefanus Harjo[5,e]

[1] Materials Sciences Research Center, Japan Atomic Energy Agency, Tokai, Naka, Ibaraki, Japan

[2] Earthquake Research Inst., Univ. of Tokyo, Bunkyo-ku, Tokyo, Japan

[3] Dept. of Architecture, Tokyo Univ. Sci., Noda, Chiba, Japan

[4] Department of Structural Engineering, Building Research Inst., Tsukuba, Ibaraki, Japan

[5] J-PARC Center, Japan Atomic Energy Agency, Tokai, Naka, Ibaraki, Japan

[a]suzuki.hiroshi07@jaea.go.jp, [b] kusunoki@eri.u-tokyo.ac.jp, [c] manabu@rs.noda.tus.ac.jp

[d] t_mukai@kenken.go.jp, [e] stefanus.harjo@j-parc.jp

Keywords: Neutron Diffraction, Reinforced Concrete, Transverse Rib, Bond Stress

Abstract. It has been demonstrated in our past studies that neutron diffraction can be an alternative method to conventional strain gauges for measuring the stress distribution along rebar embedded in concrete. The current study investigated the possibility of the bond stress evaluation using neutron diffraction in order to find a further capability of neutron diffraction for the structural engineering study on the reinforced concrete structure. Several peaks appeared in the bond stress distribution measured by neutron diffraction, showing the inhomogeneous bond variation along the embedded rebar. This result suggests that the neutron diffraction technique with high spatial resolution makes it possible to investigate local bond resistance caused by the transverse ribs. The bond stress distribution measured by the neutron diffraction technique is expected to bring detailed understanding of the bond mechanism between rebar and concrete for the reinforced concrete structure.

Introduction

The reinforced concrete (RC), which is widely utilized for various architectural and civil engineering structures, is well known as a composite structure, in which concrete with relatively low tensile strength and ductility is strengthened by reinforcements such as steel rods (rebars) with high tensile strength and/or ductility. In general, quantitative evaluation of bond resistance between rebar and surrounding concrete is important to discuss the performance of the RC structures [1-4]. In our previous studies, we have investigated the potential of the neutron diffraction technique for the stress measurement of rebar embedded in concrete as an alternative method to the conventional strain gauge. Our first relevant work was carried out using the engineering diffractometer RESA-1 in JRR-3 (Japan Research Reactor No. 3), and we demonstrated that the neutron diffraction technique can be a novel strain measurement method for rebar embedded in concrete [5, 6]. More recently, three-dimensional deformation behavior of the embedded rebar including the axial and transverse strains was successfully measured under pull-out loading using Time-of-Flight (TOF) neutron diffraction with the engineering diffractometer, TAKUMI in MLF (Materials and life Science Experimental Facility) of J-PARC (Japan Proton Accelerator Research Complex) [7]. Furthermore, it was demonstrated by some application studies using TAKUMI that the neutron diffraction technique is available to assess the bond deterioration due to rebar corrosion and crack generation in concrete [8]. As described above, our previous studies commonly discussed on the bond condition between rebar and concrete based on a change in the axial stress distributions along the embedded rebar measured by neutron diffraction. However, most of the papers investigating the bond strength for the RC structures have rather discussed on a change in the "bond stress distribution" rather than the axial stress distribution [9, 10].

The current study was, therefore, undertaken to evaluate the bond condition of the RC structure based on the bond stress distribution along the embedded rebar, in order to find further capability of neutron diffraction for the structural engineering study of the RC structure.

Experimental Procedure

RC Specimens. Figure 1 (a) shows a schematic illustration of the RC specimen used in this study. A ferritic steel deformed-bar with 9.53 mm in a nominal diameter, so-called D10 rebar, was embedded in cylindrical concrete with approximately 50 mm in diameter and 460 mm in length. The length of the bonded region is 320 mm, and an un-bonded region with 110 mm in length was artificially introduced at the loading edge of the specimen. This un-bonded region was utilized for determining the reference lattice parameter with stress-free condition. The boundary between bonded and un-bonded regions was defined as the origin of length scale (X=0). The maximum size of the mixed aggregates in the concrete was 13 mm in diameter. This specimen was cured in air for total 29 days before the neutron experiment, i.e. first 14 days at ambient condition and last 15 days in a constant temperature (20±1 °C) and humidity (60±5 RH%) room. The compressive and tensile strengths of concrete cured in air for 50 days under the same condition as the reinforced concrete specimen were measured as a reference to be 34.6 MPa and 2.79 MPa, respectively. The specimen was mounted on the pull-out loading device which consists of a hydraulic center-hole jack, a center-hole load cell and coil spring, as shown in Fig. 1 (b).

Neutron Experiment. The engineering diffractometer, TAKUMI [11], installed at BL19 in MLF of J-PARC was employed in this study. Figure 1 (c) shows the schematic layout of TAKUMI for the stress measurement of rebar embedded in concrete. This is an energy dispersive neutron engineering diffractometer, which measures energy spectrum diffracted from a sample when irradiating with white pulsed neutrons at 25 Hz generated in the mercury target. The incident neutron beam was formed into a rectangular shape with 5 mm in width and 10 mm in height by using a gauge definition slit, irradiating the reinforced concrete specimen mounted on the loading device settled on a XYZθ positioner. The specimen with the loading device was oriented 45° to the incident beam, and therefore allowed us to measure the strains of rebar in the axial and transverse directions simultaneously using both detector banks installed at ±90°. The radial collimators for 5 mm gauge width were installed in front

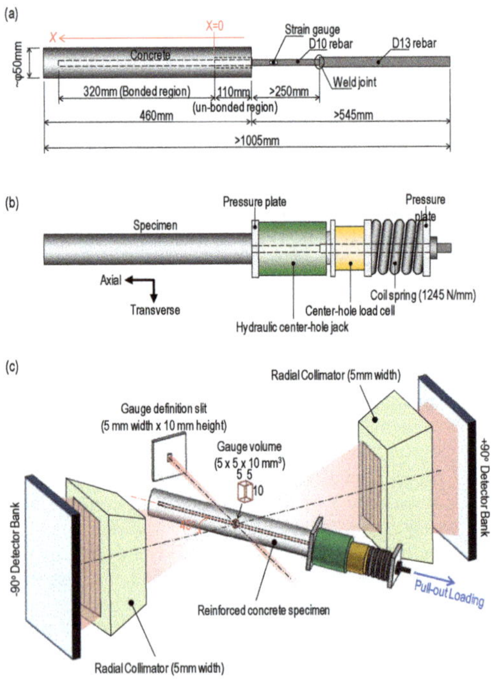

Fig. 1 Schematic illustrations of (a) sample used in this stidy, (b) load frame and (b) measurement configulation of TAKUMI.

Materials Research Proceedings **2** (2016) 25-30 doi: http://dx.doi.org/10.21741/9781945291173-5

of each detector bank, resulting in the gauge volume of $5\times5\times10$ mm^3. The diffraction patterns were measured at 5 mm intervals in minimum along rebar over the length of 250-300 mm including the first 50 mm un-bonded region under different pull-out loadings, i.e. approximately 125 MPa and 248.5 MPa. The detailed descriptions of the experimental condition including the sample preparation can be found in the reference [7].

Data Analysis. The average lattice constants, a in the axial and transverse directions were determined by multiple-fitting over 15 peaks from the 110 to 431 reflections by the Rietveld refinement software, Z-Rietveld [12]. The axial stress in a rebar axis, σ_A was calculated using the strains in the axial and transverse directions of rebar in assumption of equiaxial-stress condition [7]. On the other hand, the bond stress, τ is equivalent to the shear stress acting at the boundary between rebar and concrete, expressed by following equation,

$$\tau = \frac{A}{\pi D}\cdot\left|\frac{d\sigma_A}{dX}\right| = \frac{D}{4}\cdot\left|\frac{d\sigma_A}{dX}\right|, \quad (1)$$

where, A and D indicate a nominal cross-sectional area and a nominal diameter of rebar, respectively. According to Eq (1), the bond stress distribution can be obtained by making a differential curve of the axial stress distribution. In this study, the binomial smoothing was, at first, applied to the measured axial stress distribution to reduce the influence of data scattering, and then the bond stress distribution was obtained by plotting slopes fitted within a range of neighboring data in the smoothed axial stress distribution, as a function of the axial position of rebar.

The average bond stress was, in our previous studies [5-8], roughly calculated by Eq (1) with the slope of the axial stress distribution within a range of the anchorage region, approximately estimated by eye. In contrast, it was obtained accurately in this study normalizing the area of the bond stress diagram in the anchorage region by its length.

Results and Discussion

Figure 2 shows the axial stress distributions and corresponding bond stress distributions along the embedded rebar in concrete under two different pull-out loadings, i.e. σ_{ap}=125 MPa and 248.5 MPa. Figure 2(a) exhibits the axial stress distribution with a spatial resolution of 5 mm in minimum, showing typical stress variation for the RC structure under pull-out loading as an increase in the tensile stress towards the end of the bonded region (X=0 mm). In addition, compressive residual stresses generated due to drying shrinkage of concrete can be found in the bonded region. The bond stress distribution derived from the smoothed axial stress distribution is shown in Fig. 2(b). Lengths of the anchorage region can be predicted to be about 85 mm for σ_{ap}=125 MPa and about 145 mm for σ_{ap}=248.5 MPa, and the average bond stresses are calculated to be about 6.4 MPa and 5.9 MPa, respectively. These values are different from rough estimation in our previous study [7], but more reliable since the bond stress distribution obtained here assists to determine these values accurately. On the other hand, several peaks can be

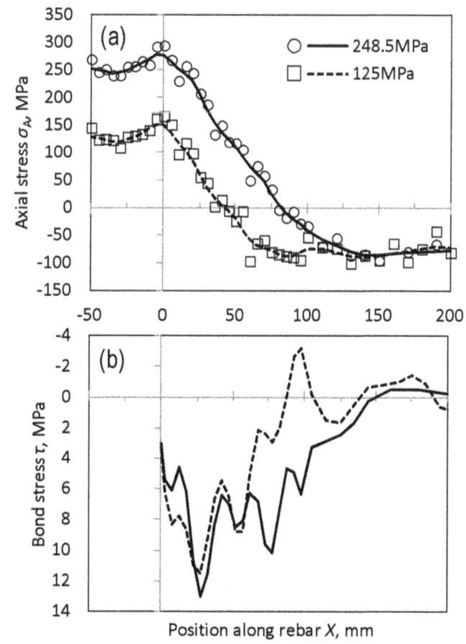

Fig. 2 (a) Axial stress and (b) bond stress distributions with 5 mm in spatial resolution. Average error bar of the measured axial stresses is ±22 MPa.

Materials Research Proceedings **2** (2016) 25-30 doi: http://dx.doi.org/10.21741/9781945291173-5

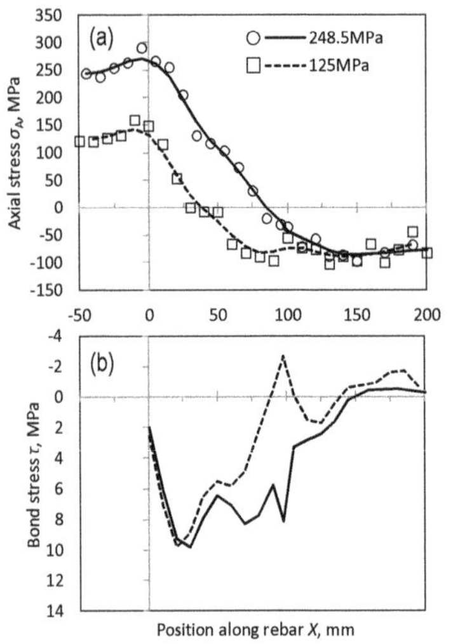

Fig. 3 (a) Axial stress and (b) bond stress distributions with 10 mm in a spatial resolution.

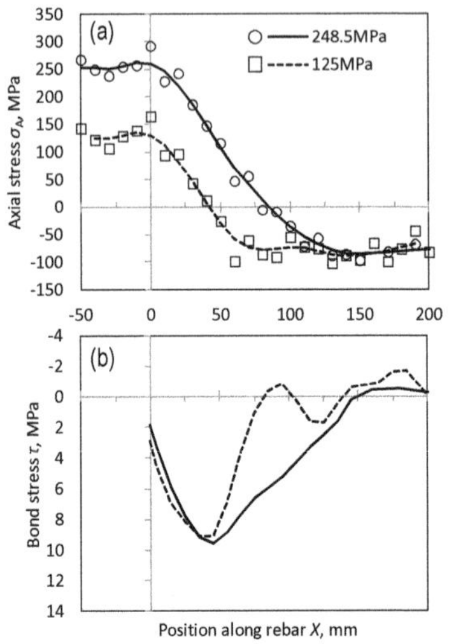

Fig. 4 (a) Axial stress and (b) bond stress distributions with 10 mm in spatial resolution, but it is 5mm offset from that in

found in the bond stress distribution with about 20 mm in intervals. Since the error bars of the bond stresses are estimated to be more than ±10 MPa, the oscillations in Fig. 2(b) is suspected to be given by the scatter in the axial stresses. However, the cyclic variation showing similar pattern regardless of the applied stresses may be rather associated with ribs regularly distributed along rebar. To know the reason why several peaks appear in the bond stress distribution, therefore, following discussions are made based on the geometrical relationship between the transverse rib and measurement positions.

Figure 3 (a) shows the axial stress distribution with 10 mm in a spatial resolution, extracted from the original axial stress distribution shown in Fig. 2(a). The bond stress distribution obtained here looks similar pattern to the variation of Fig. 2(b), but showing some broader peaks. The average bond stresses are calculated to be about 5.8 MPa for both applied stresses, which are slightly smaller than those derived from the bond stress distribution with 5 mm in a spatial resolution. On the other hand, Fig. 4 (a) shows the axial stress distribution with the same spatial resolution of 10 mm, but it is 5mm offset from that in Fig. 3(a). As shown in Fig. 4(b), the bond stress distribution derived from the axial stress distribution has no peaks, which is clearly different from the variations shown in Figs. 2(b) and 3(b). The average bond stresses are calculated to be about 5.8 MPa for σ_{ap}=125 MPa and about 5.6 MPa for σ_{ap}=248.5 MPa, which are similar value to those in former case with 10 mm in a special resolution. Comparing the average bond stresses for all cases with different special resolutions, it can be confirmed that they agree within less than 1.0 MPa, regardless of the special resolution.

The results showing different trend of the bond stress distribution can be explained by the position relationship between the transverse ribs and measurement positions, as represented in Fig. 5(a). The distance between transverse ribs of D10 rebar is approximately 6.7 mm. If the rib position coincides with the measurement position at X=10 mm, they are approximately overlapped every 20 mm when measuring the axial stress distribution with 5 mm in a spatial resolution, as shown in (i) in Fig. 5(a). Simply assuming a step-like axial stress distribution along the embedded rebar, the axial stress distribution measured with 5 mm in a spatial resolution can be represented by the red line in the bottom figure of Fig. 5(a). In this case, unchanged flat variation can be observed every 20 mm, which brings a decrease in the bond stresses at corresponding positions. This simple model can approximately demonstrate the bond stress distribution shown in Fig 2(b). On the other hand, if the stress distribution is measured with 10 mm in a spatial resolution starting from X=5 mm, the measurement positions cannot be overlapped with the rib positions at all (see (ii) in Fig. 5). In this case, the axial stress distribution fluctuates every 10 mm, showing similar trend to the case (i) in Fig. 5(a) but lower resolution, resulting in broader peaks in the bond stress distribution as shown in Fig. 3(b). If starting measurement from X=0 mm with 10 mm in a spatial resolution, on the other hand, the measurement positions can be overlapped with another ribs every 20 mm, as shown in (iii) in Fig. 5. In this case, the axial stress distribution behaves similar trend to the case (ii) in Fig. 5(a), but seems to be shifted by 5 mm. Therefore, the bond stress distribution in Fig. 4(b) shows slightly different from that in Fig. 3(b). Considering above mechanisms, neutron diffraction can investigate local bond resistance taking advantage of higher spatial resolution than the strain gauge method. According to Fig. 5(b), the horizontal bearing force, F_H on the rib face is estimated to be smaller than the friction force, f_H on a straight part of rebar because of higher axial restriction of deformation around the transverse rib by the surrounding concrete. Therefore, lower and higher bond stresses in Fig. 2(b) are considered to be derived from the friction force, f_H and the horizontal bearing force, F_H, respectively.

Summary

In the present study, the capability of the neutron diffraction technique for the structural engineering studies on the reinforced concrete structure was discussed based on the bond stress evaluation between rebar and concrete. Several peaks appear in the bond stress distribution measured by neutron diffraction, showing the inhomogeneous bond variation along the embedded rebar. The horizontal bearing force on the rib face might increase the bond resistance around the transverse ribs, increasing

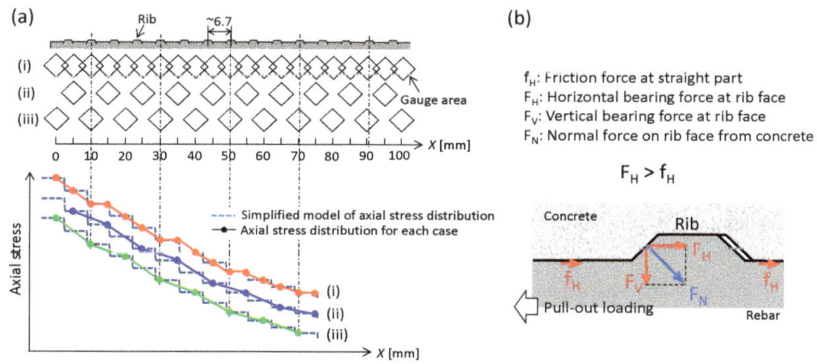

Fig. 5 (a) Schematic relationship between transverse rib and measurement positions and schamtic diagram of the axial stress distribution estimated for each spatial resolution. (b) Schamtic illustration of force working around a transverse rib.

the bond stresses locally. On the other hand, lower bond stresses are considered to be derived from weaker friction force acting at the straight part of rebar between transverse ribs. This result suggests that neutron diffraction technique with high spatial resolution can investigate local bond resistance caused by the transverse ribs. The bond stress distribution measured by the neutron diffraction technique is expected to bring detailed understanding of the bond mechanism between rebar and concrete for the reinforced concrete structures.

The absorption coefficient of concrete is approximately 0.5 cm^{-1} which is about a half of iron one [5]. For the future, to use deuterated concrete would decrease the transmission and allow higher spatial resolution and allow us to see more detailed cyclic variation associated with the ribs.

The neutron diffraction experiments were performed using the TAKUMI engineering diffractometer at the J-PARC/MLF with the approval of the J-PARC Center as Proposal No. 2012B0058.

References

[1] Y. Goto, Cracks formed in concrete around deformed tension bars, ACI Journal, 68 (1971) 244-251.

[2] A. H. Nilson, Internal measurement of bond slip, J. Am. Concr. Inst., 69 (1979) 439-441.

[3] R. Tepfers, Cracking of concrete cover along anchored deformed reinforcing bars Magazine of Concrete Research 31 (1979) 3-12.
http://dx.doi.org/10.1680/macr.1979.31.106.3

[4] M. N. S. Hadi, Bond of high strength concrete with high strength reinforcing steel, The Open Civil Engineering Journal, 2 (2008) 143-147.
http://dx.doi.org/10.2174/1874149500802010143

[5] H. Suzuki, M. Kanematsu, K. Kusunoki, Neutron diffraction studies on strain evaluation of rebar in reinforced concrete, Powder Diffraction, 24, Issue S1 (2009) S68-S71.

[6] K. Kusunoki, K. Kabayama, T. Mukai, Y. Hatanaka, H. Suzuki, A. Tasai, Experimental study on bonding action with neutron diffraction technique, 8CUEE CONFERENCE PROCEEDINGS, (2011) 701-705.

[7] H. Suzuki, K. Kusunoki, Y. Hatanaka, T. Mukai, A. Tasai, M. Kanematsu, K. Kabayama, S. Harjo, Measuring strain and stress distributions along rebar embedded in concrete using time-of-flight neutron diffraction, Measurement Science and Technology, 25 (2014) 025602.
http://dx.doi.org/10.1088/0957-0233/25/2/025602

[8] H. Suzuki, K. Kusunoki, M. Kanematsu, A. Tasai, Y. Hatanaka, N. Tsuchiya, S.C. Bae, S. Shiroishi, S. Sakurai, T. Kawasaki, S. Harjo, Application of neutron stress measurement to reinforced concrete structure, JPS Conference Proceedings, 8 (2015) 031006.
http://dx.doi.org/10.7566/jpscp.8.031006

[9] F. I. Faiyadh, Bond characteristics of oil saturated concrete, Int. J. Cement Composites and Lightweight Concrete, 7 (1985) 115-131.
http://dx.doi.org/10.1016/0262-5075(85)90067-3

[10] S. W. Lee, K. H. Tan, E. H. Yang, Bond-slip response of reinforcing bars embedded in high performance fiber reinforced cement composites, Int. J. Civil Environmental, Structural, Construction and Architectural Engineering, 9 (2015) 626-631.

[11] S. Harjo, T. Ito, K. Aizawa, H. Arima, J. Abe, A. Moriai, T. Iwahashi, T. Kamiyama, Current status of engineering materials diffractometer at J-PARC, Materials Science Forum, 681 (2011) 443-448.
http://dx.doi.org/10.4028/www.scientific.net/MSF.681.443

[12] R. Oishi, M. Yonemura, Y. Nishimaki, S. Torii, A. Hoshikawa, T. Ishigaki, T. Morishima, K. Mori, T. Kamiyama, Rietveld analysis software for J-PARC, Nuclear Instruments and Methods, A **600** (2009) 94–96.

Residual Stresses 2016: ICRS-10 Materials Research Forum LLC
Materials Research Proceedings **2** (2016) 31-36 doi: http://dx.doi.org/10.21741/9781945291173-6

Stress Analysis of the Bi-Metallic Coins – a Potential Shrink Fit Ring & Plug Standard

Scott Olsen[1, a *], Vladimir Luzin[1, b]

[1] Australian Nuclear Science and Technology Organisation, Locked Bag 2001, Kirrawee DC, NSW, 2232 Australia

[a] sol@ansto.gov.au, [b] vll@ansto.gov.au

Keywords: Residual Stress, Neutron Diffraction, Bi-Metallic Coin

Abstract. The "Shrink-fit ring and plug" system is known in the residual stress neutron community mostly due to the VAMAS Round Robin specimens, that were measured by most of the neutron residual stress instruments worldwide. This standard, however, can be challenging for some low flux instruments and considerable beamtime and efforts are required to accomplish measurements accordingly to the measurement protocol. Theoretically, the residual stress distribution is not simple, being neither plain stress nor plain strain, and essentially it is a 3D distribution with a large gradient toward the flat cut, especially for the axial component. Another "shrink-fit ring-and-plug" system is being considered, namely bi-metallic coins. With an easier zero plane stress state, they represent another potential candidate for a standard. Bi-metallic coins are in current circulation in many countries of the world. In the given study we report on an assessment of the residual stress state of 7 different bi-metallic coins measured by means of neutron diffraction to reconstruct the full stress state. The magnitudes of the stresses in the specimens were different obviously due to differences in the coinage process and materials in use. While in some cases residual stresses are weak and therefore difficult to measure accurately, in some cases stresses reach ~100 MPa. Although question in variability of the coinage process through years and within series is still debatable, tight standards and tolerances of the mint industry suggest the probability of consistency in the residual stress state making bi-metallic coin an interesting alternative to the VAMAS ring-and plug standard..

Introduction

The so-called VAMAS shrink-fit "ring-and-plug" sample [1] has been used over the last decade as a standard round-robin sample by the neutron community for the purpose of standardization and instrument calibration. The idea was so successful that all existing, newly built, or upgraded stress scanners made measurements of this sample, usually during commissioning. Collected data is commonly among the very first results, as in the case of residual stress diffractometer KOWARI at ANSTO [2]. Despite this success, there were few drawbacks with this approach:

- There were only two samples made and although they can become available within reasonable time, the associated logistics and mailing samples around the world without risk of losing them is problematic.
- Although aluminium barely activates during a normal neutron diffraction experiment (transmission tomography can be different!) moving samples that have been irradiated across borders can be problematic.
- The samples were produced in one single and unrepeated process, there is a seriously challenging problem if someone wishes to reproduce them exactly; there were some unsuccessful attempts in the past to make copies.

There are also problems of a scientific nature and related to the experimental process:

- The samples are made of aluminium alloy AA7050. Although this has the advantage of providing high penetration and low stiffness, both good for neutron strain measurements, aluminium is a weak scatterer, therefore for some residual stress diffractometers with moderate performance it is a challenging task to carry out the prescribed measurement programme in full. As a consequence incomplete data sets were collected on some instruments, while on other beamlines spatial resolution or accuracy was sacrificed to have measurements done within reasonable experimental time.

- Another drawback of AA7050 alloy is that it is anisotropic and special efforts must be made to keep it under control and separate measurements of a specially made d_0 sample (a stand-alone plug sample) should be performed slowing down the overall experiment.
- With overall dimensions 50 mm diameter and 50 mm height, the residual stress state is not simple, neither plane-strain state, nor plane stress state. In general, a gradient along the axis of the cylinder is present for strains and stresses. It is especially pronounced for the axial stress component which changes from a maximum value in the mid-height to exact zero at the top and bottom surfaces. Thus, results of the measurements, in principle, depend on the experimental setup and how measurements were executed, e.g. on the exact shape and size of the gauge volume

In an attempt to overcome these drawbacks, a much simpler standard sample can be envisaged. It should be made of a better neutron scatterer, still with sufficiently low stiffness and it should have a highly symmetric shape, like the VAMAS sample, cylindrical. If the sample is to comply with the plane stress condition, then to have a 2D shrink ring-and-plug system made of copper (or steel), a few mm in thickness and 30-40 mm in diameter would be a good candidate. Such systems are available in mass-production and they are bi-metallic coins. They are produced in almost all countries around the world from similar materials in similar size, though technologies of their production are expected to be different and undisclosed.

The aim of this study is to investigate residual stress in bi-metallic coins, using several easily obtainable candidates, and to assess how reasonably they can be used as standard specimens. The issues to investigate are

- How strong are the residual stresses in the coins? It is difficult to measure weak stresses since very small strain/stress error bars should be experimentally achieved.
- Is the material very anisotropic and should a special d_0 be a part of the measuring programme?
- How uniform is the material? Since stamping of the surface image (relief) is not uniform, it can potentially induce non-uniformity in the stress state that should be avoided.
- Although there are expectations of tight standards and tolerances in the mint industry, can the consistency of minting technology and variability in the residual stress state (through years of minting and within the same year) be addressed and satisfactorily resolved in an experimental manner?

This is a first of the kind study on bi-metallic coin residual stress since no experimental results are available up to now. Beyond the particular consideration of bi-metallic coins as a stress standard, this work brings a new characterisation of these every-day objects.

Principles of bi-metal coin production (minting)

The production process starts with punching coin blanks to be the ring and core/plug. A hole is also punched through a blank to produce a ring, whose diameter is accurately sized to fit inside the ring. The blank of the plug is also treated specially by adding a groove all the way around the edge e.g. by milling. The stamping (or double stamping) finally fixes the two parts of the assembly together and the force of the stamping plunger is adjusted to provide conditions for material of the inside edge of the ring to be pressed into the groove, locking the two parts into place (Fig. 1).

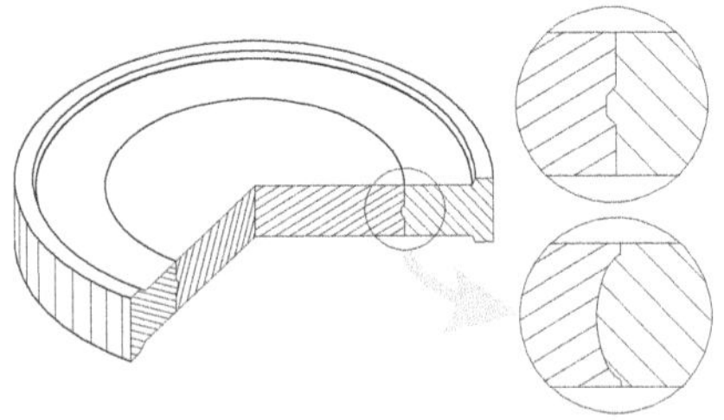

Fig. 1. Structure of a generic bi-metallic coin. A variable groove design is shown in two options.

The different materials with different elastic and plastic properties, different sizes of ring and plug, different groove designs and tolerances are

Materials Research Proceedings **2** (2016) 31-36 doi: http://dx.doi.org/10.21741/9781945291173-6

reflected in different accumulated stresses. Although the locking mechanism by groove is partially responsible for the integrity of a bi-metallic coin, the residual stress should be generally present and also play a role.

Samples and materials

A number of bi-metallic coins were selected for the current study of the residual stresses and their details are given in Table 1. All selected coins are made from copper alloy of a certain kind (with few exceptions for the ring material) and all fall into a very narrow interval of sizes: 26 – 28 mm diameter.

Experimental: principles

Due to the plane stress condition the stress state of a bi-metallic coin with outer radius b and inner

Table 1. Bi-metallic coins and their characteristics (data collected from multiple sources)

Coin	*Description*
	**UK2:** 2 pound coin (UK) Mass = 12 g; Diameter = 28.4 mm; Thickness = 2.5 mm Outer ring: Nickel-brass (76% Cu, 20% Zn and 4% Ni) Centre: Cupro-nickel (75% Cu , 25% Ni) Years of minting: 1997-present
	**CAN2:** 2 dollar coin, "toonie" (Canada) Mass = 7.30 g; Diameter = 28 mm; Thickness = 1.75 mmOuter ring: 99% Ni Inner core: aluminium bronze (92% Cu, 6% Al, 2% Ni) Years of minting: 1996–present
	**SA5:** 5 Rand coin (South Africa) Mass = 9 g; Diameter = 26 mm; Thickness = 2.8 mm Outer ring : Copper-Nickel Inner core: brass Years of minting: 2004–present
	**TW20:** 20 Yuan coin (Taiwan) Mass = 8.5 g; Diameter = 26.85 mm; Thickness = 2.15 mm Outer ring: Aluminium-bronze Inner core: copper-nickel Years of minting: 2001–present
	**MEX10:** 10 Pesos coin (Mexico) Mass = 10.329 g; Diameter = 28 mm; Thickness = 2.3 mm Outer ring Aluminium-bronze Inner core: Copper-nickel-zinc (65% Cu, 10% Ni, 25% Zn) Years of minting: 1997–present
	**Rus10:** 10 Rouble coin (Russia) Mass = 8.4 g; Diameter = 27 mm; Thickness = 2.1 mm Outer ring: Brass Inner core: Copper-nickel Years of minting: 1992–present
	**AUS5:** 5 dollar coin (Australia) Mass = 10.6 g; Diameter = 28.12 mm; Thickness = 2.6 mm Outer ring: Austenitic stainless steel Inner core : Aluminium bronze Years of minting: 1988–Present

radius a is extremely simple

$$\sigma_r^{ring}(r)=\frac{1}{(K^2-1)}\left(1-\frac{b^2}{r^2}\right),\ \sigma_h^{ring}(r)=\frac{1}{(K^2-1)}\left(1+\frac{b^2}{r^2}\right),\ \sigma_a^{ring}(r)=0; \quad (1)$$

$$\sigma_r^{plug}(r)=-p,\ \sigma_h^{plug}(r)=-p,\ \sigma_a^{plug}(r)=0;$$

(2)

where $K = b/a$ and apart from the geometric dimensions is characterized by only one parameter p, a pressure on the plug. Stress-strain relationships are also simplified to

$$\sigma_h(r)=\frac{E}{(1-\nu^2)(1-2\nu)}(\varepsilon_h+\nu\varepsilon_r);$$

$$\sigma_r(r)=\frac{E}{(1-\nu^2)(1-2\nu)}(\varepsilon_r+\nu\varepsilon_h);$$

$$\varepsilon_a=\frac{\nu}{(1-\nu)}(\varepsilon_h+\varepsilon_r).$$

(3)

In principle, for full reconstruction of the stress state, the stress measurement of the inner core is sufficient. Experimentally, the measurement programme might vary depending on assumption regarding d_0 anisotropy:

(EXP1) If a d_0 isotropy assumption is made, the experiment can be carried out non-destructively. The measurements of the central core involve detection of the two principal directions, hoop/radial and axial, from which by applying the condition of the axial component to be zero, the hoop/radial stress component and d_0 can be resolved.

(EXP2) If, however, d_0 is assumed to be anisotropic, the stresses can be resolved only in a destructive way. The experimental programme of the central core then involves detection of the two principal directions, hoop/radial and axial, each measured in loaded (intact coin) and unloaded (the central core is removed from the ring). Thus two principal elastic strains can be derived from the pair of measurement and the hoop/radial stress component can be calculated.

In both cases, due to the uniform stress state in the plug (Eq. 1, no radial dependence, only uniform pressure p), measurements with an elongated gauge volume can be used reducing overall measurement time in this way and/or improving counting statistics. With the thickness range of the coins of 1.75 to 2.6 mm, and core diameters at least 15mm, a gauge volume with dimensions $1\times1\times10$ mm^3 seems the most adequate. Thus, measurements of the two directions (in transmission and reflection geometry) are easily possible, even for moderate performance neutron stress scanners.

Experimental: neutron diffraction setup

Neutron residual stress measurements of bi-metallic coins were performed on the KOWARI neutron diffractometer at OPAL research reactor at ANSTO [2]. For all coins the Cu(311) reflection was used at 90°-geometry employing a neutron wavelength of $\lambda = 1.54$ Å. Because different alloys had slightly different lattice spacings, there was a natural variation in the exact position of the diffraction peak, but all of them they were within range of ±1°.

A gauge volume with size of $1\times1\times10$ mm^3 was consistently used since it could be positioned within each and every coin central core. Not just one, but 5 locations (exact centre, ±3 mm, ±6 mm) were systematically measured for each coin to assess uniformity and statistical variations of the material of the plug with measurement times of 3 minutes. The high flux of KOWARI, optimised for this wavelength, yielded 50 - 60 μstrain accuracy on average or ~7-8 MPa in terms of calculated stresses.

Experimental: results and data analysis

The study has been performed in several steps:

Materials Research Proceedings **2** (2016) 31-36 doi: http://dx.doi.org/10.21741/9781945291173-6

First, the attempt was made to conduct a non-destructive experiment (EXP1), i.e. the inner core part was measured in two directions, hoop/radial and axial, to assess elastic strain anisotropy. The experimental results are presented in Fig. 2. As can be seen from the diagram, three out of seven coins demonstrate tension of the core. As this is a non-physical result, but rather an artefact of the data treatment, the assumption of isotropic d_0 is not viable for the majority (if not all) of the coins.

Secondly, a full, but destructive analysis (EXP2) was performed next. The two principal directions, hoop/radial and axial, were measured in two states, intact and free, where the central core is removed from the ring. Stresses were calculated with properly measured d_0 and are shown in Fig. 3. They are quite different from those shown in Fig. 2: with d_0 taken correctly into account, all stresses are compressive, as expected.

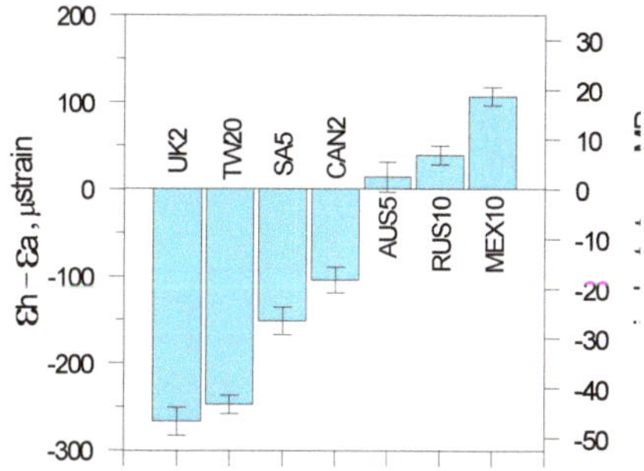

Fig. 2. Measure of the appeared anisotropy in strain and stress scale for seven bi-metal coins.

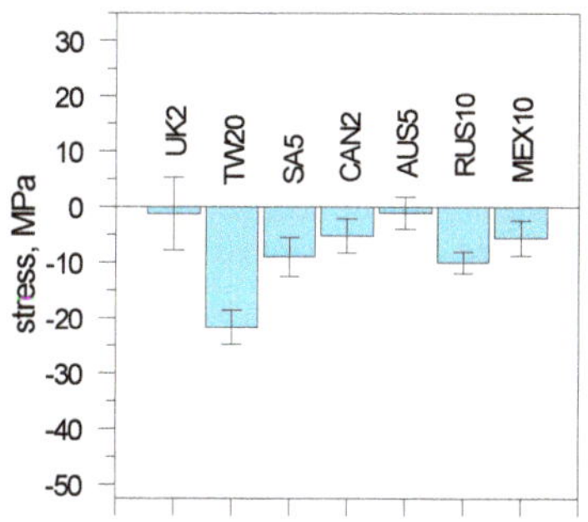

Fig. 3. Residual stress (negative pressure) for seven bi-metal coins.

Due to the significant anisotropy in coin UK2, additional measurements were made on it to assess in-plane anisotropy, uniformity and reproducibility. The dependence of the lattice d-spacing on the in-plane direction is shown in Fig. 4 (left) and indicates no in-plane anisotropy and non-uniformity, thus the cylindrical symmetry of coin is confirmed for this specimen. In comparison, the aluminium alloy AA7050 exhibits much larger degree of in-plane d0 anisotropy presented in Fig. 4 (right) inherited from the plate rolling process. To address the question of reproducibility, two UK2 coins were compared in intact conditions, from minting years 2002 and 2008, and within error bars they provided the same result. However, this coin is not the best sample to be used to check reproducibility of stress since the magnitude of stress turned out to be close to zero.

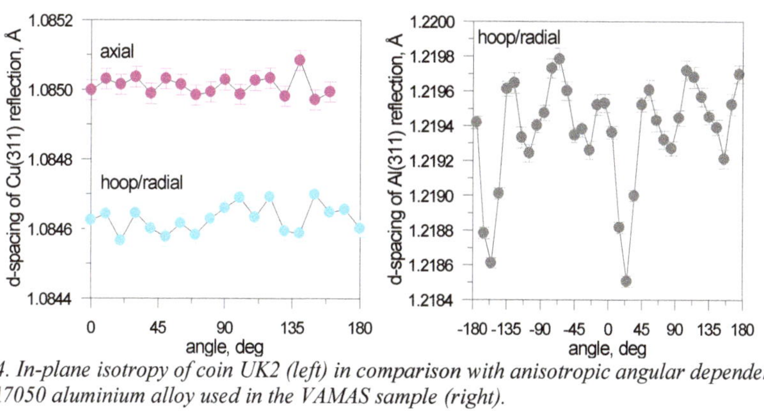

Fig. 4. In-plane isotropy of coin UK2 (left) in comparison with anisotropic angular dependence of AA7050 aluminium alloy used in the VAMAS sample (right).

Discussion

A big discrepancy between Fig. 2 and 3 clearly shows that the anisotropy in d_0 must be taken into account to obtain correct stress values and obtain physically meaningful results. Only then stress calculations result in the compressive state of the central core.

The selected set of seven coins revealed a significant variability of stresses ranging from near to zero stress in UK2 and AUS5 coins, to very low stress of ~ 5-10 MPa for coins CAN2, MEX10, SA5 and RUS10, to the highest stress of -22 ± 4 MPa for TW20 coin. Based on the results collected from this coin selection, coin TW20 can be considered the best candidate providing high stresses that can be easily measured. Considering the bi-metallic system as a whole, the stresses in the ring reach even higher values which can be demonstrated within the elastic model expressed in equations (1)-(2). The results of these calculations are given in Fig. 5.

The following considerations may suggest using the TW20 coin as a local/alternative standard:

- Stress measurements in the plug can be done extremely fast with an elongated gauge volume, e.g. 1×1×10 mm^3, achieving a good accuracy of 5-7 MPa. Only two directions are required to be measured and anisotropy of d0 is reduced to only two values. This cylindrical symmetry is more advantageous than the angular dependence of d_0 for the VAMAS sample.
- More advanced and complete measurements can be also considered as a part of the measurement programme with measurements of the ring requiring longer measurement times while using 1×1×10 mm^3 gauge volume. However, with stresses ~150 MPa, the accuracy of stress determination can be reduced to 10-15 MPa which is still within easily achievable measurement times.
- On KOWARI, the total measurement time of one coin, both central coin and outer ring, takes (approximately) one day versus one week of beamtime to measure the VAMAS sample.

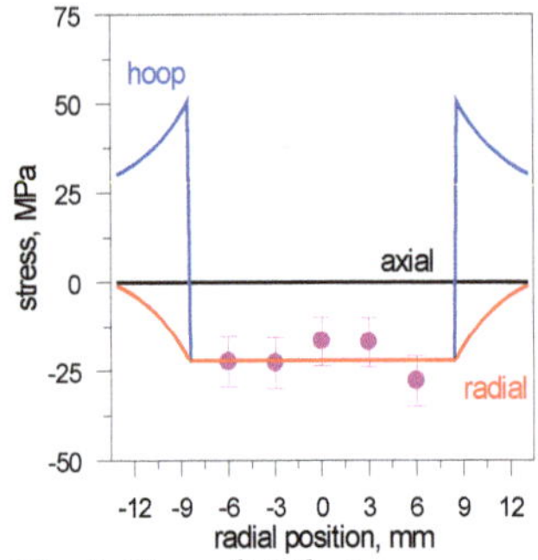

Fig. 5. Stress distribution (experimental and calculated) in coin TW20.

Overall, the material of coin TW20 (copper), its dimensions (Ø27 mm and thickness 2.1 mm), the magnitude of stress, the uniformity and in-plane isotropy of the material, makes it the best candidate for being an alternative "ring-and-plug" standard suitable for fast neutron measurements.

Summary

Several bi-metallic coins were studied using neutron diffraction and their residual stress state was determined. For the coins the stress varied between a few MPa to -22 MPa in the TW20 coin.

Due to the nature of the material and its thermo-mechanical treatment during processing, the d_0 anisotropy consistently occurs for all coins. Therefore, a ring and plug measurement, in the disassembled condition is required for a proper d_0 measurements and an overall stress analysis.

Although questions of anisotropy and uniformity and reproducibility were addressed, at least for coins of the UK2 types, a more systematic research is still required, namely for coins TW20, which so far are considered the best candidate for an alternative standard sample for fast and easy stress calibrations.

References

[1] Webster, G. A., 2000, Neutron diffraction measurements of residual stress in a shrink-fit ring and plug VAMAS Report no. 38: National Physical Laboratory.

[2] Kirstein, O., Luzin, V., and Garbe, U., 2009, The Strain-Scanning Diffractometer Kowari: Neutron News, v. 20, no. 4, p. 34-36.
http://dx.doi.org/10.1080/10448630903241175

Residual Stresses 2016: ICRS-10 Materials Research Forum LLC
Materials Research Proceedings **2** (2016) 37-42 doi: http://dx.doi.org/10.21741/9781945291173-7

Residual Stress Measurements Inside a Small Inner Diameter Hole at Low Bragg Angle Using X-Ray Diffraction Technique

M. Belassel[1,a*], J. Pineault[1,b], M. Brauss[2,c].

[1]Proto Manufacturing Limited, 2175 Solar Crescent, Oldcastle, Ontario, Canada

[2]Proto Manufacturing Inc., 12350 Universal Drive, Taylor, Michigan, 48180, USA

[a]mbelassel@protoxrd.com, [b]xrdlab@protoxrd.com, [c]mbrauss@protoxrd.com

Keywords: Residual Stress, Bolt-Hole, Inner Diameter, Low Bragg Angle, Abusive Machining

Abstract. A new X-Ray diffraction (XRD) goniometer has been specially developed to nondestructively measure residual stress (RS) on the inner diameter (ID) of small holes in metal and ceramic components. The major advantage of this novel goniometer is its ability to perform RS measurements on the ID of small holes without the need to section the component, thus maintaining the integrity of the residual stress field inherent to the component of interest. This new patented XRD goniometer has been used on a wide variety of aerospace components on features such as holes in airframe/structural components, as well as fastener/bolt holes, air holes, and confined areas on rotating and non-rotating turbine engine components. In many instances, measurements can be performed at 2θ angles much lower than the widely accepted lower limit of approximately 130°. The selection of the actual Bragg angle used depends upon the material itself, the hole ID, and the thickness of the component. Results obtained indicate that the new goniometer can be used to measure RS on the ID of small holes with good accuracy and precision even at relatively low 2θ angles in the vicinity of 100°.

Introduction

Since the discovery of RS and its measurement early in the last century, measurement techniques have been evolving steadily [1]. Many techniques have been developed thanks to industry demand and specific needs [2]. Many RS measurement techniques exist and each generally requires a different experimental setup; laboratory equipment is often suitable for small parts, whereas portable systems are generally required for oversized parts. Moreover, the geometry of the XRD goniometer may vary to facilitate access to the measurement locations of interest on a component. Both Psi and Omega modes have been widely implemented on most current XRD RS measurement instruments. Omega mode is suitable for tight geometries such as grooves as may be found in the root of a gear tooth when measuring along the groove (i.e. parallel to the teeth), whereas Psi mode in turn is more adapted to measurements in the direction normal (perpendicular) to the groove (such as the opening of the groove permits). Each geometry has its advantages and disadvantages and each may be complementary in certain instances.

In this study, RS analysis will be extended to measurements on the ID of bolt holes in the hoop direction. Such measurements are applicable for both automotive and aerospace materials; this study includes examples on a Ni-Base super-alloy. For these applications, a new XRD instrument and goniometer were designed and fabricated that operate in Psi mode [3].

Materials Research Proceedings **2** (2016) 37-42 doi: http://dx.doi.org/10.21741/9781945291173-7

Stress Measurement Technique

Basic Principles. The x-ray diffraction technique uses the distance between crystallographic planes, i.e. the d-spacing, as a strain gage in a crystalline material. When the material is in tension, the d-spacing increases and when the material is in compression, the d-spacing decreases. The presence of residual stresses in a material produces a shift in the x-ray diffraction peak angular position that is directly measured by the detector [3]. The diffraction angle 2θ is measured experimentally and the d-spacing is then calculated using Bragg's law:

$$n\lambda = 2d\sin\theta \tag{1}$$

Where: λ is the wavelength of the radiation.

Diffraction peaks collected should have a sufficient peak/background ratio and a limited noise level. These will depend on the *{hkl}* plane selected, the collection time, the size of the irradiated area and the material microstructure. Once the d-spacing is measured for unstressed d_0 and stressed d conditions, the strain is calculated using either of the following relationships:

$$\varepsilon = \frac{(d-d_0)}{d_0} \quad \text{or} \quad \varepsilon = -\frac{1}{2}\cot(\theta)\cdot\Delta(2\theta) \tag{2}$$

To improve measurement accuracy when determining Δ(2θ), the θ angle selected should be close to 90° (i.e. 2θ ~ 180°). The widely published and agreed upon low limit for 2θ is 130°; using 2θ values below this will diminish Δ(2θ) resolution and as such will result in a loss in RS measurement accuracy and a concomitant increase in experimental error. The plot in Figure 1 demonstrates the relationship between Δ(2θ) for various 2θ values. As 2θ deceases so does the slope, indicating a reduced resolution in peak separation.

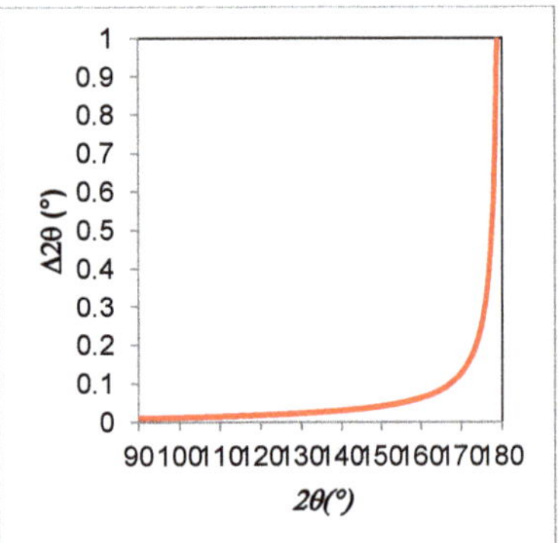

Figure 1: Variation in Δ(2θ) resolution with diffraction angle 2θ.

Sin²ψ Method. For the sin²ψ method, where a number of d-spacings are measured, stresses are calculated from an equation derived from Hooke's law for isotropic, homogeneous, fine grain materials, ψ is the angle subtended by the bisector of the incident and diffracted beam and the surface normal, and ϕ is the direction of stress measurement [4].

$$\frac{d_{\phi\psi}-d_0}{d_0} = \tfrac{1}{2}S_2^{hkl}\left(\sigma_\phi - \sigma_{33}\right)\sin^2\psi + \tfrac{1}{2}S_2^{hkl}\sigma_{33} - S^{hkl}{}_1\left(\sigma_{11}+\sigma_{22}+\sigma_{33}\right) + \tfrac{1}{2}S_2^{hkl}\tau_\phi \sin 2\psi \tag{3}$$

Where: $\varepsilon_{\phi\psi}$ is the strain at a given ψ, ϕ tilts, ½ S_2^{hkl} and S_1^{hkl} are the x-ray elastic constants of the material.

Materials and Measurement Conditions

The material of interest for this study is a Ni-base super-alloy. Three examples of surface conditions that can be found on the ID of machined holes were investigated as follows:

1. Hole drilled with no coolant.
2. Hole drilled with coolant.
3. Drilled, honed and shot peened hole.

Residual stress versus depth profiles were collected on the face of each of the machined holes using conventional XRD instrumentation and on the ID of the same holes using a new XRD instrument. The measurement conditions were as follows: 1) for the conventional instrument operating in Omega mode, the {311} diffraction plane was selected for measurements at 2θ=151.9° using $Mn_{k\alpha}$ radiation (λ=2.1034 Angstroms), and 2) for the new instrument operating in Psi mode, the {311}diffraction plane was selected for measurements at 2θ=111.3° using $Co_{k\alpha}$ radiation (λ=1.7902 Angstroms). Selection of the diffraction angle 2θ to be employed using the new instrument is predicated upon the diameter of the hole (*D*), and the thickness of the component (*W*). A geometric relationship can thus be established as seen in Eq. 4 and in Figure 2:

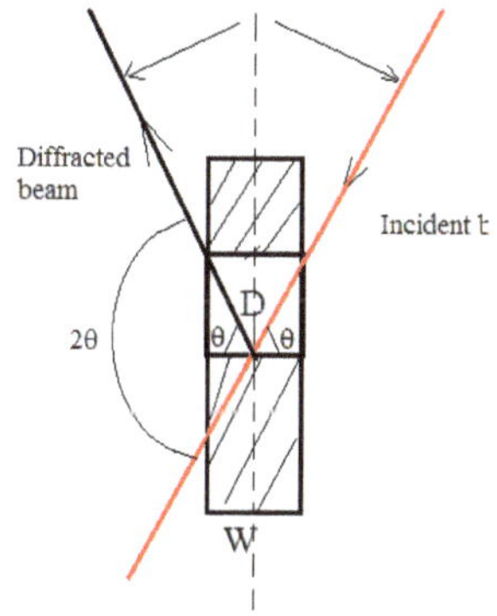

Figure 2: Determining optimal 2θ for use on the ID of a hole using new Psi mode instrument.

$$\tan(\theta_{max}) = \frac{2D}{W} \tag{4}$$

For Ni-base super-alloys where the actual collimator and the detector size are taken into account, the optimal 2θ to use is 111.3°. In this case, the ratio is: $\frac{2D}{W} = 4$, $2\theta_{max}$ = 152°,which is widely sufficient for this experiment. For other materials, the ratio may vary according to the Bragg conditions available for use and the component geometry.

In Figure 3, the RS on a Ni-base super-alloy turbine engine disk is being measured using: a) a conventional Omega mode instrument on the face of the hole, and b) using the new instrument in Psi mode on the ID of the same hole.

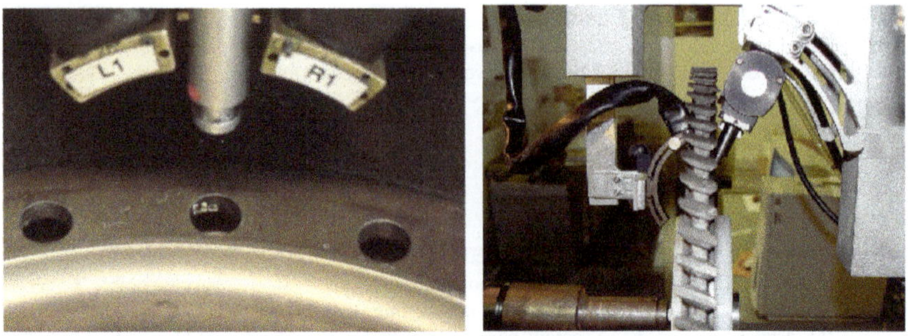

Figure 3: Measuring RS a) on the face of machined holes using a conventional XRD instrument and b) on the ID of machined holes using the new XRD instrument

Results and Discussion

Prior to performing RS measurements, the alignment of the instrument was verified using a stress free Ni powder as per ASTM E915 [5], as well as a known high stress proficiency standard.

Microstructure characterization of the different processes revealed the following:

1. The hole drilled with no coolant exhibited severely distorted grains and undesirable microstructure near the hole ID surface.
2. The hole drilled with coolant exhibited only slightly distorted grains near the hole ID surface.
3. The drilled, honed and shot peened hole showed no signs of abusively machined material.

(a) **(b)** **(c)**

Figure 4 – Hole cross sections (ID to the left) of holes machined as follows: a) drilled with no coolant, b) drilled with coolant, and c) drilled, honed and shot peened.

Materials Research Proceedings **2** (2016) 37-42 doi: http://dx.doi.org/10.21741/9781945291173-7

The RS vs. depth results obtained on both the face of the hole, and the ID of the hole, can be seen in Figure 5. The results obtained on the ID of the machined holes were found to be much more effective for detecting the relatively thin tensile stressed layers inherent to the various machining and cold working processes applied. In the case of the hole drilled with coolant, RS that appeared compressive on the face of the disk, was found to be tensile on the hole ID. In the case of the hole drilled without coolant, the magnitude of the tensile RS found on the ID was much greater when compared to tensile RS found on the face of the disk. Thus, the very steep stress gradients produced by machining demand nondestructive measurement of RS on the ID.

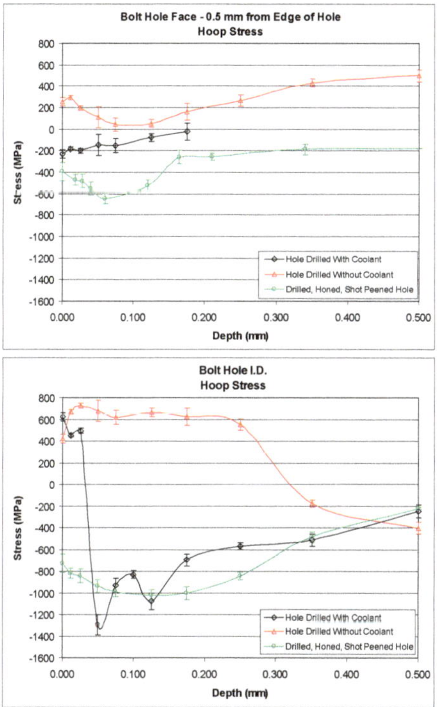

a) b)

Figure 5: Stress vs. depth profiles a) on the face of machined holes using a conventional XRD instrument, and b) on the ID of machined holes using a new XRD instrument.

Clearly, the new XRD instrument adapted for measurements on the ID surface was more effective in revealing the actual stress gradients present i.e., it was more effective in revealing the effect of each machining and cold working processes applied as well as the associated depth, sign and magnitude of undesirable tensile RS [6]. This ability is critical for the aerospace industry where accurate RS measurements are needed to help improve and optimize the machining and cold working processes applied to life limiting components so as to insure a long service life of such components.

Materials Research Proceedings **2** (2016) 37-42 doi: http://dx.doi.org/10.21741/9781945291173-7

The experimental errors associated with the 2θ angles selected should be higher for low 2θ and vice versa [1]. The magnitude of the experimental errors obtained in this study was deemed good given the magnitude of the RS present and the associated RS gradients observed [7]. The magnitude of the experimental errors were found to be dependent upon the RS levels measured, as well as the material condition and microstructural effects present in the hole that were investigated. A detailed methodology to successfully measure RS at a suitable Bragg angle (2θ) was developed prior to applying it to real components. This methodology also employs a statistical approach where each measurement is reproduced to guarantee its accuracy.

Summary

New goniometer head geometry has been developed and successfully applied to a new instrument that can be used to measure RS on the ID of small holes nondestructively. Such measurements were heretofore impossible to execute without sectioning the component. The application of this technique on real components presented many challenges that were addressed using a new, high accuracy instrument (and associated methodology) to reliably measure RS on the ID of small holes. Moreover, the new instrument provided RS results that correctly characterized the actual RS state present on the disk hole ID where a conventional instrument failed. This new instrument thus opens the way to the accurate evaluation of RS on the ID of small holes which is of great interest to both industry in general, and to the aerospace industry in particular.

References

[1] V. Hauk, “Structural and Residual Stress Analysis by Nondestructive Methods”, Elsevier, Amsterdam, 1997.

[2] J.A. Pineault, M. Belassel, M.E. Brauss, “X-Ray Diffraction Residual Stress Measurement in Failure Analysis”, Failure Analysis and Prevention, Vol. 11, ASM Metals Handbook, American Society for Metals, 2002, p. 484-497.

[3] M.Brauss, Proto Patent No. US 7,283,612 B2, issued October 16, 2007 for X-Ray Diffraction Method.

[4] Noyan I.C. and Cohen J.B., ‘Residual Stress’, Springer Verlag, 1987. *http://dx.doi.org/10.1007/978-1-4613-9570-6*

[5] ASTM E915-10, “Standard test method for verifying the alignment of x-ray diffraction instrumentation for residual stress measurement”.

[6] M. Belassel et al., “Effect of Thermal and Mechanical Loading on the Generation of Macro and Micro Stresses in Eutectoid Steel”, Proc. 4th Int. Conf. Residual Stresses, ICRS 4. Soc. Exp. Mechanics, 1994, 392-401.

[7] J. Lu et al. “Handbook of Measurement of Residual Stresses”, SEM, Published by the Fairmont Press, Inc., 1996.

Materials Research Proceedings **2** (2016) 43-48 doi: http://dx.doi.org/10.21741/9781945291173-8

Comparison of Residual Stress Measurement Techniques and Implementation Using X-Ray Diffraction

M. Belassel[1,a*], J. Pineault[1,b], N. Caratanasov[1], M. Brauss[2,c]

[1]Proto Manufacturing Limited, 2175 Solar Crescent, Oldcastle, Ontario, Canada

[2]Proto Manufacturing Inc., 12350 Universal Drive, Taylor, Michigan, 48180, USA

[a]mbelassel@protoxrd.com, [b]xrdlab@protoxrd.com, [c]mbrauss@protoxrd.com

Keywords: Residual Stress, X-Ray Diffraction, Measurement Technique, Geometry

Abstract. Regardless of the particular residual stress (RS) measurement technique being used, all are based on the same basic principles when using x-ray diffraction (XRD). Every technique has both its advantages and disadvantages, many of which are well known to engineers and scientists however, some of the important "finer points" are unfortunately not widely discussed or known by those not well versed in the subject. This paper will try to bring to light many of these commonly misunderstood issues by comparing the different techniques and attempt to illuminate the associated problems a user may encounter when measurements become challenging i.e. when RS measurements are to be performed in tight grooves or on textured materials for example. In this study, different techniques including the: Cosα technique, MET (used in Psi, Omega, or Modified Psi mode) have been evaluated and tested on a variety of materials and geometries.

Introduction

RS measurements using XRD techniques were first performed as early as 1925 [1]. Since then, measurement techniques and equipment slowly evolved in the following decades and major improvements were realized in the 1970s and the 1980s. Many RS measurement techniques based on XRD have been developed and implemented on a variety of instruments that are widely used in industry and academia today. As the use of XRD became more widespread in the 1950s, a technique called the Single Exposure Technique (SET) was commonly used however, it had very limited accuracy and did not work well on materials that did not possess a near random grain orientation distribution. Furthermore, the SET used only 1 data point, and relied on precise knowledge of the unstressed lattice spacing which was often difficult to determine experimentally with sufficient accuracy on the actual components or samples under investigation. As such, the quality of the data was very difficult to assess because non-linear relationships cannot be observed with only 2 data points. Due to the inherent limitations of the SET, scientists and engineers developed a new technique called the Multiple Exposure Technique (MET) which uses many more data points resulting in high confidence RS measurement results that do not rely on knowing the unstressed lattice spacing a priori [2]. Since then, many related standards have been developed, accepted and applied to the measurement of RS via XRD worldwide.

In this study, different techniques including the: Cosα technique, MET (used in Psi, Omega, or Modified Psi mode) will be evaluated and tested on a variety of materials and geometries. This paper will emphasize the different findings and the limitations associated with each of the techniques evaluated.

Principles

Generalized technique: these principles apply to MET when used in either Omega or Psi mode. The XRD technique uses the distance between crystallographic planes, i.e. d-spacing, as a strain gage and can only be applied to crystalline, polycrystalline and semi-crystalline materials [2]. When the material is in tension, the d-spacing increases and when the material is in compression, the d-spacing

Materials Research Proceedings **2** (2016) 43-48 doi: http://dx.doi.org/10.21741/9781945291173-8

decreases. The presence of RS in the material produces a shift in the XRD peak angular position that is directly measured by the detector [3]. For a known x-ray wavelength λ and n equal to unity, the diffraction angle 2θ is measured experimentally and the d-spacing is then calculated using Bragg's law:

$$n\lambda = 2d\sin\theta \tag{1}$$

Where λ is the wavelength of the radiation. Once the d-spacing is measured for unstressed (d_0) and stressed (d) conditions, the strain is calculated using the following relationship:

$$\varepsilon = (d - d_0)/d_0 \tag{2}$$

For the $\sin^2\psi$ method where a number of d-spacings are measured, stresses are calculated from an equation derived from Hooke's law for isotropic, homogeneous, fine grain materials:

$$\varepsilon_{\phi\psi} = \frac{1}{2}S_2(\sigma_\phi - \sigma_{33})\sin^2\psi + \frac{1}{2}S_2\sigma_{33} - S_1(\sigma_{11} + \sigma_{22} + \sigma_{33}) + \frac{1}{2}S_2\tau_\phi \sin 2\psi \tag{3}$$

Where, ½ S_2 and S_1 are the x-ray elastic constants of the material, σ_ϕ is the stress in the direction of the measurement ϕ, ψ is the angle subtended by the bisector of the incident and diffracted x-ray beam and the surface normal, and $\varepsilon_{\phi\psi}$ is the crystallographic strain at a given ψ tilt, see figure 1.

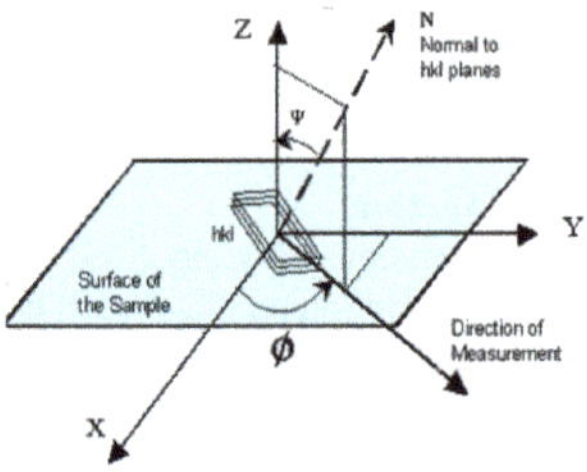

Figure 1: Definition of the axis and the direction of measurement.

The main difference between the Omega and Psi modes can be seen in the defocusing effects observed at high ψ angles. RS measurements performed in Psi mode are relatively insensitive to defocusing errors when compared to Omega mode. The increased defocusing effect and associated errors observed in Omega geometry can be mitigated by employing two detectors which can be used to minimize the magnitude of negative ψ angles employed in a RS measurement and thus minimize defocusing errors. At the same time, care must be taken in the selection of the incident x-ray beam size to limit both defocusing and beam divergence errors [2]. Moreover, Psi mode offers a constant absorption of x-rays for all ψ tilts due to the constant x-ray path whereas, in Omega mode the absorption varies with the ψ tilt angle and requires a correction. The depth of penetration is very similar for both modes except at high ψ tilt angles [4].

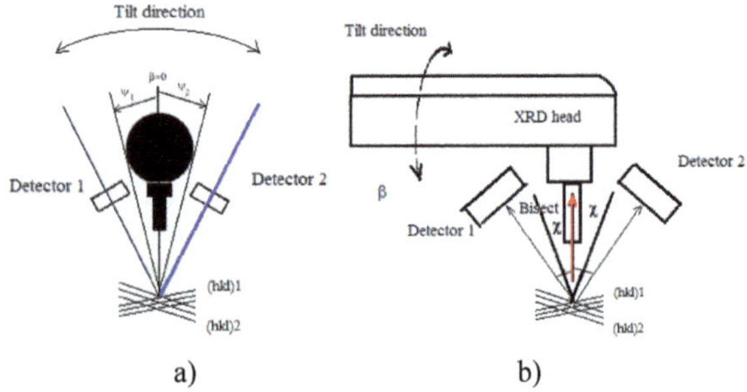

Figure 2: Goniometer geometries: a) Omega mode and b) Modified Psi mode.

Modified Psi. When working in Modified Psi mode which is a special geometry set-up convenient for vertical beam direction, the x-ray beam is perpendicular to the surface of the sample and the diffracting {hkl} planes lie in a plane with an angular offset of (180-2θ)/2 and the ψ angles for each plane vary with the inclination, see figure 2. In this case, for each normal to the {hkl} plane, the ϕ and ψ are a function of the goniometer tilt angle as shown in Eq. 4. The relationship between the β, χ, ψ, and ϕ angles are shown in Eq. 4 and Eq. 5.

$$\cos\psi = \cos\beta.\cos\chi \tag{4}$$

$$\cos\phi = \frac{\text{cin}\beta.\cos\chi}{\sin\psi} \tag{5}$$

Where β is the tilt angle and χ is the offset angle between the incident beam and the bisector of incident and diffracted beams. The derived Eq. 6 is shown as follows:

$$\begin{aligned}\varepsilon &= \frac{1}{2}S_2\left(\sigma_{11}\sin^2\beta\cos^2\chi + \sigma_{22}\sin^2\chi\right)\\ &+\frac{1}{2}S_2(\tau_{12}\sin\beta\sin 2\chi + \tau_{13}\sin 2\beta\cos^2\chi + \tau_{23}\cos\beta\sin 2\chi) + C\end{aligned} \tag{6}$$

The technique does not provide a direct measurmeent of the shear stress which is necessary when dealing with curved surfaces and non uniform processes applied to heterogenious meaterials. This techniques also requires two detectors at opposite ϕ angles [5]. The most accurate measurement for this configuration is a triaxial measurement where tilt angles ψ, ϕ can be correctly used as in the $\text{Sin}^2\psi$ method [6].

Cos(*α*) method. This method uses the diffraction ring (also called the Debye Ring) where the position of the peak at each angle α around the ring is measured [7]. This special set-up which doesn't required any change in the incident angle. Using the relationship between the derived strains in Eq. 7 and Eq. 8, one can plot the strain as a function of $\cos\alpha$ to calculate the normal stress σ_{11} and as a function of sinα to calculate the shear stress σ_{12}.

The constitutive equations are as follows:

$$\varepsilon_{\alpha 1} = \frac{1}{2}[(\varepsilon_\alpha - \varepsilon_{\pi+\alpha}) + (\varepsilon_{-\alpha} - \varepsilon_{\pi-\alpha})] = \frac{1}{2}S_2.\sigma_{11}\left[\sin^2(\psi_0 - \eta) - \sin^2(\psi_0 + \eta)\right]\cos\alpha \tag{7}$$

$$\varepsilon_{\alpha 2} = \frac{1}{2}[(\varepsilon_\alpha - \varepsilon_{\pi+\alpha}) - (\varepsilon_{-\alpha} - \varepsilon_{\pi-\alpha})] = \frac{1}{2}S_2.2\tau_{12}.(\sin 2\eta\sin\psi_0)\sin\alpha \tag{8}$$

Where ψ_0 is the inclination of the sample relative to the goniometer, 2η is the angle subtended by the incident beam and diffracted beam, and ε_α, $\varepsilon_{\pi+\alpha}$, $\varepsilon_{-\alpha}$ and $\varepsilon_{\pi-\alpha}$ are the measured strains from sectors of the ring as shown on figure 3.

When using Eq. 7 and Eq. 8, it is necessary that symmetrically opposite peaks are available on the Debye Ring. It should be noted that $\psi_0 \neq 0°$ when using this technique. It is recommended to tilt the head in such a way that one side of the Debye Ring represents ψ angles close to 0°, and the other side of the ring

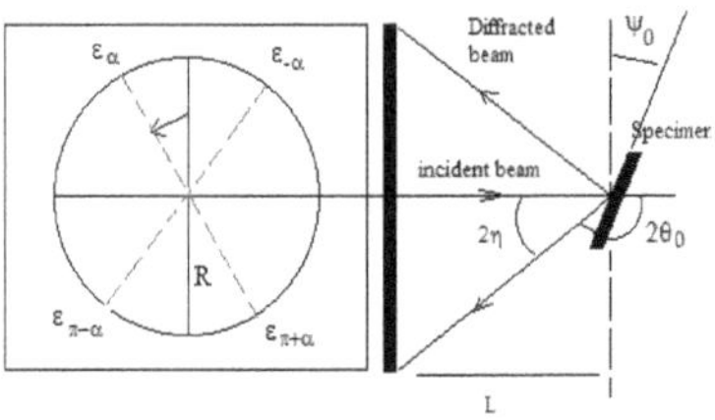

Figure 3: Principle of Cos α method using Debye Ring

Materials Research Proceedings **2** (2016) 43-48 doi: http://dx.doi.org/10.21741/9781945291173-8

represents ψ angles as high as possible to better represent the strains present. The shear stress σ_{12} provided by this technique is not of interest at this moment.

Experiments

Three samples were selected for RS measurement on the surface employing all of the aforementioned techniques. The samples were as follows: 1) a shot peened flat steel sample with a grooves that were flat at the bottom with nominal stress of -450±35 MPa, 2) an aluminum plate with low stress that exhibits some preferred orientation, and 3) a flat shot peened Ni-base alloy sample with a nominal stress of -1040±50 MPa. No measurements were performed with Psi and Modified Psi modes on this sample. The error reported is only 1Standard deviation.

In the case of the steel sample, RS measurements were collected in both the axial direction parallel to the grooves, and in the direction transverse to the grooves. The groove dimensions were as follows: Groove 1 (GR1): 2 mm wide by 1.5 mm deep, and Groove 2 (GR2): 2 mm wide by 2 mm deep. The most important parameters that determine the capability of any measurement technique and associated goniometer hardware to successfully measure RS in confined areas such as grooves are: 1) the angular opening between the incident beam and diffracted beam, and 2) the direction of ψ tilting associated with the RS measurement. RS measurements were performed employing 4 different goniometers thus using all available geometries.

All of the RS measurements on the steel and aluminum samples were performed using Cr kα (λ=2.291Å) x-radiation diffracting from the {211} plane at 156° 2θ, and the {222} plane at 157° 2θ respectively. RS measurements on the Ni-base alloy sample were performed using Cr_kβ (λ=2.08487Å) x-radiation diffracting from the {311}plane at 149° 2θ when using the Cos(α) technique, and Mn_kα (λ=2.10314Å) x-radiation diffracting from the {311}plane at 152° 2θ when using the other techniques. Table 1 lists the available angular tilt range ψ for each technique using the diffraction conditions reported above. The aperture size used in the experiment was 1 mm except the cos a where the aperture was 2 mm. A mask is applied on the top around the groves to avoid any diffraction from those surfaces.

Table 1: Goniometer capability and ranking for GR1:W:2 mm x D:1.5 mm, GR2:W:2 mm x D:2 mm

Goniometer Geometry/Mode	Technique	Angular opening (°)	Angle GR1 @ψ=0 (°)	Angle GR2 @ψ=0 (°)	Selection (Groove)
PSI	$Sin^2\psi$	24	33.7	26.5	Transverse
Omega	$Sin^2\psi$	48	33.7	26.5	Axial
Modified PSI	$Sin^2\beta$	48	33.7	26.5	Transverse
Debye Ring	cosα	48	33.7	26.5	None

As such, when measuring RS in the bottom of grooves it is desirable to have a goniometer (or 2 goniometers) that work in both Psi and Omega modes so as to enable RS measurements in two directions.

Results and Discussions

RS measurements were performed employing the 4 different goniometers. The results reported in table 2 indicate that the RS measurements in the axial direction were performed relatively easily in GR1 using all of the techniques. RS measurements performed in the transverse direction in GR1 were possible using Psi and Modified Psi with reduced accuracy due to the limited ψ tilt range available. RS measurements in the transverse direction of GR1 were not possible using Cosα and Omega because of the very limited ψ tilt range available in these geometries. Any result not within 2 standard deviation for each material is considered failed.

Materials Research Proceedings **2** (2016) 43-48 doi: http://dx.doi.org/10.21741/9781945291173-8

Table 2: RS measurement results on different head goniometer geometries and techniques.

Goniometer Geometry/Mode	Technique	Direction	Steel sample GR1 (-470±70 MPa)	Steel sample GR2 (-450±70 MPa)	Al sample Low stress <50MPa	Ni-Base alloy sample (-1040±100MPa)
PSI	$Sin^2\psi$	Axial	-524±19	-228±36*	+28±8	N/A
		Trans	-526±15*	-204±26[F]	+1±10	
OMEGA	$Sin^2\psi$	Axial	-435±8	-475±5	+12±13	-1070±50
		Trans	F	F	-21±6	
Modified PSI	$Sin^2\beta$	Axial	-443±5	-596±11 [F]	+18±12	N/A
		Trans	-386±19*	-188±46 [F]	+13±8	
Debye Ring (2D detector)	$Cos\alpha$	Axial	-485±42	-642±234 [F]	-29±31	-719±50 [F]
		Trans	F	F	-86±10[F]	

Note: * indicates results were corrected after removing bad data points, F indicates a failed measurement
N/A indicates data not available

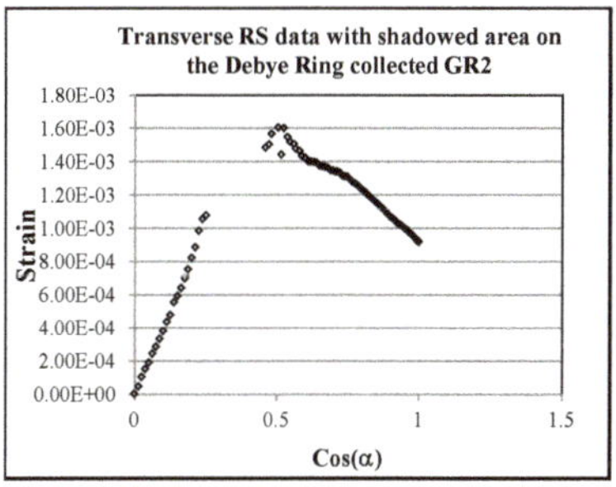

Figure 4: Strain vs. Cos α data plot observed measuring RS in the groove area

For GR2, the RS measurements failed in all directions in all Modes except using Omega Mode in the axial direction. Omega-Mode is the only one that is capable of accurately measuring RS in the bottom of a very narrow, deep groove in the axial direction. In such cases, the beam size must be selected carefully and in some instances, a mask on top of the groove may be used to obtain reliable RS results. The success of Psi-Mode in the axial direction is related to the narrow angle opening between the beam and the detector (i.e. the high back reflection {hkl} selected in this particular instance). The Psi-Mode data was corrected after removing bad points at high ψ tilt angle. For Cosα, because of the ψ tilt angular range necessary to perform a reliable RS measurement, a portion of the Debye-Ring is masked (or shadowed) and the data obtained is rendered unusable i.e., when using the Cosα technique on GR2, high ψ angles are not an option so the data missing from the Debye-Ring renders the strain calculation useless (see Figures 4 and 5). As an analogy, this is equivalent to attempting to use Psi or Omega Modes to calculate the shear stress when all of the positive (or negative) ψ tilts are not available for the calculations.

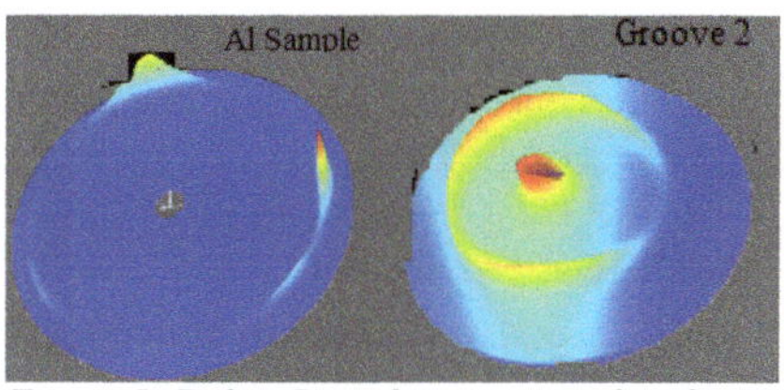

Figure 5: Debye Ring data measured in the groove 2 and Al sample

One approach to making all the above techniques successful is to remove the material around the location of interest (i.e. the bottom of the groove) so as to expose the location of interest. The disadvantage to this approach is that the part must be sacrificed at the expense of the RS measurements and as such only non-production parts would be analyzed.

For the aluminum sample, all methods were capable of producing RS results within an acceptable range however, some scatter was observed due to the texture present in the material only, no size effect was observed during the measurements. The Cosα technique had more difficulties and the measurements were repeated few times before a stress value could be calculated [8]. This is because of spotty data on the Debye Ring. This confirms the problem with the shadowing effect observed in the grooves. When looking at the slope of the data curve for a given RS value, the Cosα technique has a reduced slope when compared to the other techniques [9]. This explains the high experimental errors in the RS values.

When perform RS measurements on the Ni-Base alloy sample, the Cosα technique failed due to the weak peak intensity collected using Cr_kβ line. RS measurements using the Omega technique worked well however no measurements were performed with Psi and Modified Psi modes on this sample .

Summary

The results obtained indicate that amongst the various techniques applied to RS measurement, the MET using $\mathrm{Sin}^2\psi$ was capable of measuring RS on a wide variety of materials with different geometries and material conditions whereas the Cosα technique was limited to isotropic materials and near flat surfaces. Moreover, the Omega-Mode was the only one that was capable of measuring RS in deep grooves in the axial direction. The use of Psi-Mode may be advantageous when tight constraints related to the geometry of the test specimen become an issue. The Psi-mode goniometer is preferred when compared to the Modified Psi-mode goniometer because the latter is more limited in terms of sample geometry constraints and the scattering vector is offset from the ψ tilt plane. For successful RS measurements on a wide variety of components as well as in confined spaces and complex geometries, a combination of Psi and Omega Modes are ideal. For the Cosα technique to be viable for RS measurement, more development would be required to deal with scattered data on the Debye Ring and its various geometric limitations.

References

[1] LESTER, H. H., and ABORN, R. H. Behavior Under Stress of the Iron Crystals, Army Ordnance, v. 6, 1925, 1926, pp. 120, 200, 283, 364

[2] Noyan I.C. and Cohen J.B., 'Residual Stress', Springer Verlag, 1987. *http://dx.doi.org/10.1007/978-1-4613-9570-6*

[3] J.A. Pineault, M. Belassel, M.E. Brauss, "X-Ray Diffraction Residual Stress Measurement in Failure Analysis", Failure Analysis and Prevention, Vol. 11, ASM Metals Handbook, American Society for Metals, 2002, p. 484-497.

[4] V. Hauk, "Structural and Residual Stress Analysis by Nondestructive Methods", Elsevier, Amsterdam, 1997.

[5] EN 15305, "Non-destructive Testing - Test Method for Residual Stress analysis by X-ray Diffraction", August 2008.

[6] J. Lu et al., "Handbook of Measurement of Residual Stresses", Published by the Fairmont Press, Inc., 1996.

[7] S. Tairaet al., "A Mehod of X-Ray Microbeam Measurement of Local Stress and its Application to Fatigue Crack Growth Problems", J. Mat. Sci. Japan. 27, 1977, p.251-256.

[8] T. Kondoh et al., "X-Ray Stress Measurement for Titanium Aluminide Intermetallic Compound" Adv. In X-Ray Anal. Vol.43, 2000, p107-116.

[9] T. Sasaki et al., "Measurement of Macro- and Microstresses of Composite Materials by X-Ray Diffraction Method Using Imaging Plate", ICDD Publications, Vol. 41,1999, p479-492.

Materials Research Proceedings 2 (2016) 49-54 doi: http://dx.doi.org/10.21741/9781945291173-9

Residual Stress Field Prediction in Shot Peened Mechanical Parts with Complex Geometries

Maxime Gelineau [1, a *], Laurent Barrallier [2, b], Emmanuelle Rouhaud [3, c], Régis Kubler [2, d] and Quentin Puydt [1, e]

[1]Institut de Recherche Technologique M2P, 4 rue Augustin Fresnel, 57070 Metz – France

[2]Arts et Métiers ParisTech d'Aix-en-Provence, Laboratory of Mechanics Surface and Materials Processing (MSMP), 2 cours des Arts et Métiers, 13617 Aix-en-Provence – France

[3]Université de Technologie de Troyes, Laboratory of Mechanical Systems and concurrent Engineering (LASMIS), 12 rue Marie Curie BP 2060, 10010 Troyes – France

[a]maxime.gelineau@irt-m2p.fr, [b]laurent.barrallier@ensam.eu, [c]emmanuelle.rouhaud@utt.fr, [d]regis.kubler@ensam.eu, [e]quentin.puydt@irt-m2p.fr

Keywords: Shot Peening, Residual Stresses, Complex Geometries, Finite Element Simulation

Abstract. In order to introduce automatically the residual stresses field into a Finite Element model with complex geometry, a PYTHON code has been developed and linked to the software ABAQUS. A comparison between modelling and experiment is carried out by using X-ray diffraction analysis to determine the in-depth residual stress state of Ni-based alloy samples after shot peening.

Introduction

Most manufacturing industries perform mechanical surface treatments at the end of the manufacturing chain to reinforce relevant working parts. Shot peening is probably the most common of those processes. This treatment modifies the near surface of treated parts by introducing compressive residual stresses due to the repeated impacts of shots leading to an enhanced life.

The objective of this work is to simulate the residual stress state after shot peening, for mechanical parts with complex geometries (thin sheets, curved surfaces). It is part of an industrial collaborative project which one of the work packages consists in developing a complete simulation model of a structure made of a Ni-based alloy and submitted to cyclic loadings in service including the manufacturing process history.

Numerical Model

Introducing residual stresses in a Finite Element model. Several methods have been developed to introduce the stress-free strains into a Finite Element model [1]. Indeed, depending on the capabilities of the finite-element codes, it is possible to introduce the plastic deformation field and/or the stress field as initial conditions. Nevertheless these methods are not directly applicable if the geometry of the part is complex. They require some assumptions:

- The analytical technique used is available only if the geometry of the treated part can be reduced to a semi-infinite body. It means that the treatment is homogeneous, the treated surface is regular and the depth affected by the treatment (Δ in Fig. 1) is small (few hundred micrometers) compared to the other dimensions of the part (Δ e, e.g. for sheet of thickness e).
- The computational technique implies that the geometry has to be plane in order to introduce in a convenient way the mechanical fields into the model.

A predictive model for complex geometries. Because the geometry of the mechanical part induces a rebalancing of the stresses, a model to predict residual stresses for complex geometries build with the analytical relationships for massive and plane geometries is proposed.

Considering that residual stresses arise from the incompatibility of plastic strains between each depth, it is possible to establish analytical relationships between residual stresses $\bar{\bar{\sigma}}$ and incompatible plastic strains $\bar{\bar{\varepsilon^p}}$ in the case of an isotropic semi-infinite plane geometry [2,3] such as:

$$\bar{\bar{\varepsilon^p}}(z) = \begin{pmatrix} \varepsilon^p(z) & 0 & 0 \\ 0 & \varepsilon^p(z) & 0 \\ 0 & 0 & -2\varepsilon^p(z) \end{pmatrix} \text{ and } \bar{\bar{\sigma}} = \begin{pmatrix} \sigma(z) & 0 & 0 \\ 0 & \sigma(z) & 0 \\ 0 & 0 & 0 \end{pmatrix} \; with \; \varepsilon^p(z) = \frac{\nu-1}{E}\sigma(z). \quad (1)$$

Typical residual stresses (RS) and plastic strains (PS) profiles in shot peened massive and plane geometry are presented in Fig. 1.

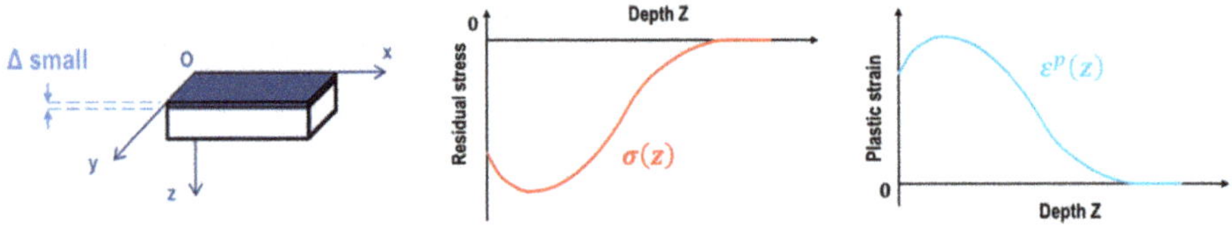

Figure 1. Typical residual stress and plastic strain profiles induced by shot peening in a semi-infinite body

Because the hypothesis of a semi-infinite body can be made for most shot peened parts on a local basis, and since it is easy to get analytically [1,2,4] or experimentally [5,6,7] the RS profile for massive and plane geometry, a simplified approach is proposed for complex geometries by using the stress-free strains computed for the semi-infinite body (with the same peening conditions).

The approach consists in introducing into the Finite Element model an unbalanced stress field as initial stresses field: this field is computed thanks to the stress-free strains due to shot peening for a semi-infinite body; then it is transferred into the geometry of the part and will lead to the residual stress field after static equilibrium.

The computation is driven by analytical relationships proposed in the literature [3]. In a local coordinate system, the initial stress tensor is given by Eq. 2:

$$\overline{\overline{\sigma^0}} = -2\mu\overline{\overline{\varepsilon^p}} \quad (2)$$

where μ is the shear modulus.

Implementation of the method. In order to generate residual stress field automatically into a mechanical part, a PYTHON code has been developed and linked to the Finite Element software ABAQUS. This code proposes an efficient way to calculate the residual stress field induced by estimating the initial stress state for each integration point of the model.

The method is divided into three main stages:

- The depth of each integration point is computed by using nodes and elements data of the surface layer through the connectivity tables of the Finite Element model.
- Using SIGINI subroutine, initial stresses corresponding to the plastic strains that would exist in a semi-infinite body are introduced locally in a local coordinate system defined with the normal vector of the surface.

Materials Research Proceedings **2** (2016) 49-54 doi: http://dx.doi.org/10.21741/9781945291173-9

- Then a static equilibrium computation is performed. After this purely elastic computation, the result is the stress field due to shot peening in the mechanical part.

Numerical Test Cases

Massive and plane geometry. The first study focuses on a cube of size 10 mm as illustrated in Fig. 2. Only the top of the part is considered as a shot peened area. A random RS profile for the semi-infinite body (SIB) is used. The shot peened part is a Ni-based alloy (Inconel 718 DA) with an elastic isotropic behavior (elastic constants: E = 210 GPa, μ = 81 GPa; mesh size: $1\times1\times10^{-2}$ mm^3 minimum).

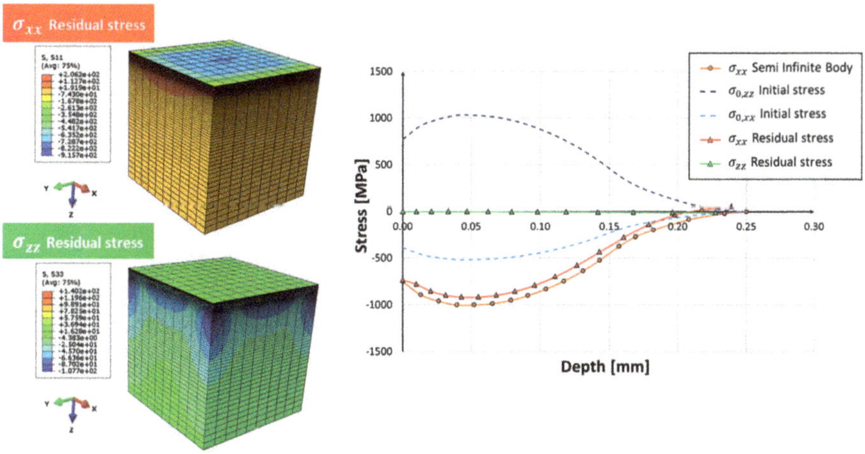

Figure 2. Comparison between the residual stresses within a semi-infinite body (orange line), the initial stresses introduced into the Finite Element model (dashed lines) and the residual stresses computed in the middle of the cube (red and green lines)

It can be observed that the RS profile computed for the Finite Element model (so-called σ_{xx} Residual stress in Fig. 2) is very close to the one of the SIB. It means the geometry is thick enough to be considered as a semi-infinite body. Hence the result from this preliminary study shows a good agreement with the theoretical expected results.

Complex test cases. The method has been also applied to several cases to ensure the numerical tool works when the geometry and/or the peening conditions become complex. Consequently the objective is to look at the initial stresses field within the part. We have to check if the initial stress gradient through the thickness is correctly described.

For instance, as illustrated in Fig. 3, we used three random peening conditions (it means three different RS profiles within a SIB). One area is considered as non-peened. We can notice that the initial stresses field introduced into the mechanical part is accurately described, especially in the sharp corner where the border through the thickness is equidistant to both surfaces of areas 3 and 4.

Besides, we performed successfully the same kind of computation for several other Finite Element models, with a more complex geometry: curved shapes (convex and concave), smooth and notched fatigue test specimen, etc.

For example, the numerical method has been proven successful in the case of a spring of 10 mm diameter. A random RS profile for the SIB is used. The entire surface of the part is considered as shot-peened, except the ends. Fig. 4 shows the results computed after the equilibrium computation.

Materials Research Proceedings **2** (2016) 49-54 doi: http://dx.doi.org/10.21741/9781945291173-9

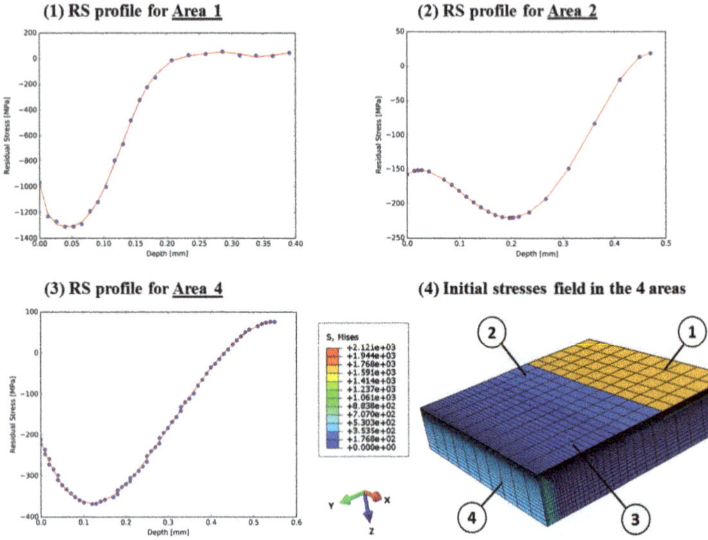

Figure 3. Residual stress profiles within a semi-infinite body for three peening conditions (1,2,3) and preview of the associated initial stresses field introduced into the FE model (4)

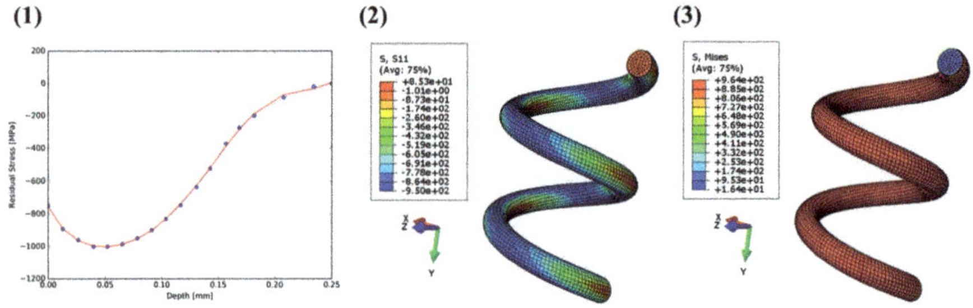

Figure 4. Residual stress profile within a semi-infinite body (1), preview of the associated residual stresses field σ_{xx} (2) and the Von Mises stress distribution (3)

However, the RS field become sometimes complex: it differs from the one that would be found in a semi-infinite body due to the geometry. Thus, even if the treatment is considered homogeneous and the surface regular, the global equilibrium of the structure may change the RS field. Hence we need experimental data to understand what occurs after the rebalancing of the stresses.

Modelling vs experiment

Experimental procedure. A Ni-based alloy (Inconel 718 DA) thin plate (75×25×5 mm^3) is homogeneously peened with cast steel shot S130 on one surface, with the following settings: F22-23A Almen intensity, 200% coverage. Then the study of residual stress state is performed.

Besides, the modelling approach is based on the knowledge of the RS profile through the thickness of a SIB. As we mentioned before, the hypothesis of SIB for a 10 mm thickness sample seems reasonable. Thus the previous surface treatment is carried out on a thicker Inconel 718 DA plate (25×19×10 mm^3).

Comparison of residual stress states. A characteristic residual stress profile after shot peening is achieved by using X-ray diffraction analysis. Material removal correction is evaluated using relationships derived from Moore and Evans method [8]. The main test parameters were as follows:

- χ goniometer (SEIFERT XRD 3000 PTS),
- Mn Kα radiation (30 kV and 20 mA),
- Rear Cr filter in front of P.S.D.,
- 2 Φ angles,
- 11 χ angles,
- Collimator giving a spot of 1.5 mm diameter,
- Family of diffraction planes: {311} at 2Θ = 150°.

Fig. 5 shows the experimental and modelling results of the residual stress state after shot peening. The 2 Φ angles equal to 0° and 90° represent respectively the transverse (TD) and longitudinal direction (LD) of the plate. The SIB profile (green line) is a polynomial interpolation between experimental data for the 10 mm sample. This curve is used to compute the initial stresses field. Then this field is applied to the Finite Element model of the 5 mm plate.

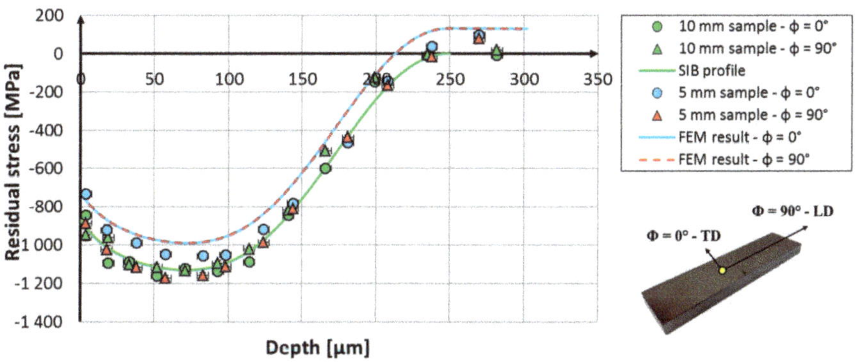

Figure 5. Distribution of the residual stress below the surface of Inconel 718 DA alloy

Residual stress profiles after equilibrium (FEM results) in both directions TD and LD are respectively presented with red dashed line and blue line, whereas experimental values are given by blue circles and red triangles. As a matter of fact, the Finite Element approach underestimates the compressive residual stresses. However the gap with the SIB profile is reasonable: it remains below 200 MPa.

It means the 5 mm plate could be considered as a SIB too, and the present approach is suitable to simulate the residual stress state after shot peening.

Conclusion

A method that enables to predict the residual stress field due to shot peening knowing the stress-free strains into a semi-infinite body has been presented. An experimental validation has been performed on a thin plate by using X-ray diffraction analysis. According to these preliminary results, the proposed method is efficient.

Materials Research Proceedings **2** (2016) 49-54 doi: http://dx.doi.org/10.21741/9781945291173-9

Nevertheless, this work needs to be completed with other experimental measurements in order to test the semi-infinite body hypothesis. What makes this hypothesis valid in terms of geometries is not clearly defined yet. Thus the perspective is:

- To evaluate the deformation field of thin shot peened plates by using three-dimensional coordinate measuring machine (MMC). It will provide the profile height which is also a characteristic parameter to link with the rebalancing of the stresses.
- To lead the experimental validation with curved surfaces.
- To define the scope of validity of the method.

Acknowledgments

This research work is part of the collaborative CONDOR project coordinated by the Institut de Recherche Technologique Matériaux Métallurgie et Procédés. It is a pleasure to acknowledge all the industrial partners: Airbus Helicopters, ARCELORMITTAL, AREVA, MISTRAS group SA, PSA Peugeot Citroën and SAFRAN.

References

[1] P. Renaud, Modélisation numérique du grenaillage des pièces initialement cémentées ou carbonitrurées, Thèse, Arts et Métiers ParisTech, 2011.

[2] F. Cochennec, Simulation numérique du grenaillage de mise en forme pour une intégration produit-procédés en conception mécanique, Thèse, Université de Technologie de Troyes, 2009.

[3] E. Rouhaud, A. Milley, J. Lu, Introduction of residual stress fields in finite element three-dimensional structures, Proceedings of the 5th International Conference on Residual Stresses (ICRS5), Linköping Suède, 16-18 juin 1997.

[4] I. Lillamand, « Evolution d'une couche grenaillée sous sollicitations thermiques et mécaniques, cas de la fatigue oligocyclique », Thèse de l'Ecole Nationale Supérieure d'Arts et Métiers (1999), 218 p.

[5] M. Choumad, « Analyse des contraintes résiduelles générées lors du grenaillage : approches analytique, numérique et expérimentale des impacts de billes », Thèse de l'Université de Technologie de Troyes (2007), Spécialité Systèmes Mécaniques et Matériaux, 167 p.

[6] J. Lu, « Traitements de surface mécaniques – Principes », Techniques de l'Ingénieur, Article m1190 (2006), 18 p.

[7] P.J. Withers, H.K.D.H. Bhadeshia, « Residual stress – Measurement techniques », Materials Science and Technology (2011), Vol. 17 (pp. 355-365). *http://dx.doi.org/10.1179/026708301101509980*

[8] L. Castex, « Redistribution des contraintes dans une plaque après enlèvement de matière », Réunion des 10 et 11 octobre 1984 à Ai-en-Provence (ENSAM), Groupement Français pour l'analyse des contraintes par diffractométrie X.

Comparison of Two X-Ray Residual Stress Measurement Methods: Sin^2 ψ and Cos α, Through the Determination of a Martensitic Steel X-Ray Elastic Constant

D. Delbergue[1,2,a], D. Texier[1,b], M. Lévesque[2,c], P. Bocher[1,d *]

[1]Laboratoire d'Optimisation des Procédés de Fabrication en Aérospatiale (LOPFA), École de technologie supérieure (ÉTS), 1100 Notre-Dame Ouest, Montréal, QC H3C 1K3, Canada

[2]Laboratory for Multi-Scale Mechanics (LM2), École Polytechnique de Montréal, C.P. 6079, Succ. Centre-ville, Montréal, QC H3C 3A7, Canada

[a]dorian.delbergue.1@ens.etsmtl.ca, [b]damien.texier@etsmtl.ca, [c]martin.levesque@polymtl.ca, [d]philippe.bocher@etsmtl.ca

Keywords: X-Ray Diffraction, Residual Stresses, X-Ray Elastic Constant, Single Exposure

Abstract. X-ray diffraction technique for residual stresses measurement is usually associated to the sin^2 ψ method, a method based on the interception of the diffraction cone and line detectors. To overcome this loss of information, the cos α method is an alternative method which uses a single exposure to collect the entire diffraction cone via a 2D detector. The present paper compares both sin^2 ψ and cos α methods, through the X-ray elastic constant (XEC) determination of a quenched and tempered martensitic steel. The full-cone measurement method demonstrates a smaller scatter and a better repeatability of the measurements. This latter point is of considerable interest since larger scatter in XEC may result in large variation in residual stress values, especially at high stress levels.

Introduction

Aerospace, automotive, and power industries are constantly looking for means to improve fatigue life. Many surface-treatment processes, such as laser peening or shot peening, are known to improve fatigue life due to introduction of near-surface compressive residual stresses. However, the relationship between processes and fatigue life improvement is not fully understood. These surface-treatment processes introduce a residual stress profile in a shallow region beneath the surface which may delay the crack nucleation and/or propagation. Therefore, residual stresses are more and more accounted for fatigue life prediction models. Consequently, it is of major concern to assess residual stresses with the best possible accuracy. Many characterization techniques, such as hole-drilling, X-ray diffraction (XRD), or ultrasonics, have been developed during the last century to measure them.

X-ray diffraction technique has been widely used in industries and laboratories for residual stresses measurement. The XRD technique is based on the measurement of the crystallographic lattice deformation. According to Bragg conditions, a change in lattice spacing induce a shift of the diffracted X-ray angle [1]. Then, the strain and stress can be calculated from the peak shift. For decades, the traditional sin^2 ψ method has been employed for stress calculation [2]. The dedicated diffractometer mainly use line detectors, most commonly Position Sensitive Scintillation Detectors (PSSD), in order to capture the scattered X-rays. In the case of stress-free isotropic polycrystalline material, X-rays are diffracted in all directions by the grains, giving form to a cone. The line detectors only capture a limited part of the scattered X-rays corresponding to two diffraction cone radii. In the case of plane-stress state, a minimum of two sample orientations are required for strain measurements [2].

In 1978, the cos α method was developed in Japan for stress calculation [3]. This method, also called the single exposure method, allows stress calculation by capturing the resulting diffraction

Materials Research Proceedings **2** (2016) 55-60 doi: http://dx.doi.org/10.21741/9781945291173-10

cone of a single incident X-ray beam via a 2D detector. The intersection of the diffraction cone with the area detector provides a ring named Debyer-Scherrer ring [1]. The introduction of new 2D photosensitive detectors, such as image plate (IP), and the development of powerful calculation algorithms, bring the cos α method up to date [4]. Since few years, portable apparatus equipped with IP allow quick residual stress measurements using the cos α method.

The two presented XRD methods measure the elastic strain for a chosen diffraction peak corresponding to specific planes of atoms in the crystal structure. This family of planes does not necessarily have the same elastic constants than the bulk material; the latter being the expression of the average elastic behavior of all plane families. Therefore, the X-ray elastic constants (XEC) determination is needed for the chosen peak of the material subjected to X-ray residual stress measurement [5]. The XEC are determined by measuring the lattice strain of given crystallographic planes using a diffractometer while the specimen is subjected to constant axial loading. A wrong XEC value may underestimate the residual stress state, leading to non-conservative fatigue predictions [6, 7].

The present work aims at measuring the XEC with both methods on a quenched and tempered martensitic steel, and test their accuracy and repeatability.

Stress analysis

For both apparatuses, the scattered X-rays are recorded for a specific family of planes {hkl}. When the diffraction peaks have been fitted and the corresponding Bragg's angles determined, the d-spacing (also called the lattice spacing) $d^{\{hkl\}}$ can be calculated using the Bragg's law. The strain is then obtained by Eq. 1:

$$\varepsilon_{\varphi\psi}^{\{hkl\}} = \frac{d_{\varphi\psi}^{\{hkl\}} - d_0^{\{hkl\}}}{d_0^{\{hkl\}}}. \tag{1}$$

where $\varepsilon_{\varphi\psi}^{\{hkl\}}$ is the measured strain for the {hkl} planes for a specimen orientation referred by the φ and ψ angles, and $d_0^{\{hkl\}}$ is the d-spacing for an unstressed specimen, usually powder. The stress tensor is considered biaxial in the irradiated layer of material (which is about 5 µm thick in the case of steel).

With the $\sin^2\psi$ method, the traditional equation for X-ray stress measurement is given as [8]:

$$\varepsilon_{\varphi\psi}^{\{hkl\}} = \frac{1}{2} S_2^{\{hkl\}}\, \sigma_\varphi \sin^2\psi + S_1^{\{hkl\}}(\sigma_{11} + \sigma_{22}). \tag{2}$$

where $1/2S_2^{\{hkl\}} = \left(1+v^{\{hkl\}}\right)/E^{\{hkl\}}$ and $S_1^{\{hkl\}} = -v^{\{hkl\}}/E^{\{hkl\}}$ are the XEC for the family of planes {hkl}, σ_{11} and σ_{22} are the stress tensor components, σ_φ is the stress in the φ direction, φ and Ψ angles are defined as the in-plane direction, and the angle between specimen normal and diffraction plane normal, respectively, as shown in Fig. 1. In Eq. 2, the strain $\varepsilon_{\varphi\psi}^{\{hkl\}}$ appears to be a linear function of $\sin^2\psi$. The differentiation of Eq. 2 with respect to $\sin^2\psi$ and the isolation of σ_φ lead to Eq. 3:

$$\sigma_\varphi = \frac{E^{\{hkl\}}}{1+v^{\{hkl\}}} \frac{\partial \varepsilon_{\varphi\psi}^{\{hkl\}}}{\partial \sin^2\psi} = \frac{1}{1/2S_2^{\{hkl\}}} \frac{\partial \varepsilon_{\varphi\psi}^{\{hkl\}}}{\partial \sin^2\psi}. \tag{3}$$

Eq. 3 shows that for the $\sin^2\psi$ method the stress σ_φ is function of the XEC, $1/2S_2^{\{hkl\}}$, and the linear regression between $\varepsilon_{\varphi\psi}^{\{hkl\}}$ and $\sin^2\psi$. Therefore, the $\sin^2\psi$ method implies the measurement of strains for different ψ angles provided by the tilt of the machine. By changing ψ_0, the angle between specimen normal and incident X-ray beam, the ψ angle will subsequently change as they are fixed together by the Bragg's angle.

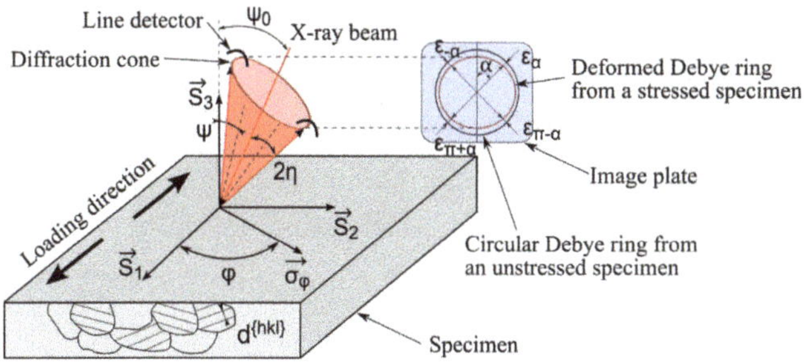

Fig. 1: Schematic representation of a specimen and grains, specimen S and measurement σ_φ *directions, diffraction cone, line detector and image plate.*

In the cos α method, the strain is determined by comparing the recorded Debye ring with the perfectly circular ring from an unstressed specimen. In fact, the strain along the normal to the diffraction planes varies with the ψ angle [1]. As the strain affects peak location, the Debye ring is not circular for specimen under stress. Eq. 4 expresses $\bar{\varepsilon}_\alpha^{\{hkl\}}$ the parameter used to calculate the stress [9]:

$$\bar{\varepsilon}_\alpha^{\{hkl\}} = \frac{1}{2}[(\varepsilon_\alpha^{\{hkl\}} - \varepsilon_{\pi+\alpha}^{\{hkl\}}) + (\varepsilon_{-\alpha}^{\{hkl\}} - \varepsilon_{\pi-\alpha}^{\{hkl\}})]. \tag{4}$$

where $\varepsilon_\alpha^{\{hkl\}}$, $\varepsilon_{\pi+\alpha}^{\{hkl\}}$, $\varepsilon_{-\alpha}^{\{hkl\}}$, and $\varepsilon_{\pi-\alpha}^{\{hkl\}}$ are strains calculated at four points located at 90° on the Debye ring (as depicted in Fig. 1). The stress is calculated by varying α from 0° to 90° in order to cover the whole ring and is a linear function of the regression between $\bar{\varepsilon}_\alpha^{\{hkl\}}$ and cos α. As a consequence, Eq. 5 is the equation providing the stress for the cos α method:

$$\sigma_\varphi = \frac{E^{\{hkl\}}}{1+\nu^{\{hkl\}}} \frac{1}{\sin 2\eta \sin 2\psi_0} \frac{\partial \bar{\varepsilon}_\alpha^{\{hkl\}}}{\partial \cos\alpha} = \frac{1}{1/2 S_2^{\{hkl\}}} \frac{1}{\sin 2\eta \sin 2\psi_0} \frac{\partial \bar{\varepsilon}_\alpha^{\{hkl\}}}{\partial \cos\alpha}. \tag{5}$$

where 2η is the Debye ring semi-angle.

The data are obtained for a φ angle equal to zero to estimate the uniaxial stress.

Experiment

The comparison between the two residual stress calculation methods was performed while measuring the X-ray elastic constant of a martensitic steel. The micro-tensile specimen geometry is presented in Fig. 2c). Specimen was extracted from a rolled block and contour shapes were machined with a CNC machine. Then 1 mm thick specimen was sliced using a precision cutting machine Struers Secotom-50. It was then manually polished using SiC papers (up to grade 1200) and a jig specially designed for thin specimens, allowing keeping parallelism between the two main faces. The specimen was electropolished as suggested by the ASTM standard [10]. The final specimen thickness was 0.798 mm.

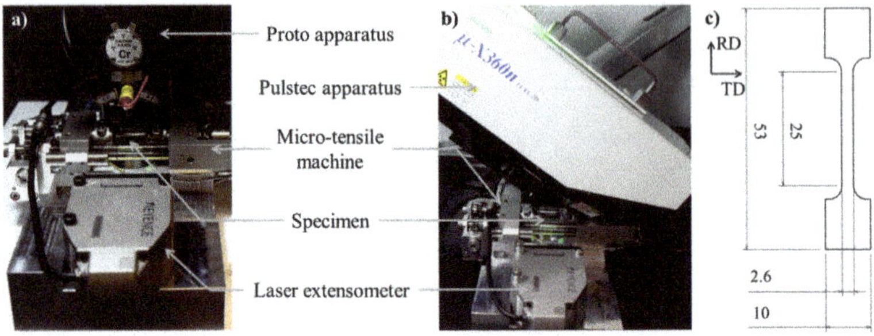

*Fig. 2: Experimental setups: micro-tensile machine and optical extensometer paired up with: a) the Proto iXRD apparatus, **b)** the Pulstec μ-X360 apparatus, and **c)** specimen geometry presentation (dimension are given in mm).*

A Proto iXRD diffractometer was used to collect the diffraction peaks for the stress calculation via the $\sin^2\psi$ method, while the entire Debye ring was captured by a Pulstec μ-X360 apparatus for the $\cos\alpha$ method. Both machines were equipped with Cr-tube (λ = 2.291 Å). The Kα doublet from {211} planes family were used due to their high Bragg's angle (2θ = 156°), providing better accuracy on strain measurement. The Pulstec machine was set so that the ψ_0 angle is equal to 33.2° to capture the diffraction cone in a measurement of 30 s. The Proto apparatus was limited to seven different incident angles between ±13° (*i.e.* $-23.6° \leq \psi \leq 23.6°$), so the detectors do not hit the micro-tensile machine. 30 exposures of 0.25 s were used for each one of the seven incident angles, resulting in a measurement time of 53 s. All diffraction peaks were fitted using Gaussian peak fitting. In order to determine the most accurate XEC, a 0.5x3 mm^2 rectangular aperture was chosen for Proto iXRD and a 1 mm diameter aperture for Pulstec μ-X360, so the larger volume of diffracting material was used for each apparatus. The stresses were first assessed using $1/2\,S_{2_{Macro}}$ = 6.50 E^{-6} MPa^{-1} calculated from the macroscopic values. The XRD measurements were repeated six times for each loading condition to test equipment repeatability.

A 5 kN micro-tensile machine, manufactured by Kammrath & Weiss GmbH, was paired up with the diffractometers in order to load the specimen in tension. Nine loading conditions were chosen from 0% to 70% of the yield strength, which has been previously determined by three macroscopic tensile tests.

Micro-tensile tests were conducted under displacement control (displacement rate of 3 $\mu m.s^{-1}$) and the displacement was stopped for the various loading dwells. The specimen elongation/contraction was continuously recorded by a Keyence LS-7030M optical extensometer during the whole experiment to assess the macroscopic strain. The experimental setup with the micro-tensile machine and the laser extensometer is presented in Fig. 2a) & b) when combined with the two XRD apparatus.

Discussion and Results

Fig. 3 presents the nominal stress-strain evolution applied during the experiment using Pulstec diffractometer. The plateaus correspond to the XRD measurement periods, *i.e.* the six measurement repetitions. The nominal stress and strain were found to be constant during these measurement periods. The difference in applied stress between the beginning of 1st measurement and the end of the 6th is less than 1MPa in average, except for the highest loading condition where the load decrease by 4.3MPa. This larger difference is due to a slightly sliding of the specimen in the grip for this specific condition. Nevertheless, the difference in load does not appear to be the main parameter affecting the stress measurement in this case as the lowest stress value was not measured for the 6th repetition.

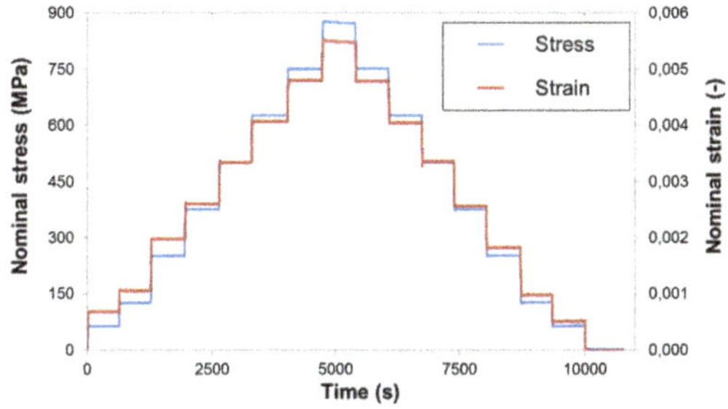

Fig. 3: Nominal stress and nominal strain evolution during the measurements with Pulstec apparatus.

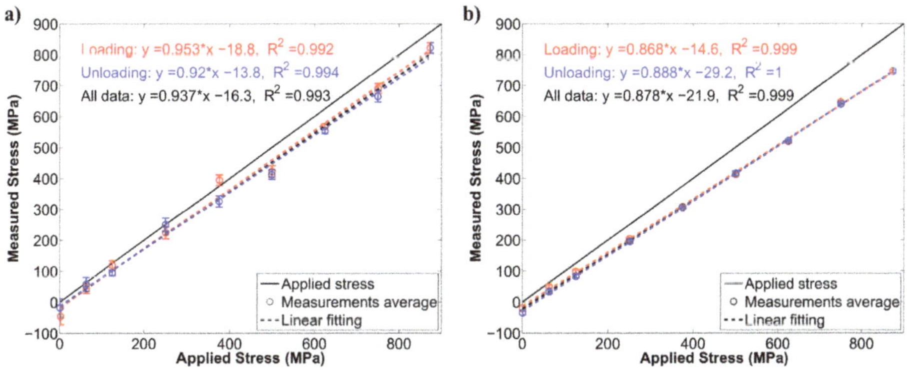

*Fig. 4: Presentation of measured stresses calculated with a) the $sin^2\psi$ method (via Proto iXRD) and **b)** the cos α method (via Pulstec μ-X360), for the different loading conditions.*

Average stresses are plotted versus applied load, and are presented in Fig. 4 for the two stress calculation methods, therefore for the two diffractometers. The error bars correspond to a 95% confidence interval calculated using the Student's law. For both graphs, the stresses measured during loading and unloading are depicted as red and blue plots, respectively. The straight solid line represents a reference line in the hypothetical case where the XEC values can be estimated with the macroscopic mechanical constants of the material. As depicted in Fig. 4a) & b), the various plots are not aligned with this reference line showing that $1/2S_2^{\{211\}}$ is different from the macroscopic XEC and has to be calculated to make the data slope fit with the reference line, as proposed by Munsi et al. [11]. The correction coefficients were found to be 1.067 for the measurements done with $\sin^2\psi$ method and 1.139 for the measurements done with the cos α method. This lead to two XEC for the {211} family of planes: $1/2S_2^{\{211\}}{}_{\sin^2\psi} = 6.93 \pm 0.32$ MPa^{-1} and $1/2S_2^{\{211\}}{}_{\cos\alpha} = 7.40 \pm 0.1$ MPa^{-1}. Errors are calculated using a 95% confidence interval for the linear regression coefficients.

As exposed earlier in the Stress Analysis section, the XEC should be the same for both stress calculation methods. Consequently, the XEC previously determined should not differ from one measurement method to the next. Fig. 4 shows that the discrepancy of the results is lower with the cos α method, demonstrating a higher repeatability of this measurement. This may be due to the fact that in the current study the 2D detector captures 35 times more diffraction peaks, in a single

Materials Research Proceedings **2** (2016) 55-60 doi: http://dx.doi.org/10.21741/9781945291173-10

exposure, than the two line detectors. This provides more statistical data for the stress calculation. For $\sin^2\psi$ method (Fig. 4a)), the large difference in fitting curves for the loading and unloading conditions may show a low accuracy. This may be explained by a too small ψ angle scan for the $\sin^2\psi$ method, even if the volume irradiated was twice bigger for the Proto diffractometer.

Conclusions

The stresses measured by X-ray diffraction are in linear relation with the X-ray elastic constant. Therefore, a wrong XEC value may introduce a large bias in the measured stress. The use of a micro-tensile machine to stress the specimen at given macroscopic loads permits to compare the $\sin^2\psi$ and the cos α methods for stress calculation, while accessing the XEC of the studied material. Depending on the methods used for the stress calculation, a large difference in material intrinsic constant can be found. In the present study, the cos α method via the 2D detector has shown a better measurement repeatability than the $\sin^2\psi$ method via the line detectors.

Acknowledgements

The authors would like to thank the research funding from the CRIAQ project MANU508 supported by the Natural Sciences and Engineering Research Council of Canada (NSERC).

References

[1] B.D. Cullity, Elements of X-ray diffraction, Addison-Wesley Publishing Company, Inc., 1956.

[2] I.C. Noyan, J.B. Cohen, Residual stress measurement by diffraction and interpretation, Springer-Verlag, New York, 1987.

[3] S. Taira, K. Tanaka, T. Yamasaki, A method of X-ray microbeam measurement of local stress and its application to fatigue crack growth problems, Journal of the Society of Materials Science, Japan, 27 (1978), pp. 251-256 (in Japanese).
http://dx.doi.org/10.2472/jsms.27.251

[4] K. Hiratsuka, T. Sasaki, K. Seki, Y. Hirose, Development of measuring system for stress by means of image plate for laboratory X-ray experiment, JCPDS - International Centre for Diffraction Data, 46 (2003), pp. 61-67.

[5] P.S. Prevey, A method of determining the elastic properties of alloys in selected crystallographic directions for X-ray diffraction residual stress measurement, Adv. in X-ray Analysis, 20 (1977), pp. 345-354.
http://dx.doi.org/10.1007/978-1-4613-9981-0_30

[6] V. Savaria, F. Bridier, P. Bocher, Measuring in-depth residual stress gradients: the challeng of induction hardened parts, in: S.J.B. Kurz, E.J. Mittemeijer, B. Scholtes (Eds.), International Conference on Residual Stresses 9 (ICRS 9), 2014, Garmisch-Partenkirchen, Germany.

[7] V. Savaria, F. Bridier, P. Bocher, Predicting the effects of material properties gradient and residual stresses on the bending fatigue strength of induction hardened aeronautical gears, International Journal of Fatigue, 85 (2016), pp. 70-84.
http://dx.doi.org/10.1016/j.ijfatigue.2015.12.004

[8] P.S. Prevéy, X-ray diffraction residual stress technique, in: R.E. Whan (Ed.) Materials Characterization, ASM International, 1986, pp. 380-392.

[9] K. Tanaka, Y. Akiniwa, Diffraction Measurements of Residual Macrostress and Microstress Using X-Rays, Synchrotron and Neutrons, JSME International Journal Series A, 47 (2004), pp. 252-263.
http://dx.doi.org/10.1299/jsmea.47.252

[10] ASTM, E1426-98 Standard Test Method for Determining the Effective Elastic Parameter for X-Ray Diffraction Measurements of Residual Stress, 2009.

[11] A.S.M.Y. Munsi, A.J. Waddell, C.A. Walker, A Method for Determining X-ray Elastic Constants for the Measurement of Residual Stress, Strain, 39 (2003), pp. 3-10.
http://dx.doi.org/10.1046/j.1475-1305.2003.00044.x

Residual Stresses 2016: ICRS-10 Materials Research Forum LLC
Materials Research Proceedings **2** (2016) 61-66 doi: http://dx.doi.org/10.21741/9781945291173-11

Residual Stress Measurement of Ti-Metal Samples by Means of XRD with Ti and Cu Radiation

Lasse Suominen[1, a *], Theo Rickert[2, b], Sebastian Send[3, c]

[1]Stresstech Oy, Tikkutehtaantie 1, 40800 Vaajakoski, Finland

[2]American Stress Technologies, Inc., 540 Alpha Drive, Pittsburgh PA 15238-2912, USA

[3]Stresstech GmbH, Konnwiese 18, D-56477 Rennerod, Germany

[a] lasse.suominen@stresstech.com, [b] theo.rickert@astresstech.com, [c] sebastian.send@stresstech.de

Keywords: X-Ray Elastic Constant Measurement (XEC), X-Ray Diffraction (XRD), Residual Stress, Titanium Alloys, Ti-Radiation, Cu-Radiation

Abstract: The use of titanium in structural components has been growing for years, especially in highly demanding applications in the aerospace industry where quality control is essential. One of the critical properties is fatigue strength, which is strongly affected by residual stresses. Residual stresses may be an unavoidable by-product of the manufacturing process or intentionally imparted through processes like shot peening. X-ray diffraction (XRD) is commonly used for residual stress measurement. Yet titanium alloys are more difficult to measure than other metals like steels and aluminium alloys. Many titanium alloys have a two-phase microstructure, one of them being hexagonal alpha titanium. Also, the commonly used Cu-radiation generates very strong fluorescence, which results in a low penetration depth. Both reduces measurement quality. Ti-radiation is much less frequently used. It benefits from very low fluorescence and a higher penetration depth. This paper compares XRD residual stress measurements using Cu- and Ti- radiation on two titanium grades: Grade 2 and Grade 5 (Ti6Al4V). The samples from both are in the rolled condition and have been shot peened. The x-ray elastic constants were determined by XRD with Cu- and Ti-radiation and samples and residual stress depth distributions were measured up to 0.5 mm depth.

Introduction

Titanium alloys are widely used in aerospace structures due to their excellent strength to weight ratio. For aerospace applications reliability is critical from which follows that all factors affecting the performance of the components have to be studied and controlled extremely well. One important factor is residual stress, which is always present due to the manufacturing processes and which affects fatigue strength significantly. Titanium alloys are challenging because their machining is difficult due to their high chemical reactivity, poor thermal conductivity, high strength that is maintained at elevated temperatures and low modulus of elasticity [1]. The same factors make welding challenging. Since productivity is always an important issue it is always attempted to minimize the need for machining and welding [2, 3].

For the evaluation of titanium components, reliable, fast and easy residual stress measurement is a key issue. There are several methods for measuring residual stresses nondestructively and destructively. This paper deals with X-ray diffraction, using X-ray instruments specifically designed for residual stress measurement. The measurement accuracy depends on a number of factors, such as grain size, preferred orientation, accessibility of measurement location etc. The X-ray wavelength is given by the X-ray tube chosen. It is preferable to select it so that the 2θ – angle of the diffraction peak analyzed is the highest possible that has sufficient intensity. However, some combinations of material and radiation cause significant fluorescence, which affects the net intensity level and also the effective measurement depth [3]. X-ray diffraction does not measure stress directly, but strain.

Materials Research Proceedings **2** (2016) 61-66 doi: http://dx.doi.org/10.21741/9781945291173-11

The stresses have to be calculated using elastic constants [4]. While X-ray elastic constants (XEC) are the best choice, they are often not available and macroscopic elastic constants determined by mechanical testing are commonly used. It is also possible to calculate elastic constants based on single crystal properties.

CuKα (copper) is the most commonly used radiation to measure titanium alloys. It creates very strong fluorescence. A newer alternative is TiKα radiation (titanium), which has very low fluorescence and nearly twice the penetration depth [5]. The aim of this work is to compare the performance of both of these wavelengths in residual stress measurement. The X-ray elastic constants are also measured with both wavelengths and compared to macroscopic and calculated values.

Experimental procedure

The material used in this work are two commercially available, 3 mm thick, titanium plates, one from commercially pure titanium (Grade 2) and one from two-phase alloy Ti6Al4V (Grade 5). The main phase in both samples is the hexagonal α-phase. Ti6Al4V has a maximum of 10% cubic β-phase [3]. Typical values for some mechanical properties of these alloys are shown in Table 1.

Table 1. Material properties of Grade 2 and 5 titanium alloys [6].

Grade	Ultimate strength [MPa]	Yield strength [MPa]	Elastic modulus [GPa]	Poisson's ratio
Grade 2	241	172	103	0.32
Grade 5	896	827	114	0.32

The samples were shot peened to minimize grain size and preferred orientation effects (Fig. 1).

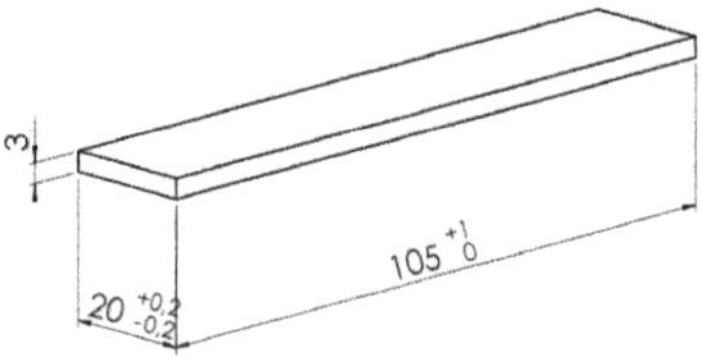

Figure 1. Sample dimensions [mm].

They were loaded on a four-point bending device for the XEC measurements. The bending load was controlled by a calibrated load cell and the load stresses were calculated using Eq. (1):

$$\sigma = \frac{3Fa}{bh^2} \quad (1)$$

in which σ = stress
F = applied force
a = support span
b = width of the specimen
h = thickness of the specimen

Five load steps were used: for Grade 2: 50, 211, 372, 533 and 695 N (corresponding stresses 17, 72, 128, 183 and 283 MPa); for Grade 5: 102, 434, 766, 1099 and 1431 N (stresses 44, 187, 330, 473 and 616 MPa).

X-ray measurements. The X-ray measurements were made using the modified-ψ measurement mode on an Xstress G3 equipped with Mythen detectors. Maximum χ - tilts were ±45°, with four tilts on both sides. The diffraction peak for the family of {213} planes was used with CuKα (Bragg angle 2θ = 139.3°), that for the family of {110} planes with TiKα (Bragg angle 2θ = 137.4°). For Grade 5, the XEC was also measured on a sample not shot peened. This was not possible for the Grade 2 sample due to its large grain size.

The XEC measurements were performed using the method described in the EN15305 [7]. To determine $\varepsilon_{\phi\psi}$, the stress σ^A is calculated for each applied load.

$$\varepsilon_{\phi\psi} = \frac{1}{2}S_2^{hkl} sin^2\theta(\sigma_{11}^R - \sigma_{33}^R - \sigma_{11}^A) sin^2\chi + \frac{1}{2}S_2^{hkl} sin^2\theta\tau_{13}^R sin2\chi + K] \quad (2)$$

where

$$K = S_1^{hkl} Tr(\sigma^R) + \frac{1}{2}S_2^{hkl}(sin^2\theta\sigma_{33}^R + cos^2\theta\, \sigma_{\phi+\frac{\pi}{2}}^A) \quad (3)$$

This represents an elliptical function

$$\varepsilon_{\phi\psi} = a{*}sin^2\chi + b{*}sin2\chi + c \quad (4)$$

where

$$a = \frac{1}{2}S_2^{hkl} sin^2\theta(\sigma_{11}^R - \sigma_{33}^R - \sigma_{11}^A) \quad (5)$$

in which the XEC

$$\frac{1}{2}S_2^{hkl} = \frac{1+\nu}{E} \quad (6)$$

can be evaluated by the least square method. The superscript R with the stress parameters refers to residual stresses, the superscript A to applied stress.

Residual stress depth profiles were measured on the shot peened surfaces. Electropolishing was used to remove material for the various depths. The measurements with CuKα and TiKα were made at exactly the same locations and depths.

Results and analysis

X-ray elastic constant measurements were made starting from high load to low load and back up again. This was performed five times, once for every combination of X-ray radiation, surface condition and material. Typical measurement data are shown in Fig. 2.

The XEC results are summarized in Table 2, which also includes XECs calculated from Young's moduli and Poisson's ratios determined by mechanical testing (macroscopic), XECs predicted from single crystal properties (after Kröner) and some literature data.

All elastic constant values are in the same range, except that the Grade 5 TiKα results are lower by about 20%. There is a systematic trend that plastic deformation (shot peening) reduces the value of the XEC. The same trend has been found by Bahadur et.al. [10] in austenitic steels. The differences between the (213) CuKα and the (110) TiKα values are significant. Some variability in the Grade 5 XECs could be due to its two-phase microstructure. Another, quite probable factor is preferred orientation, which can also affect peak intensity. For a randomly oriented microstructure Imax should decrease moderately as function of sin2ψ or may be nearly constant. With TiKα radiation Imax decreases clearly more than expected and with CuKα it even increases significantly – for the shot peened (Fig. 3) as well as the not shot peened surfaces.

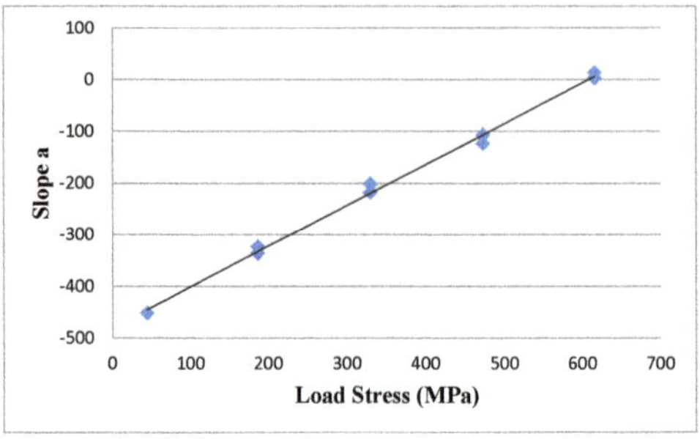

Figure 2. XEC graph for CuKα and Grade 5, measuring from high load to low and back to high. Slope 10.8 ±0.26 10^{-6} MPa^{-1}.*

*Table 2. Measured XEC values ($1/2S_2$), 10^{-6} MPa^{-1} in italic. Average error of measured XEC's is 0.3*10^{-6} MPa^{-1}*

	Grade 2	Grade 5	Grade 5 (not shot peened)	α-phase, Calculated, Kröner	α-phase, XEC [8]
XEC CuKα (213)	*10.4*	*10.7*	*11.3*	11.7	11.90 **
XEC TiKα (110)	*11.1*	*8.7*	*9.2*	12.0	11.98
Macroscopic*	12.8	11.6			
XEC (213) [9]		10.9/11.9			

*calculated from values in Table 1.
** lattice {302}

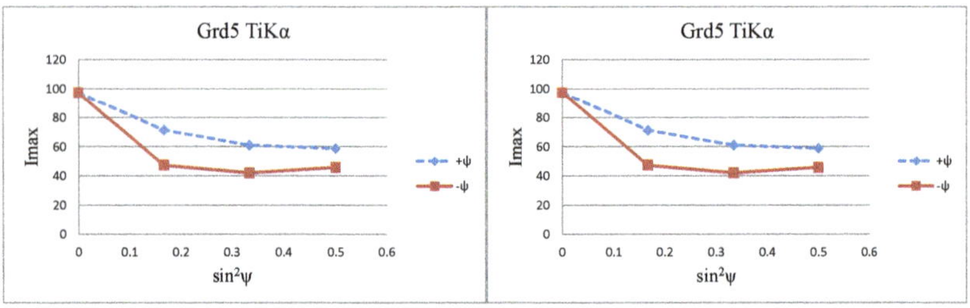

Figure 3. Maximum intensity of diffraction peaks as function of tilt angle for Grade 5 with Tikα (110) and CuKα (213). Examples are from shot peened surfaces.

Residual stress depth distribution. The Grade 5 depth distributions were measured up to 0.5 mm depth and Grade 2 to 0.3 mm depth (Fig. 4). In Grade 2, the grain size became too large at that depth. Stresses were calculated using an XEC of 11.0*10^{-6} MPa^{-1}. The measurements with CuKα and TiKα were made at exactly the same locations, in the longitudinal direction of the samples (Fig. 1). In the

Materials Research Proceedings **2** (2016) 61-66 doi: http://dx.doi.org/10.21741/9781945291173-11

CuKα measurements Mythen detectors were used due to their better sensitivity and a threshold energy capability to decrease the fluorescence radiation [11].

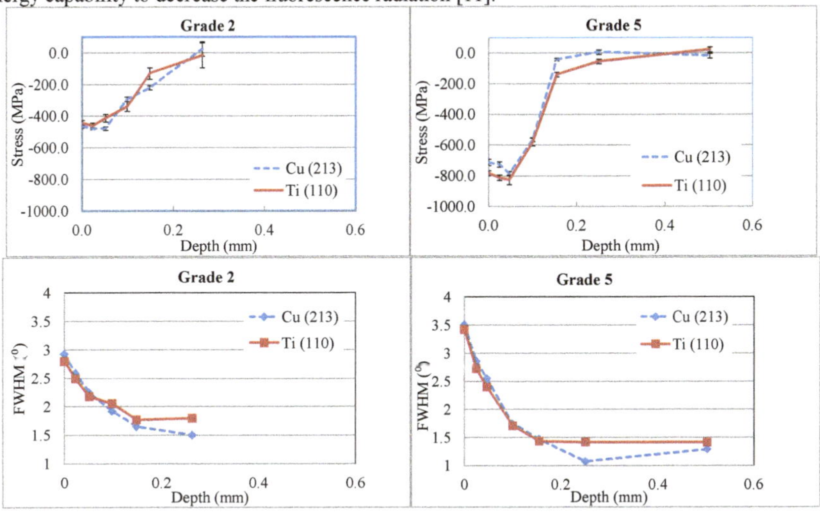

Figure 4. Depth distribution of residual stresses and full width half maximum (FWHM) of shot peened Grade 2 and Grade 5 samples measured by CuKα ($2\theta = 139.3°$) and TiKα ($2\theta = 137.4°$).

The CuKα and TiKα measurement results are very close to each other. There is a small difference in the stress levels of Grade 5 (Fig. 4) measured by CuKα and TiKα, where the CuKα values are slightly lower. This correlates with the higher XEC value for CuKα (10.7 vs. 8.7) but the XEC difference is much bigger. Shot peening deformed the material up to about 0.2mm depth, a little bit deeper for Grade 2 than Grade 5. On average, the TiKα measurements have higher error bars compared to CuKα. In the measurements shown in Figure 4, the average error bars were ±15 to ±19 MPa for Grade 5 and ±17 to ±35 MPa for Grade 2. These higher error bars may be related to the multiplicity factor, which is 24 for the {213} peak (CuKα) and only 6 for {110} (TiKα). The higher the multiplicity factor the less sensitive the measurement should be to preferred orientation.

Conclusions

The X-ray elastic constants measured with CuKα radiation for the (213) lattice planes agree reasonably well with values found in literature. Also the unalloyed titanium sample gives similar values for the (110) planes. But the Grade 5 (Ti6Al4V) XEC values measured on (110) planes using TiKα are clearly smaller than other measured and reported values. Slightly larger residual stresses in the (110) measurement support this difference only to a small degree. There is no clear explanation for these low XEC values. It may be related to preferred orientation or the second phase.

The residual stress results with Cu- and Ti-radiations are comparable. Both work equally well provided that the material's strong fluorescence with Cu-radiation can be properly addressed in the data analysis. The benefit of Ti-radiation is its greater penetration depth and the shorter measurement times. On the downside, its diffraction peak has a lower multiplicity factor, which makes the measurement more sensitive to texture.

Materials Research Proceedings **2** (2016) 61-66 doi: http://dx.doi.org/10.21741/9781945291173-11

References

[1] M. Rahman, Y.S. Wong and A.R. Zareena, Machinability of titanium alloys. JSME Ser C 2003; 46(1): 107–115.
http://dx.doi.org/10.1299/jsmec.46.107

[2] D.M. Dimitrov et.al., Residual stress evaluation of a high performance machined pre-formed Ti6Al4V part. Procedia CIRP 45 (2016) 155 – 158.
http://dx.doi.org/10.1016/j.procir.2016.02.049

[3] G. Bruno, B.D. Dunn, Surface and bulk residual stress in Ti6Al4V welded aerospace tanks, Transactions of the ASME Vol. 126, Aug 2004
http://dx.doi.org/10.1115/1.1763932

[4] M.A. Iadicola, T.H. Gnäupel-Herold, Effective X-ray elastic constant measurement for in situ stress measurement of biaxially strained AA5754-O. Materials Science and Engineering A 545 (2012) 168–175
http://dx.doi.org/10.1016/j.msea.2012.02.100

[5] V. Hauk, Structural and Residual Stress Analysis by Nondestructive Methods, Elsevier, 1997 112.

[6] Information on http://www.rtiintl.com/Titanium/RTI-Titanium-Alloy-Guide.pdf

[7] EN 15305:2008 Non-destructive Testing - Test Method for Residual Stress analysis by X-ray Diffraction

[8] V. Hauk, Structural and Residual Stress Analysis by Nondestructive Methods, Elsevier, 1997 135.

[9] I.C. Noyan, J.B. Cohen, Residual Stress Measurement by Diffraction and Interpretation, Springer-Verlag, 1987

[10] A. Bahadur, B. Ravi Kumar and S. Ghosh Chowdhury, Evaluation of changes in X-ray elastic constants and residual stress as a function of cold rolling of austenitic steels, Materials Science and Technology March 2004 Vol. 20
http://dx.doi.org/10.1179/026708304225012170

[11] D. Šišak Jung, L. Suominen, J. Parantainen and C. Hoermann, MYTHEN Detector – Perspectives in Residual Stress Measurements (2014) Adv. Mat. Res. 996, 203-208

Residual Stresses 2016: ICRS-10 Materials Research Forum LLC
Materials Research Proceedings **2** (2016) 67-72 doi: http://dx.doi.org/10.21741/9781945291173-12

Residual Stresses in Uniaxial Cyclic Loaded Pearlitic Lamellar Graphite Iron

Mattias Lundberg[1, a, *], Jonas Saarimäki[1, b], Ru Lin Peng[1, c], Johan J. Moverare[1, d]

[1]Department of Management and Engineering, Division of Engineering Materials,

Linköping University, SE-581 83 Linköping, Sweden

[a]mattias.lundberg@liu.se, [b]jonas.saarimaki@liu.se, [c]ru.peng@liu.se, [d]johan.moverare@liu.se

Keywords: Residual Stress, XRD, Lamellar Graphite Iron

Abstract. The mechanisms behind residual stress generation have been a topic of interest for quite some time since it is well-known that residual stresses can benefit the fatigue life of components. We have studied the residual stresses in lamellar graphite iron generated by fatigue damage. Cylindrical test specimens, with close to zero residual stresses of fully pearlitic lamellar graphite iron, manufactured and subjected to uniaxial load controlled cyclic loading, have been investigated. The load conditions used were: pure tension, pure compression, and alternating tension/compression over one thousand cycles. Measurements were performed using a four-circle goniometer Seifert X-ray machine equipped with a linear sensitive detector and a Cr-tube. Evaluation of the residual stresses were conducted using the $sin^2\psi$-method on the α-Fe {211} diffraction peak together with material removal technique to obtain depth profiles.

Introduction

It is well known that compressive residual stresses (RS) at the surface of a specimen prolongs the fatigue life. The associated increase in strain hardening of the surface, can be equally important for the specimen fatigue life since the combination of strain and compressive RS at the surface of the sample inhibits crack nucleation and propagation.

Steels, aluminium, titanium, and nickel alloys are some of the metallic materials in which work hardening at the surface can result in compressive RS and increased fatigue strength of the material [1,2]. These homogenous metallic material withstand considerable plastic deformation before final fracture compared to cast iron, especially lamellar graphite iron (LGI) which has an inhomogeneous microstructure. Lamellar graphite iron plasticises already at tensile loads ~ 40 MPa [3]. Graphite acts as notches all over the specimen volume and is the reason for its poor tensile strength.

Relaxation of near surface RS due to cyclic loading are a well-known phenomenon. Local plastic deformations (microcracks) as well as pinning and un-pinning of dislocations are believed to be a source of stress relaxation [2,4–6]. Residual stress relaxation mechanisms due to cyclic loading are affected by the initial magnitude of the RS, the gradient of RS, degree of cold work, cyclic loading and material response to cyclic loading [1]. In a multiphase material, such as pearlitic LGI, the different response of the phases to cyclic loading can render a measurable shift in RS in one of the phases. In this study, we have three phases (ferrite, cementite, lamellar graphite) but only the stresses in the ferritic phase, which has the largest volume fraction in the material, are measured with laboratory X-rays, since diffraction peaks for the other phases can not be obtained.

In multiphase materials, such as duplex steels, the different phases are known to have different amounts of RS and often also different signs on the stresses [7–9]. During cooling and solidification, the differences in volume contraction between phases gives a thermal mismatch, resulting in RS. The different thermal coefficients of the phases in cast iron (ferrite, cementite and graphite) are the source of RS in untreated material. Surface treatments and cyclic loading also result in plasticity mismatch between the different phases, causing large variations in RS between the phases which can also have different signs. The effects of these RS in a LGI under cyclic loading have not been thoroughly investigated. In this paper, the RS generated in a pearlitic LGI under cyclic loading have been investigated to fill in the knowledge gap in RS progression in cast iron.

Experimental setup

Four cylindrical specimens, with a 6.3 mm diameter of over the 25 mm gauge length, of pearlitic LGI were manufactured. The specimens were manually ground using 2400 and 4000 grit SiC paper to lower the machining effects. Afterwards, the specimens were stress relieve annealed at 600 °C for one hour and cooled slowly in air to room temperature. Surface oxides were gently removed manually with 4000 grit SiC paper. The gauge length was electrolytic polished in A2-solution to achieve a stress free surface.

Uniaxial testing was performed on three test samples in an Intron 8801 servo hydraulic test machine. The fourth sample was used as reference. A sinusoidal load cycle was used in pure tension (between 20 MPa and 200 MPa), pure compression (between -20 MPa to -200 MPa), and alternating tension/compression (± 150 MPa, starting in tension). The material ultimate tensile strength is 250 MPa. Each specimen was run for 1000 cycles.

X-ray measurements were performed using a four-circle goniometer Seifert X-ray machine, equipped with a linear sensitive detector and a Cr-tube. Evaluation of RS were conducted using the $sin^2\psi$-method [10] with the α-Fe {211} diffraction peak, at $2\theta \approx 156.5°$. A ø 1 mm collimator was used to minimize the effects of specimen curvature. Diffraction data was obtained from four different measuring points, A–D, ~ 90° rotation of the specimen between each point. At every depth all diffraction data was superimposed to obtain a good average value of the RS. Peak position was calculated using a double pseudo-Voigt curve fit.

Electron backscatter diffraction (EBSD) for grain size determination was performed using an OXFORD detector in a Hitatchi SU-70 field emission gun scanning electron microscope (FEG-SEM). The specimen was tilted to 70° with a working distance of 20 mm, acceleration voltage of 15 kV and a step size of 0.75 µm were used. The HKL software Channel 5 was used to evaluate the EBSD measurements. The angle mismatch between two neighbouring points of 10° or higher was set as definition of a grain boundary.

Results

The biaxial stresses for the four samples were calculated from the superimposed diffraction data (SDD), results depicted in Fig. 1. No significant differences can be observed between the samples.

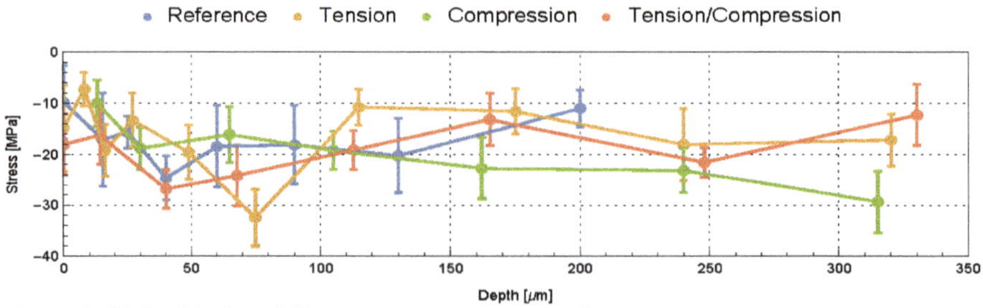

Figure 1: SDD of the four different specimens investigated.

Residual stress fluctuations and standard deviation for the reference sample are shown in Fig. 2 (a). Small tensile residual stresses for the pure tensile loaded specimen are shown in Fig. 2 (b) at some of the measured positions. The RS in the compression and tension/compression specimens in Fig. 2 (c) and (d) only show three measuring points where the RS are tensile.

Materials Research Proceedings **2** (2016) 67-72 doi: http://dx.doi.org/10.21741/9781945291173-12

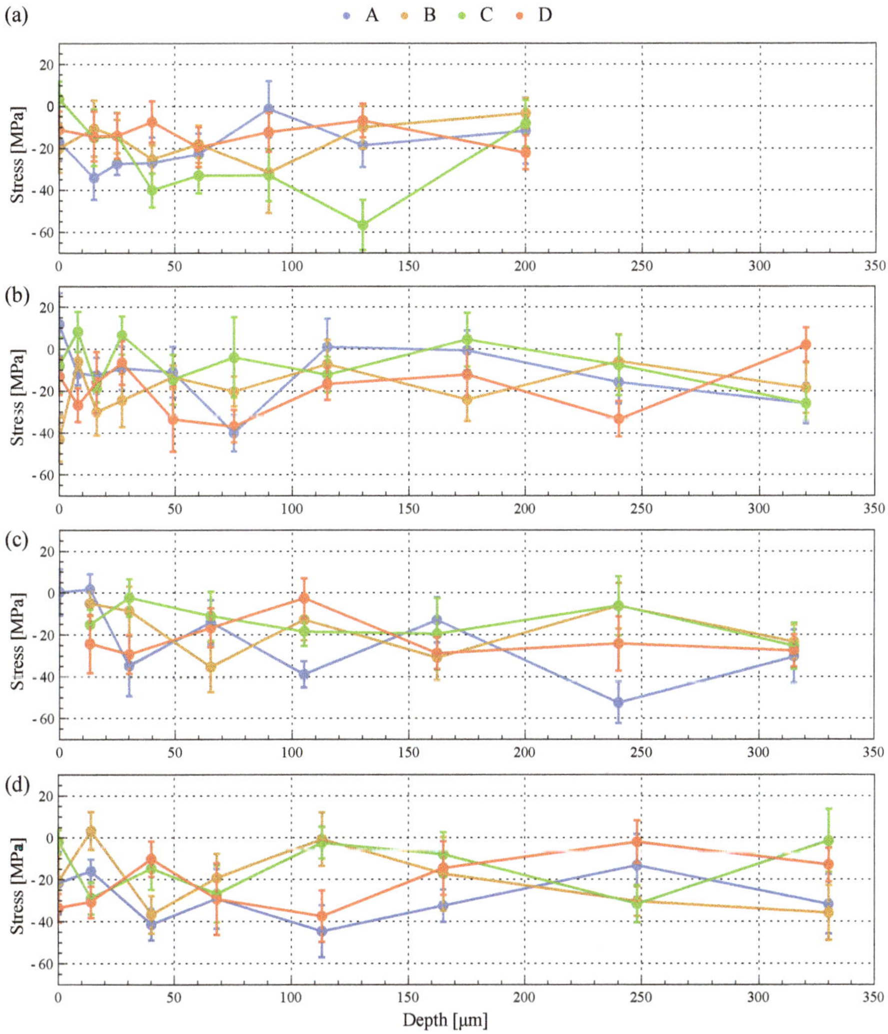

Figure 2: Residual stress profile for the four points in the (a) reference-, (b) pure tension-, (c) pure compression-, and (d) tension/compression- specimen.

All samples show an increase in Full Width at Half Maximum (FWHM) for SDD with increasing depth. No significant changes between the test specimens can be seen in Fig. 3 when comparing the average FWHM for SDD.

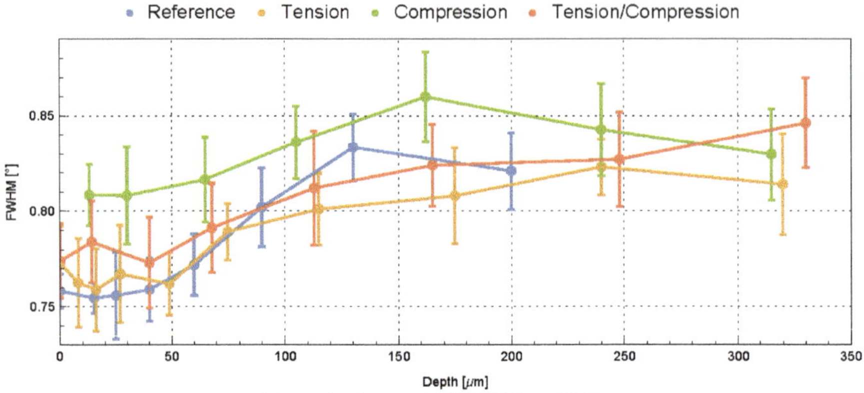

Figure 3: Averaged FWHM from SDD.

Grain size measurements conducted from EBSD-mapping of a representative area can be seen in Fig. 4. There are a few grains with a diameter larger than 200 μm and a fairly even distribution of grains in the range between 50–160 μm. Average grain size were calculated to be 100 μm with a standard deviation of 35 μm. The grain size measurement plot includes a lot of noise which results in a vast amount of "grains" in the 2–5 μm range. This noise comes from inclusions found in the matrix.

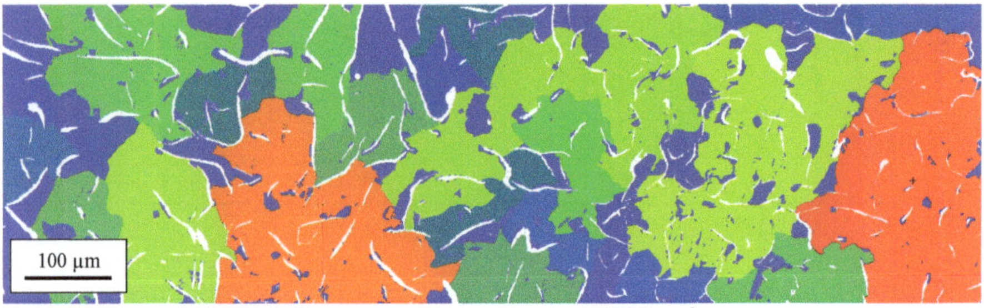

Figure 4. Grain size measurement using EBSD technique.

Discussion

Biaxial stresses for all four samples were calculated from the SDD. No significant differences can be observed between the samples. Since the accuracy of the x-ray diffraction measurements yields an uncertainty of ± 20 MPa [11] as well as the overlapping standard deviations seen in Fig. 1. The reference specimen should not show any RS, nevertheless all specimens show compressive RS. To validate the accuracy of the equipment secondary reference measurements were conducted on stress free iron powder (1.5 ± 2.5 MPa). Since the linear fits in Fig. 5 are in the range of -15 – -20 MPa and are nearly horizontal, as well as the secondary reference measurements on stress free iron powder showed zero stresses, thermal mismatch between the phases is the most probable explanation for the measured RS. The purely compressive loaded specimen indicates an increase in compressive stress with increasing depth.

Extracting diffraction data from more than one eutectic cell in cast iron requires a large enough diffraction volume i.e., large enough collimator. Convex surfaces make the use of a large collimator inappropriate due to the curvature effect giving rise to pseudo strains [12]. Using a small collimator

Materials Research Proceedings **2** (2016) 67-72 doi: http://dx.doi.org/10.21741/9781945291173-12

gives a small diffraction volume which can be effected by the heterogeneous strain in and around the eutectic cell structure. Fig. 2 shows the data obtained using a ø 1 mm collimator measuring four points evenly distributed around the cylindrical test specimen, thus resulting in diffraction data from

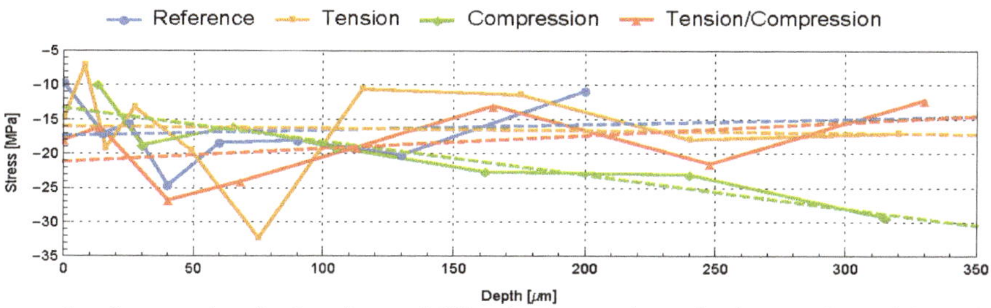

more than four eutectic cells. Superimposed diffraction data can be used to increase the statistics and to obtain the average RS in- and outside of the eutectic cells, seen in Fig. 1

Figure 5. Linear fits showing the change of RS with depth from SDD.

From the EBSD-mapping and the corresponding grain size measurement, the average grain size of 100 µm results in more than the 50 individual grains within the diffraction volume needed to obtain good RS measurements. Fluctuations seen in Fig. 2 are most likely caused by thermal mismatch between the phases and the complex nature of the materials solidification since more than 50 individual grains are within the diffraction volume for every measurement, position A–D.

The FMHW distribution at each position A–D and corresponding depth varies non-conventionally and the averaged FWHM data varied between the points A–D. The averaged FWHM from SDD in Fig. 3 is therefore used to illustrate FWHM behaviour. All specimens show an increase in FWHM with depth as a result of incomplete stress relieve annealing or compositional gradient. A slightly higher dislocation density is most likely the cause of the higher FWHM values exhibited by the pure compression specimen.

The material will be strained under the influence of a mechanical load. The different macroscopic behaviour of LGI under tension and compression is related to its lamellar graphite showing notch effect under tension. If the load distribution over the whole cross-section is uniform and there is no surface effect, little RS would be expected after a low plastic loading either in compression or tension. The lower yield strength of the ferrite than the cementite can generate detectable RS under a certain degree of plastic deformation. RS will be different depending on the loading mode given that the effect of machine hardening had been removed and the specimen was stress relieve annealed. However, only a small intimation of the pure compression specimen can be seen in Fig. 5. The lack of this intimation for the other specimens is presumably due to stress relaxation mechanisms. It is well known that stress relaxation mainly occurs during the first cycles. One thousand cycles were applied to ensure a more homogenous strain field in the specimen, because of the anisotropic behaviour of the different phases. Due to the stress relaxation mechanisms, implementing one to three cycles would probably be more suited to detect large RS differences. Stress relaxation and its effects on fatigue life in cast iron need to be further investigated.

Conclusions

In this paper, the residual stresses in a pearlitic lamellar graphite iron after cyclic loading have been investigated to better understand the role of the different phases and their load capacity under influence of loadings. Four cylindrical test specimen of fully pearlitic lamellar graphite iron were normalized followed by testing and residual stress measurement conducted with laboratory X-rays.

- There are no significant changes between the reference, pure tension, pure compression, and alternating tension/compression specimens in stresses or FWHM.
- Superimposed diffraction data (SDD) gives a more reliable result.
- Due to the stress relaxation mechanisms, implementing one to three cycles would likely be more suited to detect large residual stress differences.

Acknowledgments
Agora Materiae and AFM Strategic Faculty Grant SFO-MAT-LiU#2009-00971 at Linköping University are acknowledged. Scania CV AB for the material and Viktor Norman for operational help with the MTS machine.

References
[1] H. Wohlfahrt, Shot peening and residual stresses, in: Sagamore Army Mater. Res. Conf. Proc., 1981: pp. 71–92.
[2] I. Nikitin, M. Besel, Correlation between residual stress and plastic strain amplitude during low cycle fatigue of mechanically surface treated austenitic stainless steel AISI 304 and ferritic-pearlitic steel SAE 1045, Mater. Sci. Eng. A. 491 (2008) 297–303.
http://dx.doi.org/10.1016/j.msea.2008.03.034
[3] M. Lundberg, M. Calmunger, R.L. Peng, In-situ SEM / EBSD Study of Deformation and Fracture Behaviour of Flake Cast Iron, in: 13th Int. Conf. Fract., 2013: pp. S12–038.
[4] W.Z. Zhuang, G.R. Halford, Investigation of residual stress relaxation under cyclic load, Int. J. Fatigue. 23, Supple (2001) 31–37.
http://dx.doi.org/10.1016/S0142-1123(01)00132-3
[5] J.D. Almer, J.B. Cohen, B. Moran, The effects of residual macrostresses and microstresses on fatigue crack initiation, Mater. Sci. Eng. A. 284 (2000) 268–279.
http://dx.doi.org/10.1016/S0921-5093(99)00779-0
[6] K. Zhan, C.H. Jiang, V. Ji, Residual Stress Relaxation of Shot Peened Deformation Surface Layer on S30432 Austenite Steel under Applied Loading, Mater. Trans. 53 (2012) 1578–1581.
http://dx.doi.org/10.2320/matertrans.M2012111
[7] N. Jia, R.L. Peng, Y.D. Wang, G.C. Chai, S. Johansson, G. Wang, P.K. Liaw, Interactions between the phase stress and the grain-orientation-dependent stress in duplex stainless steel during deformation, Acta Mater. 54 (2006) 3907–3916.
http://dx.doi.org/10.1016/j.actamat.2006.04.019
[8] M.L. Martinez-Perez, F.J. Mompean, J. Ruiz-Hervias, C.R. Borlado, J.M. Atienza, M. Garcia-Hernandez, M. Elices, J. Gil-Sevillano, R.L. Peng, T. Buslaps, Residual stress profiling in the ferrite and cementite phases of cold-drawn steel rods by synchrotron X-ray and neutron diffraction, Acta Mater. 52 (2004) 5303–5313.
http://dx.doi.org/10.1016/j.actamat.2004.07.036
[9] J.J. Moverare, M. Odén, Deformation behaviour of a prestrained duplex stainless steel, Mater. Sci. Eng. A. 337 (2002) 25–38.
http://dx.doi.org/10.1016/S0921-5093(02)00022-9
[10] I.C. Noyan, J.B. Cohan, Residual stress - Measurement by Diffraction and Interpretation, 1987.
[11] P.J. Withers, H.K.D.H. Bhadeshia, Residual stress. Part 1–measurement techniques, Mater. Sci. Technol. 17 (2001) 355–365.
http://dx.doi.org/10.1179/026708301101509980
[12] I.C. Noyan, J.B. Cohan, Residual stress - Measurement by Diffraction and Interpretation, 1987.

Residual Stresses 2016: ICRS-10 Materials Research Forum LLC
Materials Research Proceedings **2** (2016) 73-78 doi: http://dx.doi.org/10.21741/9781945291173-13

3D Residual Stresses in Selective Laser Melted Hastelloy X

Jonas Saarimäki[†, 1, a, *], Mattias Lundberg[†, 1, b], Johan J. Moverare[1, c], Håkan Brodin[2, d]

[1]Department of Management and Engineering, Division of Engineering materials, Linköping University, Linköping, SE-58183, Sweden

[2]Siemens Industrial Turbomachinery AB, Finspång, SE-612 31, Sweden

[†]Joint first author

[a]Jonas.saarimaki@liu.se, [b]mattias.lundberg@liu.se, [c]johan.moverare@liu.se, [d]hakan.brodin@siemens.com

Keyword: Triaxial Stress, SLM, HX

Abstract. 3D residual stresses in as manufactured EOS NickelAlloy HX, produced by laser powder bed additive manufacturing, are analysed on the surface closest to the build-plate. Due to the severe thermal gradient produced during the melting and solidification process, profound amounts of thermal strains are generated. Which can result in unwanted geometrical distortion and effect the mechanical properties of the manufactured component. Measurements were performed using a four-circle goniometer Seifert X-ray machine, equipped with a linear sensitive detector and a Cr-tube. Evaluation of the residual stresses was conducted using $sin^2\psi$ method of the Ni {220} diffraction peak, together with material removal technique to obtain in-depth profiles. An analysis of the material is reported. The analysis reveals unwanted residual stresses, and a complicated non-uniform grain structure containing large grains with multiple low angle grain boundaries together with nano-sized grains. Grains are to a large extent, not equiaxed, but rather elongated.

Introduction

Additive manufacturing, free form fabrication, rapid prototyping and 3D-printing are some of the different designations for processes where components can be built to finished or near-finished shape without machining a block of material or casting material in a mould [1–3]. The processes were primarily developed for simpler materials, such as thermoset plastics and plaster. The lasers equipment originally used could only melt materials with low melting points, for instance brass, and was not powerful enough to completely melt steel. Therefore, this manufacturing method could not meet the requirements for parts subjected to high stress levels or elevated temperatures, e.g., superalloys [4]. With time, the process control was improved and more powerful lasers were developed. With the higher input possible from a more powerful laser it is possible to create a microstructure with a low amount of porosity and no internal defects such as solidification cracks or poor bonding [5].

Free-form fabrication of superalloys is gaining increased interest from the industry, since the available range of alloys is growing. Today, alloys for selective laser melting (SLM) include aluminium, titanium, tool steel, stainless steel and heat resistant materials of cobalt- and nickel-base. In the case of melting of metal powders, the dominating manufacturing process is laser melting, often denoted selective laser melting, direct laser metal sintering (DMLS) or LaserCUSING. All of these names are trademarks for different companies manufacturing equipment for laser melting.

The laser melting manufacturing process can briefly be described as a layer-by layer process, where powder is distributed on a powder bed, see Fig. 1. Firstly, a powder distributer travels over the powder bed cavity contained by the build chamber walls and build plate. Molten and solidified powder constitutes the component surrounded by un-molten powder. Secondly, a laser beam melts the powder layer and creates a new slice of solid material in the component. Thirdly, a ram lowers the build platform and the process is repeated until a finished geometry is formed. After finalisation, the remaining loose powder is removed and the component is cut off from the build platform.

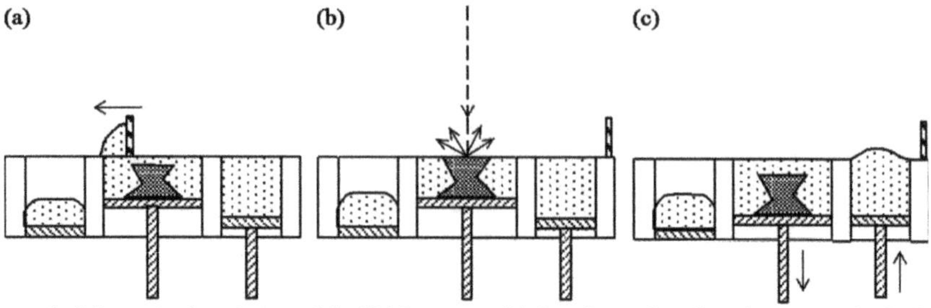

Figure 1. Schematic description of the SLM process. (a) Powder is distributed on a powder bed, the build platform. (b) The powder is melted by a laser beam and a slice of solid metal is formed. (c) The powder bed is lowered and the process is repeated until a finished component is formed.

Although selective laser melting allows manufacturing of complex geometries, it comes with drawbacks compared to the conventional manufacturing technologies. The temperature gradient and consequent plastic deformation leads to residual stresses and deformation due to the locally focused energy input [6]. Residual stresses can influence the geometrical accuracy and mechanical strength as well as contribute to crack initiation. Previous research has been conducted using methods such as the crack compliance method which is not suited for near surface stresses [7], hole drilling which requires large dimensional sizes and smooth surfaces to be effective [8]. In this study, x-ray diffraction was used to measure surface and bulk stresses using material removal technique.

The purpose of this study is to examine residual stress levels in as manufactured SLM material. No post processing or heat treatments were done prior to testing since gas atomised EOS NickelAlloy HX powder is used for manufacturing e.g., gas turbine burners used in the as manufactured state.

Experimental details

The material used in the current study is manufactured from powder EOS NickelAlloy HX. In literature Hastelloy X can be identified as Alloy X, when not available from the original manufacturer. The powder material is gas atomized and sieved to a fraction (10–45 μm) suitable for the SLM process. After manufacturing no post-processing, such as heat treatment or hot isostatic pressing, was conducted. The nominal composition in wt.% of EOS NickelAlloy HX is shown in Table 1. During the SLM manufacturing the test specimen was attached to the building platform via area A in Fig. 2. After manufacturing, the test specimen was removed from the platform by wire electro discharge machining. The typical microstructure of the laser melted material after manufacturing is shown in Fig. 3-4, where the building direction is indicated by the arrows. Previous work by Brodin *et al.* [9] on alloy X has shown that material bulk properties meet or exceed the properties of both hot-rolled and cast Hastelloy X in heat treated condition.

Table 1. Chemical composition, EOS NickelAlloy HX [wt.%].

Ni	Cr	Fe	Mo	W	Co	C	Si
Bal.	20.5–23.0	17.0–20.0	8.0–10.0	0.2–1.0	0.5–2.5	≤0.1	≤0.1
Mn	S	P	B	Se	Cu	Al	Ti
≤0.1	≤0.03	≤0.04	≤0.01	≤0.005	≤0.5	≤0.5	≤0.15

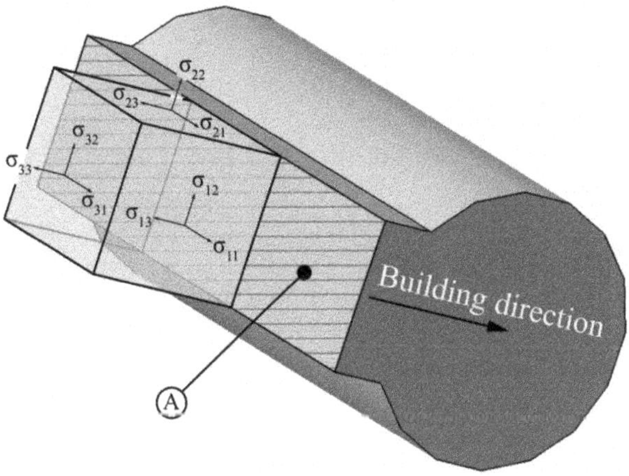

Figure 2. Sample geometry and stress components where A denotes build plate attachment area and the arrow the building direction.

Triaxial X-ray measurements [10] were performed using a four-circle goniometer Seifert X-ray machine, equipped with a linear sensitive detector, Cr-tube and a 2 mm spot collimator. The {220} diffraction peak of γ-phase was used for the stress measurement with the $sin^2\psi$-method. Values off the x-ray elastic constants used were calculated using the Young's modulus (190 GPa) and Poison's ratio (0.31) based on the work by Saarimäki [11] resulting in: $s_1 = -1.63\times10^{-6}$ and $\frac{1}{2}s_2 = 6.89 \times 10^{-6}$ MPa. Which differ from the experimental work on single crystal Hastelloy X [12]. Raw powder from the distributor was used and analysed for two ϕ angles (D1 = 0° and D2 = 90°) together with 13 ψ angles, evenly distributed between ± 60° to determine the unstressed lattice spacing, d_0. No corrections were made considering the stress redistribution due to material removal. The orientation imaging map (OIM) in Fig. 4 was obtained using electron backscatter diffraction (EBSD) with a step size of 1.5 µm and 15 kV in a Hitachi SU70 FEG analytical scanning electron microscope (SEM).

Results

The EBSD analysis revealed a complicated non-uniform grain structure containing large grains with several low angle grain boundaries together with nano-sized grains. Grains are to a large extent, not equiaxed, but rather elongated, depicted in Fig. 3-4.

Stress free lattice spacing, d_0, determination [13] of the gas atomized powder is shown in Fig. 5. The stress free lattice spacing d_0 was calculated by fitting two linear models and computing the intersect point, resulting in d_0 = 1.2728 Å.

The directional grains could cause a crystallographic texture as well as anisotropy leading to the non-linear behaviour seen in Fig. 6.

Initial surface measurements in Fig. 7 reveal large tensile stresses (σ_{11} = 660 MPa, σ_{22} = 740 MPa, σ_{33} = 755 MPa). A steep gradient is evident from the surface to a depth of ~20 µm, with the lowest residual stresses obtained in the principal directions (σ_{11}= -70 MPa, σ_{22} = 30 MPa,

σ_{33} = 100 MPa). From 20 – 45 µm, the stress in the σ_{11} direction increase greatly. After this depth the changes in stresses reduces and are fairly stable in value. At 445 µm depth, the residual stresses in the principal directions are as follows: σ_{11}= 400 MPa, σ_{22} = 90 MPa and σ_{33} = 85 MPa.

σ_{12} shows compressive stresses at the surface which approaches zero stress level at 445 µm. Little or no shear stresses were calculated for σ_{13} and σ_{23}.

Materials Research Proceedings **2** (2016) 73-78 doi: http://dx.doi.org/10.21741/9781945291173-13

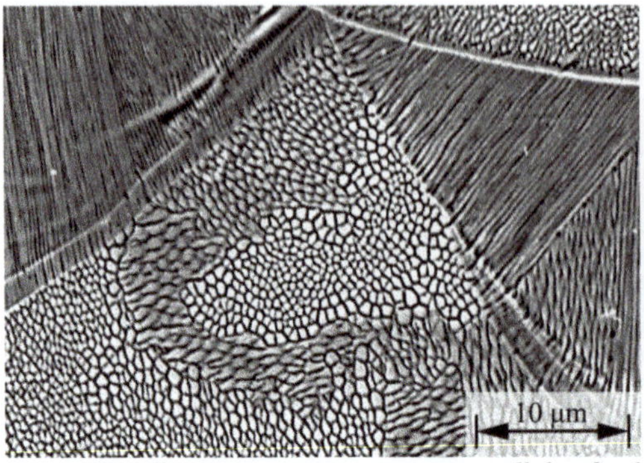

Figure 3. Bulk microstructure comprising cellular dendrites. Figure 4. Orientation imaging map.

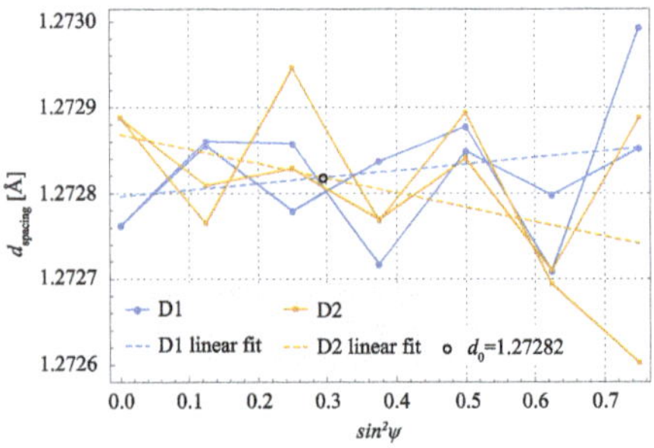

Figure 5. d_0 determination.

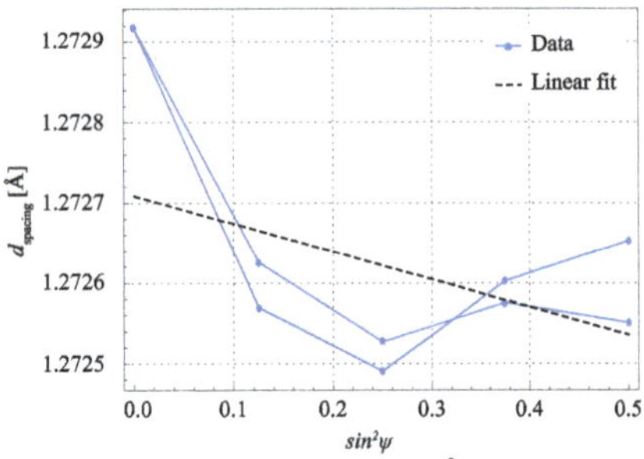

Figure 6. $d_{spacing}$ vs. $sin^2\psi$ at 45 μm and $\phi = 45°$.

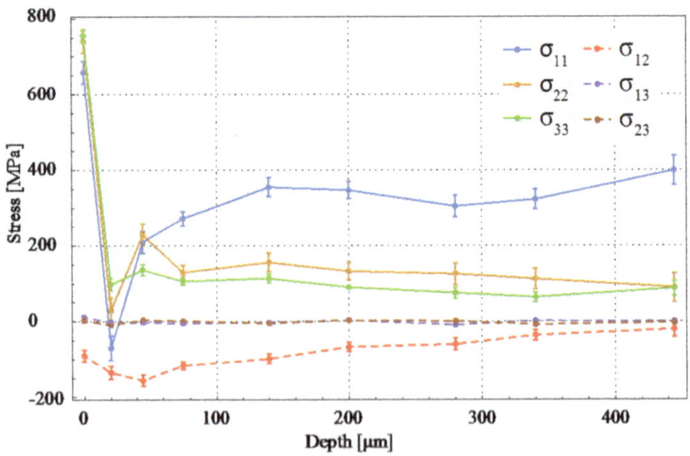

Figure 7. Residual stress profile.

Discussion

Electron backscatter diffraction together with an OIM was used to analyse the microstructure in Fig. 3-4. The analysis revealed a non-uniform grain structure containing large elongated (not equiaxed) grains with many low angle grain boundaries together with nano-sized grains. The grains are oriented directionally, parallel to the building direction giving this unconventional microstructure. The directional grains could cause a crystallographic texture as well as anisotropy leading to the non-linear behaviour in Fig. 6.

The gas atomised powder used to determine the stress free lattice spacing d_0 was not completely stress free, as seen in Fig. 5. The D1 linear fit has a positive slope and the D2 linear fit has a negative slope. The presence of residual stresses is likely due to the rapid cooling from molten to solid state since the powder is gas atomised. The assumption to use the gas atomised powder for d_0 determination is reasonable since both the powder and the alloy are free from precipitates. Furthermore, the cellular dendritic SLM material and the dendrites in the gas atomised powder particles have similar primary dendrite arm spacing's.

The residual stresses in the specimen are significant since warping of the specimen is seen just by looking at it. The high surface residual stress values (σ_{11}= 660 MPa, σ_{22}= 740 MPa, σ_{33}= 755 MPa) shown in Fig. 7 are above the yield stress but lower than the tensile strength (676 MPa in the relevant direction, i.e., horizontal), as reported by Brodin and Saarimäki [14]. Electric discharge machining, can change the microstructure to a depth of 5 – 10 μm. This could generate the steep stress gradient observed at the first measured points. Within the measured volume, all stresses except σ_{11} decrease from 145 μm to a depth of 440 μm at which the stresses are fairly stable. Even though the residual stresses should approach zero and eventually become compressive in the bulk, the stresses extend to a greater depth than investigated here. The high σ_{11}stress level is responsible for the specimen deformation caused by the layer by layer build process during solidification and cooling.

Biaxial stress state assumes that there is a linear relationship between $d_{spacing}$ and $sin^2\psi$ [15]. However, a biaxial stress state cannot be assumed in this case since in depth and shear components are $\neq 0$ which results in the poor linear fit in Fig. 6. Thus, the triaxial stress calculations performed are believed to better reflect the actual stress state in the specimen.

Due to the lack of data regarding the x-ray elastic constants (XEC) s_1 and $\frac{1}{2}s_2$, they were calculated using the Young's modulus (bulk parameter) and v. However, these calculated values could render non-negligible errors. Using the XEC for single crystal Hastelloy X [12] or Inconel 718 [16] would generate a shift in measured stresses of approximately 30 %. Hence, accurate XEC need to be determined.

Conclusions

Residual stress measurements and microstructural analysis was conducted on SLM material from EOS NickelAlloy HX powder. We show that the microstructure is fine-grained and the grains are elongated along the build direction. It is clear that the grains are allowed to grow over several layers during the building process. Electric discharge machining locally changes the microstructure resulting in the steep decline in residual stress from the surface (0 – 20 μm).

All stresses except σ_{11} decrease from 145 μm to a depth of 440 μm where σ_{11} is still increasing. Assuming a biaxial stress state is misleading since the SLM process induce large out of plane residual stresses. Accurate XEC need to be determined due to the lack of data regarding the x-ray elastic constants (XEC) s_1 and $\frac{1}{2}s_2$ for SLM EOS NickelAlloy HX.

Acknowledgements

The authors would like to thank Agora Materiae, graduate school, Faculty grant SFO-MAT-LiU#2009-00971, and the project teams at Linköping University, Siemens Industrial Turbomachinery AB for valuable discussions.

Materials Research Proceedings **2** (2016) 73-78 doi: http://dx.doi.org/10.21741/9781945291173-13

References

[1] J. Chu, S. Engelbrecht, G. Graf, D.W. Rosen, A Comparison of Synthesis Methods for Cellular Structures with Application to Additive Manufacturing, Rapid Prototyp. J. 16 (2010) 275–283. *http://dx.doi.org/10.1108/13552541080000501*

[2] D.W. Rosen, Computer-Aided Design for Additive Manufacturing of Cellular Structures, Comput. Aided. Des. Appl. 4 (2007) 585–594. doi:10.1080/16864360.2007.10738493. *http://dx.doi.org/10.1080/16864360.2007.10738493*

[3] G. Marchelli, R. Prabhakar, D. Storti, M. Ganter, The guide to glass 3D printing: developments, methods, diagnostics and results, Rapid Prototyp. J. 17 (2011) 187–194. *http://dx.doi.org/10.1108/13552541111124761*

[4] S. Das, J.J. Beaman, M. Wohlert, D.L. Bourell, Direct laser freeform fabrication of high performance metal components The authors, Rapid Prototyp. J. 4 (1998) 112–117. *http://dx.doi.org/10.1108/13552549810222939*

[5] M. Agarwala, D. Bourell, J. Beaman, H. Marcus, J. Barlow, Direct selective laser sintering of metals, Rapid Prototyp. J. 1 (1995) 26–36. doi:10.1108/13552549510078113. *http://dx.doi.org/10.1108/13552549510078113*

[6] P.J. Withers, H.K.D.H. Bhadeshia, Residual stress. Part 2 – Nature and origins, Mater. Sci. Technol. 17 (2001) 366–375. doi:10.1179/026708301101510087. *http://dx.doi.org/10.1179/026708301101510087*

[7] P. Mercelis, J.-P. Kruth, Residual stresses in selective laser sintering and selective laser melting, Rapid Prototyp. J. 12 (2006) 254–265. doi:10.1108/13552540610707013. *http://dx.doi.org/10.1108/13552540610707013*

[8] C. Casavola, S.L. Campanelli, C. Pappalettere, Preliminary investigation on distribution of residual stress generated by the selective laser melting process, J. Strain Anal. Eng. Des. 44 (2009) 93–104. doi:10.1243/03093247JSA464. *http://dx.doi.org/10.1243/03093247JSA464*

[9] H. Brodin, O. Andersson, S. Johansson, Mechanical Behaviour and Microstructure Correlation in a Selective Laser Melted Superalloy, in: ASME Turbo Expo 2013 Turbine Tech. Conf. Expo., San Antonio, Texas, USA, 2013. doi:10.1115/GT2013-95878. *http://dx.doi.org/10.1115/GT2013-95878*

[10] H. Dölle, The influence of multiaxial stress states, stress gradients and elastic anisotropy on the evaluation of (Residual) stresses by X-rays, J. Appl. Crystallogr. 12 (1979) 489–501. *http://dx.doi.org/10.1107/S0021889879013169*

[11] J. Saarimäki, The mechanical properties of lattice truss structures with load- bearing shells made of selectively laser melted Hastelloy X™, KTH Royal Institute of Technology, 2011. *http://urn.kb.se/resolve?urn=urn:nbn:se:kth:diva-41320.*

[12] H.A. Canistraro, E.H. Jordan, S. Shixiang, L.H. Favrow, F.A. Reed, Elastic Constants of Single Crystal Hastelloy X at Elevated Temperatures, Trans. ASME. 120 (1998) 242–247. doi:10.1115/1.2812350. *http://dx.doi.org/10.1115/1.2812350*

[13] V.M. Hauk, R.W.M. Oudelhoven, G.J.H. Vaessen, The state of residual stress in the near surface region of homogeneous and heterogeneous materials after grinding, Metall. Trans. A. 13 (1982) 1239–1244. doi:10.1007/BF02645507. *http://dx.doi.org/10.1007/BF02645507*

[14] H. Brodin, J. Saarimäki, Mechanical properties of lattice truss structures made of a selective laser melted superalloy, in: 13th Int. Conf. Fract., Beijing, China, 2013: pp. 1–10. *http://urn.kb.se/resolve?urn=urn:nbn:se:liu:diva-95433.*

[15] I.C. Noyan, J.B. Cohen, Residual stress Measurement by diffraction and Interpretation, 1987.

[16] P.S. Prevey, A method of determening the elastic properties of alloys in selected crystallographic directions for x-ray diffraction residual stress measurement, Adv. X-Ray Anal. 20 (1977) 345–354. *http://dx.doi.org/10.1007/978-1-4613-9981-0_30*

Materials Research Proceedings **2** (2016) 79-84 doi: http://dx.doi.org/10.21741/9781945291173-14

Numerical Simulation of Residual Stresses Induced by Weld Repair in a Stainless Steel Pipe Considering the Influence of an Initial Fabrication Weld

Gervasio Salerno[1,a *], Chris Bennett[1,b], Wei Sun[1,c] and Adib A. Becker[1,d]

[1]Gas Turbine and Transmissions Research Group, Faculty of Engineering, The University of Nottingham, University Park, Nottingham, NG7 2RD, UK

[a]epxgs3@nottingham.ac.uk, [b]chris.bennett@nottingham.ac.uk, [c]w.sun@nottingham.ac.uk, [d]a.a.becker@nottingham.ac.uk

Keywords: Finite Element Modelling, Weld Repair, Residual Stress

Abstract. This work presents the application of a finite element (FE) model developed to simulate the repair process in the case of components with a pre-existing stress state. The approach is tested in the case of a repair of a laser beam weld in a stainless steel pipe with the region of repair located in the heat affected zone of the original weld. The area of the repair is removed and refilled testing different approaches in terms of the number, and direction of the repair passes. The comparison between the refilling procedures is presented with the aim of evaluating the effects on the final residual stress distribution.

Introduction

Weld repair is commonly adopted to restore the integrity of a weld when defects are detected. According to the size and position of the defect, different repair geometries may be used, including shallow or deep and short or long excavations, resulting in different residual stress distributions. In the literature, only a few works are available where the attention was focused on the numerical simulation of weld repair. On the one hand the benefits of a 2D simulation in computational welding mechanics have been highlighted by Dong et al. [1], Brown et al. [2], Kim et al. [3] and Limpus et al. [4]: easier modelling procedure and smaller models which require lower computing power. On the other hand, Feng et al. [5] as well as Bonnaud and Gunnars [6] pointed out that only 3D simulations can capture effects due to a repair, for instance the details at start/stop locations. Brust and Rudland [7] discussed the merits and limitation of 2D versus 3D solutions including several examples of their simulations: a partial circumferential pipe repair, an industrial weld fabrication and ship structure weld analyses.

In this work, the repair of a laser beam weld in a stainless steel pipe is simulated by means of a numerical FE approach developed by Salerno et al [8] adopting 3D simulations. The area of the repair is removed and refilled testing different strategies in terms of the number and direction of the repair passes. The comparison between the refilling procedures is presented highlighting the effects on the final residual stress distribution.

Case study

A laser beam welding (LBW) was used to join two pipes made of SUS316 stainless steel to create a whole unique pipe with length 800 mm. The work from Deng and Kiyoshima [9] was used as a reference for the simulation of the fabrication weld. The inside diameter of the pipe was 200 mm, and the wall thickness 10 mm. As a small defect was assumed to appear at the boundary of the HAZ and the fusion zone, the repair procedure was carried out by removing a slot 5 mm deep through the thickness, 5 mm wide and covering an arc of 40°, located as shown in Fig1.

Materials Research Proceedings **2** (2016) 79-84 doi: http://dx.doi.org/10.21741/9781945291173-14

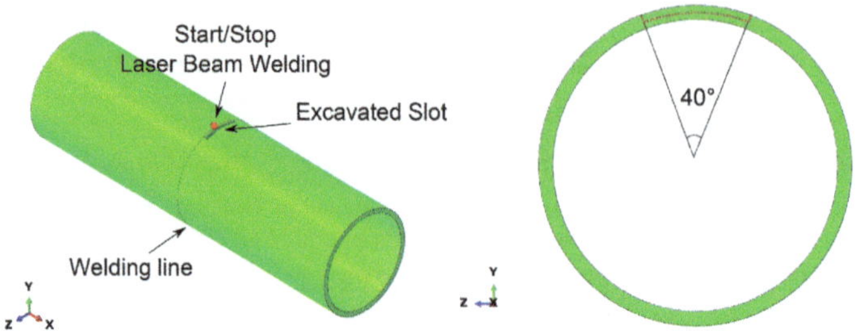

Fig. 1 Fabrication weld and excavated slot

To obtain full penetration through the wall thickness, the power of the LBW and the speed were set as 12 kW and 1000mm/min, respectively as by Deng and Kiyoshima [9]. It was assumed that the slot was refilled using TIG welding. The weld power for the TIG process was set with a trial/error approach to ensure the elements simulating the filler being deposited reached the melting temperature. Different refilling approaches were investigated as described in Table 1. The terms Same Direction (SD) and Opposite Direction (OD) refer to the direction of the original LBW. In the case of two passes, these were assumed to be on top of each other, with the first one filling half the depth of the slot.

Table 1 Refilling procedure investigated

Case A	One pass - Same Direction (SD)
Case B	One pass - Opposite Direction (OD)
Case C	Two passes OD + OD
Case D	Two passes OD + SD
Case E	Two passes SD + SD
Case F	Two passes SD + OD

FE model

The computational procedure developed by Salerno et al. [8] was adopted in the present study using the commercial FE software Abaqus. A view of half mesh and cross section is shown in Fig. 2.

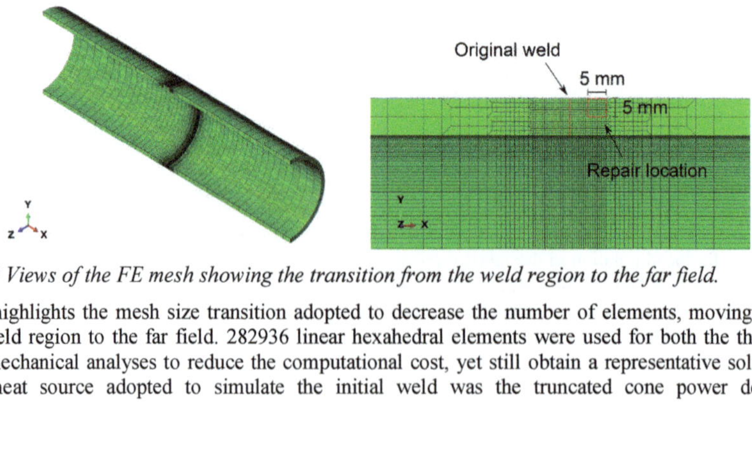

Fig. 2 Views of the FE mesh showing the transition from the weld region to the far field.

This highlights the mesh size transition adopted to decrease the number of elements, moving from the weld region to the far field. 282936 linear hexahedral elements were used for both the thermal and mechanical analyses to reduce the computational cost, yet still obtain a representative solution. The heat source adopted to simulate the initial weld was the truncated cone power density

Materials Research Proceedings **2** (2016) 79-84 doi: http://dx.doi.org/10.21741/9781945291173-14

distribution which is commonly chosen to simulate beam welding processes while the double ellipsoid was selected for the simulation of the TIG weld repair [10]. The geometric parameters of the heat sources were chosen to produce a plausible weld pool in both cases as shown in Figs. 3 and 4.

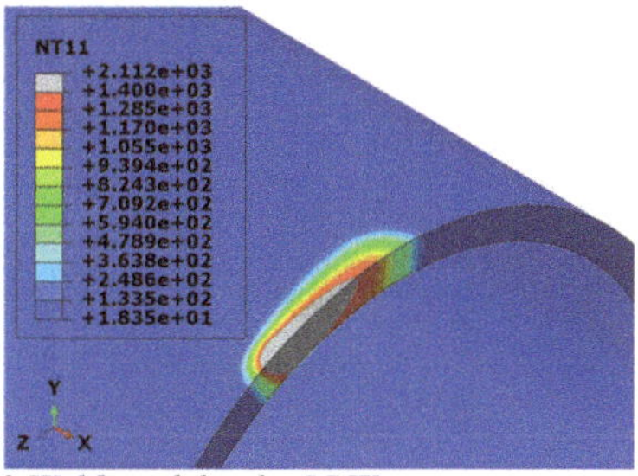

Fig. 3 Weld pool for the LBW

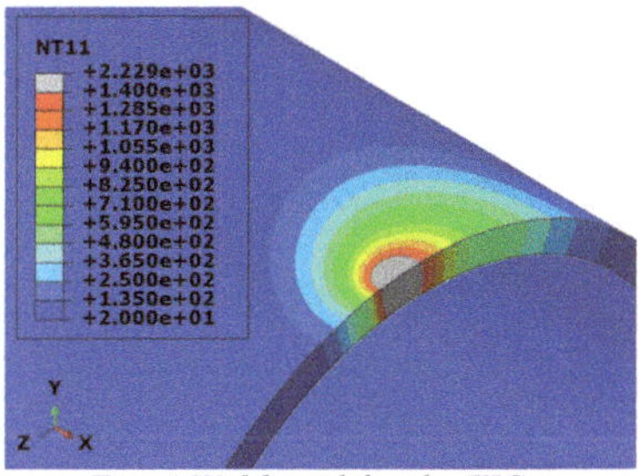

Fig. 4 Weld pool for the TIG

In the mechanical analyses, the elastic behavior was modelled using the isotropic Hooke's law while the von Mises yield criterion was adopted to determine the onset of plastic behavior. Strain hardening was taken into account using a linear isotropic hardening law. The annealing temperature option was used to simulate the loss of the hardening memory of the material. This was assumed to be 800°C for SUS316 stainless steel [11]. When the temperature of a material point is higher than the annealing temperature, the equivalent plastic strain is reset to zero. If the temperature of a material point falls below the annealing temperature at a subsequent point in time, the material point can work harden again. Microstructural changes were not considered in the analyses. The boundary conditions were assigned in three nodes located on the one end of the pipe to avoid rigid body motions, constraining node 1 along X,Y,Z, node 2 along X,Y and node 3 along X.

Results and discussion

Numerical predictions for the LBW were verified by means of the numerical results from Deng and Kiyoshima [8] both in terms of thermal and mechanical analyses. In Fig.5 hoop and axial residual stresses due to the LBW and the repair procedures investigated are presented on the paths shown. The effects of the repair on the initial residual stress are evident, causing a redistribution in both the axial and hoop stress. The repair approaches investigated induce visible differences on the axial stress (Fig.5a). The highest tensile stresses in the axial distribution are regulated by the direction of the repair pass both for one and two passes. These appear approximately at the boundary of the slot being refilled (70° and 110°) due to the restraint imposed by the surrounding material. For the same reason, lower local maxima of axial residual stress are visible at the starting location when the material is deposited in two passes. These are not present when the deposition is simulated in a single pass. In the case of two passes, the direction of the second pass governs the final distribution of the axial stress which appears to be comparable to the one obtained with a single pass in the same direction of the second one. The hoop stress is not as affected by the repair strategy adopted as the axial stress. Trends on the two axial paths considered are comparable for all the cases analyzed both for the axial and hoop (Figs.5a and 5b). Along the axial path on the O.D. the hoop stress becomes compressive because of the repair pass while it stays tensile for the path on the I.D. As highlighted in Table 2, the maximum and minimum axial and hoop stresses are not significantly influenced by the repair approach adopted. The highest axial stresses are in the case of two passes when the second pass is in the same direction of the LBW. The highest hoop stress is not due to the repair.

Materials Research Proceedings **2** (2016) 79-84 doi: http://dx.doi.org/10.21741/9781945291173-14

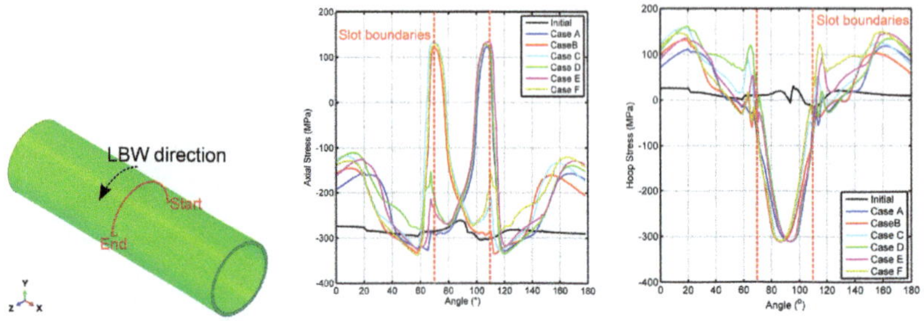

(a – Semi-circumferential path located at the centreline of the repair weld)

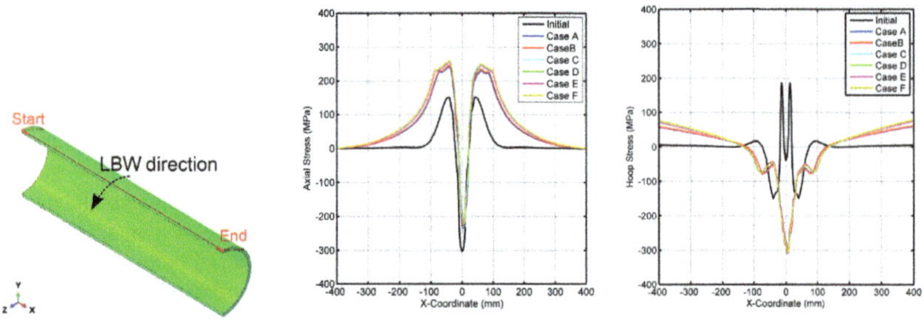

(b – Axial path located at the start of the initial weld)

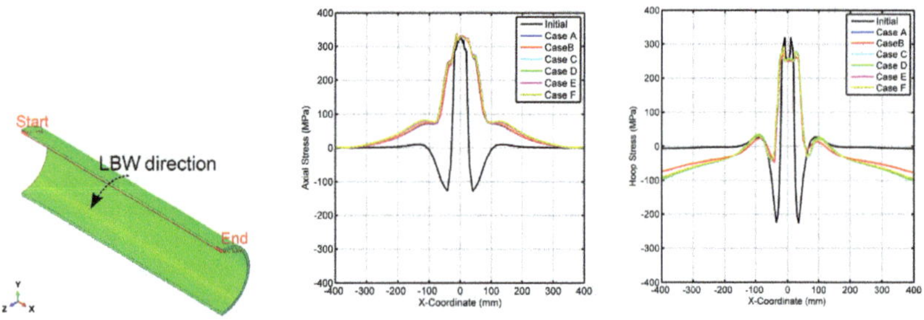

(c– Axial path located at the start of the initial weld)

Fig. 5 Axial and Hoop stress on highlighted paths. O.D. (a-b) and I.D. (c)

Table 2 Maximum and minimum axial and hoop stresses in MPa for the cases studied.

	Axial - Max	Axial - Min	Hoop - Max	Hoop - Min
LBW	350	-314	354	-257
Case A	410	-353	354	-320
Case B	409	-353	354	-320
Case C	435	-361	354	-321
Case D	425	-360	354	-320
Case E	439	-361	354	-321
Case F	422	-361	354	-321

Materials Research Proceedings **2** (2016) 79-84 doi: http://dx.doi.org/10.21741/9781945291173-14

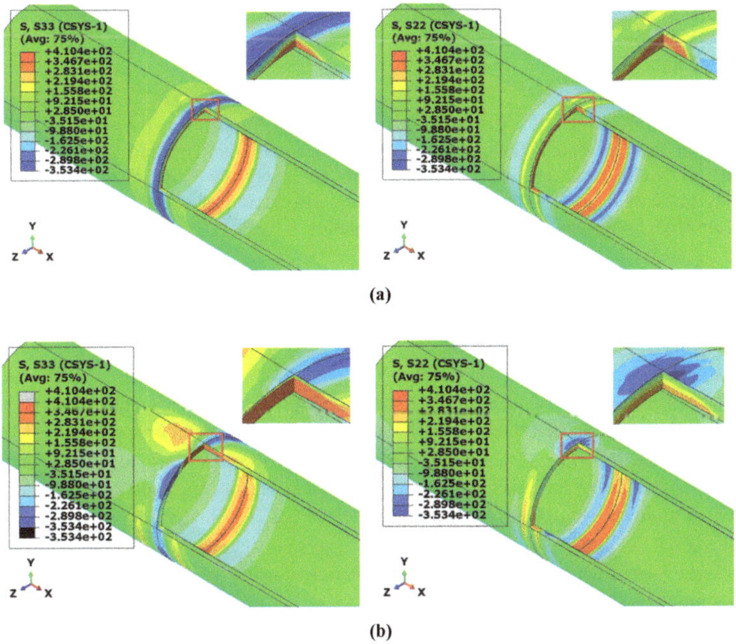

Fig. 6 Axial (left) and Hoop (right) residual stress in the pipe (Mpa): Initial residual stress (a) - Case A (b)

Fig. 6 shows the distribution of the axial and hoop stress after the LBW and the case A. A quarter of the pipe was hidden for convenience to evaluate the distribution of the stress across the thickness. Tensile regions appear in the axial stress distribution because of the repair with highly compressive areas located at the start/stop area while a compressive stress is visible in the hoop distribution in the repair region. However the field results do not highlight relevant differences for the repair strategies analyzed.

Conclusions and future work

- The initial residual stress field redistributes because of the repair procedure whatever repair strategy is adopted.
- The number of passes and the deposition sequence simulated do not appear to have a significant effect on the magnitude of residual stress, both for the axial and the hoop stress. The numerical model suggests there is no relevant benefit of using a repair strategy or another in terms of lowering the maximum tensile stress.
- The distribution of the hoop stress is less affected by the number of passes and their direction compared to the axial stress.
- In the case of two passes the second pass governs the distribution of the final axial stress. This appears very similar to the one obtained with a single pass in the same direction as the second one. Therefore, if the aim of the analysis is to obtain a quicker, yet still representative solution, it is sufficient to simulate the material deposition in a single step with the deposition direction of the final pass.

Future work will include the investigation of different slot geometries and refilling strategies, welding power for the repair procedure and the microstructural evolution driven by the weld passes on the material properties.

Acknowledgments
The authors wish to thank Rolls-Royce plc for their financial support of the research, which was carried out at the University Technology Centre in Gas Turbine Transmission Systems at The University of Nottingham. The views expressed in this paper are those of the authors and not necessarily those of Rolls-Royce plc. The authors are also grateful for access to the University of Nottingham High Performance Computing (HPC) Facility.

References

[1] Dong, P. and Hong, J.K. and Bouchard, P.J., Analysis of residual stresses at weld repairs, International journal of pressure vessels and piping 82 (2005) 258-269. http://dx.doi.org/10.1016/j.ijpvp.2004.08.004

[2] Brown, T.B. and Dauda, T.A. and Truman, C.E. and Smith, D.J. and Memhard, D. and Pfeier, W., Predictions and measurements of residual stress in repair welds in plates, International journal of pressure vessels and piping 83 (2006) 809-818. http://dx.doi.org/10.1016/j.ijpvp.2006.08.012

[3] Kim, K.S., Lee, H.J., Lee, B.S. Jung, I.C. and Park, K.S. Residual stress analysis of an Overlay weld and a repair weld on the dissimilar Butt weld. Nuclear Engineering and Design 239.12 (2009): 2771-2777. http://dx.doi.org/10.1016/j.nucengdes.2009.08.022

[4] Limpus, C. R., et al. Effect of size of butt weld repairs on weld overlay residual stresses. ASME 2007 Pressure Vessels and Piping Conference. American Society of Mechanical Engineers, 2007. http://dx.doi.org/10.1115/pvp2007-26636

[5] Feng, Z. and Wang, X.L. and Spooner, S. and Goodwin, G.M. and Maziasz, P.J. and Hubbard, C.R. and Zacharia, T., A finite element model for residual stress in repair welds, Technical Report, Oak Ridge National Lab., TN (United States), 1996. http://dx.doi.org/10.2172/244602

[6] Bonnaud, E., and J. Gunnars. Three Dimensional Weld Residual Stresses Simulations of Start/Stop and Weld Repair Effects. Procedia Engineering 130 (2015): 531-543. http://dx.doi.org/10.1016/j.proeng.2015.12.260

[7] Brust, F. W., and Rudland, D. L. Three dimensional aspects of computational weld modelling. ASME 2008 Pressure Vessels and Piping Conference. American Society of Mechanical Engineers, 2008. http://dx.doi.org/10.1115/pvp2008-61558

[8] Salerno, G., Bennett, C.J., Sun, W. and Becker, A.A., FE modelling strategies for weld repair in pre-stressed thin components: submitted to Journal of Strain Analysis for Engineering Design (2016). http://dx.doi.org/10.1177/0309324716666436

[9] Deng, D., and Kiyoshima, S. Numerical simulation of residual stresses induced by laser beam welding in a SUS316 stainless steel pipe with considering initial residual stress influences. Nuclear Engineering and Design 240.4 (2010): 688-696. http://dx.doi.org/10.1016/j.nucengdes.2009.11.049

[10] Lacki, P., and Konrad, A. Numerical simulation of the electron beam welding process. Computers & Structures 89.11 (2011): 977-985.

[11] Ogawa, K. and Deng, D. Influences of material properties and FEM elements on welding residual stress analysis. Symposium of Welded Structure, Osaka, Japan. Vol. 4. 2006. http://dx.doi.org/10.1016/j.compstruc.2011.01.016

Materials Research Proceedings **2** (2016) 85-90 doi: http://dx.doi.org/10.21741/9781945291173-15

Analysis of Residual Stress Relaxation Under Mechanical Cyclic Loading of Shot-Peened TRIP780 Steel

Clément Mauduit[1,2*], Régis Kubler[2], Laurent Barrallier[2], Sophie Berveiller[3], Quentin Puydt[1], Martine Monin[4], Bastien Weber[5]

[1]IRT M2P, 4 Rue Augustin Fresnel, 57078 Metz, France

[2]MSMP, Arts et Métiers ParisTech, 2 Cours des Arts et Métiers, 13617 Aix en Provence, France

[3]LEM3, Arts et Métiers ParisTech, 4 Rue Augustin Fresnel, 57070 Metz, France

[4]PSA, 18 Rue des Fauvelles, 92250 La Garenne Colombes, France

[5]ArcelorMittal, Voie Romaine, 57280 Maizières-lès-Metz, France

Keywords: TRIP Steel, Shot-Peening, Residual Stresses, Relaxation

Abstract. Shot-peening is a common mechanical surface treatment used in automotive and aeronautical industries to enhance life duration of mechanical parts by introducing compressive residual stresses. TRIP 780 steel fatigue type specimens are shot-peened and loaded under cyclic bending. The martensite phase transformation and the evolution of residual stresses in the different phases are determined by XRD at different numbers of cycles. A FEM model is also proposed to better understand the evolution of residual stresses in the first cycles.

Introduction

Most surface treatments aim at enhancing life duration of mechanical parts by introducing compressive residual stresses. Shot-peening (SP) process generates a compressive residual stress by deformation of the surface which is accommodated by elastic residual strains. However, those induced stresses may not be stable during the life of a part submitted to thermal and mechanical fatigue. This study deals with the analysis of mechanical stress relaxation and microstructure evolutions of a TRIP-aided steel, TRIP780 [1, 2], after shot peening and subsequent cyclic mechanical loading. This steel exhibits a multiphase microstructure (ferrite, bainite, and austenite) where each mechanical constituent participates to the global behavior [3, 4]. Austenite is a metastable phase which can transform into martensite under a thermomechanical loading.

To reach this goal, the mechanisms responsible for relaxation (plasticity, phase transformation) have to be identified. The first part presents the experimental set-up of the performed analysis and the obtained results are presented and discussed. In a second part, a FEM model using a phenomenological approach is presented in order to understand the redistribution of mechanical fields (macroscopic stress and plastic strain) during the first cycles.

Experimental methods

TRIP780 steel exhibits a multiphase microstructure (austenite, bainite and ferrite). Its chemical composition is given in Table 1. The volume fraction of initial austenite is about 13% with 70 % of ferrite and 17 % of bainite. The macroscopic yield strength of TRIP780 steel provided by ArcelorMittal is 560 MPa.

Standard fatigue type specimens cut in the transverse direction (TD) of a 2 mm thick TRIP780 steel sheet (Fig. 1) are submitted to cyclic bending.

C	Mn	Si	Fe
0.21	1.6	1.6	Bal.

Table 1 Chemical composition of TRIP780 steel in % wt (iron balanced)

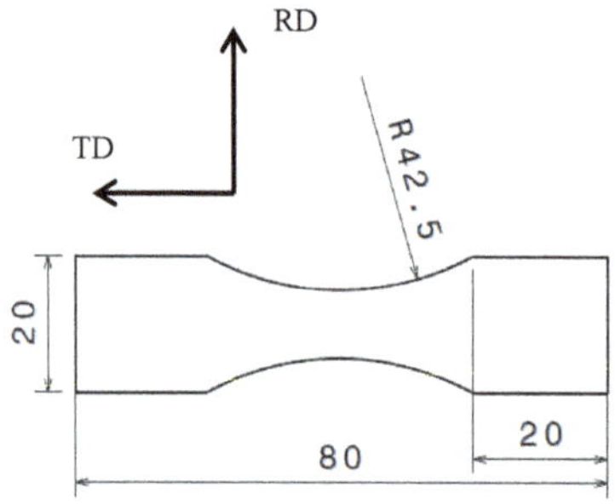

Figure 1 Specimen geometry (a). Thickness 2 mm. Dimensions in mm. Bending test bench (b)

Specimens are shot-peened in a turbine Wheelabrator machine successively on each face. The Almen intensity is F19A. The shot-peening conditions are as following: 400 μm diameter conditioned cut wire shots- hardness700 HV, coverage 230%. The resulting Ra roughness is 4.5 μm.

The specimens are submitted to cyclic reverse bending (R=-1) with an imposed curvature which leads to pure bending in the area of interest. The bending test bench was developed for this study by the academic and industrial partners (Fig. 1b).The residual stresses are determined at 3 interrupted states (initial, 1000 and half-life cycles) with three targeted loading intensities (585, 600 and 650 MPa at the surface). Bending loads are calculated and calibrated from a pure elastic behavior.

Residual stresses are determined in austenite (FCC) and ferritic constituents (CC and BCC) using X ray diffraction (XRD) analysis with $\sin^2\psi$ method [5]. A D500 Siemens XRD goniometer is used to determined stress in ferritic constituents (ferrite, bainite and martensite) with Cr radiation and Vanadium filter on the {211} planes with 40kV and 30 mA. The austenite fraction determination is carried out on the same apparatus following the ASTM standards [6]. When more than 5% of austenite is detected, residual stresses in the austenitic phase are determined with an Elphyse Set-X machine using Mn radiation with Cr filter on the {311} planes. Residual stresses profiles are obtained by electropolishing in the transverse direction. Since the specimens are 2 mm thick, the determined stresses are corrected taking into account the layer removal [7].

Experimental results

After shot-peening, the retained austenite has transformed to martensite and a gradient of martensite and residual stresses in the specimen is determined. Figure 2 presents the austenite fraction profile after SP. Associated stress profiles of ferritic and austenitic phases are displayed in Figure 3. The determined stress in each constituent φ represents the quantity $\sigma_{xx}^{\varphi} - \sigma_{zz}^{\varphi}$, with the macroscopic value σ_{zz}^{Macro} being equal to zero for a uniform shot peening treatment.

With 13% austenite in the as-received material, we assume that austenite is transformed in the first 200 μm. At surface, the assumption can be made that austenite is totally transformed. The shot-peened specimens show a maximum stress in compression of -600 MPa in ferritic constituent and -600 MPa in austenite phase too. As the detection threshold of austenite is 5%, residual stresses determination is carried out from 70 μm

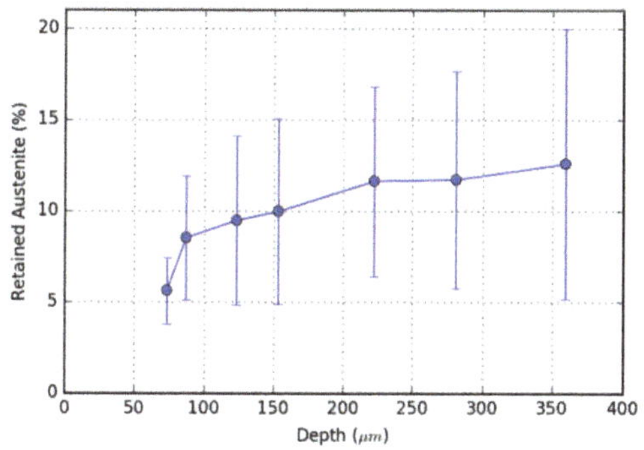

Figure 2 Initial retained austenite profile after SP

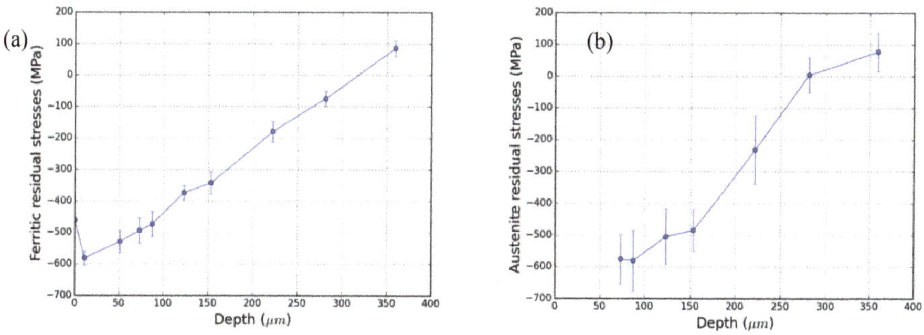

Figure 3 Corrected residual stress profiles after SP (a) in ferritic constituents, (b) in austenite in the transverse direction

In the following part, the focus is made on the macroscopic stress $\sigma_{macro}(z)$ at each depth z in the material, resulting from the averaging of the stresses in austenite σ_{xx}^{aust} and ferritic σ_{xx}^{ferr} constituents by a mixture law:

$$\sigma_{xx}^{Macro}(z) = f^{aust}.\sigma_{xx}^{aust}(z) + (1 - f^{aust})\sigma_{x}^{ferr}(z) \quad (1)$$

where f^{aust} is the austenite volume fraction determined by XRD and x is the measurement direction (TD in this case).

During the cyclic tests, SP specimens are submitted to an imposed curvature at 3 loads (585, 600 and 650 MPa at the surface). In Figure 4, residual macroscopic stresses after 0, 1000 and half-life cycles under the 3 loads are displayed. For the highest loading of 650 MPa (Figure 4a), residual stresses relax mainly in the first thousand cycles. The relaxation still takes place between 1000 and half-life cycles. For 600 and 585 MPa (Figure 4b and 4c respectively), relaxation in the subsurface appears to be more important than at the surface. For each profile except 650 MPa, residual stresses reach a level of -50 MPa in subsurface (after 100 µm depth), and reach -300 MPa at the surface. For 600 MPa and 585 MPa, profiles cross at 300 µm at the value of 0 MPa. The relaxation behavior seems to be different in the first 50-100 µm than in the rest of the sample. It corresponds with the zone where austenite has been transformed by SP.

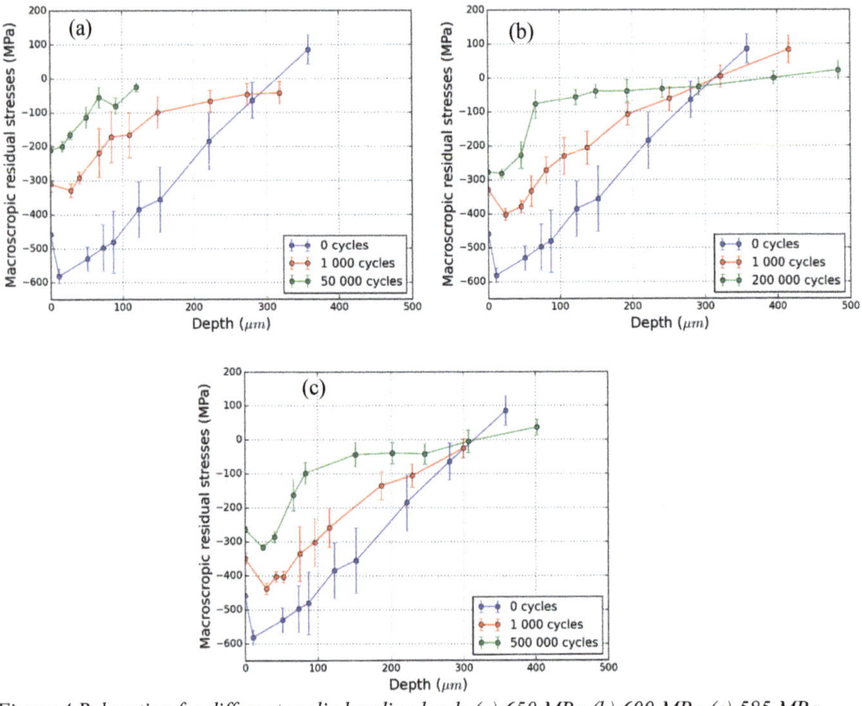

Figure 4 Relaxation for different cyclic bending loads (a) 650 MPa (b) 600 MPa (c) 585 MPa (R=-1).

Simulation

In this part, finite element (FE) simulations are carried out in order to understand the role of the material behavior and of the structural effect on the relaxation of a shot peened specimen. The simulations are performed with Abaqus 6.13/Standard FE software.

The first step is to model the initial mechanical state after shot-peening. Besides compressive residual stresses, shot-peening generates an important hardening at the samples surface. As a first approximation, the residual stresses are generated by the fictitious thermal method [9]. It consists in introducing fictitious dilatation coefficients α in a partitioned part, apply a fictive thermal load ΔT which generate residual stresses due to incompatible plastic strains. Plastic strain values associated with the targeted residual stresses are obtained by equation (2) (semi-infinite body hypothesis) [8]:

Materials Research Proceedings **2** (2016) 85-90 doi: http://dx.doi.org/10.21741/9781945291173-15

$$\underline{\underline{\epsilon}}^{p}(z) = \frac{\nu - 1}{E}\sigma_{res}(z) \qquad (2)$$

In the hypothesis of an orthotropic residual stress state, dilatation coefficients are given by the following equation [9]:

$$\underline{\underline{\alpha}}(z) = -\frac{2\mu + C + K(z)}{2\mu.\Delta\theta}.\ \underline{\underline{\epsilon}}^{p}(z) \qquad (3)$$

$$\text{with } K(z) = \left|\frac{\sigma_y}{3\epsilon^{p}_{xx}(z)}\right| \qquad (4)$$

and C is kinematic hardening slope, σ_y the yield strength, and $\Delta\theta$ the imposed temperature change.

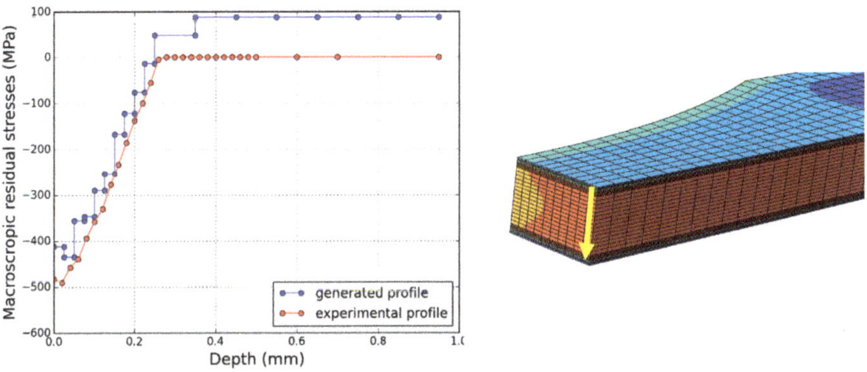

Figure 5 Simulated macroscopic residual stresses profile after shot peening and sample geometry and mesh (symmetry simplification).

The fictitious thermal method is applied to a FE model of the specimen to generate the residual macroscopic stress profile (figure 5) based on experimental data. The generated stress is slightly different from the experimental one. Indeed, the experimental profile is not self-balanced.

This method generates only the required plastic strain to generate the residual stress profile. They are about 0.3-0.4% which is far below the real plastic strain in shot-peened material (until 30-50%). The aim of this method is not to initiate the hardening internal variables. Todo so, the yield strength σ_y, could be change with depth in order to model some hardening effects.

Mechanical relaxation is separated in two parts [10]. Static relaxation occurs in the first bending cycles due to macroscopic strain by the superposition of residual stresses and loading. This phenomenon is modelled by a cyclic hardening law on the first 5 cycles of loading. Some numerical investigations are also presented for a material modelled with standard phenomenological constitutive laws with different hardenings.

The second part is the dynamic relaxation which needs another description of the damage mechanisms occurring at high number of cycles.

Materials Research Proceedings **2** (2016) 85-90 doi: http://dx.doi.org/10.21741/9781945291173-15

Conclusions

The initial state of shot-peened TRIP steel and the stress relaxation under 3 reverse bending loads were investigated in this study. This experimental work is followed by FE simulations which aim to dissociate static and cyclic relaxation.

References

[1] Cooman, B. D. (2004), 'Structure and properties relationship in TRIP steels containing carbide-free bainite, Current Opinion in Solid State and Materials Science **8**, 285 - 303. http://dx.doi.org/10.1016/j.cossms.2004.10.002

[2] Bleck, W. (2002), 'Using the TRIP effect-the dawn of a promising group of cold formable steels', TRIP aided high strength ferrous alloys, 13--22.

[3] Fu, P. & Jiang, C. (2014), 'Residual stress relaxation and micro-structural development of the surface layer of 18CrNiMo7-6 steel after shot peening during isothermal annealing', Materials and Design **56**, 1034--1038. http://dx.doi.org/10.1016/j.matdes.2013.12.011

[4] Ackermann, S.; Kulawinski, D.; Henkel, S. & Biermann, H. (2014), 'Biaxial in-phase and out-of-phase cyclic deformation and fatigue behavior of an austenitic TRIP steel', International Journal of Fatigue **67**(0), 123 - 133. http://dx.doi.org/10.1016/j.ijfatigue.2014.02.007

[5] EN, NF15305 Avril 2009. Essais non-destructifs: méthode d'essai pour l'analyse des contraintes résiduelles par diffraction des rayons X. AFNOR.

[6] ASTM, E. (2003). 975-03. Standard Practice for X-Ray Determination of Retained Austenite in Steel with Near Random Crystallographic Orientation.ASTM Internacional.

[7] Moore, M. G., & Evans, W. P. (1958). Mathematical correction for stress in removed layers in X-ray diffraction residual stress analysis (No. 580035). SAE Technical Paper.

[8] Zarka, J., Frelat, J., Inglebert, G., & Kasmai- Navidi, P. (1990). A new approach to inelastic analyses of structures.

[9] Ahdad, F.; Desvignes, M.; Castex, L.; Journet, B.; Pons, G.; Ericsson, T. & AHMAD, M. (1995), 'Preision de la duree de vie d'une structure precontrainte: application au grenaillage', Materiaux et techniques **83**(10-11), 13--18.

[10] Holzapfel, H., Schulze, V., Vöhringer, O., & Macherauch, E. (1996). Stability and relaxation behaviour of shot peening induced residual stresses in AISI 4140 during bending fatigue. In Proc. 6th Int. Conf. on Shot Peening, San Francisco (pp. 413-423).

Materials Research Proceedings 2 (2016) 91-96 doi: http://dx.doi.org/10.21741/9781945291173-16

Validation of XRD Stress Analyses Combining *in-situ* Tests and Integrated Peak Processing

B. Voillot[1-2,a*], R. Billardon[2,c], J.L. Lebrun[3], F. Hild[1,b]

[1]LMT-Cachan, ENS Cachan/CNRS/Université Paris-Saclay, 61 avenue du Président Wilson 94235 Cachan, France

[2]Safran Landing Systems, Safran group, Inovel Parc Sud, 7 rue Général Valérie André 78140 Velizy-Villacoublay, France

[3]Université Paris-Est, Institut de Recherche en Constructibilité, ESTP, 94230 Cachan, France

[a]voillot@lmt.ens-cachan.fr, [b]hild@lmt.ens-cachan.fr, [c]rene.billardon@safrangroup.com

Keywords: X-ray Diffraction, Ti5553 Alloy, *in-situ* Tensile Test, Peak Correlation, Integrated Methods

Abstract. The surface integrity has a significant effect on the fatigue life of structural components (e.g., landing gears). To estimate sub-surface residual stresses, X-ray diffraction (XRD) is applied to a Ti5553 alloy sample in an *in-situ* tensile test. This material is challenging since it is made of α and β phases of different proportions, shapes and scales ranging from submicrometre to millimetre sizes. Therefore stress variations occur between grains. For millimetric probed volumes, the studied microstructure leads to shallow and noisy diffraction signals. It is shown that combing a new integrated method of diffraction peak registration and *in-situ* XRD measurements in tensile tests allows qualitative and quantitative results of residual stress analyses to be obtained for Ti5553 alloy samples.

Introduction

Aircraft landing systems are not only subjected to high dynamic load levels upon landing, but also to fatigue cycles during landing, take-off and taxiing. The structural parts of landing systems are made of alloys with high specific strength. The continuous development of such materials recently led to the introduction of forged and machined parts made of titanium alloys with high ultimate tensile strength [1]. The fatigue life of structures depends on intrinsic (i.e., bulk) and extrinsic (i.e., surface) properties [2,3]. Consequently, chemical and mechanical treatments are performed as part of the manufacturing process to enhance the resistance to corrosion and fatigue of all structural components (in particular parts made of high strength titanium alloys). Compressive residual stresses are created on the sub-surface of forged and machined parts. These residual stresses have to be evaluated to quantify their influence on the fatigue properties of structures. The most common non-destructive evaluation technique used to determine residual stresses is X-ray diffraction (XRD).

The aim of this work is to prove the feasibility of XRD for challenging stress analyses. Qualitative, quantitative and reliable estimation of stresses in coupons made of high strength two-phase titanium alloy (i.e., Ti5553) are needed. *In-situ* tensile tests are carried out to compare stress estimations obtained via XRD measurements with other techniques (e.g., applied stress data using load cells). To post-process XRD measurements two in-house codes have been developed. The images shot during the experiment have been used to check that the probed volume has not moved during the experiment thanks to digital image correlation (DIC [4]).

XRD stress analysis on two-phase titanium alloy

One of the challenges of XRD analyses of two-phase alloys such as Ti5553 is the microstructure complexity [5,6], see Figure 1. Ti5553 is a metastable β-alloy made of body-centred cubic (BCC)

matrix and hexagonal close-packed (HCP) inclusions (i.e., α-phase) with different volume fractions, shapes and scales of grains ranging from sub-micrometre (i.e., lamellae) to millimetre size (i.e., β-grains). These dimensions are to be compared to the 1-mm in diameter X-ray spot that is used herein. The probed volume by the X-ray beam spreads over a few mm^2 and penetrates down to 3 μm under the surface. The consequence is the heterogeneity of the material concerned with XRD measurements, especially the variability in number and shape of α- and β-grains within the probed volume. These grains may experience small variations in lattice parameters or local stresses, which result in small strain variations. Consequently, XRD peaks are shallow, which make XRD analyses challenging. In the present work, only stresses in the α-phase will be estimated. The use of {213} planes corresponds to Bragg's angle 2θ = 140.6°. The elastic parameters chosen for the present analyses are linked to the α-phase for an isotropic elastic behaviour in a two-phase titanium alloy [7] (i.e., $\frac{1}{2}S_2 = \frac{(1+\nu)}{E} = 11.9 \times 10^{-6}\mathrm{MPa}^{-1}$, where ν and E are respectively Poisson's ratio and Young's modulus for the α-phase).

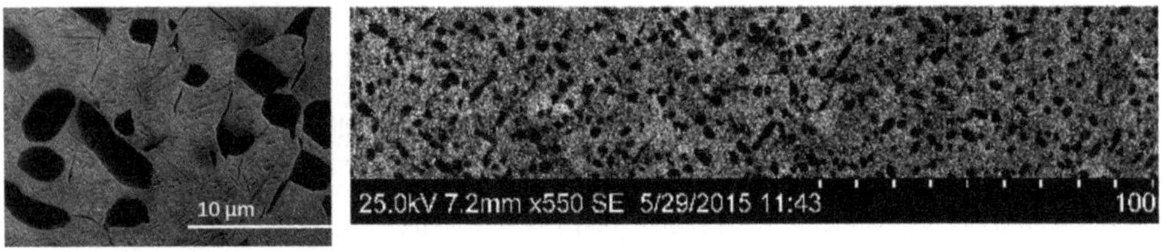

Figure 1: SEM pictures of Ti5553 alloy from a bogie of a landing gear. Sub-micrometre size α-lamellae (left) and micrometre size α-nodules (right) embedded in β-matrix

When a material is loaded (as herein) or exhibits residual stresses (e.g., landing gears after shot-peening), the crystal lattices deform. XRD enables variations of interreticular distances to be measured by analysing peaks due to diffraction phenomena in crystalline materials. The link between diffraction peak angle and interreticular distances is given by Bragg's law. Elastic stresses are then evaluated by estimating peak shifts with the chosen elastic law. Post-processing is carried out with the so-called $\phi\psi$ isotropic elastic law [8].

Morphological variations in two-phase materials induce different stresses within grains. Therefore the stresses differ within the considered volume, thereby inducing micro-shifts of diffraction peaks in the diffractogram. For millimetre-size probed volumes, such microstructures create shallow and noisy diffraction peaks. The post-processing steps need special care to be exercised [8,9]. Results of stress evaluations in such two-phase alloys thus need to be validated with independent measurements. It is proposed to use an *in-situ* tensile test for which the applied global forces are measured by the load cells and can be linked to stresses measured via XRD. In the present case, the stress states may slightly vary between two grains of the same coupon [5,6]. This is an additional issue for the present work since locations of XRD measurements are not perfectly reproducible and some uncertainties are expected.

In-situ tensile test for XRD stress analyses

A peak correlation method for XRD measurements is applied and compared with a new approach that is based on integrated algorithms [10], which are used in DIC techniques [4]. The *in-situ* XRD measurements are carried out in the centre of the top face of a dogbone flat coupon made of Ti5553 alloy (Figure 2). After electropolishing, a speckle pattern is sprayed on the top surface to enable images to be registered via DIC. Two opposite actuators are used so that the monitored zone is motionless (this hypothesis was checked with DIC).

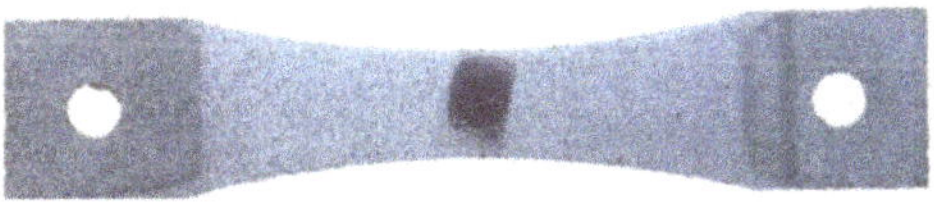

Figure 2: Ti5553 speckled coupon with an unpainted area to enable for XRD measurement. Sample dimensions: 50 × 6 × 0.5 mm³

Figure 3 illustrates the *in-situ* setup consisting of the miniature testing machine within the XRD goniometer.

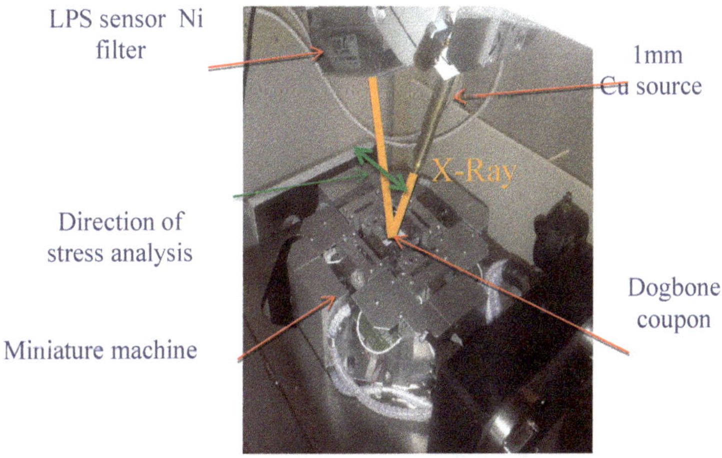

Figure 3: Dogbone Ti5553 coupon in miniature testing machine.
Parameters for XRD measurements: Cu source, 1-mm collimator, linear XRD sensor, Ni filter

The experimental procedure consists of loading the coupon at various load levels (Figure 4). The central area of specimen is left without speckle to allow for XRD measurements. Knowing transverse sections everywhere enables the mean applied stresses to be estimated, in particular in the centre of the coupon with the knowledge of measured forces. Both XRD signals and image acquisitions are carried out for each loading step. Forces and corresponding stresses vary within the range of residual stresses induced by machining (≈ 200 MPa) or shot peening (≈ 700 MPa), which are typical extrinsic states at the end of the manufacturing process of the studied alloy. The maximum applied stress corresponds to maximum load that can be applied by the selected testing machine. The two load levels at the middle of the experiment are longer than the other ones to assess the repeatability of the measurements, which turned out to be very good. All stress analyses via XRD are carried out in the loading direction.

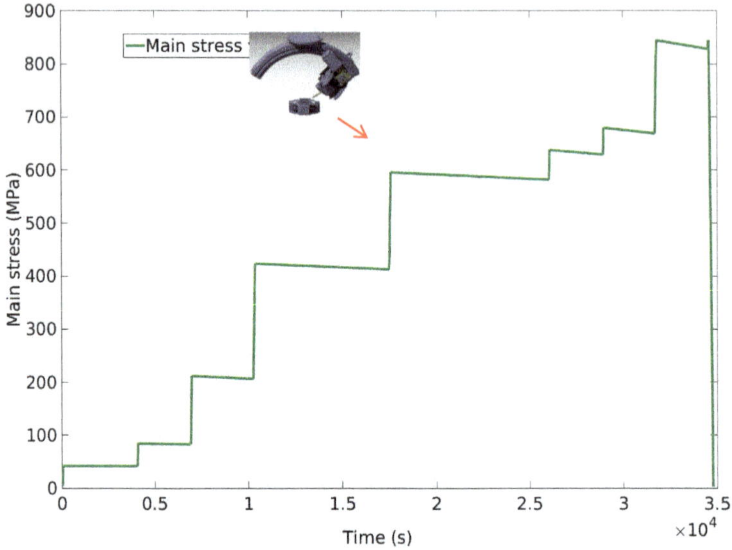

Figure 4: Load history for a Ti5553 coupon and XRD measurements

Stress evaluations

XRD diffraction peaks are registered then modelled by a Gaussian profile coupled with isotropic elasticity. It allows the peak shift to be related to stresses thanks to an integrated approach. The latter is of great interest in two-phase alloys since peaks are flat and noisy, and the determination of peak positions is not always accurate [10]. Let f denote the measured diffractogram, and g a reference signal (here modelled as Gaussian). An integrated approach consists of minimizing the sum of squared differences ρ^2

$$\rho(x) = \frac{f(x) - I_b - Dx}{\Delta I} - g\left(\frac{x - x_0(\sigma_{\phi\phi}, \sigma_{\phi 3}, \varepsilon_0)}{\sigma_0}\right)$$

over the *whole* measurement data points x *and* diffractograms with respect to the peak height ΔI, the mean background level I_b and its spatial drift D, the peak width σ_0, and the three mechanical unknowns $\sigma_{\phi\phi}, \sigma_{\phi 3}, \varepsilon_0$ via a Gauss-Newton scheme. In this formulation, the peak position x_0 is directly parameterised with the normal stress $\sigma_{\phi\phi}$, the shear stress $\sigma_{\phi 3}$, and ε_0 that is a term combining the trace of stress tensor and the reference angle [10]. The elastic and isotropic law used to model stresses once lattice strains $\varepsilon_{\phi\psi}$ are measured via XRD in the direction $\phi\psi$ reads

$$\varepsilon_{\phi\psi} = \frac{1}{2} S_2 \sigma_{\phi\phi} sin^2\psi + S_1 tr(\sigma) + \frac{1}{2} S_2 \sigma_{\phi 3} sin2\psi$$

where S_1 and S_2 are the elastic parameters of the α-phase for the considered {213} plane family.

Figure 5 displays applied stress estimations in the analysed zone (Figure 2) for different load levels (Figure 4). The first estimation, which is referred to as digital signal correlation (DSC), uses peak registration as commonly performed in commercial codes (e.g., *StressDiff*). The results given by the integrated method, i.e., integrated digital signal correlation (IDSC), which consists of directly seeking the stresses within the peak registration procedure (i.e., the peak shift is parameterised with the unknown stresses), are also shown.

The fact that the registration residuals for DSC and IDSC are very close validates the integrated approach and also the isotropic elastic model [8]. Further, the stresses can also be estimated with the force levels measured by the load cells of the testing machine. Very similar results are obtained with the three different approaches. Part of the observed differences is due to geometric defects induced by electropolishing, which was not uniform. Figure 5 also reports the standard uncertainties of stress estimations for each load and for DSC and IDSC algorithms (green markers) and compares them to stress evaluations with the load data (black markers). The standard resolutions for both methods are similar and small (≈ 15MPa) compared to stress estimation levels. It is an additional validation for the registration of XRD peaks and the hypothesis of isotropic elasticity. Root mean square errors between stress estimations by DSC or IDSC and load data are higher. They include errors in evaluating the real section geometry. The mean value of these errors is 38 MPa, which remains small given all the challenges that had to be addressed.

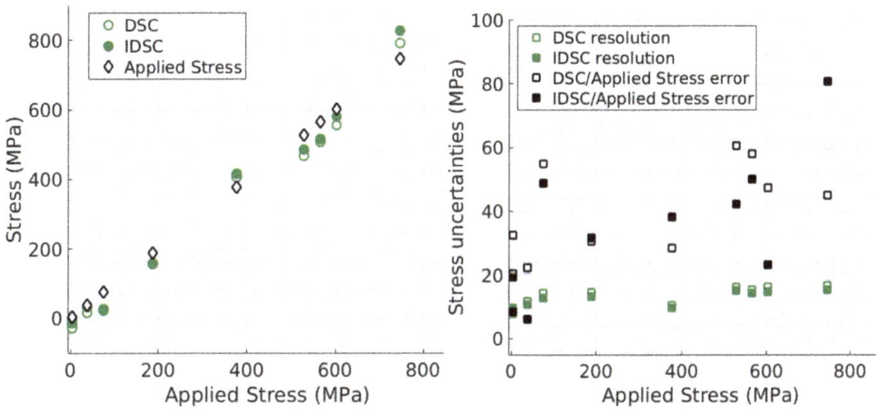

Figure 5: Estimation of stresses with different methods (left) and corresponding standard stress uncertainties (right). For each stress evaluation 13 ψ positions were considered [11]

Conclusion

A new procedure has been used to estimate stresses via XRD analyses using isotropic elasticity. An *in-situ* tensile test has been carried out to validate both experimental methodologies and an in-house integrated signal registration procedure. It has been applied to high strength Ti5553 alloy exhibiting shallow and noisy diffraction peaks.

References

[1] R. R. Boyer, R. D. Briggs, The use of β titanium alloys in the aerospace industry, J. Mater. Eng. Perform., 14 (2005) 681-685. http://dx.doi.org/10.1361/105994905X75448

[2] E. C. Reed, J. A. Viens, The influence of surface residual stress on fatigue limit of titanium, J. Eng. Ind., 82 (1960) 76-78. http://dx.doi.org/10.1115/1.3663004

[3] A. Cox, J.P. Villain-Chastre, S. Turner, M. Jackson, The effect of finish milling on the surface

integrity and surface microstructure in Ti-5Al-5Mo-5V-3Cr, in: Proc. 13th World Conference on Titanium, 45, 2016. http://dx.doi.org/10.1002/9781119296126.ch45

[4] S. Roux, F. Hild, Stress intensity factor measurements from digital image correlation: post-processing and integrated approaches, Int. J. Fract., 140 (2006) 141-157. http://dx.doi.org/10.1007/s10704-006-6631-2

[5] M.E. Fitzpatrick, A.T. Fry, P. Holdway, F.A. Kandil, J. Shackleton, L. Suominen, Determination of residual stresses by X-ray diffraction, in: Measurement Good Practice Guide No. 52, 2005

[6] A. Bhattacharjee, V.K. Varma, S.V. Kamat, A.K. Gogia, S. Bhargava, Influence of beta grain size on tensile behavior and ductile fracture toughness of titanium alloy Ti–10V–2Fe–3Al, Metall. Mat. Trans. 37A (2006) 1423-1433. http://dx.doi.org/10.1007/s11661-006-0087-x

[7] S. Fréour, D. Gloaguen, M. François, R. Guillen, E. Girard, et al. Determination of the macroscopic elastic constants of a phase embedded in a multiphase polycrystal - application to the beta-phase of Ti17 titanium based alloy, Materials Science Forum, Trans Tech Publications Inc., 404-407 (2002) 723-728

[8] J. Lu, Handbook of measurement of residual stresses, Society for Experimental Mechanics, 1996

[9] F. Lefebvre, M. François, J. Cacot, C. Hemery, P. Le-Bec, E. Baumhauer, D. Bouscaud, T. Bergey, D. Blaize, D. Gloaguen, J. L. Lebrun, A. Cosson, R. Kubler, Y. Cheynet, E. Daniel, H. Michaud, J.C. Monvoisin, P. Blanchet, P. Allain, Y. Mrini, J. M. Sprauel, P. Goudeau, P. Barbarin, C. Charles, J.M. Le Roux, W. Seiler, C. Fischer, L. Desmas, A. Ouakka, M.J. Moya, Y. Bordiec, External reference samples for residual stress analysis by X-ray diffraction, Mat. Sci. Forum 681 (2011) 215-222. http://dx.doi.org/10.4028/www.scientific.net/MSF.681.215

[10] B. Voillot, F. Hild, J.L. Lebrun, R. Billardon, Evaluation of residual stresses due to mechanical treatment of Ti5553 alloy via XRD, in: Proc. 13th World Conference on Titanium, 268, 2016. http://dx.doi.org/10.1002/9781119296126.ch268

[11] EN 15305:2008, Non-destructive testing - test method for residual stress analysis by X-ray diffraction, 2008

Residual Stresses 2016: ICRS-10 Materials Research Forum LLC
Materials Research Proceedings **2** (2016) 97-102 doi: http://dx.doi.org/10.21741/9781945291173-17

Residual Stresses in Modelling Fatigue Lifetime of Gas Nitrided Iron-Based Alloys

H. Weil[1, 2, a], L. Barrallier[1,b], S. Jégou[1,c], A. Courleux[2,d], G. Beck[2,e]

[1]MSMP Laboratory - Arts et Métiers ParisTech, 2 cours des Arts et Métiers, 13617 Aix en Provence, France

[2]Hispano-Suiza - Groupe SAFRAN, 18 boulevard Louis Seguin, 92707 Colombes, France

[a]hadrien.weil@ensam.fr, [b]laurent.barrallier@ensam.eu, [c]sébastien.jégou@ensam.eu, [d]alice.courleux@hispano-suiza-sa.com, [e]guillaume.beck@hispano-suiza-sa.com

Keywords: Residual Stress, Nitriding, Fatigue, Rupture, Hardness, Crossland, Dang Van

Abstract. The objective of this work is to propose a model abled to optimize the mechanical properties from gaseous nitriding of steels (hardness, residual stresses) for a given loading and fatigue lifetime. Multiaxial fatigue criteria are used and the present study focuses on the influence of residual stresses. Starting from a surface loading profile, the theoretical residual stress in-depth profile can be calculated aiming infinite fatigue lifetime. The model allows then optimization of the nitriding parameters.

Introduction

Compressive residual stresses provide the material with a better fatigue strength [1]. Fatigue lifetime calculations can take into account residual stresses thanks to multiaxial fatigue criteria [2]. Gaseous nitriding generates significant improvements in mechanical properties of the surface of steel-made parts, such as hardness and the development of high compressive residual stresses [3-5].

Residual stresses found origins through the volume change resulting from the phase transformations during nitriding [6]. For a given steel composition, the generation of residual stresses is influenced by the temperature, time, and ammonia dissociation rate of nitriding [7].

The present work aims to consider the development of residual stresses from nitriding to the calculation of the fatigue lifetime of nitrided parts. A model is proposed to calculate the theoretical residual stress in-depth profile depending on the loading and required durability of a workpiece.

Modelling

Multiaxial fatigue criteria. Fatigue criteria allow taking into account the mechanical characteristics of treated materials. The interest has focused on criteria expressed in terms of hydrostatic pressure P_H since it is an addition of the loading and residual stresses. The Dang Van [9] criterion was used, in order to predict the initiation of material failure.

The Dang Van criterion, E_{DV}, is used as shown in (Eq. 4) [9].

$$E_{DV} = \max_{t \in T}\left(\frac{\tau(t)+\alpha_{DV}P_H(t)}{\beta_{DV}}\right) \tag{4}$$

The hydrostatic pressure $P_H(t)$ is an addition of applied stresses $\sigma_{app}(t)$ and residual stresses $\sigma_{res}(t)$:

$$P_H(t) = \frac{1}{3}tr(\sigma_{app}(t) + \sigma_{res}(t)) \tag{2}$$

Materials Research Proceedings **2** (2016) 97-102 doi: http://dx.doi.org/10.21741/9781945291173-17

The slope, α_{DV}, of the line is calculated with respect to the fatigue properties of the non nitrided core material [2]. The coefficient β_{DV} takes into account the hardness *HV(Z)* at a given depth *Z* induced by nitriding [1] :

$$\beta_C(z) = \beta_0 \left(\frac{HV(Z)}{HV_0}\right)^n \tag{3}$$

Where the coefficients β_0 and HV_0 are a constant and hardness for the core material. The coefficient n is a work-hardening coefficient.

The shear stress $\tau(t)$ is equal to the Tresca criterion. Thus, it is calculated with the eigenvalues of the stress tensor given in (Eq. 5).

$$\tau(t) = \frac{1}{2}\max(|\sigma_1(t) - \sigma_2(t)|, |\sigma_2(t) - \sigma_3(t)|, |\sigma_3(t) - \sigma_1(t)|\,) \tag{5}$$

This fatigue criterion can be used, on one hand, to calculate the fatigue lifetime by setting the mechanical properties (residual stress, hardness) of the nitrided layers, or on the other hand, the optimisation of the mechanical properties can be achieved by setting the required fatigue lifetime.

Calculation of a fatigue lifetime. The purpose is to determine a fatigue lifetime in-depth profile for a nitrided layer and for a given failure probability. At each depth, the number of cycles N such that *E(N) = 1 is determined (Fig. 1)*. The hydrostatic pressure and applied stresses are functions of the depth, whereas the coefficients of criterion also depend on the number of cycles N [2]. At each depth, $\alpha(N)$ and $\beta(N)$ must then be determined for each new number of cycles N.

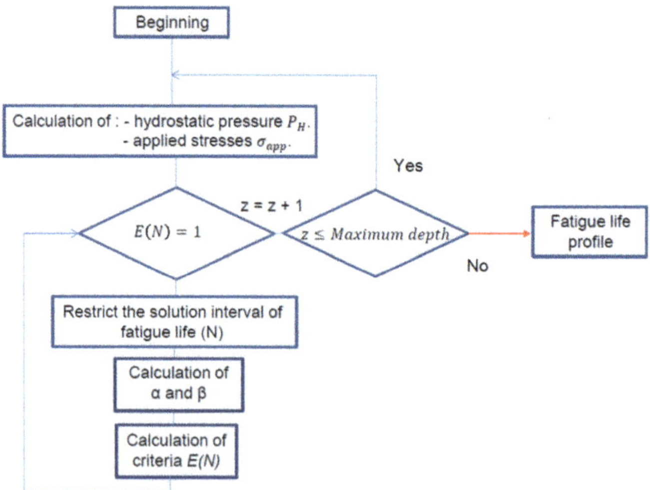

Figure 1. Determination of the fatigue lifetime.

Influence of compressive residual stresses

In order to show the influence of the residual stress and hardness on the fatigue strength, calculation of the fatigue lifetime on a tooth flank was performed with and without nitriding.

Materials Research Proceedings **2** (2016) 97-102 doi: http://dx.doi.org/10.21741/9781945291173-17

Input data. The present work deals with gaseous nitriding of a carbon iron-based alloy. Material and nitriding parameters remain confidential. The residual stress σ_{res} and hardness HV in-depth profiles used in order to point out the effect of the mechanical properties provided by nitriding on the fatigue lifetime of a carbon iron-based alloy are given in Fig. 2.a and 2.b respectively. The residual stress is obtained by X-ray diffraction in the ferrite. A Siemens D500 diffractometer is used with Chromium radiation K_{α} on the diffraction plane in ferrite $\{2\ 1\ 1\}$. In order to determine the mean residual stress $\sigma_{res\,xx} - \sigma_{res\,zz}$, the $sin^2\omega$ method is used and Electro-chemical polishing provides each depth. The hardness is measured with a microdurometer.

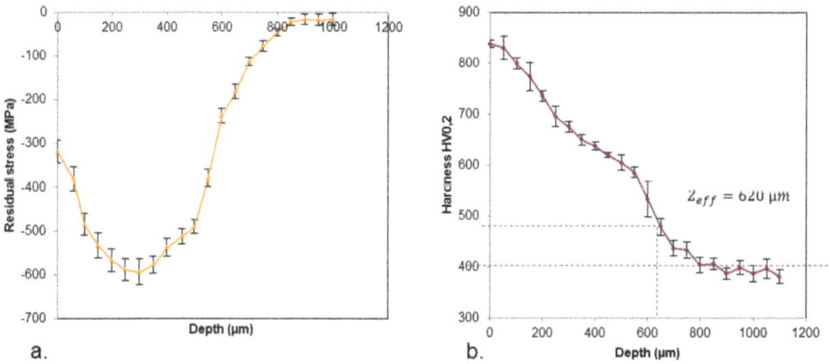

Figure 2: Residual stress in ferrite obtained by X-ray diffraction (a.) and Hardness obtained by microdurometer with effective depth Z_{eff} (b.) in-depth profiles after nitriding of a carbon iron-based alloy.

The shear stress $\tau(t)$ and hydrostatic pressure P_H are determined according to the residual stress in-depth profile from nitriding (Fig. 2.a) and the shear stress from simulation of the contact pressure on a tooth flank (Fig. 3).

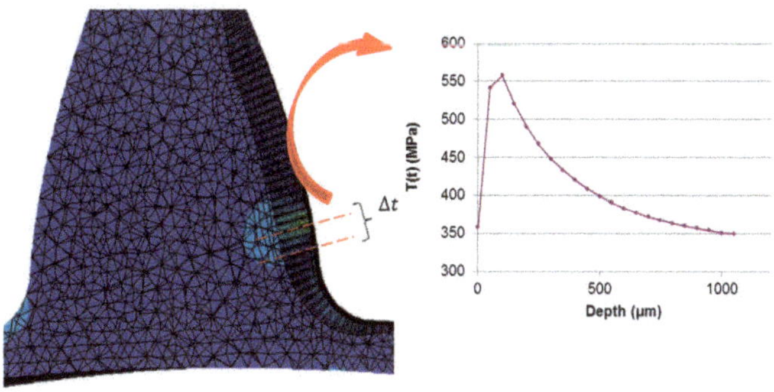

Figure 3: Shear stress profiles from simulation of the contact pressure on a tooth flank at a given interval of time dt.

Materials Research Proceedings **2** (2016) 97-102 doi: http://dx.doi.org/10.21741/9781945291173-17

Results. The comparison between the loading before (as round marks) and after (as square marks) nitrding is given in Fig. 4 for a durability of 10^7 cycles and 10^{-6} % failure probability.

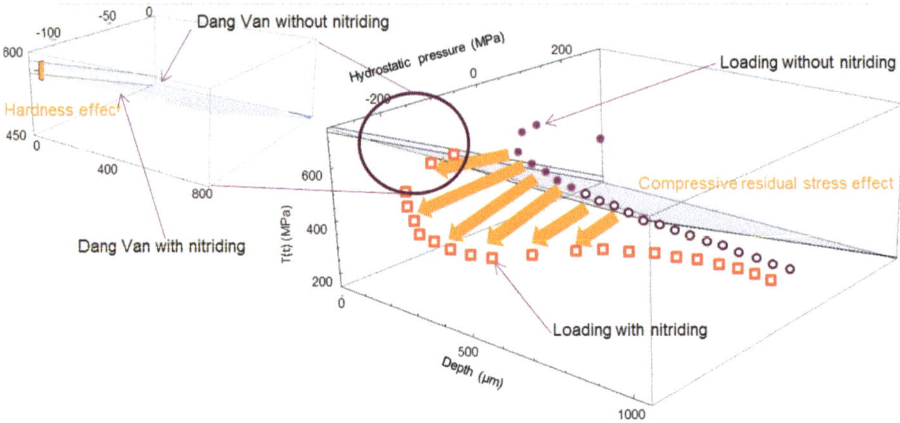

Figure 4: Loading with (red square) and without (purple round) nitriding according to the Dang Van criterion (purple surface) considering 10^7 cycles and a 10^{-6} % probability of failure. Filled marks are located above the Dang Van surface.

The boundary surface (shown in purple in Fig. 4) separates the rupture zone (above) from the non-rupture zone (underneath). The loading points without nitriding are above the fatigue limit surface between 0 and 250 µm depth. Therefore, the probability of failure close to the surface is not negligible for a steel without nitriding. The necessity of nitriding is given as empty square points, all being below the fatigue limit surface. This result is explained by the decrease of the hydrostatic pressure due to the compressive residual stress effect provided by nitriding.

Increased hardness through the nitrided layer also has a positive effect on the fatigue strength of the material. It shifts up the fatigue boundary surface and increases the non-rupture zone. The hardness effect is lower than for residual stresses as given in Fig. 4. However it is important to point out the significant influence of hardness on increasing the yield strength [2].

Optimization of residual stresses: the minimum requirement for gears

Thanks to previous multiaxial fatigue criteria, the theoretical residual stress and hardness in-depth profiles can be calculated for a given load, probability of failure, and core hardness. The profiles are calculated independently of one another. An example is given in Fig. 5 and by considering the previous loading on the flank of a gear tooth. The durability is equal to 10^7 cycles, the probability of failure to 10^{-6} % and the core hardness to 400 $HV_{0,2}$.

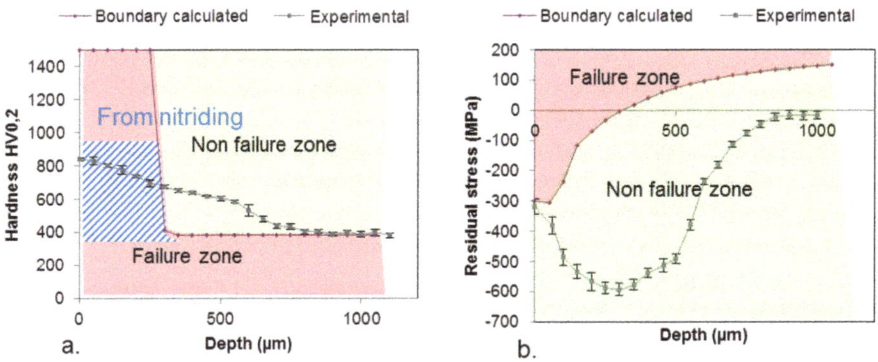

Figure 5: theoretical and experimental hardness (a.) and residual stress (b.) in-depth profiles based on the Dang Van's criterion for 10^7 cycles and a 10^{-6} % probability of failure for a flank gear tooth (the applied load is given in Fig. 2).

As illustrated in Fig. 5.a, the calculated stress and hardness profiles are a boundary between a rupture and non-rupture zone. To improve the mechanical properties, a compressive residual stress must be sustained up to 350 µm in depth when considering a constant hardness of 400 $HV_{0,2}$. Moreover, compressive residual stresses of the order of -310 MPa are needed to avoid crack initiation at the extreme surface due to the hertz pressure on the flank [11]. Tensile stresses due to the rebalancing in sublayers may theoretically be permitted, but must be limited, and even more avoided, such that rupture is prevented. Without any residual stresses, the boundary is more important than the 1000 $HV_{0,2}$ of hardness provided by nitriding (Fig. 5.b). In the case of residual stress, the experimental curve is located in the non-rupture zone. However, the experimental hardness is not sufficient against the rupture.

Conclusion

Nitriding has benefit properties for the improvement of the fatigue lifetime of the gears. This improvement includes the residual stress and increased hardness in nitrided layers. Multiaxial fatigue criteria were used to firstly predict the fatigue lifetime based on mechanical properties improvements given by nitriding, and secondly predict the ideal residual stress and hardness in-depth profiles for a given durability and probability of failure of nitride steel-made parts. Although the effect of hardness on the lifetime is not as significant as for residual stresses, the present model will allow to carefully optimize the nitriding parameters for a given fatigue lifetime. Specimens were sized to validate the fatigue lifetime prediction and the results will come soon. The influence of the relaxation of residual stress during the mechanical tests will also be studied.

References

[1] M.A. Terres, S. Ben Mohamed, H. Sidhom. Influence of ion nitriding on fatigue strength of low-alloy (42CrMo4) steel: Experimental characterization and predictive approach, International Journal of Fatigue. **32**, 1795-1804 (2010). http://dx.doi.org/10.1016/j.ijfatigue.2010.04.004

[2] M. Chaussumier. PhD Thesis, Ecole Nationale Supérieure d'Arts et Métiers, France (2000).

Materials Research Proceedings **2** (2016) 97-102 doi: http://dx.doi.org/10.21741/9781945291173-17

[3] L. Barrallier, J. Barralis. On origin of residual stress generated by nitriding treatment on alloy steels, Proc. 4th Int. Conf. Residual Stress, Baltimore. **43**, 629-642 (1994).

[4] L. Barrallier. Classical nitriding of heat treatable steel, in: Thermochemical Surface Engineering of Steels, Woodhead Publishing, 2015, pp. 392-411. http://dx.doi.org/10.1533/9780857096524.3.393

[5] S. Jégou, L. Barrallier, R. Kubler. Phase transformations and induced volume changes in a nitride ternary Fe-3%Cr-0.345%C alloy, Acta Materialia. **58**, 2666-2676 (2010). http://dx.doi.org/10.1016/j.actamat.2009.12.053

[6] L. Barrralier. PhD Thesis, Ecole Nationale Supérieure d'Arts et Métiers, France (1992).

[7] L. Castex, J. Barralis et J.C. Chaize. Etude de la tenue en fatigue de l'acier 32CDV13 nitruré, Mémoires et Etudes Scientifiques Revue de Métallurgie. **84**, 13-23 (1987).

[8] B. Crossland. Effect of large hydrostatic pressures on the torsional fatigue strength of an alloy steel, Fatigue of Metals-International Conférence-SEP LONDON, 138-149 (1956).

[9] O. Message, K. Dang Van, B. Griveau. On a new multiaxial fatigue criterion : theory and application. Biaxial and Multiaxial Fatigue, Mechanical Engineering Publications, 1989, pp. 479-496.

[10] B. Weber. PhD Thesis, Institut National des Sciences Appliquées de Lyon, France (1999).

[11] K.L. Johnson, Contact mechanics, Cambridge University Press, Cambridge, 1985. http://dx.doi.org/10.1017/CBO9781139171731

Residual Stresses 2016: ICRS-10 Materials Research Forum LLC
Materials Research Proceedings **2** (2016) 103-108 doi: http://dx.doi.org/10.21741/9781945291173-18

Comparison of X-Ray Residual Stress Measurement Values by Cos α Method and Sin² Ψ Method

Ami Kohri [1, a*], Yasuhiro Takaku [1,b] and Masashi Nakashiro [1,c]

[1] IHI Inspection & Instrumentation Co., Ltd. (IIC), 6-17, Fukuura 2-chome, Kanazawa-ku, Yokohama-city, Kanagawa 236-0004 Japan

[a] a_kohri@iic.ihi.co.jp, [b] y_takaku@iic.ihi.co.jp, [c] m_nakashiro@iic.ihi.co.jp

Keywords: X-Ray Diffraction, Residual Stress, Cos α Method, Sin² Ψ Method

Abstract. In recent years, Sasaki et al. has developed the X-ray residual stress measurement device (μ-X360) using the "cos α" method [1, 2]. μ-X360 has been used widely in Japan. μ-X360 uses a 2D detector, and the measurement can be done by only single irradiation of X-ray. Accordingly the device is small and light, and stress measurement is conducted in a short time. For these reasons, μ - X360 is expected to be applied to on-site stress measurement of the structures. Generally the "$\sin^2 \psi$" method has been used for the X-ray residual stress measurement. We, IIC, have been providing field measurement services using the portable X-ray stress measuring device (Stresstech Inc. Xstress 3000) with the "$\sin^2 \psi$" method. There are many measurement results with the "$\sin^2 \psi$" method, but there are few reports about the measurement with the "cos α" method. Therefore, we measured various samples by the "cos α" method and compared with the measured values with the "$\sin^2 \psi$" method, and clarified the characteristics of the results with the "cos α" method.

Introduction

The residual stress measurement has been used for quality management of machine parts, soundness and safety evaluation of structural members. There are various residual stress measurement techniques such as sectioning method, center hole drilling method, X-ray diffraction (XRD) method, neutron diffraction method and so on. In these techniques, XRD method has been adopted widely, because it is non-destructive, inexpensive and relatively simple.

The $\sin^2 \psi$ method has been generally applied in the residual stress measurement by XRD method. We have been providing residual stress measurement services not only in our laboratory but also in the field using Xstress 3000 with $\sin^2 \psi$ method.

On the other hand, Sasaki et al. developed the portable device, μ-X360 suitable for the field measurement by using the cos α method. This device is expected to be applied to the field measurements more and more.

The measurement performance of "$\sin^2 \psi$" method is excellent and effective. However, the "$\sin^2 \psi$" method requires a goniometer of the device to irradiate X-rays from some different angles to a specimen to be measured. Therefore, the device cannot be applied in some cases because of the restriction of the device.

In case of the "cos α" method, goniometer is not necessary in the device. Therefore, the device using the "cos α" method is smaller and lighter than the device using the "$\sin^2 \psi$" method. Consequently the "cos α" method can be used in various cases and for various targets.

However, it is necessary to compare the data obtained from the "$\sin^2 \psi$" and the "cos α" methods in order to evaluate the measured residual stresses correctly in the actual structure, because these two methods have different measurement principles.

This paper describes the effect of the measurement principles on the residual stresses measured in various loading and material conditions.

Materials Research Proceedings **2** (2016) 103-108 doi: http://dx.doi.org/10.21741/9781945291173-18

Measurement method and test specimen.
We used measurement devices, X-stress 3000 with $\sin^2 \psi$ method and μ-X360n (Pulstec Industrial Co., Ltd) with cos α method in this experiment. Figure 1 shows the X-rays residual stress measurement devices. We measured powder samples (Fe, SUS, Ni, Al), and high compressive stress standard samples (Fe, SUS, IN718). Figure 2 shows the powder samples, and Figure 3 shows the high compressive stress standard samples. In addition, we measured SM490 and SUS316 plates by four-point bending tests and Figure 4 shows the test setup for X-ray stress measurement. Table 1 shows measurement conditions.

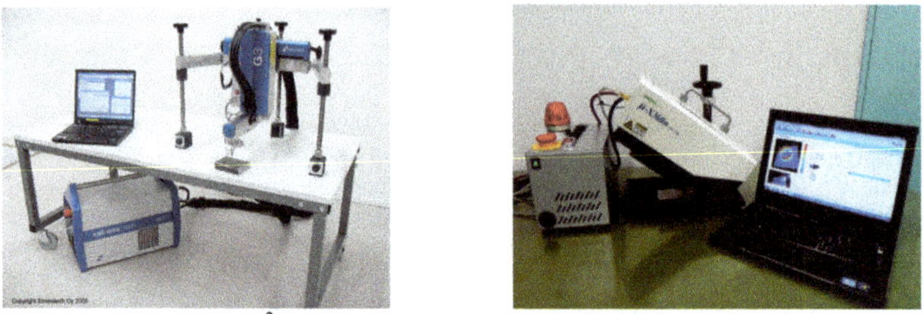

(a) X-stress 3000 with $\sin^2 \psi$ method (b) μ-X360n with cos α method

Fig.1 X-rays residual measurement devices.

Fig.2 Powder samples

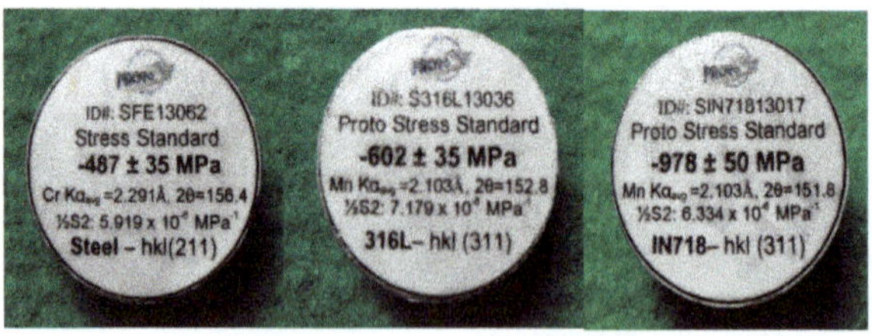

Fig.3 High compressive stress standard samples

Materials Research Proceedings **2** (2016) 103-108 doi: http://dx.doi.org/10.21741/9781945291173-18

Table 1 Stress measurement conditions

(a) For the powder samples.

Specimen	αFe	γFe	Ni	Al	
Mesurement method	cosα				
Characteristic X-ray	Cr -Kα	Cr -Kβ		Cr -Kα	
Diffraction plane	211	311	311	311	222
Diffraction angle[deg]	156.4	148.5	158.3	139.0	156.1
X-ray tube voltage[kV]	30				
X-ray tube current[mA]	1				
X-ray slit	φ2.0mm				

(b) For the high compressive stress standard samples and four point bending tests.

Specimen	Fe		SUS316L		IN718	
Mesurement method	cosα	$\sin^2\psi$	cosα	$\sin^2\psi$	cosα	$\sin^2\psi$
Characteristic X-ray	Cr-Kα		Cr-Kβ	Mn-Kα	Cr-Kβ	Mn-Kα
Diffraction plane	211		311	311	311	311
Diffraction angle[deg]	156.4		148.5	152.3	150.9	156.0
X-ray tube voltage[kV]	30					
X-ray tube current[mA]	1.0	6.7	1.0	6.7	1.0	6.7
X-ray slit	φ2.0mm					

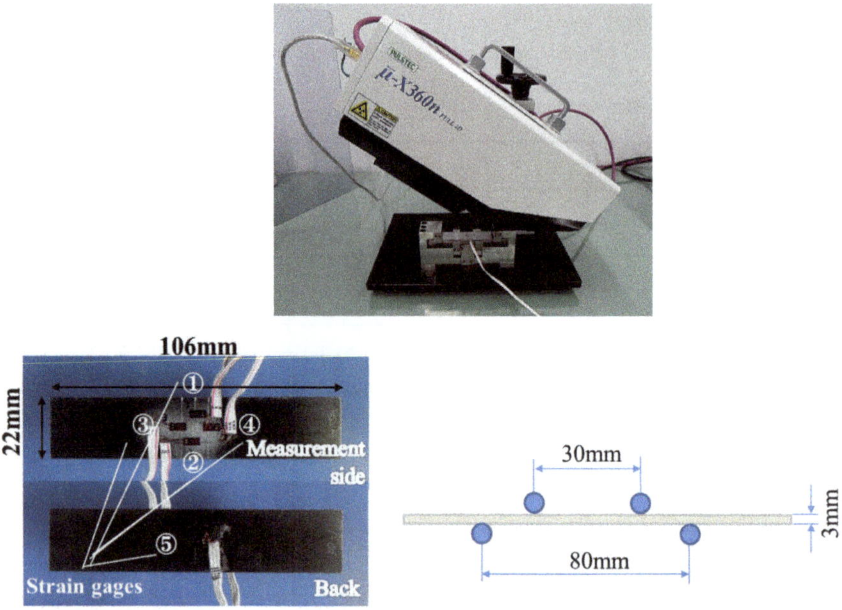

Fig.4 Experimental setup for X-ray stress measurement in the four-point bending testing.

Measurement results of the powder samples.
Figure 5 shows stress measurement results of the powder samples (αFe, γFe, Ni, Al). The gray scale in Figure 5 shows measurement range of $\sin^2\psi$ method. The measurement result of αFe by cos α method was 0±1.7 MPa. All measurement results of other material powder samples by cos α method were within ± 20MPa. These results were equivalent to the measurement values with $\sin^2\psi$ method. The measurement accuracy of the cos α method was satisfactory.

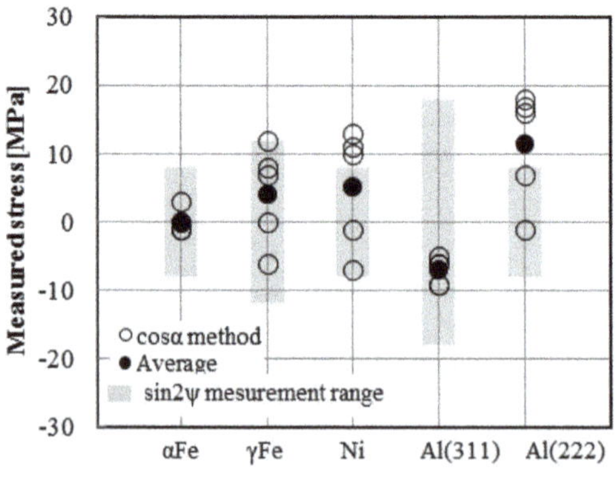

Fig.5 Stress measurements results of powder samples.

Materials Research Proceedings **2** (2016) 103-108 doi: http://dx.doi.org/10.21741/9781945291173-18

Measurement results of the high compressive stress standard specimens.
Figure 6 shows the results of measurements of the high stress standard specimens (Fe, SUS316L, IN718). The gray scale in Figure 6 expresses an official nominal level. The measured values for the high stress standard specimen were within the ranges of the nominal values in both the cos α and the $\sin^2\psi$ methods. The measurement values with the cos α and the $\sin^2\psi$ methods were almost equal.

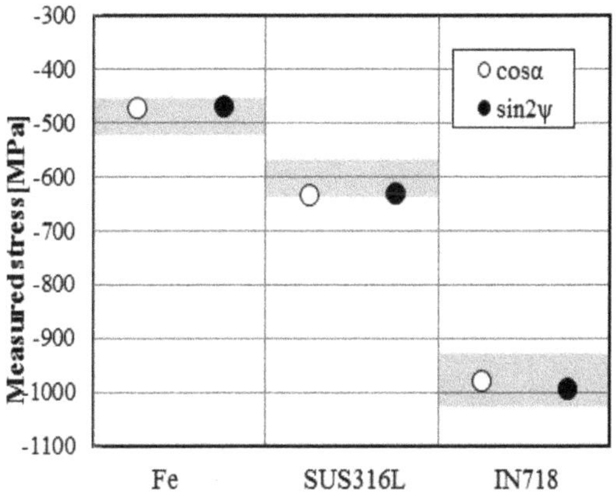

Fig.6 Residual stresses measured using cos α and $\sin^2\psi$ methods for the surface of the shot peened specimens of Fe, SUS316L, and Inconel718.

Measurement results of the four point bending tests.
Figure 7 shows the measurement results of the four point bending test of SM490. Specimens were measured without electropolishing the surface. The measured data by cos α and $\sin^2\psi$ methods showed a linear relationship with the strain gage data during the test. In the measurement of SM490, measured values by cos α and $\sin^2\psi$ methods were almost equal.

Figure 8 shows the measurement results of the four point bending test of SUS316 plate. Both the results of the cos α and $\sin^2\psi$ methods show linear relationship, a slope of 1.0, with the measured stress by strain gage. However, there was a difference of 80 MPa between $\sin^2\psi$ and cos α methods. We consider that this difference may be induced because the characteristic X-ray of Mn-Kα was used for the measurement with $\sin^2\psi$ method ,while the characteristic X-ray of Cr-Kβ was used for the measurement with cos α method. And furthermore, there is a possibility that the measured data were affected by the difference of the measurement area of each method.

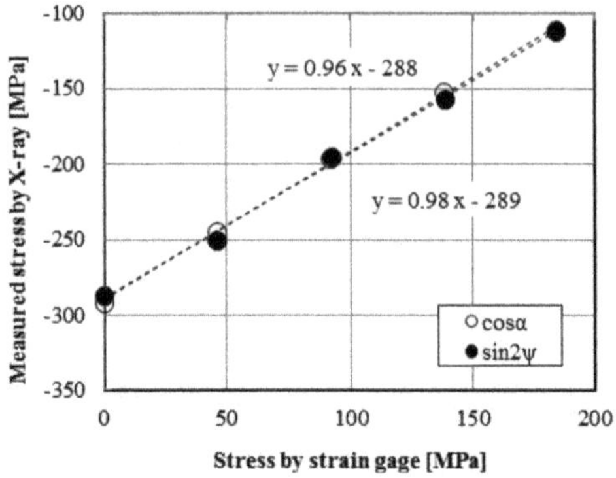

Fig.7 Measurement results of the four-point bending test of SM490 test piece.

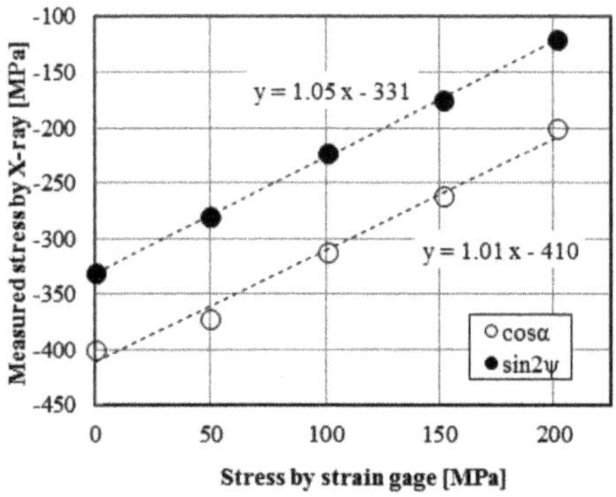

Fig.8 Measurement results of the four-point bending test of SUS316 test piece.

Summary

We measured stresses in various loading and materials conditions by cos α and $\sin^2 \psi$ methods and compared measurement values. As a result, in various powder samples and high stress standard test specimens, the measurement values with the cos α and the $\sin^2 \psi$ methods were almost equal. Also, in the four point bending test of SM490, measured values by cos α and $\sin^2\psi$ methods were almost equal. However, in case of the four point bending test of SUS316, the measured values by cos α and $\sin^2 \psi$ methods had a slight difference because of the difference in each measurement condition. Both cos α and $\sin^2 \psi$ methods are effective as the residual stress measurement technique, but we must be careful about measurement condition when we manage the aging change of stress precisely.

References

[1] Sasaki, T and Hirose, Y, J. Soc. Mat. Sci., Japan 44, 504, (1995) 1138-1143

[2] Miyazaki, T and Sasaki, T, J. Appl. Cryst. (2016). 49, 426–432. http://dx.doi.org/10.1107/S1600576716000492

Residual Stresses 2016: ICRS-10 Materials Research Forum LLC
Materials Research Proceedings **2** (2016) 109-114 doi: http://dx.doi.org/10.21741/9781945291173-19

Residual Stress Influence on the Flexural Buckling of Welded I-Girders

Benjamin Launert[1,a *], Michael Rhode[2,b], Arne Kromm[2,c], Hartmut Pasternak[1,d] and Thomas Kannengiesser[2,e]

[1]Chair of Steel and Timber Structures, Brandenburg University of Technology (BTU), Konrad-Wachsmann-Allee 2, 03046 Cottbus, Germany

[2]Federal Institute for Materials Research and Testing (BAM), Department 9.4 – Weld Mechanics, Unter den Eichen 87, 12205 Berlin, Germany

[a]benjamin.launert@b-tu.de, [b]michael.rhode@bam.de, [c]arne.kromm@bam.de, [d]hartmut.pasternak@b-tu.de, [e]thomas.kannengiesser@bam.de

Keywords: Stability Design, Finite Element Method, 2-D Welding Simulation, Inherent Strain, PDA

Abstract. The nonlinear analysis became a common tool to precisely assess the load-bearing behavior of steel beam and column members. The failure level is significantly influenced by different types of imperfections, among geometric also structural imperfections (residual stresses). Here are still gaps in the knowledge. Nowadays, 3-D welding simulation developed to a level where it could provide reliable estimation of weld-induced distortion and residual stresses. Nevertheless, modelling and computational effort are still in a less practicable range. In this study a simplified procedure to implement residual welding stresses in continuous large scale members is proposed and the influence on the ultimate limit state of slender members in compression is evaluated for two common structural steel grades. The results showed significant improvements in the utilization of load bearing capacity compared with simplified design methods. The comparatively general approach in this study offers potential for future optimization.

Introduction

Welded I-girders are used in different applications in steel construction (e.g. industrial buildings, bridges) due to either dimensions and/or efficiency through customizable plate thicknesses, shapes and/or materials. Many standards, including Eurocode 3 (EC 3) permit the use of non-linear finite element analysis (FEA) for the design of steel structures. The development in this field allows performing "experiments" in computing software instead of the laboratory or expensive in-situ experiments. Still, the implementation of imperfections (including residual stresses due to weld manufacturing) remains a major task in the performance of such analysis. Unlike geometrical imperfections, residual stresses (RS) are not standardized. For that reason, more or less founded simplified distribution functions as proposed by the Swedish design code BSK 99 [1] are used. Limitations on the applicability of this model are not reported except for the plate thicknesses which should not exceed 40 mm. However, the influence of RS seemed somewhat overestimated for many cases comparing conventional structural steel S355 and S690QL. As in the EC 3, direct proportionality of the tensile and compressive RS (the latter are the main interest in this study) and the yield strength is assumed. An opposite effect was recently noticed in [2,3] showing relatively reduced RS in high strength steel. A significant overestimation of RS was generally noticed if the chords are narrow. A lower limit is missing. Additionally, the correlation with the plate thicknesses only is doubted since no record is taken of actual welding parameters. Hence, a new approach was required. This contribution (as part of a national research project [2]) presents an improved general

Materials Research Proceedings **2** (2016) 109-114 doi: http://dx.doi.org/10.21741/9781945291173-19

procedure to access the load-bearing capacity numerically under realistic consideration of welding effects (with the focus on RS).

2-D Thermo-elasto-plastic analysis

The geometry is shown in Fig. 1 with the chord width (b_f) 150 mm and the web height (h_w) 220 mm. Plate thicknesses were 25 mm (t_f) and 15 mm (t_w) and base materials were conventional S355 and S690QL. The weld type is a T-joint with subsequently welded two-sided single-layer fillets. A reference model (3-D simulation) was presented in [3]. The calculation time was several days. Thus, a simplified modelling procedure is absolutely required in practice. A simplified 2-D modelling procedure in Ansys (16.2) was developed. The results are shown Fig. 1. Verification is presented in Fig. 2.

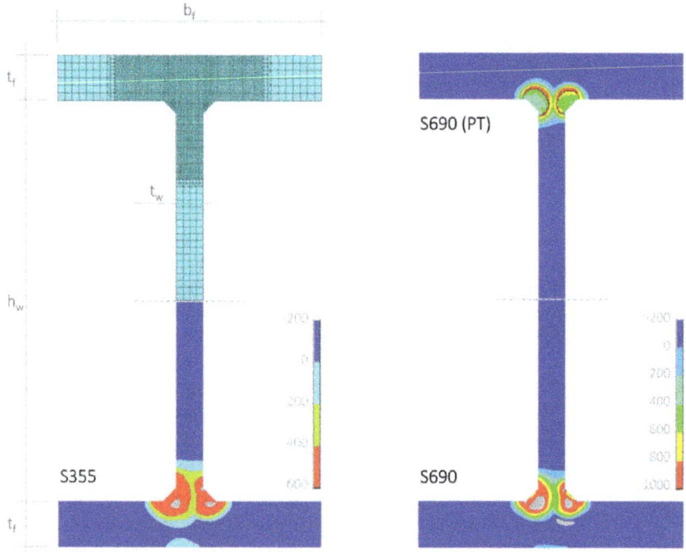

Fig. 1: Meshed 2-D model and calculated RS (in MPa) for S355 and S690QL (with/without PT)

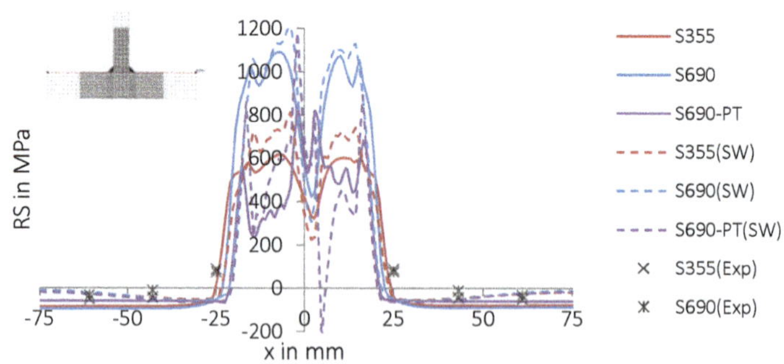

Fig. 2: RS distribution compared to [3], SW-Simufact, Exp-Experiment, exemplary for the chord

Elements types were plane 55 for the thermal analysis and plane 182 for the mechanics. A generalized plane strain condition was assumed. The element edge length close to the weld was 1 mm. A model using symmetry condition about the web middle axis has shown identical stresses compared with those determined from the full section model (which is due to the web height in this example). Nevertheless, the computation effort was comparatively small also for the full section model (less than half an hour). The temperatures were more or less independent of the grade and mostly depend on welding parameters. The particular yield strengths (mean values were 463,9 MPa for S355 and 831,3 MPa for S690QL) were used for scaling of the stress strain behaviour. The hardening behaviour was simplified (a constant hardening modulus of 1000 MPa was assumed at all temperature levels). Further data (Young's modulus, Poisson ratio and thermal expansion coefficient) were taken as done in [3]. A good match of (longitudinal) RS was achieved. The peaks are slightly underestimated. An additional simplified modelling procedure (PT) was used for consideration of transformation effects during heating and cooling in S690QL. The results were comparable as well even with their simplifications.

Plasticity-based analysis

The following part is based on the assumption that RS are caused by incompatible permanent strains (so called inherent strains [4,5]) induced by either temperature gradients and/or phase transformations. This analysis is therefore sometimes referred as plasticity-based analysis (PDA) [6]. The actual analysis is of elastic type though. The procedure is that initial strains are applied to the full-structure FE-model and that "reactions" in terms of deformations and stresses are calculated by conducting a linear elastic analysis. Inherent strain (taken as the accumulated plastic strain) in a weld joint is produced only in a limited region near the weld. However, RS (or elastic strain) due to inherent strain is produced in the entire weldment. Their relations are expressed by elastic response equations. This is the idea behind PDA.

Applications were carried out mainly to control deformations in large welded structures using equivalent forces and moments (known as Equivalent Load Method). Numerical procedures incorporating inherent strain directly are limited. The coupling with a subsequent capacity analysis is of interest in structural design but has not been realized either. The plastic strain distribution can be calculated easily using simplified 2-D modelling approach. The distribution in T-joints obtained from comparative 2-D and 3-D models was however found quite different except of those associated with the longitudinal direction [6]. Thus, 2-D model may not be applicable in the prediction of angular distortion in T-joints but in the prediction of longitudinal RS and longitudinal buckling which strongly depends upon the distribution of longitudinal cumulative plastic strain [6]. A direct transfer (as proposed in [6]) seemed unnecessary. A modified procedure is presented and detailed in subsequent paragraphs.

1. *2-D thermal elastic-plastic analysis (TEPA) to calculate the distribution of (longitudinal) cumulative plastic strain,*
2. *Definition of a simplified distribution pattern,*
3. *Creation of a suitable mesh applicable to the criteria of PDA and capacity analysis and*
4. *Elastic analysis of the full-structure elastic FE-model imposing initial strain to reproduce RS (and deformations).*

The distribution of plastic strain is initially needed from TEPA (1.). This is shown in Fig. 3 a) in case of S690. It was noticed that the plastic strain pattern is typically of somewhat simple shape. For that reason, a simplified distribution pattern was derived (2.). According to [5], the inherent strain zone of a single fillet weld may be generally simplified by three partial ellipses and the fillet weld metal. This approach was simplified to ensure a better applicability of the model. An ellipsoid (radii are indicated by r_x and r_y) with the center in the application point of the weld torch was used (see Fig. 3 a) as well). From the distribution function it is known that the maximum values are

Materials Research Proceedings **2** (2016) 109-114 doi: http://dx.doi.org/10.21741/9781945291173-19

approximately equal to the yield strain (ε_y) within the first ellipsoid. Values then drop linearly to zero to the border of the second ellipsoid. Elsewhere no initial strains are applied. A section of the distribution at the top surface of the chord is given in comparison with reference data in Fig. 3 b). A reasonable agreement was observed.

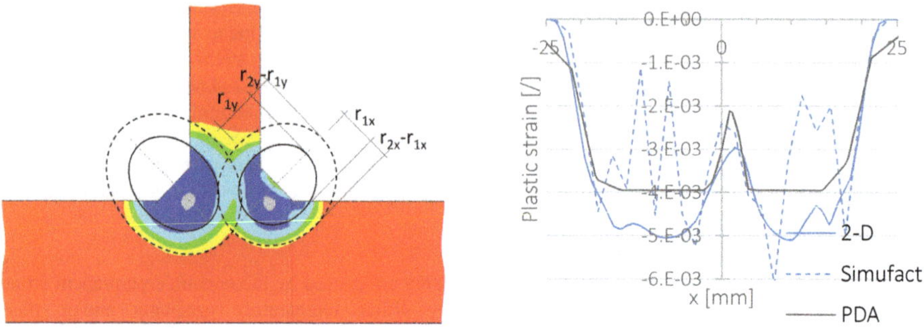

Fig. 3: Calculated plastic strain distribution compared to proposed simplified method (exemplary for S690QL)

The implementation in the global (3-D) FE-model was realized using the "inistate" command (alternative ways are possible, e.g. by the definition of artificial expansion coefficients to create equivalent thermal strains as proposed in [6]). A certain restriction is that PDA requires a fine mesh to ensure that initial strains are applied precisely. It is therefore necessary to provide a suitable mesh applicable to the criteria of PDA and capacity analysis (3.). This is, to a certain degree, always a compromise. An adapted 2-D mesh was generated with a reduction in element numbers from 6480 to 1092. This mesh was extruded in longitudinal direction (to different length) with element lengths of 15 mm. The element type has then been changed to solid 185. Finally, a linear elastic analysis was conducted (4.). It was shown that this procedure is capable of creating realistic RS (and also deformations) which can be directly utilized in a subsequent capacity analysis. Applicability in terms of deformations was exemplified but only verified qualitatively yet. Hence, only the RS are used in section 4.

Load bearing capacity

The final step was a nonlinear analysis of the load-bearing behaviour. The static system is a simply supported beam under pure compression. The investigations are given exemplary for weak-axis buckling. The cross section is the same as given in section 2. The material behaviour is equal to the one from the TEPA except that values are only needed for room temperature. Deviations from the ideal shape (so called geometrical imperfections) were taken as the first eigenmode determined from linear buckling analysis (LBA). The maximum value was scaled to L/1000 (which is the maximum permitted manufacturing tolerance). This is a conservative assumption, but represents the state-of-art. The PDA could be used for future deformation analysis. For the consideration of RS, three approaches were compared:

Case 1) Without RS,

Case 2) Direct definition of RS based on BSK 99 [1] and

Case 3) Indirect generation of RS using PDA (definition of initial strains by inistate).

Materials Research Proceedings **2** (2016) 109-114 doi: http://dx.doi.org/10.21741/9781945291173-19

EC 3 and other standards use buckling curves. Load influencing factors are covered by the definition of so called reduction factors χ. Depending on section type, manufacture, failure and/or direction of failure and in parts material, different buckling curves are distinguished. The reduction factor χ is given by Eq. 1

$$\chi = \frac{N_R}{N_{pl}}. \quad (1)$$

The slenderness $\bar{\lambda}$ is varied by the length l and is given by Eq. 2

$$\bar{\lambda} = \sqrt{\frac{N_{pl}}{N_{cr}}} \quad \text{,where} \quad (2)$$

$$N_{pl} = A \cdot f_y, N_{cr} = \frac{\pi^2 \cdot E \cdot I}{l^2}.$$

Accordingly, the length l to be used in the analysis for certain slenderness $\bar{\lambda}$ can be given by Eq. 3

$$l = \sqrt{\bar{\lambda}^2 \cdot \frac{\pi^2 \cdot E \cdot I}{A \cdot f_y}}. \quad (3)$$

The lengths l were calculated using the cross-sectional properties without consideration of the weld seams. The slenderness range relevant for RS was found close to 1,0 [7]. For that reason, the slenderness $\bar{\lambda}$ was varied in steps of 0,2 from 0,6 to 1,4. Solid element type 185 was used. As for the welding, this was necessary due to the local plastic strain distribution for case 3). Works are ongoing to make the method less sensitive to the mesh and to reduce the element level in terms of shell (or even beam) element types suitable in this context. Solid model was not necessary for cases 1) and 2), but has been used for comparative purpose. The calculated results are given in Fig. 4. The values are compared with the buckling curves stated by the EC 3. For the investigated case, curve "c" applies for comparison.

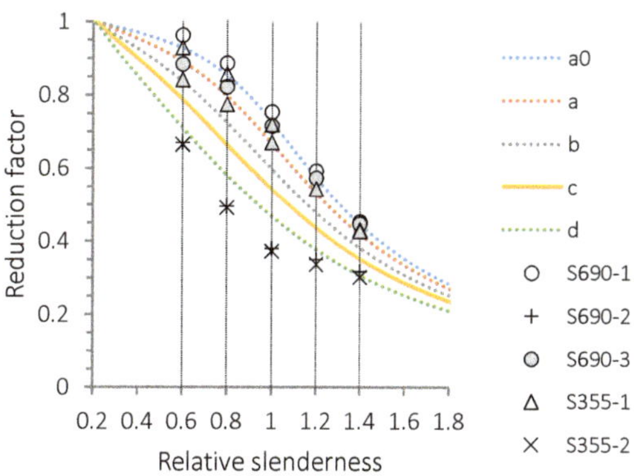

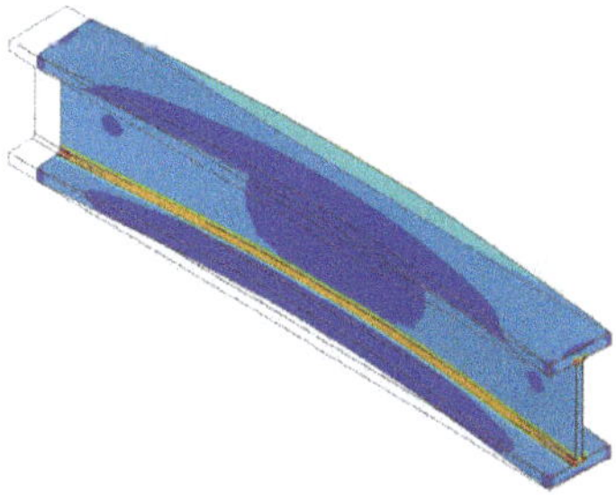

Fig. 4: Results in comparison with European buckling curves (EC 3)

The comparison of results without RS shows that there are minor differences with an increasing yield strength difference though values are normalized. With RS, results differ significantly. The RS model by [1] led to highly unrealistic compressive RS (see Introduction) causing a significant underestimation of the ultimate load. Instead, the RS generated through inherent strain (PDA) equalled realistic RS. The comparison of materials shows generally favourable behavior of the S690, which is explained by relatively decreased compressive RS. On average, the buckling curve is raised by one line indicated by curve "b" for the S355 and curve "a" for the S690. With respect to EC 3, large reserves are identified.

Summary

An improved method for the capacity accessment of welded girders was presented. The modelling using 3-D solid elements in welding simulation is often not required in case of constant cross sections through the lengths. Nearly identical results are generated using 2-D solid models encompassing the assumption of generalized plane strain. The observed computation times demonstrated the applicability in practice. The calculated plastic strain distribution was then simplified for application to the total girder. A first approach was presented using two ellipsoids for each fillet. This resulted in good agreement with reference data. A further standardization of these variables could be helpful. The method is however, to some extent, restricted by the necessary local discretization density, which is needed in similar form also for the capacity model. The transferability to analogeous shell or beam models requires further research. Nevertheless, the proposed method is a significant improvement in comparison with existent simplified models. A particular advantage is the wide applicability apart from the herein presented example (easy adaption to different cross sectional shapes). The method appeared principally also suitable to take into account weld-induced deformation of large structures. The capacity analysis was eventually on an example for weak-axis buckling of a centrally loaded compression member. Results showed the superior buckling behavior of high strength steel S690. The assignment of a buckling curve was approximately one curve higher compared to that of conventional S355.

References

[1] N.N., Swedish Design Rules for Steel Structures, Boverket - Swedish National Board of Housing, Building and Planning, Karlskrona, 2003.

[2] H. Pasternak, T. Kannengiesser et al., IGF 18104 BG - Enhancement of Load Capacity of Welded High-Strength I-Shape Sections Using Improved Design Models for the Consideration of Residual Stress (in German), 04/2014-04/2016, Project description, 2013.

[3] H. Pasternak, B. Launert, T. Kannengiesser and M. Rhode, Advanced residual stress assessment of plate girders through welding simulation, submitted to Procedia Engineering (2016), 12th International Conference "Modern Building Materials, Structures and Techniques", May 26th – 27th, 2016, Vilnius.

[4] Y. Ueda, M. G. Yuan, M. G., The Characteristics of the Source of Welding Residual Stress (Inherent Strain) and Its Application to Measurement and Prediction, Trans. JWRI 20 (1991), No. 2, 119-127.

[5] Y. Ueda, N.-X., Measuring Methods of Three-Dimensional Residual Stresses with Aid of Distribution Function of Inherent Strains (Report 3): Distributions of Residual stresses and Inherent Strains in Fillet Welds, Trans. JWRI 24 (1995), No. 2, 123-130.

[6] G. Jung, Plasticity based distortion analysis for fillet welded thin plate t-joints, PhD-thesis, Ohio State University, 2003.

[7] H. Pasternak, B. Launert and T. Krausche, Welding of girders with thick plates – Fabrication, measurement and simulation, Journal of Constructional Steel Research 115 (2015), 407-416. http://dx.doi.org/10.1016/j.jcsr.2015.08.037

Materials Research Proceedings 2 (2016) 115-120 doi: http://dx.doi.org/10.21741/9781945291173-20

Measurement of Highly Non-uniform Residual Stress Fields in Thin Plate Using a New Side Cut Destructive Method

H. K. Kim[a] *, M. J. Pavier[b] and A. Shterenlikht[c]

Mech Eng Dept, University of Bristol, UK

[a]HK.kim@bristol.ac.uk, [b]Martyn.Pavier@bristol.ac.uk, [c]mexas@bristol.ac.uk

Keywords: Non-uniform Residual Stress, Mechanical Strain Relaxation Technique, Digital Image Correlation, Thin Friction Stir Welding Plate, Plasticity Effect

Abstract. A recently proposed Mechanical Strain Relaxation (MSR) technique for the measurement of residual stress in thin plates is presented. The measurement of residual stress involves making a single straight cut and collecting the relaxation displacement data from the side surface. In this work the method was applied to an Aluminium friction stir welded (FSW) specimen. The cut was introduced with a wire electrical discharge machining. The displacements were recorded with a 3D digital image correlation (DIC) method. The measured FSW residual stress profile agreed well with that measured by Energy Dispersive X-ray Diffraction (EDXRD). It was observed that the amount of plastic strain, caused by stress redistribution during the relaxation process, strongly depends on the direction of the propagation of the cut. In particular, if a cut is propagated along the thickness of a plate, then the effect of plastic flow on the measured residual stress is negligible. Another attractive feature of the method is that it is relatively insensitive to random experimental noise.

Introduction

Over the last two decades a wide range of mechanical strain relaxation (MSR) techniques have been developed and used to measure residual stresses in a wide variety of engineering components. Well known examples of MSR techniques are: hole drilling [1], slitting [2] and contour method [3]. Recently, new technique [4] has been proposed and it follows the general scheme of Mechanical Strain Relaxation (MSR) techniques, i.e. measuring the deformation due to removal of some stressed material in a body and then that measured relaxation is converted to stress.

The main target of the proposed method is that the shape of in-plane residual stress field is highly non-uniform, including discontinuities. Other popular techniques, such as e.g. hole drilling, cannot be used in such cases because they assume stress uniformity within hole diameter [1] and many holes are required to measure residual stress field in relatively large measurement area, leading to complexity of the experiment [5].

During cutting residual stress is redistributed and this might cause plastic deformation where the magnitude of residual stress is close to yield [2]. However, it is shown that this plastic strain on relaxation can be mitigated by choosing the correct propagation of the cut [4]. For this is reason, the technique is attractive in practical application because the measurement can be conducted regardless of the propagation of the cut. It is not always possible for many techniques, for example the slitting method, the propagation of the cut is very important because non-uniform RS can only be measure along depth [2].

In this work, RS field in 3 mm thick FSW plates have been measured by the proposed method. The application is ideal for the technique because the specimen is very thin and the stress field is non-uniform. The techniques involves a single cut at the midsection of the longitudinal direction of the plate, propagated either from the front to the rear surface, called 'side cut', or from the top to bottom surface, called 'top cut' to observe the difference. 3D DIC were used to measure relaxation displacement on the half width of side surface. Then the residual stress field was reconstructed from the measured relaxation via analytical model.

Materials Research Proceedings **2** (2016) 115-120 doi: http://dx.doi.org/10.21741/9781945291173-20

Analytical model

The analytical model for the technique is based on the elastic solution for the problem of self-equilibrated loading at the one end of 2D semi-infinite strip of width 2c [6], see Fig. 1(a). This analytical model will be employed to reconstruct residual stress field in thin Friction stir welded (FSW) plate from the measured displacement fields on the side surface from 3D DIC, see Fig. 1(b). Other application of the elastic problem to the measurement of residual stress was investigated and described in [4]. Brief overview is given here.

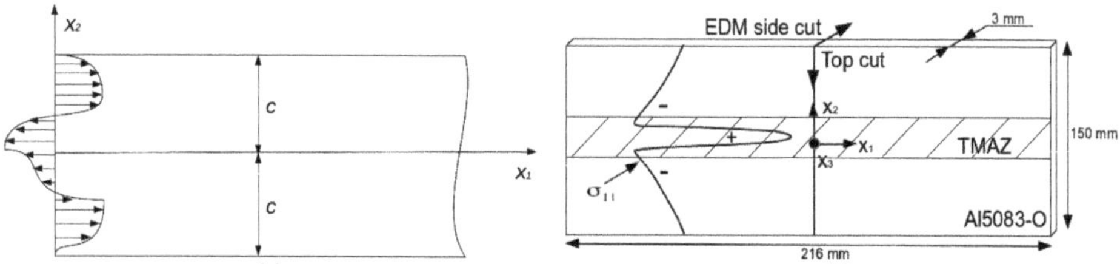

(a) (b)

Fig. 1 : (a) The problem geometry for a 2D semi-infinite rectangular strip, of width 2c, with arbitrary self-equilibrated end load [4]. The boundary conditions are: $\pm \boldsymbol{c} : \boldsymbol{\sigma_{22}} = \boldsymbol{\sigma_{12}} = \mathbf{0}$ *and* $\boldsymbol{x_1} = \mathbf{0} :$ $\int_{-c}^{c} \boldsymbol{\sigma_{22}} \boldsymbol{dx_2} = \int_{-c}^{c} \boldsymbol{\sigma_{12}} \boldsymbol{dx_2} = \mathbf{0}$ *and (b) The FSW specimen used in this work. The specimen is cut with wire EDM on the* $\boldsymbol{x_1} = 0$*, the symmetry plane.*

The solution uses the two even, f, and odd, g, stress functions which satisfy the bi-harmonic equation:

$$f = e^{-\frac{\gamma x_1}{c}}\left(\xi \cos\frac{\gamma x_2}{c} + \frac{\gamma x_2}{c}\sin\frac{\gamma x_2}{c}\right); \qquad g = e^{-\frac{\phi x_1}{c}}\left(\psi \sin\frac{\phi x_2}{c} + \frac{\phi x_2}{c}\cos\frac{\phi x_2}{c}\right) \tag{1}$$

The boundary conditions, see Fig. 1(a), lead to the following constraints:

$$\sin 2\gamma + 2\gamma = 0; \qquad \xi = -\gamma \tan\gamma; \qquad \sin 2\phi - 2\phi = 0; \qquad \psi = -\phi / \tan\phi \tag{2}$$

which have infinite number of solutions for dimensionless γ, ξ, ϕ, ψ. A complete stress function, suitable for any arbitrary self-equilibrated stress on the boundary, is constructed as follows:

$$\theta = \sum_{i=1}^{\infty} a_i Re(f_i) + b_i Im(f_i) + c_i Re(g_i) + d_i Im(g_i) \tag{3}$$

Using plane stress and small strain assumptions, the surface in-plane displacements can be expressed as follows:

$$Eu_1 = \sum_{i=1}^{\infty} a_i \int Re\left(\frac{\partial^2 f_i}{\partial x_2^2} - v\frac{\partial^2 f_i}{\partial x_1^2}\right)dx_1 + b_i \int Im\left(\frac{\partial^2 f_i}{\partial x_2^2} - v\frac{\partial^2 f_i}{\partial x_1^2}\right)dx_1$$
$$+ c_i \int Re\left(\frac{\partial^2 g_i}{\partial x_2^2} - v\frac{\partial^2 g_i}{\partial x_1^2}\right)dx_1 + d_i \int Im\left(\frac{\partial^2 g_i}{\partial x_2^2} - v\frac{\partial^2 g_i}{\partial x_1^2}\right)dx_1 \tag{4}$$

Materials Research Proceedings **2** (2016) 115-120 doi: http://dx.doi.org/10.21741/9781945291173-20

$$\mathrm{Eu}_2 = \sum_{i=1}^{\infty} a_i \int Re\left(\frac{\partial^2 f_i}{\partial x_1^2} - v\frac{\partial^2 f_i}{\partial x_2^2}\right)dx_2 + \; b_i \int Im\left(\frac{\partial^2 f_i}{\partial x_1^2} - v\frac{\partial^2 f_i}{\partial x_2^2}\right)dx_2$$

$$+\, \mathrm{c_i} \int Re\left(\frac{\partial^2 g_i}{\partial x_1^2} - v\frac{\partial^2 g_i}{\partial x_2^2}\right)dx_2 + \; d_i \int Im\left(\frac{\partial^2 g_i}{\partial x_1^2} - v\frac{\partial^2 g_i}{\partial x_2^2}\right)dx_2$$

The unknown coefficients in the series expansion of the stress function, $\mathrm{a_i}, b_i, c_i, d_i$, are found from the solution of the standard linear least square (LLS) problem:

$$\min_x \left|\left|\boldsymbol{Ax} - \boldsymbol{u}\right|\right|_2 \tag{5}$$

where **x** is the vector of $4n$ unknown series coefficients, **u** is a vector of $2m$ displacement values, **A** is a $2m \times 4n$ matrix of integral functions taken at the locations of the measurement points. To improve LLS stability, the number of measured displacement points, m, should be considerably higher than the number of terms in the series expansion, n. In practice, hundreds or thousands of data points are needed, which means the method requires a full-field measurement technique.

Experiment

A 216 mm × 150 mm × 3 mm plate was manufactured from aluminium alloy 5083-O, $\sigma_Y =$ 145 MPa, UTS = 290 MPa. The specimen was welded under force control, with a spindle speed of 400 rpm and feed-rate of 200 mm/min using a shoulder and pin diameter of 22 and 5 mm, respectively [7]. 3D DIC were used to measure u_1 and u_2 relaxation displacements due to the cut.

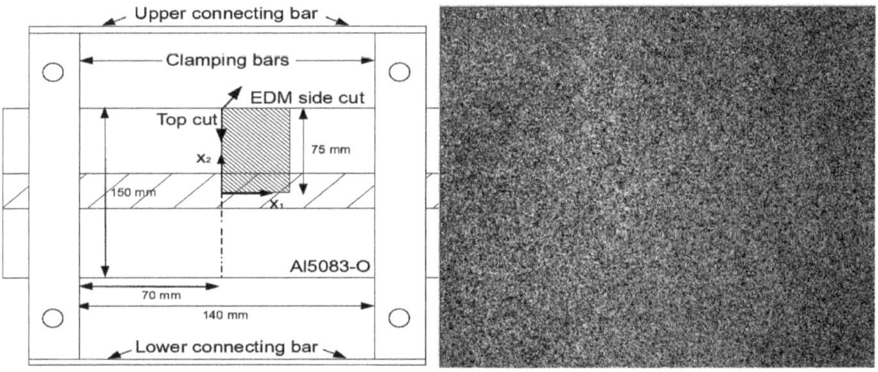

(a) (b)

Fig. 2: Layout of the specimen and the clamps. EDM wire cut cut was done along the dashed line. The fine hatched area indicates the surface fragment where the DIC data was collected. (b) Surface pattern created with matt black and white spray paints.

Clamping was introduced to avoid rigid body movement of the specimen during the cut. The clamping rig was made of four Aluminium bars ($\sigma_Y = 503$ MPa, UTS = 572 MPa) and top and bottom bars were bolted at 70 mm away from the midsection of the specimen where EDM cutting is carried out, see Fig. 2(a). The clamping bars are connected rigidly to keep the two halves of the cut specimen together after
the EDM cut.

Materials Research Proceedings **2** (2016) 115-120 doi: http://dx.doi.org/10.21741/9781945291173-20

Dantec Dynamics 3D DIC system [8] was used to measure the relaxation. The system consists of two 12-bit 2448 × 2050 pixel CCD cameras with zoom lenses and LED light illumination. Taking specimen symmetry into account, only half width of the specimen was recorded to achieve high displacement resolution. The field of view was 76 × 63 mm^2, and the DIC stereo parameters were: the angle of 18.87° and baseline of 116.34 mm. The size of the pixel was 31 μm/pixel. The estimated displacement accuracy was between 0.31 and 0.62 μ/pixel, based on the accuracy of sub-pixel resolution algorithms, typically between 0.02 to 0.01 pixel [9]. 35 and 25 pixel sizes of subset and step size were chosen, giving around 0.78 mm data point spacing.

To achieve the optimum accuracy of the DIC, suitable surface preparation is critical. The surface should have uniformly distributed suitable size of the random speckle pattern. Both black and white speckle size of at least 3-5 pixels is typically recommended to maintain acceptable noise level [10]. Matt black/white spray paints was found most effective in this work, see Fig 3(b). Additionally, the spray paint pattern can also meet the requirement of the preservation of the surface during the EDM cutting.

Two different propagations of the cut was applied to see the difference in reconstructed residual stress field due to plastic strain produced by different way of stress redistribution: (1) cutting from the front to rear face, called as side cut, and (2) cutting from the top to bottom surface, called as top cut. The EDM wire diameter was 0.25 mm and the speed of the wire was 18.75 mm/min for the top cut and 0.7 mm/min for the side cut.

Results

u_1 and u_2 surface relaxation displacement fields side cut were obtained from 3D DIC, see Fig. 3. There are few spots where the subsets could not manage to correlate maybe due to surface pattern damage. Unexpected high magnitude of u_2 relaxation at the far away from the cut has been observed 3(a). This might be the error caused by rigid-body rotation during the cut or out of focus in outer region of the lenses.

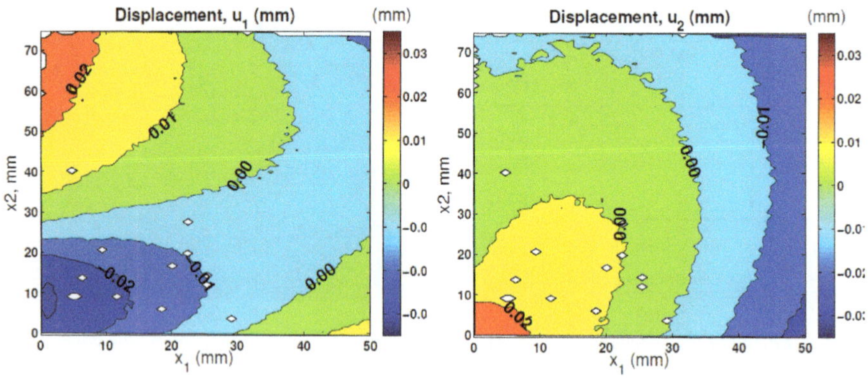

(a) (b)

Fig. 3: u_1 (a) and u_2 (b) relaxation displacement results from the front cut. 3D DIC subset and step size are 35 and 25 pixels. The displacement field was collected up to x_1 = 50 mm from the cut, x_1 = 0 mm.

Fig. 4(a) shows the u_1 relaxation displacement, obtained from the side and top cut in the region of near the cut. Significant local distortion very close to the cut is seen in the top cut u_1 field. The magnitude of u_1 is much lower in the top cut than in the side cut. This might be due to the plastic strain induced during the top cut [4], which leads to large errors in the reconstructed residual stress fields, see Fig. 4(b).

The half width of the residual stress fields reconstructed from measured relaxation displacement fields are compared against EDXRD result, see 4(b). 15 series limits and the displacement data fields from around $x_1 = 0$ to 5 mm were used to reconstruct residual stress field to minimise the analytical uncertainty caused by data far away from the cut [4]. Large tail at the tip in the reconstructed residual stress, see Fig. 4(b), has been observed due to the limitation of accurate displacement measurement of DIC at an edge. Also, due to high uncertainty of $u_2$displacement field from DIC, u_2 displacement field was set to zero and it gives more reliable reconstructed residual stress. As a result, although there is oscillation, the reconstructed residual stress field from the front cut agrees very well with EDXRD result, but the result from top cut has large discrepancy with the others due to the distortion near the cut in measured relaxation displacement fields, see Fig. 4(a). It is noted that the EDXRD error bars are taken from [7] and the error bars in the measurement using this method was simulated by adding 5% random error to the DIC measured displacement field.

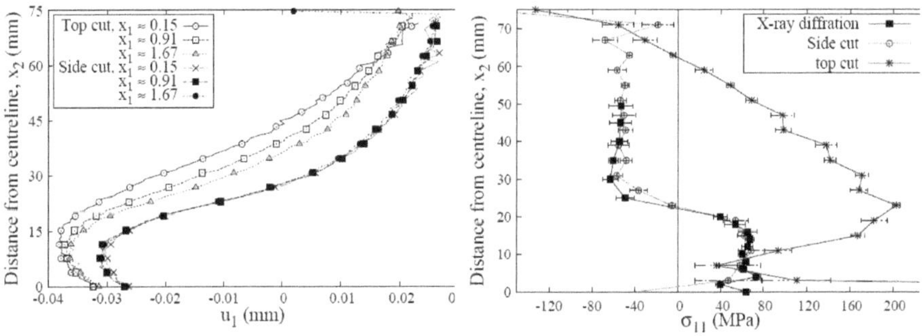

(a) (b)

Fig. 4: (a) u_1 relaxation displacement profiles near the cutting edge. The magnitude of u_1 near the cut is lower in the top cut than that in the side cut. (b) Residual stress fields, reconstructed from measured relaxation displacement fields from the side and top cut are compared against EDXRD result [7].

Concluding remarks

Residual stress profile, induced by FSW in thin Al5083-O plate was measured by the new side cut non-uniform residual stress technique. The technique consists of measuring full-field the relaxation displacements due to the cut at midsection of the welding direction of the plate by 3D DIC and reconstructing the residual stress field along the cut from the measured relaxation displacements using analytical model.

Two different cutting propagations were applied to see the plastic flow caused by different behaviour of stress redistribution during the EDM cutting process. In this example, the top cut removed the layer of material with uniform stress, leading to the stress redistribution in the remaining ligament. On the other hand, the side cut removed the layer of material with self-equilibrated stress, leading to minimum stress redistribution. As a result, the residual stress field, reconstructed from the side cut had much better agreement with that from EDXRD than that from the top cut.

There are many errors can be added in the relaxation displacement measurement and cause the distortion in the reconstructed residual stress fields. In this example, although the half the width of residual stress field were successfully reconstructed, instability of the reconstruction was shown as oscillation in the residual stress field. The possible causes of instability of the technique could be the limited collection of relaxation displacement data and the rigid body rotation during the cut. Thus, theses errors should be controlled to achieve precise experimental result. Improving 3D DIC measurement system is one of the technical solution. higher resolution of the CCD camera can achieve desire spatial resolution with increased size of the field-of-view.

The clamping is the one of the major issue in the experiment. the relaxation due to the is highly influenced by the clamping, leading to error in reconstruction of residual stress. No clamping is ideal, but this easily cause rigid-body motion. Thus, the optimum clamping strategy must be investigated. the minimum clamping, very far away from the cut, is normally recommended so that the relaxation of the residual stress cannot be affected.

Acknowledgements

The work was supported by Bristol Alumni Foundation and carried out using the computational facilities of the Advanced Computing Research Centre, University of Bristol. The authors are indebted to Dr. Graeme Horne for providing the specimens and EDXRD data.

References

[1] G. S. Schajer, Measurement of non-uniform residual stresses using the hole-drilling method, Part I - Stress Calculation Procedures, J. Eng. Mater-T ASME, 110(4):338–343, 1988. http://dx.doi.org/10.1115/1.3226059

[2] W. Cheng and I. Finnie, Residual Stress Measurement and the Slitting Method, Springer, New York, 2007.

[3] M. B. Prime, Cross-sectional mapping of residual stresses by measuring the surface contour after a cut, J. Eng. Mater-T ASME, 123:162–168, 2001. http://dx.doi.org/10.1115/1.1345526

[4] H. K. Kim, H. E. Coules, M. J. Pavier, and A. Shterenlikht. Measurement of highly non-uniform residual stress fields with reduced plastic error, Exp. Mech., 55:1211–1224, 2015. http://dx.doi.org/10.1007/s11340-015-0025-1

[5] W. Xu, J. Liu, and H. Zhu, Analysis of residual stresses in thick aluminum friction stir welded butt joints. Material Design, 32(4):2000–2005, April 2011. http://dx.doi.org/10.1016/j.matdes.2010.11.062

[6] S. Timoshenko and J. N. Goodier, Theory of Elasticity, McGraw-Hill, 3rd edition, 1970.

[7] G Horne, M. J. Peel, and D. J. Smith. Measurement of elastic follow-up for combined applied and residual stresses. In Proceedings of the ASME Pressure Vessels and Piping, *PVP2013*, 14-18 July 2013. http://dx.doi.org/10.1115/pvp2013-97624

[8] Dantec Dynamics, ISTRA 4D Software Manual Q-400 system, 2012.

[9] B. Pan, H. Xie, B. Xu, and F. Dai. Performance of sub-pixel registration algorithms in digital image correlation, Measurement Science and Technology, 17(6):1615–1621, June 2006. http://dx.doi.org/10.1088/0957-0233/17/6/045

[10] P. Reu. Hidden Components of DIC: Calibration and Shape Function - Part 1. Exp. Techniques, 36(2):3–5, March 2012. http://dx.doi.org/10.1111/j.1747-1567.2012.00821.x

Materials Research Proceedings 2 (2016) 121-126 doi: http://dx.doi.org/10.21741/9781945291173-21

Evaluation of Residual Stress in the Hot Forming Process Using Hole Drilling

M. Jalali Azizpour[1, a*] and Hamed Fattahi[2,b]

[1,2] Department of mechanics, Ahvaz Branch, Islamic Azad University, Ahvaz, Iran

[a]mjalali@iauahvaz.ac.ir, [b]hfattahi13@yahoo.com

Keywords: Hot Forming, Residual Stress, Drilling, FEM, Aluminum Alloy

Abstract. In the present study, it was attempted to perform the residual stress in the aluminum sheet A.A5083 via hot metal forming process by gas according to the Finite Element Method. The existence of the residual stress in the industrial pieces results in changing the reaction of structure against the applied load. The understanding of the effect of residual stress on the structural behavior is required and the design of the sensitive pieces regardless of the residual stress can result in financial and fatality damages. The aim of the present study was to apply an efficient approach to determine the stresses caused during the forming process with the hole drilling method. Thus, the sheet of 5083 aluminum alloy was formed by that method in an imperfect conical mold and the mentioned parameters were evaluated. At first, the section was analyzed initially and after preparing the mapping, it was modeled in the analysis software and its reliability was evaluated. The finite element analysis method of the hot metal forming process by the gas dependent on the temperature was evaluated. To simulate the forming process, ABAQUS/Explicit as the finite element software was used. At the end, the results obtained by the residual stress via the finite element method were compared with the results of the experimental tests of the piece.

Introduction

The hot metal forming process by gas blow is one of the latest methods for forming the sheet. Meanwhile, this method is considered as the most important super plastic forming methods. the hot forming by gas has been originated from super plastic forming and the hot blow forming. some of the advantages of this method against other forming methods are as follow: use in high temperatures, ability to achieve high length changes, ability to construct complicated forms made by materials, ability to increase the forming speed and reduce the cost of mould [1, 2].Sorgente[3] evaluated the possibility of reducing forming time via controlling the pressure profile in a part of pieces which was under the greatest strain at the end of forming process. The hot forming is a usable method in industry that increases the formability of aluminum and magnesium. Despite this fact, this method is applied for some special alloys such as aa5083 super plastic alloy with controllable features for the hot forming [4]. the residual stress in the industrial pieces results in changing the structure reaction against the applied load .thus, the recognition, determination of residual stress and evaluation of effect of residual stress on the structure behavior are necessary and the design of sensitive pieces without considering the residual stress may results in some financial and physical damages [5]. the pieces forming by the super plastic forming method has so many advantages. the forming by the super plastic forming method has so many advantages. the ability to form the pieces with complicated forms by a single mould and with no need to mandrel is one of them because the forming power is applied by gas. Some of them are: ability to form the pieces with complicated forms by a single mould and with no need to mandrel because the forming power is applied by gas. Despite that, this method has some disadvantages such as need to high temperature or power to ensure from complete forming of material inside the mould[6,7]. yu wu evaluated the properties of deformation in the thin sheet of magnesium alloy. he experimentally showed that the mean strain rate 10-2 was available and he also reported about the uniform distribution across the piece with a dramatic reduction of forming time [8].meanwhile, the material behavior before applying the super plastic forming is one of the main items that must be considered because the particles size should be fine to prevent from increasing the costs of alloying and transformation processes before the recrystallization [9]. The main objective of the present study was to determine the residual stress experimentally and numerically in an imperfect cone made by gas pressure in a temperature close to melting temperature of AA5083 aluminum sheet and also the distribution of sheet thickness after the forming process.

The hot metal forming by gas blow

In the hot metal forming by the gas or the fast plastic forming, the aluminum sheet is heated to the forming temperature (400-510°C). The heated sheet is put between two molds (steel or ceramic) manufactured by the sheet itself and then it is dragged to the forming tool by the air or gas pressures (nitrogen or argon) in the dorsal surface of the sheet. The forming pressure in one controlled method is constantly increased from the ambient pressure to the ultimate pressure of forming in the range of 17-34 bars or higher.

During the first several seconds to the first minute, as the pressure raises, the sheet matches itself with the tool level. In order to begin the stretching of sheet after the first period of pressuring, the pressure is raised faster. With regard to the size and the complexity of forming bodies, the forming usually is completed in 2 to 12 minutes that is considerably faster than the super plastic forming.

Finite element model for sheet forming

The shell elements are the sheets in the space and they have no shear or flexural stiffness matrix. Thus, only the non-zero stress components in the shell are the components that are parallel with the middle surface of shell. The warm forming by gas includes the modeling of pieces by the complicated geometry. Thus, the correct and exact definition of surface in the complicated mould and also the high quality finite element mesh across it is required. Often, the surfaces of mould are irregular morphologically.

The mechanic properties of AA5083 (based on the temperature used in simulation) were obtained knowing the the physical and thermal properties ofAA5083 and plotted in Fig.1 [7,10].

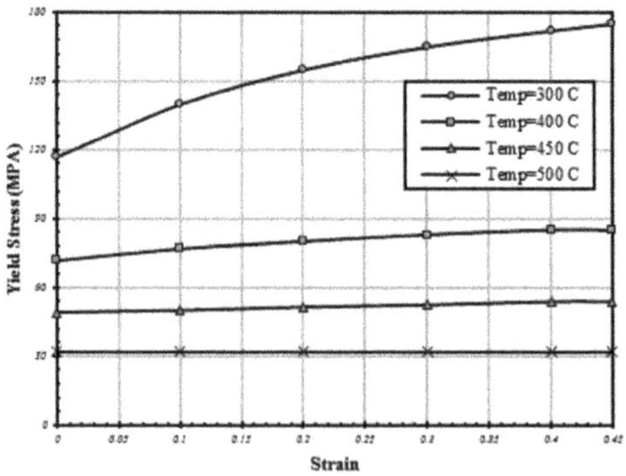

Fig. 1: Diagram of AA5083 stress changes based on temperature [7]

The forming sheet is defined as one deformable piece and the mould is modeled as the analytical rigid pieces. In the software analysis, the opening dimensions of the mould and sheet are in compatible with the conditions of the practical test. The circular sheet ´s diameter is 150mm. the

Materials Research Proceedings **2** (2016) 121-126 doi: http://dx.doi.org/10.21741/9781945291173-21

external diameter of the mould and the sheet holder is 150mm and their internal diameter is 115mm.Meanwhile, the radius of the opening edge is 5mm. For the mould and the sheet holder, the Reference Point is determined to apply the mechanic and thermal boundary conditions for the analytical rigid components. Table 1 indicates the loading and thermal conditions.

In order to measure the residual stress in the depth by Central Drilling Method, Strain Gauge installed on the lower plane was applied. Fig. 2 indicates the place of strain gauge schematically. The strain gauge ´s data are the average strains across the gage length. Thus, in order to extract the data in the Finite Element Method, the average of node ´s data in the strain gauge is applied.

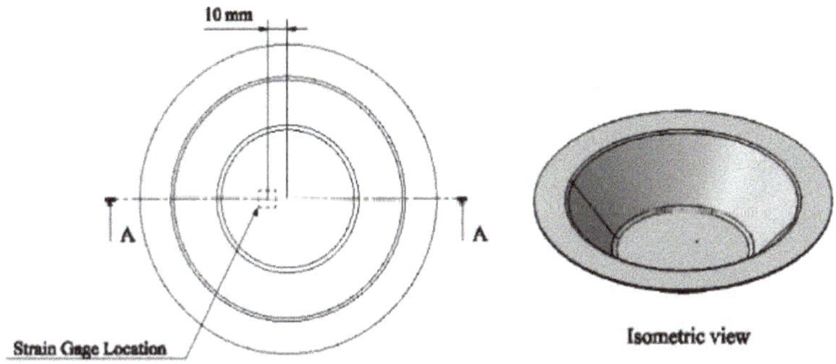

Fig.2: Place of the central drilling in the specimens after forming

Experimental test

In order to evaluate the inductive residual stress practically in the hot forming process by gas blow, the Central Drilling Method was used (according to ASTM E-837-92).

In this method, the drilling process by the hand drill was used to release the residual stresses caused by the hot metal forming by gas.

Table 1: Warm forming conditions in the tested speciments

Forming Temperature (ºC)	Forming Pressure (bar)	Forming Time (s)
500	5	10

First, the place of the hole was marked and then the place of drill was adjusted on the intended point by the universal tripod tool. Fig. 3 indicates the tripod tool and the drill mounted on the work piece(according to ASTM E-837-92).

Materials Research Proceedings **2** (2016) 121-126 doi: http://dx.doi.org/10.21741/9781945291173-21

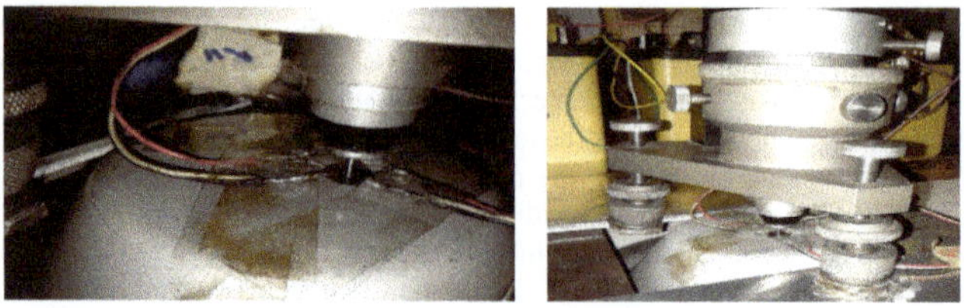

Fig.3: Preparation of piece to measure the residual stress

Simulation and experimental results of tested specimen

The results extracted by ABAQUS software include the distribution of displacement and stress caused by the simulation of warm metal forming. Figures 4, 5 indicate the sheet ´s displacement and the contour of basic stresses of the sheet under the forming process respectively.

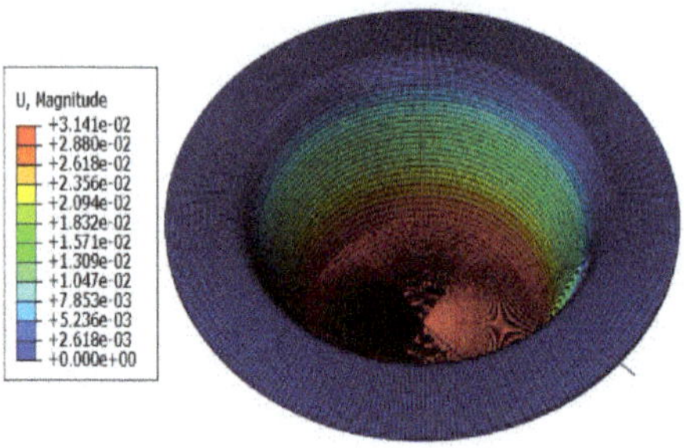

(a)Displacement contour

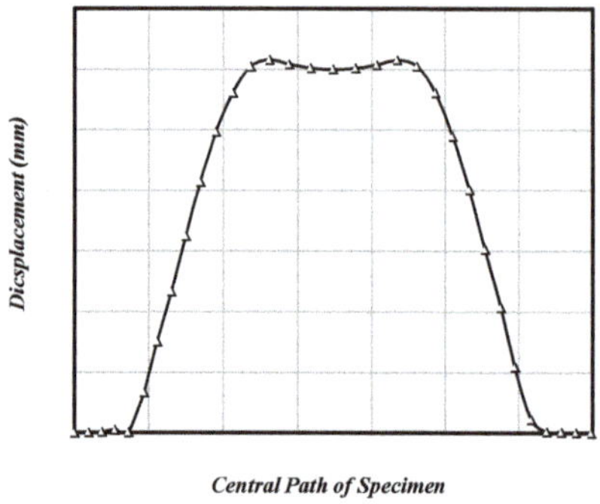

(b) Displacement plot

Fig.4: sheet´s displacement under the warm forming process

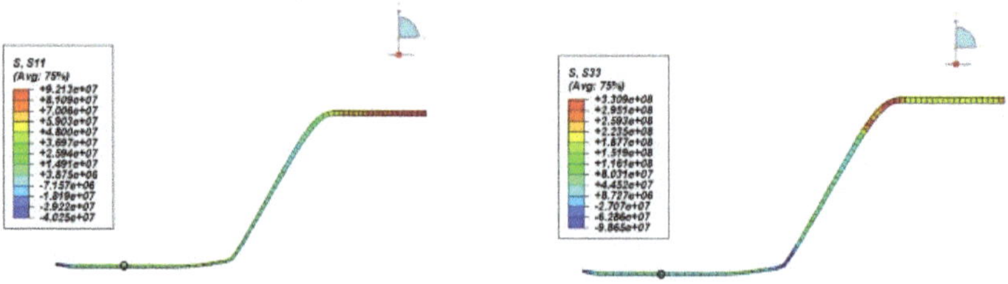

Fig.5: Contour of principal stresses (Pa)

Materials Research Proceedings **2** (2016) 121-126 doi: http://dx.doi.org/10.21741/9781945291173-21

In the experimental test, after making the hole, the released strain was recorded for every removed layer (fig.6). With regard to the strains obtained by the experimental test, the residual stress in the piece (equations (2) - (4) in ASTM-E837-92 Standard) was computed.

$$\sigma_{min}, \sigma_{max} = \frac{\varepsilon_3 + \varepsilon_1}{4\bar{A}} \pm \frac{\sqrt{(\varepsilon_3 - \varepsilon_1)^2 + (\varepsilon_3 + \varepsilon_1 - 2\varepsilon_2)^2}}{4\bar{B}} \quad (2)$$

$$\bar{A} = -\frac{1+\nu}{2E} \times \bar{a} \quad (3)$$

$$\bar{B} = -\frac{1}{2E} \times \bar{b} \quad (4)$$

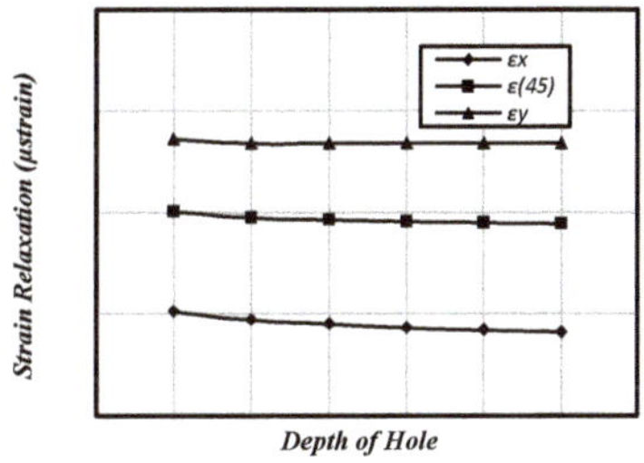

Fig.6: Results of the released strain in the experimental test

Comparison of experimental & numerical simulation results

Due to the innate and computational errors in the measuring methods of the residual stress (central drilling) and also the errors of finite element method, it isn't possible to predict the exact value of the residual stress. In this section, the results of tests were evaluated and compared. In fig.7, the stress values obtained by the Finite Element Method and the experimental test were compared.

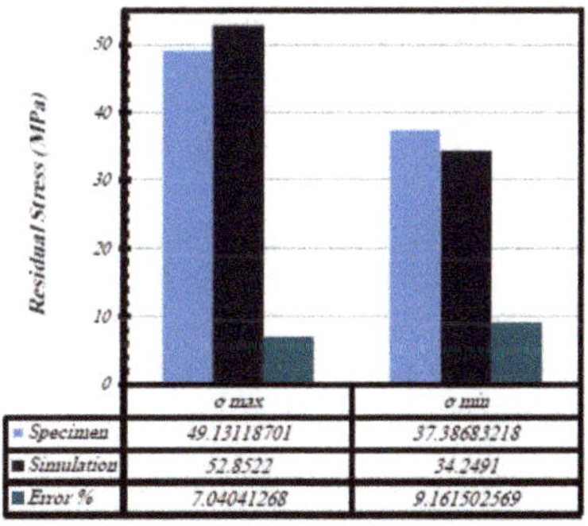

	σ max	σ min
Specimen	49.13118701	37.38683218
Simulation	52.8522	34.2491
Error %	7.04041268	9.161502569

Fig.7: Comparison of residual stress by the simulation and experimental test

As shown in Fig.7, the maximum error in the finite element computations and the experimental test for the maximum stress in the work piece is 7%. In this figure, a very good conformity is seen between these two measuring method. Center hole drilling is measuring method the other is computaional.

Hambli et al [11] studied on the optimization of pressure algorithm in sheet hot forming method by gas blow .They reported their numerical and experimental results about the maximum height of cone with the standard deviation 12%.

The following reasons can be considered as the error source in the measuring method of residual stress by the central drilling.

1. In this method, the released strains are low that is one of the main error resources in the Central Drilling Method.
2. Making the hole causes the stress concentration on the gauge.
3. The computations of Central Drilling Method based on the plane stress.
4. Error in the strain reading.

Summary & conclusion

In the present study, the inductive stresses in the new method of hot metal forming via the aluminum alloy by the gas blow was evaluated by the practical test and the finite element analysis. The finite element analysis of hot metal forming by gas in the Abaqus software indicates that there is a reasonable compatibility between the results obtained by the experimental test and the finite element analysis based on the conditions of experimental test. Also, the results of finite element simulation can be used as a reliable method for the conceivability evaluations, determination of conditions and effective parameters of the process.

References

[1] S.Toros,F.Ozturk and I.Kacar, "Review of Warm Forming of Aluminum–Magnesium Alloys", *Journal of Materials Processing Technology, 207 (1-3),* Oct.2008, pp. 1–1. http://dx.doi.org/10.1016/j.jmatprotec.2008.03.057

[2]F.Shehata,M.J.Painter, andR.Pearce,"Warm Forming of Aluminum/Magnesium Alloy Sheet", *Journal of Mechanical Work Technology, 2 (3),* Dec.1978, pp. 279-291. http://dx.doi.org/10.1016/0378-3804(78)90023-2

[3]D.Sorgente,L. Tricarico, "Pressure Profile Designing in Superplastic Forming Based on the Strain Rate and on Post-forming Properties", *Journal of Materials Engineering and Performance,* Vol 23, 2014,pp. 2025–2033. http://dx.doi.org/10.1007/s11665-014-1027-3

[4]J.Roesler,H. Harders and M.Baeker, "Mechanical behavior of engineering materials", Springer, New York,2007.

[5]D.Wilson, "Aluminum versus Steel in the Family Car-the Formability Factor", *Journal of Mechanical Work Technology*, 16 (3), June.1988, pp. 257-277. http://dx.doi.org/10.1016/0378-3804(88)90055-1

[6]Tang, J.S., Fuh, Y.K., Lee, S., 2014, "Superplastic forming process applied to aero-industrial strakelet:"wrinkling, thickness, and microstructure analysis", *Journal of Advance Manufacturing Technology*,

[7]H.Yu Wu, T. M.Da, "Modified male die rapid gas blow forming of fine-grained Mg alloy AZ31B thin sheet", *Journal of Advance Manufacturing Technology*,2015.

[8] M.Bakhshi-Jooybari, A.Gorji, and M.Elyasi, "Developments in sheet hydroforming for complex industrial parts",*Metal Forming*,Vol 3,2012, pp. 55-84. http://dx.doi.org/10.5772/48142

[9] A.H.Van Den Boogaard, J.Hu ′etink, "Modelling of aluminium sheet forming at elevated temperatures", *Materials Processing and Design: Modeling*, Simulation and Applications, Proceedings of the 8th International Conference on Numerical Methods in Industrial Forming Processes, NUMIFORM,2004.

[10] P.F.Bariani, "Hot Stamping of AA5083 Aluminum Alloy Sheets", *Journal of Manufacturing Technology*, 62 (1), April 2013, pp. 251-254.

[11] R.Hambli, A. Potiron, F.Guerin and B. Dumon, "Numerical Pressure Prediction Algorithm of Superplastic Forming Procedd Using 2D and 3D Modles", *Journal of Materials Processing Technology*, 112, October, 2001, pp. 83–90. http://dx.doi.org/10.1016/S0924-0136(01)00549-0

Residual Stresses 2016: ICRS-10 Materials Research Forum LLC
Materials Research Proceedings **2** (2016) 127-131 doi: http://dx.doi.org/10.21741/9781945291173-22

Through Thickness Residual Stress and Microstructural Mapping of AA7085-T7452 Die Forging

P. S. Baburamani*[a], K. F. Walker [b], P. K. Sharp [c], J. Niclis [d] and A. Shekhter [e]

Aerospace Division, Defence Science and Technology Group, Fishermans Bend, Vic 3207

*Corresponding author: E-mail: [a]pud.baburamani@dsto.defence.gov.au; [b]kevin.walker@dsto.defence.gov.au; [c]khan.sharp@dsto.defence.gov.au; [d]james.niclis@dsto.defence.gov.au; [e]alex.shekhter@dsto.defence.gov.au;

Keywords: Forging, Slitting, Microstructure, Through-thickness, Residual Stress, Hardness

Abstract. Significant weight savings and reductions in manufacturing costs have motivated the drive towards the use of unitised structures such as large die forgings in modern aircraft. However, detrimental bulk tensile residual stresses, present in these thick forgings can impact the durability and structural integrity of the primary aerospace structure. Residual stresses can lead to distortion during component machining, and part rejection. In addition, failure to account for the residual stress effects in the fatigue crack growth characteristics of the component in design and in performance could impact the structural integrity. Therefore, there is a need to measure and analyse the residual stresses present in the forged components, and examine the residual stress-microstructure, including grain structure/flow relationship for the forging, if such a relationship could be developed. The preliminary results of this experimental program have shown a link between the measured residual stress and hardness, along with changes in the grain structures through the thickness.

Introduction

Structural unitisation of components is being actively pursued by the aerospace industry with the primary aim of weight reduction and cost savings. Large aluminium forgings and thick section rolled products make a significant contribution towards unitisation of critical, primary airframe components, such as wing-carry-through bulkheads. Inherent with the thick section wrought product forms, and particularly in die forged components, is the built-in bulk residual stresses on a macro scale [1], arising from the thermal gradients in the thermo-mechanical processing and the post-solution treatment rapid quenching. The presence of process related bulk residual stresses results in: a) part distortion during machining which is a cost driver and b) high degree of scatter and/or variability in fatigue crack growth rate [2], which is a performance driver affecting the durability and damage tolerance of the component. The residual stresses in an aluminium 7050-T7452 forged block that caused distortion during machining was reported to be a major cost driver for a commercial aircraft manufacturer, as a result of excess manufacturing time, schedule overruns, and part rejection and scrapped material [3].

The surface and subsurface compressive residual stresses in an as-quenched (before stress relief treatment) aluminium alloy forgings have been reported to be in the range -200 MPa – 250 MPa [4]. The mechanical stress relief treatment by cold compression (forgings) or stretching (rolled plate) that follows, aims to relieve (or reduce) the severity of the bulk residual stresses existed in the as-quenched state in the component. However, cold compression stress relief in die forged components do not always fully relieve the complex 3 D bulk residual stresses, and the unrelieved residual stresses may be of sufficient magnitude to cause distortion in machined components [4-6]. Moreover, the three dimensional residual stress distribution in the stress relieved aluminium die forging varied

Materials Research Proceedings **2** (2016) 127-132 doi: http://dx.doi.org/10.21741/9781945291173-22

with the position, and the geometry of the forging, as a result of varying thermal gradients and the plastic flow in the three orthogonal directions [4].

As a first step towards a better understanding of the residual stresses in thick sections, a preliminary experimental program was carried out to measure the residual stresses present in the open die forged, mechanically stress relieved, thick AA7085-T7452 aluminium (Al-Zn-Mg-Cu-Zr) alloy, and examine if a residual stress-microstructure relationship for the forged material, can be developed. The objective of the work reported here was to measure the residual stress profile using X-Ray diffraction and by the slitting method on samples cut from various depths and transverse positions, and along the through-thickness, to correlate the results with the grain flow, microstructure and hardness.

Experimental Methods and Materials

AA7085-T7452 aluminium alloy open die forging, which had been compression stress relieved and, measuring 1750 mm (length) x 1200 mm (width) x 152 mm (thickness) was used in this study. The samples B1, B2, B3, B4 and B5 (80 mm x 25 mm x 10 mm) were taken from five locations at the centre along the through thickness direction of the open die forging and were used in the x-ray diffraction residual stress measurements and, for hardness measurements and microstructural examination (See Fig. 1). The specimens were without any surface irregularities, and had a smooth surface finish, close to N6.

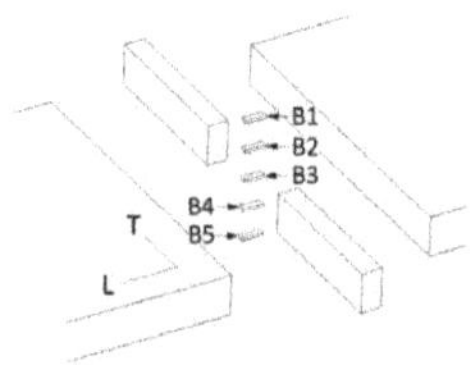

Figure 1 Specimen cut up diagram for B1, B2, B3, B4 and B5 from the open die forging

The microstructures were characterised using standard metallographic procedures in the longitudinal and short transverse surfaces of the samples. Hardness measurements were taken at five locations in the sample using a Rockwell hardness tester and on the B-scale using 100 kg load. X-Ray diffraction (XRD) was performed using a Proto iXRD Combo with a Cobalt X-Ray source and using a 2 mm circular aperture or 1.0 x 5 mm, or 2.0 x 5 mm aperture. In addition, four larger samples of 80 mm x 25 mm x 25 mm, identified B1/2, B2/3, B3/4 and B4/5 were sent to Hill Engineering, California for residual stress measurements by the slitting method [7], to obtain through-thickness residual stress distribution at those locations. Sample B1/2 was located between B1 and B2, and B2/3 was located between B2 and B3 and so forth. These larger samples were thought to be relatively more representative of the through-thickness locations in the forging for that heat treated temper. The material elastic properties used by Hill Engineering in the residual stress calculations were; Young's modulus = 71.7 GPa and the Poisson's ratio, $\nu = 0.33$ [7]. Prior to the slitting measurements, the back side of the rectangular specimens was skimmed using EDM wire to provide a flat surface to mount the strain gage. The specimens were slit at the approximate mid-length using a wire EDM along the sample thickness direction. A single strain gage was mounted to the back face of the specimens, behind the slit. The measured strain versus slit depth was used to calculate the residual stress [7].

Results and Discussion

Hardness. The through-thickness hardness was measured in the longitudinal and the short transverse surfaces of the specimens B1, B2, B3 and B5 (Fig. 1), and the mean values are shown in Table 1. The specimen locations x (B1, B2, etc.) relative to the forging thickness t (152 mm), x/t are also indicated

Materials Research Proceedings **2** (2016) 127-132 doi: http://dx.doi.org/10.21741/9781945291173-22

in the table. The standard deviations for the hardness in the longitudinal surface were 0.3 to 0.9 and for the short transverse direction 0.1 to 0.6.

Table 1 Rockwell hardness (HRB) in the longitudinal and short transverse surfaces

Specimen	Location, x/t	HRB Mean-Longitudinal	HRB Mean-Short transverse
B1	0.03	86.0	87.6
B2	0.27	84.8	85.0
B3	0.50	83.5	80.7
B5	0.97	87.4	87.4

The hardness reached its maximum values near the top (B1) and bottom (B5) surface locations in the forging, in both directions, and as we approach the centre of the forging, the hardness is seen to decrease, reaching a minimum value at the centre (B3), in both the longitudinal and short transverse surfaces (Table 1). This trend may be attributed to less deformation and slow cooling at the centre in the thick section forging, leading to lower work hardening and lower flow stress as suggested by Robinson et al. [8], compared to the top and bottom surfaces which can absorb more deformation and cool faster. This trend is more strongly observed in the short transverse direction (see Figure 2). Also note the differences in the hardness results for B3 which is at the centre (core) of the forging; the short transverse hardness is lower than the longitudinal hardness, an indication of the inhomogeneity present in the thick forging. Zhang et al. [5], examined the through-thickness hardness distribution in the centre of the specimen in a 121 mm thick 7050-T7452 forging and a minimum value of hardness occurred at one end of the specimen near the surface, corresponding to a low value of tensile or compressive residual stress and a maximum hardness at the centre of the specimen. These results differ from the trends observed in the present work, the sample size and the extent of deformation and the thermal gradients could determine the actual trend observed. The through-thickness hardness and through thick residual stress appeared to follow the same trend in their work [5], and a similar trend is also seen in this work (Fig. 2). Figure 2 compares the near-surface residual stresses measured by the slitting method and by X-ray diffraction with the Rockwell hardness measured in the short transverse and longitudinal surfaces, respectively. The hardness measured in the short transverse faces of B1, B2, B3 and B5 appear to nearly relate to the through thickness residual stresses measured near the surface on samples B1/2, B2/3, B3/4 and B4/5 by the slitting method (Fig. 2).

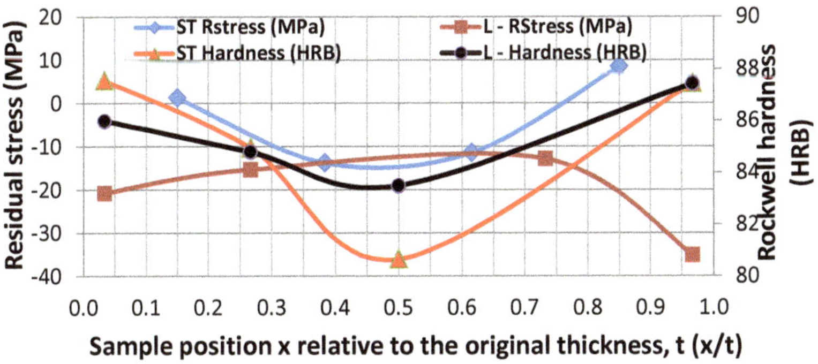

Figure 2 Residual stress and hardness in the longitudinal and short transverse surfaces

Materials Research Proceedings **2** (2016) 127-132 doi: http://dx.doi.org/10.21741/9781945291173-22

Through-Thickness Residual Stresses. Fig. 2 also shows the distribution of residual stresses at the centre of the longitudinal surface of the specimens measured by x-ray diffraction and the residual stress close to the surface at a depth of 0.03 mm (30 μm) measured by the slitting method and, the average hardness values measured on the same longitudinal and short transverse surfaces, respectively. Overall, the measured XRD residual stresses were low (< 35 MPa) and compressive in all the sample surfaces, and in through the thickness of the forging. The residual stresses are more compressive (20 to 35 MPa), near the top and the bottom surfaces and less compressive (12 to 15 MPa) in the interior regions of the samples from the forging. The mean error or uncertainty in the XRD residual stress measurements was 7.5 MPa and the range was 4 to 12 MPa. This kind of variations in the residual stresses measured has also been reported in the literature [5]. The longitudinal residual stresses in the cold compressed forging were reported to vary with the position along the length of the forging due to multiple compression and overlap [3], during post-quench cold compression stress relief [5].

The relatively low values of residual stresses are possibly due to the original forging (and hence the samples) was solution treated, rapidly cooled and cold compression stress relieved at ~1.5% to 3% plastic strain, and subsequently age hardened, and also possibly due to the samples being put through a number of cutting operations to produce the smaller sample sizes used in these experiments. It is more than likely; each cutting or machining operation would relieve some of the residual stresses [9].

However, these observations are similar to those of Robinson et al. [10]. In their work [10], 7050 aluminium forging was investigated. The surface residual stress was compressive and around 40 MPa, after overageing. The residual stresses had dropped from about -200 to -220 MPa in the as cold water quenched state to -50 to -70 MPa after cold compression (stress relief), and to -40 to - 50 MPa after overageing. Figure 2 also includes hardness to show how it compares with the residual stress measurements. It can be seen that higher residual stresses (XRD) near the top and bottom ends (-20 MPa and -35 MPa) aligns with higher hardness results (86.0 and 87.4 HRB). As the residual stresses show a decrease towards the core inner regions of the forging (-12 and -15 MPa), the hardness numbers also decrease (84.8 and 83.5 HRB) on the longitudinal surfaces.

The uncertainties in the residual stress measurements by the slitting method were 0.8 to 1.2 [7]. The residual stresses and the hardness follow similar trends as those observed for the longitudinal surface hardness and XRD surface residual stresses. Note that in Figure 2, all the residual stresses are compressive, for XRD, and the residual stresses are tensile near the top (B1/2) and bottom surfaces (B4/5) [1.3 MPa and 8.5 MPa] and, compressive in the inner regions in the slitting method. The observed trends in Figure 2 for the measured residual stresses (slitting) and the short transverse surface hardness are more closely aligned suggesting a potential intrinsic relationship between the two parameters. Zhang et al [5] commented that the through thickness residual stress profiles and the spatial variation in stress distribution are likely to have been caused by the non-uniform plastic deformation during the stress relief treatment.

Through-Thickness Microstructures. The through-thickness microstructures were assessed from samples B1, B2, B3, B4 and B5 taken from different locations from the surfaces, interiors and the core of the forging in the short transverse (through the 152 mm thickness) direction. These surfaces correspond to the direction in which the slitting operations were performed in samples B1/2, B2/3, B3/4 and B4/5. The representative microstructures, as marked, and the residual stress distribution obtained from the slitting method [Fig. 3 (d)], are shown in Figure 3.

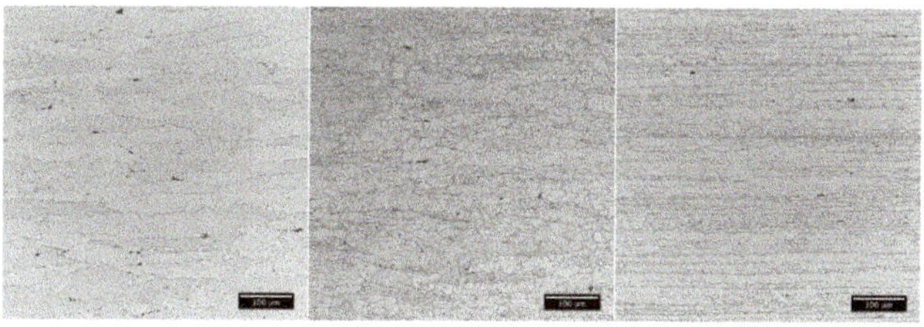

(a) B1 (b) B2 (c) B3

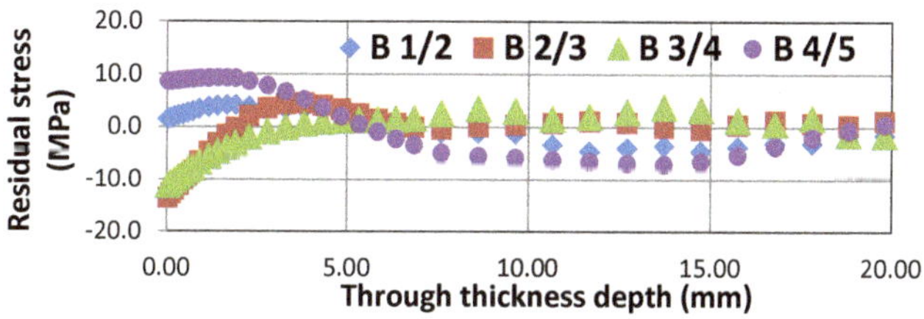

(d)

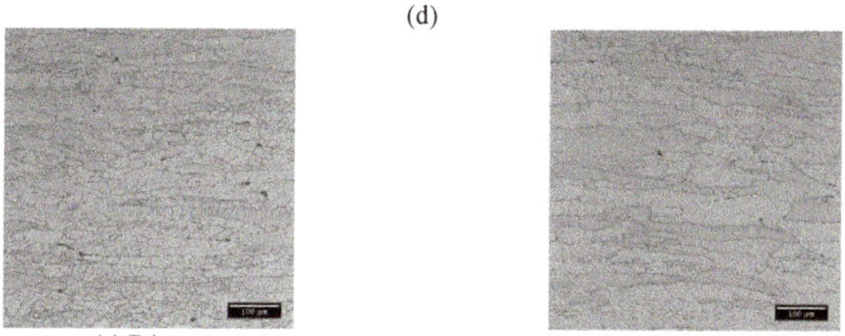

(e) B4 (f) B5

Figure 3 Through-thickness residual stresses measured by the slitting method (d) and the microstructures of specimens in the short transverse direction; (a) B1, (b) B2, (c) B3, (e) B4 and (f) B5

The microstructures showed a dual grain structure consisting of elongated large grains, and smaller, polygonised grains in un-recrystallised and partially recrystallised forms. For example, close to the surface, B1 and B5, where more deformation is likely during mechanical working and during compressive stress relief, large, elongated and worked grain structure with very few polygonised grains are shown, [Fig. 3 (a) and 3 (f)]. In B2, there is a mixture of elongated grains and large polygonised or equi-axed grains that appear to be partially recrystallised [Fig. 3 (b)]. In B3, which is at the core there is less deformation during the thermo-mechanical processing stage and it is likely to retain more heat compared to the top and bottom surfaces because it cools slowly, there are elongated grains that do not appear to be severely worked and with a large number of small sub grains [Fig. 3 (c)]. B4 is an extension of B3 but showing more worked structure and larger sub grains [Fig. 3 (e)], possibly because it receives more deformation than B3. Variations and inhomogeneous grain distributions through the samples at various positions along the through-thickness of the forging are evident in these micrographs, in line with the heterogeneous grain distribution observed in aluminium alloy forgings reported in the literature [10]. The microstructures in the thick section die forging have evolved during the thermo-mechanical/metallurgical processing, and the nature of the grain structure is dependent mainly on the amount of working the various regions in the through-thickness face are subject to, and the thermal gradients. The larger (coarser) grains near the top (B1, Fig. 3 a) and bottom (B5, Figure 3 f) surfaces have resulted in slightly higher hardness values (see Table 1 and Fig. 2). In general, any mutual dependency between grain distribution and size, and the residual stresses is likely to be governed by the thermal gradients and deformation inhomogeneities during mechanical working and compressive deformation stress relief, and the strain hardening behaviour of the material in the production of thick section forged products and components.

Conclusion

The preliminary experimental work and the results reported in this paper have identified the nature and extent of non-uniform contours in the residual stresses in the thick section forged product and, the inhomogeneous microstructural features present in the forging. In addition, the results present another dimension in the inhomogeneity space in terms of grain structure and size, as well as variations in hardness and, potentially indicative of varying strength and fatigue properties in the final component. The observed trends in Fig. 2 for the measured residual stresses by the slitting method and the surface hardness are more closely aligned, suggesting a potential intrinsic relationship between the two parameters. Slitting method provides through the thickness profile of residual stress with depth, the surface and the interiors of the specimens provide different type of contours, which confirms the inhomogeneous distribution of residual stresses in thick section aluminium alloy products. Results from XRD and slitting methods showed that relatively higher residual stresses at the top and bottom of the samples were aligned with higher hardness values and interior and core regions showed lower residual stresses and lower hardnesses. The work hardening capability of the section through the thickness and the thermal and deformation gradients are thought to be one of the factors responsible for the observed results. The preliminary experiments and the results have provided an opportunity to explore these aspects further, with the use of more representative larger samples to start with.

References

[1] G. S. Wilson, A. F. Grandt Jr., R. J. Bucci, R. W. Schultz, *Int. J. Fatigue*, 31 (2009) 1286-1299. http://dx.doi.org/10.1016/j.ijfatigue.2009.02.023

[2] R. W. Bush, R. J. Bucci, P. E. Magnusen, G. W. and Kuhlman, *ASTM STP 1189,* Ravinder Chona (Ed.), American Society for Testing and Materials, Philadelphia, 1993 568-589.

[3] D. M. Walker and R. Y. Hom, *Advanced Materials and Processes*, 160 6 (2002), 57-60.

[4] J. S. Robinson, S., Hossain, C. E. Truman, A. M. Paradowska, D. J. Hughes, R. C. Wimpory, M. E. Fox, *Materials Science and Engineering A*, 527 (2010) 2603-2612. http://dx.doi.org/10.1016/j.msea.2009.12.022

[5] Z. Zhang, Y. Yang, L. Li, B. Chen, H. Tian, *Materials Science and Engineering A*, 644 (2015) 61-68. http://dx.doi.org/10.1016/j.msea.2015.07.018

[6] M. B. Prime, and M. R. Hill, *Scripta Materialia,* 46 (2002) 77-82. http://dx.doi.org/10.1016/S1359-6462(01)01201-5

[7] A. T. DeWald, Measurement report, Hill Engineering, LLC, California, 2016, 6p.

[8] J. S. Robinson, C. E. Truman, S. Hossain, R. Wimpory, *Materials Science Forum*, 571-572 (2008) 45-50. http://dx.doi.org/10.4028/www.scientific.net/MSF.571-572.45

[9] Altenkirch, J., Steuwer, A., Peel, M., Richards, D. G., Withers, P. J., The effect of tensioning and sectioning on residual stresses in aluminium AA7749 friction stir welds, *Materials Science and Engineering A*, 488 (2008) 16-24. http://dx.doi.org/10.1016/j.msea.2007.10.055

[10] J. S. Robinson, P. J. Tiernan and J. F. Kelleher, *Materials Science and Technology*, 31 4 (2015) 409-417. http://dx.doi.org/10.1179/1743284714Y.0000000571

Validation of Neutron Diffraction and the Incremental Deep Hole Drilling Residual Stress Measurements of a High Strength T Butt Weld Test Piece Using the Contour Residual Stress Measurement Technique

Glen Sloan[1, a *], Xavier Ficquet[2, b], Douglas Cave[2, c], Karim Serasli[2,d], Ed Kingston[2, e], Valerie Linton[3, f]

[1]School of Mechanical Engineering, University of Adelaide, North Terrace, Adelaide, Australia

[2]Veqter Ltd, Unicorn Business Park, Whitby Road, Bristol, BS4-4EX, UK

[3]Energy Pipelines CRC, Wollongong, Australia

[a]glen.sloan@bigpond.com.au, [b]xavier.ficquet@veqter.co.uk, [c]douglas.cave@veqter.co.uk, [d]karim.serasli.veqter.co.uk, [e]ed.kingston@veqter.co.uk, [f]valerie.linton@epcrc.com.au

Keywords: Welding, Residual Stress, Neutron Diffraction, Contour Method, Stress Corrosion Cracking, Corrosion Fatigue, T Butt

Abstract. Off shore ring stiffened cylindrical structures are subjected to high stresses caused by static or variable external hydrostatic pressure and residual stresses due to the fabrication and welding processes. The use of numerical simulation allows a straightforward calculation of the stresses induced by the hydrostatic pressure in the structure. However it is more intricate to determine the residual stresses resulting from the progressive manufacture of components and assembly using a multi-pass welding processes. This paper presents the work carried out to ascertain the residual stresses present within a high strength welded T-plate test piece, representative of part of a ring stiffened cylindrical structure, by the contour method to determine the longitudinal residual stresses across the width of the test piece. This test piece had previously been subjected to neutron strain scanning in a pattern across the test piece and also the incremental Deep Hole Drilling (iDHD) and, the Deep Hole Drilling (DHD) processes to validate the neutron results for the longitudinal and transverse components of residual stress in the weld toe and centreline of the T butt weld. Therefore the work on this test piece is unique as three methods of residual stress measurement have been used. The results of these measurements are presented and compared to highlight agreements and discrepancies in the measured residual stresses by each method.

Introduction

Ring stiffened cylindrical structures operate in some of the worst environmental conditions particularly when they are used in naval platforms. These naval platforms will be subject to high stress low cycle fatigue due to change in depth. They are manufactured from high strength quenched and tempered steels and therefore are susceptible to welding induced defects such as Hydrogen Assisted Cold Cracking (HACC) for which the presence of tensile residual stresses is considered one of the driving forces for the crack development. These platforms also operate in environments where the microstructure of the quenched and tempered steels coupled with tensile residual stresses can result in Stress Corrosion Cracking (SCC) which can then initiate failure by corrosion fatigue. Therefore a comprehensive knowledge of residual stress profiles is required not just at the weld but any location where there is susceptibility to any of these failure modes. This knowledge of residual stresses becomes even more critical when proposing alternate welding details such as pass placement as per [1]. Fracture Mechanics assessment standards such as BS7910 do not address alternate pass placement strategies as discussed in [1], [2]. A research program to investigate these factors was

Materials Research Proceedings **2** (2016) 133-138 doi: http://dx.doi.org/10.21741/9781945291173-23

implemented by the University of Adelaide with details of the work program presented elsewhere [2], [3].The test piece discussed in this paper had previously had strain measurements made using neutron diffraction (ND) on the Kowari strain scanner at the Bragg Institute, ANSTO, Sydney, Australia. The scanning pattern for this ND work is presented in [1] which is a sister paper for this conference. While comprehensive results were achieved across the base plate limited strain scanning results for the longitudinal direction were obtained in the fusion zone of the weld to the base plate and in the weld itself due to excessive beam path lengths as discussed in [1] and [2]. To partially overcome this two sets of iDHD/DHD measurements were undertaken as described in [1] and [2]. While these measurements gave additional information there were still critical gaps in knowledge, principally on the weld caps for sides 1 and 2 where it was not practical to undertake iDHD/DHD measurements on the entire weld cap. While more extensive DHD measurements could be undertaken the test piece would end up with numerous holes which could preclude use of other nondestructive or destructive techniques. Therefore it was decided to acquire additional residual stress information via use of the contour method.

Experimental Methodology

As detailed in [1] the longitudinal direction for plate rolling, curvature shaping and welding all coincided in the test piece which is standard practice for ring stiffened cylindrical structures. In addition, the longitudinal stress profiles across all areas of the base plate and web are the ones that show the most dramatic change but for which it is hardest to gather information by the ND method particularly in the region under the T butt weld. Therefore the decision was made to undertake sample contouring by cutting transversely across the test plate and therefore causing relaxation in the longitudinal direction. The contouring residual stress work was undertaken by Veqter Ltd using the methods described on their web site [4]. The cut was undertaken using the electro discharge method at a location greater than 30mm from the closest pre-existing DHD site. The relaxed profile was measured using CMM on a grid profile of 0.5 mm. As detailed in [1] the relatively short weld length enabled each weld pass to be undertaken without stops or starts along the length of the weld which would have involved other activities such as grinding. Therefore there is consistency in residual stress profiles along the length. In addition the welders practiced extensively achieving constant heat input levels [2]. The heat input levels for the passes was 1.7 ± 0.1 kJ/mm. Therefore correlation between residual stress levels along the length of the test piece determined by the different methods was assured.

Results

A pictorial representation of the results are presented in Fig 1. The areas of the base plate away from the region of the web and the T butt weld show tensile regions towards the concave surface of the base plate with a zigzag pattern of tensile and compressive zones through the base plate. These extend up to the HAZ and fusion zones of the T butt weld noting that the test piece had reduced weld caps flush with the faces of the abutting web stiffener. The contour plot clearly shows that the residual stress of the first pass of the capping run side 2 i.e. at the intersection of the base plate and the web, see Fig 2 is compressive. This is not the case for side 1 as the equivalent weld pass has low tensile residual stresses. Compressive residual stresses are present on both sides of the weld for the capping passes up to the last capping pass on both sides where they are tensile. For these passes the residual stresses are up to 0.25 of the nominal minimum yield strength of the weld metal. The highest tensile stresses have occurred towards the center of the fusion zone in the base plate and at the root region of the full penetration T butt weld. The residual stresses in the web away from the weld region quickly drop to low compressive values. The residual stress profiles to be addressed in the discussion section are indicated by the red lines shown on the cross sectional plot in Fig.1.

Materials Research Proceedings **2** (2016) 133-138 doi: http://dx.doi.org/10.21741/9781945291173-23

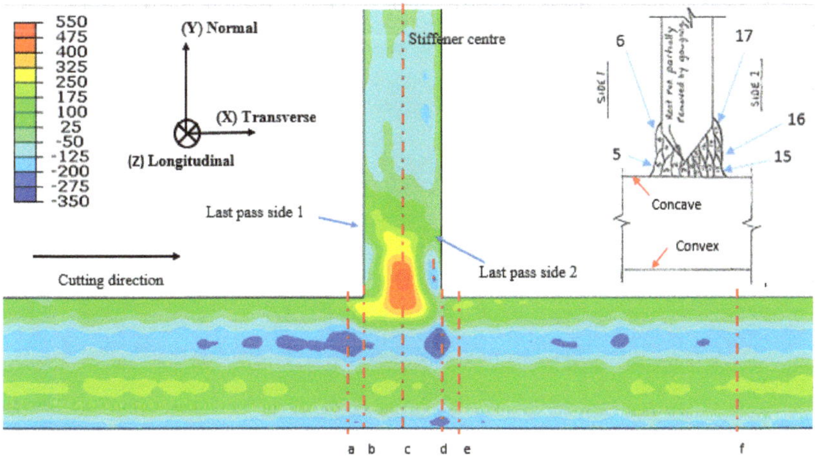

Fig. 1 Output from the FEA analysis of undoing the measured surface profile. The residual stress profile generated is for the longitudinal direction relative to the welding. The insert shows pass placements and the dashed red lines, a to f, are through thickness profiles which will be discussed.

Discussion

Throughout the following discussion section the datum for the position of each measurement by all methods will be changed to that which was used in [2] for the neutron diffraction testing. The initial areas of discussion will effectively be a sanity check for the comparison of the contour results against the neutron diffraction results for areas away from the T butt weld itself. These will then be followed by the more complicated areas in way the weld.

The first location of interest is the through thickness residual stress profile at 90 mm from the centerline of the web on side 2 of the weld, line f Fig.1. In Fig 2 comparative plots of the longitudinal residual stress profiles determined by the neutron diffraction technique with those from the contour method. In addition a plot of the neutron diffraction results from Pearce [3] on a curved plate segment shaped to the identical radius, and in fact from the same shaped plate used for the base plate of this test piece has been superimposed on these results together with the theoretical longitudinal residual stress profile for the shaped base plate calculated by the formulas given in Kendrick [5]. As can be seen in Fig 2 there is correlation between the results as they all show the zigzag pattern through the base plate for a shaped plate to a radius with tensile residual stresses towards the concave surface and compressive residual stresses towards the convex surface. Also the magnitude of the near surface residual stresses is less than that of the through thickness residual stress peaks and therefore corresponds with the results of Pearce [3] and calculations from Kendrick [5]. The kink upwards in the longitudinal residual stresses near the convex surface determined by the contour method cannot be explained at this stage as it is well below any surface induced profile from activities such as shot peening. The errors for the residual stresses determined by ND for all plots are given in [1] and for the DHD/iDHD are 25 MPa maximum.

Materials Research Proceedings **2** (2016) 133-138 doi: http://dx.doi.org/10.21741/9781945291173-23

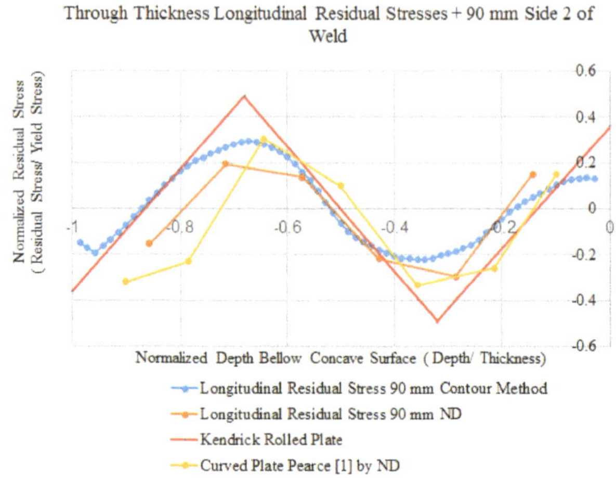

Fig. 2 Plots of longitudinal residual stress through the base plate by various methods at 90 mm from centerline of the web member. All closely match the theoretical profile developed by Kendrick [5]

The next locations of interest are the through base plate profiles 5 mm either side of the face of the abutting web member, lines a and e Fig 1. These locations are clear of the HAZ and Fusion Zones of the weld. The residual stress profiles obtained by the contour method are shown in Fig. 3 with the results from the neutron diffraction testing down these two paths superimposed. While at first glance there appears to be little change from the profile shown in Fig. 2 there are subtle differences. The through thickness profiles show a zigzag pattern starting with tensile residual stresses near the concave surface and compressive residual stresses near the concave surface of the base plate similar to those determined at the +90 through base plate location side 2 of the weld the magnitudes of the peaks have changed. The plots from contouring now have the magnitude of the peak stress near the concave surface greater than the through thickness peak. In addition the levels of tensile stresses are reducing rapidly at the surface whereas the profile at ± 20 mm (not shown) do not reflect this and therefore establishes that the welding has affected only a small distance either side of the abutting member.

The profiles through the centerline of the web as determined by ND, iDHD/DHD and by Contour Method are given in Fig 4. A sanity check on the contour results with respect to possible significant variation across the equivalent gauge volume, i.e.5 mm, of the DHD measurements was undertaken and showed that there was insignificant variation in the base plate and minor variation in the weld region. The variation in the weld region was equivalent to the apparent variation in residual stresses by the two methods at the weld root region. While this indicates that there is variation between the two destructive methods in the base plate there is commonality in the weld region. The main thing that is shown is that the two methods have established that there is a high tensile residual stress levels in the region of the centerline of the web/weld. While the neutron diffraction profile in the base plate does not directly correspond with those from the other two methods it follows the general trend. While no longitudinal residual stresses in the weld region could be obtained by this method along this line the transverse and the normal strains were determined and were tensile [2]. The ND full cap individual result [1], [2] corresponds closely with the DHD and Contouring results increasing confidence in the profiles obtained.

Materials Research Proceedings **2** (2016) 133-138 doi: http://dx.doi.org/10.21741/9781945291173-23

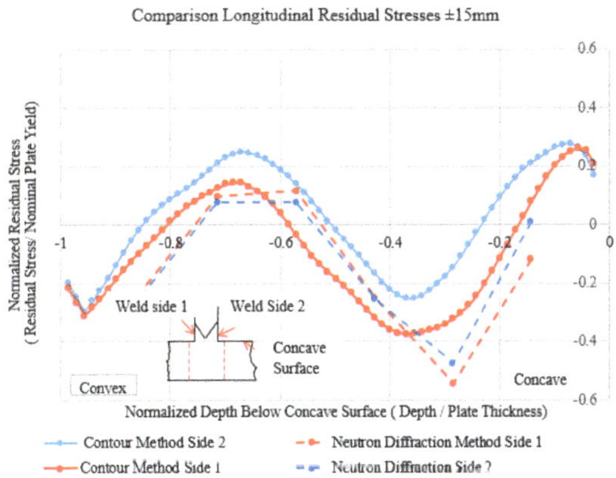

Fig. 3 Comparison plot of longitudinal residual stress profiles through the base plate at ± 15mm from the centerline of the web. Profiles show the slight alteration from Fig 2 due to the effect of the weld.

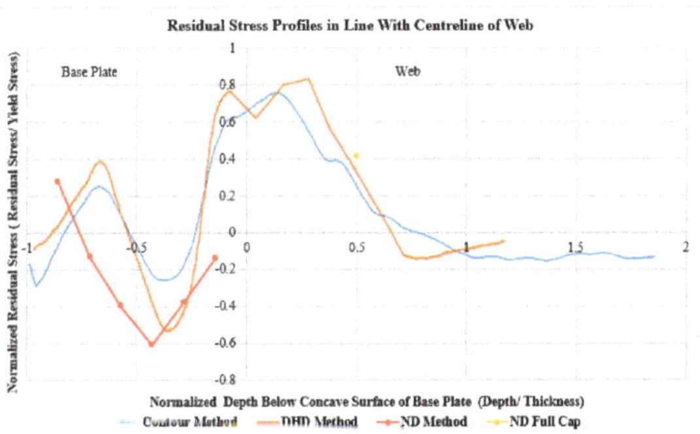

Fig.4 Comparative plots of longitudinal residual stresses through the base plate and web centerline determined by 3 methods.

Turning to the results presented in Fig 5 for the weld toe region side 2 the results from all methods show that there is a major drop in the longitudinal residual stress magnitude at the weld toe and in the subsequent passes of the weld cap. Of particular interest is that the single ND result in the weld cap corresponds closely with the equivalent profile obtained by the contouring method both for the 8 mm and 10mm line. All profiles show a drop in the residual stresses at the weld toe region down to low tensile or compressive. This confirms that the alternate pass placement strategy has achieved its aims namely reducing the risk of SCC and HACC at the weld toe regions where defects are subjected to high stress low cycle fatigue. As noted in [1] the results for the DHD test centered 11mm from the centered of the web and therefore 1 mm of the face of the web are more closely aligned to the ND results for + 15mm from the center of the web, 5mm away from the face. They still show a drop in magnitude at the weld toe. The residual stresses at the weld toe on side 1, line b Fig1 (not shown) have been reduced to low magnitude tensile stresses and therefore reducing risk of HACC and SCC.

Materials Research Proceedings **2** (2016) 133-138 doi: http://dx.doi.org/10.21741/9781945291173-23

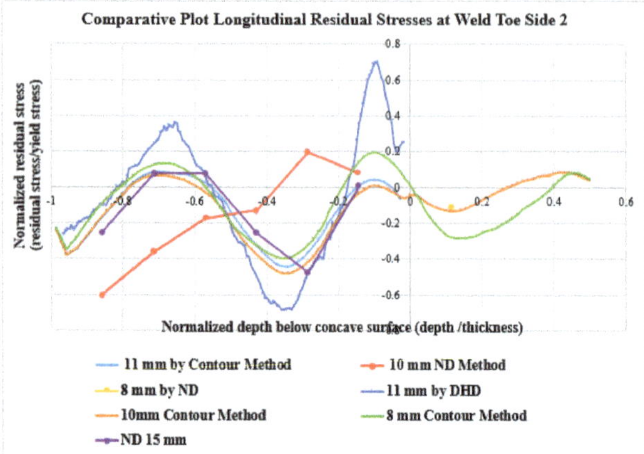

Fig.5 Comparative Plots of longitudinal residual stresses through weld toe region side 2 of weld determined by 3 methods.

Summary

The additional longitudinal residual stress information obtained by the cutting method has enhanced the understanding of what is the impact of the specific pass placement strategy shown in Fig 1 for a T butt weld. The results obtained by the contour method confirm those previously obtained by the ND and DHD methods in that the passes in the capping runs towards the base plate and abutting weld member toes are altered to have compressive residual stresses. The same applies to the base plate at the weld toe side 2. Therefore the pass placement strategy progressing away from the base plate reduces the risk of SCC at the weld toe region and in the weld caps. Similarly the risk of formation of HACC defects in the weld cap and weld toe region will also be reduced as the residual stresses are reduced. The high tensile residual stress region has been moved to the center of the weld. Therefore any defects forming in this region will be embedded away from the weld caps and are more easily detected by non destructive testing.

Reference

[1] G. W. Sloan, V. M. Linton, O. Kirstein, X. Ficquet, E. Kingston, I Brown Residual Stress Measurement in a High Strength T Butt Weld Specimen by the Neutron Diffraction and Deep Hole Drilling Techniques, ICRS 10 Sydney 2016.

[2] G. W. Sloan, "The influence of welding variables on the residual stress profile around T butt and MMAW butt welds in thick section high strength steel", Master's Thesis, University of Adelaide 2016.

[3] S. Pearce, "Breakdown of residual stresses in highly restrained thick section welds", Doctoral thesis, University of Adelaide, 2016.

[4] Information on http://www.veqter.co.uk

[5] S. Kendrick, "Structural design of submarine pressure vessels", NCRE report R483, March 1964.

Materials Research Proceedings 2 (2016) 139-144 doi: http://dx.doi.org/10.21741/9781945291173-24

Cementite Residual Stress Analysis in Gas-nitrided Low Alloy Steels

L. Barrallier[1,a*], S. Goekjian[1,b], F. Guittonneau[1,c], S. Jégou[1,d]

[1] MSMP Laboratory, Arts et Métiers ParisTech, 2 cours des Arts et Métiers, F-13617 Aix-en-Provence, France

[a] laurent.barrallier@ensam.eu, [b] sarah.goekjian@ensam.eu, [c] fabrice.guittonneau@ensam.eu, [d] sebastien.jegou@ensam.eu

Keywords: Nitriding, Residual Stress, X-ray Diffraction, EBSD

Abstract. This paper deals with the measurement of residual stresses in cementite after gas-nitriding of a 33CrMoV12-9 steel. During nitriding, precipitation of nanometric alloying elements nitrides and cementite at grain boundaries occurs leading to an increase of superficial hardness and providing compressive residual stresses in the surface layer. The stress state in the ferritic matrix has generally been measured to characterize the mechanical behaviour of the nitrided case while the other phases are not taken into account. In order to better understand the mechanical behaviour (e.g. fatigue life and localization of cracks initiation) of heterogeneous material such as in case of nitrided surfaces, the nature (sign, level) of residual stresses (or pseudo-macro-stresses) of the present phases can be calculated from measurements using X-ray diffraction to select the considered phase. Due to a low volume fraction of cementite through a nitrided case, an approach based on X-ray and electron backscattered diffractions (XRD and EBSD respectively) is proposed to perform stress measurements in cementite. An optimization of the surface preparation (by mechanical and/or chemical polishing techniques) prior to EBSD analysis was performed in order to minimize deformation induced by surface preparation. Pseudo-macro-stresses were calculated in tempered martensite and cementite. Results are compared to local residual stress measurements carried out by a cross-correlation method using EBSD patterns.

Introduction

Gaseous nitriding is a thermochemical surface treatment used to improve surface hardness and generate compressive residual stresses in mechanical steel parts for aeronautics or high performance automotive gearboxes [1]. It involves the diffusion of nitrogen from the surface to the core of the material using an ammonia and nitrogen and/or hydrogen gas mixture. The present work deals with low carbon alloyed steels characterized by a tempered martensite microstructure ($(Cr,Fe)_{23}C_6/(Cr,Fe)_7C_3$ carbides in ferrite). The temperature of gaseous nitriding (between 480-580 °C) is chosen below the eutectoid transformation and tempering temperatures (ferrite to austenite in Fe-N binary system) in order to prevent the core material from mechanical properties alterations, but high enough to activate the diffusion of nitrogen [2]. Co-diffusion of carbon occurs during nitriding and contributes to a complex microstructure gradient characterized by tempered carbides transformed into chromium incoherent nitrides (Cr,Mo)N and grain boundary cementite $(Fe,Mo)_3C$. Semi-coherent chromium nitrides also form from the solid solution of chromium into ferrite [3]. Nano-scale (Cr,Mo)N precipitates provide the hardness increase after nitriding [4]. Close to the surface, iron nitrides (γ'-Fe_4N_{1-x}, ε-$Fe_{2-3}N$) generally develop as a compound layer depending on the nitrogen potential [5]. Fig. 1 shows a typical nitriding layer of a 33CrMoV12-9 steel nitrided at 520 °C during 100 h. Precipitation and transformation of phases also generate specific volume changes (in regard to the density of the initial and formed phases) in the nitriding layer leading to the generation of compressive residual stresses at the nitriding temperature and during cooling [3, 6] [7]. Only residual stresses in ferrite have usually been determined using X-ray diffraction technics. A few papers give the macroscopic stress state using mechanical methods (hole drilling…) [6],8,9]. Due to the multiphase character of nitriding layers, residual stresses depend on the considered crystalline phase (pseudo-macro-stresses) and are different to macro-stresses. The residual stress state is a substantial input for the fatigue resistance of nitrided parts, especially in the case of the initiation of high cycle fatigue (HCF) cracks [10]. Whereas macro-stresses drive the crack propagation, micro-stresses govern the local damage due to heterogeneities at the microstructure scale. The aim of this paper is the determination of residual stresses in the grain boundary cementite in order to take them into account in HCF criteria.

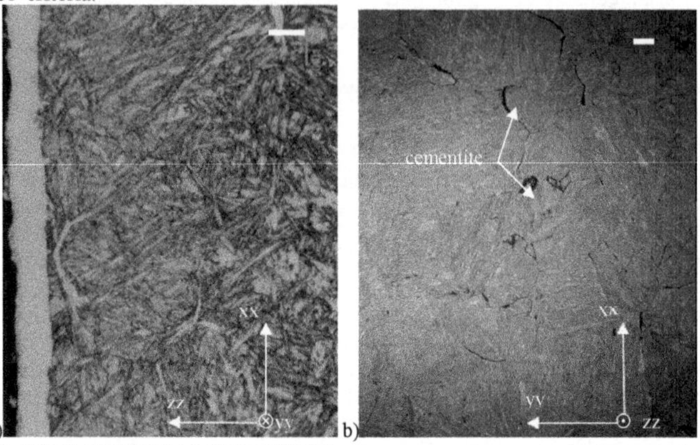

a) b)

Fig. 1: Nitrided 33CrMoV12-9 @ 520 °C/100 h: a) cross-section optical micrograph showing compound and diffusion layers, b) scanning electron micrograph (SEM) @ 50 µm depth and parallel to the nitrided surface: grain boundary cementite.

Diffraction stress analysis

X-Ray diffraction. Stress analysis can be carried out by X-ray diffraction using crystalline planes as a strain gauge. The $\sin^2\psi$ method is usually chosen in case of polycrystalline and multiphase materials. $\sigma_{xx}^{\phi_i} - \sigma_{zz}^{\phi_i}$ residual stress are only accessible in case of a multiphase (phases ϕ_i) material without unstressed reference (axis are defined in Fig. 1). X-ray elastic constants (XEC) of considered phases must be determined from elastic constants of a single crystal and the orientation distribution function (ODF) of polycrystalline aggregates in case of textured sample. Fig. 2 gives the evolution of the Young modulus $E_{\{hkl\}}$ of Fe_3C orthorhombic single crystal in spherical coordinates (data from [11]). The elastic anisotropy of Fe_3C cementite is highly marked, $E_{\{011\}} = 319$ GPa and $E_{\{110\}} = 65$ GPa. XEC are calculated using homogenization methods based on direct Kröner-Eshelby approach for un-textured materials and polycrystalline Kröner-Eshelby approach using ODF for textured materials.

Electron back scattering diffraction. EBSD was used to determine crystal orientations using SEM electron diffraction pattern (Kikuchi pattern). The gage volume is given by the operating conditions such as the voltage, tilt angle… and nature of the material (electron absorption effect). Orientation maps can be performed at the scale of the microstructure with a 0.1-1 µm spatial resolution. Quantitative stress analysis is based on the measurement of the misorientation of Kikuchi bands. Maurice and al. developed a method using 3D Hough transform [12] while Wilkinson and al. compared Kikuchi patterns with a reference taken from an identical crystal orientation [13]. Based on a cross-correlation method, these technics allow to determine tensorial residual stress maps (assuming σ_zz^(ϕ_i)=0) for a given phase and orientation. For more details see [14,15]. Only the single crystal elastic constants of the considered phase are needed to perform stress analysis. Such stress mapping actually gives the stress gradient through a single crystal (or grain) and not the absolute residual stresses. It implies that stresses between grains cannot be evaluated.

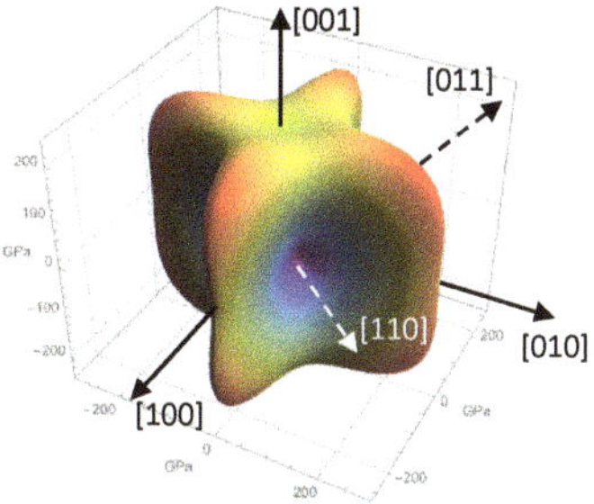

Fig. 2: Young modulus of Fe_3C single crystal (data from [11]).

Experiments

Material. Sample (15×15×6 mm^3) was made from oil quenched (austenitization @ 920 °C/1 h) and tempered (2 h @ 620 °C) 33CrMoV12-9 steel. Gas nitriding was performed at 520 °C during 30 h and $K_N = 4.3$ $atm^{-1/2}$. The preparation of the sample for stress analyses consists in cutting the sample parallel to the surface at a 50 μm depth. The obtained surface was mechanically polished (220, 500, 1200, 2000, 4000 mesh SiC paper and 1 μm diamond solution), and several finishing preparations (vibrating, ionic and electrolytic polishing) were explored in order to reduce surface plastic deformation and thus optimize the acquisition of Kikuchi patterns.

Surface finishing preparations. Vibrating polishing was performed using a Buehler® VibroMet 2 vibratory device during 16 h with a polishing solution made with 50 mL of colloidal alumina OP AN 0.05 and 200 mL of L PS4 lubricant followed by utrasonic cleaning of the sample during 30 min in pure ethanol. Ionic polishing was performed on a 2×15×6 mm^3 sample using a JEOL® IB-09010CP cross-section polisher operating at 4.5 kV with argon flux during 20 h. Electropolishing was performed on a Struers® Electropol device operating at 15 V and 0.5 A using a Struers A2 electrolyte during 30 s.

Tab. 1: X-ray diffraction parameters used for residual stress and texture analyses.

	Phase	Texture analysis	Stress analysis	
Device	Seifert®			Siemens®
Mount type	Bragg-	4 circles	χ	Ω
Phase	-	Fe_3C	Fe_3C	α-Fe
Diffracting peak	-	{121}/{210}	{051}	{211}
Wavelength	Cr-Kα			
Voltage [kV]/intensity	40/30			
2θ step [°2θ]	0.03	0.1	0.05	0.032
Time / peak [s]	2000	200	7200	300
Angles [°]	38-141 °2θ	α (0-40°) and β (0-360°) with	9 ψ	13 ψ

Observation and EBSD acquisition. Light micrographs were obtained using a Leica® Aristomet device coupled with a Nikon® digital acquisition system. Surface was etched with a 10 % Nital solution. SEM observations were carried out on a JEOL® 7001F device operating at 15 kV using backscattered electrons. For EBSD analysis, sample was tilted at 70° using hkl® device. Basic EBSD analysis and stress analysis (cross correlation method) were performed using Channel 5 and CrossCourt® softwares respectively.

X-ray diffraction. Residual stress and texture analyses of cementite were performed on Seifert® XRD 3003 PTS and Siemens® D500 devices using position sensitive detector (PSD). Acquisition parameters are given in Tab. 1. Fig. 3 gives the X-ray diffraction spectrum of the analysed surface. The volume fraction of cementite was estimated to 4.6 ± 0.66 % from Rietveld analysis and Maud software [16].

Results

Surface preparation. Several quality parameters were considered to select the best surface finishing preparation. The IR criterion gives the indexing rate of Kikuchi patterns in the maps. The band contrast (BC) is extracted from the Kikuchi pattern analysis and represents the contrast of the Kikuchi bands (the higher, the better the Kikuchi pattern is). The mean angular deviation (MAD) is obtained from Hough transform of Kikuchi patterns and characterizes the misorientation between the fitted solution and measured band orientations (the lower, the better the solution is). Tab. 2 indicates that the best Kikuchi patterns, on both ferrite and cementite, were obtained using ionic polishing.

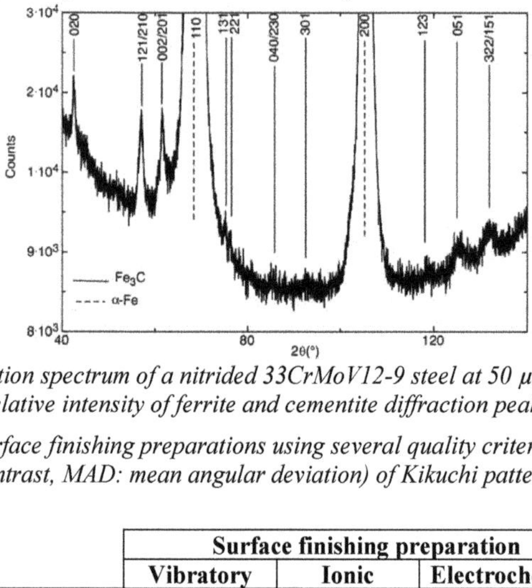

Fig. 3: X-ray diffraction spectrum of a nitrided 33CrMoV12-9 steel at 50 μm below the surface. Relative intensity of ferrite and cementite diffraction peaks.

Tab. 2: Quality of surface finishing preparations using several quality criteria (IR: indexing rate quality, BC: band contrast, MAD: mean angular deviation) of Kikuchi patterns for ferrite and cementite.

		Surface finishing preparation				
		Vibratory		Ionic		Electrochemical
Criterion	Phase	Aver.	Best	Aver.	Best	
IR	α-Fe	8.3	11.5	41.8	52.5	27.3
	Fe_3C	5.0	12.7	14.9	25.4	10.6
BC	α-Fe	65.0	105.0	80.1	91.9	73.7
	Fe_3C	67.0	102.0	63.7	68.0	50.6
MAD	α-Fe	0.75	0.67	0.66	0.59	0.70
	Fe_3C	0.84	0.77	0.83	0.77	0.91

EBSD analysis. Fig. 4 gives several maps from EBSD analyses of grain boundary cementite clusters and ferrite (Fig. 4a). The BC map (Fig. 4b) indicates polycrystalline cementite with a mean grain size around 0.73 ± 0.4 μm. The distribution of the grain size is given in Fig. 4f. The grain size of tempered martensite is close to 10 μm. Fig. 4c and Fig. 4e give the local misorientation between pixels and grains respectively. These qualitative criteria indicate that cementite is more strained than ferrite. The Euler angles map (Fig. 4d) and inverse pole figures (IPF) (Fig. 4g to 4i) give the orientations of cementite and ferrite grains.

Materials Research Proceedings **2** (2016) 133-144 doi: http://dx.doi.org/10.21741/9781945291173-24

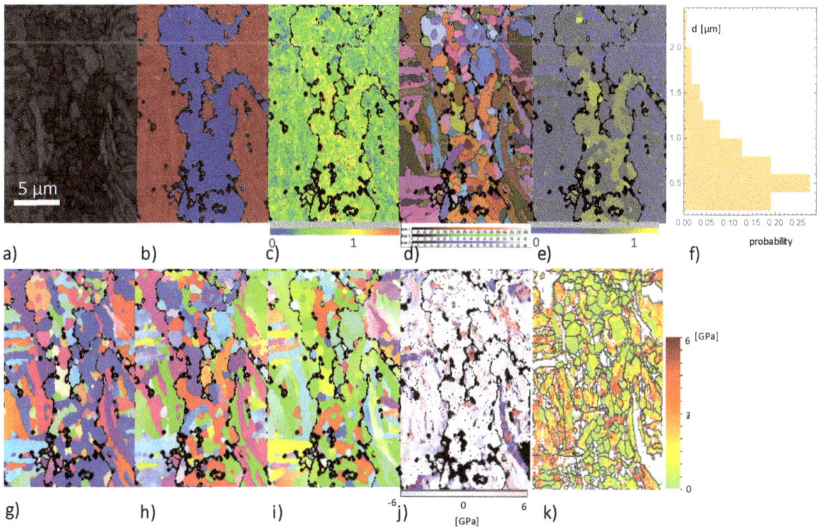

Fig. 4: Maps from EBSD analysis of grain boundary cementite clusters and ferrite (same scale on every map): a) band contrast, b) phases: red = α-Fe, blue = cementite, c) mean average disorientation between pixels d) Euler angles, e) mean average disorientation between grains, f) Fe_3C grain size distribution, g) IPFx, h) IPFy, i) IPFz, j) hydrostatic pressure, k) von Mises stress.

Texture. The EBSD Euler angles map indicates a local orientation of cementite at grain boundaries. Fig. 5a shows the inverse pole figures along X, Y and Z axes from EBSD analysis where the <101> direction remains perpendicular to the surface. The local anisotropy of cementite can be linked to the local ferrite grains anisotropy and the Bagaryatskii orientation relationship: $(100)_{Fe_3C} \parallel (0\bar{1}1)_\alpha, (010)_{Fe_3C} \parallel (0\bar{1}\bar{1})_\alpha, (001)_{Fe_3C} \parallel (211)_\alpha$ [17]. Because of a bigger gage volume and un-textured ferrite matrix, no texture is observed for cementite using X-ray diffraction according to the average contribution of the local textured cementite at grain boundaries (Fig. 5b). Consequently, XEC of cementite were calculated assuming an isotropic behaviour (as the ferrite phase).

Residual stresses. X-ray residual stress analyses were performed using ½S_({211})=5.83 〖10〗^(-6) MPa^(-1) and ½S_({051})=6.22 10-6 MPa-1 for □-Fe and Fe3C r espectively, according to calculations from single crystal elastic constants and the Kröner-Eshelby invariant method. Diffraction peaks were fitted using a pseudo-Voigt function and linear background. Residual stresses of both phases were considered as axisymmetric, so that stress analysis was performed in one direction [1]. It follows that compressive stresses of 〖(σ〗_xx^(α-Fe)-σ_zz^(α-Fe))=-1325±1 MPa without shear stresses and 〖(σ〗_xx^(Fe_3 C)-σ_zz^(Fe_3 C))=-619±92 MPa without significant shear stress were calculated for ferrite and cementite respectively. The local residual stress states from EBSD analysis in tempered martensite and cementite are given in Fig. 4j for the hydrostatic pressure and Fig. 4k for von Mises stresses. The cross-correlation method globally gives high stresses in absolute value (up to 6 GPa). The quality of the Kikuchi patterns was found very low generating non negligible correlation errors. The local (inside a selected orientation) stresses in tempered martensite appear higher than in cementite, that is in agreement with X-ray diffraction

Materials Research Proceedings **2** (2016) 133-144 doi: http://dx.doi.org/10.21741/9781945291173-24

measurements. Stress heterogeneities seem to be more important in ferrite due to the native microstructure (martensite laths).

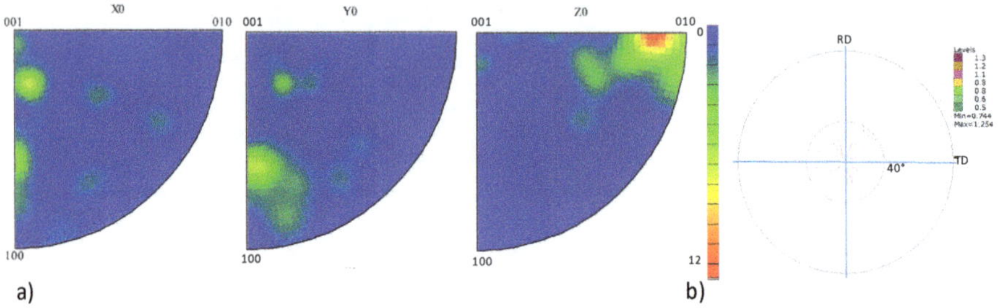

Fig. 5: Cementite texture analysis: a) IPF at the scale of grain boundaries from EBSD analysis, b) pole figure at the macroscopic scale using {121}/{210} diffracting planes (X-ray diffraction).

Conclusion

Residual stress analyses were performed in a nitrided layer using a local approach based on EBSD analysis and a pseudo-macroscopic approach based on X-ray diffraction both in the ferrite matrix and cementite at grain boundaries. Studied surface must be very well prepared using an optimized ionic surface polishing method for good quality EBSD analysis. Pseudo-macro-stresses were found compressive in both ferrite and cementite and the level of stress in cementite was twice less than in tempered martensite. Local stress analysis by EBSD is in agreements and shows strong mechanical heterogeneities between grains.

References

[1] L. Barrallier, Classical Nitriding of Heat Treatable Steel, in E.J. Mittemejier, M.A.J. Somers (Eds.), Thermochemical Surface Engineering of Steels, Elsevier Ed., 2014, pp. 392-410.

[2] D. Pye, Practical nitriding and ferritic nitrocarburizing, ASM International, 2003.

[3] L. Barrallier, PhD thesis, Arts et Métiers ParisTech, 1992.

[4] J.N. Locquet, PhD thesis, Arts et Métiers ParisTech, 1998.

[5] G. Fallot, PhD thesis, Arts et Métiers ParisTech, 2015.

[6] M.A. Terres, S.B. Mohamed, H. Sidhom, Int. J. of Fatigue, 32 (2010) 1795-1804.

[7] S. Jégou, PhD thesis, Arts et Métiers ParisTech, 2009.

[8] C. Mansilla, V. Ocelík, J.T.M. de Hosson, Mat. Sc. and Eng.: A, 636 (2015) 476-483. http://dx.doi.org/10.1016/j.msea.2015.04.023

[9] V. Goret, PhD thesis, Arts et Métiers ParisTech, 2006.

[10] M. Chaussumier, PhD thesis, Arts et Métiers ParisTech, 2000.

[11] C.C. Jiang, S. Srinivasan, A. Caro, S. Maloy, Journal of Applied Physics, 103 (2008).

[12] C. Maurice, R. Fortunier, Journal of Microscopy, 230 (2008) 520-529. http://dx.doi.org/10.1111/j.1365-2818.2008.02045.x

[13] A. Wilkinson, G. Meaden, D. Dingley, Ultramicroscopy, 1106 (2005) 307-313.

[14] A. Wilkinson, T. Britton, Materialstoday, 15, n°19 (2012).

[15] T. Britton, A. Wilkinson, Ultramicroscopy, 111, (2011) 1395-1404. http://dx.doi.org/10.1016/j.ultramic.2011.05.007

[16] L. Lutterotti, S. Matthies, H. Wenk, A. Schultz, J.J. Richardson, J. Appl. Phys., 81 (1997) 594-600 http://dx.doi.org/10.1063/1.364220

[17] Y. A. Bagaryatskii, Dokl. Akad. Nauk. SSSR, 73 (1950) 1161–1164.

Residual Stress Measurement in a High Strength T Butt Weld Specimen by the Neutron Diffraction and Deep Hole Drilling Techniques

G. W. Sloan[1 a *], V. M. Linton[2,b], O. Kirstein[3,6,c], X. Ficquet[4,d], E. Kingston[5,e]

[1]School of Mechanical Engineering, University of Adelaide, North Terrace, Adelaide, Australia

[2]Energy Pipelines CRC, Wollongong, Australia

[3]Instrument Technology Division, European Spallation Source, Sweden

[4]Veqter Ltd, Unicorn Business Park, Whitby Road, Bristol, BS4-4EX, UK

[5]Veqter Ltd, Unicorn Business Park, Whitby Road, Bristol, BS4-4EX, UK

[6]School of Mechanical Engineering, University of Newcastle, Callaghan, Newcastle, Australia

[a *]glen.sloan@bigpond.com.au, [b]valerie.linton@epcrc.com.au, [c]oliver.kirstein@esss.se, xavier.ficquet@veqter.co.uk, ed.kingston@veqter.co.uk

Keywords: Residual Stress, High Strength Steel, Quenched and Tempered, Neutron Diffraction, Deep Hole Drilling, T Butt Welds

Abstract. While defect assessment standards such as BS 7910 "Guide to methods of assessing the acceptability of flaws in metallic structures" [1] present residual stress profiles for T butt welds in thick sections they inherently assume that the weld pass placement strategy is such that the last weld pass is always on the base plate. There is limited information on what the residual stress profiles are if alternative pass placement strategies are used in high strength weldments and structures. Similarly there is limited information on what are the residual stress profiles under the body of the weld and the superimposition on the balancing residual stresses in the base plate away from the weld on pre-existing residual stresses. This paper presents neutron strain scanning work on a high strength, curved base plate, double sided T Butt weld undertaken such that the pass placement in each layer progressed away from the base plate towards the abutting member to determine the potential of reduction in residual stresses at the base plate weld toes. Key results were validated by the Deep Hole Drilling Technique. The results obtained show that a major reduction in the magnitude of the residual stresses occurred at the weld toes of the base plate with a corresponding increase in the center of the weld. Weld toe through thickness profiles were reduced from those of BS 7910. Balancing residual stresses in the base plate were minimal. Discussion includes comparison to results obtained in similar test plates welded with a different pass placement strategy.

Introduction

Ring stiffened cylindrical structures for the offshore industry and naval platforms are usually fabricated from high strength steels with butt welds and T butt welds. The fabrication method involves manufacturing plate cans from thick plates with the longitudinal direction of initial rolling of the plate coinciding with that for shape rolling into a curve. Depending on the diameter 3 or more plates will be needed to be welded together to form a cylindrical can. The ring frames are also fabricated from high strength steel and are welded into the can via double sided T butt welds undertaken in the 1G down hand orientation. The same also applies for rigid bulkheads in the structures to resist buckling of the structures when subjected to sea pressure. These types of structures operate in environments where Stress Corrosion Cracking (SCC) is a frequent problem for

Materials Research Proceedings 2 (2016) 145-150 doi: http://dx.doi.org/10.21741/9781945291173-25

crack initiation in service as is Hydrogen Assisted Cold Cracking (HACC) during manufacture. In the case of naval platforms they are also subjected to a high stress low cycle fatigue regime throughout their service life. Therefore an understanding of residual stress profiles is essential for determining service life and structural integrity safety regimes for the welds and the structural components. While standards such as BS 7910 [1] present residual stress profiles for T butt welds they are limited to the weld toe region only and they assume that the last weld pass is on the base plate with the residual stress level being at tensile yield. No profiles are given for the region under the weld or in the base plates away from the welds and how much the existing profiles in the components are altered by adjacent welding. Hence there is a huge gap in knowledge of residual stress profiles throughout a fabricated ring stiffened cylindrical structure. In 2004 the University of Adelaide and its research partners, ISIS Rutherford Appleton Laboratories, Australian Nuclear Science and Testing Organization, ASC Pty Ltd, and the Commonwealth of Australia, implemented a program to address these gaps [2], [3]. The program was structured to look at the progressive build up or superimposition of residual stress profiles throughout the fabrication process for these types of structures and addressed both current and potential future weld pass placement strategies for the T butt welds. The areas of primary interest were the base plate both beneath the weld and away from the weld region and also the weld. The material used throughout the entire program was BIS 812 EMA quenched and tempered plate with a minimum yield strength of 690 MPa The primary method of determining the subsurface residual stress profiles was neutron diffraction (ND) as it was nondestructive and could be repeated with the destructive techniques Deep Hole Drilling (DHD), [4], and the Contouring technique being used to validate key profiles and also capture additional residual stress information where it was not possible to successfully neutron strain scan due to beam path length issues. This paper addresses the ND and DHD and incremental Deep Hole Drilling (iDHD), [4], results obtained on one of the T butt test pieces looking at the alternate weld pass placement strategy of progressing the weld passes in each layer from the base plate up to the web member. The results are extracted from [3].The welding for these test pieces was undertaken in the 1G orientation using the MMAW process at a heat input of 1.7 kJ/mm with preheat of 120^0C and outgassing at the same temperature to reduce risk of HACC. The weld consumable for the test pieces looking at this pass placement strategy was Atom Arc E120 consumables which produces welds of 690 MPa yield strength or greater for this heat input.

Experimental Methodology

In order to maximize the longitudinal strain scan results in and under the weld this test piece was fabricated with a reduced number of layers by omitting the normal capping runs to give a slight taper as experience with the larger T butt weld test piece of the series with full caps [3] had shown that with full caps the beam path lengths were excessive for successful longitudinal strain measurement in this region. The pass placement strategy for full cap and reduced cap are shown in Fig 1. This was preferred over cutting back the sample post welding to reduce the beam path length in the longitudinal direction due to concerns over relaxation of stresses. The test piece was fabricated with the lineup of longitudinal directions as shown in Fig 1. The test piece was fabricated in 2008 when the scanning table for the Kowari strain scanner was limited to 10kg only and hence the test piece dimensions were limited in length to comply with this requirement while maintaining thickness of base plate and web member to other test pieces in the series [3]. The neutron strain scanning was undertaken on the Kowari strain scanner at the Bragg Institute, ANSTO, Sydney in 2010.

The plan for the neutron strain scanning is shown in Figure 2. As can be seen in this figure the through thickness scan lines concentrated on the area adjacent to and below the weld utilizing the reduced beam path lengths for longitudinal strain scans in both the weld and the zone immediately underneath the abutting web/weld. The methodology for undertaking DHD and iDHD residual stress measurements is given in [4] with the iDHD method being used in the plastic deformation regions of the heat affected zone and weld metal in addition to the DHD method, [4]. The subsequent locations

Materials Research Proceedings **2** (2016) 145-150 doi: http://dx.doi.org/10.21741/9781945291173-25

for the DHD/iDHD tests are shown also in Fig 2. These were located at the 11 mm from the centerline of the web, so as to avoid the initial 1.5mm gun drilled hole drill piece breaking if it was located at exactly the weld toe and the second set of DHD/iDHD measurements were undertaken at the centerline of the web. The general separation rule between DHD sites is 6 x the diameter so as to avoid relaxation of residual stress at subsequent sites. As the two sets of DHD/iDHD measurements were undertaken several months apart the separation between the two DHD sites had to be 6 x 5mm, i.e. 30 mm, as the diameters of the final EDM core is 5mm, and therefore the second DHD site was centered further along the weld.

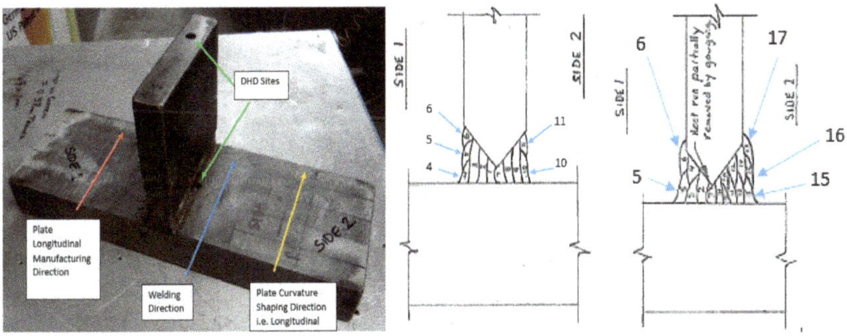

Fig 1 showing a) the cumulative line up of the longitudinal directions of the main manufacturing steps of the test piece and b) the weld pass placement strategy of the test piece and c) that for the larger test piece previously neutron strain scanned reflecting a standard weld cap. Note the reduction in layers in b) but the same pass placement strategy as c).

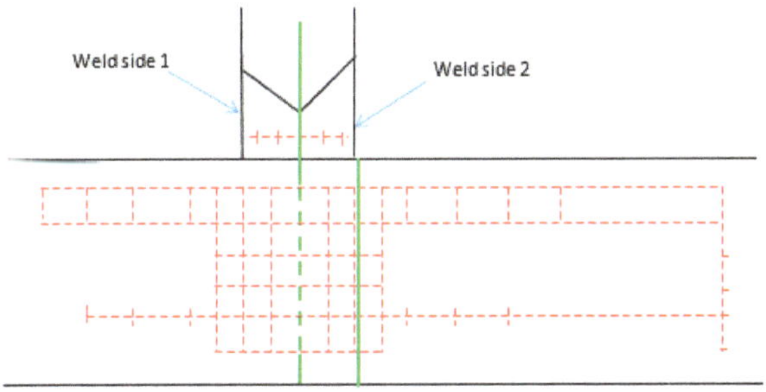

Fig.2 Plan for residual stress measurements undertaken on T butt weld test piece. The red grid represents the Neutron diffraction testing plan with strain measurements undertaken at the intersections of the transverse and vertical line. The two green lines are the locations of the centers of the DHD cores.

Results

The residual stress profiles determined by ND at the weld toes on side 1 and side 2, Fig. 3, are completely different as those on side 1 show a zigzag pattern with reduced magnitude close to the weld and then largely compressive at a depth of 10 mm below the concave surface before reverting to

Materials Research Proceedings **2** (2016) 145-150 doi: http://dx.doi.org/10.21741/9781945291173-25

tensile at a depth below the concave surface of 20 mm. The zigzag pattern for side 1 does not reflect that of a curved plate [2], [3] and therefore the original residual stress profile in the curved base plate has been over ridden. This is reinforced by the through thickness profiles for Side 2 in which there is no zigzag pattern and the 3 profiles follow the same pattern. For both side 1 and side 2 the profiles indicate that the residual stresses near the weld toe will be low tensile or low compressive stresses. This is supported by the result for the only location on the scan line across the weld where all 3 strains were captured was the one at side 2 of the weld pass closest to the weld cap surface. The residual stresses at this location were -84.5 MPa, -160 MPa, and 19.1 MPa for the longitudinal transverse and normal directions respectively.

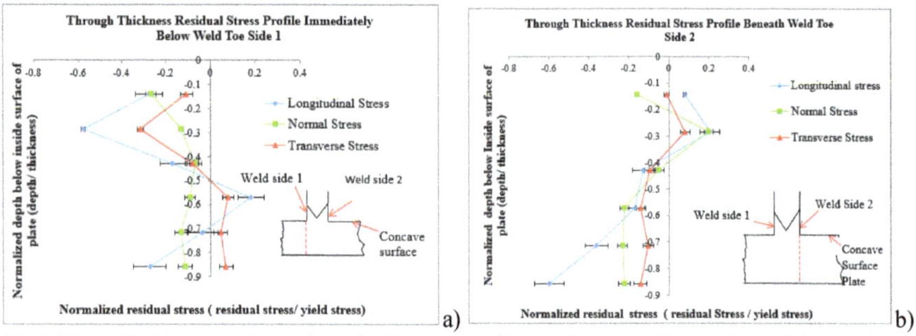

Fig 3 Through thickness plots of residual stress profiles in the base plate at the weld toes a) side 1 and b) side 2 determined by ND

The through thickness profiles for the residual stresses determined by ND at the centerline of the web are presented in Fig. 4. They are radically different from the profiles presented in Fig. 3 in that they show substantial compressive residual stress values near mid thickness of the base plate changing to lower compressive values closer to the concave surface and to tensile values towards the convex surface. No residual stress results could be determined along this line in the weld and its heat affected zone, fusion zone and in the weld due to the beam path length in the longitudinal direction.

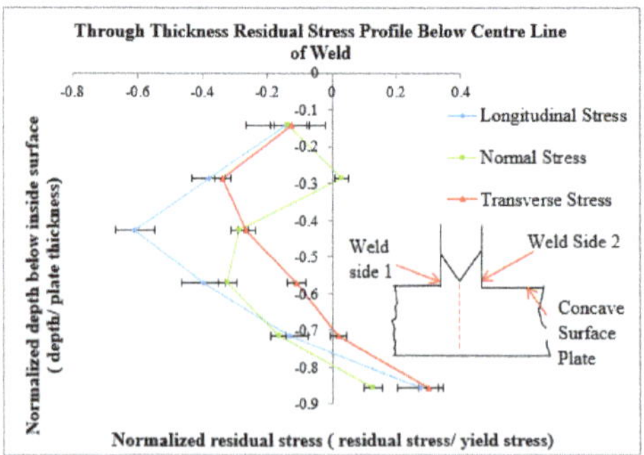

Fig. 4 Through thickness plots of residual stress profiles in the base plate at centerline of web determined by ND.

The through thickness plots for the residual stress profiles obtained by iDHD/ DHD in the base plate and web are shown in Fig 5. The longitudinal and transverse residual stress profiles at the weld toe side 2 show a rapid reduction in the magnitude of the stress levels close to the weld on the concave side of the base plate, which is opposite to that for the un-welded base plate, and with a zigzag equilibrium
balancing pattern through thickness. Those through the centerline of the web while showing a zigzag pattern through the web indicated high tensile stresses in the weld and beginning of the web.

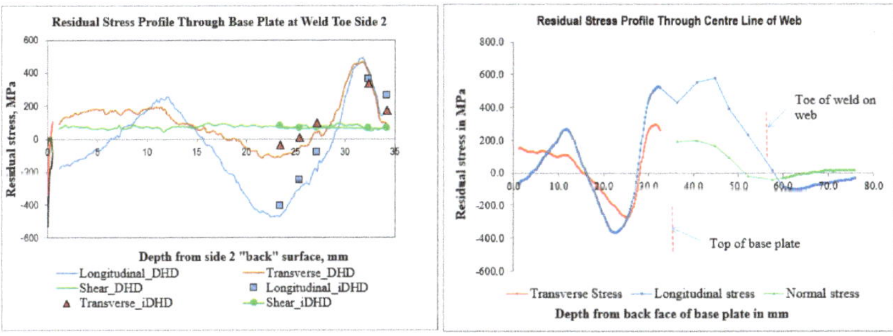

Fig. 5 Through thickness plots of the results of the DHD tests undertaken at a) the toe of the weld side 2 and b) through the centerline of the weld/ web.

Discussion

The comparative plots of the through thickness residual stress profiles for ND and IDHD/DHD methods are presented in Figure 6. Looking first at those through the base plate and centerline of the web in Fig 6a) there appears to be good correlation for the results in that they are generally following the same profile though the ND results appear to be slightly out of phase with those determined by DHD. The differences reflect one of the limitations of ND in that due to time considerations for strain scanning, particularly for the longitudinal direction, and due to beam path length issues, no longitudinal strains were captured near the centerline in the weld. Since a more detailed profile was not achieved for the ND measurements there was a need to use alternative methods such as DHD/ iDHD.

Turning to the residual stress profiles at the weld toe region there is a distinct difference in the shapes of the profiles determined by ND for 10 mm from the centerline of the web i.e. directly under the weld toe to those determined by DHD centered 11 mm from the centerline of the web. There is closer correlation between the DHD profiles and the ND profiles measured + 15 mm from the centerline of the web both with respect to magnitudes of residual stress and to profile shape. The key contributing factors to this are the rapidly changing residual stresses induced by the welding moving away from the weld toe as well as the differences in gauge volume and position of the centers of gauge volume. The DHD is based on a 5 mm diameter core removed between + 8.5 mm and + 13.5 mm whereas the ND measurements are based on a gauge volume from + 8 mm to 12 mm and hence the gauge volume for the ND is more closely aligned to the ND results centered on plus 15 mm.

Notwithstanding this all the results at the weld toe indicate that there is a distinct drop in the magnitude of the residual stress profiles at the weld toe with the one ND result achieved in the weld cap indicating compressive residual stresses. It is apparent that equilibrium balancing through thickness results in the high tensile residual stress levels found in the weld and mid thickness of the

Materials Research Proceedings **2** (2016) 145-150 doi: http://dx.doi.org/10.21741/9781945291173-25

base plate. As detailed in [3] the transverse strain profile across the weld/ web member 4mm above the concave surface of the base plate varied from -681 micro strain in the weld cap side 2 to plateau at + 1245 micro strain mid thickness before decreasing to -253 micro strain in the weld cap side 1. The normal strain profile varied from 421 micro strain in the weld cap side 2 to reverse plateau at approximately zero micro strain mid thickness and then increase to 597 micro strain in the weld cap side 1. This strain profile supports what is evident in Fig 6. The reduction of the tensile residual stresses at the weld toes side 1 and side 2 indicates that the alternate pass placement strategy shown in Fig 1 will reduce the risk of HACC defects forming as well as SCC when in service. This is a major achievement as crack/ defect propagating from the weld toes through the base plate can lead to catastrophic failure of the structure. As crack propagation can lead to catastrophic failure the reduction in tensile residual stresses by the alternate pass placement strategy at the weld toes regions needs to be validated via further testing using an alternative method such as contouring the results of which are subject of a sister paper for this conference.

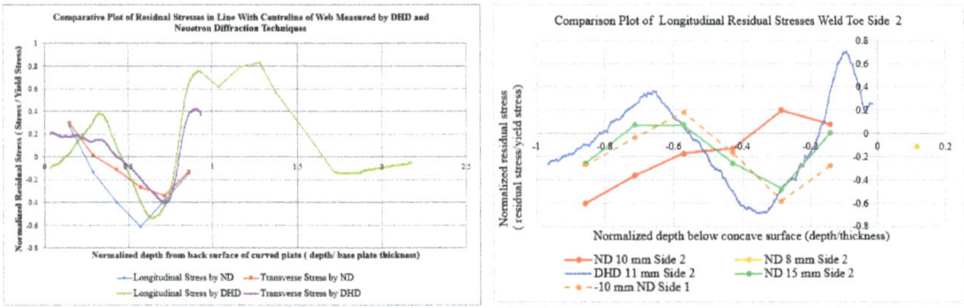

Fig 6. Comparative plots of the longitudinal residual stress profiles obtained by the ND and DHD at the centerline of the web on the left and at the weld toe side 2 on the right.

Summary

The residual stress profiles for the alternate pass placement strategy, from the ND and DHD testing undertaken on this test piece, clearly show that there is a reduction is the residual stress magnitude at the weld profile when compared with the yield strength values given in BS 7910. Further work needs to be undertaken by a 3rd method such as contouring to establish the rate of change of the residual stress profiles away from the weld toe and to demonstrate the advantages of the alternate pass placement strategy of progressing weld passes from the base plate to the web to give confidence before implementing this strategy on full scale structures.

References

[1] BS 7910 "Guidance on methods of assessing the acceptability of flaws in metallic structures", 1999 with amendments 2000, 2009.

[2] S. Pearce "Breakdown of residual stresses in highly restrained thick section steel welds", Doctoral thesis, University of Adelaide 2009

[3] G. Sloan "The influence of welding variables on the residual stress profiles around T butt and MMAW butt welds in thick section high strength steel", Master's thesis, University of Adelaide 2015

[4] Information on http://www.veqter.co.uk

Residual Stresses 2016: ICRS-10 Materials Research Forum LLC
Materials Research Proceedings **2** (2016) 151-156 doi: http://dx.doi.org/10.21741/9781945291173-26

The Residual Stress Effect on the Shape Memory Polymers

A. Kallel[1,a*], M. Lamraoui[2,b], J. Fitoussi[2,c], A. Tcharkhtchi[2,d],

[1] Léonard de Vinci Pôle universitaire, Research center, Paris la Défense, France

[2] Laboratoire PIMM, UMR 8006 CNRS, Arts et Métiers ParisTech, Paris 75013, France

[a]achraf.kallel@devinci.fr, [b]menad.lamraoui@ensam.eu, [c]joseph.fitoussi@ensam.eu, [d]abbas.tcharkhtchi@ensam.eu

Keywords: Shape Memory Polymers, Residual Stress Effect, Driving Force, Blend PCL /SBS

Abstract. The current paper presents an experimental study of the residual stress role to recover polymer blend to its original position after deformation. In this study, we use a polymer blend of 40% PolyCaproLactone (PCL) and 60% Styrene-Butadiene-Styrene (SBS). The Shape Memory Polymer (SMP) is able to storing a permanent macroscopic after passing through a temporary one; then under an external stimulus; they can recover their initial shapes. The recovery rate depends on the mechanical property of the mixture and also of the number of the loading cycles. Indeed, one multiplying the number of mechanical loading cycle; generally tensile test; the recovery rate does not change in a linear manner and therefore the residual stress is not added in integer from one cycle to another. In this work we studied the mechanism of the shape as well as property memory effect of the blend under study during a shape memory cycle, to establish a relationship between the "shape memory effect" and the "properties memory effect". Finally, we measure the "Driving Force" responsible for the shape memory effect by an original method. This study allows establishing a relationship between the rate of recovery and residual stresses introduced into the polymer during its deformation.

Introduction

Shape memory polymers (SMPs) have the capacity of changing their shapes from a temporary shape to a permanent shape upon an external stimulus. The stimulus could be provided by thermal [1, 2], magnetic or electric sources [3]. In the case of thermal stimulus, the cycle process consists of first a deformation at high temperature, relaxation, fixing and finally recovery [4-6]. SMPs are useful in many domains as medical and biological devices [8, 9], smart textile [10, 11], tubing and actuating materials and packaging components [12-14]. The capacity of SMPs to change shape is generally explained by the presence of at least two different phases with different mechanical properties (hard and soft). Recently different new concepts regarding to certain aspects related to shape memory effect were shown [4-6]. We can consider as concept the Property Memory Effect (PME) in which stress-strain tensile tests and both mono and multi-frequency DMA (Dynamic Mechanical Analyzer) tests on virgin and 100% recovered samples of PU (rubber) revealed that the polymer at the end of the shape memory tests regains 100% of its initial form without regaining some of its physical properties like the glass transition temperature, the tensile modulus, the heat expansion coefficient and the free volume fraction. After recovery tests on a stretched sample, the majority of the polymers regain only a certain percentage of their initial shape. The Degree of Shape Memory Effect (DSME) is defined as the percentage of the residual shape. This denotes the Partial Shape Memory Effect (PSME) concept. The successive cyclic tensile tests on partial shape memory polymers (PSMP) show that the DSME increases with the increase of the number of cycles. This test also showed the shape capacity of polymer increases by the number of cycles. This new concept of shape memory effect is

Materials Research Proceedings **2** (2016) 151-156 doi: http://dx.doi.org/10.21741/9781945291173-26

of very important value. We can in a legitimate way suppose that a polymer without shape memory effect can be transformed into a shape memory polymer.

Another new concept concerns the driving force responsible of recovery phenomenon. It has been shown that during shape memory cycling tests, and during fixing test the polymer is cooled down rapidly in order to fix the shape obtained at the end of the test. In this period the residual stresses will be locked in the structure. This residual stress is indeed the origin of driving force responsible for the shape memory effect [4-6]. During recovery, at a temperature higher than the glass transition of soft segments, these stresses are released and the SMP regains its initial shape.

In the present work, in order to study the origin of this driving force "Driving force", the relationship between the recovery rate and the residual stresses, introduced into the polymer during its deformation, has been determined.

Experimental

Materials. A mixture of polycaprolactone (PCL) (40%) and Styrene-Butadiene-Styrene (SBS) (60%) is used. This blend shows almost a total shape memory effect. We choose this percentage because it gives a good combination of shape recovery ratio and shape fixing ratio [18]. The PCL that was used for this study is PCL CAPA 860 supplied by Perstrop Limited UK with a density of $1.14\ g/cm^3$ of density and its Tg and Tm are respectively -50°C and 60°C. The used Styrene-Butadiene-Styrene (SBS); provided by the company Sinopec Group, is an amorphous copolymer (Tg =-80°C) between the polybutadiene rubber (PB) and thermoplastic polystyrene (PS).

Determinations of melting point and glass transition temperature. A TA instruments DSC-Q10 V9.0 Build 275 type Differential Scanning Calorimetry has been used to determine the transition temperature of polymers. Specimens have been cooled with nitrogen from room temperature to -80 °C (Minimal temperature that the machine can achieve) and then heated until 80 °C at a constant rate of 3 °C / min.

Pure PCL presents a melting point of 59,2 °C [18]. Pure SBS is an amorphous polymer without any melting point. The blend shows a melting point of 57,6 °C which is almost the same as the pure PCL (Fig. 1). In order to determine the glass transition temperature, dynamical mechanical thermal analysis (DMTA) tests has been carried out using a DMA Q800 TA Instruments. A 1Hz frequency has been performed between -100°C to 0°C with a heating rate of 3°C/min,

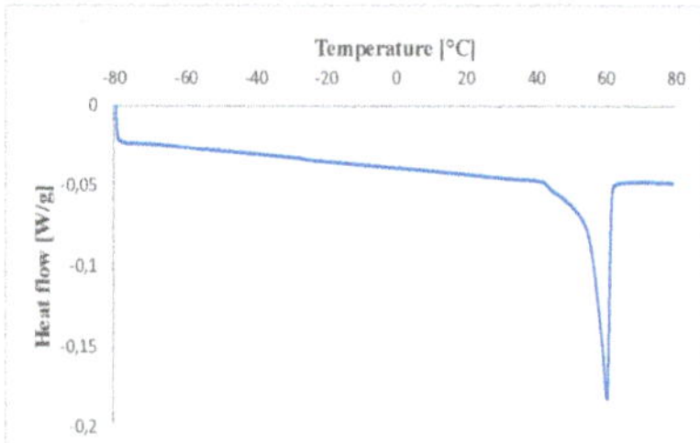

Fig. 1: DSC Results of: (a) pure PCL, (b) pure SBS and (c) PCL (40%) /SBS (60%) blend.

These curves in Fig.2 show that the transition temperature of the SBS and PCL are -82°C and -42°C, respectively. These results reinforce the fact that the SBS and the PCL are not miscible because their Tg remain the same in the blend. Fig. 3 showed the MEB images of a virgin simple (right image) and a simple after one SM cycle (left image). The bright region was PCL phase and the dark region was SBS phase. That shows the intrinsic immiscibility between the SBS, the soft phase, and the PCL the hard phase. This confirms DMA results showing two different transition temperatures: the lowest corresponds to the soft phase (SBS) and the highest corresponds to the hard phase (PCL).

Materials Research Proceedings **2** (2016) 151-156 doi: http://dx.doi.org/10.21741/9781945291173-26

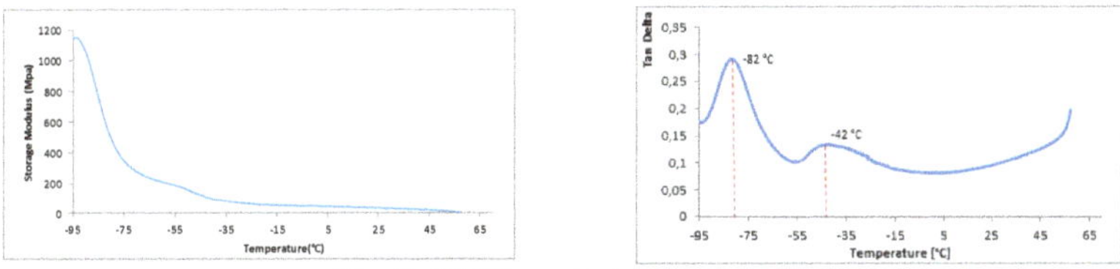

Fig. 2: DMTA analysis of virgin sample

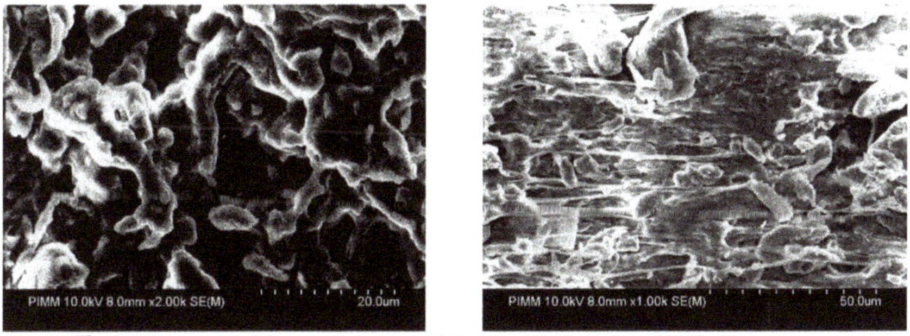

Fig. 3: The MEB images of the cross-section of the blend SBS 60% and PCL 40%.

Tensile and shape memory test. The stress-strain tests have been done by a uniaxial-tensile loading Instron 5566-type machine with a load cell of 1 kN at room temperature, according to the standard NF ISO 6239 [17].

Shape memory test was carried according to the following steps: (1) heating the sample up to 50°C and holding for 10 minutes; (2) extending to the strain of 100% with a speed of 6 mm/min (tensile test) (3) fixing this deformation by cooling down the sample to the room temperature using a fan for 10 min (4) unloading the sample to zero stress and then recording the strain (5) heating up the sample to above the glass transition temperature of the blend and then recording the strain (recovery test). During this step, the length of the sample decreases and thanks to the shape memory effect, the sample recovers all its substantial original shape. It is important to indicate that the recovery test cannot practically be performed isothermally because during heating, the oven from room temperature up to 50°C and before reaching this temperature, the sample begins its deformation for recovering its initial shape. So recovery test was carried out in ramp (constant temperature rate) condition.

For multi-cycle tests, successive shape memory cycles were performed on the same sample. At the end of the first recovery cycle test, the sample is subjected to another shape memory cycle, i.e. tensile, fixing and recovery. This cycle was repeated a several times.

Results and discussion

Thermal effects on the mechanical and the shape memory properties. The results of tensile tests on the SBS/PCL blend are shown in Figure 5 and table 2. These results clearly show the influence of temperature on mechanical properties. The stress decreases with increasing temperature. We can note that the thresholds which separate elastic and plastic areas are at 8% of strain (ε). The strain energy can be stored in the polymer structure, which subsequently will be released under the effect of thermal stimulation. This stored energy is the source of the driving force that will bring the sample to its original shape.

Materials Research Proceedings **2** (2016) 151-156 doi: http://dx.doi.org/10.21741/9781945291173-26

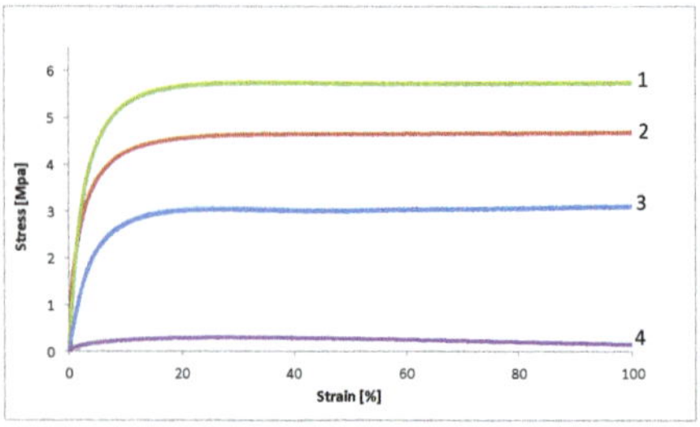

Fig. 5: Tensile tests of SBS / PCL blend at different temperatures: (1) 25°C, (2) 35°C, (3) 50°C, (4) 60°C

Table 1: Young modulus values and stress at 100% of strain at different temperature

T(°C)	Modulus [MPa]	Stress at 100% strain [MPa]
25	174	5.7
35	103	4.7
50	66	3.1
60	7	0.3

Fig. 5 shows that the blend is less rigid at high temperature. The curve's initial slope (Young Modulus) as well as the sample's stiffness is inversely proportional to the temperature. 50°C will be used as a reference temperature for the entire tensile test. The recovery test is carried out in a thermal chamber for a range of temperature after holding at 100% of deformation. The recovery rate and the fixing rate are respectively:

$$R_r(\%) = \frac{L_u - L_f}{L_u - L_i} * 100 \quad (1), \text{ and } R_f(\%) = \frac{L_u}{L_m} * 100 \quad (2)$$

Where L_i is the initial sample length, L_m the length after 100% of deformation, L_u the length after tensile test without stress (jaws are release) and L_f is the t length after recovery.

To calculate the recovery rate, w used L_u, the length after tensile te: without stress, and not $L_{m,}$ the lengt after 100% of deformation, becaus when we release the jaws th sample's length changes. The fixin rate is not equal to 100%.

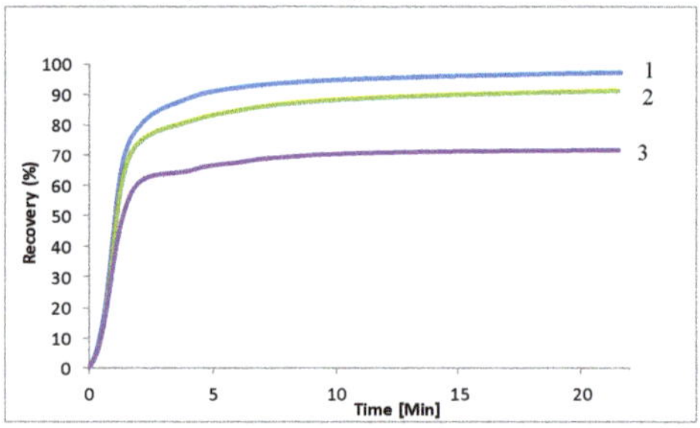

Fig. 6: Recovery rate at different temperature: (1) 75°C, (2) 65°C, (3) 55°C

Table 2: Recovery tests at different temperature

Temperature [°C]	Rr [%]	Rf [%]
75	97,2	92,9
65	91,1	94,1
55	71,9	93,4

Molecular mobility increases by rising the temperature. For our blend, 100% recovery is reached in about 20 min with 75 °C while for 55 °C the polymer reached only 71,9% of recovery. For this blend, the SBS chains, the soft phase, ensure the recovery.

Multi-cycle effects on mechanical and shape memory properties. Multi-cycle tests were performed to study their effects on the mechanical and shape memory properties. For these assessments, we will use 50°C for tensile test and 55°C for recovery test.

The residual stress stored in the sample after each tensile test, need a thermal stimulus to be released. The amount of the internal energy stored during the cycle is more important when the tensile test is made at -30°C than at 23°C. Therefore, for sample which the tensile test is made at -30°C, the recovery starts at low temperature and can reach 55% at the ambient temperature.

Materials Research Proceedings **2** (2016) 151-156 doi: http://dx.doi.org/10.21741/9781945291173-26

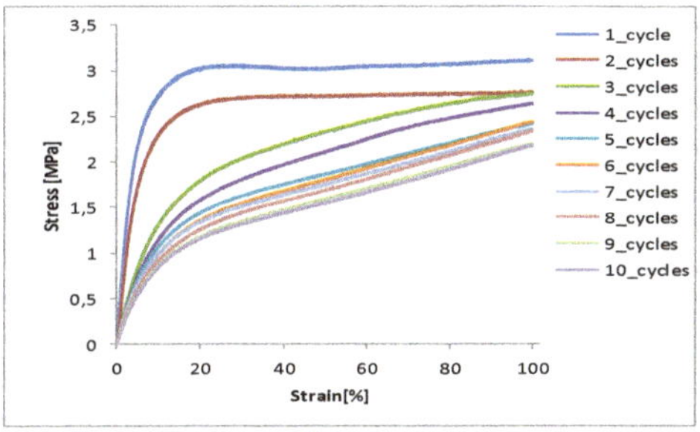

Fig. 7: Tensile test for different cycles

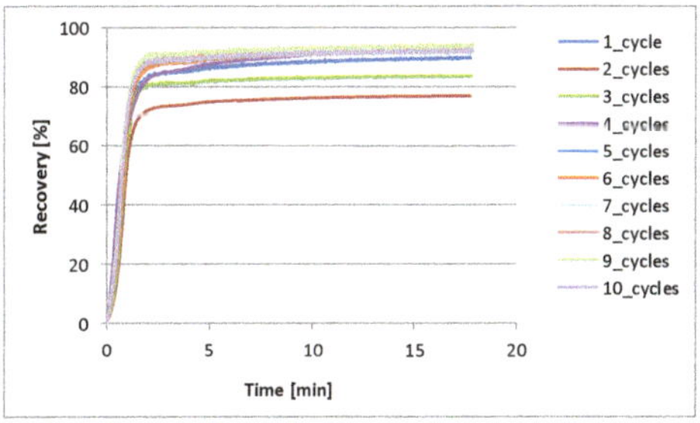

Fig. 8: Recovery rate for different cycles.

Tensile tests were carried out on a sample after shape memory cycles. Results are given in figure 8. We can notice that the specimen's stress decreases from 3.1 Mpa (cycle1) to 2.1 Mpa after 10 shape memory cycles. The Young modulus undergoes a significant decrease because it passes from 65.6 Mpa in the first cycle to 12.1 Mpa in the tenth cycle.

Table 3: Maximal stress and Modulus for each cycle

Strain cycles	Stress [MPa]	Modulus [MPa]
1	3,1	65,6
2	2,7	46,8
3	2,7	20,5
4	2,6	17,6
5	2,4	17,6
6	2,4	15,2
7	2,3	14,9
8	2,3	13,7
9	2,1	12,3
10	2,1	12,1

Table 4: Fixing rate and Recovery rate at different cycles

Cycles	Rr [%]	Rf [%]
1	89,7	94,6
2	76,9	94
3	83,5	93,7
4	90,7	92,2
5	92,2	88,8
6	92	89,1
7	92,4	88,1
8	93,1	87,8
9	93,7	87,7
10	92	88,1

Conclusion

After a shape memory cycle; the soft material, more sensitive to the deformation, is going to store some strain energy which will be afterward a source for the driving force to return to the initial shape. Whereas the hard phase, during the tensile test, resist to the deformation of the soft chains and this resistance is going to create residual stress in the hard phase which will be released by an external stimulus (heating). The direct measure of the residual stress by diffraction for our example will be very difficult to see impossible because the mixture is not crystalline and does not present any arrays as the case of the steel or other metals. We remark that the blend regains 90% of its initial shape after the first memory test and that the recovery rate varies in a random way until the fifth cycle from which it stabilizes around 92 % until tenth cycles. The fixing rate decrease from 94% in the first cycle to 88% in the fifth one. Then it keeps this rate until the tenth cycle. We can notice that even when the sample regains almost its initial shape after each cycle, it does not have the same mechanicals properties. We can notice that the sample does not remain the same viscoelastic properties even when it regains almost its initial shape after each cycle.

References

[1] Lendlein A, Kelch S. Shape-memory polymers. Angew Chem Int Ed 2002;41:2034–57. http://dx.doi.org/10.1002/1521-3773(20020617)41:12<2034::AID-ANIE2034>3.0.CO;2-M

[2] Behl M, Lendlein A. Shape-memory polymers. Materials Today 2007;10:20–8. http://dx.doi.org/10.1016/S1369-7021(07)70047-0

[3] Leng J, Lu H, Liu Y, Du S. Conductive nanoparticles in electro activated shape memory polymer sensor and actuator. Paper presented at the proceedings of SPIE – the international society for optical engineering. Nanosensors and microsensors for bio-systems, San Diego, CA, United States, vol. 6931; 2008. p. 693109. http://dx.doi.org/10.1117/12.775743

[4] Tcharkhtchi A. and al, Some New Concepts of Shape Memory Effect of Polymers. Polymers 2014, 6, 1144-1163. http://dx.doi.org/10.3390/polym6041144

[5] Farzaneh, S., Fitoussi, J., Lucas, A., Bocquet, M., Tcharkhtchi, A., "Shape memory effect and properties memory effect of polyurethane", Journal of Applied Polymer Science, Volume 128, Issue 5, pages 3240–3249, 2013. http://dx.doi.org/10.1002/app.38530

[6] S. Abdallah-Elhirtsi, J. Fitoussi, B. J. Rashmi, K. Prashantha, S. Farzaneh, M. F. Lacrampe, P. Krawczak, A.Tcharkhtchi, "Study of partial shape memory effect of polymers by multicycle tests", , Polymer Composites, 36 (6), pp. 1145-1151 (2015). http://dx.doi.org/10.1002/pc.23472

[7] T. Tadaki and al. Shape Memory Alloys. Annual Review of Materials Science. Vol. 18: 25-45 (1988). http://dx.doi.org/10.1146/annurev.ms.18.080188.000325

[8] Lendlein A, Langer R. Biodegradable, elastic shape-memory polymers for potential biomedical applications. Science 2002;96:1673–6. http://dx.doi.org/10.1126/science.1066102

[9] Wache HM, Tartakowska DJ, Hentrich A, Wagner MH. Development of a polymer stent with shape memory effect as a drug delivery system. J Mat. Sci: Mater Med 2004;14:109-12. http://dx.doi.org/10.1023/A:1022007510352

[10] Hu JL, Ding XM, Tao XM. Shape memory polymers and their applications to smart textile products. J China Textile Univ 2002;19:89–93. Application textile

[11] Russel DA, Hayashi S, Yamada T. The potential use of memory film in clothing. Paper presented at the techtextil symposium-new protective textiles (through textile technology index database), April 1999. Application textile

[12] J. Karger-Kocsis. Recent advances in shape memory polymers and composites: a review. J Mater Sci (2008) 43:254–269. http://dx.doi.org/10.1007/s10853-007-2176-7

[13] T. Xie. Recent advances in polymer shape memory. Polymer 52 (2011) 4985-5000. http://dx.doi.org/10.1016/j.polymer.2011.08.003

[14] J. Hu. Recent advances in shape–memory polymers: Structure, mechanism, functionality, modeling and applications. Progress in Polymer Science 37 (2012) 1720– 1763. http://dx.doi.org/10.1016/j.progpolymsci.2012.06.001

[15] Lin JR, Chen LW. Study on shape-memory behavior of polyether-based polyurethanes. I. Influence of the hard-segment content. J Appl Polym Sci 1998;69:1563–74. http://dx.doi.org/10.1002/(SICI)1097-4628(19980822)69:8<1563::AID-APP11>3.0.CO;2-W

[16] Lin JR, Chen LW. Study on shape-memory behavior of polyether-based polyurethanes. II. Influence of soft-segment molecular weight. J Appl Polym Sci 1998;69:1575–86. http://dx.doi.org/10.1002/(SICI)1097-4628(19980822)69:8<1575::AID-APP12>3.0.CO;2-U

[17] NF ISO 6239:1986. Plastics Determination of Tensile Properties by Use of Small Specimens. Association Française de Normalisation (AFNOR): Seine-Saint-Denis, France, 1986.

[18] Zhang H., Wang H., Zhong W. and Du Q. A novel type of shape memory polymer blend and the shape memory mechanism. Polymer 50 (2009) 1596-1601. http://dx.doi.org/10.1016/j.polymer.2009.01.011

Materials Research Proceedings 2 (2016) 157-162 doi: http://dx.doi.org/10.21741/9781945291173-27

Residual Stress Measurements of Alumina-Zirconia Ceramics by Time-of-Flight Neutron Diffraction

Kunyang Fan[1], Jesus Ruiz-Hervias[1a*], Jonas Gurauskis[2], C. Baudin[2]

[1] Materials Science Department, Universidad Politécnica de Madrid, E.T.S.I. Caminos, Canales y Puertos, Profesor Aranguren 3, E-28040 Madrid, Spain

[2] Instituto de Cerámica y Vidrio, CSIC, Kelsen 5, E-28049 Madrid, Spain

[a]jesus.ruiz@upm.es

Keywords: Ceramic Composites, Residual Stress, Neutron Diffraction, Tape Casting

Abstract. Neutron strain scanning and Rietveld analysis were used to study the residual stresses in Al_2O_3/Y-TZP ceramic composites fabricated by different green processing techniques (a novel tape casting and conventional slip casting) and with different zirconia content. The results show that the residual stresses in zirconia particulates are tensile and the ones in alumina matrix are compressive, with almost flat through-thickness residual stress profiles in all bulk samples. The residual stresses for both phases were mainly dependent on the zirconia content, irrespective of the measurement direction and the fabrication process.

Introduction

Alumina-zirconia ceramics have received considerable attention in both engineering and academic fields due to their improved mechanical properties when compared with pure alumina ceramics [1-4]. It has been proposed that the residual stresses due to thermal and elastic mismatches between alumina and zirconia could contribute to toughening and enhance the structural performance of these ceramic composites [5, 6].

Numerous works have been done on residual stresses analysis in Al_2O_3-ZrO_2 composites [6-10] and a wide range of results were obtained depending on fabrication routes, composition design and measurement techniques. In this study Al_2O_3/Y-TZP (alumina/tetragonal ZrO_2 stabilized with 3 mol% Y_2O_3) composites were investigated. Samples were prepared by a novel tape casting route [11, 12] for stacking green ceramic tapes made from high solid content of water-based slurries at room temperature and using low pressures. The conventional slip casting [13] was taken as reference for comparison. The effect of the Y-TZP content on the Al_2O_3/Y-TZP composites was also studied.

Time-of-flight (TOF) neutron diffraction and Rietveld analysis have been applied for the non-destructive determination of the residual stresses in Al_2O_3/Y-TZP bulk samples. Through-thickness residual stress profiles corresponding to both phases, i.e. alumina and zirconia, were obtained, and compared to available data in the literature and to estimations by theoretical models. The effects of Y-TZP content and the new processing method on the microstructure and residual stresses of the ceramic composites were established.

Experimental

Sample preparation and characterization

Monoliths of Al_2O_3/Y-TZP composites with 5 and 40 vol. % of Y-TZP were obtained using high-purity α-Al_2O_3 and polycrystalline tetragonal zirconia stabilized with 3 mol. % Y_2O_3, named as Y-TZP, as the starting powders. Two kinds of green processing methods were used for each composition: the novel tape casting and conventional slip casting. The studied specimens were named as A-5YTZP(slip), A-5YTZP(tape), A-40YTZP(slip) and A-40YTZP(tape), in order to

Materials Research Proceedings **2** (2016) 157-162 doi: http://dx.doi.org/10.21741/9781945291173-27

describe compositions and fabrication techniques. Subsequent sintering was carried out at a maximum temperature of 1500°C, with a dwell time of 2 h (heating and cooling rates of 5°C/min).

High-density (relative density>98.5 %) sintered samples were obtained for each studied Al_2O_3/Y-TZP composite. Characteristic microstructures were observed by scanning electron microscopy on diamond polished and chemically etched (85% H_3PO_4, 7 min at 200°C) sample surfaces. The microstructure is very similar for both fabrication techniques, and depends on the zirconia content. Representative scanning electron micrographs of A-5YTZP and A-40YTZP composites are presented in Fig. 1, as a function of Y-TZP reinforcement content. A dense microstructure was observed in all the studied materials, where the alumina matrix (in dark grey) and zirconia particulates (in light grey) were generally well-dispersed. The grain size of Al_2O_3 matrix was clearly decreased, as zirconia content increased from 5 to 40 vol. %.

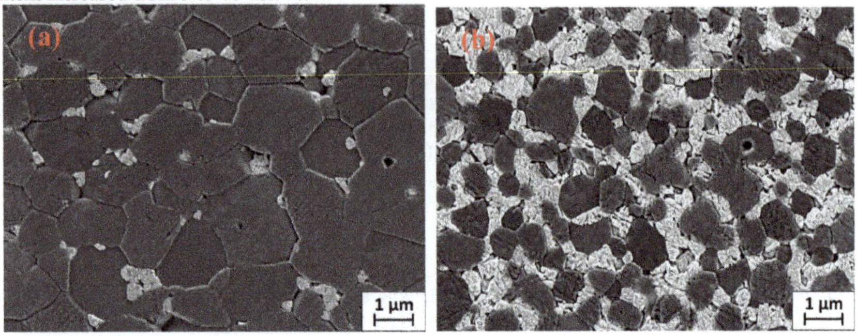

Fig. 1. SEM micrographs of polished and chemically etched surfaces of the studied ceramic composites: (a) A-5YTZP; (b) A-40YTZP. Al_2O_3 grains appear with dark grey color and Y-TZP particulates are in light grey color.

Residual stress measurement

The time-of-flight neutron diffraction technique ENGIN-X (ISIS, UK) was used for residual strain scanning in each ceramic composite sample (dimensions of 20 mm×20 mm×5 mm). With two detector banks centered on Bragg angles of $2\theta_B = \pm 90$ degrees, the ENGIN-X instrument allows simultaneous measurements of strain in two principal stress component directions. In this case, the in-plane direction, parallel to the larger plane of the samples (20 mm×20 mm), and the normal direction, perpendicular to it, were chosen. The measurement gauge volume was set to 15×1×1 mm^3. Through-thickness strain scanning was carried out along the sample thickness in 0.4 mm steps. The stress-free reference lattice parameters of α-Al_2O_3 and Y-TZP were obtained by measuring both the α-Al_2O_3 and Y-TZP starting powders.

TOF diffraction data were analyzed by Rietveld refinement of the complete spectrum, using the TOPAS-Academic V5 program [14]. The instrument diffraction profile was modeled by using a convolution of a pseudo-Voigt function with two back-to-back exponentials. Monoclinic zirconia was detected in the Y-TZP powder but not in the composites. Consequently, all the studied A/Y-TZP composites were analyzed by using a two-phase model consisting of the hexagonal α-Al_2O_3 phase and tetragonal Y-TZP phase. The space groups and initial atomic structure information used in refinements are taken from [15] [16].

The lattice parameters of each phase were obtained after refinement. The average elastic strain representative of a phase, named mean phase strain, was determined from the change of the average lattice parameters. Mean phase strains for both phases were respectively calculated by averaging the strain over the unit cell, as:

Materials Research Proceedings **2** (2016) 157-162 doi: http://dx.doi.org/10.21741/9781945291173-27

$$\bar{\varepsilon} = (2\varepsilon_a + \varepsilon_c)/3 \tag{1}$$

where ε_a and ε_c are the strains along lattice axis (*a* and *c*), calculated as $\varepsilon_a = (a - a_0)/a_0$, $\varepsilon_c = (c - c_0)/c_0$, respectively. The lattice parameters *a* ($a=b$) and *c* were obtained from bulk samples, and a_0 and c_0 are the stress-free lattice parameters measured from the starting powders.

Taking into account the fabrication process and specimen symmetry, the strains measured in the normal and in-plane directions are considered the principal strains ε_{ii}, i=1, 2, 3, as $\varepsilon_{11} = \varepsilon_{22} = \varepsilon_{In-plane}$ and $\varepsilon_{33} = \varepsilon_{Normal}$. Thus, mean phase stresses for both alumina and zirconia phases were calculated along the in-plane and normal directions using Hooke's law:

$$\sigma_{In-plane} = \frac{E}{1+\nu}\bar{\varepsilon}_{In-plane} + \frac{E\nu}{(1+\nu)(1-2\nu)}\left(2\bar{\varepsilon}_{In-plane} + \bar{\varepsilon}_{Normal}\right) \tag{2}$$

$$\sigma_{Normal} = \frac{E}{1+\nu}\bar{\varepsilon}_{Normal} + \frac{E\nu}{(1+\nu)(1-2\nu)}\left(2\bar{\varepsilon}_{In-plane} + \bar{\varepsilon}_{Normal}\right) \tag{3}$$

where $\bar{\varepsilon}$ corresponds to the calculated mean phase strain for both Al_2O_3 and Y-TZP phases, given by Eq. 1, and E and ν are the bulk elastic constants of Al_2O_3 (E=400 GPa and ν=0.22) and Y-TZP (E=210 GPa and ν=0.31) [17].

Results and discussion

Fig. 2 shows a representative diffraction pattern for a A-40YTZP (tape) composite, including measured (blue line) and calculated profile (red line, overlapping the blue observed profiles). The difference plot between the observed and calculated intensities was shown in grey line below the spectrum. Good fitting was achieved for all reference powders and bulk samples, with the weighted residual error R_{wp} ranging from 4% to 10%.

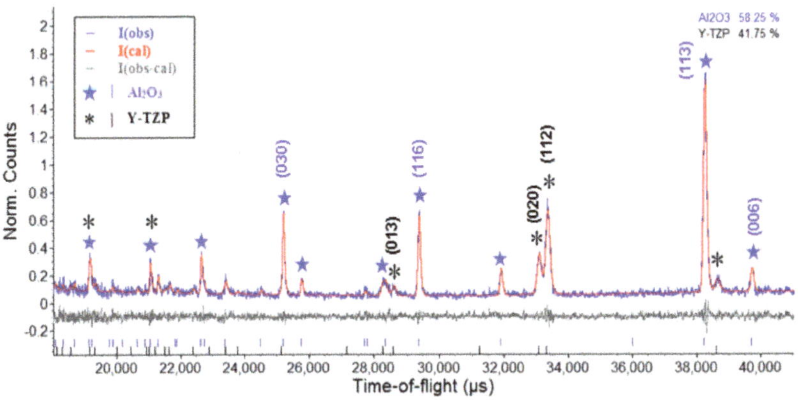

Fig. 2. Representative TOF neutron diffraction pattern analyzed with Rietveld refinement for a A-40YTZP (tape) sample. Individual peaks of Al_2O_3 (★) and Y-TZP (*) were identified at the bottom of the profile with blue and black tick marks, respectively.

Materials Research Proceedings **2** (2016) 157-162 doi: http://dx.doi.org/10.21741/9781945291173-27

Phase composition corresponding to the measured gauge volumes was evaluated by Rietveld refinement, as shown in the upper right of the profile fitting window (Fig. 2). Almost constant Y-TZP contents were recorded at different scanning positions of the same sample, with values very close to the nominal ones.

Through-thickness residual stress profiles of the alumina matrix and the zirconia particulates are shown in Fig. 3 and Fig. 4, for the slip casting and tape casting Al_2O_3/Y-TZP samples, respectively. For all the studied Al_2O_3/Y-TZP composites, compressive stresses were found in Al_2O_3 matrix and tensile ones were in the Y-TZP particles, This is due to the lower thermal expansion coefficient of Al_2O_3 matrix ($\alpha_{A,25-1000℃}$= 8.6×10^{-6} °C^{-1}) [18] as compared with Y-TZP particles ($\alpha_{YTZP,25-1000℃}$= 10.8×10^{-6} °C^{-1}) [19]. Almost flat stress profiles were obtained for both phases in each specimen, which indicates the homogenous distribution of residual stress inside the samples. No obvious orientation effects were discovered, according to the similar stress behaviors in normal and in-plane directions.

In addition, no significant differences can be observed between samples with the same composition made by conventional slip casting (Fig. 3) or by tape casting routes (Fig. 4). This seems to indicate that the low pressure (18 MPa) used for stacking tapes in the novel tape casting green process did not produce additional macro-residual stresses after sintering. Consequently, the micro-residual stresses in both phases were mainly induced by thermal and elastic mismatch between phases, during cooling from the sintering temperature to room temperature. Irrespective of the fabrication process (slip casting or tape casting), as Y-TZP content increased from 5 vol. % to 40 vol. %, tensile stresses in Y-TZP particles decreased from an average value of 730±50 MPa in A-5YTZP composites to approximately 400±30 MPa in A-40YTZP composites. On the contrary, the compressive stresses in Al_2O_3 matrix were increased (in absolute value) from an average value of -70±12 MPa in A-5YTZP composites to approximately -320±12 MPa in A-40YTZP composites. As the stress profiles are rather flat, the stresses were averaged along the sample thickness, and the reported error is the standard deviation.

The residual stresses were estimated with the modified Eshelby model proposed by Taya et al. [20]. In the model, the isotropic average stress fields in the Al_2O_3 matrix and Y-TZP particulate can be theoretically estimated from the volume fraction, elastic modulus, Poisson's ratio and thermal expansion coefficients of the phases. The average residual stresses obtained in both phases are the following:

- A-5YTZP composites: -39 MPa in alumina and 741 MPa in zirconia
- A-40YTZP composites: -326 MPa in alumina and 489 MPa in zirconia

The average residual stresses calculated from the model agree quite reasonably with the ones obtained from the neutron diffraction experiments.

The values of $f_A\sigma_A + f_{YTZP}\sigma_{YTZP}$ were also calculated to check the static equilibrium condition [21], where f_A, f_{YTZP} are the volume fractions of Al_2O_3 matrix and Y-TZP in composites; and σ_A, σ_{YTZP} are the measured mean phase stress of Al_2O_3 matrix and Y-TZP particulates, respectively. As can be seen in Figs. 3 and 4, those values are close to zero in all cases.

Summary and conclusions

Through-thickness residual strain scanning was carried out in Al_2O_3/Y-TZP ceramic composites with different zirconia content (5 and 40 vol.%), by means of time-of-flight neutron diffraction. The whole spectra were analyzed with the Rietveld method. The obtained results showed that alumina matrix is under compression and Y-TZP particulates are under tension in the composites, due to thermal expansion mismatch between the matrix and the particles. Almost flat through-thickness mean phase residual stress profiles were obtained in both phases. The novel tape casting process does not produce a noticeable change on the residual stress state when compared to the traditional slip casting route. Residual stresses for both phases were mainly dependent on the Y-TZP content in composites.

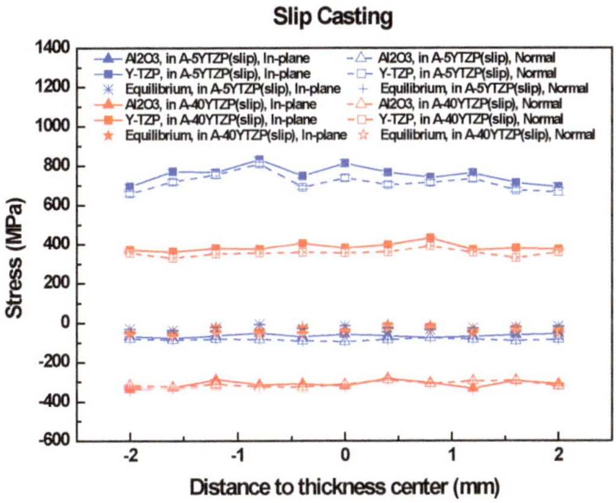

Fig. 3. Mean phase stress profiles for Al_2O_3 (Δ) and Y-TZP (□) phases through the thickness of A/Y-TZP (slip casting) bulk samples, both in the in-plane (solid line) and normal (dash line) directions. The values of $f_A\sigma_A + f_{YTZP}\sigma_{YTZP}$ were presented without line.

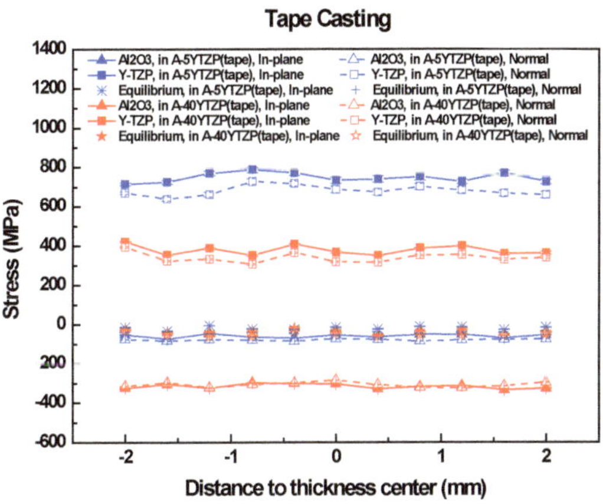

Fig. 4. Mean phase stress profiles for Al_2O_3 (Δ) and Y-TZP (□) phases through the thickness of A/Y-TZP (tape casting) bulk samples, both in the in-plane (solid line) and normal (dash line) directions. The values of $f_A\sigma_A + f_{YTZP}\sigma_{YTZP}$ were presented without line.

Acknowledgements:
This research has been supported by the Spanish Ministry of Education through grants BIA2011-26486 and BIA2014-53314-R. EU-NIM3 funding under the program 'Access to ISIS Neutrons' is acknowledged for the measurements at ISIS ENGIN-X. One of the authors (Kunyang Fan) would like to thank China Scholarship Council (CSC) for the grant to do a Ph.D in Spain.

References
[1] K. Kageyama, Y. Harada, H. Kato, Preparation and mechanical properties of alumina-zirconia composites with agglomerated structures using pre-sintered powder, Mater. Trans., JIM. 44 (2003) 1571-1576.
[2] F. Gutierrez-Mora, D. Singh, N. Chen, et al., Fracture of composite alumina/yttria-stabilized zirconia joints, J. Eur. Ceram. Soc. 26 (2006) 961-965. http://dx.doi.org/10.1016/j.jeurceramsoc.2004.12.035

[3] W. Tuan, R. Chen, T. Wang, et al., Mechanical properties of Al_2O_3/ZrO_2 composites, J. Eur. Ceram. Soc. 22 (2002) 2827-2833. http://dx.doi.org/10.1016/S0955-2219(02)00043-2
[4] A. Nevarez-Rascon, A. Aguilar-Elguezabal, E. Orrantia, et al., On the wide range of mechanical properties of ZTA and ATZ based dental ceramic composites by varying the Al_2O_3 and ZrO_2 content, Int. J. Refract. Met. Hard Mater. 27 (2009) 962-970. http://dx.doi.org/10.1016/j.ijrmhm.2009.06.001
[5] K.B. Alexander, P.F. Becher, X.L. Wang, et al., Internal Stresses and the Martensite Start Temperature in Alumina-Zirconia Composites: Effects of Composition and Microstructure, J. Am. Ceram. Soc. 78 (1995) 291-296. http://dx.doi.org/10.1111/j.1151-2916.1995.tb08799.x
[6] C. Baudín, J. Gurauskis, A.J. Sánchez-Herencia, et al., Indentation Damage and Residual Stress Field in Alumina-Y_2O_3-Stabilized Zirconia Composites, J. Am. Ceram. Soc. 92 (2009) 152-160. http://dx.doi.org/10.1111/j.1551-2916.2008.02813.x
[7] X.L. Wang, C.R. Hubbard, K.B. Alexander, et al., Neutron Diffraction Measurements of the Residual Stresses in Al_2O_3-ZrO_2 (CeO_2) Ceramic Composites, J. Am. Ceram. Soc. 77 (1994) 1569-1575. http://dx.doi.org/10.1111/j.1151-2916.1994.tb09758.x
[8] T. Adachi, T. Sekino, T. Nakayama, et al., Measurement of microscopic stress distribution of multilayered composite by X-ray stress analysis, Mater. Lett. 57 (2003) 3057-3062. http://dx.doi.org/10.1016/S0167-577X(02)01436-2
[9] G. De Portu, L. Micele, Y. Sekiguchi, et al., Measurement of residual stress distributions in Al_2O_3/3Y-TZP multilayered composites by fluorescence and Raman microprobe piezo-spectroscopy, Acta Mater. 53 (2005) 1511-1520. http://dx.doi.org/10.1016/j.actamat.2004.12.003
[10] J. Ruiz-Hervías, G. Bruno, J. Gurauskis, et al., Neutron diffraction investigation for possible anisotropy within monolithic Al_2O_3/Y-TZP composites fabricated by stacking together cast tapes, Scripta Mater. 54 (2006) 1133-1137. http://dx.doi.org/10.1016/j.scriptamat.2005.11.061
[11] J. Gurauskis, A. Sanchez-Herencia, C. Baudin, Joining green ceramic tapes made from water-based slurries by applying low pressures at ambient temperature, J. Eur. Ceram. Soc. 25 (2005) 3403-3411. http://dx.doi.org/10.1016/j.jeurceramsoc.2004.09.008
[12] J. Gurauskis, A.J. Sánchez-Herencia, C. Baudín, Al_2O_3/Y-TZP and Y-TZP materials fabricated by stacking layers obtained by aqueous tape casting, J. Eur. Ceram. Soc. 26 (2006) 1489-1496. http://dx.doi.org/10.1016/j.jeurceramsoc.2005.02.013
[13] A. Tsetsekou, C. Agrafiotis, A. Milias, Optimization of the rheological properties of alumina slurries for ceramic processing applications Part I: Slip-casting, J. Eur. Ceram. Soc. 21 (2001) 363-373. http://dx.doi.org/10.1016/S0955-2219(00)00185-0
[14] A. Coelho, TOPAS-Academic V5, Coelho Software, Brisbane, Australia. http://www.topas-academic.net/ (2012).
[15] J. Lewis, D. Schwarzenbach, H. Flack, Electric field gradients and charge density in corundum, α-Al_2O_3, Acta Crystallogr. Sec. A. 38 (1982) 733-739. http://dx.doi.org/10.1107/S0567739482001478
[16] M. Yashima, S. Sasaki, M. Kakihana, et al., Oxygen-induced structural change of the tetragonal phase around the tetragonal–cubic phase boundary in ZrO_2–$YO_{1.5}$ solid solutions, Acta Crystallogr., Sect. B: Struct. Sci. 50 (1994) 663-672. http://dx.doi.org/10.1107/S0108768194006257
[17] W. Pabst, G. Ticha, E. Gregorova, Effective elastic properties of alumina-zirconia composite ceramics-Part 3. Calculation of elastic moduli of polycrystalline alumina and zirconia from monocrystal data, Ceramics- Silikaty. 48 (2004) 41-48.
[18] T. D, Thermal expansion data. III. Sesquioxides, M_2O_3 with the corundum.and the A-, B- and C-M_2O_3 structures, Br. Ceram. Trans. J. 83 (1984) 92–98.
[19] H. Schubert, Anisotropic Thermal Expansion Coefficients of Y_2O_3-Stabilized Tetragonal Zirconia, J. Am. Ceram. Soc. 69 (1986) 270-271. http://dx.doi.org/10.1111/j.1151-2916.1986.tb07424.x
[20] M. Taya, S. Hayashi, A.S. Kobayashi, et al., Toughening of a Particulate-Reinforced Ceramic-Matrix Composite by Thermal Residual Stress, J. Am. Ceram. Soc. 73 (1990) 1382-1391. http://dx.doi.org/10.1111/j.1151-2916.1990.tb05209.x
[21] V. Hauk, Structural and residual stress analysis by nondestructive methods: Evaluation-Application-Assessment, Elsevier1997.

Residual Stresses 2016: ICRS-10 Materials Research Forum LLC
Materials Research Proceedings **2** (2016) 163-168 doi: http://dx.doi.org/10.21741/9781945291173-28

Combining Sectioning Method and X-Ray Diffraction for Evaluation of Residual Stresses in Welded High Strength Steel Components

Arne Kromm[1,a*], Michael Rhode[1,b], Benjamin Launert[2,c], Jonny Dixneit[1,d], Thomas Kannengiesser[1,e] and Hartmut Pasternak[2,f]

[1]Bundesanstalt für Materialforschung und -prüfung (BAM), Unter den Eichen 87, 12205 Berlin, Germany

[2]Chair of Steel and Timber Structures, Brandenburg University of Technology (BTU), Konrad-Wachsmann-Allee 2, 03046 Cottbus, Germany

[a]arne.kromm@bam.de, [b]michael.rhode@bam.de, [c]benjamin.launert@b-tu.de, [d]jonny.dixneit@bam.de, [e]thomas.kannengiesser@bam.de, [f]hartmut.pasternak@b-tu.de

Keywords: Weld Residual Stress, Sectioning Method, X-Ray Diffraction

Abstract. Residual stresses and distortions in welded I-girders for steel construction are relevant when evaluating the stability of steel beams and column members. The application of high strength steels allows smaller wall thicknesses compared to conventional steels. Therefore, the risk of buckling has to be considered carefully. Due to the lack of knowledge concerning the residual stresses present after welding in high strength steel components conservative assumptions of their level and distribution is typically applied. In this study I-girders made of steels showing strengths of 355 MPa and 690 MPa were welded with varying heat input. Due to the dimension of the I-girders and the complex geometry the accessibility for residual stress measurement using X-ray diffraction was limited. Therefore, saw cutting accompanied by strain gauge measurement has been used to produce smaller sections appropriate to apply X-ray diffraction. The stress relaxation measured by strain gauges has been added to residual stresses determined by X-ray diffraction to obtain the original stress level and distribution before sectioning. The combination of both techniques can produce robust residual stress values. From practical point of view afford for strain gauge application can be limited to a number of measuring positions solely to record the global amount of stress relaxation. X-ray diffraction can be applied after sectioning to determine the residual stresses with sufficient spatial resolution.

Introduction

This work is part of an ongoing research project on recommendations for the implementation of welding imperfections in the numerical analysis of welded thick-walled I-girders. One focus is put on welding residual stresses and their impact on the stability of specific components. Simulation approaches for capacity design are often quite conservative due to lack of knowledge concerning important input parameters [1]. For example, welding residual stresses are often assumed to be tensile up to the yield strength in the heat affected zone (HAZ). This is not the case especially when welding high strength steels which show a stress reducing effect due to solid state phase transformations [2-3]. The resulting extent of equilibrium compressive residual stresses is important for the ultimate load. But experimental proofs of spatially resolved residual stresses for welded thick-walled I-girders are difficult to obtain due the limited accessibility of such structures. Especially the HAZ adjacent to the weld restricts local measurements as it is covered by the parallel chords even in large scale I-girders. The sectioning method is an established technique to determine residual stresses in larger structures. Strain gauges are applied on the surface of the specimen. Sectioning by saw cutting leads to stress/strain relaxations which are recorded by strain gauges locally. Applying a large

Materials Research Proceedings **2** (2016) 163-168 doi: http://dx.doi.org/10.21741/9781945291173-28

number of strain gauges the stress relaxation can be calculated accurately. The spatial resolution is in principle limited by the geometry of the strain gauge. In the present work the sectioning method was combined with residual stress determination by X-ray diffraction (XRD). For this purpose the I-girders were cut just to the size until the areas of interest were accessible by XRD. The final residual stresses were determined by simply summing up the stresses found by sectioning and subsequent XRD.

Experimental

Welding. Full-scale specimens welded at a steelwork company were used. The high strength steels S355J2+N and S690QL (minimum yield strength: 355 MPa, respectively 690 MPa) were welded by gas metal arc welding (GMAW). The plate thicknesses were 20 mm for the chords and 10 mm for the web plate. Fillet welding was performed using strength matching filler materials in a single layer. Important parameters are shown in Table 1. Two heat inputs were applied varied mainly by the welding speed.

Table 1: Welding parameters

Sample	Material	Current [A]	Voltage [V]	Welding speed [cm/min]	Heat input [kJ/cm]
355L	S355J2+N	328	33	64	10.1 (low)
355H		329	33	40	16.3 (high)
690L	S690QL	343	33	64	10.6 (low)
690H		338	33	40	16.7 (high)

Sectioning. Sectioning was performed using a conventional band saw. The sections are indicated in Fig. 1. The cutting strategy was chosen from preliminary tests taking into account the strain gauge response and also literature recommendations [4]. The strains were recorded longitudinally and transversely to the welding direction. From these strains the stresses in longitudinal direction were calculated.

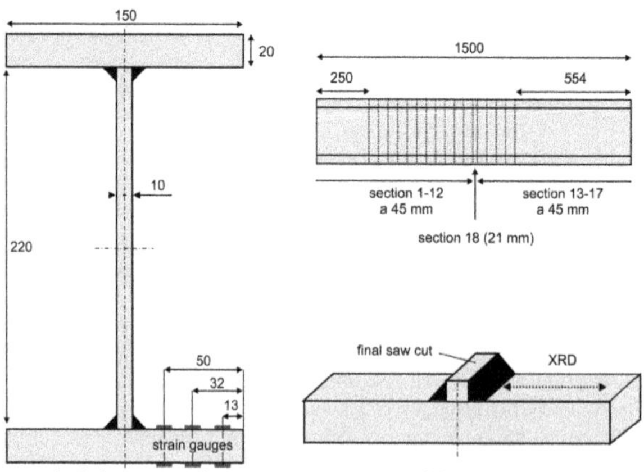

Figure 1: Schematic of sectioning, position of the strain gauges and measuring position for XRD

Materials Research Proceedings **2** (2016) 163-168 doi: http://dx.doi.org/10.21741/9781945291173-28

The first section had a length of 250 mm. From cut number two on the sections had lengths of 45 mm. Cut number twelve was located directly adjacent to the strain gauges. The second half of the beam was then cut beginning from the other side. Section number 13 had a length of 554 mm. Sections 14 to 17 showed again lengths of 45 mm. The resulting section number 18 which carried the strain gauges then resulted in a length of 21 mm. From this last section the chords were cut from the fillet and used for final residual stress measurement by XRD as shown in Fig 1.

X-ray diffraction. The residual stresses in longitudinal direction were determined using the $\sin^2\psi$-method. Measuring and evaluation parameters are given in Table 2. The measurements were performed along a line starting adjacent to the weld covering the HAZ and the base material. Up to 19 points were chosen along a distance of 50 mm (see Fig 1). The measurement was made at the welded side and also at back side of the chords at similar positions.

Table 2: Measuring and evaluation parameters for residual stress analysis by XRD

Focus	2 mm
Radiation	CrKα
Diffraction line	Ferrite: 211
E{211} and ν{211}	220.000 MPa and 0.28

Results and Discussion

Longitudinal residual stresses determined from the strain gauge response at the upper (weld) and lower side of the chords are shown in Fig. 2 and Fig. 3. The closer the strain gauge was located relative to the weld the higher the stresses were. Up to 100 MPa were found adjacent to the weld independent from the material or the heat input. Away from the weld the stresses turned into compression. The mid and the outer position showed about -50 MPa. The back side of the chords was characterized by quite constant compressive residual stresses between approximately -50 MPa to -100 MPa. With higher heat input (Fig. 2, right and Fig. 3, right) the stresses tended to show slightly higher absolute values. In the following the stresses obtained at three fixed positons by strain gauges were added to the spatially resolved stresses obtained from XRD. Ranges were defined for each strain gauge - indicated in Fig. 2 and Fig. 3 by the dashed lines - in which the stresses from the strain gauges were assumed to be constant.

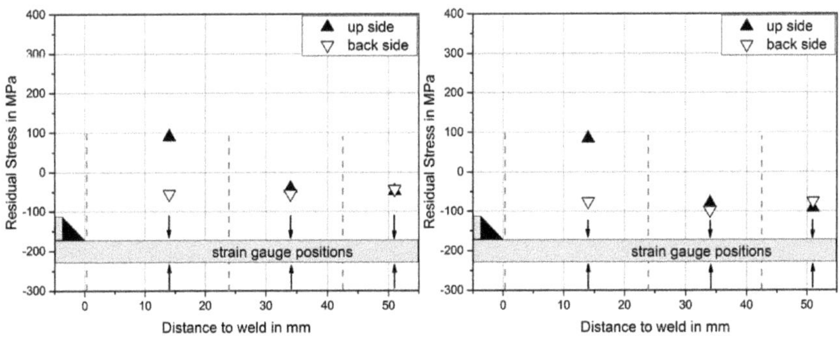

Figure 2: Longitudinal residual stresses determined by strain gauges during sectioning of samples 355L (left) and 355H (right)

Materials Research Proceedings **2** (2016) 163-168 doi: http://dx.doi.org/10.21741/9781945291173-28

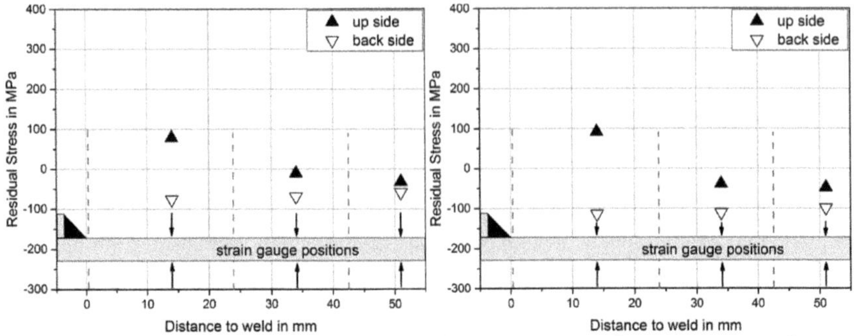

Figure 3: Longitudinal residual stresses determined by strain gauges during sectioning of samples 690L (left) and 690H (right)

Residual stresses determined from XRD after sectioning are shown for S355 in Fig. 4. The highest stresses were found close to the weld at the upper side of the chord. For the low heat input up to 300 MPa were registered. At high heat input up to 200 MPa remained after sectioning adjacent to the weld. Within the first 10 mm a gradient was present in the HAZ. Here the stresses turned into compression. Approximately -100 MPa were left in the base material. The back side of the chords showed stresses near 0 MPa in the range near the weld. The rest was characterized by stresses between 0 MPa and -100 MPa, which may stem from initial fabrication stresses present in the plates before welding. Note, that the plates were not subjected to stress relieve heat treatment prior to welding. No cutting influence is expected due to waterjet-cutting of the plates. The XRD results are always affected by the surface preparation which was critical at some points. This applied also in quality to the high strength steel S690 shown Fig. 5. The chord back side showed the same stress distribution also in quantity. Differences to S355 appeared at the upper side of the chord. Adjacent to the weld metal the tensile stress showed a dip. The stress was decreased here between 50 MPa and 150 MPa dependent on the heat input. Low heat input formed lower stresses at all. In the base material the stresses were around 0 MPa.

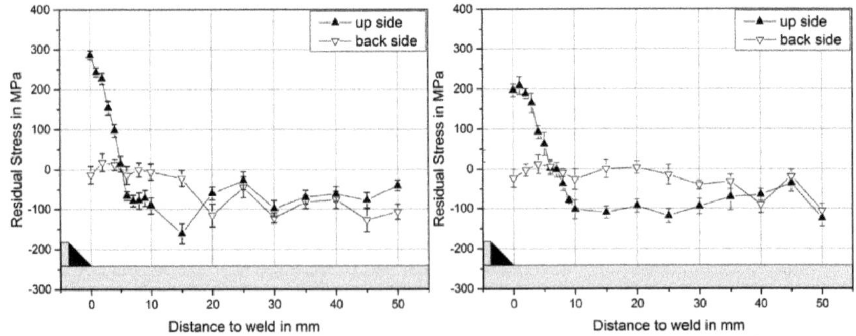

Figure 4: Longitudinal residual stresses determined by XRD after sectioning of samples 355L (left) and 355H (right)

Materials Research Proceedings **2** (2016) 163-168 doi: http://dx.doi.org/10.21741/9781945291173-28

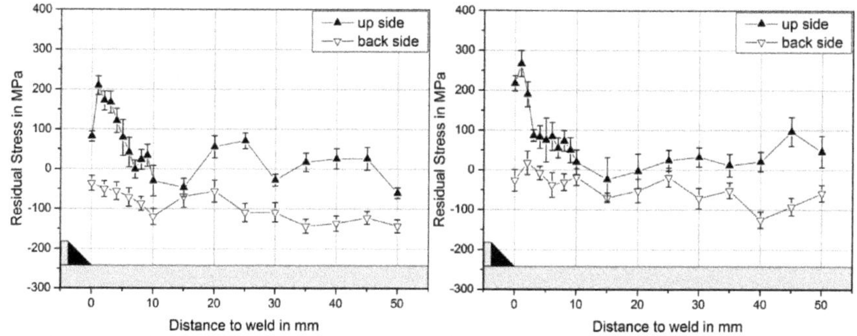

Figure 5: Longitudinal residual stresses determined by XRD after sectioning of samples 690L (left) and 690H (right)

After superposition of the stresses from sectioning and XRD the differences between HAZ and base material became higher. The highest tensile stresses up the yield strength were obtained for S355 welded with low heat input adjacent to the weld (see Fig. 6). The stresses remained tensile within a distance of 10 mm to the weld. With higher distance to the HAZ the stresses changed into compression up to -200 MPa. The stresses at the back side are in principle in compression showing the same absolute values as on the upper side. Also the back side of S690 was completely in compression up to -200 MPa (see Fig. 7). The upper side showed around 0 MPa far from the weld and HAZ. In the HAZ the highest tensile stresses were found around 2 mm away from the weld metal. In the transition to the weld the stresses were decreased from 300 MPa to about 150 MPa at low heat input. For the high heat input this decrease was less, from 360 MPa to 310 MPa. Related to the yield strength of the materials the higher stresses, up to the yield point, were present in the mild steel S355. The high strength steel S690 was stressed just to 52 % of the yield strength. As the geometry (restraint) of the samples was similar in each case the main reason was assumed to be the phase transformation behavior. Stress release by martensite formation is known to be effective in high strength steels [2-3].

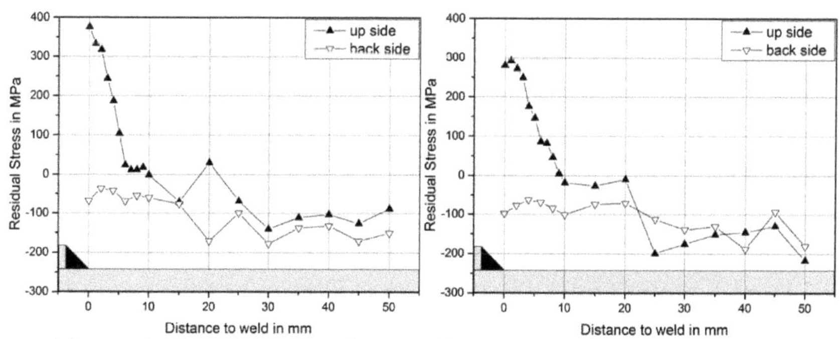

Figure 6: Longitudinal residual stresses determined from strain gauges and XRD of samples 355L (left) and 355H (right)

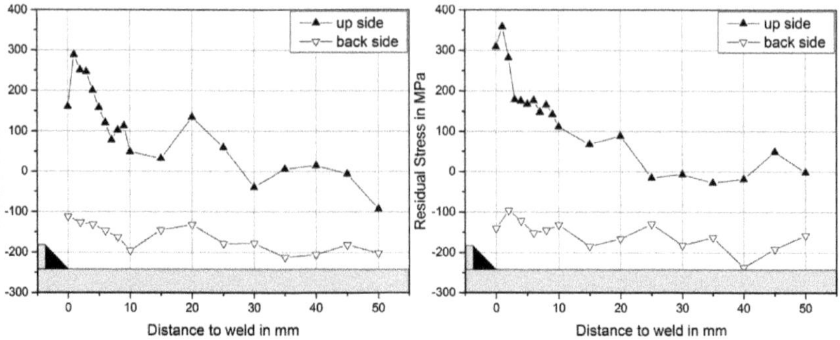

Figure 7: Longitudinal residual stresses determined from strain gauges and XRD of samples 690L (left) and 690H (right)

Summary

Residual stresses were determined on welded I-girders made of S355 mild and S690 high strength steel. A combination of the sectioning method and XRD was applied. The residual stresses released during sectioning showed a similar level independent from the material and the heat input applied. XRD has shown that the local residual stresses present in the HAZ are less affected by the heat input but were characteristic for the type of steel welded due to the impact of the solid state phase transformation. Residual stresses in the mild steel S355 reached the yield point, while the stresses in the high strength steel S690 just showed about half of that. Extensive use of strain gauges is just necessary to record the global stress relaxation by sectioning in order to have sufficient access by XRD. The latter can be applied then to ensure residual stress determination with high spatial resolution.

Acknowledgements

This IGF-project (IGF-project-no. 18104 BG) of the Research Association for Steel Application (FOSTA) was supported by the Federal Ministry for Economic Affairs and Energy by the AiF as part of the program for support of the cooperative industrial research (IGF) on the basis of a decision by the German Bundestag. Sincere thanks are given for this support and to the representing companies actively involved in the project board.

References

[1] H. Pasternak, B. Launert, T. Krausche, Welding of Girders with Thick Plates - Fabrication, measurement and simulation, Journal of Constructional Steel Research 115 (2015) 407-416. http://dx.doi.org/10.1016/j.jcsr.2015.08.037

[2] T. Nitschke-Pagel, H. Wohlfahrt, Residual Stresses in Welded Joints - Sources and Consequences, Mater. Sci. Forum 404-407 (2002) 215-226. http://dx.doi.org/10.4028/www.scientific.net/MSF.404-407.215

[3] J. Hensel, T. Nitschke-Pagel, K. Dilger, S. Schönborn, Effects of Residual Stresses on the Fatigue Performance of Welded Steels with Longitudinal Stiffeners, Materials Science Forum 768-769 (2014) 636-643. http://dx.doi.org/10.4028/www.scientific.net/MSF.768-769.636

[4] N. Tebedge, G. Alpsten, L. Tall: Residual-stress Measurement by the Sectioning Method, Experimental Mechanics, 13 (1973) 88–96. http://dx.doi.org/10.1007/BF02322389

Residual Stresses 2016: ICRS-10 Materials Research Forum LLC
Materials Research Proceedings **2** (2016) 169-174 doi: http://dx.doi.org/10.21741/9781945291173-29

Influence of Weld Repair by Gouging on the Residual Stresses in High Strength Steels

Arne Kromm[1,a *], René Schasse[2,b], Ping Xu[1,c], Tobias Mente[1,d] and Thomas Kannengiesser[1,e]

[1]Bundesanstalt für Materialforschung und -prüfung (BAM), Unter den Eichen 87, 12205 Berlin, Germany

[2]Schweißtechnische Lehranstalt Magdeburg GmbH, An der Sülze 7, 39179 Barleben, Germany

[a]arne.kromm@bam.de, [b]schasse@sl-magdeburg.de, [c]ping.xu@bam.de, [d]tobias.mente@bam.de, [e]thomas.kannengiesser@bam.de

Keywords: Repair Welding, Weld Residual Stress, Carbon Arc-Air Gouging

Abstract. Carbon arc-air gouging is a common technology when repairing defects in welded structures. Often this technique is applied in repeated cycles even on the same location of the joint. Due to the multiple heat input by gouging and subsequent re-welding, the residual stresses are strongly influenced. This can become crucial when microstructure and mechanical properties are adversely affected by multiple weld reparations. Knowledge about the relation of gouging and residual stresses is scarce but important when high strength steels, which are sensitive to residual stresses, are processed. The present study shows the effect of repair welding on a high strength steel structural element. The weld and the heat affected zone were subjected to multiple thermal cycles by gouging and subsequent repair welding. The residual stresses were determined by X-ray diffraction at different positions along the joint. The results showed that the residual stress level has increased by the repair cycles. This is most pronounced for the heat affected zone. Adapted welding procedures may prevent detrimental residual stress distributions.

Introduction

During fabrication of steel components, non-destructive testing for quality control is frequently applied. Emerging failures are often treated locally by repair welding. In this case, multiple repair cycles may be required to completely remove the failure. Standards do not limit the number of such repair cycles. Therefore, it is unknown how the number of repair cycles influences the mechanical properties and residual stresses. These stresses become important when high strength steels are applied, which are more sensitive to residual stresses due to their limited ductility.

For the removal of weld defects, carbon arc-air gouging is a commonly used method during fabrication. The advantage is the excellent failure finding ability due to the good accessibility even in the edges of a component. Carbon arc-air gouging uses a copper covered graphite electrode. Typical diameters are 6, 8 or even 10 mm. A positive polarity is used for the electrode when gouging steel. The specific current load reaches from 10.5 A/mm² for a 6 mm diameter electrode to 7.5 A/mm² for a 10 mm diameter electrode. The molten material is blown out of the weld by an air stream. The gouging depth is controlled by the angle of approach which is usually between 30° and 45° [1-2].

Residual stresses in repair welds are not reported in the literature up to now. Similar behavior may be expected from slit welds which are used to evaluate the cold cracking behavior of welds. The notch created by gouging is comparable to the slit used in specific tests [3-7]. Published data reveals that the residual stresses in such tests are influenced by slit characteristics, i.e. length [3-4] and position in the sample [8-9], and the stiffness (restraint intensity) of the test pieces [5]. Nevertheless, the number of gouging and respective repair welding cycles has not been investigated yet. Therefore, the present work focuses on the residual stress development considering multiple repair cycles.

Materials Research Proceedings **2** (2016) 169-174 doi: http://dx.doi.org/10.21741/9781945291173-29

Experimental

The sample for residual stress analysis was adopted from a typical detail in a real bridge construction (Fig. 1, left). The lower flange joint of the cross girders of an open bridge deck and the orthotropic plates are static sensitive joints. In service, the load acts transverse to the weld. The weld is under tensile stress. In case of railway bridges, this type of joint is therefore checked 100% by non-destructive testing.

In this study, the test sample consisted of two base plates acting as the lower flange of the main beam, namely the cross girder. The transverse stiffener, located on the root side of the weld, simulated the cross-girder web (Fig. 1, right).

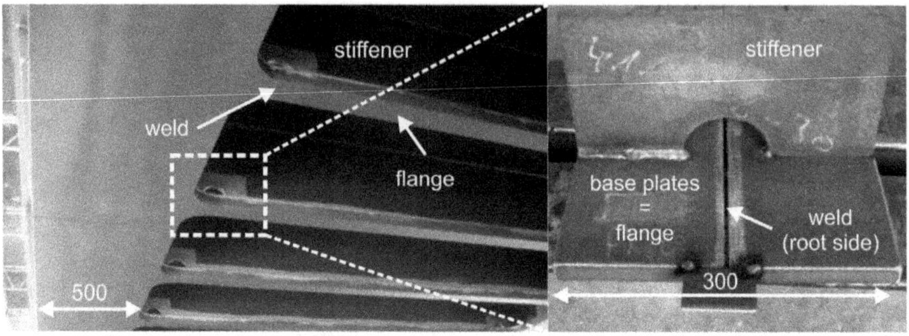

Figure 1: Weld detail in a bridge construction – cross girder joint (left), sample simulating the cross girder joint consisting of two base plates and a transverse stiffener on the root side (right)

Welding was performed according to typical parameters used during fabrication as indicated in Table 1. Two base materials with a thickness of 20 mm (base plates) and 10 mm (stiffener) were chosen. The high strength steels S355J2+N and S690QL (minimum yield strength: 355 MPa respectively 690 MPa) were welded using gas metal arc welding (GMAW). Matching filler materials G3Si and G69 4 M Mn3Ni1CrMo were used with 1.2 mm wire diameter, in accordance to ISO 14341 and ISO 16834-A.

Table 1: Welding parameters

Layer	Preheat/interpass temperature [°C]	Current [A]	Voltage [V]	Welding speed [cm/min]	Heat input [kJ/cm]
Root	120	250-260	28-29	35	12.45
Intermediate	200	300-310	31-32	42	13.73
Top	200	280-290	29-30	27	18.68

A carbon electrode with a diameter of 6 mm at a current of 380-400 A was applied for gouging in each case. It was performed over a length of 200 mm located on the top surface of the weld according to Fig. 3. During each repair cycle the weld seam was not removed completely. A small edge of the original weld seam with extent of 1-2 mm was left. After the initial welding cycle and between, the joints were allowed to cool down to ambient temperature. Gouging and subsequent repair welding was repeated up to five times. The depth of the gouged region was about 7 mm.

Materials Research Proceedings 2 (2016) 169-174 doi: http://dx.doi.org/10.21741/9781945291173-29

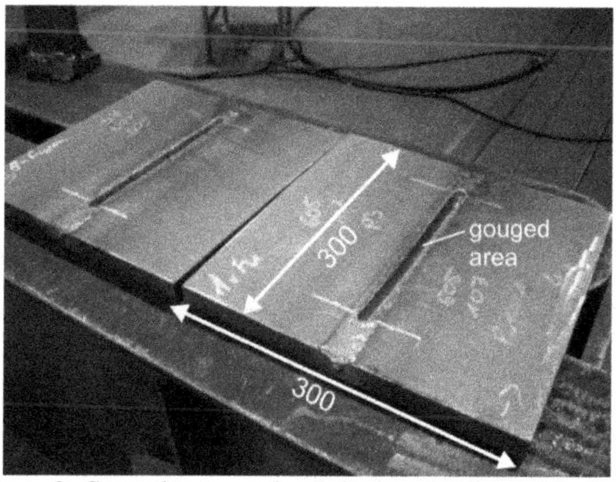

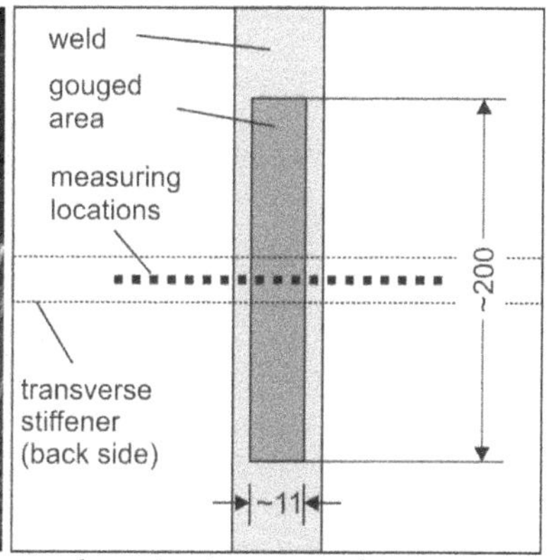

Figure 2: Gouged test samples (left), location for residual stress determination on sample top surface indicated by black dotted line (right)

The residual stresses were determined by X-ray diffraction using the sin²ψ-method [10]. Measuring and evaluation parameters are given in Table 2. The measurements were performed along the line located in the center on the top surface of the joint covering the weld, HAZ and base material. 25 points were chosen along a distance of 48 mm. Fig. 2 (right) shows the location and arrangement of the residual stress measurement. There are single locations in the weld and also HAZ where the residual stresses could not be determined due to microstructural reasons (nonlinear d vs. sin²ψ distributions).

Table 2: Measuring and evaluation parameters for residual stress analysis

Measuring mode	sin²ψ
Focus	2 mm
Radiation	CrKα
Diffraction line	Ferrite: 211
E{211}	211.000 MPa
ν{211}	0.3

Results and Discussion

Residual stresses found for the as welded condition in the samples of steel S355 (see Fig. 5 left) showed higher values for the longitudinal direction as compared to the transverse one. This accounts for the weld metal itself and also for the heat affected zone (HAZ) in the base material. The highest peak values may reach the yield point of the steel. The transverse direction is characterized only by moderate tensile stress levels not above 150 MPa. The stress level and distribution is mainly influenced by the restrained shrinkage of the weld and HAZ.

The S690 shows comparable residual stress characteristics (Fig. 5 right), although the highest stress peaks, which were found only in the HAZ, had values up to 250 MPa. The weld metal showed lower stresses, which were caused by the phase transformation to bainite and martensite [11]. Even compressive residual stresses may appear.

Materials Research Proceedings **2** (2016) 169-174 doi: http://dx.doi.org/10.21741/9781945291173-29

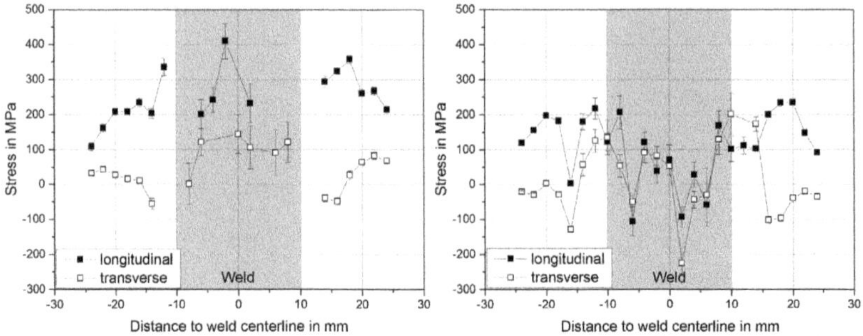

Figure 5: Longitudinal and transverse residual stresses in the as welded condition of S355 (left) and S690 (right)

After the first repair cycle the residual stresses in S355 were shifted to higher values, especially in the HAZ and the base material. The weld metal showed more or less the same characteristics as found for the as welded condition. In the HAZ particularly, the transverse residual stresses were affected by the repeated heat input due to the deposition of the weld metal. The same trend was observed for the higher strength material S690 (Fig. 6, right). While the weld metal was less affected, the HAZ and the base material showed elevated tensile residual stresses around 400 MPa. The stress distribution appeared asymmetric, due to manual welding and gouging. The main reason for the residual stress rise after the first repair cycle was the increase of the shrinkage restraint due to the notch resulting from gouging. Applying a number of repair cycles to the same weld led to even higher stress levels which reached the yield strength, particularly in case of S355 (Fig. 7, left). The weld metal of S690 was less affected, due to the stress reduction caused by the phase transformation, (Fig. 7, right). Nevertheless, the HAZ showed increased residual stress levels with repeated gouging.

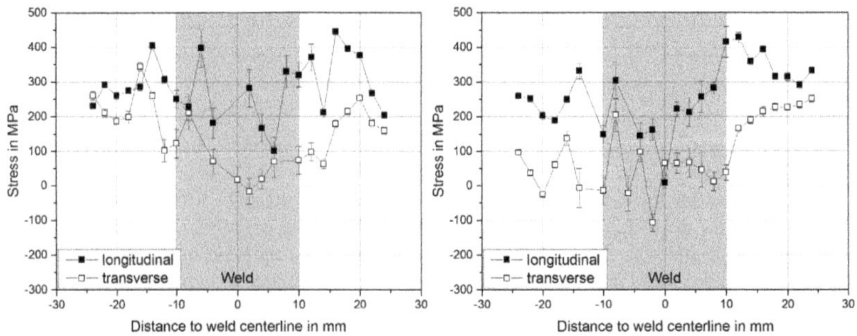

Figure 6: Longitudinal and transverse residual stresses after the 1st repair cycle of S355 (left) and S690 (left)

Materials Research Proceedings **2** (2016) 169-174 doi: http://dx.doi.org/10.21741/9781945291173-29

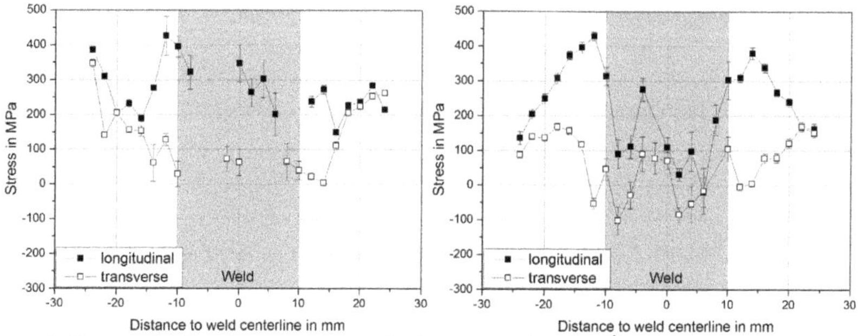

Figure 7: Longitudinal and transverse residual stresses after the 5th repair cycle of S355 (left) and S690 (right)

Adapting the welding procedure allowed control of the residual stress levels found after gouging and repeated deposition of the weld metal in the notch. Applying a Pilgrim-step sequence during welding (three steps) resulted in moderate stress levels longitudinal as well as transverse to the welding direction (Fig. 8). Even lower stresses than in the initial as welded condition were obtained. This applied for S355 as well as for S690. With shorter welds the shrinkage length was less. Therefore, longitudinal residual stresses were lower compared to the conventional welding procedure.

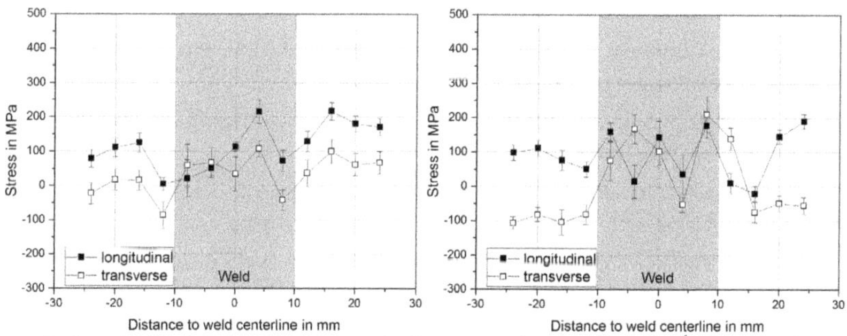

Figure 8: Longitudinal and transverse residual stresses after the 1st (left) 5th repair cycle (right) of S690 after applying back-step-repair welding.

Summary

Residual stresses were determined on welded samples of S355 and S690 high strength steel. The sample geometry represented a common detail used in bridge constructions. The residual stress state was analyzed by X-ray diffraction for the as welded condition and for subsequent first and fifth repair cycle. The residual stresses significantly increased after the first repair cycle and continued to increase with further weld repair cycles. The stress increase was more pronounced for the longitudinal direction. S355 was affected in both weld metal and HAZ, but the higher strength steel S690 was only influenced in the HAZ. The reason was the stress reduction due to solid state phase transformation acting in the weld metal, which was more prominent in case of S690. Elevated tensile residual stresses in the HAZ can become critical when the mechanical properties are degraded.

Materials Research Proceedings **2** (2016) 169-174 doi: http://dx.doi.org/10.21741/9781945291173-20

Applying a Pilgrim-step sequence during the repair welding procedure is an appropriate way to reduce the residual stresses significantly.

References

[1] V.S. Miltyutin, The effect of arc-air cutting on the quality of metal, Welding Production 26 (1979) 44-46.

[2] Doshchechkina et al., Changes in the Structure of the surface zone in the air-cutting of metals, Welding Production 27 (1980) 40-42.

[3] K. Satoh, Y. Ueda, H. Kihara, Recent Trend of Researches on Restrain Stresses and Strains for Weld Cracking, Transactions of JWRI (1972) 53-68.

[4] K. Satoh, An Analytical Approach to the Problem of Restraint Intensity in Slit Weld. Transactions of JWRI (1972) 69-76.

[5] C. Schwenk, T. Kannengiesser, M. Rethmeier, Restraint Conditions and Welding Residual Stresses in Self-Restrained Cold Cracking Test, in: S.A. David, T. DebRoy, J.N. DuPont, T. Koseki, H.B. Smartt (Eds.), Trends in Welding Research. Proceedings of the 8th International Conference, 2009, pp. 766-773.

[6] Y. Ueda, K. Fukuda, Y.C. Kim, R. Koki, Characteristics of Restraint Stress-Strain of Slit Weld in a Finite Rectangular Plate and the Significance of Restraint Intensities as a Dynamical Measure, Transaction of JWRI 11 (1982) 105-113.

[7] Y. Ueda, K. Fukuda, I. Nishimura, H. Iiyama, N. Chiba, Cracking in welded corner joints. Metal Construction and British Welding Journal 1 (1984) 30-34.

[8] R.D. Stout, R. Vasudevan, W. Pense, A Field Weldability Test for Pipeline Steels, Welding Journal 55 (1976) 89-94.

[9] R. Vasudevan, R.D. Stout, W. Pense, A Field Weldability Test for Pipeline Steels - Part II, Welding Journal 59 (1980) 76-84.

[10] E. Macherauch and P. Müller, Das $\sin^2\psi$ - Verfahren der röntgenographischen Spannungsmessung, Z. angew. Physik 13 (1961) 305-312 (in German).

[11] T. Nitschke-Pagel, H. Wohlfahrt, Residual Stresses in Welded Joints - Sources and Consequences, Mater. Sci. Forum 404-407 (2002) 215-226. http://dx.doi.org/10.4028/www.scientific.net/MSF.404-407.215

Residual Stresses 2016: ICRS-10 Materials Research Forum LLC
Materials Research Proceedings **2** (2016) 175-180 doi: http://dx.doi.org/10.21741/9781945291173-30

Residual Stress States After Piezo Peening Treatment at Cryogenic and Elevated Temperatures Predicted by FEM Using Suitable Material Models

Alexander Klumpp[1,a*], Mehmet Tamam[1,b], Franziska Lienert[1,c], Stefan Dietrich[1,d], Jens Gibmeier[1,e] and Volker Schulze[1,f]

[1]Karlsruhe Institute of Technology, Kaiserstraße 12, 76131 Karlsruhe, Germany

[a]alexander.klumpp@kit.edu, [b]mehmet.tamam@student.kit.edu, [c]franziska.lienert@kit.edu, [d]stefan.dietrich@kit.edu, [e]jens.gibmeier@kit.edu, [f]volker.schulze@kit.edu

Keywords: Mechanical Surface Treatment, Machine Hammer Peening, Piezo Peening, Residual Stresses, Finite Element Simulation, Temperature Variation, Material Modeling

Abstract. Piezo peening is a recently developed mechanical surface treatment and belongs to machine hammer peening technologies. It has proven suitable to generate a wide range of compressive residual stress profiles and penetration depths depending on the parameters chosen for the process. By this means, greatly enhanced fatigue behavior could be achieved. In this study, the residual stress states after modified piezo peening treatments were determined experimentally and by 3D finite element (FE) simulation. Low alloy steel AISI 4140 was treated at ambient, cryogenic and elevated temperatures. Residual stresses were determined experimentally using the $\sin^2(\psi)$ method combined with subsequent electrolytic surface layer removal. The FE simulation makes use of a material model, which is capable of describing strain-rate and temperature dependent material behavior as well as the Bauschinger effect and allows for the emulation of surface layer removal for proper residual stress determination. Thus, the applicability of appropriate material modeling to predict experimentally determined residual stress profiles could be demonstrated.

Introduction

Due to the generation of smooth surfaces together with compressive residual stresses and work hardening, machine hammer peening (MHP) has become a crucial process step, e.g. in the fabrication of molds and dies. Mostly utilizing electromagnetic, pneumatic and hydraulic transducers, today's MHP processes allow for the generation of specific surface characteristics [1].

A recently developed MHP technology utilizing a piezo-electric power transducer is piezo peening [2]. It has been applied to the quenched and tempered steel AISI4140, where the fatigue strength could be greatly improved. This was found to be mainly due to the introduction of near-surface compressive residual stress fields. It was shown that residual stress profiles can widely be varied depending on the applied process parameters [2]. During the last decades, computational mechanics such as the finite element (FE) method has been applied extensively to understand process-property-relationships. Therefore an approach towards the FE simulation of piezo peening has been presented in a recent publication [3], showing good agreement between numerical and experimental results. Since strain-rate and temperature dependent material modeling has been applied in simulation, it is particularly interesting to investigate the influences of temperature variation upon flow stresses and residual stress profiles. On the one hand, potential influences of cryogenic and elevated temperatures on the residual stress profiles after piezo peening are explored experimentally, since the effect of temperature on residual stress generation has not yet been investigated for this process. Furthermore, tendencies regarding residual stress maxima and penetration depths in experiment and simulation are compared, such that the obtained results serve as validation for the applied material model.

Materials Research Proceedings **2** (2016) 175-180 doi: http://dx.doi.org/10.21741/9781945291173-30

Process Description and Experimental Setup

A schematic drawing of the utilized piezo peening device [2] is shown in Fig. 1 (left). Specimens are mounted to a linear x-y-slide to be peened by the spherical hammer head. The latter is driven by the piezo actuator with a specific frequency (f), stepover distance (s) and stroke (h), generating "impulsive regular" [4] deformation of the specimen surface. By means of the x-y-slide, the work piece surface can be treated using different patterns, such as meanders. The soft bearing on top is used to control the contact properties. The hammer head is lifted off the specimen surface after each stroke. In Fig. 1 (right), the experimental setup for piezo peening at cryogenic temperatures is shown. Specimens were cooled down to -180 °C using liquid N_2 flowing through a brass block specimen holder. For piezo peening at elevated temperatures, the same block was heated by heating rods, thus achieving temperatures of +200 °C. In each case a two-point temperature controller was used.

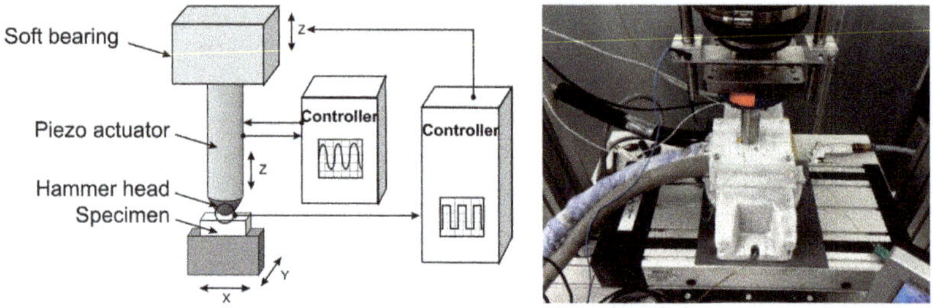

Figure 1: Piezo peening process (schematic, left) [2] and test bench for cryogenic peening at KIT (right)

Low alloy steel AISI 4140 with hardness of 430 HV1 was used for the investigations. The chemical composition is shown in Table 1. The material was austenitized at 850 °C for 20 minutes, oil-quenched and then tempered at 450 °C for 120 minutes. Afterwards, it was furnace-cooled to room temperature.

Table 1: Chemical composition of AISI 4140

Chemical composition (wt.-%)						
Fe	C	Cr	Mo	Mn	Si	Ni
Base	0.425	1.011	0.222	0.803	0.252	0.101

Flat specimens with a thickness of 4 mm (Fig. 2) were used throughout the experimental investigations. piezo peening was carried out on an area of 10*10 mm^2. Near-surface residual stresses were evaluated in the center of the peened area, using Cr-Kα radiation on {211}-α-ferrite diffraction line at 2θ = 156.4°. X-ray stress analysis was carried out according to the $\sin^2(\psi)$ method [5], using E{211} = 220 GPa and ν{211} = 0.28 as Young's modulus and Poisson's ratio, respectively. Residual stress depth profiles were determined by incremental electrolytic layer removal. Comparability of experiment and simulation was established by FE layer removal simulation. A detailed description of the experimental setup and residual stress measurements can be found in [2].

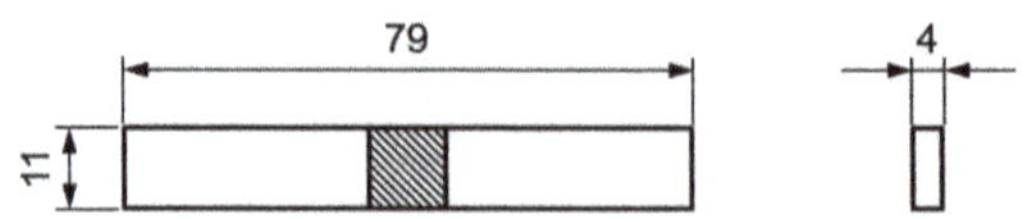

Figure 2: Specimen geometry with peened area (hatched)

Finite Element Simulation

The commercially available equation solver Abaqus/Explicit was used for stress computation due to its suitability for nonlinear, dynamic problems. FE geometry and mesh were modeled using parameterization. Material data were implemented as user material subroutine (VUMAT) Fortran77 code. A 2*2*2 mm^3 sized cuboid consisting of approx. 860,000 hexahedron elements of type C3D8R was used as simulation model and is shown in Fig. 3 (left).

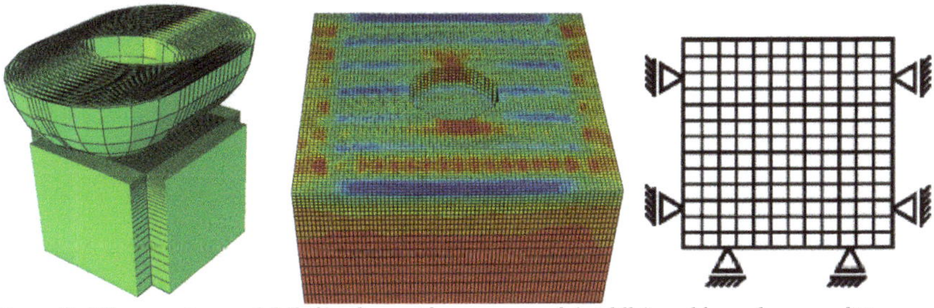

Figure 3: FE geometry model (left), element layer removal (middle) and boundary conditions (right)

For proper resolution, mesh refinement was introduced towards the impact surface. Semi-infinite elements of type CIN3D8 were used at the laterals to dampen the stress waves induced by the impact of the hammer head on the surface. Hammer heads were modeled as rigid half-spheres (Fig. 3, left), which is valid if the yield strength ratio of hammer head and work piece exceeds a value of 2.0 [6]. Eight consecutive impacts were simulated on 1.8 mm length, using an imposed sinusoidal hammer head movement and an individual hammer head model for each stroke. General contact with a friction coefficient of μ = 0.4 was defined between hammer head and work piece. A post-computation Python script was used to evaluate the residual stress state. Residual stresses were determined in a circular area; cf. Fig. 3 (middle). Stress redistribution due to electrolytic layer removal was realized by successively setting Young's modulus to zero in near-surface layers. After each layer, the mean value of superficial residual stresses within the area was determined. By this means, comparability between experiment and simulation was established. Floating boundary conditions as illustrated in Fig. 3 (right) were chosen to impede horizontal movement of the laterals and vertical movement of the underside. Furthermore, variable mass scaling, as described in [3], was applied.

The physically based, elastic-viscoplastic material model for BCC materials presented in [3] was used to describe the material behavior in the present strain, strain-rate and temperature range. Changes in strain-rate and temperature mainly affect the yield strength, whereas hardening curves remain similar in a wide range of temperature and strain rate [7]. This effect is due to thermally activated deformation [8] and allows for modeling a viscoplastic "overstress" independently of work hardening [7]. Combined nonlinear isotropic and kinematic hardening as proposed by [9] was applied to account for the Bauschinger effect. It was shown in literature that including kinematic hardening leads to a more realistic description of simulations of shot peening and machine hammer peening [3, 9, 10].

The framework of small strains was adopted, thus assuming additive decomposition of the strain tensor in elastic and plastic parts. The consistent viscoplasticity approach [11] was used together with the von Mises yield criterion, which reads

$$f \equiv J_2\,(\boldsymbol{\sigma}' - \xi) - R - \sigma_G - \sigma^* = 0, \tag{1}$$

Materials Research Proceedings **2** (2016) 175-180 doi: http://dx.doi.org/10.21741/9781945291173-30

where ξ and R denote the kinematic and the isotropic hardening variables, respectively. Two additive nonlinear kinematic variables and one nonlinear cyclic isotropic variable as proposed by [12] and [13], respectively, were applied. A full description of the constitutive equations can be found in [7] and [3]. The variables σ_G and σ^* denote the athermal and thermal parts of the flow stress, respectively [8]. The former slightly depends on temperature, whereas the latter shows strong dependence on temperature and strain-rate. The constitutive formula developed by [8] was used to reproduce the experimentally obtained flow stresses. As can be seen from Fig. 4, the thermal flow stress parts can be increased by higher strain-rates, and even more pronounced, by lower temperatures. Three temperatures were chosen for the investigations: cryogenic (-180 °C); ambient (20 °C) and warm (200 °C). Young's modulus was determined a priori and set to 225, 210 and 194 GPa, respectively. Thus, flow stresses during piezo peening at varying temperatures, where strain-rates about 100/s apply at a peening frequency of 500 Hz, are adequately reproduced. The full constitutive parameter set can be found in [3].

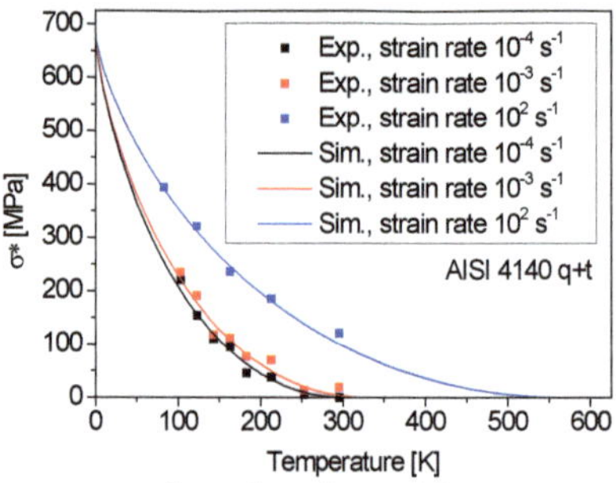

Figure 4: Strain-rate and temperature dependent thermal flow stress part [3]

Results and Discussion

Experimental standard process parameters are summarized in Table 2. Besides the temperature, either stepover distance or stroke was varied while all other process parameters remained unchanged throughout the investigations.

Table 2: Standard process parameters

Frequency	Hammer head material	Hammer head diameter	Stroke	Feed rate
500 Hz	WC-Co	5 mm	36 µm	50 mm/s

Fig. 5 shows transverse (left) and longitudinal (right) residual stress profiles determined experimentally after piezo peening at different temperatures and stepover distances with 36 µm stroke. At all temperatures, residual stress penetration depths increase by reducing stepover distance, which is a usually observed effect [2]. Furthermore, three effects seem to apply: A slight increase in penetration depth could possibly be assumed for cryogenic temperatures, regardless of stepover distance and measurement direction. The effect of stepover distance on residual stress penetration depth seems to be more pronounced at cryogenic and less pronounced at elevated temperatures. Furthermore, cryogenic temperatures appear to shift residual stress maxima below the surface, while maximum residual stresses can be found on the surface after warm peening. This effect can probably be explained by an increasing dominance of Hertzian pressure caused by the cooling to cryogenic temperatures, which is typical for hard work pieces [14].

Materials Research Proceedings **2** (2016) 175-180 doi: http://dx.doi.org/10.21741/9781945291173-30

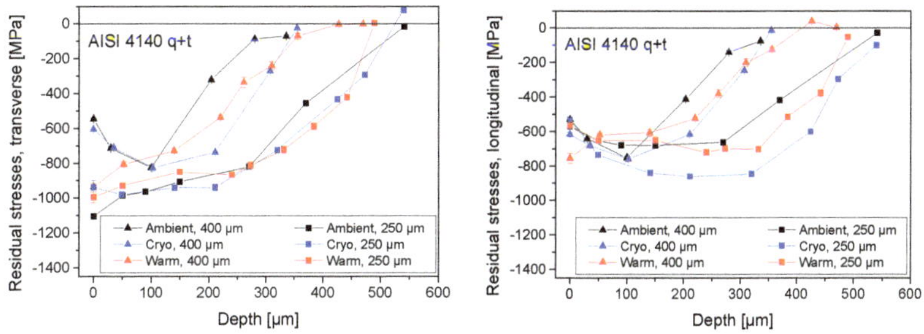

Figure 5: Experimentally determined transverse (left) and longitudinal (right) residual stress depth distributions for different temperatures and stepover distances

The question arises whether such effects can be reproduced by the strain-rate and temperature dependent FE simulation. A comparison of experimentally and numerically determined residual stress profiles with varying stroke (36 and 48 µm) and stepover distance (100 and 250 µm) is shown in Fig. 6 (left). For the sake of clarity, only transverse residual stresses are considered here.

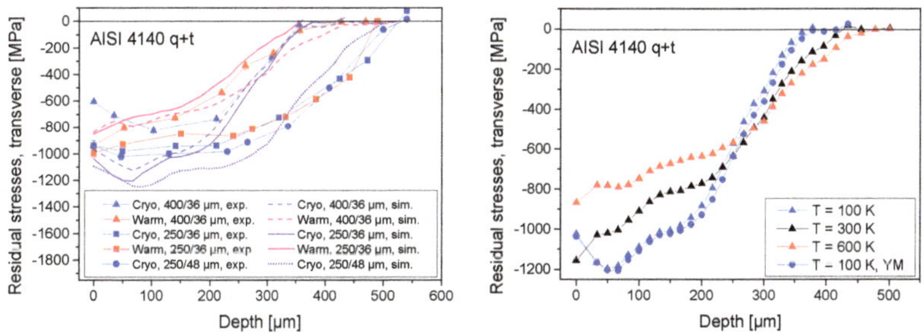

Figure 6: Comparison of residual stress depth distributions (left) and simulation properties (right)

As can be seen from Fig. 6 (left), generally satisfactory agreement was achieved between simulation and experiment. However, a certain discrepancy regarding the penetration depths could be seen for 250 µm stepover distance. Cryogenic peening with 48 µm stroke hardly affects residual stresses induced by the peening experiment, which is probably caused by the limited rigidity of the test bench. Despite the underestimation of residual stress penetration depths, the experimentally observed shift of residual stress maxima below the surface at cryogenic temperatures can also be recognized in the simulations. Contradictory to the experiment, numerically obtained residual stress penetration depths after warm peening increase for each stepover distance. However, this effect was shown to be an intrinsic property of strain-rate dependent material models [15] and is illustrated in Fig. 6 (right): Higher temperatures and lower strain-rates lead to lower maximum residual stresses and higher residual stress penetration depths and vice versa. Young's modulus, varied from 225 to 210 GPa (denoted by "YM" in Fig. 6, right), has only limited influence upon residual stress generation.

An adequate constitutive description of flow stresses is decisive for any simulation of residual stress generation. Yet, the prevailing deformation mechanisms are expected to be more complex than modeled in this investigation. For instance, the process was modeled adiabatically, which may not be entirely valid, especially for larger work pieces. Those remaining issues in material modeling for simulations of MHP processes are therefore still subject to thorough investigations of ongoing work.

Summary
Experimentally and numerically obtained residual stress profiles after piezo peening at temperatures between -180 and +200 °C were presented in this paper. Agreement between simulation and experiment for different parameters is found applying a strain-rate and temperature dependent material model in the simulations. Contrary tendencies concerning residual stress penetration depths were found, but may be attributed to the material model definition. More attention should be devoted to a real process description, taking into account the prevailing deformation mechanisms and process conditions.

Acknowledgement
Results have been derived within the international ERANET-CORNET research project *HaPTec* "Further qualification of machine hammer peening technology for industrial use".

References
[1] V. Schulze, F. Bleicher, P. Groche, Y. B. Guo, Y. S. Pyun, Surface Modification by Machine Hammer Peening and Burnishing, submitted to Proc. CIRP (2016). http://dx.doi.org/10.1016/j.cirp.2016.05.005
[2] F. Lienert, J. Hoffmeister, V. Schulze, Residual Stress Depth Distribution after Piezo Peening of Quenched and Tempered AISI 4140, MSF, vol. 768-769 (2013), pp. 526–533. http://dx.doi.org/10.4028/www.scientific.net/MSF.768-769.526
[3] A. Klumpp, F. Lienert, S. Dietrich, V. Schulze, Residual Stresses after Piezo Peening Treatment predicted by FEM Simulation, in Proc. IDE (2015), pp. 105–115.
[4] A. Klumpp, J. Hoffmeister, V. Schulze, Mechanical Surface Treatments, in Proc. ICSP 12 (2014), pp. 12–24.
[5] E. Macherauch, P. Müller, Das sin2 psi-Verfahren der röntgenographischen Spannungsmessung, Zeitschrift für angewandte Physik, vol. 13, no. 7 (1961), p. 305.
[6] K.-I. Mori, K. Osakada, N. Matsuoka, Rigid-plastic finite element simulation of peening process with plastically deforming shot, JSME Int. J. (A), vol. 39, no. 3 (1996), pp. 306–312.
[7] G. Z. Voyiadjis, F. H. Abed, A coupled temperature and strain rate dependent yield function for dynamic deformations of bcc metals, Int. J. Plas., vol. 22, no. 8 (2006), pp. 1398–1431. http://dx.doi.org/10.1016/j.ijplas.2005.10.005
[8] U. F. Kocks, A. S. Argon, M. F. Ashby, in: Progress in materials science, Thermodynamics and Kinetics of Slip, vol. 19 (1975), pp. 110–170.
[9] M. Klemenz, V. Schulze, O. Vöhringer, D. Löhe, Finite Element Simulation of the Residual Stress States after Shot Peening, MSF, vol. 524-525 (2006), pp. 349–354. http://dx.doi.org/10.4028/www.scientific.net/MSF.524-525.349
[10] E. Rouhaud, A. Ouakka, C. Ould, J. L. Chaboche, M. François, Finite elements model of shot peening, effects of constitutive laws of the material, in Proc. ICSP 9 (2005), pp. 107–112.
[11] A. Carosio, K. Willam, G. Etse, On the consistency of viscoplastic formulations, Int. J. of Solids and Structures, vol. 37, no. 48-50 (2000), pp. 7349–7369. http://dx.doi.org/10.1016/S0020-7683(00)00202-X
[12] J.-L. Chaboche, Time-independent constitutive theories for cyclic plasticity, Int. J. Plasticity, vol. 2, no. 2 (1986), pp. 149-188. http://dx.doi.org/10.1016/0749-6419(86)90010-0
[13] D. Lee, F. Zaverl, A generalized strain rate dependent constitutive equation for anisotropic metals, Acta Metallurgica, vol. 26, no. 11 (1978), pp. 1771–1780. http://dx.doi.org/10.1016/0001-6160(78)90088-3
[14] V. Schulze, Modern mechanical surface treatment: states, stability, effects, first ed., John Wiley & Sons, Weinheim, 2006.
[15] S. A. Meguid, G. Shagal, J. C. Stranart, 3D FE analysis of peening of strain-rate sensitive materials using multiple impingement model, Int. J. of Impact Eng., vol. 27, no. 2 (2002), pp. 119–134. http://dx.doi.org/10.1016/S0734-743X(01)00043-4

Residual Stress Analysis in Injection Moulded Polycarbonate Samples

Arnaud Magnier[1a*], Berthold Scholtes[1b], Thomas Niendorf[1c]

[1]Institute of Materials Engineering, University of Kassel, Germany

[a]magnier@uni-kassel.de, [b]scholtes@uni-kassel.de, [c]niendorf@uni-kassel.de

Keywords: Residual Stress, Plastics, Polycarbonate, Hole Drilling Method, Ring-core Method, Injection Moulding

Abstract. The current paper presents results of residual stress measurements in injection moulded polycarbonate samples, which have been processed in various ways to introduce different residual stress states. The hole drilling as well as the ring-core method were used and methodological developments as compared to measurement procedures applied on metallic samples are outlined. In this context the time dependent viscoelastic behaviour of the investigated material as well as temperature fluctuations during testing are of high importance. It is demonstrated that manufacturing parameters, i.e. mould temperature and injection rate, have a significant impact on the resulting residual stress states. A frame made of aluminium was used to induce pronounced tensile residual stresses in the sample by preventing shrinkage. Holes of different diameters were drilled in order to get information at different depths from the surface.

Introduction

Injection moulding is a widely used manufacturing process for components made of plastics. The process comprises three stages: filling, packing and cooling. Each stage affects the materials properties of the moulded product and contributes to the formation of residual stresses. Many investigations deal with their analysis by testing and/or simulation [1]. As residual stresses can cause warpage and shrinkage, they are mostly undesired. However, it was also demonstrated that they can have positive effects on the materials properties. Compressive residual stresses improve the notched Izod impact strength [2] as well as slow down crack propagation [3]. Nevertheless, residual stresses are always superimposed to external stresses, and, thus, it is of crucial importance to predict them. Although simulation methods have been steadily improved, predictions have to be confirmed by measurements. The hole drilling method enables measuring depth gradients of plane stress states locally. Thereby, the hole drilling method can be used for samples with a minimum curvature of 3.5 times the diameter of the hole for an uni-axial stress state [4], while a minimal curvature of 6 times the diameter is recommended for multi-axial stress states [5]. However, plastics are challenging materials for mechanical stress measurement methods, as they are highly influenced by temperature fluctuations and behave like a visco-elastic material [6, 7]. In the current work it will be shown that injection rate and mould temperature significantly influence the absolute values of residual stresses in polycarbonate samples. Due to a hybrid structure made of polycarbonate and aluminium, which has been designed for the current work, it is demonstrated that the hole drilling method is appropriate to measure compressive as well as tensile residual stresses in the polycarbonate. By drilling holes with diameters of 2 mm and 4 mm and milling a circumferential slot with an inner diameter of 14 mm, it is revealed that the results are reproducible and eventually information about residual stresses is obtained at different depths following various treatments.

Materials and experimental set up

Four samples with the dimensions of 60 mm × 155 mm × 4 mm were prepared for the current work. All samples were injection moulded using Polycarbonate (Makrolon 2805). The injection moulding

Materials Research Proceedings **2** (2016) 181-186 doi: http://dx.doi.org/10.21741/9781945291173-31

parameters are given in Table 1. The material is characterized by a Young's modulus of about 2.7 GPa and a measured Poisson´s ratio of 0.4.

Table 1: Injection moulding parameters of the different samples

	Sample 1	Sample 2	Sample 3	Sample 4
Injection Rate (mm/s)	1000	100	100	100
Mould Temperature (°C)	30	80	80	80
Packing Pressure (MPa)	50	50	50	50
Melt Temperature (°C)	290	290	290	290
Additional Information			Annealed	With aluminium frame (Fig. 1)

Process parameters of sample 1 were employed to manufacture a sample showing high residual stresses. A high injection rate of about 1000 mm/s, which should promote high flow-induced residual stresses [1], and a low mould temperature of about 30°C were chosen. For the other samples, the mould temperature was increased to 80°C and the injection rate reduced to 100 mm/s. The third sample was annealed for 6 hours at 110°C after moulding in order to release residual stresses. As the thermal stresses were expected to be the main mechanism inducing residual stresses [1], compressive residual stresses were expected near the surface. In order to generate tensile residual stresses near the surface, the fourth sample was moulded in an aluminium frame, which prevents shrinkage of the polycarbonate in the longitudinal direction. Thus, tensile residual stresses in the longitudinal direction were expected at point A in Fig. 1. At the outer parts of the sample (Point B in Fig. 1) shrinkage in the transversal direction is also prevented, as the width of polycarbonate, i.e. the sample dimension in transversal direction, is bigger at the outer part than in the middle of the sample. Thus, tensile residual stresses in the transversal direction are expected at point B. Because of the equilibrium of residual stresses, compressive residual stresses are then expected in the transversal direction in the middle of the sample (Point A in Fig. 1).

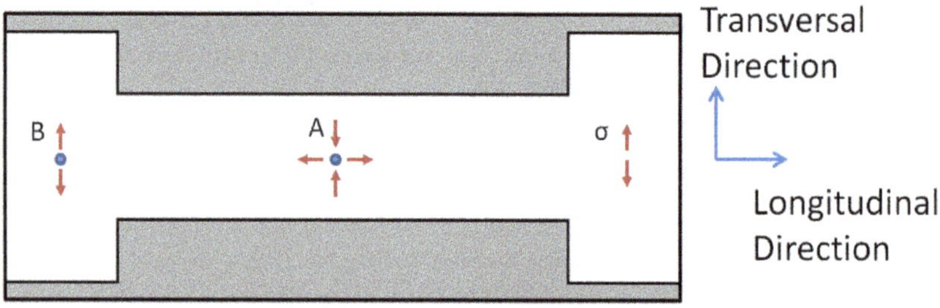

Figure 1: Polycarbonate sample (white) with an aluminium frame (grey). The thickness of the frame is 4 mm. The red arrows highlight the expected orientation of the residual stresses. See text for details.

In order to measure residual stresses, the hole drilling method was applied. Measurements were carried out at point A in Fig. 1, using the standard procedure [9], and following the recommendations in [6, 7, 8]. Calibration coefficients correlating residual stresses and released strains were calculated by Finite Element Analyses (FEA). The reader is referred to the aforementioned studies for details including guidelines detailing strain gauge installation and analysis of the time dependant measured strain signals. The signal is in fact strongly influenced by temperature fluctuations and the visco-elastic behaviour of the polycarbonate sample during and after drilling. In order to address these

Materials Research Proceedings **2** (2016) 181-186 doi: http://dx.doi.org/10.21741/9781945291173-31

issues the authors developed a procedure that minimizes effects. Findings will be reported elsewhere [6]. For the hole drilling method, two measurements were made for each sample: one with a 2 mm diameter blind hole at depth increments of 0.04 mm, and the second one with a 4 mm diameter blind hole at depth increments of about 0.08 mm. The ring-core method was also applied. The internal diameter of the circumferential slot was 14 mm and the external diameter was about 17 mm. The depth increments were 0.2 mm. Those three experiments were conducted complementary on each of the tested samples to gain information about residual stresses at different depths.

Experimental Results

The results of the residual stress measurements and the corresponding released strains are shown in Figs. 2 and 3. In Fig. 2 the results are arranged according to the measurement technique applied. The graphs a and b depict results obtained by the hole drilling method with a 2 mm diameter hole; the graphs c and d show respective data obtained by the hole drilling method with a 4 mm diameter hole; the graphs e and f finally depict results from the ring-core method. The curves in Fig. 2 are designated with the production parameters of the specimen (mould temperature and injection rate) or special treatments (annealed or with aluminium frame). For the purpose of clarity, the released strains are only highlighted by the maximum measured strain from the strain gauge rosette, and only the principal residual stresses are shown. For sample 4 both principal stresses are shown in red in Fig. 2. The samples 1, 2 and 3 almost show rotationally symmetric stress fields; therefore, only one principal stress is depicted in Fig. 2. The small difference between maximum and minimum residual stress is shown in Fig. 3. The maximum principal stress is exactly in the longitudinal direction (cf. Fig. 1). In Fig. 3 results are summarized differently. Graphs a and b highlight residual stress profiles of sample 4, which was moulded in an aluminium frame; graphs c and d depict data for the sample with a mould temperature of 30 °C; graphs e and f show residual stresses of the annealed sample.

Fig. 2b reveals that all samples have a compressive residual stress maximum at a depth of approximately 0.3 mm. This can also be found in Fig. 2d. As expected, the annealed sample shows the lowest amount of residual stress. If not annealed the sample shows a maximum compressive residual stress of about -3.3 MPa. Reducing the mould temperature to 30 °C and increasing the injection rate to 1000 mm/s results in a 40% higher compressive stress of about -4.6 MPa. The sample with the aluminium frame shows the highest absolute value of residual stresses. It reaches compressive stresses of about – 6 MPa and an average tensile stress of about 3 MPa, which are in the transversal and longitudinal direction, respectively (cf. Fig. 1). Similar residual stress profiles with constant tensile stress in the middle of the sample are reported in literature [1]. The result of the ring-core method shown in Fig. 2f seems to be less accurate. This can be attributed to the geometry of the model used in FEA to calculate the calibration coefficients. The sample was actually modelled as a 60 mm diameter disc. Modelling based on a 200 mm diameter disc provides slightly differing coefficients and, thus, affects results especially at the depth of 3.4 mm, where the result was so much diverging that it could not be shown. However, this difference is not sufficient for explaining the divergence at the depth of 2.4 mm, where the absolute value of the residual stress is strongly increasing. As this increase is also seen for the stress free sample, this result doesn´t seem to be reliable. Concerning sample 4, its aluminium frame wasn´t modelled in the simulation. This shouldn´t influence the results obtained by the hole drilling method significantly, as the hole diameter is small, but surely influences the ring-core measurement having a much larger measuring diameter. Consequently, in Fig. 2f the ring-core measurement was only evaluated up to a depth of 2 mm, as from here on errors occurred due to new residual stress equilibrium and the weak adhesion between polycarbonate and aluminium. Still, Fig. 3 shows a good agreement between the measurements made employing the hole drilling method with a 2 mm diameter and 4 mm diameter independent of the condition of the tested sample.

Materials Research Proceedings **2** (2016) 181-186 doi: http://dx.doi.org/10.21741/9781945291173-31

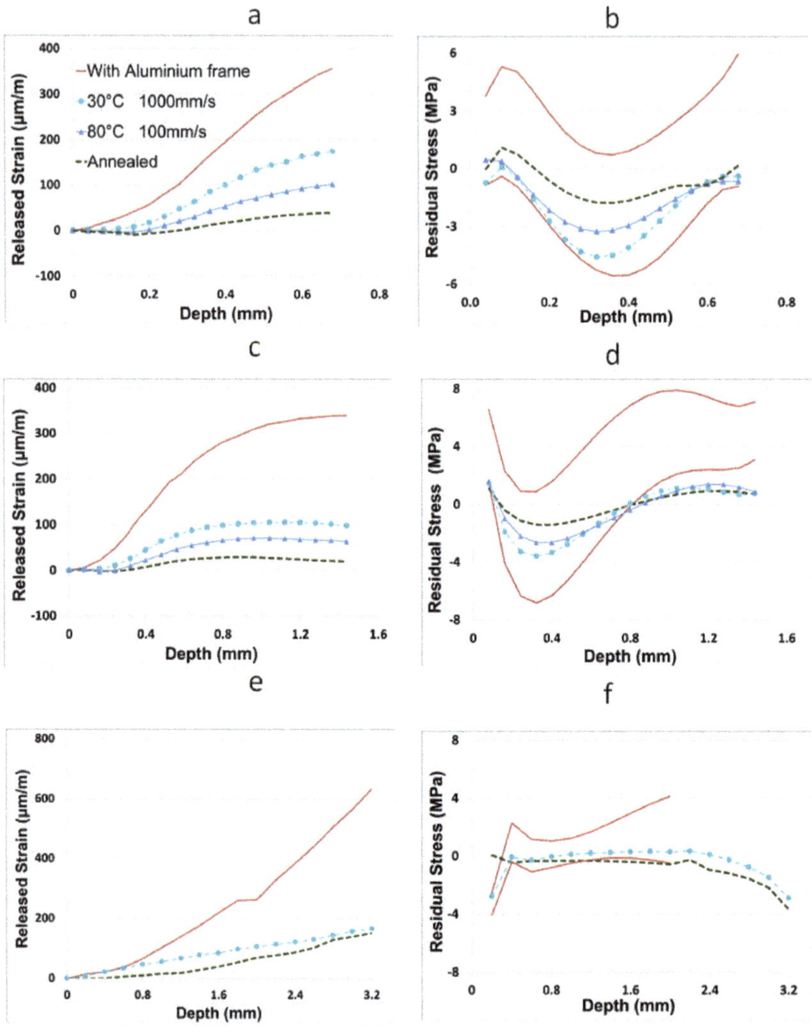

Figure 2: Released strain and measured residual stress applying the hole drilling method with a 2 mm hole (a and b), a 4 mm hole (c and d) and applying the ring-core method (e and f). Due to rotationally symmetric stresses, only one principal stress value is plotted for samples manufactured without the aluminium frame

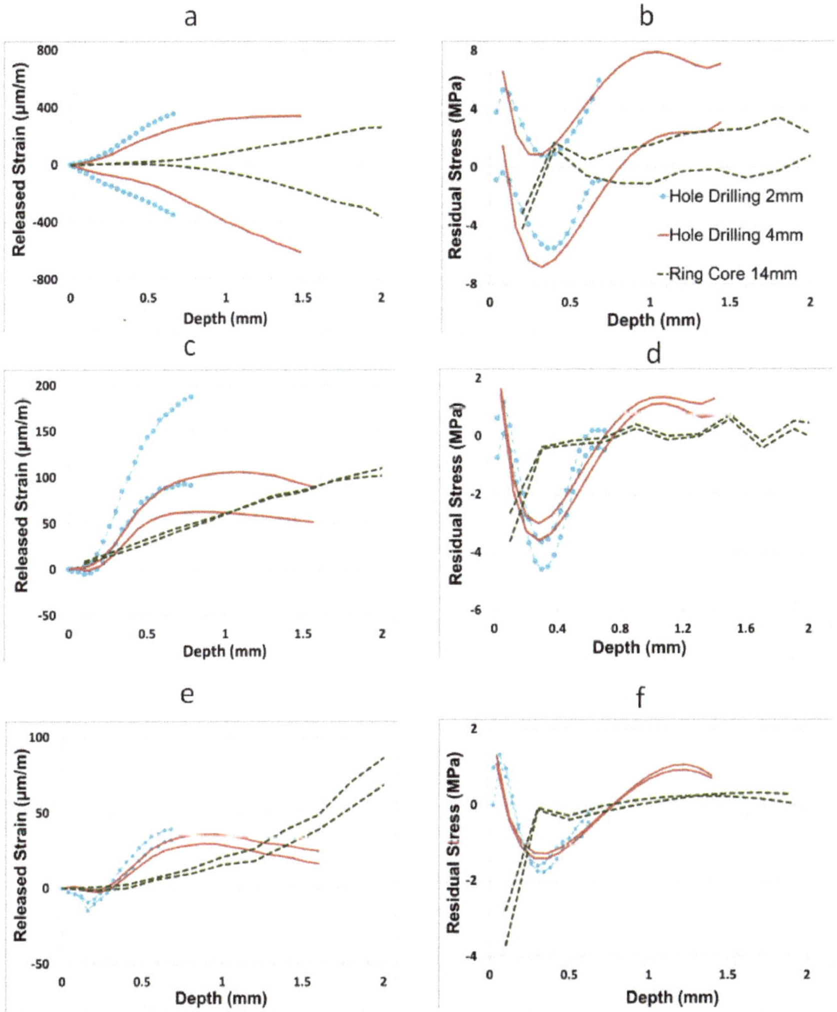

Figure 3: Measured maximum and minimum released strain and principal residual stress in polycarbonate samples: samples moulded into an aluminium frame (a and b), with a mould temperature of 30°C (c and d) and post-moulding annealed (e and f)

Conclusion

Residual stress profiles were characterized on polycarbonate by applying the hole drilling and the ring-core methods. The measurements employing the hole drilling method revealed the expected trend of the residual stress profiles. Varying the injection moulding parameters, different maximum amounts of compressive residual stresses were introduced and measured near the surface. It was also possible to set and determine tensile residual stresses within the whole depth of a sample by moulding into an aluminium frame. The results of the hole drilling method could be obtained with two different hole diameters. The results using the ring-core method were less satisfactory and may be strongly affected by the geometry of the sample, including the aluminium frame. In summary, the results presented prove that reliable residual stress analyses in plastics can be carried out provided that appropriate measurement parameters are applied. Thereby, temperature fluctuations and the visco-elastic material behaviour have to be considered adequately.

Acknowledgements
The authors gratefully acknowledge sample manufacturing and support by "Institute of Materials Engineering – Polymer Technology", at the university of Kassel.

References

[1] A. Guevara-Morales, U. Figueroa-Lopez, Residual stresses in injection moulded products, Journal Material Science 49 (2014) 4399-4415. http://dx.doi.org/10.1007/s10853-014-8170-y

[2] P. So and L.J. Broutman, Residual stresses in polymers and their effect on mechanical behaviour, Polymer engineering and science 16 (1976) 785-791. http://dx.doi.org/10.1002/pen.760161202

[3] A. Siegmann, A. Buchman, S. Kenig, Residual stresses in polymers.II. Their effect on mechanical behavior, Polymer engineering and science 21 (1981) 997-1002. http://dx.doi.org/10.1002/pen.760211503

[4] Garcia Sobolevski, E. (2007), Residual stress analysis of components with real geometries using the incremental hole drilling technique and a differential evaluation method, PhD thesis, Universität Kassel

[5] G.S. Shajer (2013), Pratical Residual Stress measurement methods, Wiley. http://dx.doi.org/10.1002/9781118402832

[6] A. Magnier, B. Scholtes, T. Niendorf, Aspects concerning the analysis of residual stress in plastic materials using the hole drilling method, (In preparation).

[7] A. Magnier, A. Nau, B. Scholtes, Some Aspects of the application of the Hole Drilling Method on Plastic Materials, in: S. Bossuyt et al. (eds.), Residual Stress, Thermomechanics and Infrared Imaging, Hybrid Techniques and Inverse Problems, Vol. 9, Conf. Proceedings of the Society for Experimental Mechanics Series, Springer Int. Publishing (2016) 371-380. http://dx.doi.org/10.1007/978-3-319-21765-9_45

[8] A. Nau, B. Scholtes, M. Rohleder, J.P. Nobre, Application of the Hole Drilling Method for Residual Stress Analyses in Components made of Polycarbonate, Journal of Plastics Technology 7 (2011) 66-85.

[9] ASTM E837-13, Standard Test Method for Determining Residual Stresses by the Hole-Drilling Strain-Gage Method, ASTM International. (2013)

Residual Stresses 2016: ICRS-10 Materials Research Forum LLC
Materials Research Proceedings **2** (2016) 187-192 doi: http://dx.doi.org/10.21741/9781945291173-32

Local Residual Stress Analysis on Deep Drawn Cups by Means of the Incremental Hole-Drilling Method

Simone Schuster [1,a*], Jan Pagenkopf [1,2,b] and Jens Gibmeier [1,c]

[1] Karlsruhe Institute of Technology (KIT), Institute for Applied Materials (IAM), Kaiserstr. 12, D-76131 Karlsruhe, Germany

[2] Fraunhofer Institute for Mechanics of Materials IWM, Wöhlerstr. 11, D-79108 Freiburg, Germany

[a] simone.schuster@kit.edu, [b] jan.pagenkopf@iwm.fraunhofer.de, [c] jens.gibmeier@kit.edu

Keywords: Incremental Hole-drilling Method, Crystallographic Texture, Multiple Case-Specific Calibration Functions, Deep Drawing

Abstract. In addition to residual stresses sheet metal forming induces characteristic crystallographic texture, hence, the material behavior is anisotropic. In general, the standard evaluation procedures of residual stress analysis techniques are limited to isotropic material states. In the present paper deep drawn steel cups of dual phase steel DP600 are analyzed by using a recently proposed calibration approach for residual stress analysis by means of the incremental hole-drilling method for highly textured material states. It is based on the differential method, which is enhanced with four case specific calibration functions. The multiple case specific calibration functions are determined by means of finite element simulations using the orientation distribution function (ODF) in combination with Hill's assumption and single crystal elastic constants of iron to calculate the effective elasticity tensor to account for elastic anisotropy. Supplementary, the deep drawing process is simulated using a finite element model based on the Hill48 yield criterion. Finally, the comparison shows that the numerical results are in satisfactory agreement to the experimental data.

Introduction

Standard methods of residual stress measurement techniques are restricted to isotropic material states. However, forming processes like e.g. rolling or deep drawing cause preferred orientations of the grains due to the limited possibilities of gliding. Crystallographic textures oftentimes result in anisotropic material behavior.

The standard approach of X-ray diffraction stress analysis according to the $\sin^2\psi$-method [1] is no longer applicable in case of textured material states, since the 2θ-$\sin^2\psi$ distributions are strongly non-linear. A remedy is the application of special measurement strategies like e.g. the crystallite group method [2] and stress factors [3], where the texture of the sample must be known a priori. These measurement strategies are elaborate and time-consuming and the knowledge of the stress free lattice parameter D_0 is required. Furthermore, formed components often obtain complex geometries and can be large (e.g. A-/B-/C-pillar, cowl). Shadowing effects can occur due to the complex shape of the sample. Owing to a limited installation space of the X-ray diffractometer the samples have to be cut and stress redistributions must be considered during the residual stress calculation. Since the penetration depth of conventionally generated X-rays is limited to a few microns, layers of the material must be removed by means of e.g. electrochemical polishing to determine residual stress depth distributions.

The incremental hole-drilling method has great potential, since it is versatile and fast compared to X-ray diffraction. Standard stress calculation methods (e.g. integral method [4], differential method [5]) use calibration data, which is based on isotropic material behavior. Significant errors in stress calculation can occur, if conventional calibration data is applied to residual stress analysis on

Materials Research Proceedings **2** (2016) 187-192 doi: http://dx.doi.org/10.21741/9781945291173-32

strongly textured materials [6]. Thus, we proposed a new calibration approach based on the differential method [7]. Four case-specific calibration functions must be determined numerically to account for the anisotropic elastic material properties. The elastic constants of the textured sample must be known beforehand for the calibration. A model assumption considering the interactions of the grain boundaries e.g. Voigt, Reuss, Eshelby/Kröner or Hill [8] can be used to calculate the effective elasticity tensor by means of the orientation distribution function (ODF) and single crystal elastic constants. In the present study residual stress depth distributions are determined at different positions of a deep drawn cylindrical cup made of dual phase steel DP600. They are compared to results obtained from finite element (FE) simulations of the deep drawing process.

Experimental investigation

A deep drawn steel cup with a nominal diameter of 100 mm and a drawing ratio of 1.8, which is the relationship between the blank diameter prior to the drawing operation and the punch diameter, was analyzed. Here, a cold rolled dual phase steel sheet DP600 (ferritic-pearlitic microstructure, model sample taken out during the rolling process prior to the final heat treatment) with a thickness of 1 mm was deep drawn with a blank holder force of 180 kN. The deep drawing operation leads to characteristic earing at the edge of the cup caused by the planar anisotropy of the material. Three different measuring locations were defined as can be seen in Fig. 1. A measuring location below an ear (path I) and one below a trough (path II) are located at the half of the maximum cup height. The rolling direction (RD) of the steel sheet corresponds to path II. The third measuring position is located in RD at the radius between cup wall and bottom. Stress components in drawing direction (DD) and tangential direction (TD) were determined. The chemical composition of the dual phase steel is shown in Table 1.

Table 1: Chemical composition of DP600 in weight-%

C	Si	Mn	P	S	B	Al	Cr	Mo	Ni
0.093	0.29	1.663	0.011	<0.001	0.0002	0.042	0.341	0.007	0.047

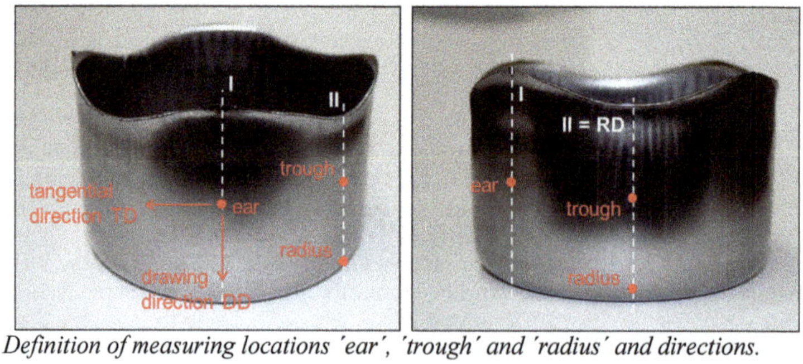

Fig. 1: Definition of measuring locations 'ear', 'trough' and 'radius' and directions.

X-ray diffraction texture analyses were carried out using a diffractometer of type XRD 3003 PTS from Seifert. A pin hole collimator with a nominal diameter of 1 mm and CoKα radiation was used. Incomplete pole figures for the lattice planes of type {110}, {200}, {211} and {220} were measured using a β-range of 0°..65° and a α-range of -170°..170° each in a step size of 5°. Fig. 2 illustrates the pole figures {100}, {110} and {111}, which were recalculated from the ODF. The rolling direction of the cold rolled steel sheet prior to the deep drawing operation is still visible for the ′trough′ and ′radius′ location and points into the drawing direction. In contrast, the maximum intensity in the pole figures at the measuring position below the ear is turned 45° with respect to the drawing direction.

Materials Research Proceedings **2** (2016) 187-192 doi: http://dx.doi.org/10.21741/9781945291173-32

The ODF, Hill's model assumption [8] and the single crystal elastic constants of iron (C_{11} = 230 GPa, C_{12} = 135 GPa and C_{44} = 117 GPa [9]) were used to determine the elasticity tensor C_{ijkl}. The crystallographic texture leads to a Young's modulus ratio E_{max}/E_{min} within the measuring plane of 1.09 in case of the trough and the radius and 1.03 in case of the ear.

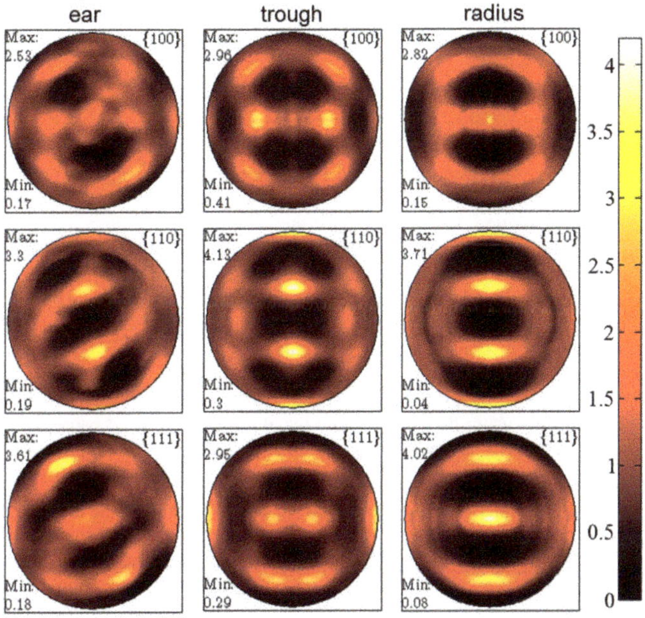

Fig. 2: Pole figures of type {100}, {110} and {111} for the three measuring locations 'ear', 'trough' and 'radius'.

A hole-drilling device RS200 from Vishay Measurement Group was used for the hole-drilling experiments. TiN coated end mills with a nominal diameter of 0.8 mm were used. Accordingly, strain gage rosettes of type EA-11-031RE-120 from Vishay Measurement Group were applied on the sample. The residual stress depth profiles were calculated using the new calibration approach based on the differential method and multiple case-specific calibration functions considering the elastic anisotropy, which we recently proposed in [7]. The following steps have to be conducted:

(i) Texture measurement and determination of the ODF

(ii) Calculation of the elasticity tensor C_{ijkl}

(iii) Determination of four case-specific calibration functions (two FE simulations required)

(iv) Residual stress calculation

Basically, the residual stress components in the two perpendicular directions of the rosette can be determined with this approach. The FE model shown in [7] was used for the determination of the case-specific calibration functions. The elastic anisotropy was considered by means of the elasticity tensor. Additionally, the FE model accounts for the small component's thickness. An external load was applied to the outside surfaces of the model to induce an in-plane calibration stress. Furthermore, the direction of calibration stress was aligned to the orientation of the elastic constants in drawing

Materials Research Proceedings **2** (2016) 187-192 doi: http://dx.doi.org/10.21741/9781945291173-32

and tangential direction of the cup. Drilling of the hole was realized by a stepwise deletion of the elements. An integral strain relaxation over the strain gage area was calculated for each drilling step.

Complementary, XRD stress analyses by means of the $\sin^2\psi$-method were carried out. Due to the limited installation space of the stationary diffractometer the sample must be cut, which causes complex redistributions of the original residual stresses. Furthermore, non-linear 2θ-$\sin^2\psi$ distributions were obtained. Thus, by this means no reliable determination of the residual stress state on the analyzed cup was suitable.

Finite element simulation of the deep drawing process

The numerical simulation of the deep drawing process was performed using the finite element software package Abaqus. First, the forming of the cylindrical cup was simulated with an explicit time integration scheme. An elasto-plastic model with isotropic hardening was used to describe the material behavior. The comparison between the punch forces measured during the forming process and predicted by the simulation was used to adjust the flow curve of the material for high strains. In order to account for the plastic anisotropy of the sheet, the Hill48 [10] yield function was applied. The anisotropy parameters were determined to give the best fit to both the Lankford parameters (r-values) and yield stresses evaluated from tensile tests in three directions (0°, 45° and 90° to rolling direction). Due to the material and specimen symmetry, only one quarter of the blank was modeled. The sheet was meshed with brick elements with reduced integration (C3D8R) using six elements over the sheet thickness. The tools were modeled as analytical rigid bodies and the friction coefficient was set to $\mu = 0.09$ for all contact pairs. After the forming step, an elastic springback simulation was performed using the implicit time integration scheme of Abaqus/Standard to obtain the residual stresses in the final component.

Results and discussion

Figure 3 shows the numerically determined residual stress distribution for the components in drawing (a) and tangential (b) direction of the deep drawn cup. The measuring locations are marked by black dots. The FE model is appropriate to describe the characteristic earing of the cup. Steep gradients for the stress in drawing direction can be detected on the outside surface of the cup. An area with maximum tensile stresses of more than 1000 MPa is located approx. 15 mm below the trough (marked by a black X). Actually, two belts of high tensile stresses (>500 MPa) along the outer surface can be observed. In contrast, the stress distribution of the component in tangential direction (3b) shows lower lateral gradients. Only at the edges of the cup very high tangential stresses (approx. 1000 MPa) can be detected. Again, an area with high tensile stresses of approx. 500 MPa is located below the trough. A comparison of the numerically and experimentally determined stress depth distributions for the three measuring locations is illustrated in Fig. 4. The hole-drilling experiments were evaluated up to a drilling depth of 0.4 mm. It is worth noting that the wall thickness of the cup is approx. 1 mm. The measurement accuracy of the incremental hole-drilling method is ±25 MPa for standard applications. It is slightly higher in case of textured material states due to the higher experimental effort. In general, a good qualitative and quantitative agreement between the numerical and experimental results was achieved as indicated by the plots in Fig. 4. At the measuring location ′ear′ the residual stresses in drawing and tangential direction are almost identical and constant over depth. The numerically determined residual stress profile in tangential direction is in very good accordance to the experimental result. For the stress component in drawing direction a gap with an average deviation of approx. 67 MPa exist. In case of the measuring location ′trough′ the residual stress in drawing direction is slightly overestimated by the FE model. Whereas, the difference of the stress component in tangential direction is higher (approx. 218 MPa) especially near to the surface. The tensile residual stress distributions at the measuring location ′radius′ are lowest of all measuring positions. The numerical results are conforming with the experimental results with an accuracy better than 60 MPa.

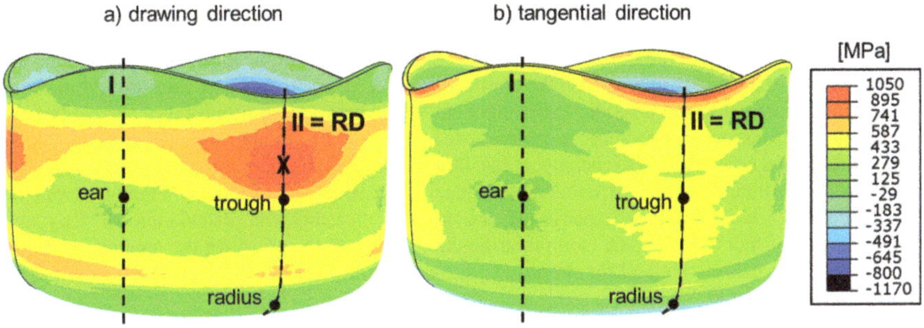

Fig. 3: Distribution of the stress components in drawing (a) and tangential direction (b) calculated by means of FE simulation.

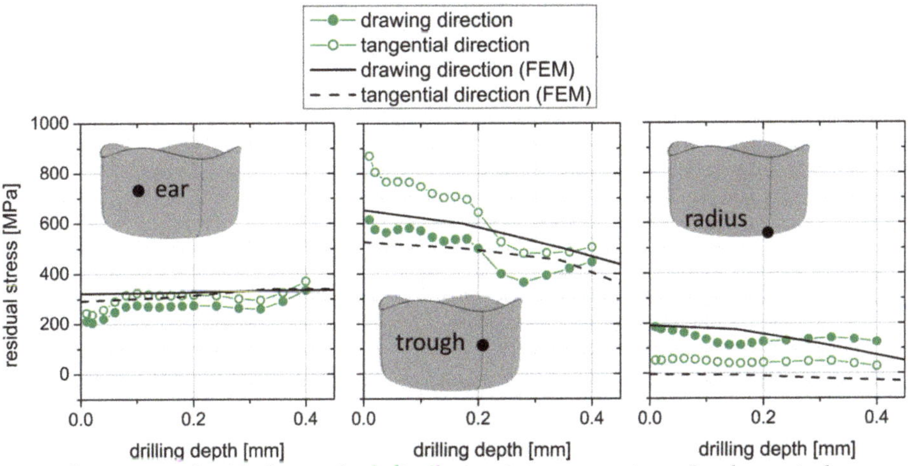

Fig. 4: Comparison of residual stress depth distributions between experimental and numerical results.

Despite the large lateral residual stress gradients especially for the measuring location ´trough´ (as indicated by Fig. 3) a good agreement between experiment and FE model could be achieved. Hence, with recourse to the preliminary studies in [6, 7] we can conclude that the new calibration approach should always be applied in case of textured material states regardless of the additional effort. Deviations between experiment and FE model can be explained by (i) the simplifying assumptions in the FE model (no kinematic hardening, lack of data of the flow curve for high strains), (ii) large residual stress gradients near the measuring locations and (iii) high residual stresses with respect to the materials yield strength (i.e. measuring position ´trough´), which can cause plastic deformations in the vicinity of the drilled hole and thus, an overestimation of the residual stress magnitude.

Conclusion

We demonstrated a reliable residual stress analysis by means of the incremental hole-drilling method on a deep drawn cup with a strong local anisotropy using an evaluation approach based on multiple case-specific calibration functions, which we proposed in [7]. The knowledge of the effective

Materials Research Proceedings **2** (2016) 187-192 doi: http://dx.doi.org/10.21741/9781945291173-32

elasticity tensor C_{ijkl} is required. It can be calculated using the ODF (texture measurement) in combination with single crystal elastic constants of the material and a model assumption of e.g. Hill [8]. The case-specific calibration functions can also account for the component's thickness. Since the wall thickness of the cup is thinner than the application limit of the incremental hole-drilling method, it is highly recommended to consider the component's thickness for the provision of the calibration data. The deep drawing operation was simulated using a FE model. Although, we had no experimental data for the yield behavior for large deformations, the numerical results are in satisfactory accordance to the experimental residual stress depth distributions. Finally, it can be concluded that the additional expense of the texture measurement, calculation of C_{ijkl} and determination of four case-specific calibration functions is worth the effort for the application of the incremental hole drilling method for strongly textured material states as in the present case the deep drawn DP600-steel cup.

Acknowledgement
The Graduate School 1483 "Process chains in production: Modelling, interactions and assessments of process zones" and the German research foundation (DFG) are gratefully acknowledged for funding. Further, we thank Prof. M. Liewald, P. Schmid and his team at the Institute for Metal Forming Technology in Stuttgart for the realization and support during the deep drawing experiments.

References

[1] E. Macherauch, P. Müller, Das $\sin^2\psi$-Verfahren der röntgenographischen Spannungsmessung, Z angew. Physik, 13 (1961) 340-345.

[2] V. Hauk, G. Vaessen, Eigenspannungen in Kristallitgruppen texturierter Stähle, Z. Metallkde., 76 (1985) 102-107.

[3] H. Behnken , V. Hauk, Berechnung der röntgenographischen Spannungsfaktoren texturierter Werkstoffe – Vergleich mit experimentellen Ergebnissen, Z. Metallkde., 82 (1991) 151-158.

[4] G. S. Schajer, Measurement of non-uniform residual stresses using the hole-drilling method. Part I - Stress calculation procedures, J. Eng. Mater. Technol., 110 (1988) 338–343. http://dx.doi.org/10.1115/1.3226059

[5] T. Schwarz, H. Kockelmann, Die Bohrlochmethode - ein für viele Anwendungsbereiche optimales Verfahren zur experimentellen Ermittlung von Eigenspannungen, Messtech. Briefe, 29 (1993) 33–38.

[6] S. Schuster, J. Gibmeier, Residual stress analysis of strongly textured materials by means of the incremental hole-drilling method – Survey on the application limits, Mater. Test., 56 (2014) 915-922. http://dx.doi.org/10.3139/120.110651

[7] S. Schuster, J. Gibmeier, Incremental hole-drilling for residual stress analysis of strongly textured material states – A new calibration approach, Exp. Mech., 56 (2016) 369-380. http://dx.doi.org/10.1007/s11340-015-0104-3

[8] R. Hill, The elastic behaviour of a crystalline aggregate, Proc. Phys. Soc. A, 65 (1952) 349-354. http://dx.doi.org/10.1088/0370-1298/65/5/307

[9] Landolt-Börnstein, Group III Condensed Matter, Table 3, Cubic system. Elements, 29a:14, Springer, 1992.

[10] R. Hill, A theory of the yielding and plastic flow of anisotropic metals, Proc. R. Soc. Lond. A, 193 (1948) 281-297. http://dx.doi.org/10.1098/rspa.1948.0045

Residual Stresses 2016: ICRS-10 Materials Research Forum LLC
Materials Research Proceedings **2** (2016) 193-198 doi: http://dx.doi.org/10.21741/9781945291173-33

Bending Fatigue Behavior of Blast Cleaned Grey Cast Iron

Maqsood Ahmad[1,a*], Ru Lin Peng[2,b], Mathias König[3,c], Sten Johansson[2,d]

[1]Base Engine & Materials Technology, Volvo Group, SE-40508, Gothenburg, Sweden

[2]Dept of Management and Engineering, Linköping University, SE-581 83 Linköping, Sweden

[3]Materials Technology for Basic engine, Scania CV, SE-15187, Södertälje, Sweden

[a]maqsood.ahmad@volvo.com, [b]ru.peng@liu.se, [c]mathias.konig@scania.com, [d]sten.johansson@liu.se

Keywords: Cast Iron, Blast Cleaning, Machining, Bending Fatigue, Residual Stress

Abstract. This paper presents a detailed study on the effect of an industrial blast cleaning process on the fatigue behavior of a grey cast iron with regard to the residual stresses and microstructural changes induced by the process. A comparison was also made to the effect of a machining operation which removed the casting skin layer. The blast cleaning process was found to greatly improve the fatigue resistance in both the low and high cycle regimes with a 75% increase in the fatigue limit. X-ray diffraction measurements and scanning electron microscopic analyses showed that the improvement was mainly attributed to compressive residual stresses in a surface layer up to 800 μm in thickness in the blast cleaned specimens. The machining also gave better fatigue performance with a 30% increase in the fatigue limit, which was ascribed to the removal of the weaker casting skin layer.

Introduction

Blast cleaning is a standard industrial process used in foundries to remove mould residues and oxide scales from the surface of castings. During the process, accelerated hard particles/shots impinge on the cast surface. The impact not only gives the desired cleaning effect but also induces surface plastic deformation and residual stresses, see for example [1], in the same way as the widely used shot peening process for surface strengthening of metallic components. However, being a cleaning process, a high blasting force is often used and little attention is paid to the control of process parameters such as blasting intensity and coverage.

The fatigue behaviour of structural components is sensitive to their surface condition. A blast cleaning process that significantly alters the surface integrity is important for the performance of cast irons subjected to cyclic loading. However, published research work on fatigue of cast irons with casting skin after clean blasting or shot peening are very limited, see however [2-4]. Benefits of blasting [2] or gentle shot peening [3] to the bending fatigue resistance of as-cast compacted graphite irons were reported. In [4] a gentle shot peening was also found to increase the bending fatigue limit of an as-cast grey iron. More publications on machined cast irons can be found, which show that the effect of shot peening varies with the peening parameters and materials. In general, the bending fatigue resistance of machined cast irons can be improved to certain degrees by shot peening, see for example [5] on grey cast iron, [3] on compacted graphite iron, [5,6]on ductile iron and [7] on austempered cast iron. On the other hand, shot peening using high peening intensity may not be beneficial; it was observed in [8] to lower the axial fatigue strength of a machined grey cast iron and in [6] to have no effect on the bending fatigue life of a machined ductile iron. The current work investigated the effect of an industrial clean blasting process on the bending fatigue behaviour of an as-cast grey cast iron. Comparison with machining that removed the casting skin was also made. Four point bending fatigue testing was performed and the results were analysed with regard to

Materials Research Proceedings **2** (2016) 193-198 doi: http://dx.doi.org/10.21741/9781945291173-33

residual stresses measured by X-Ray diffraction and microstructural changes studied using a scanning electron microscope.

Experimental Details

The material used in the study is a grey iron grade VIG290/205 equivalent to Mo-Cr-Cu alloyed SS-EN-1561-GJL 300. Smooth specimens (Fig. 1) for bending fatigue test were manufactured from cast plates in three different conditions: as-cast, machined and as-blast cleaned. For the as-cast condition, the casting surfaces were gently cleaned using a steel brush. For the blast cleaned condition, a typical process used for in-house cleaning components of heavy duty diesel engines was employed. Steel shots of S390 were used and the blast intensity measured on Almen strips placed in the centre location of the cast plates was averaged to 0.18 mmC. For the machined condition, the casting skin was removed by milling, leaving a surface with Ra 3.2 μm in roughness.

Four-point bending fatigue testing was performed in a MTS hydraulic load frame with a stress ratio of 0.1 and frequency of 25 Hz. Residual stresses were determined on the Ferrite-211 planes using the standard $\sin^2\psi$ method with Cr-K_α radiation and 9 ψ-angles ranging from -55° to +55°. The X-ray elastic constant for stress calculation is $5.81 \cdot 10^{-6}$ MPa^{-1}. Layer removal by electrolytic polishing was combined for measuring residual stresses below the surface. Microstructural analyses on longitudinal cross-sections were carried out in a Field Emission Scanning Electron Microscope.

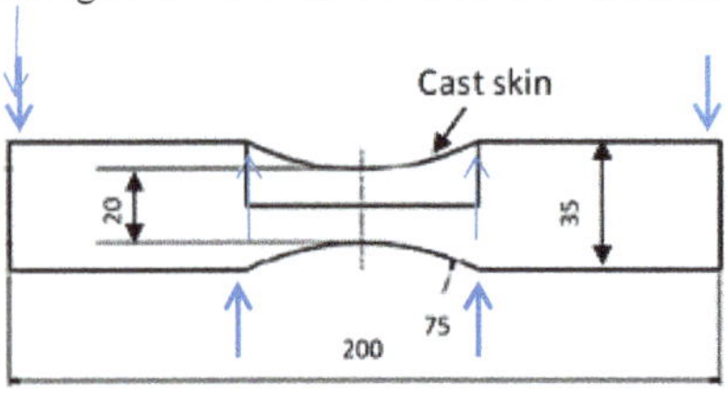

Fig. 1 Geometry of the fatigue specimens with casting skin. Specimen thickness was 15 mm. The gauge section of the machined specimens was reduced to 15 mm (width) by 10 mm (thickness) but the radius was kept unchanged. The arrows indicate positions for applying the bending load.

Results and Discussion

SEM and XRD analyses. As indicated in Fig. 2a, a casting skin, typically 300 μm thick, was observed in as-cast specimens. The casting skin contained an outer scale of iron and silicon oxides and a subsurface layer of pearlite with flaky inclusions. EDS analysis revealed that the flaky inclusions were essentially lamellar graphite with Si oxides in the edges. Polygonal grains of mainly ferritic phase were also observed in the casting skin layer (Fig. 2b). The microstructure below the casting skin, as shown in Fig. 2a, consisted of a matrix of essential pearlite and type A graphite.

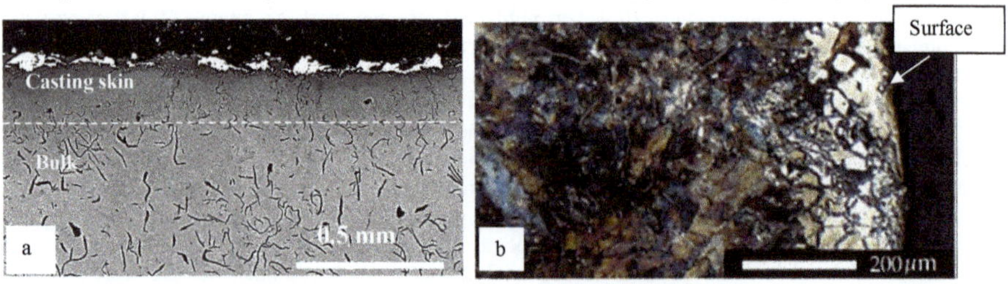

Figure 2. Longitudinal cross-section showing (a) the casting skin with oxide scale (SEM image) and (b) ferrite grains (light grains) in the casting skin (optical micrograph).

Impact of the blasting shots removed the outer oxide scale (Fig. 3a) and also induced heavy plastic deformation in a surface layer (Fig.3b and c). Protrusions formed at the surface in association with reorientation of flaky inclusions (Fig. 3b). However, a comparison of Fig. 3a with Fig. 2a indicated

Materials Research Proceedings **2** (2016) 193-198 doi: http://dx.doi.org/10.21741/9781945291173-33

similar surface roughness for the as-cast surface and blast cleaned surface. Closer examination of the surface layer under high magnifications dislocated the formation of massive micro cracks in embedded oxides, which did not seem to propagate into the surrounding matrix, as shown in Fig. 3c. This is in contrast to one of our previous studies where microcracks were common in machined specimens of the same material after shot peening using 0.16 mmC intensity [8]. That the casting skin layer better withstood the impact load could be attributed to a damping effect of the oxide scale and better ductility of the casting skin layer containing polygonal ferrite-grains.

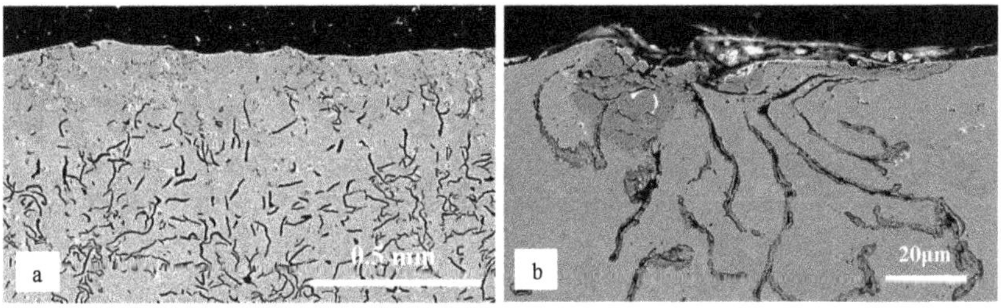

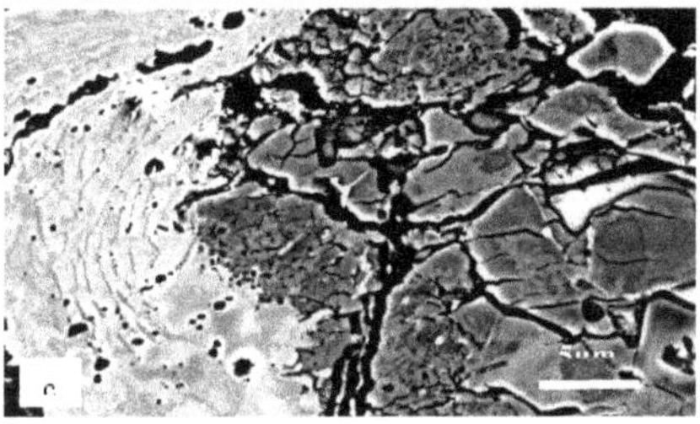

Figure 3. SEM images from a longitudinal cross-section of a blast cleaned specimen.

(a) Overview of the near surface region, (b) protrusion related to flaky inclusions and (c) shattered oxides (dark grey particles) in subsurface

Hardness measured in the matrix of the casting skin layer was plotted in Fig. 4a for both the as-cast and blast cleaned specimens. Due to the inhomogeneous microstructure, the hardness values were scattered. Nevertheless, hardness values in the casting skin were in general lower that the averaged bulk hardness plotted at 750 μm depth (they were actually measured at and beyond 600 μm depth) due to the partial pearlitic microstructure with ferrite grains. Strain hardening from the blast cleaning process was obvious; increased hardness was observed in a surface layer of about 300 μm after blast cleaning.

Residual stress profiles from the XRD measurements are presented in Fig. 4b-d together with corresponding diffraction peak width measured in FWHM (Full Width at Half Maximum intensity). Very low thermal stresses from the casting process were found (Fig. 4b). Somewhat lower FWHM in the surface layer was also observed, which could be explained by the presence of polygonal ferrite grains. Such grains are expected to show a smaller diffraction peak width that the thin ferritic lamellar in a pearlite. The blast cleaning induced compressive residual stresses to a large depth, about 800 μm, as shown in Fig. 4c, and the FWHM profile indicated the penetration of plastic deformation to at least 500 μm below surface. At the surface the longitudinal stress was similar to the transverse stress, about -300 MPa, but at the subsurface the peak longitudinal stress, -420 MPa was in lower magnitude than the peak transverse stress, -500 MPa. The much larger affected depth than normal shot peening was ascribed to the high blasting intensity and the large media size. Actually, a comparable affected depth was reported in [9] for heavy peening of machined specimens of the same material using 0.16 mmC peening intensity and an ever larger depth was reported in [6] for peening of a machined nodular cast iron using ~0.2 mmC peening intensity. The machining employed in this study was shown to induce also compressive residual stresses, which, however, only extended to a very shallow depth, about 120 μm, see Fig. 4d.

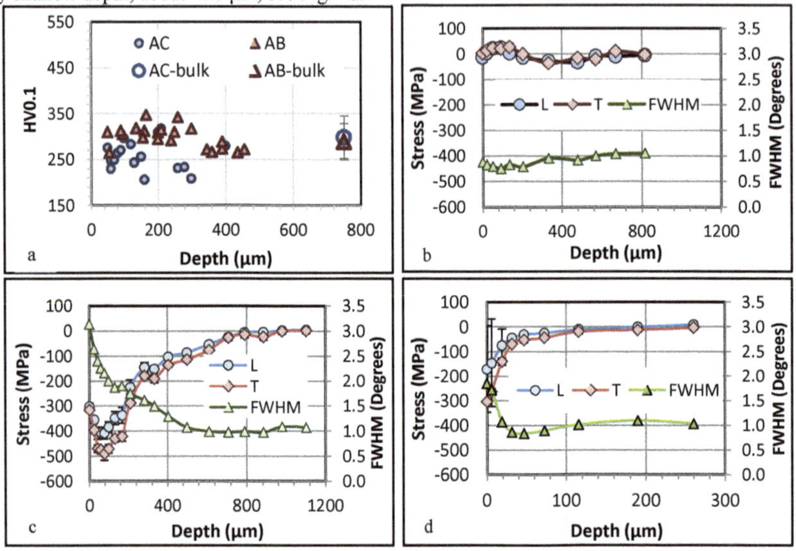

Fig. 4 (a) Hardness profile in as-cast (AC) and as blasted (AB) specimen. (b), (c) and (d) are residual stresses measured along the longitudinal (L) and transverse (T) direction in as-cast (b), as blasted (c), as machined (d) specimens, respectively.

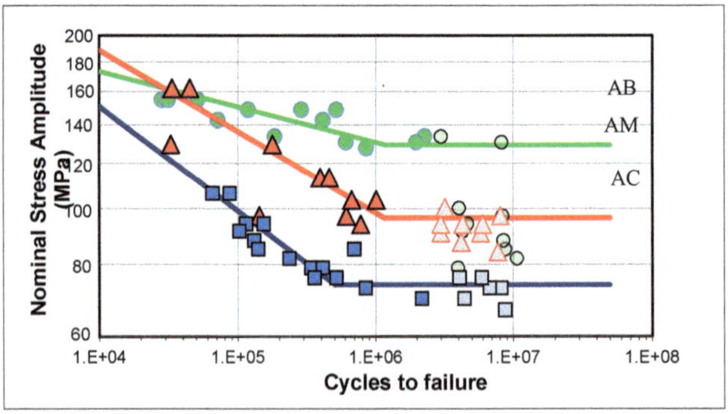

Figure 5. S-N curves for as-cast (AC), as-clean blasted (AB) and as-machined (AM). Filled symbols: failed, open symbols: runout.

Fatigue tests. Results from the bending fatigue tests are presented in Fig. 5 as S-N curves and the derived fatigue limits are given in Table 1. It can be seen that the blast cleaned specimens outperformed the as-cast specimens with better fatigue resistance in both the low cycle fatigue and high cycle fatigue regimes. The fatigue limit was increased by almost 75%, from 73.8±5.7 MPa for the as-cast to 129.1±3.9 MPa after blast cleaning. The machining also shifted the whole S-N curve upwards, though the improvement was not as significant as the blast cleaning operation: the fatigue limit was 96.4±6.4 MPa, giving a 30% increase.

Table 1 Fatigue limit with standard deviation derived from the data in Fig. 5 by Dixon method

Specimens	As-cast	Blasted	Machined
Fatigue limit in MPa	73.8±5.7	129.1±3.9	96.4±6.4
Change in %	-	74.8	30.6

Efforts to locate fatigue initiation points on the fracture surfaces gave no results due to its coarse surface topography. Therefore the analysis below on fatigue crack initiation was based on examination of longitudinal cross-sections of fatigued specimens. For the as-cast specimens, secondary fatigue cracks were observed to form in the surface oxide scale, see e.g. Fig.6 (a) or below the oxide scale, which then propagated inwards along flaky inclusions into the matrix. For the blasted specimens, microcracks (Fig. 6b) were observed to grow from graphite often in the regions 600 μm to 900 μm below the surface but very few microcracks could be found in the casting skin layer. This indicated that the compressive residual stresses effectively supressed fatigue crack initiation in the casting skin. The fatigue damage and growth process started instead deeper in the specimens where low compressive or even tensile stresses prevailed (Fig. 4c). Outward growth of the cracks towards the surface was also retarded by the compressive residual stresses. As a result, a better fatigue resistance was obtained. Strain hardening could also have contributed to the improvement, the effect was however considered to be minor, as the depth with obvious hardness improvement was much smaller than the depth of compressive stresses (Fig. 4a).

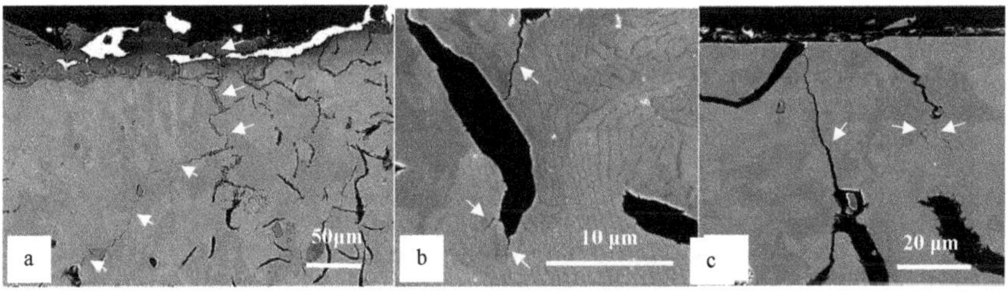

Figure 6 Surface crack iniation and inward propagation in an as-cast specimen (a), subsurface microcracks in a blast cleaned specimen (b), and near surface cracking in a machined specimen (c), indicated by white arrows.

High intensity shot peening applied on machined cast irons were found to be non-beneficial or even to give negative effect on fatigue performance due to increased surface roughness and other damages such as microcracks in subsurface [8,10]. Although the blast cleaning process employed here had similar intensity as the shot peening for machined specimens of the same cast iron in [8], due to the ductile casting skin no obvious damage to the surface layer was induced. This could have ensured the benefit of surface enhancement from the blasting.

For the machined specimens, near surface secondary fatigue cracks were observed in association with graphite inclusions, see Fig. 6c. Compressive residual stresses from the machining dropped sharply to very low values within a distance of 40 μm from the surface (Fig. 4d) and did not seem to prevent surface/near surface crack initiation from graphite inclusions. However, as the weak casting skin was removed, the surface layer exposed to high applied stresses had better strength, which was considered to be the main reason for the observed improvement in fatigue resistance. Machining also improved the surface roughness, however, due to the notch effect of graphite inclusion in grey cast irons, such effect may not be obvious.

Summary

Four-point bending fatigue tests were performed on a grey cast iron in three different surface conditions: as-cast with casting skin, blast-cleaned and machined (milling). The main results are summarised below:

The blast cleaning process investigated in this study was found to greatly improve the fatigue performance of as-cast specimens in both the low and high fatigue life regimes, with an increase of 75% in the fatigue limit. The improvement was attributed to that the blasting induced compressive residual stresses to a depth of 800 μm and obvious hardening to a depth of more than 300 μm but at the same time resulted in little damage in the form of increased surface roughness or microcracking.

The machined specimens showed better fatigue performance than the as-cast specimens, which was mainly due to the removal of the soft casting skin with a brittle outer oxide scale prone to fatigue crack initiation. The fatigue limit was increased by 30%.

The current work shows that in spite of the high peening intensity blast cleaning processes may effectively increase the fatigue properties of as-cast irons with casting skins.

Acknowledgement

Financial supports from VINNOVA – Sweden's Innovation Agency through the program FFI - Strategic Vehicle Research and Innovation are acknowledged.

Reference

[1] S.S. Mroz, G.M. Goodrich, Quantification of Shot Blast Cleaning Effects on Cast Iron as-cast Surfaces, Transactions of the American Foundry Society, Vol. 114. 114 (2006) 493-506.

[2] S. Boonmee, D. Stefanescu, Effect of casting skin on fatigue properties of CG iron, International Journal of Metalcasting. 7 (2013) 15-26. http://dx.doi.org/10.1007/BF03355550

[3] R. Lin Peng, T. Vuoristo, D. Bäckström, M. Ahmad, M. Lundberg, S. Johansson, Fatigue strength of shot peened compacted graphite iron, ICSP-12, Goslar, Germany (2014).

[4] B. Kaouache, D. Bäckström, M. Ahmad, T. Vuoristo, S. Johansson, R. Lin Peng, To increase fatigue strength of grey iron by shot peening, ICEAF-IV, Skiathos, Greece (2015).

[5] M. Wollmann, S. Koda, L. Wagner, The Influence of Shot Peening on the Fatigue Behavior of Gray Cast Iron and Ductile Iron, in Shot Peening and Other Mechanical Surface Treatments - Science and Applications, L. Wagner Ed. Self-publishing (2014).

[6] S. Bagherifard, I. Fernandez-Pariente, R. Ghelichi, M. Guagliano, Effect of severe shot peening on microstructure and fatigue strength of cast iron, Int. J. Fatigue. (2013).

[7] A. Zammit, M. Mhaede, M. Grech, S. Abela, L. Wagner, Influence of shot peening on the fatigue life of Cu–Ni austempered ductile iron, Mater. Sci. and Eng. A 545 (2012) 78-85. http://dx.doi.org/10.1016/j.msea.2012.02.092

[8] M. Lundberg, R. Lin Peng, M. Ahmad, D. Bäckström, T. Vuoristo, S. Johansson, Fatigue strength of machined and shot peened grey cast iron, Adv. Mater. Res. 891-892 (2014) 30-35. http://dx.doi.org/10.4028/www.scientific.net/AMR.891-892.30

[9] M. Lundberg, R.L. Peng, M. Ahmad, T. Vuoristo, D. Bäckström, S. Johansson, Residual stresses in shot peened Grey and compact iron, HTM J. Heat Treat. Mater. 69 (2014) 38-45. http://dx.doi.org/10.3139/105.110207

[10] Y. Uematsu, T. Kakiuchi, K. Tokaji, K. Nishigaki, M. Ogasawara, Effects of shot peening on fatigue behavior in high speed steel and cast iron with spheroidal vanadium carbides dispersed within martensitic-matrix microstructure, Mater. Sci. and Eng. A 561 (2013) 386-393. http://dx.doi.org/10.1016/j.msea.2012.10.045

Residual Stresses 2016: ICRS-10 Materials Research Forum LLC
Materials Research Proceedings **2** (2016) 199-204 doi: http://dx.doi.org/10.21741/9781945291173-34

Elastic and Elastic-Plastic Behaviour of a Crack in a Residual Stress Field

Guiyi Wu[a], Chris Aird[b], David Smith and Martyn Pavier[c*]

Department of Mechanical Engineering, University of Bristol
Queen's Building, University Walk, Bristol BS8 1TR, UK

[a]guiyiwu@outlook.com, [b]chris.aird@edf-energy.com, [c]martyn.pavier@bristol.ac.uk

Keywords: Fracture Mechanics, Residual Stress, Finite Element Analysis, Strip Yield Model, Plasticity

Abstract. The behaviour of a crack in the centre of a plate subject to a far-field stress has been studied where the plate contains an initial residual stress. Elastic and elastic-plastic conditions have been considered. For elastic conditions a series of analyses based on stress intensity factor solutions have been developed to calculate the state of opening and the stress intensity factor for cracks of different lengths relative to the size of the residual stress field and different magnitudes of applied stress relative to the magnitude of the residual stress. For elastic-plastic conditions a strip yield model has been used to develop a similar set of analyses. The results of these analyses compare closely with those of finite element modelling.

Introduction

In an engineering component containing a crack, residual stresses interact with stresses generated by applied loading in a complex manner. Typically, numerical techniques are required to predict the likelihood of fracture, providing results that are particular to the set of conditions considered. The work we describe here represents an attempt to understand the generic behaviour of the crack for a straightforward geometry, loading and residual stress distribution that we hope will provide insight into the results of more involved analyses.

Terada [1] provided an early analysis of the effect of residual stress on a crack. He proposed a residual stress distribution to represent a butt weld between two plates and then developed a method to evaluate the stress intensity factor assuming elastic behaviour. Chell and Ewing [2] discussed the effect of plasticity on fracture when residual stresses exist while Labeas and Diamantakos [3] used finite element analysis to explore the stress intensity factor for a crack of varying length embedded in a residual stress field.

In this paper we investigate the behaviour of a crack in an infinite two dimensional plate with an initial residual stress distribution under superimposed uniaxial tension. First we describe our method for generating a residual stress distribution using a stress function. Next we use existing stress intensity solutions to explore the linear elastic behaviour of the crack as the length of the crack is altered and the magnitude of the applied stress is increased. We then use the strip yield model to investigate the elastic-plastic behaviour of the crack with increasing applied stress, for one size of the crack.

Residual Stress

For the analytical work we a residual stress distribution chosen as representative of the residual stresses produced by a butt weld in a plate, as shown in Fig. 1.

Materials Research Proceedings **2** (2016) 199-204 doi: http://dx.doi.org/10.21741/9781945291173-34

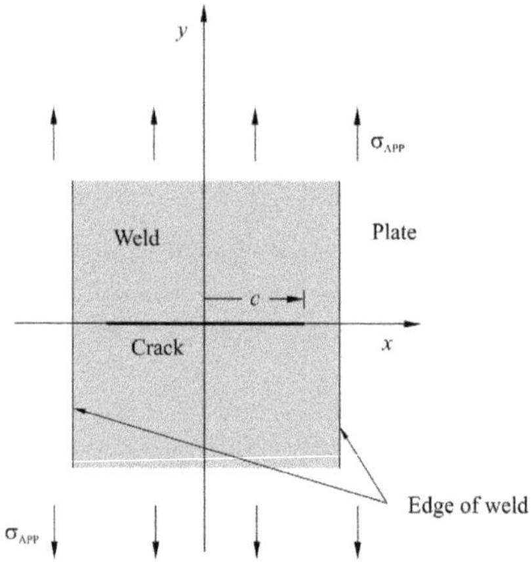

Fig. 1 Geometry of model.

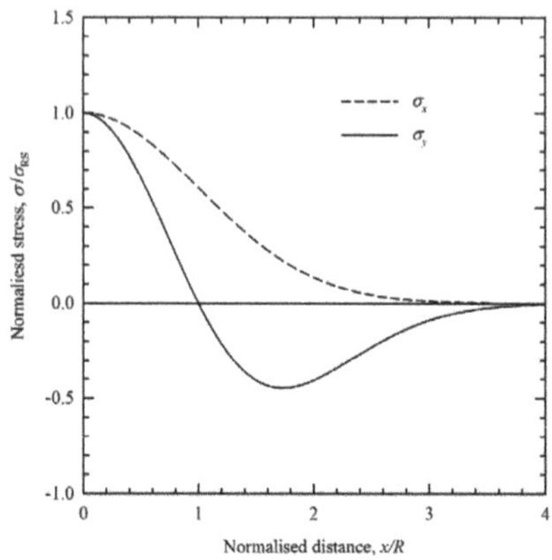

Fig. 2 Normalised residual stress components σ_{xx} and σ_{yy} versus x at $y = 0$.

Our analytical results will be compared with those of finite element modelling using ABAQUS 6.11, for which a residual stress distribution in equilibrium must be generated. This is achieved by defining the stress function

$$\phi = -\sigma_{\text{RS}} R^2 \exp\left[-\frac{x^2}{2R^2} - \frac{y^2}{2R^2}\right] \tag{1}$$

where R characterises the size of the residual stress distribution and will be of the order of half the width of the wled in Fig. 1. Stress components are calculated from the stress function by

$$\sigma_{xx} = \frac{\partial^2 \phi}{\partial y^2}, \sigma_{yy} = \frac{\partial^2 \phi}{\partial x^2}, \sigma_{xy} = \frac{\partial^2 \phi}{\partial x \partial y} \tag{2}$$

For $y = 0$, this gives the same distribution for σ_{yy} as proposed by Terada [1]:

$$\sigma_{yy} = \sigma_{\text{RS}} \left[1 - \frac{x^2}{R^2}\right] \exp\left[-\frac{x^2}{2R^2}\right] \tag{3}$$

These stress components were used to input an initial stress distribution using the SIGINI routine. The graph of Fig. 2 plots the normalised stresses σ_{xx} and σ_{yy} versus x for $y = 0$.

Elastic behaviour

In this section the elastic behaviour is studied for a crack in a residual stress field subjected to additional uniaxial applied load in the y direction as defined in Fig. 1. Depending on the length of the crack, the level of applied load and the magnitude of the residual stresses the crack may be closed, partially open or fully open.

Fig. 3(a) shows the behaviour of the crack for positive σ_{RS}, that is when the residual stress is tensile at the centre. The half-length of the crack c is normalised with respect to the size of the tensile region of the residual stress field R and the applied stress σ_{APP} is normalised with respect to σ_{RS}.

For all sizes of crack, the crack is always closed if the applied stress is compressive and greater in magnitude than the value of the residual stress at $x = 0$. As the magnitude of the compressive applied stress is reduced, the crack opens first at $x = 0$ when the sum of the applied and residual stresses equal zero. The boundary between the fully closed and partially open regions of crack behaviour, the line *AB*, is given by $\sigma_{APP} = -\sigma_{RS}$.

Once the crack is partially open, increasing the magnitude of the applied stress will eventually cause the crack to open completely. The magnitude of applied stress to open the crack fully depends on the length of the crack. For crack half-lengths less than $\sqrt{5}R$ the boundary between the fully open and partially open regions is the line *AC*. This line is given by the condition that the sum of the stress intensity factor at the tip due to the applied stress K_{APP} and the stress intensity factor due to the residual stress K_{RS} is equal to zero. K_{RS} is derived by integration of the expression for the stress intensity factor due to pairs of splitting forces applied to the crack surface presented in Tada, Paris and Irwin [4].

For crack half-lengths greater than $\sqrt{5}R$, if the applied stress is high enough that the crack is fully open and the applied stress is then reduced the crack closes first at some point along the crack surface. The position of this point depends on the crack length and is calculated by integration of the expression for crack opening due to pairs of splitting forces applied to the crack surface [4]. The boundary between the fully open and partially open regions is the line *CD*. For crack half-lengths greater than $\sqrt{5}R$ a small regime of crack behaviour exists as shown in Fig. 3(a) where the crack is partially open: open at the tips and in the centre but closed between these points. This regime is bounded by the two lines *CD* and *CE*

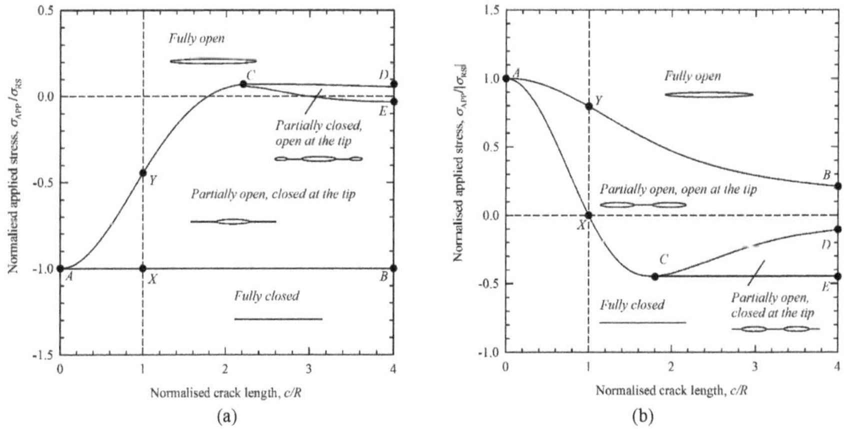

Fig. 3 Elastic crack behaviour map for (a) tensile residual stress and (b) compressive residual stress.

The behaviour of the crack for negative σ_{RS}, that is when the residual stress is compressive at the centre, is shown in Fig. 3(b). The crack is now fully closed provided the applied load is less than a certain value that depends on the crack length. When the crack half-length is less than $\sqrt{3}R$ the crack opens first at the tip. Points on the line *AC* are given by the condition that the sum of the residual stress and the applied stress equals zero at the crack tip. For crack half-lengths greater than $\sqrt{3}R$, the crack becomes open at the tip when the stress intensity at the tip, $x = c$, is zero. Points on the line CD are given by simultaneous solution of the equations

$$\begin{aligned} K^{a}_{\mathrm{APP}} + K^{a}_{\mathrm{RS}} &= 0 \\ K^{c}_{\mathrm{APP}} + K^{c}_{\mathrm{RS}} &= 0 \end{aligned} \tag{4}$$

K^{a}_{APP} and K^{c}_{APP} are the stress intensity factors for twin cracks with crack tips located at $x = \pm a$ and $x = \pm c$ due to a uniform applied stress [4].

For all sizes of crack, the crack is always fully open provided the applied stress is high enough. If the crack is fully open and the applied stress is then reduced the crack closes first at the centre. The applied stress at which closure occurs depends on the length of the crack. The boundary between the fully open and partially open regions is the line AB. This line is given by the condition that the sum of the opening at the centre of the crack due to applied stress δ_{APP} and the opening due to the residual stress δ_{RS} is equal to zero.

In addition to the behaviour of the crack for different applied stresses, stress intensity factors may also be calculated. For example, Fig. 4 shows the stress intensity for a crack of half-length $c = R$ versus the applied stress for three cases: no residual stress, tensile and compressive residual stress. For the residual stress cases, these results correspond to the vertical dashed line in Fig. 3(a) and (b) with $c/R = 1$.

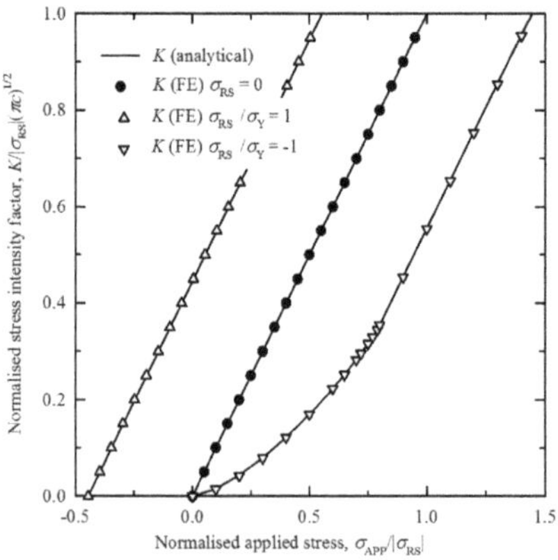

Fig. 4 Elastic normalised stress intensity factor $K/|\sigma_{RS}|\sqrt{\pi c}$ *versus normalised applied stress* $\sigma_{APP}/|\sigma_{RS}|$ *for zero, tensile and compressive residual stress.*

The finite element results shown in Fig. 4 for comparison were evaluated using the ABAQUS 6.11 finite element system. Calculations of stress intensity factors used the JEDI code [5].

Elastic-plastic behaviour

We now use the strip yield model of Dugdale [6] and Barenblatt [7] to examine the elastic-plastic behaviour of a crack in a residual stress field. We will not present revised crack behaviour maps since in general these maps are similar to the maps for elastic behaviour of Fig. 3. Instead we choose one crack length defined by $c = R$ and study the behaviour of the crack as the applied load is increased from an initial state where the crack is fully closed.

Materials Research Proceedings **2** (2016) 199-204 doi: http://dx.doi.org/10.21741/9781945291173-34

Whne the applied load is sufficiently high that the crack opens at the tip, a yielded zone of length ρ forms ahead of the tip. The size of this yielded zone is calculated using the condition that the total stress intensity factor for an extended crack of half-length $c+\rho$ is zero. When residual stresses act in addition to the applied stress, the total stress intensity factor K_{TOT} is calculated as the sum of the stress intensity factors due to the applies stress K_{APP}, the residual stress K_{RS} and stresses applied to the crack tip to represent the yield zone K_{Y} . Therefore

$$K_{\mathrm{TOT}} = K_{\mathrm{APP}} + K_{\mathrm{RS}} + K_{\mathrm{Y}} = 0 \tag{5}$$

Solution of Eq. (5) gives the size of the plastic zone ρ. The effective stress intensity factor is evaluated using the technique of Burdekin and Stone [8], which requires the crack opening displacement to be calculated at $x=c$. The total crack opening displacement δ_{TOT} is evaluated as the sum of the crack opening due to the applied stress δ_{APP} , the residual stress δ_{RS} and the stresses in the yielded zone δ_{Y}. That is

$$\delta_{\mathrm{TOT}} = \delta_{\mathrm{APP}} + \delta_{\mathrm{RS}} + \delta_{\mathrm{Y}} \tag{6}$$

Finally the effective stress intensity factor is obtained by

$$K_{\mathrm{EFF}} = \sqrt{E'\sigma_{\mathrm{Y}}\delta_{\mathrm{TOT}}} \tag{7}$$

Fig. 5 compares the results of the strip yield model for tensile residual stress and compressive residual stress, where the magnitude of the residual stress is equal to the yield stress, that is $\sigma_{\mathrm{Y}}/|\sigma_{\mathrm{RS}}|=1$. Fig. 5 also shows the results for the standard strip yield model, the case with no residual stress. The figure also shows two dimensional plane stress finite element results obtained for K_{EFF} using the calculation

$$K_{\mathrm{EFF}} = \sqrt{E'J} \tag{8}$$

where J is the value of the J-integral evaluated using the JEDI procedure [5]. Very good agreement is obtained even for applied stresses approaching the yield stress where large plastic zones develop and small scale yielding conditions certainly do not exist.

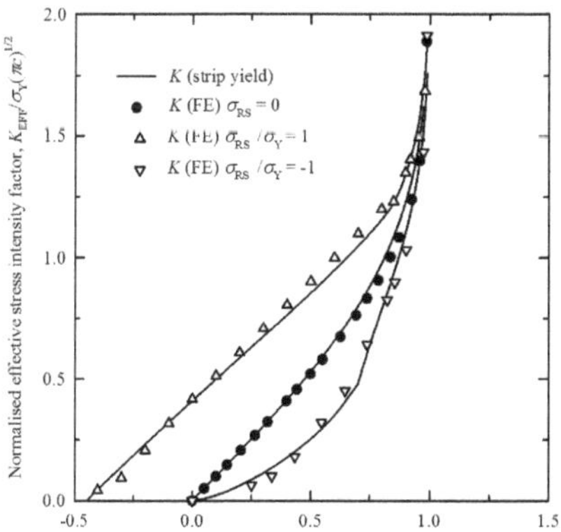

Fig. 5 Elastic-plastic normalised effective stress intensity factor $K_{\mathrm{EFF}}/|\sigma_{RS}|\sqrt{\pi c}$ versus normalised applied stress $\sigma_{APP}/|\sigma_{RS}|$ for zero, tensile and compressive residual stress.

Materials Research Proceedings **2** (2016) 199-204 doi: http://dx.doi.org/10.21741/9781945291173-34

Conclusions

Analytical and finite element methods have been used to map the elastic behaviour of a crack in a residual stress field. Depending on the length of the crack and the magnitude of the superimposed applied stress the crack may be fully open, partially open and open at the tip, partially open but closed at the tip or fully closed. Of course, a non-zero stress intensity factor only develops when the crack is open at the tip.

Effective stress intensity factors for an elastic-perfectly plastic crack in a residual stress field under plane stress conditions have been calculated using strip yield and finite element methods with good agreement, even for high levels of applied stress approaching the yield stress. The crack opening behaviour is similar to that for an elastic crack.

The results that have been presented were based on the distribution of residual stress of Eq. (3). Different distributions will give different results, but not markedly different since the calculations depend on a weighted average of the didtribution.

References

[1] Terada H. An analysis of the stress intensity factor of a crack perpendicular to the welding bead, Engng. Fracture Mech., 8 (1976), 441-111. http://dx.doi.org/10.1016/0013-7944(76)90024-2

[2] Chell GG and Ewing DJF. The role of thermal and residual stresses in linear elastic and post yield fracture mechanics. Int. J. Fracture, 13 (1977), 467-479

[3] Labeas G and Diamantakos I. Numerical investigation of through crack behaviour under welding residual stresses. Engineering Fracture Mechanics, 76 (2009), 1691-1702. http://dx.doi.org/10.1016/j.engfracmech.2009.03.006

[4] Tada H, Paris PC, Irwin GR (2000). The stress analysis of cracks handbook (3rd edition), ASME Press, New York. http://dx.doi.org/10.1115/1.801535

[5] Beardsmore DW (2008). A code for the calculation of J for cracks inserted in initial strain fields for the role of J and Q in the predictions of crack extension and fracture. Proc. ASME 2008 Pressure Vessels and Piping Conference, Chicago, USA, Paper No. PVP2008-61169, 955-966. http://dx.doi.org/10.1115/pvp2008-61169

[6] Dugdale DS (1960). Yielding of steel sheets containing slits. J. Mech. Phys. Solids, 8, 100-104. http://dx.doi.org/10.1016/0022-5096(60)90013-2

[7] Barrenblatt DS (1962). The mathematical theory of equilibrium cracks in brittle fracture. Adv. Appl. Mech. 7, 55-125. http://dx.doi.org/10.1016/s0065-2156(08)70121-2

[8] Burdekin FM and Stone DEW (1966). The crack opening displacement approach to fracture mechanics in yielding materials, J. Strain Analysis for Eng. Design, 1, 145-153. http://dx.doi.org/10.1243/03093247V012145

Residual Stresses 2016: ICRS-10 Materials Research Forum LLC
Materials Research Proceedings **2** (2016) 205-210 doi: http://dx.doi.org/10.21741/9781945291173-35

Multi-axial Analyses of Welding Stresses in High-Strength Steel Welds

Dirk Schroepfer[1,a*], Kerstin Flohr[1,b], Arne Kromm[1,c] and Thomas Kannengiesser[1,d]

[1]Bundesanstalt für Materialforschung und –prüfung (BAM), 9.4 Division - Weld Mechanics, Unter den Eichen 87, 12205 Berlin, GERMANY

[a]dirk.schroepfer@bam.de, [b]kerstin.flohr@bam.de, [c]arne.kromm@bam.de, [d]thomas.kannengiesser@bam.de

Keywords: High-Strength Steel, Welding, Reaction Stress, X-Ray Diffraction

Abstract. Today's efforts for lightweight design result in a growing application of high-strength structural steels from 960 MPa. In welded structures of these steels increased demands regarding component safety and a high elastic ratio should be considered. Hence, the prevention of an evolution of high weld-induced tensile residual stresses is required. Recent studies showed that component related restraint conditions of welds are able to elevate welding induced stresses to critical values, depending on material characteristics, the welding process and parameters. This work involves multi-axial welding loads as a consequence of the superposition of local residual stresses, global reaction stresses and moments, varying the welding parameters under different restraint conditions. The global welding loads are measured via GMA-weld tests in a special testing facility and via a DIC(Digital Image Correlation)-system in a slot weld. Local transverse residual stresses were analysed by means of X-ray diffraction. The application of a less amount of weld runs due to a modified welding parameters and welds seam configurations revealed as a beneficial approach to reduce welding loads in high-strength steels.

Introduction

In modern steel constructions, e.g. mobile cranes, high-strength steels are required to ensure low self-weights. Besides higher ratios of lifting capacity to total weight of such constructions, higher energy and cost efficiencies are provided [1]. A series of high-strength base materials and filler metals was developed by the steel producers, recently. S960QL is one of the most important steel grades for the mobile crane industry. However, the design of welded structures and the welding process become more challenging with increasing material strength due to higher requirements for the component safety, higher elastic ratios and closer technological boundaries [2], for instance the cooling time from 800 °C to 500 °C ($\Delta t_{8/5}$-cooling time). Besides ensuring metallurgical requirements by keeping a defined working range of heat control, high tensile residual stresses should be avoided. These stresses are contributors for crack initiation and may be detrimental for the lifetime, load capacity and component safety [3].

Several recent numerical and experimental studies revealed that loads and stresses due to welding are affected by an interaction of the welding parameters, the applied materials and the restraint, which is subjected to the design of the weld [4–8]. Regarding the influence of a hindered shrinkage of a weldment, the residual stress (σ^{rs}) formation should be differentiated into a local and a global scope. Following Fig. 1a, local restraint stresses σ_{loc} develop due to inhomogeneous volume changes in the weld and HAZ. They are subjected to the material behaviour and welding process. If the weld is externally restrained, these local stresses are superimposed globally by normal reaction stresses σ_y due to the hindered lateral shrinkage Δl of the component; see Fig. 1b. As a rule, component related restraints are accompanied by a restrained angular distortion $\Delta\beta$, which causes bending moments M_x while welding and cooling. Considering a multi-axial load analysis of welded components, the resulting bending stress σ_{Mx} should be additionally superimposed; see Fig. 1c. Existing analyses involved the interaction between heat control and stress build-up in welds under global restraint. Lower working temperatures (preheat and interpass temperature $T_{p/i}$) and heat inputs significantly decrease the resulting reaction forces [4]. The modification of process parameters at modern power sources enables the weldability of narrow weld seams considering the seam configuration [9].

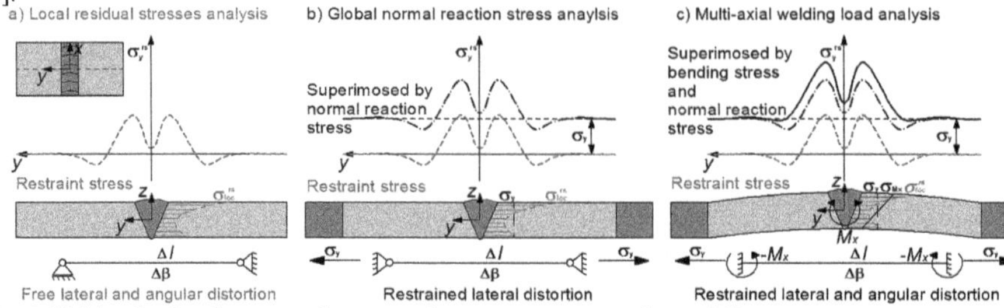

Fig. 1. Welding load and stress analyses considering restraints according to [4].

A reduced overall heat input and amount of filler metal for the same weld is possible, which may provide beneficial effects regarding distortion and residual stresses. However, an advanced understanding for stress prediction and optimization in high-strength steel component welds considering multiaxial loads is missing. Hence, this research focusses on the influence of the weld seam configuration and welding process parameters on the global reaction forces and bending moments in welded components.

Experimental

Test Material. Plates with a thickness of 20 mm of high-strength quenched and tempered fine grained structural steel S960QL were welded with a similar high-strength solid wire according to ISO 16834-A [10]. The chemical composition and mechanical properties are given in Table 1.

Table 1. Chemical compositions (FES, Fe balanced) and mechanical properties of the test materials.

Element in %	C	Si	Mn	Cr	Mo	Ni	V	Nb	Ti
Base material (S960QL, EN 10025-6 [11])	0.16	0.32	1.24	0.20	0.61	0.05	0.040	0.015	0.003
Filler material (G 89 6 M21 Mn4Ni2CrMo [10])	0.08	0.87	1.84	0.35	0.57	2.23	0.005	-	0.070
Property	$R_{p0.2}$ in MPa		R_m in MPa		A_5 in %	Av in J at -40 °C			HV10
Base material (mechanical testing)	1035		1050		17	96			332±8
Filler material (producer testing report)	938		980		15	62			354±7

Table 2. Welding parameters and variation of seam configuration and restraint condition.

Weld preparation	Welding current [A]	Welding voltage [V]	Welding speed [mm/min]	Wire feed speed [m/min]	Welding parameters	Working temperature $T_{p/i}$ [°C]	Heat input [kJ/mm]
V groove, 45°	265 ± 10	26.5 ± 0.1	330	8.7	Conv.	100	1.3
V groove, 30°	320 ± 10	28.7 ± 0.1	400 to 420	11	Mod.		

Test no.			1	2	3	4	5	6
Groove angle	α	[°]	45	30	45	30	45	30
Welding parameters	-	-	Conv.	Mod.	Conv.	Mod.	Conv.	Mod.
Restraint intensity	R_{Fy}	[kN/(mm·mm)]	0 [a]	0 [a]	3 [b]	3 [b]	10 [c]	10 [c]

Welding parameters of the power source: Conv.: transitional arc, Mod.: modified spray arc;
[a] Free shrinking weld test, [b] Restrained weld test in the 2-MN-testing facility, [c] Self restrained weld test (slot weld).

Welding Parameters. All welds were performed with automated GMA multilayer welding. The welding parameters are shown in Table 2. The welding parameters and seam preparation by means of groove angle were varied. The 45° V-butt joints were welded with the conventional and the 30° groove angles with the modified welding parameters. Considering technological and mechanical requirements of the welds, cooling times were measured for the applied heat control ($\Delta t_{8/5}$ = 6 s to 8 s). $\Delta t_{8/5}$-times and tested properties of the welds are according to the specifications.

Weld Tests and Dimensions. For a variation of the restraint conditions, free shrinking, externally restrained and self-restrained weld tests were performed; see Fig. 2 for dimensions and build-up sequence. The external restraint in weld tests no. 3 and 4 was conducted using a special in-house developed 2-MN-testing facility [4,7]. In this test setup, a shrinkage restraint and synchronous measurement of forces and stresses of the weld were achieved; see Fig. 2b and 3b.

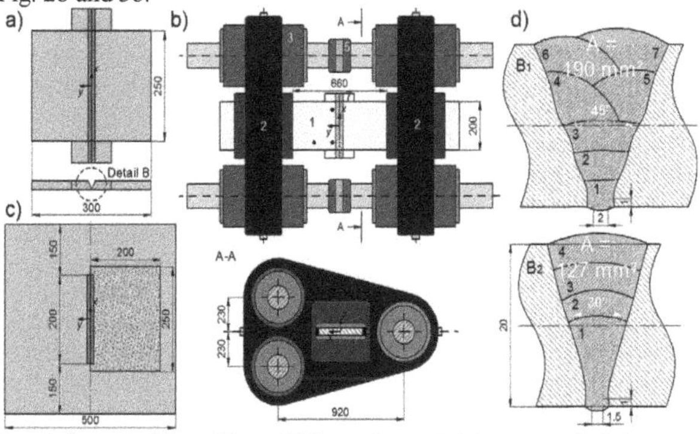

Fig. 2. Specimen dimensions for the free shrinking (a), externally restrained at 2-MN-testing facility setup (b) and self-restrained weld tests (c); d) build-up sequences for groove angles 45° and 30°

(1 – specimen / themocouples; 2 – test desks / hydraulic clamping (500 bar); 3 – hydraulic cylinder (200 bar); 4 – piston rod, 5 – load cell; 6 – weld seam, start and end tabs; 7 – DIC area).

The self-restraint in weld tests no. 5 and 6 was achieved by a 200 mm long slot in squared plate; see Fig. 2c and 3a. For a quantification of different restraint conditions, the restraint intensity in weld transverse direction R_{Fy} was introduced by Satoh [12]. It is the components stiffness towards the weld seam related to the seam length. For simple butt joints it can be estimated according to [12]. The values, given in Tab. 2, were experimentally and analytically determined for each setup. They are in good agreement with typical values for mobile crane components [4].

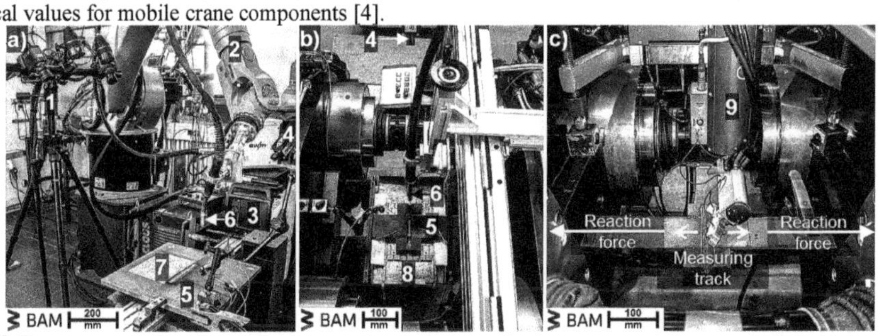

Fig. 3. Test setup: a) self-restrained weld test, b) 2-MN-testing facility, c) X-ray diffraction under restraint

(1 – DIC-system, 2 – welding robot, 3 – welding machine, 4 – two colour pyrometer, 5 – specimen with attached thermocouples, 6 – GMAW-torch, 7 – analysed area (DIC), 8 – heating mat, 9 – X-ray diffractometer).

Measurements. Type-K thermocouples were attached 10 mm adjacent to the weld to monitor the working temperature. The $\Delta t_{8/5}$-cooling times were determined at the weld surface using a two colour pyrometer (measuring range: 350 °C to 1300 °C). At the testing facility, reaction forces were measured using load cells at each piston rod of the 3d-hydraulic cylinder system; see Fig. 2b and 3b. In the slot welds, strain fields near the weld seam were measured using a DIC-system while welding and cooling; see Fig. 3a. In the welds, local transverse residual stresses were determined via X-ray diffraction using the $\sin^2\psi$-method on the top surface after cooling to ambient temperature.

Results

Global Reaction Stresses. Fig. 4a shows the reaction force $F_y(t)$, temperature $T(t)$ and bending moment $M_x(t)$ for test no. 3 in the 2-MN-testing facility, cp. Table 4. Preheating of the tack welded specimen initiates compressive forces. Hence, at the beginning of the root weld, the reaction force is F_y = -40 kN. The bending moment remains more or less zero. While root welding, the already solidified inserted weld metal generates transversal shrinking forces and a rising of the bending moment. Both increase further to a first maximum of F_y = 110 kN and M_x = 0.8 kNm during cooling to T_i = 100 °C. A reduction of force and moment is obvious while welding of the second layer due to a local heat input combined with stress relief. Subsequent cooling to T_i leads to a new continuous increasing of the reaction force and bending moment, the next weld run to a force and moment reduction. This evolution of the forces and moments was detected for every weld sequence. The amplitude of the reaction force increases with every weld run as a result of the welding heat input. The amplitude of the bending moment decreases and the mean level increases slightly with each weld run due to the increasing height of the weld. However, cooling of weld run four shows an increased moment build-up with a maximum of over M_x = 1 kNm. This is caused by the welding of the first layer above the neutral axis of the specimen in combination with an asymmetric insertion. Weld run five almost relieves this high bending moment. Subsequent cooling to ambient temperature after completion of the weld leads to a bending moment build-up of M_x = 0.95 kNm and to a maximum reaction force of $F_{y,end}$ = 404 kN. In Fig. 4b, the bending moments of the weld tests no. 3 and 4 are shown. Exceeding the neutral axis while root welding, a more symmetrical insertion of the weld metal and the reduced weld seam volume lead to reduced bending moments for the modified weld [13]. During welding, reduced amplitudes of the bending moment and after cooling to ambient temperature only M_x = 0.1 kNm occur. Fig. 4c presents a comparison of the reaction stress build-up $\sigma_y(t)$ of both welding processes of test no. 3 and 4. $\sigma_y(t)$ was determined using $F_y(t)$ and the actual load bearing layer height $H_{layer}(t)$. As the two graphs exhibit the same tendency, a reduction of total welding time by 40 % using the modified welding process is obvious. Further quantitative differences reveal in a decreased stress level and 10 % lower end reaction stresses $\sigma_{y,end}$ after cooling to ambient temperature in the modified weld. This is caused by a reduced amount of necessary weld runs and the lower overall heat input due to the decreased weld seam volume.

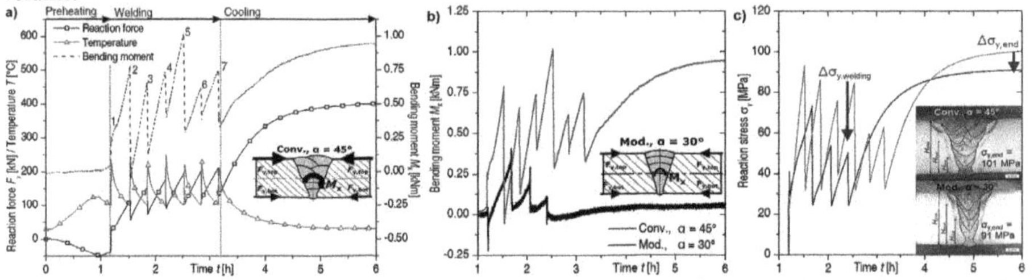

Fig. 4. a) Reaction force $F_y(t)$, temperature $T(t)$ and bending moment $M_x(t)$ for weld test no. 3; comparison of bending moments (b) and reaction stresses (c) for two different weld parameters at the 2-MN-testing facility.

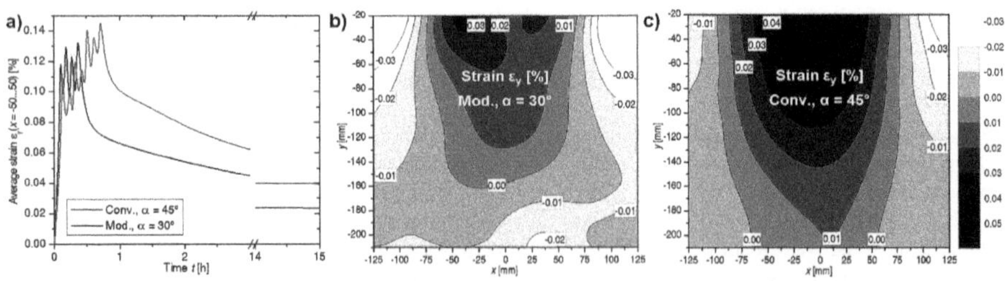

Fig. 5. Strains at the measured area of self-restrained specimens (test no. 5 and 6): a) average strains $\varepsilon_{y,avg}(t)$ 10 mm adjacent to the weld edge while preheating, welding and cooling for the different weld processes; Strain fields $\varepsilon_y(x,y)$ after cooling to ambient temperature for the modified (b) and the conventional weld (c).

Global strain fields. Fig. 5a shows the $\varepsilon_{y,avg}(t)$-graphs measured with the DIC-system 10 mm adjacent to the weld edge of both weld processes at the slot welds. A decreased strain build-up while welding is revealed for the modified weld. As the graphs are qualitatively equal, the three additionally needed weld runs cause an increased expansion of about $\Delta\varepsilon_y$ = 0.02 % at the top of the specimen towards the weld seam, which cannot be compensated during temperature equalization. Hence, the global strain field $\varepsilon_y(x,y)$ after subsequent cooling to ambient temperature of the modified weld (Fig. 5b) shows significantly reduced values compared to the conventional weld (Fig. 5c).

Local Residual Stresses. Fig. 6a and b show transversal residual stress distributions measured at three different restraint conditions across the weld ($x = 0$ mm). The graphs exhibit typical residual stress distributions for steels with an undergoing phase transformation and a broader stress profile for the conventional welds [8]. At the weld metal of the modified welds, higher tensile residual stresses of the amount of 60 % of $R_{p0.2}$ of the weld metal are obvious, which are due to a deeper weld penetration and almost independent of the restraint condition. However, the maximum tensile residual stresses of the conventional welds ($\Delta\sigma_{y,WM}$) rises with increasing restraint intensity up to 70 % of $R_{p0.2}$. In the area remote to the weld, residual stresses of the restrained welds are increased due to reaction stresses ($\Delta\sigma_{y,BM}$), which is also pronounced for the conventional welds; see also Fig. 4. Note that, the base material surface was blast cleaned and reveals residual stresses of about -160 MPa. Since almost no bending moments occur in the modified welds, the bending stresses at the HAZ are similar.

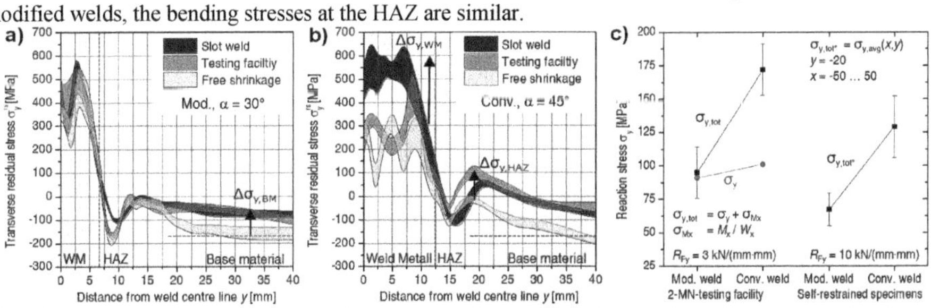

Fig. 6. Transverse residual stress distributions $\sigma^{rs}_y(y)$ at different restraint conditions for modified (a) and conventional welds (b); c) reaction stress vs. welding process for restrained weld tests (calculation: $\sigma_{y,tot}$ – reaction forces measured at 2-MN-testing facility, $\sigma_{y,tot*}$ – strain field measured with DIC-system)

The conventional welds show a stress increase in the HAZ due to restrained shrinkage and angular distortion ($\Delta\sigma_{y,HAZ}$). Therein, higher tensile residual stresses exhibit at the 2-MN-testing facility weld, despite a higher restraint intensity R_{Fy} of the slot weld. The specimens welded and measured in the testing facility reveal also higher total reaction stresses at the top of the welds, calculated by a superposition of global normal stresses σ_y and bending stresses σ_{Mx}; see Fig. 6c. The obvious bending stress increase is a result of the high bending restraint R_{Mx} of the 2-MN-testing facility [14]. It is approx. 20 times as high as in the slot welds due to the comparatively long levers to the two left-hand piston rods; see Fig. 2b.

Summary

The present work shows the effect of the seam configuration due to a modified welding process on welding loads and stresses in butt-joints of high-strength steel components. 20 mm thick plates were multilayer-GMA-welded under different defined restraint conditions. The groove angle of the joints was varied between 30° and 45° with adapted welding parameters. Occurring reaction stresses, moments and strain fields close to the seam of the restrained welds were observed in-situ while welding and cooling. Local transverse residual stresses were analysed by means of X-ray diffraction at all welds. From this work, the following conclusions can be drawn:

- While welding, decreased reaction stress amplitudes and a 10 % lower end reaction stress were found for the modified weld with a reduced weld seam volume.
- In highly self-restrained specimens, the smaller groove angle leads to decreased strains adjacent to the weld by approx. 70 %.
- A high effect was found for the bending moment in specimens welded under high bending restraints due to the 2-MN-testing facility. The additional consideration of resulting bending stresses reveals over 80 % higher total reaction stresses at the top of the conventional welds.

Materials Research Proceedings **2** (2016) 205-210 doi: http://dx.doi.org/10.21741/9781945291173-35

- The local analyses of stress superposition at the weld area surface mainly reveals a higher effect of the restraint condition on the local residual stresses in the conventional welds, especially elevated tensile residual stresses in the critical HAZ due to increased bending stresses.
- Based on recent studies [4,6,7] this work involves a multi-axial analysis of welding stresses regarding local restraint stresses, global normal reaction stresses and bending stresses as a consequence of a lateral and angular restraint. All measurement methods are in good agreement.
- The present study accomplishes a first quantification of the beneficial effects of reduced groove angles due to modified processes for component related welds of high-strength steel S960QL.

Acknowledgements
The studies were funded by the AiF-project IGF-Nr. 17978 N / FOSTA P1011. Sincere thanks are given for this support and to the representing companies actively involved in the project board.

References
[1] H.-P. Guenther, ed., Use and Application of High-Performance Steels for Steel Structures, IABSE IVBH, Zürich, 2005.

[2] R. Rauch, S. Kapl, et al., High Strength Low Alloy Steel Weldments with Accommodated Qualities to the Base Metal, BHM Berg- und Hüttenm. Monatshefte. 157 (2012) 102–107. http://dx.doi.org/10.1007/s00501-012-0060-5

[3] J. Lu, Prestress Engineering of Structural Material: A Global Design Approach to the Residual Stress Problem, in: Handb. Residual Stress Deform. Steel, 2002: pp. 11–26.

[4] D. Schroepfer, A. Kromm, T. Kannengiesser, Improving Welding Stresses by Filler Metal and Heat Control Selection in Component-related Butt Joints of High-strength Steel, Weld. World. 59 (2015) 455–464. http://dx.doi.org/10.1007/s40194-014-0219-7

[5] D. Schroepfer, T. Kannengiesser, Stress Build-up in HSLA Steel Welds due to Material Behaviour, J. Mater. Process. Technol. 227 (2016) 49–58. http://dx.doi.org/10.1016/j.jmatprotec.2015.08.003

[6] M. Rhode, A. Kromm, T. Kannengiesser, Residual Stresses in Multi-Layer Component Welds, in: Trends Weld. Res. Proc. 9th Int. Conf., ASM International, ISBN: 1-62708-998-8, Chicago, Illinois, USA, 2012: pp. 48–54.

[7] T. Lausch, T. Kannengiesser, M. Schmitz-Niederau, Multi-Axial Load Analysis of Thick-Walled Component Welds made of 13CrMoV9-10, J. Mater. Process. Technol. 213 (2013) 1234–1240. http://dx.doi.org/10.1016/j.jmatprotec.2013.01.008

[8] T. Nitschke-Pagel, H. Wohlfahrt, The Generation of Residual Stresses due to Joining Processes, in: V. Hauk, H. Hougardy, E. Macherauch (Eds.), Residual Stress. - Meas. Calc. Eval., DGM Informationsgesellschaft mbH, ISBN: 3-88355-169-4, 1991: pp. 121–133.

[9] J. Chen, C. Schwenk, et al., Predicting the Influence of Groove Angle on Heat Transfer and Fluid Flow for New Gas Metal Arc Welding Processes, Int. J. Heat Mass Transf. 55 (2011) 102–111.

[10] EN ISO 16834: Welding consumables – Wire electrodes, wires, rods and deposits for gas shielded arc welding of high strength steels – Classification, 2012.

[11] EN 10025-6: Hot rolled products of structural steels – Pt. 6: Technical delivery conditions for flat products of high yield strength structural steels in the quenched and tempered conditions, 2011.

[12] K. Satoh, Y. Ueda, S. Matsui, M. Natsume, T. Terasaki, K. Fukuda, M. Tsuji, Japanese Studies on Structural Restraint Severity in Relation to Weld Cracking, Weld. World. 15 (1977) 155–189.

[13] K. Masubuchi, Analysis of Welded Structures: Residual Stresses, Distortion, and their Consequences, Pergamon Press, 1980.

[14] K. Satoh, S. Matsui, Effect of Intensity of Bending Restraint on Weldcracking in Multipass Weld, Trans. Japan Weld. Soc. 8 (1977) 42–49. Japan Welding Society, 1977.

Residual Stresses 2016: ICRS-10 Materials Research Forum LLC
Materials Research Proceedings **2** (2016) 211-216 doi: http://dx.doi.org/10.21741/9781945291173-36

Evaluation of Residual Stress by X-Ray Diffraction and Correlative Stress Modelling

Kumar, Sandeep[1, a], Crivoi, Alexandru [1, a], Tan, Ming Jen, [2, b,*] Tai, Anna[3, c], Marinescu, Iulian[3, c]

[1]Rolls-Royce@NTU Corporate Lab., Singapore

[2]Sch. of Mech. & Aerospace Engg., Nanyang Technological University, Singapore

[3]Rolls-Royce Singapore Pte. Ltd., Singapore

kumar.s@ntu.edu.sg [a], acrivoi@ntu.edu.sg [a], mjtan@ntu.edu.sg [b,*],

anna.tai@rolls-royce.com [c], Iulian.Marinescu2@rolls-royce.com [c]

Keywords: Residual Stress, TIG Welding, X-Ray Diffraction, Numerical Simulation, Aerospace Material

Abstract. Residual stress is an unavoidable problem which occurs during any manufacturing process or during repair when it is unwanted or uncontrolled, and becomes a limitation to the service life of a component. Welding is one of the common repair methods used on gas turbine engine components that develops high residual stress, and uncontrolled residual stress may appear in several manufacturing processes which involves uneven distribution of heat, mainly in a localised manner (i.e. at the cutting/welding tip/zone). This study is mainly focused on the estimation of the surface residual stress after welding on aerospace material; here X-Ray diffraction (XRD) is used for analysis since it is an accurate NDT method used for measuring residual stress. X-ray diffraction is made a semi-destructive method by removing material using electropolishing to micron level for analysis of sub-surface stresses, as X-ray diffraction method has low penetration depth. Residual stress measurement is carried out at surface level and correlated with numerical simulation of residual stress due to welding.

Introduction

Most of the manufacturing and repair processes generate residual stresses in components. Residual stress is critical and needs to be controlled. In most cases, understanding the distribution versus the depth of the residual stress is of interest in various industries. To acquire this information, the most commonly used methods are X-ray diffraction with layer removal and central hole drilling [1]. Residual stresses in welded joints primarily develop due to differential weld thermal cycle (significantly varying temperature) experienced by the weld metal and region closed to fusion boundary i.e. heat affected zone. Type and magnitude of the residual stresses vary continuously during different stages of welding. The stresses in the weld joints develop due to typical nature of welding process i.e. localized heating and cooling leading to varying volumetric expansion and contraction of metal around the weld zone (Fig. 1 (a),(b),(c)) [2].

The objective of this paper is to measure the residual stress development at surface level to understand the significant impact on the performance of the component in aerospace materials. X-ray diffraction with layer removal can be used for the measurement of residual stress in the sub-surface level which will help to understand others problems e.g. failure. The results obtained from the semi-destructive method at surface level will be correlated with stress modelling and also used to comprehend the residual stress distribution [3]. Finite element model of the specimen is developed, and the welding simulation is performed to model the resulting stress fields.

Materials Research Proceedings **2** (2016) 211-216 doi: http://dx.doi.org/10.21741/9781945291173-36

Experimental methods

X-ray diffraction is the most accurate and best developed method of quantifying mainly the surface residual stress for weld repair analysis, but for X-ray diffraction the penetration depth and resolution is of lower magnitude and surface residual stress is generally misleading on the weld bead region due to surface deformed layer in the material. In order to determine residual stress, the strain in the crystal lattice must be measured with reference to a precisely known reference orientation from a sample surface. X-ray can also be used for sub-surface measurement by removing material (incremental etching) making it a semi-destructive technique [4].

As a result, it is normally recommended that surface and subsurface stresses are measured to fully understand and characterize the residual stress fields that are developed. Surface residual stress measurement is quite straight forward, but for the subsurface residual stress X-ray diffraction becomes a semi-destructive technique. X-ray has the tendency to be absorbed when passed through a material, so naturally that will affect the intensity, and in-turn the intensity affects the penetration depth, so material needs to be removed to measure residual stress at sub-surface level [5, 6].

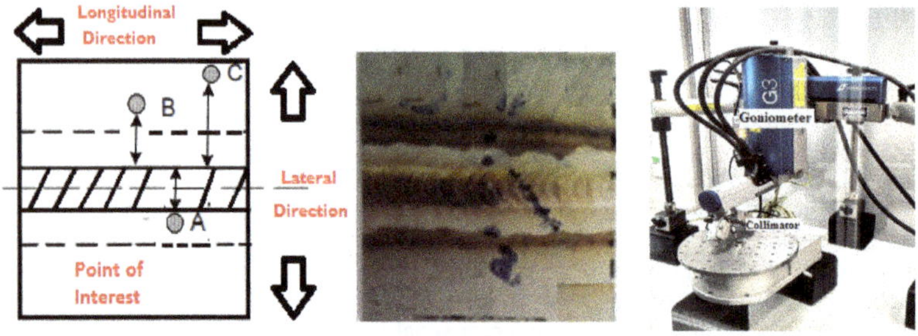

Figure 1. (a)Point of interest on welded specimen,(b) Ti-6Al-4V Coupon (c)Experimental setup for residual stress measurements.

Xstress 3000 is the equipment shown in Fig. 1(c) used for residual stress analysis, this equipment is specifically used for stress measurement on simple and complex geometries. The equipment is required to be calibrated with stress-free samples and known residual stress specimens before measurements on the specimens (Table 1) [7].

The test coupons shown in Fig. 1(b) are made up of Ti-6Al-4V, the machined specimen shown in Fig. 1(a) is a flat plate of 50 mm x 50 mm x 3.42 mm dimensions. Two specimens of diffrent thicknesses are seperatly flat-bead welded with similar Tungeston Inert Gas welding specifications for residual stress analysis (Table 2). Figure 1(a) indicates the points of interest where the residual stress is required to be analyzed at surface level. Point measurement was carried on the weld-bead, HAZ and base-metal region, with a 3mm collimator, which helps to focus the X-ray for accurate measurement. Residual stress is measured in both lateral and longitudinal direction, generally the peak stress concentration area is in longitudinal direction along the weld-bead and heat affected zone.

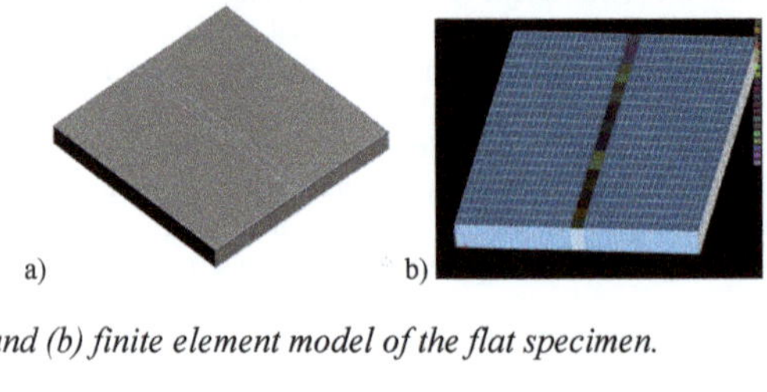

Figure 2. (a) CAD and (b) finite element model of the flat specimen.

Materials Research Proceedings **2** (2016) 211-216 doi: http://dx.doi.org/10.21741/9781945291173-36

Table 1: Xstress 3000 experimental measurements

{hkl}	-	{213} (used for copper tube)
Psi angles/Tilt oscillation	-	0, -25.8, -38.0, 0, 25.8, 38.0 / ± **3**
Exposure time	-	70 sec
Measuring Mode	-	Modified χ
X-ray Voltage	-	30 kv
X-ray current	-	10 mA
Collimator distance	-	12.7mm
Post processing-Calculation		
Background	-	Parabolic
Fit	-	Gauss fit

Modeling

The finite element model of the specimen has been developed for simulations (Fig. 2). The CAD of the plate (Fig. 2(a)) has been created using the geometry module of Salome [9]. The plate volume has been partitioned into three parts by two section planes to highlight the volume under the welding zone. Simple hexagonal mesh has been created using uniform wire discretization method with N=15 in the mesh module of Salome. The welding area width is set to 3mm, and the welding area mesh faces are sorted into 20 groups along the welding direction. To simulate the welding process, the moving heat source method is applied to the welding area faces (colored in Fig. 2). Since the total welding time is estimated to be around 20 seconds, each highlighted mesh group receives a heat flux pulse for the duration of 1 second consecutively. As the welding process moves in the lateral direction, the i-th mesh group receives a heat flux boundary condition between the (i-1)-th and i-th second. Additionally, a more advanced method called here "moving heat source with a trace" is applied here for comparison. This method leaves the residual heat flux as a boundary condition for the i-th face after the i-th second, exponentially decaying with time. The purpose of the method is to simulate the filler metal deposition on the base plate, which continues to transfer the heat to the specimen after the welding head has moved away from the point. The thermal and structural solutions of the model are generated using the Code-Aster software package and post-processed in Salome [10].

Results and Discussion

Initially X-ray diffraction is set to the required specifications, and calibration is carried out with stress free and known stress samples. Then followed by some preliminary analysis on magnitude of residual stress based on specimens with different thicknesses.

	Thickness of Ti6/4 Welded Specimens (50mmx50mm)	Longitudinal direction	Lateral direction
Residual stress is measured on either sides of weld-bead in the HAZ of specimens	1mm	High tensile stress	Low tensile stress
	3.5mm	Medium tensile stress	Low compressive stress

Table 2: Magnitude change in residual stress.

The specimens used for the analysis were welded in similar conditions, i.e. residual stress measured along the HAZ of specimens is tabulated, this will enable understanding the stress range due to

Materials Research Proceedings **2** (2016) 211-216 doi: http://dx.doi.org/10.21741/9781945291173-36

varying thicknes of the specimens. The thickness of the specimen or the component has an effect on the magnitude of the residual stress due to welding (Table 2).

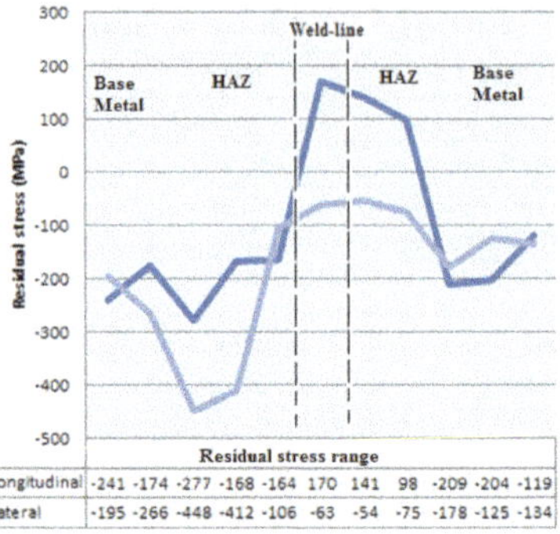

Figure 3: Surface residual stress distribution of TIG welded specimen

Residual stress is measured at different data points on the weld-bead, along HAZ and base metal region. Figure 3 shows that residual stress distribution for TIG welded specimen of 3.5 mm thickness. XRD equipment measures residual stress at surface level to a depth of less than 10 micron and close to 5 micron when using a copper tube. Figure 3 shows residual stress is completely distributed in the compressive stress region towards the base metal and HAZ region for both the longitudinal and lateral stress, while along the weld bead along the longitudinal direction stress is distributed in the high tensile region compared to lateral stress. Another observation is that the stress is linearly increasing (towards tensile region) along the weld-bead in both the directions. The maximum stress is around 450 MPa compressive along the lateral direction close to the HAZ. The stress distribution is not symmetric on either sides of the weld because of minor shift in the consistancy of welding parameters, eg:welding speed, uneven weld etc.

a) 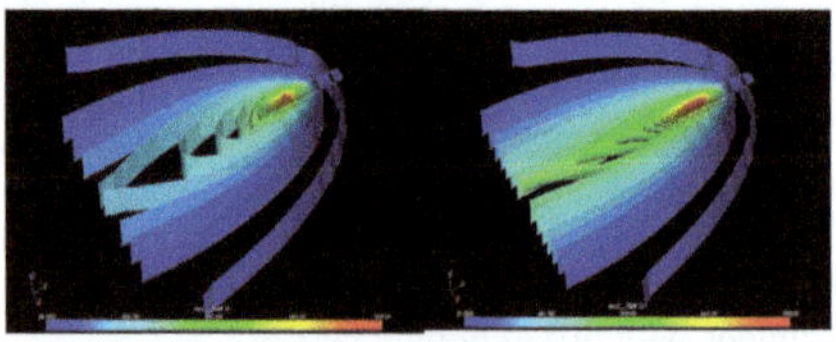 b)

Figure 4. Isothermal surfaces on the specimen during the simulated welding process. (a) Simple moving heat source method. (b) Moving heat source with a trace.

Figure 4 shows the comparison between the simulated isothermal surfaces during the welding process using two different methods. Figure 4(b) corresponds to the thermal field solution generated from the "moving heat source with a trace" boundary conditions. The isothermal surfaces in Fig. 4(b) have higher elongation and are assumed to better match the additional thermal effects from the filler material deposition than the conventional moving heat source model solution shown in Fig. 4(a).

Materials Research Proceedings **2** (2016) 211-216 doi: http://dx.doi.org/10.21741/9781945291173-36

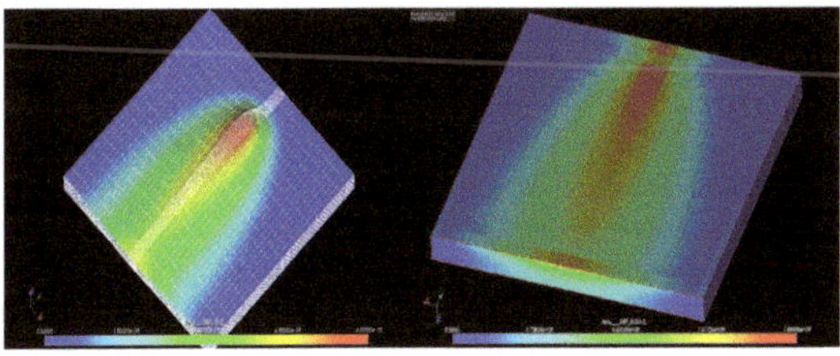

a) b)

Figure 5. (a) Displacement field map with deformed surface plot and (b) thermal stress field during the simulated welding process.

Figure 5 illustrates the thermo-mechanical effects during the welding process. Local heating of the plate causes the significant displacements occur around the weld line due to the thermal expansion of the plate material. The displacements are shown in Fig. 5(a) with a scale factor of 50. In turn, displacements are linked to the thermal stresses appearing around the heated zones as result of the elastic response (see Fig. 5(b)).

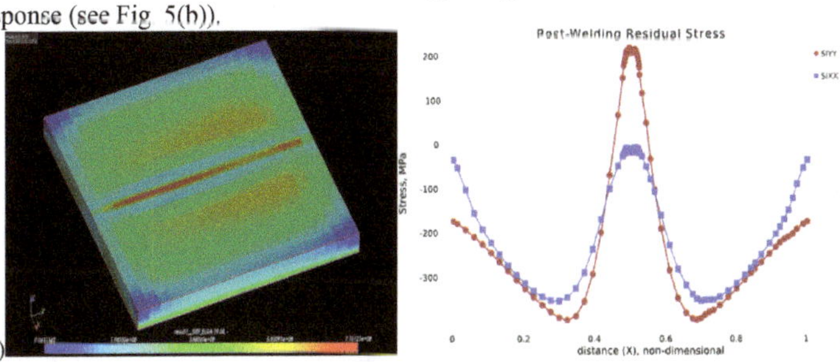

a) b)

Figure 6. (a) Post-welding von Mises residual stress field. (b) Plots of the longitudinal (SIYY) and lateral (SIXX) principal stress components of the residual stress tensor. The plot uses data points along the line perpendicular to the weld line.

Finally, the simulated welding process reaches the edge of the plate after *20 s*, and the plate starts to cool naturally. Then, after another *20 s* period of heat dissipation and natural convection, the thermal gradients inside the plate smoothen out, and the residual strains and stresses caused by elasto-plastic effects show up. Figure 6(a) shows the von Mises stress field plot after *20 s* of welding and *20 s* of natural convective cooling. Peak values are concentrated along the thin weld line zone. The last diagram shown in Fig. 6(b) illustrates the dependency of the stress components on the position in the lateral direction. Longitudinal stress reaches the peak tensile values in the vicinity of the weld line and peak compressive values in the HAZ. Lateral stress has very low absolute values near the edges and near the weld line, but reaches high compressive values in the HAZ (please refer to Fig. 6(b)).

Conclusion

It is concluded that the numerical analysis of weld residual stress appears to have a good agreement with the XRD measurement. The agreement was based on the stress graph from experimental and numerical analysis, and the range of the stress values in the three regions of the weld specimen are matched as a part of the preliminary validation study. The stresses measured in both longitudinal and transverse direction seems to have a fair correlation with modelling, considering the diffrence in experimental measurements due to the minor shift in welding consistency of the specimens; peak tensile stress is observed along the weld-bead, it can be seen to decrease and shift to the compressive region between the heat affected zone (HAZ) and the base metal region, peak compressive stress is observed near the HAZ along the longitudinal direction for both modelling and experimental measurements. Thus the comparison between measured stress profiles and model predictions at the surface level has been established with good agreement.

Future studies will mainly focus on a flat plate with minor amounts of material removal and comparing them with specimen thickness for sub-surface residual stress measurements by X-ray diffraction, and the respective correlation to sub-surface measurements by stress modelling.

Acknowledgements

This work was conducted within the Rolls-Royce@NTU Corporate Lab with support from the National Research Foundation (NRF) Singapore under the Corp Lab@University scheme.

References

[1] Warren, A. W., & Guo, Y. B. Characteristics of residual stress profiles in hard turned vs. Ground surfaces with and without a white layer. *ASME. J.Manuf. Sci. Eng.* (2009), 131 (4). http://dx.doi.org/10.1115/1.3159046

[2] IDC Technologies. Residual stress in welded joints. *IDC technical references Lecture 21*. (2003) U.S.A.

[3] Cammet, J. Quality assurance of shot peening by automated surface and subsurface residual stress measurement. *Lambda Research-The Shot Peener*, (2001) *15*(3), 7-8, U.S.A.

[4] Fitzpatrick, M. E., Fry, A. T., Holdway, P., Kandil, F.A., Shackleton, J., & Suominen, L. Determination of Residual stresses by X-ray diffraction. *National Physical Laboratory-Measurement Good Practice Guide No. 52*. (2005), U.K.

[5] Suzuki, T. Masaaki, S. Hatsuhiko, O.T.N. Muneyuki, I.Y.T. Hiroshi, S. & Atsushi, M. Residual Stress Measurement of Welding Area by Neutron Diffraction Method. *Nippon Steel Technical Report* 100, (2011), Japan.

[6] Kumar, S., Tan, M.J, Wong, B.S, & Weeks, N. Post Weld Heat Treatment surface residual stress measurements using X-ray diffraction, *Emerging Technologies in Non-Destructive Testing VI-CRC Press*. Dec. (2015), 563 -567, Belgium. http://dx.doi.org/10.1201/b19381-93

[7] Stresstech Group, V1.5, . *Xtronic Guide -Stresstech Guide*, Finland, (2013).

[8] Modenesi, P.J. Apolinario, E.R. & Pereira, I.M. TIG welding with single-component fluxes. *Journal of Materials Processing Technology*, (2000), *99*(1), 260-265. http://dx.doi.org/10.1016/S0924-0136(99)00435-5

[9] Salome, open source platform for CAE problems: www.salome-platform.org/, (2016).

[10] Code_Aster, open source - general FEA software: www.code-aster.org/, *(2016)*.

The Relationship between X-Ray Stress Measured Value and Applied Stress in Elastic/Plastic Deformation Region in Tensile Testing

Masashi Nakashiro[1a*], Yasuhiro Takaku[2b], Yukinori Mitani[2c], and Ami Kohri [2d]

[1]IHI Inspection & Instrumentation Co., Ltd. (IIC), 25-3, Minami-Ohi 6-chome, Shinagawa-ku, Tokyo 140-0013 Japan

[2] IHI Inspection & Instrumentation Co., Ltd. (IIC), 6-17, Fukuura 2-chome, Kanazawa-ku, Yokohama-city, Kanagawa 236-0004 Japan

[a] m_nakashiro@iic.ihi.co.jp, [b] y_takaku@iic.ihi.co.jp,

[c] y_mitani@iic.ihi.co.jp, [d] a_kohri@iic.ihi.co.jp

Keywords: X-Ray Residual Measurement, $Sin^2\Psi$, True Stress, Elastic Strain, Plastic Strain

Abstract. Recently in Japan, the 2D detector X-ray stress measurement equipment has developed, and requests of on-site measuring the stress on structural components are increasing. On the other hand, high residual stress, which exceeds its tensile strength stress, is often observed on welded construction parts, and sometimes it causes defects or damage. We report a result of clarified characteristics of X-ray diffraction stress value (XRD) in which exceeds the yield point in the plastic strain region on steel (SM490) and stainless steel (SUS316). For studying the correlation with XRD measurement results and the stress value, we installed flat plate test specimens in the tensile test machine and cyclically loaded to the plastic strain region to simulate the welding structure under load. The XRD measurement values indicated a high correlation with the actual stress value. These results confirm that the XRD measurement values give the actual stress condition of the material under load, irrespective of elastic-plastic deformation. XRD measurement is not only applicable for measuring internal stress but also measuring combination of internal and external stress on a material. Consequently, XRD measurement can also be used to measure the loaded stress conditions of a structure (this stress is also called dead load or static load). Moreover, this outcome expands function of XRD from a specialized measuring method for residual stress to various structural conditions.

1. Introduction

To evaluate the validity of result by X-ray diffraction (XRD) measured residual stress, it is often used comparison with mechanical stress by 4-bend testing. This method is also used as measurement of X-ray elastic constants[1)]. X-ray itself measures stress which dwells on fairly surface, so the residual stress value measured by XRD and mechanical stress value by 4-bend testing shows a high correlation when the specimen is in elastic region. However, measuring welded, heat treated, peened, or other surface treated parts are common demands in field measurement. Usually, there are yield point exceed high residual stress in these processed parts. The tensile test machine is used to produce the uniform deformation in the sample to evaluate the plastic deformation by XRD. Also, it is known that there are difference between the XRD measured value and the mechanical stress value in the plastic region. This difference occurs by influences of surface effect and phase stress[2),3)].

This research describes the correlation between XRD measured stress and mechanical stress from the elastic to plastic strain range. We calculated nominal stress value by XRD measured stress value and load value of tensile test and true stress value by specimen elongation.

Materials Research Proceedings **2** (2016) 217-222 doi: http://dx.doi.org/10.21741/9781945291173-37

2. Experimental Materials and Method

2.1 Experimental Materials

For testing, we chose the carbon steel (JIS: SM490) specimen which has a yield point and the stainless steel (JIS: SUS316) which does not have a yield point. We determined elastic region of the carbon steel as up to the yield point and the stainless steel as up to 0.2% of proof stress. Table 1 shows the sample material chemical compositions, mechanical properties and manufacturing process. These material's grain size are small more than ASTM E112 Plate IV No.5 and suit XRD stress measurement. Fig.1 shows the shape and dimensions of specimen.

Table 1 Material properties of test sample by mill sheet

Material (JIS)	Chemical compositions								Mechanical properties		
	C	Si	Mn	P	S	Ni	Cr	Mo	Yield strength/ 0.2% proof strength	Tensile strength	Elongation
	Wt%								(MPa)		(%)
Carbon steel (SM490)	0.17	0.33	1.37	0.014	0.003				418	541	26
Stainless steel (SUS316)	0.04	0.46	0.84	0.028	0.001	10.06	16.08	2.09	337	601	56

SM490: Hot rolling plate, SUS316: Solution treatment

We determined the gauge length of extensometer as 50mm. For measuring by XRD, influence of residual stress which occurred while fabrication process of specimens should be removed. Therefore, we electrolytic polished the center parts of specimens, where to XRD measuring, for 6mm diameter circle with depth of 0.15mm.

2.2 Experimental method

For mechanical test, we used 100kN screw load type tensile test machine. The measurement items were load-sell of tensile machine, XRD equipment, strain gauge, and extension meter. The strains were measured by strain gauges in elastic region and extensometer in plastic strain. The applied stress were calculated by the load divided by specimen cross area. Fig.2 shows the experimental status on tensile test machine. For measuring the stress in elastic region, we stopped the cross head of the testing machine when it gains the predefined stress point. Then we measured the stress by XRD with keeping cross head position and strain by strain gauge. For measuring the stress in plastic deformation area, we stopped the cross head when the load of testing machine was stabilized. Then we measured the stress by XDR, elongation and load. For loading methods, we used incremental load method and cyclically load method.

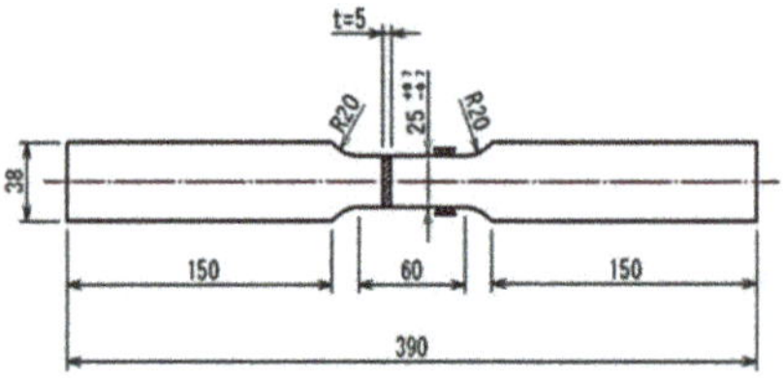

Fig. 1 Test specimen shape and dimension

1) Elastic strain: $\varepsilon_e=\sigma/E$ Plastic deformation: $\varepsilon_p= \varepsilon-\varepsilon_e= \varepsilon-\sigma/E$

2) Nominal stress: $\sigma_n = P/S_o$ P: Load Value (N) , S_o:Initial cross section area of specimen (mm^2)

3) True stress: $\sigma_t = \sigma_n \cdot (1+ \varepsilon)$, σ_t is valid until the tensile strength of maximum load point σ_B.

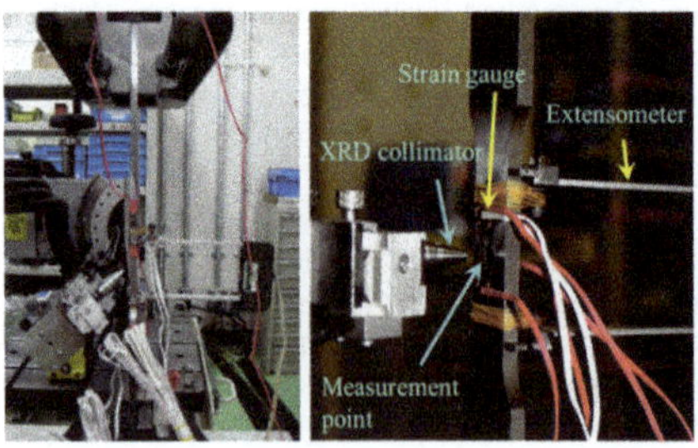

Fig. 2 Experimental setup for X-ray stress measurement in tensile testing

The measureable range of extension meter is up to 25mm deformation in 50mm gauge length and maximum strain is 0.5mm/mm. SM490 specimens were measurable until just before the fracture point. For SUS316 specimens, we set the range as 50% strain because the fracture elongation was more than 55 %. Fig.3 shows the schematic diagram of specimen stress-strain curves, uniaxial plane strain model as stress model, and plastic deformation value. The nominal stress and the true stress were calculated by the following equations.

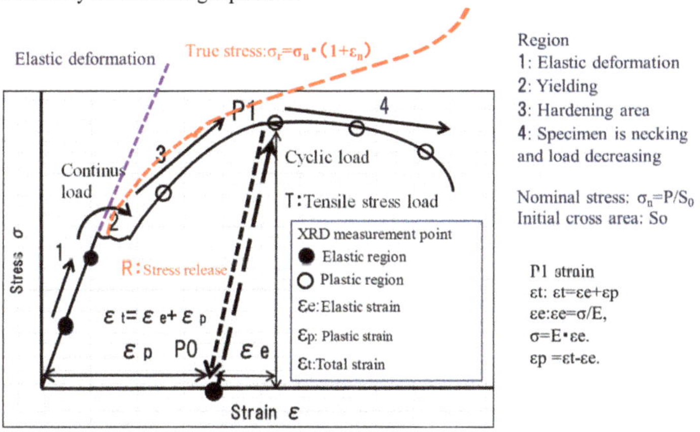

Fig. 3 The schematic diagram of specimen stress-strain curves

Sin$^2\psi$ method is used for XRD measurement, and sin$^2\psi$ method is used for ψ goniometer. The applied equipment was Stresstech Oy., X3000. Table 2 shows the measuring specification.

Table 2 Conditions for X-ray stress measurement

Material (JIS)	Carbon steel (JIS: SM490)	Stainless steel (JIS: SUS316)
Characteristic X-ray	Cr-Kα	Mn-Kα
Diffraction line, hkl	211	311
Diffraction angle (deg.)	156.4	152.3
Tube voltage (kV)	30	
Tube current (mA)	6.7	
Collimator spot size	(X-ray slit) φ3.0 mm	

3. Experimental Results and Discussion

3.1 Measurement Results of Incremental Load Process

Fig.4 shows the experimental results of the relationship between XRD measured stress value and strain at each interrupting position in the continuous incremental load test. Stress was measured by XRD after a while from the load tester stopped and load is stable. Since it takes a little while that the load to specimen gets stable when the load tester stopped. The XRD measured stress value was compared with the nominal stress and true stress value. The result showed both SM490 and SUS316 decreased their cross-section areas of specimens when the strain or elongation is increased. Nominal stress values increased a little and drew a smooth curve in plastic region. However, calculated true stress values in effective cross-section area during measurement increased monotonically. Likewise, XRD measured stress value increased monotonically, and this changing rate corresponds with changing rate of true stress.

Materials Research Proceedings **2** (2016) 217-222 doi: http://dx.doi.org/10.21741/9781945291173-37

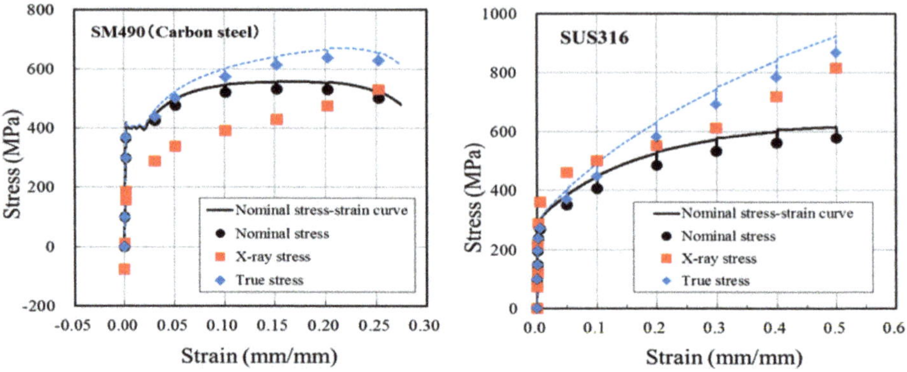

Fig. 4 Stress-strain curve for continuous tensile loading

Stainless steel does not show clear yield point, so we determined its elastic region as up to 150MPa. Fig.5 shows the relationship between XRD measured stress value and the true stress value for each measuring point in elastic and plastic regions. In the case of SM490, the difference between σ_t andσ_x is about 180MPa. There is the influence of specimen setting condition such as bending moment and specimen surface initial stress conditions. But in this figure, XRD measured stress value and the true stress value shows deep correlation, and the coefficient factor is approximately 1.

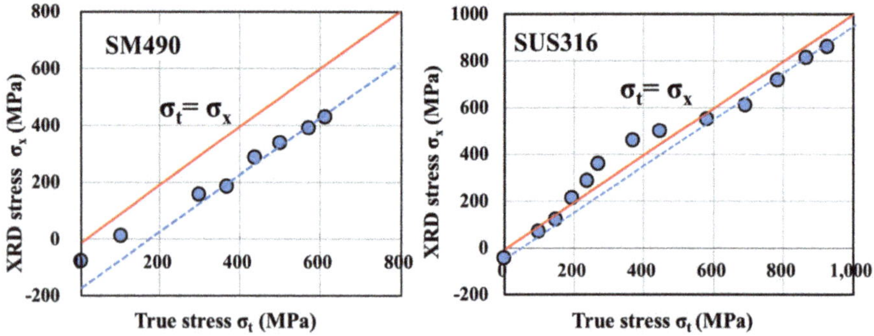

Fig. 5 Relationship between X-ray stress and true stress in continuous tensile lording

On the other hand, case of SUS316, XRD measured stress value was higher than the true stress value in the stress level from 200 MPa of proof stress to 400 MPa. We considered that these areas are large work hardening and it is because there was stress distribution difference between both inside and surface of the specimen in the plastic region. Therefore, strain value on the surface of specimen was bigger than its inside when incrementally loaded, so the XRD measured value was higher.

In any case, carbon steel and stainless steel together have the good correlation for the actual-stress and the XRD stress value to the large plastic deformation.

3.2 Measurement Results of Cyclic Load Process

In this cyclic load measurement, we stopped tensile test machine and measured the stress by XRD at predefined stress point in elastic region and predefined strain value point in plastic region. After

Materials Research Proceedings **2** (2016) 217-222 doi: http://dx.doi.org/10.21741/9781945291173-37

measuring, the load was released to level 0, and we measured XRD stress again. Then the specimen was re-loaded again until next release point, and we measured XRD stress value. This testing flow was repeated for each measuring points. Fig.6 shows the measurement results by XRD stress and applied load value on each extensometer elongation for each measured points.

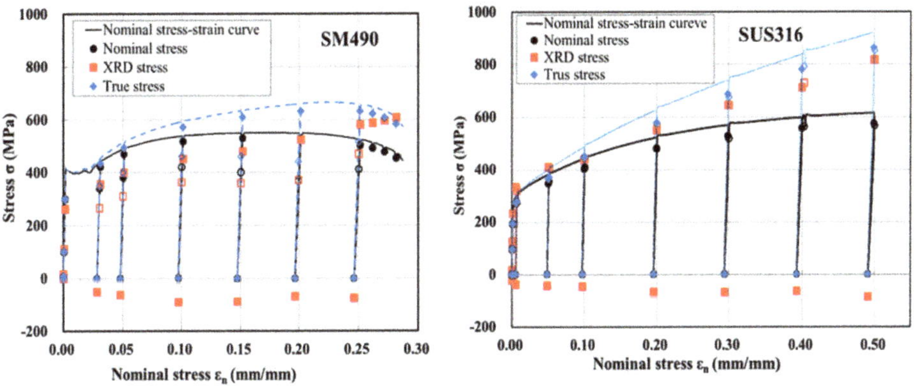

Fig. 6 Stress-strain curve for cyclic tensile loading

Both SM490 and SUS316 showed compressed stress value when loading was released in plastic region. We regarded it as Bauschinger effect by the cause of the stress distribution difference in the specimen cross sections. In the case of SM490, the true stress value over the tensile strength σ_B is reference data. Since the strain stress on SM490 ingenerates localized neck when the stress exceeds σ_B, but calculated true stress value from measured localized neck area, the true stress value shall be increased monotonically like XRD measured stress value. Fig.7 shows the relationship between $\sin^2\psi$ and d for XRD measurement results of SUS316 each plastic deformation. Irrespective of the plastic deformation, the both relation shows good linearity correlation. This figure shows the measurement effectivity was confirmed.

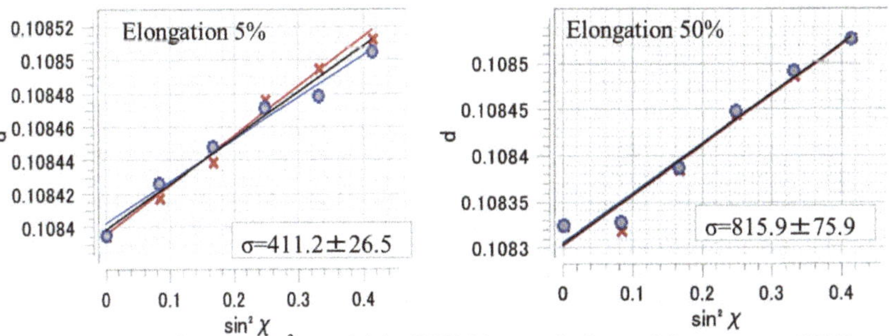

Fig. 7 Relationship between $\sin^2\psi$ and d for SUS316 on each elastic deformation and XRD stress results

Fig.8 shows the relationship between XRD measured stress value and the true stress value from these measuring results. There are high correlations with XRD stress and the true stress, as same as the case of incremental load process. In the case of SM 490, the difference between σt andσx is about

Materials Research Proceedings **2** (2016) 217-222 doi: http://dx.doi.org/10.21741/9781945291173-37

100MPa. In the case of SUS316, XRD measured stress value was higher than the true stress value on low plastic region near the proof stress as same as incremental load process. However, the stress distribution difference on inside of specimen which occurred by repeating the test got smaller, and it showed deep linear relationship with true stress alongside of incremental load process.

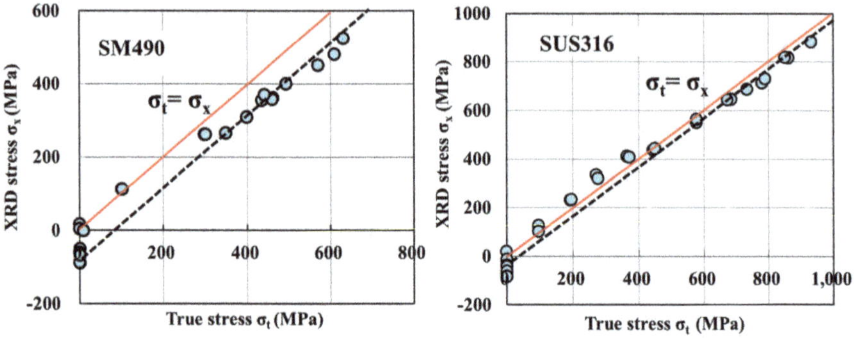

Fig. 8 Relationship between X-ray stress and true strain in cyclic tensile lording

4. Conclusion

1) We examined the relationship between the external load stress and XRD measured stress. XRD measured stress can be always shown the real stress value on the material regardless of elastic and plastic deformation.

2) XRD measured values are shown summation total stress value of the internal and external load stress.

3) XRD measurement method is a highly reliable method to evaluate the absolute stress on a material.

4) XRD measured value at the surface neighborhood of material is influenced by its surface condition.

5) The residual stress values vary depending on the measurement methods, and XRD stress values are sometimes different from the strain gauge measured values on the field measurement.

6) It is important to understand the properties of each measurement method well.

7) It is difficult to measure the absolute stress value of components as correctly on site. We have to use suitable and various measurement methods to measure highly reliable value.

References

[1] JSMS-SD-5-02, "Standard method for X-ray stress measurement'', The Society of Materials Science, Japan (2002).

[2] S. Taira, K. Hayashi and N. Urakawa, "X-ray Measurement and Analysis of Residual Stress Induced in Polycrystalline Aluminum by Uniaxial Plastic Deformation", Journal of the Japan Institute of Metals, Vol.35, No.3, pp189-196(1971).

[3] K. Kojima, "Stress Measurement for X-ray", Journal of the Society of Materials Science, Japan, Vol.47, No.6, pp.502-509(1952).

Residual Stresses 2016: ICRS-10 Materials Research Forum LLC
Materials Research Proceedings **2** (2016) 223-228 doi: http://dx.doi.org/10.21741/9781945291173-38

Influence of Heat Control on Residual Stresses in Low Transformation Temperature (LTT) Large Scale Welds

Jonny Dixneit[1, a *], Arne Kromm[1, b], Mirko Boin[2, c], Thomas Kannengiesser[1, d] and Jens Gibmeier[3, e]

[1]Bundesanstalt für Materialforschung und –prüfung (BAM), Unter den Eichen 87, 12205 Berlin, Germany

[2]Helmholtz-Zentrum Berlin für Materialien und Energie GmbH, Hahn-Meitner-Platz 1, 14109 Berlin, Germany

[3]Karlsruhe Institute of Technology, IAM, Engelbert-Arnold-Straße 4, 76131 Karlsruhe, Germany

[a]jonny.dixneit@bam.de, [b]arne.kromm@bam.de, [c]boin@helmholtz-berlin.de, [d]thomas.kannengiesser@bam.de, [e]jens.gibmeier@kit.edu

Keywords: LTT, Welding Residual Stress, Phase Transformation, Interpass Temperature.

Abstract. The current paper presents residual stress analyses of large scale LTT (Low Transformation Temperature) welds. LTT filler materials are specially designed for residual stress engineering by means of an adjusted martensite phase transformation. Controlling the level of mostly detrimental residual stresses already during the welding process would be highly attractive as time and cost consuming post processing may be prevented. In large scale welds the residual stress state is influenced by the heat control (e.g. interpass temperature) during welding. Therefore, welding residual stresses are studied here putting the focus on the influence of welding process parameters while joining heavy steel sections with a thickness of 25 mm. The residual stress state was determined at the top surface using X-ray diffraction as well as in the bulk by neutron diffraction. The results show that control of the interpass temperature is vital for the residual stresses present in the joints. This accounts for the top surface but is most pronounced for the bulk of the welds. While high interpass temperatures are appropriate to induce compressive residual stresses in the weld metal, low interpass temperatures favor unwanted tensile residual stresses instead.

Introduction

Despite the well-known benefit of post-weld heat treatments on fatigue life due to altering welding residual stresses, such processes often lead to additional production costs. Novel Low Transformation Temperature (LTT-) filler materials are specially designed for controlling weld residual stresses by means of an adjusted martensite formation already during welding and thus its application may minimize costs. The volume expansion due to phase transformation counteracts the contraction due to shrinkage of the joint. Tensile residual stresses are reduced or even beneficial compressive residual stresses are formed. The strength of these filler materials makes them potentially applicable to high-strength steels as well as to repair works in existing steel structures. The influence of varying phase transformation temperatures on the residual stresses for different weld geometries was the aim of most investigations up to now. Shiga et al. [1] and Francis et al. [2] confirmed that compressive residual stresses can be achieved, when welding LTT-alloys with varying martensit start (M_S-) temperatures. The focus was also on the influence of the welding parameters. High interpass temperatures can delay the martensite formation during multi-pass welding. It follows that the whole weld metal undergoes the phase transformation not before cooling to ambient temperature after deposition of the last bead. On the other hand, low interpass temperatures allow repeated martensite formation in the already deposited layers. Both alters the residual stresses found in a welded joint [3]. But the final residual stresses are also influenced by the restraint conditions of the structure. That means the weld is restrained by the adjacent base material (BM) as well as by the stiffness of the whole weld construction. Zenitani et al. [4] already confirmed the influence of varying restraint intensities regarding residual stresses in LTT-joints. Therefore, the present study was focused on the evaluation of the residual stresses in thick walled LTT-joints showing a high restraint concerning the impact of the interpass temperature.

Experimental

Materials and Welding. Based on already published chemical compositions two LTT alloys were prepared. Some characteristics are presented in Table 2 and Table 1. Both fillers show the main alloying elements chromium and nickel or manganese respectively. Two different interpass temperatures were applied. Butt welding was carried out in eleven runs by Pulsed Gas Metals Arc Welding (P-GMAW) using LTT flux cored wires. The weld samples (250 mm × 200 mm) exhibit a thickness of 25 mm each and were prepared with an opening angle of 45°. The high strength steel S960QL was used as a base material. The specimens were heated to preheat temperature using a furnace. Preheat and interpass temperatures were controlled by thermocouples on various positions in the heat affected zone (HAZ). The welding parameters are shown in Table 3.

Table 1: chemical composition of the base and filler material (fcw)

Chemical composition (wt.-%), *measured on pure (weld-) samples using dilatometry tests									M_S^*
	C	Si	Mn	Cr	Ni	Mo	B	Fe	(°C)
base material	0.17	0.29	0.82	0.51	0.91	0.51	0.0002	Bal.	419
LTT CrNi	0.04	0.38	0.62	12.2	4.80	0.005	-	Bal.	232
LTT CrMn	0.08	0.25	6.40	11	0.02	0.041	-	Bal.	123

Table 2: mechanical properties of base and filler materials

	$R_{p0,2}$ in MPa	R_m in MPa	reference
base material	985	1018	-
LTT CrNi	900	1123	[5]
LTT CrMn	-	1074	

Table 3: welding parameters

wire diameter	1.6 mm
arc voltage (RMS value)	29 V
welding current (RMS value)	377 A
welding speed	520 mm/min
preheat- and interpass temperature	50 °C and 200 °C
$t_{8/5}$-time (averaged),	7 s and 11 s
shielding gas	M13, 1 % O_2

X-ray diffraction. The residual stress measurement was carried out midways on the top surface of the samples using the $\sin^2\psi$ technique [6]. Residual stresses were determined in longitudinal as well as in tranverse direction of the weld line using the parameters shown in Table 4.

Table 4: measuring and evaluation parameter for X-ray diffraction

Measuring position	weld, HAZ, BM (top surface, midway)
radiation	CrK_α
X-ray spot size	2 mm
Exposure time	5 s

Materials Research Proceedings **2** (2016) 223-228 doi: http://dx.doi.org/10.21741/9781945291173-38

Ψ-range	0...±45° in 7 steps
Diffraction line	ferrite/martensite: 211
E {211}	220.000 MPa
ν {211}	0.28

Neutron diffraction. Stress measurements were performed at the instrument E3 at the BERII facility of the HZB Berlin, Germany [7]. The setup is shown in Fig. 1 (left). As a multi-axial stress state was expected in the welded joints, strain measurements were necessary in the three principle directions covering the longitudinal, transversal and weld depth direction (normal). Using a neutron wavelength of approximately 0.147 nm the α/α′ -Fe {211} diffraction peak located at approximately 2Θ = 78° was utilized for strain measurement. The exposure time varied between 10 min to 30 min. From lattice strains in principle directions the stresses were calculated. A gauge volume of 2x2x2 mm^3 was chosen for the transversal as well as weld depth direction while in longitudinal direction the gauge volume was adapted to 2x2x10 mm^3 to achieve a sufficient spatial resolution and acceptable exposure time. The measuring positions were realized midway along the specimens. Three measuring lines in distances of 4, 12.5 and 21 mm to the plate surface were chosen as indicated in Fig. 1 (right). The measuring range included in total 15 measuring points located in the weld and on both sides of the HAZ, while the BM was covered single sided only. Weld distortion was not balanced for the measuring lines. As the LTT weld metals are higher alloyed compared to the base material the unstrained lattice parameter a_0 was determined from appropriate reference specimens. For this purpose a thin comb was cut from each sample representing the weld cross section including the HAZ and BM. The prongs of the comb were prepared by Electrical Discharge Machining to reduce thermal effects to the material.

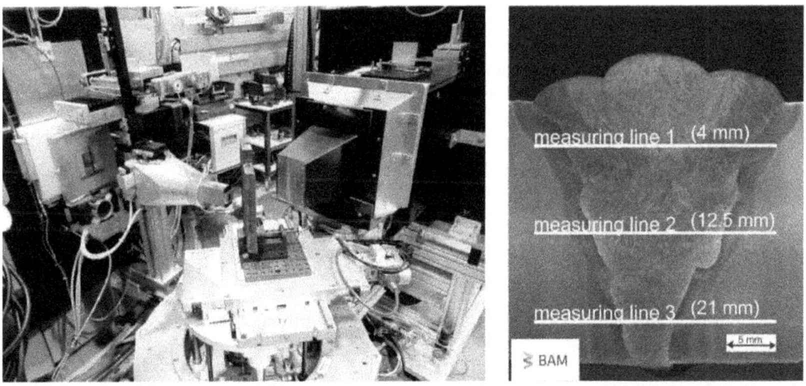

Fig. 1: setup of the neutron diffraction experiment at the instrument E3@BER II (left) and assignment of the measuring lines to the joints cross section (right)

Results and Discussion

Residual Stresses – Surface. Residual Stresses found in the surface of the welds are presented in Fig. 2 and Fig. 3. The longitudinal stress distribution of all joints is quite similar in quality. Moreover, the residual stress level in the LTT CrMn welds is almost identical except for a single tensile peak value in the left hand side weld metal at the lower interpass temperature. In case of the CrNi filler the residual stress level is slightly decreased when applying the higher interpass temperature. In this case the weld is characterized mainly by comparatively low tensile residual stresses up to about 300 MPa, while the lower interpass temperature shows peak values around

Materials Research Proceedings **2** (2016) 223-228 doi: http://dx.doi.org/10.21741/9781945291173-38

500 MPa. These stress levels are also to be found in the HAZ. In any case the transition to the BM shows a steep residual stress gradient. In this area the stresses change from tension into compression. Adjacent to the HAZ the residual stresses are decreased to compressive values between -200 MPa and -300 MPa given by the BM condition before welding (sand blasted). Some high measurement errors are due to microstructural reasons.

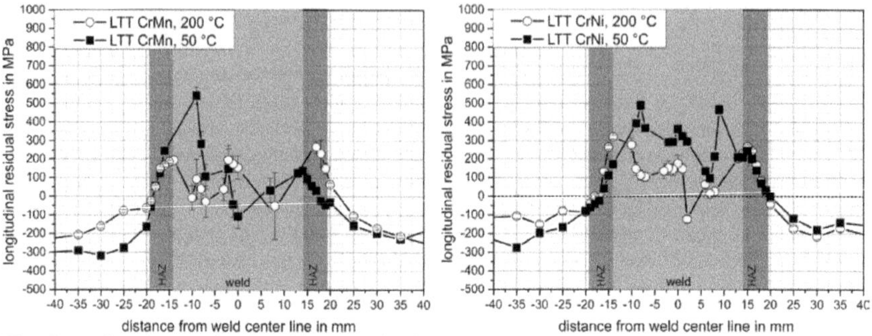

Fig. 2: surface longitudinal welding residual stresses for the CrMn-filler (left) and the CrNi-filler (right) induced during welding using different interpass temperatures

In contrast there are some differences to be observed in transversal direction. The weld metals are in tension up to about 800 MPa and the region showing tensile residual stresses is wider when using the high interpass temperature. Additionally, the residual stresses exhibit several maxima in the weld center. On the other hand, with lower interpass temperature single stress peaks appear in the center without dips. The HAZ shows low compressive residual stresses on the level of the BM up to about -300 MPa. Independent from the interpass temperature applied the longitudinal residual stresses are lower than the transverse ones. This is consistent to results found in other LTT joints [8]. Higher interpass temperatures lead to slightly decreased stress levels in longitudinal direction. But in transversal direction the high tensile residual stresses are limited to a smaller zone in the weld metal center in case of lower interpass temperatures.

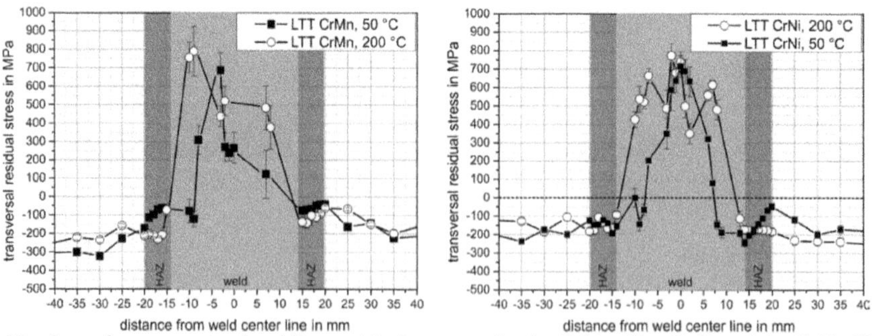

Fig. 3: surface transversal welding residual stresses for the CrMn-filler (left) and the CrNi-filler (right) induced during welding using different interpass temperatures

Residual Stresses – Bulk. Longitudinal as well as transversal residual stresses are shown for the LTT CrNi in Fig. 4. The LTT CrMn exhibits the similar residual stress distributions in quality and is therefore not shown. In case of an interpass temperature of 200 °C there are compressive longitudinal stresses detectable through the bulk. They are lost in case of the root (Fig. 1 line 3), where tensile stresses balance up to about 380 MPa in the weld metal. Similar to the top surface high tensile stresses are present in the HAZ independent from the depth. The stress magnitudes are higher than on the surface (max. 700 MPa). Also contrary to the surface in transversal direction there are quite low tensile residual stresses of up to about 350 MPa detectable in the weld metal only. The HAZ varies between tension and compression from 4000 MPa to -300 MPa. In case of an interpass temperature of 50 °C the longitudinal stress level is shifted completely into tension reaching the level of the surface stresses. Longitudinal stress peaks are also detectable through the material bulk in the HAZ up to 900 MPa. In transversal direction the welding residual stress distribution is less affected by lower interpass temperatures. The stress level shows low tensile stress up to 400 MPa in the weld and 430 MPa in the HAZ. High tensile stress levels as found at the surface are not detectable. In the normal direction the stress distribution of the LTT alloys is comparable to the transversal welding residual stresses and therefore not shown.

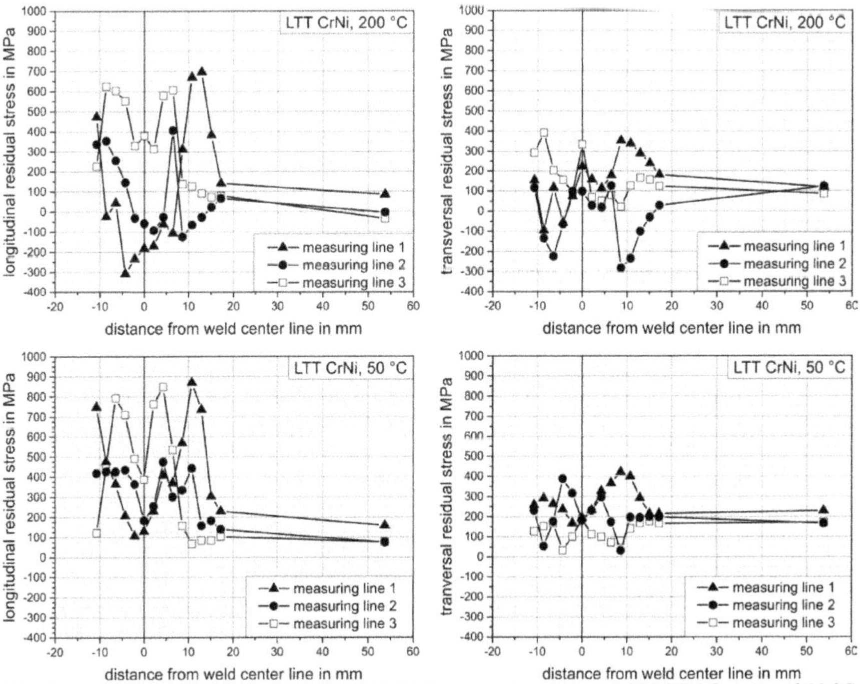

Fig. 4: residual stress in bulk for the LTT CrNi, interpass temperature 200 °C (above) and 50 °C (below)

Summary and Conclusions

The present work dealt with the residual stresses in high strength heavy plate joints welded with LTT fillers applying different interpass temperatures. The residual stresses were determined using X-ray (surface residual stresses) and neutron diffraction (bulk residual stresses). The following conclusions can be drawn:

- Independent from the interpass temperature residual stresses in longitudinal direction are lower than in transverse direction. This applies for the surface as well as for the bulk of the welds.
- Welding residual stresses in the bulk are lower compared to stresses found on the surface.
- Compressive residual stresses as a result of the martensite formation are determined in the bulk weld metal, only.
- Independent from the LTT filler used high interpass temperatures are beneficial to reduce the stresses mainly in longitudinal direction.
- Lower interpass temperatures tend to narrow the tensile zone in the weld metal but they also prevent the formation of compressive residual stresses.

Acknowledgements
The authors would like to especially acknowledge the German Research Foundation (DFG) for funding parts of this work (KA 1807/4-1 | GI 376/4-1), Lincoln Electric Europe for provision of welding consumables as well as EWM AG for support.

References

[1] C. Shiga; H.Y. Yasuda; K. Hiraoka; H. Suzuki: Effect of Ms Temperature on Residual Stress in Welded Joints of High-Strength Steels. *Welding in the World*, Vol. 54 (2010), No. 3-4, pp. R71–R79

[2] J.A. Francis; S. Kundu; H.K.D.H. Bhadeshia; H.J. Stone; R.B. Rogge; P.J. Withers; L. Karlsson: The Effects of Filler Metal Transformation Temperature on Residual Stresses in a High Strength Steel Weld Intervening Transformations. Journal of Pressure Vessel Technology vol. 131 (2009), pp. 1–15. https://doi.org/10.1115/1.3122036

[3] H. Dai; R.J. Moat; P.J. Withers: Modelling the Interpass Temperature Effect on Residual Stress in Low Transformation Temperature Stainless Steel Welds. In: *ASME 2011 Pressure Vessels and Piping Conference: Volume 6, Parts A and B*, ASME, 2011, pp. 1451–1458. https://doi.org/10.1115/pvp2011-57329

[4] S. Zenitani; N. Hayakawa; J. Yamamoto; K. Hiraoka; Y. Morikage; T. Kubo; K. Yasuda; K. Amano: Development of new low transformation temperature welding consumable to prevent cold cracking in high strength steel welds. *Science and Technology of Welding and Joining*, Vol. 12 (2007), No. 6, pp. 516–522

[5] S. Vanrostenberghe, M. Clarin, Y. Shin, B. Droesbeke, V. Van der Mee, M. Doré, G. Marquis, J. Parantainen, T. Kannengießer ZB (2015) Improving the fatigue life of high strength steel welded structures by post weld treatments and specific filler material. doi: 10.2777/705266

[6] P. Müller, E. Macherauch: Das $\sin^2\psi$-Verfahren der röntgenographischen Spannungsmessung*. *Zeitschrift für angewandte Physik*. Vol. 13 (1961), pp. 306-312 [in german]

[7] T. Poeste, R. C. Wimpory, R. Schneider: The new and upgraded neutron instruments for material science at HMI – current activities in cooperation with industry. Materials Science Forum, Vol. 524-525 (2006), pp. 223-228. https://doi.org/10.4028/www.scientific.net/MSF.524-525.223

[8] A. Kromm; T. Kannengiesser; J. Altenkirch; J. Gibmeier: Residual Stresses in Multilayer Welds with Different Martensitic Transformation Temperatures Analyzed by High-Energy Synchrotron Diffraction. *Materials Science Forum*, Vol. 681 (2011), pp. 37–42. https://doi.org/10.4028/www.scientific.net/MSF.681.37

Residual Stress Analysis in Girth-welded Ferritic and Austenitic Steel Pipes Using Neutron and X-Ray Diffraction

Nico Hempel[1, a *], Jeffrey R. Bunn[2,b], Thomas Nitschke-Pagel[1,c], E. Andrew Payzant[2,d] and Klaus Dilger[1,e]

[1]TU Braunschweig, Institute of Joining and Welding, Langer Kamp 8, D-38106 Braunschweig, Germany

[2]Oak Ridge National Laboratory, 1 Bethel Valley Road, Oak Ridge, Tennessee 37831, USA

[a]n.hempel@tu-braunschweig.de, [b]bunnjr@ornl.gov, [c]t.pagel@tu-braunschweig.de, [d]payzanta@ornl.gov, [e]k.dilger@tu-braunschweig.de

Keywords: Welding, Residual Stress, Tubes and Pipes, Ferritic Steel, Austenitic Steel, Neutron Diffraction, X-Ray Diffraction

Abstract. This paper is dedicated to the thorough experimental analysis of the residual stresses in the vicinity of tubular welds and the mechanisms involved in their formation. Pipes made of a ferritic-pearlitic structural steel and an austenitic stainless steel are each investigated in this study. The pipes feature a similar geometry and are welded with two passes and comparable parameters. Residual strain mappings are carried out using X-ray and neutron diffraction. The combined use of both techniques permits both near-surface and through-wall analyses of the residual stresses. The findings allow for a consistent interpretation of the mechanisms accounting for the formation of the residual stress fields due to the welding process. Since the results are similar for both materials, it can be concluded that residual stresses induced by phase transformations, which can occur in the structural steel, play a minor role in this regard.

Introduction

Current fatigue design standards and recommendations, like the ones given by the International Institute of Welding (IIW) [1], are based on the assumption of yield strength magnitude tensile residual stresses if the actual residual stress state is unknown. This postulate reflects uncertainties about the initial residual stress state after welding, which may depend on numerous parameters, as well as about the possible relaxation of residual stresses, which occurs when the static or cyclic yield strength is exceeded locally. Therefore, research concerning both the development and the relaxation of residual stresses is needed to improve the generalized approach given in [1].

Experimental and numerical analyses of welding residual stresses hardly ever show that the conservative assumptions made in the IIW recommendations hold. In girth-welded pipes, the pipe geometry and the heat input have been identified as the governing factors for the residual stress development [2,3], apart from material parameters. Under suitable conditions, pipe wall bending can occur, leading to compressive axial residual stresses at the weld toe of girth welds, but also to tensile residual stresses at the weld root, which was shown by other authors, see e.g. [2,3], and in previous studies on ferritic-pearlitic and austenitic steel pipes using X-ray diffraction [4-6]. In this work, these will be supplemented by neutron diffraction measurements and the results obtained from the two different steels will be compared. The thorough experimental analysis of the residual stress state after welding will serve as a basis for the investigation of residual stress relaxation under loading and the validation of numerical simulations, both of which are subject of future work.

Experimental work

Sample preparation. Pipes of the ferritic-pearlitic structural steel S355J2H+N and of the austenitic stainless steel X6CrNiTi18-10 were used for the experiments. The yield stress of the base materials is 355 MPa and 223 MPa, respectively. The tubular specimens of 200 mm length were machined on all surfaces, resulting in outer diameters of 100.5 mm and 114 mm and wall thicknesses of 7.75 mm and 7.5 mm for the structural steel and the austenitic steel, respectively. A v-shaped groove was introduced at half-length as a weld preparation. Before welding, the specimens of the structural steel were stress relieved thermally at 600 °C for 30 minutes and cooled uniformly at about 1 °C/min.

Metal active gas (MAG) welding was performed in flat position using a rotary table. The filler metal, ISO 14341-A-G 4Si1 for the structural steel and ISO 14343-A-G 19 9 NbSi for the austenitic steel, was applied in two passes, which were started at the same point and were welded in the same direction. The nominal energy inputs were similar for the structural and the austenitic steel, with 8.6 kJ/cm and 9.1 kJ/cm for the root pass and 11.8 kJ/cm and 11.2 kJ/cm for the second pass, respectively. Each pass was welded at room temperature.

Residual Stress Analysis. The residual stresses in the welded samples were determined using X-ray diffraction (XRD) and neutron diffraction (ND) in order to analyze the stress state both at the surface and within the pipe wall. The measurements were taken at points along lines perpendicular to the welding direction at $\varphi = 90°$, where φ is the circumferential angle marking the welding direction and the start/stop location at $\varphi = 0°$. As previous work revealed virtually axisymmetric residual stress states [4-6], $\varphi = 90°$ is considered to be representative. Due to the symmetry, measurements were only performed on one side of the weld centerline up to a distance of 60 mm. The coordinate x specifies the axial distance of a certain point from the weld centerline.

The hoop and axial residual stresses on the surfaces of the pipes were determined by XRD. The inner surfaces were only accessible after sectioning the tubes, the released stresses being monitored by strain gauge measurements. For details concerning the XRD measurements, please refer to [4-6].

The ND measurements for the residual stress analysis within the pipe wall were carried out at the 2nd Generation Neutron Residual Stress Mapping Facility (NRSF2) beam line [7] of the High Flux Isotope Reactor (HFIR), Oak Ridge National Laboratory. For the ferritic pipes, the diffraction of neutrons with a wavelength of $\lambda \approx 1.72$ Å on the {211} lattice planes was analyzed, whereas $\lambda \approx 1.54$ Å was used for the {311} lattice planes of austenite, in order to attain a diffraction angle of $2\theta \approx 90°$. The planes were chosen due to their similarity to the expected bulk stress/strain response. In addition, these reflections are reported to be the least-sensitive to intergranular strains [8]. The stress-free lattice spacing was determined in the base metal (BM) for both materials as well as in the weld metal (WM) and the heat-affected zone (HAZ) of the ferritic-pearlitic samples to take composition gradients into account. At each axial position, measurements were taken at five points that were evenly distributed across the wall thickness. At every point, the strain was determined in hoop, axial and radial direction, assuming that these are the principal directions, in order to compute the multiaxial residual stress state. The cubical measurement volume had an edge length of 2 mm.

For both XRD and ND, the elastic constants $E^{\{211\},\alpha\text{-Fe}} = 210500$ N/mm^2 and $\nu^{\{211\},\alpha\text{-Fe}} = 0.28$ were used for ferrite and $E^{\{311\},\gamma\text{-Fe}} = 195700$ N/mm^2 and $\nu^{\{311\},\gamma\text{-Fe}} = 0.366$ for austenite.

Results

Fig. 1 shows the hoop residual stress on the outer surface for both a ferritic and an austenitic sample as determined by XRD. Except for the weld seam, where tensile residual stresses occur in the ferritic sample of S355J2H+N and compressive residual stresses can be found in the austenitic sample of X6CrNiTi18-10, the results agree well on a qualitative level. Close to the weld seam, a small tensile maximum of about 70 MPa or 200 MPa can be seen, respectively, whereas the residual stresses are in the compressive regime at a distance of 10 to 40 mm from the weld centerline. Here, maximum compressive values of -250 MPa were determined for the ferritic sample and -550 MPa for the austenitic sample. At larger distances, the residual stresses in both samples approach 0 MPa.

The residual stresses on the inner surface after sectioning can be taken from Fig. 2. Again, the distributions are qualitatively similar and mainly differ in the maximum values. In the weld seam,

Materials Research Proceedings **2** (2016) 229-234 doi: http://dx.doi.org/10.21741/9781945291173-39

moderate tensile residual stresses were found that are higher in the ferritic sample. At x = -7 mm, a maximum of almost 500 MPa or 800 MPa is reached for the ferritic or austenitic sample, respectively. With increasing distance from the weld, the residual stresses become compressive, reaching maximum values of -200 MPa and -700 MPa, and finally converge to 0 MPa. The residual stress state was affected by the sectioning process. Strain gauge measurements on the inner surface near the weld revealed a decrease of the tensile residual stresses of about 100 MPa for a ferritic pipe and 25 MPa for an austenitic one. A more detailed examination of that effect can be found in [5,6].

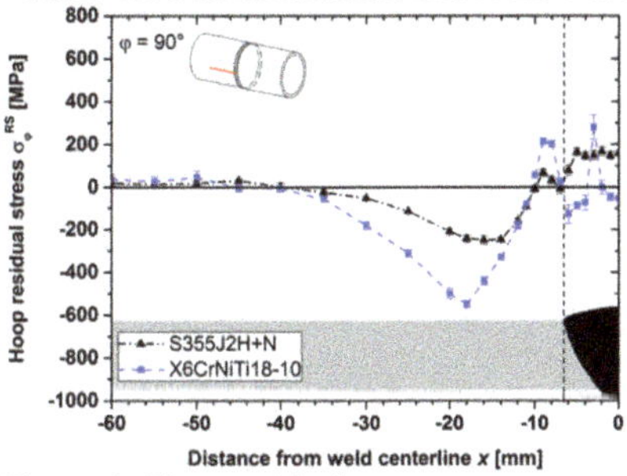

Figure 1: Hoop residual stresses on outer surface of ferritic and austenitic samples

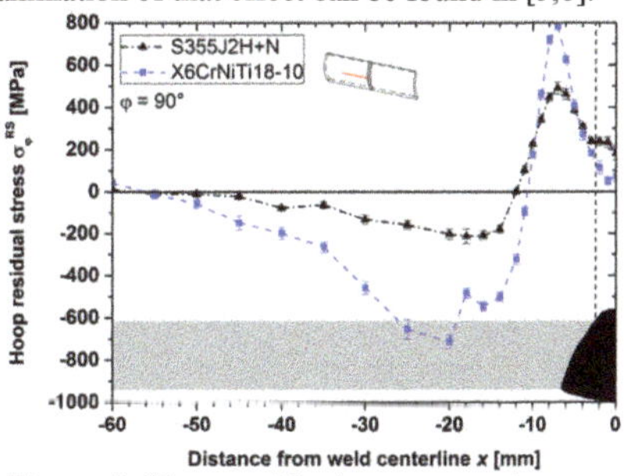

Figure 2: Hoop residual stresses on inner surface of ferritic and austenitic samples

The results of the neutron diffraction experiments for the hoop stress component are presented as contour plots in Fig. 3 for a ferritic pipe and in Fig. 4 for an austenitic one. Each diagram displays the cross section of the whole pipe wall; the white areas reflect the fact that the center of the fully immersed measurement volume had to be at least 1.5 mm away from the surfaces. The center positions of the measurement volume at each measuring point are also given. In the ferritic sample, the sign of the residual stress mainly depends on the axial distance from the weld. In the weld metal, the heat-affected zone and the adjacent base metal, the residual stresses are tensile for about $|x| < 14$ mm. For $|x| > 14$ mm, they are compressive, reaching a maximum at about 25 mm and fading with increasing distance from the weld. Across the pipe wall, the highest absolute values can be found near the inner surface, which holds both in the tensile and the compressive area. In the austenitic sample, a hoop residual stress distribution similar to the one in the ferritic pipe was found, both qualitatively and quantitatively. Distinct differences are the higher tensile residual stresses in the vicinity of the weld, where up to 380 MPa instead of 280 MPa are reached, and the presence of an area of tensile residual stress near the outer surface for $|x| > 25$ mm.

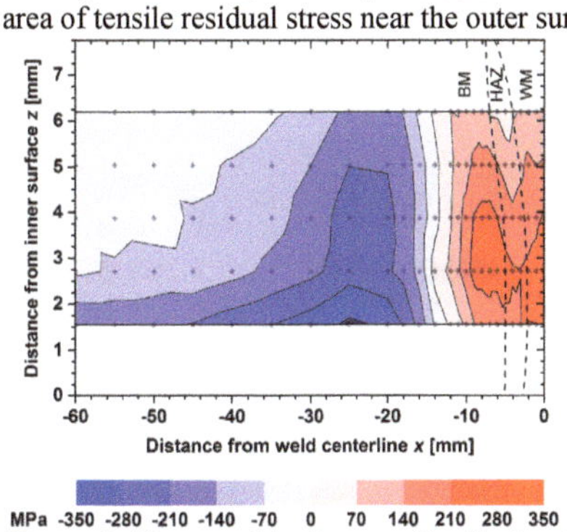

Figure 3: Hoop residual stress distribution in the pipe wall, structural steel S355J2H+N

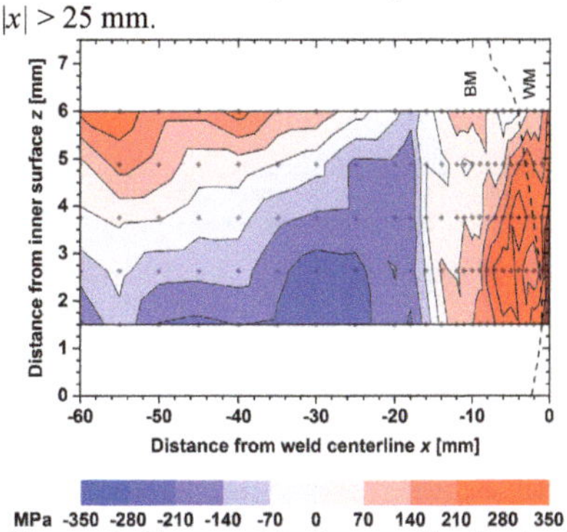

Figure 4: Hoop residual stress distribution in the pipe wall, austenitic steel X6CrNiTi18-10

Materials Research Proceedings **2** (2016) 229-234 doi: http://dx.doi.org/10.21741/9781945291173-39

The axial residual stress on the outer surface is given in Fig. 5. Qualitatively, there is a good agreement for both materials within $|x| < 30$ mm. In the vicinity of the weld, the residual stresses are compressive, taking moderate values of up to -200 MPa in the weld seam and peaking to -400 MPa or -600 MPa near the weld toe, depending on the material. With increasing distance, the residual stresses then reach a relatively small tensile maximum and decrease again, exhibiting a different behavior for the two materials. While the residual stress is almost 0 MPa at $x = -60$ mm in the ferritic sample, it asymptotically approaches a value of -450 MPa in the austenitic pipe.

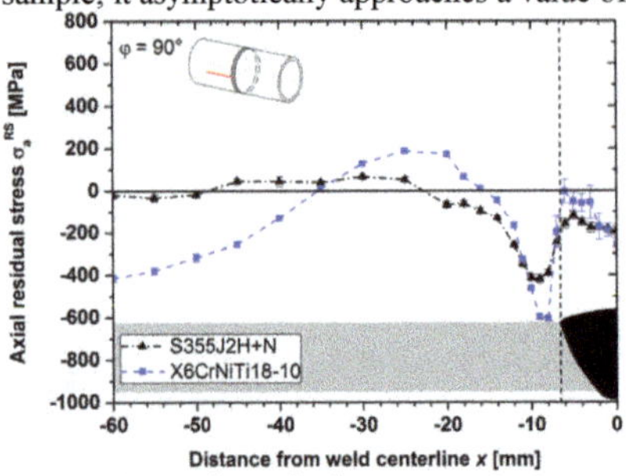

Figure 5: Axial residual stresses on outer surface of ferritic and austenitic samples

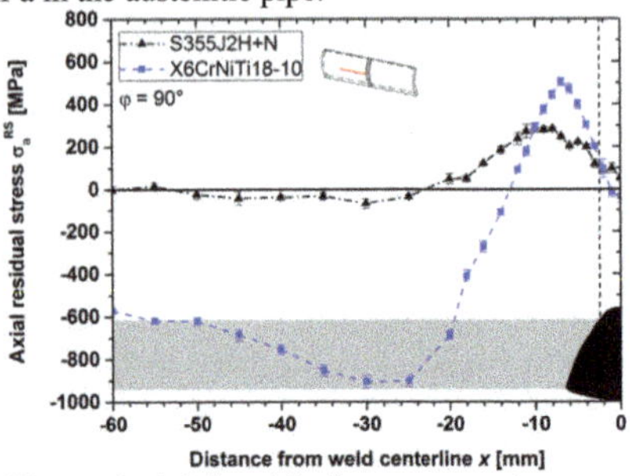

Figure 6: Axial residual stresses on inner surface of ferritic and austenitic samples

The axial residual stress on the inner surface, obtained after sectioning, shows a behavior converse to that on the outer surface, see Fig. 6, featuring small tensile stresses in the weld, a tensile maximum at $x = -6\ldots-10$ mm, and compressive residual stresses with a maximum at about $x = -30$ mm. While the latter amounts to -70 MPa in the ferritic sample, it reaches -900 MPa in the austenitic pipe. Moreover, far away from the weld, compressive stresses of -550 MPa are found in the X6CrNiTi18-10 pipe, compared to about 0 MPa in the S355J2H+N sample. Through strain gauge measurements, the residual stress release near the weld due to sectioning was found to be about 125 MPa for the ferritic pipe and 50 MPa for the austenitic sample.

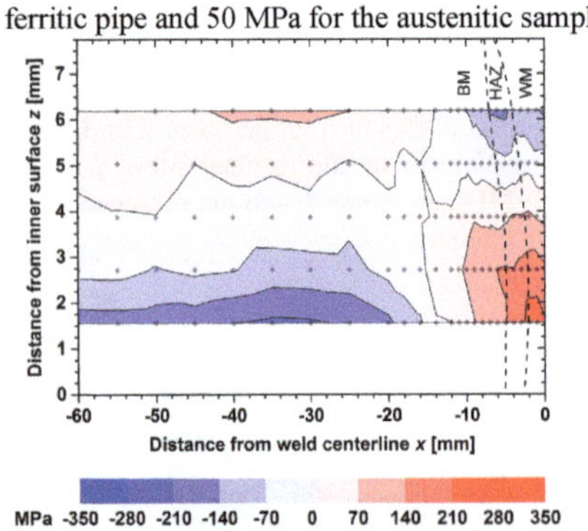

Figure 7: Axial residual stress distribution in the pipe wall, structural steel S355J2H+N

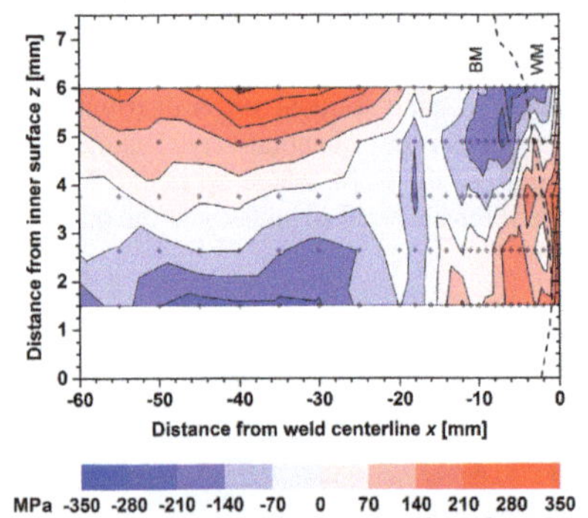

Figure 8: Axial residual stress distribution in the pipe wall, austenitic steel X6CrNiTi18-10

Neutron diffraction results for the axial residual stress component are shown in Fig. 7 for a ferritic sample and in Fig. 8 for an austenitic sample. In both cases, compressive residual stresses are found near the outer surface in the weld metal and its vicinity, reaching maximum values of -160 MPa and -240 MPa, respectively, near the weld toe. Conversely, the area around the weld root is characterized by tensile residual stresses extending to the center of the wall in the ferritic sample and to the outer surface of the austenitic sample at the weld centerline. The maximum tensile residual

Materials Research Proceedings **2** (2016) 229-234 doi: http://dx.doi.org/10.21741/9781945291173-39

stresses are 240 MPa and 200 MPa, respectively. At $|x| > 25$ mm, the residual stress state is reversed, i.e. the residual stresses are tensile near the outer surface and compressive near the inner surface with maxima occurring at x = -30…-40 mm. Here, values of 90 MPa and -220 MPa are reached in the ferritic sample, as well as 370 MPa and -240 MPa in the austenitic sample. At even larger distances, the residual stress values fade slightly, but are still significant at x = -60 mm.

Discussion

Based on XRD results, the governing mechanism responsible for the residual stress formation in the girth-welded pipes investigated here has already been discussed in detail for the ferritic-pearlitic pipes [4,5] and for the austenitic pipes [6]. It was shown that in both cases, the circumferential contraction of the weld and the highly heated areas in its vicinity during cooling is constrained by the adjacent material, thus causing tensile hoop residual stress in the former and compressive hoop residual stress in the latter areas, as can be seen in Fig. 1 to 4. Due to the self-constraining tubular geometry, the contraction of the weld also leads to necking of the pipe and thereby to pipe wall bending, which accounts for the axial residual stress state featuring inverse signs on the inner and outer surfaces, see Fig. 5 to 8. In general, the ND results confirm the findings obtained by XRD and thus the interpretation of the formation mechanisms of the residual stress fields. However, there are also slight deviations from the XRD results that will be discussed in the following paragraphs.

Both in hoop and axial directions, the XRD measurements on the inner surface of the ferritic and austenitic pipes show residual stress maxima that are at least 7 mm from the weld centerline, whereas the residual stresses in the weld metal are relatively low. The ND measurements, however, reveal that the highest values occur in the weld metal or at the notch between the weld root and the base metal. This important difference can be ascribed to the sectioning that was necessary to perform the XRD measurements on the inner surface, thereby altering the residual stress state, which was also shown at selected points by strain gauge measurements. It is also possible that the curvature of the weld reinforcement influences the XRD measurements, especially at the weld root. Moreover, the surface of the weld reinforcement is subject to a lower constraint as there is no base material in the immediate vicinity, which might also affect the residual stresses on the surface of the weld seam. The last two points can also be used as an explanation why compressive hoop residual stresses are found in the weld seam on the outer surface of an austenitic pipe, see Fig. 1, whereas tensile residual stresses occur in this region almost across the whole wall thickness, see Fig. 4.

A comparison of the residual stress states in the ferritic and austenitic samples yields a good qualitative agreement. There are, however, significant differences: First, the absolute values of the residual stresses in the austenitic samples are often higher than in the structural steel and by far surpass the yield stress of 223 MPa. This effect is most pronounced in all XRD results, i.e. on the surface. Second, at large distances from the weld, the residual stresses on or near the surfaces of the austenitic steel can approach values significantly different from 0 MPa, which does not occur in the ferritic samples. These stresses can be both tensile, see Fig. 4, and compressive, see Fig. 5 and 6.

Both the higher values and the residual stresses far away from the weld can be explained by the fact that the austenitic pipes, unlike the ferritic samples, have not been heat-treated before welding. Machining leads to significant near-surface plastic deformation, thus causing both a work hardening effect and introducing machining residual stresses. Work hardening is limited to areas 200 – 300 µm below the surface [6] and is therefore most distinct in the XRD results, where residual stress values of up to 900 MPa occur. The ND results reveal that the residual stresses in the pipe wall are much lower, see Fig. 4 and 8, which confirms the findings in [6]. Since they can still exceed the yield stress to a lesser extent, it can be inferred that work hardening effects not related to the machining process also play a role in the residual stress formation in the austenitic pipes.

Residual stresses resulting from machining, or introduced by the fabrication process in general, extend farther below the surface than the work hardening effect [6]. This explains why they can affect the results of ND measurements to a larger extent, as reflected for instance by the relatively high hoop residual stresses near the outer surface at a large distance from the weld, shown in Fig. 4. The fact that tensile stresses can be found here in contrast to the XRD measurements on the surface, see Fig. 1 and 2, can be explained by the considerable variation of the machining residual stresses in different specimens. It was found that they can be tensile, compressive or even negligibly small, presumably depending on the condition of the cutting tool and the machining parameters.

The fact that the results are similar for both materials, apart from machining effects, reveals that phase transformations, which can occur in the ferritic-pearlitic steel but not in the austenitic steel, do not contribute substantially to the formation of the residual stress field. This is due to the relatively long cooling time from 800 °C to 500 °C in the HAZ of about 18 seconds [5], leading to high transformation temperatures and thus to a dominant effect of thermal contraction.

Summary and Conclusions

In this study, the residual stress state in girth-welded pipes of ferritic-pearlitic and austenitic steel has been thoroughly investigated using both X-ray and neutron diffraction. It was shown that the results of neutron diffraction measurements mostly agree well with the X-ray measurements on the surface. Nevertheless, they also contain complementary information, especially regarding the residual stresses near the inner surface of the pipes, which is only accessible for X-ray measurements after sectioning and thus altering the residual stress state. Also, neutron diffraction confirmed that the residual stresses on the surface of the austenitic pipes are significantly affected by the machining process during sample preparation, resulting in very high residual stress values.

From the good qualitative agreement of the residual stress states in pipes of similar geometry but different types of steels it can be inferred that phase transformations play only a minor role in the formation of the residual stress fields in the girth-welded pipes investigated here.

Acknowledgement

We would like to thank the German Federation of Industrial Research Associations (AiF) for their financial support of the research project IGF No. 17.619N. This project was carried out under the auspices of AiF and financed within the budget of the Federal Ministry for Economic Affairs and Energy through the program to promote joint industrial research and development (IGF).

The authors also thank BASF SE, Germany, for the allocation of X6CrNiTi18-10 pipes.

A portion of this research used resources at the High Flux Isotope Reactor, a DOE Office of Science User Facility operated by the Oak Ridge National Laboratory. The assistance of P.A. Cornwell with the strain mapping experiments is gratefully acknowledged.

References

[1] A.F. Hobbacher, Recommendations for Fatigue Design of Welded Joints and Components, second ed., Springer, Heidelberg, 2016. http://dx.doi.org/10.1007/978-3-319-23757-2

[2] P.J. Bouchard, Validated residual stress profiles for fracture assessments of stainless steel pipe girth welds, Int. J. Press. Ves. Pip. 84 (2007) 195–222. http://dx.doi.org/10.1016/j.ijpvp.2006.10.006

[3] S. Song, P. Dong, J. Zhang, A full-field residual stress estimation scheme for fitness-for-service assessment of pipe girth welds: Part 1, Int. J. Press. Ves. Pip. 126-127 (2015) 58–70. http://dx.doi.org/10.1016/j.ijpvp.2015.01.002

[4] N. Hempel, T. Nitschke-Pagel, K. Dilger, Residual stresses in multi-pass butt-welded tubular joints, Adv. Mat. Res. 996 (2014) 488-493.

[5] N. Hempel, T. Nitschke-Pagel, K. Dilger, Residual stresses in multi-pass butt-welded ferritic-pearlitic steel pipes, Weld. World 59 (2015) No. 4 555-563.

[6] N. Hempel, T. Nitschke-Pagel, K. Dilger, Study on the near-surface residual stress state in butt-welded pipes of austenitic steel using X-ray diffraction, Weld. World (accepted 19/08/2016, DOI 10.1007/s40194-016-0378-9).

[7] W. Woo, V. Em, C.R. Hubbard, H.J. Lee, K.S. Park, Residual stress determination in a dissimilar weld overlay pipe by neutron diffraction, Mat. Sci. Eng. A 528 (2011) 8021-8027. http://dx.doi.org/10.1016/j.msea.2011.07.059

[8] M.T. Hutchings, P.J. Withers, T.M. Holden, T. Lorentzen, Introduction to the Characterization of Residual Stress by Neutron Diffraction, CRC Press, Boca Raton, 2005.

Materials Research Proceedings **2** (2016) 235-240 doi: http://dx.doi.org/10.21741/9781945291173-40

Effect of Plasticity on Residual Stresses Obtained by the Incremental Hole-drilling Method with 3D FEM Modelling

Evy Van Puymbroeck[1, a *], Wim Nagy[1,b] and Hans De Backer[1,c]

[1]Ghent University, Department of Civil Engineering EA15, Technologiepark 904, 9052 Ghent, Belgium

[a]Evy.VanPuymbroeck@UGent.be, [b]Wim.Nagy@UGent.be, [c]Hans.DeBacker@UGent.be

Keywords: Residual Welding Stresses, Incremental Hole-drilling Method, FEM, Plasticity Effects

Abstract. The incremental hole-drilling method is used to determine high residual welding stresses in an orthotropic bridge deck. When comparing the measurement results with a theoretical residual stress distribution of an orthotropic steel deck, a large difference in sign and magnitude of the residual stress values is observed. These measurement results are presented in another paper [1]. The test method, specified in ASTM E837-13a only applies when the material behavior is linear-elastic. Relaxed-plastic strain can be detected in the region of the bored hole for the evaluation of high residual welding stresses. This plastic behavior can result in a significant error of the residual stresses. In this paper, the hole-drilling procedure is simulated and the effect of plasticity on the determination of residual stresses is studied with three-dimensional finite element modelling. The calculation software Siemens NX 9.0 is used to simulate the hole-drilling procedure with both linear-elastic and elastic-plastic material behavior. First, a 3D model is set up for uniform in-depth residual stress fields with a linear-elastic material behavior to determine the calibration coefficients. The same model is used to determine similar calibration coefficients but this time with a simplistic model of material plasticity. The effect of plasticity on the uniform in-depth residual stresses is determined. The residual stresses obtained under the assumption that the material behavior is linear-elastic are an overestimation. In future research, residual strains for non-uniform in-depth residual stresses can also be studied with similar models. This will result in a more accurate determination of the residual weld stresses present in bridge constructions.

Introduction

The incremental hole-drilling method is widely used for measuring residual stresses in steel components. Plastic relaxed strain can be induced in the region of the borehole. Since this test method only applies when the material behavior remains linear-elastic, significant errors can be introduced by this plastic behavior during the determination of residual stresses [2]. In this paper, the effect of plasticity on the determination of residual stresses for uniform in-depth residual stresses is studied.

Incremental hole-drilling method

The residual stresses introduced by a welding operation can be evaluated making use of the incremental hole-drilling technique. A small hole is drilled through the center of a strain gauge rosette into the test material (Fig. 1) [3]. The surface material of the test specimen has to be exposed by drilling only through the material of the strain gauge rosette. This drilling depth is called zero depth. The initial uncertainty of the separation of the cutter from the outer surface by the strain gauge rosette and coating can be disregarded by establishing this zero depth [4].

The relieved surface strains caused by the introduction of a hole in a series of small steps are recorded. Residual stresses are calculated according ASTM E837-13a [3] taking into account the measured strains and calibration coefficients. Reliable measurements are only achieved by limiting

Materials Research Proceedings **2** (2016) 235-240 doi: http://dx.doi.org/10.21741/9781945291173-40

the residual stresses to 80% of the material's yield stress in order to take into account that the test method only applies when material behavior is linear-elastic [3]. However, the level of residual stress can be comparable with the material yield stress and then, the stress concentration due to the introduction of the hole produces zones in which the elastic limit of the material is reached. The plastic region arises at the lower circumference of the hole and when the hole depth is increased, it spreads towards the strain gauges located on the strain gauge rosette.

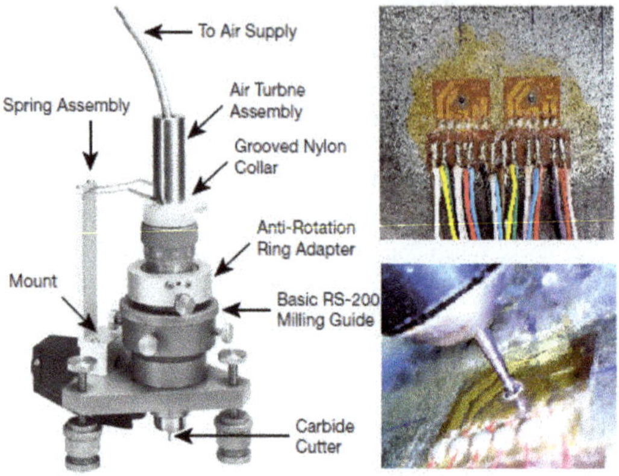

Fig. 1. Hole-drilling machine (left) and hole-drilled strain gauge rosette (right)

A type A strain gauge rosette which follows the Rendler and Vigness [5] geometry is considered in this paper to determine the effect of plasticity on the determination of residual stresses with the incremental hole-drilling method. This pattern is available in several different sizes and is recommended for general-purpose use [3]. The strain gauge rosette type CEA-06-062UL-120 is used for the finite element modelling. A schematic representation of the geometry of a strain gauge rosette used for hole drilling is shown in Fig. 2(a) and a detail of one strain gauge is shown in Fig. 2(b). The dimensions of a strain gauge for the considered strain gauge rosette can be found in Table 1. The hole diameter of this strain gauge rosette is 2mm while the maximum hole-drilling depth is 1mm.

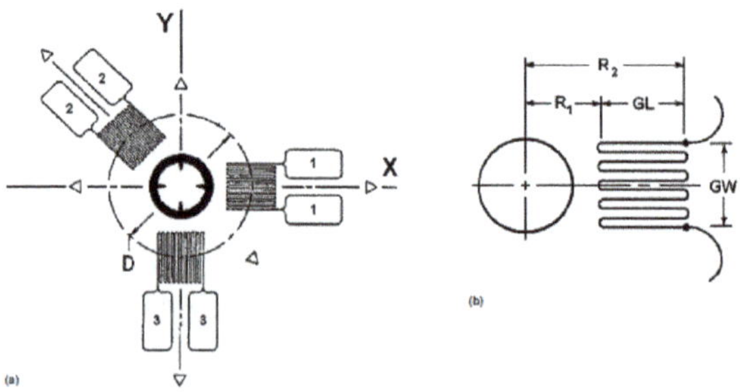

Fig. 2. Schematic geometry of strain gauge rosette (a) and detail of one strain gauge (b) [2]

Materials Research Proceedings **2** (2016) 235-240 doi: http://dx.doi.org/10.21741/9781945291173-40

Table 1. Strain gauge rosette dimensions in mm

Rosette Type	D	GL	GW	R_1	R_2
CEA-06-062UL-120	5.13	1.59	1.59	1.77	3.36

3D finite element model

A three dimensional finite element model is set up in Siemens NX 9.0 to evaluate the plastic relaxed strain readings of the considered strain gauge rosette. The geometry of the strain gauge rosette is exactly modelled by defining element nodes corresponding with the number and length of grid lines and their spacing. The strain gauge rosette's material and its bonding to the test surface is not taken into account. It is assumed that the measured strains by the strain gauge rosette are identical to the relaxed strains of the test specimen [6]. Nodal displacements for both ends of every grid line are simulated and they are used to calculate the radial strains.

The hole-drilling procedure is simulated by deactivating elements with the element birth/death feature of the Nastran Solver in NX. Element layers are incrementally deactivated in a number of steps corresponding to the hole-depth increment specified in ASTM E837-13a. For each step, strains are simulated which are used to calculate residual stresses according to the ASTM standard. Only radial strains 1 and 3 (Fig. 2 (a)) will be considered for the evaluation of the residual stresses since the only interest are the normal x- and y-stresses. The used mesh for the model is shown in Fig. 3 and it contains 436410 three-dimensional elements.

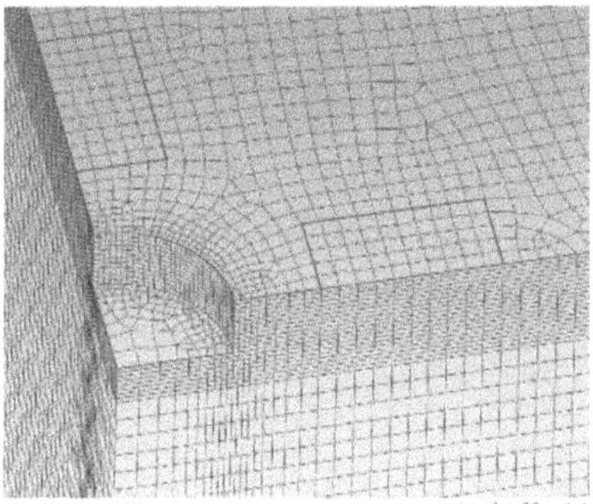

Fig. 3. Mesh of finite element model with drilled hole

Geometry, constraints, material properties and loading conditions. The hole-drilling simulation of a flat plate with a thickness of 6mm is performed. The effect of increasing the plate thickness is studied and it has no influence (less than 1%) on the results. Due to symmetry, only a quarter of the total plate is modelled. Therefore, symmetry constraints are applied on the sides of the model where the hole will be situated which means that the displacements along x- and y-axes are zero. The displacement of the bottom surface along the z-direction is set to zero. The model width is chosen 15 times the hole diameter and the thickness is larger than the rosette's mean diameter according to the minimum thickness for thick plates recommended by ASTM E837-13a. The steel grade S235 is used as material for the modelling. The presence of a residual stress field is simulated by imposing a uniform pressure distribution of 50MPa on the sides not containing the drilled hole.

In order to make the distinction between linear-elastic and elastic-plastic material behavior, a different stress-strain curve is specified. For linear-elastic material behavior, a linear stress-strain curve is specified until the yield stress is reached (Fig. 4). Elastic-plastic material behavior is specified with bilinear stress-strain curve (Fig. 5) and isotropic strain hardening plasticity model. A simplistic model of material plasticity has been used.

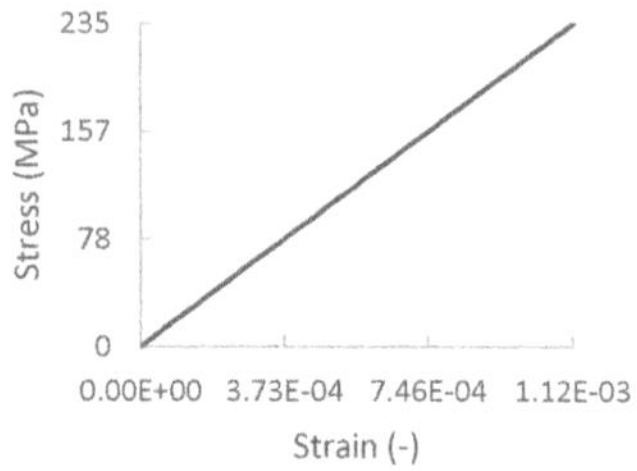

Fig. 4. Linear-elastic stress-strain behavior

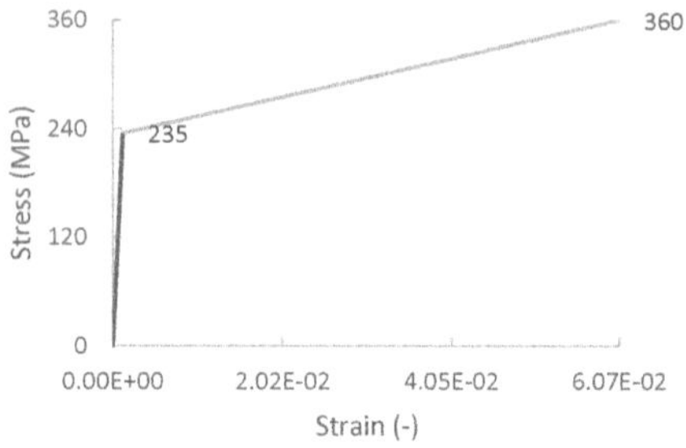

Fig. 5. Elastic-plastic stress-strain behavior

Residual stress calculation. The principal residual stresses σ_x and σ_y can be calculated from the measured relaxed strains ϵ_1 and ϵ_3, the calibration coefficients a and b, the material's Young's modulus E and Poisson's ratio ν. The following relationships apply:

$$\sigma_x = -\frac{E(\epsilon_1+\epsilon_3)}{2a(1+\nu)} - \frac{E(\epsilon_1-\epsilon_3)}{2b} \qquad \sigma_y = -\frac{E(\epsilon_1+\epsilon_3)}{2a(1+\nu)} + \frac{E(\epsilon_1-\epsilon_3)}{2b} \tag{1}$$

The calibration coefficients are calculated with finite element modelling by applying a known uniform stress distribution on the sides not containing the hole and evaluating the relaxed strains [2]. These imposed pressures are assumed to be equal to the principal residual stresses before a hole was drilled [7].

Determination of calibration coefficients

Linear-elastic material properties. First, a test piece with uniform in-depth residual stress field is considered in order to numerically calibrate the hole-drilling process and determine the calibration coefficients. A linear-elastic material law is used and the uniform residual stress field is split up into hydrostatic ($\sigma_x = \sigma_y$) and deviatoric ($\sigma_x = -\sigma_y$) residual stress fields. The model with hydrostatic residual stress field is used to determine the calibration coefficient a and the model with deviatoric residual stress field results in calibration coefficient b. Both models are evaluated for ten depth-step increments of 0.1mm until a final hole depth of 1mm is reached. The relaxed strains for each depth-step are evaluated and the calibration coefficients are determined. In Table 2 the calibration coefficients determined with the 3D-model are shown. The results for elastic-plastic material behavior are also already mentioned.

When a comparison is made between the calibration coefficients obtained by the 3D model with linear-elastic material behavior and the coefficients specified in ASTM, the results should be the same. However, the calculated coefficients from the 3D model with linear-elastic material behavior are on average 4% smaller than the theoretical ones tabulated in ASTM. The difference becomes larger when the hole depth increases and is caused by the location of the strain gauge rosette on the surface. The relaxed strain at deeper hole depth increments has less contribution to the strain response since the distance to the strain gauge rosette increases when reaching deeper hole depths. The determination of the calibration coefficients is dependent on geometric variables including plate thickness, hole diameter and drilling depth. Errors in strain measurement, hole depth and geometry cause also a difference between calculated and theoretical calibration coefficients. The level of uniform residual stress applied in the model also influences the calibration coefficients. Aoh and Wei

[8] have shown that calibration coefficient *a* becomes ill conditioned faster than calibration coefficient *b* which can also can be noticed in Table 2.

When a comparison is made with calibration coefficients obtained from other studies [8], similar trends and error rates were observed in the experimental calibration by different research works. It is therefore concluded that the coefficients calculated with the specified 3D FEM model are in agreement with the conventional coefficients of the incremental hole-drilling method. Therefore, this model will be used to determine the effect of plasticity on residual stresses obtained by the incremental hole-drilling method.

Table 2. Calibration coefficients

	ASTM E837-13a (linear-elastic material behavior)		NX9 Linear-elastic material behavior		NX9 Elastic-plastic material behavior	
Hole depth [mm]	Calibration coefficient *a*	Calibration coefficient *b*	Calibration coefficient *a*	Calibration coefficient *b*	Calibration coefficient *a*	Calibration coefficient *b*
0.1	0.015	0.028	0.016	0.028	0.016	0.027
0.2	0.036	0.066	0.037	0.065	0.036	0.062
0.3	0.059	0.110	0.058	0.108	0.056	0.102
0.4	0.083	0.155	0.079	0.153	0.076	0.144
0.5	0.104	0.200	0.097	0.197	0.094	0.186
0.6	0.124	0.242	0.113	0.238	0.109	0.225
0.7	0.140	0.280	0.125	0.276	0.122	0.260
0.8	0.154	0.314	0.135	0.309	0.131	0.292
0.9	0.165	0.343	0.143	0.338	0.139	0.319
1	0.173	0.370	0.148	0.363	0.144	0.341

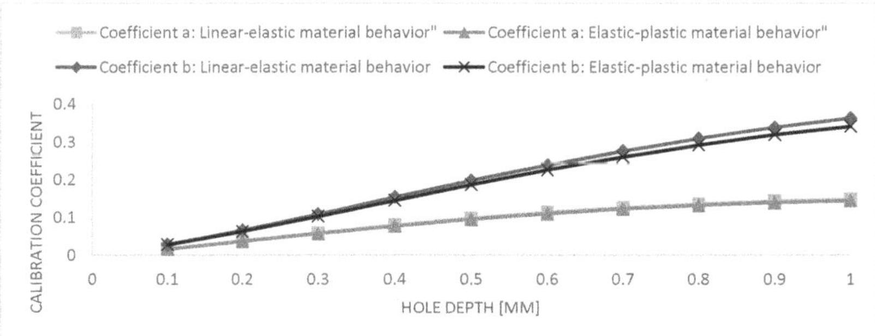

Fig. 6. Comparison between linear-elastic and elastic-plastic material behavior

Elastic-plastic material properties. A comparison is made between the calibration coefficients obtained with linear-elastic and elastic-plastic material properties and the results are shown in Fig. 6. It shows that calibration coefficients are both smaller for the elastic-plastic material behavior. The difference between linear-elastic and elastic-plastic values also increases when the hole depth is increased. A smaller calibration coefficient means that the residual strains and their corresponding residual stresses are lower than the theoretical ones. Taking the effect of plasticity into account, the residual stresses are lower than expected. This means that the residual stresses obtained under the assumption that the material behavior is linear-elastic are an overestimation. For calibration coefficient a, the mean error rate between linear-elastic and elastic-plastic material behavior is 2.74% while the mean error rate for calibration coefficient b is equal to 5.37%. The effect of plasticity is larger for deviatoric residual stress fields.

Conclusion

The hole-drilling procedure is simulated for a uniform in-depth residual stress field and the effect of plasticity is determined based on calibration coefficients. Lower residual stresses are present when elastic-plastic material behavior is assumed. The plasticity introduces relaxation of the material and by ignoring this, the residual stresses will be overestimated by specifying linear-elastic material behavior.

References

[1] Van Puymbroeck, E., Nagy, W. and De Backer, H., Determination of residual stresses with the incremental hole-drilling method to improbe the design of critical components in bridge construction, 16th International Conference and Exhibition : Structural Faults & Repair, 2016.

[2] Beghini, M., Bertini, L. and Santus, C., A procedure for evaluating high residual stresses using the blind hole drilling method including the effect of plasticity, The Journal of Strain Analysis for Engineering Design, Vol. 45(4), 2010. http://dx.doi.org/10.1243/03093247JSA579

[3] ASTM E837-13a, Standard Test Method for Determining Residual Stresses by the Hole-Drilling Strain-Gage Method, ASTM International, 2015.

[4] Vishay Measurements Group, Model RS-200 Milling Guide Instruction Manual Version 2.0, Micro-Measurements, USA, 2011.

[5] Rendler, N.J. and Vigness, I., Hole-drilling strain-gage method of measuring residual stresses, Experimental Mechanics, Volume 6, Issue 12, pp 577-586, 1966. http://dx.doi.org/10.1007/BF02326825

[6] Nau, A., Scholtes, B., Evaluation of the High-Speed Drilling Technique for the Incremental Hole-Drilling Method, *Experimental Mechanics*, 53, 531-542, 2013. http://dx.doi.org/10.1007/s11340-012-9641-1

[7] Beghini, M., Bertini, L., Recent Advances in the Hole Drilling Method for Residual Stress Measurement, *Journal of Materials Engineering and Performance*, ASM International, 7, 163-172, 1998. http://dx.doi.org/10.1361/105994998770347882

[8] Aoh, J.-N., Wei, C.-S., On the Improvement of Calibration Coefficients for Hole-Drilling Integral Method: Part I-Analysis of Calibration Coefficients Obtained by a 3-D FEM Model, *Journal of Engineering Materials and Technology*, ASME, Vol. 124, 250-258, 2002. http://dx.doi.org/10.1115/1.1416685

Residual Stresses 2016: ICRS-10 Materials Research Forum LLC
Materials Research Proceedings **2** (2016) 241-246 doi: http://dx.doi.org/10.21741/9781945291173-41

Ultrasonic Non-destructive Testing and in Situ Regulation of Residual Stress

Chunguang Xu[1, a*], Haibing Tian[2,b], Wentao Song[3,c] and Jianfeng Song[4,d]

[1,2,3,4] School of Mechanical Engineering, Beijing Institute of Technology, Beijing, China

[a]xucg@bit.edu.cn, [b]738427973@qq.com, [c]18810328150@163.com, [d]15827349393@163.com

Keywords: Residual Stress, Ultrasonic, Non-Destructive Testing, Petroleum Pipeline

Abstract. Based on acoustoelasticity theory, the relationship between ultrasonic and stress in mechanical component is researched. Further, the principle, method and technology of exciting ultrasonic critically refracted longitudinal wave is analyzed. In order to ensure that test results are exact and realization of quantity traceability, the calibration technology of ultrasonic testing residual stress is studied. At last, the ultrasonic stress testing and calibration system is established. This technology now is widely used in testing residual stress in high-speed rail and petroleum pipeline.

Introduction

Residual stress greatly influences the service performance of practical engineering components, notably their strength, fatigue life, and dimensional stability. The fatigue life of metallic materials can be extended by the near-surface macroscopic compressive residual stresses that retard fatigue crack initiation and crack growth. Thus, it is important to know the magnitudes of these stresses, and to account for them during the design process. However, the "locked-in" character of residual stresses makes them challenging to measure, because there are no external loads that can be manipulated.

The engineering properties of materials and structural components, notably fatigue life, distortion, dimensional stability, corrosion resistance, and brittle fracture, can be considerably influenced by residual stresses[1]. Such effects usually lead to considerable outlays in the repair and restoration of parts, equipment, and structures. Accordingly, residual stress analysis is an essential stage in the design of parts and structural elements, and in the estimation of their reliability under actual service conditions[2].

Measurement of the actual stress in a structure or component is of great importance in all areas of engineering[3]. Non-destructive methods offer the obvious advantage of specimen preservation, and are particularly useful for production quality control and measurement of detailed calibrations on representative specimen material to valuable specimens. However, these methods commonly require provide the required computational data. A promising non-destructive method for subsurface stress measurement uses an acoustic transducer as an ultrasonic strain gauge. In this approach, ultrasonic measurements are taken before and after the specimen is subjected to stress[4].

Principle of residual stress testing

The testing theory of acoustoelasticity

Acoustoelasticity theory is one of the main bases of ultrasonic stress testing. Acoustoelasticity theory is based on the finite deformation of continuum mechanics to study the relationship between the elastic solid stress state and the macroscopic elastic wave velocity. Based on the four basic assumptions of acoustoelasticity, the elastic wave equation (acoustoelasticity equation) in stress medium under initial coordinates can be obtained [5].

Materials Research Proceedings **2** (2016) 241-246 doi: http://dx.doi.org/10.21741/9781945291173-41

$$\frac{\partial}{\partial X_J}\left[\left(\delta_{IK}t^i_{JL}+C_{IJKL}\right)\frac{\partial u_K}{\partial X_L}\right]=\rho^i\frac{\partial^2 u_I}{\partial t^2} \qquad (1)$$

Where δ_{IK} is Kronecker delta function, ρ^i is the density of the solid in the loading condition, u_I is the dynamic displacement, X_J is the particle position vector, C_{IJKL} is the equivalent stiffness, which depends on the material constant and the initial displacement field and t^i_{JL} is the Cauchy stress shown in the initial coordinates under the solid loading state.

In the case of homogeneous deformation and the solid is isotropic, Eq. (1) can be analytically expressed. Therefore, the equation for the ultrasonic propagation velocity and stress in solid can established in Cartesian coordinates [6]. For the longitudinal wave which propagates along the stress direction:

$$\rho_0 V^2=\lambda+2\mu+\frac{\sigma}{3\lambda+2\mu}\left[\frac{\lambda+\mu}{\mu}(4\lambda+10\mu+4m)+\lambda+2l\right] \qquad (2)$$

In Eq. (2), λ and μ are the Lame elastic constants; l,m,n are the Murnaghan elastic constants; the elastic constants of different materials are shown in Table 1 [7]. ρ_0 is the density of the solid before deformation; σ is the stress applied in one direction (tensile stress is positive and compressive stress is negative); V is the velocities of the longitudinal wave.

Table 1. Lame and Murnaghan constants of the materials, unit [GPa].

Material	λ	μ	l	m	n
Steel (1045)	120	79	-179	-496	-628
Aluminium (6061)	62	26	-201	-305	-300
Copper (99.9%)	104	46	-542±30	-372±5	-401±5

The testing principle of ultrasonic LCR wave method

When a longitudinal wave propagates from a medium in which the wave velocity is slower to a medium in which the wave velocity is faster, according to the Snell law, there is an incidence angle that makes the refraction angle of the longitudinal wave equal to 90°. A longitudinal wave with a refraction angle equal to 90° is called the critically refracted longitudinal wave (L_{CR} wave). The angle of incidence is the first critical angle.

We obtain the relationship between the longitudinal wave velocity that propagates along the stress direction and the stress, as shown in Eq. (2). In the actual detection, the distance between the transmitting and receiving transducer is fixed, and we can reflect the change of the sound velocity by calculating the change of the sound time and then determine the acoustic elastic effect. From Eq. (2), we can obtain the relationship between the stress variation and the changing time of sound propagation:

$$d\sigma = K\cdot dt \qquad (3)$$

$$K=\frac{-2V_0(3\lambda+2\mu)}{(\frac{4\lambda+10\mu+4m}{\mu}+\frac{2l-3\lambda-10\mu-4m}{\lambda+2\mu})L} \tag{4}$$

Where, K is stress coefficient of measured component, the unit is MPa/ns; Δt is the time variation under the condition of stress; L is the distance between the transmitting and receiving transducer, V_0 is the longitudinal wave velocity under the condition of zero stress.

System of residual stress testing

Schematic diagram of the testing system needs to meet the requirements are shown in Figure 1. It mainly includes specialized ultrasonic transducer, ultrasonic transceiver, temperature sensor and transmitter, automatic scan device, trigger and data collector, portable industrial control computer and corresponding algorithm software, calibration block, etc. The index of residual stress test system as follows: Detection range $\pm\sigma_S$ (σ_S is the yield strength of tested material); test resolution is ±20MPa; temperature range: 0°~40°; test depth: 0.5~10mm.

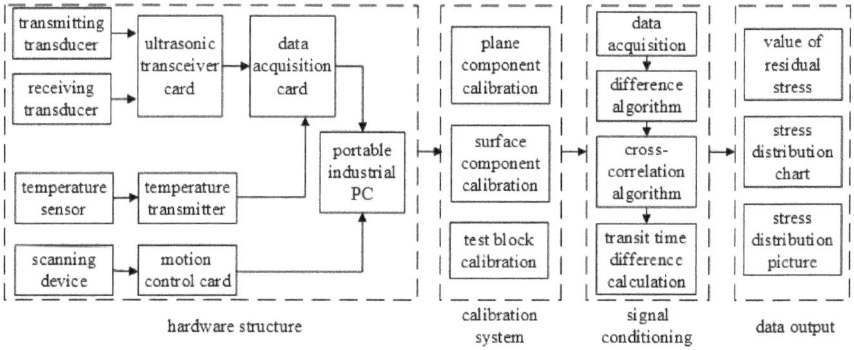

Fig. 1 Schematic diagram of residual stress testing system

Experiment studies

Residual stress testing for pipeline weld joint

Using the ultrasonic testing system to test residual stress of China's East-West pipeline, and evaluate the dangerous areas in pipeline, as shown in Fig. 2. The material of pipeline is X70 steel, and welding procedure is manual arc welding. We tested residual stress around straight weld joint in a section of pipeline. In order to verify the accuracy of the test results, a hydrostatic test has been carried out. From the Fig. 3, it is observed that the blasting place is consistent with the dangerous area evaluated by ultrasonic method.

Materials Research Proceedings **2** (2016) 241-246 doi: http://dx.doi.org/10.21741/9781945291173-41

Fig. 2 Testing point location

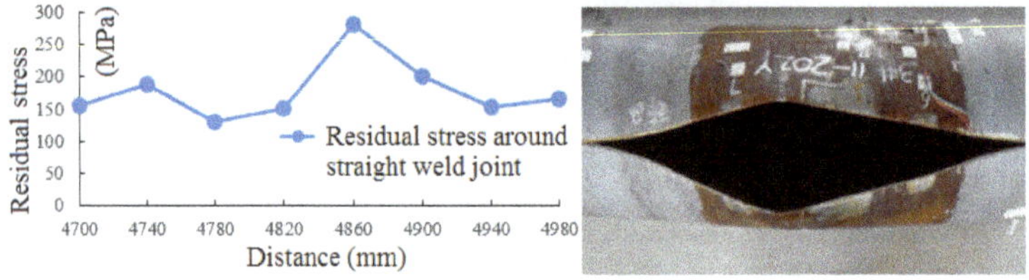

Fig. 3 Hydrostatic experimental verification

Residual stress distribution of steel pipe inner surface

The subject in this experiment is a steel pipe whose diameter is 105 mm, wall thickness is 20mm. The material of pipe is 685 steel, and tested length is 4 m. We test eight angles such as 0°, 45°, 90°, 135°, 180°, 225°, 270° and 315°. Every time, ultrasonic testing probe fixed in auto creeping mechanism moves along one of eight direction. Test residual stress while ultrasonic testing probe moves 40mm. Finally, the residual stress distribution of steel pipe inner surface is shown as in Fig. 4.

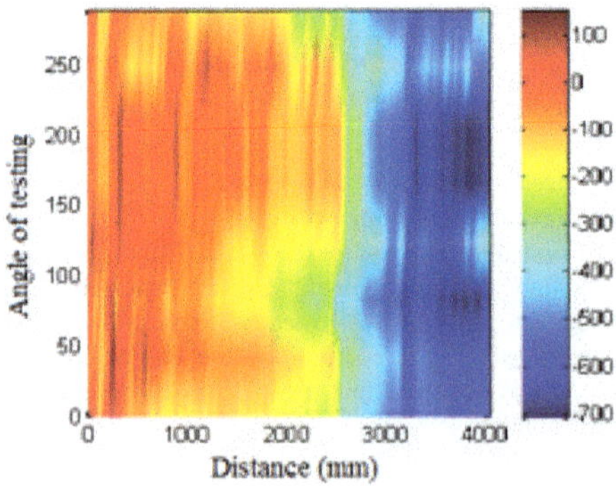

Fig. 4 Residual stress distribution of inner surface

Residual stress testing for rail

According to acoustoelasticity theory, we successfully accomplished the experiment of the residual stress of rail detection.

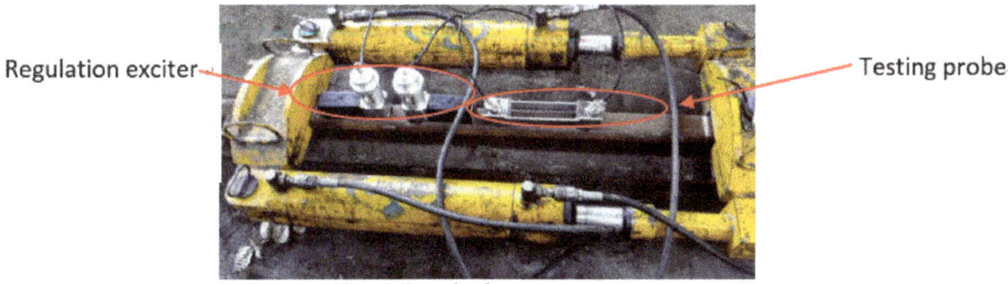

Fig. 5 Residual stress testing

The detection sensor used in this testing is shown in Fig. 5, which functions in the mode of pitch and catch. This detection sensor is composed of sound wedge and two ultrasonic transducer. Also, the distance of the inspiring and receiving ultrasonic transducer can be adjusted. Similarly, the angle of the ultrasonic transducer can also be adjusted. This will ensure the good coupling of the sensor and the sample and improve the detection accuracy.

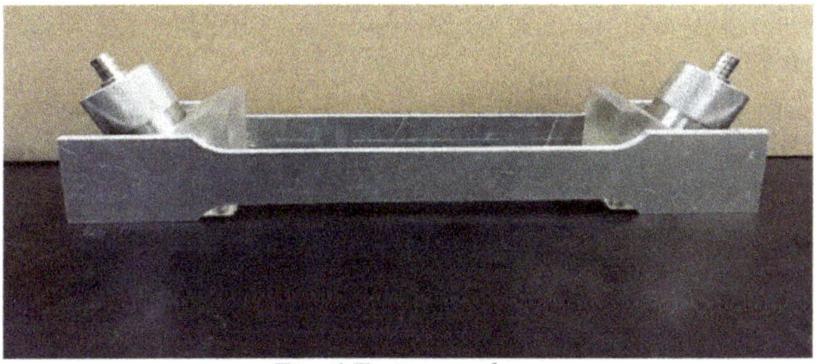

Fig. 6 Testing probe

As a rule, if the residual stress is in a state of compression, the rail is safe. Otherwise, it's in unsafe working condition. If the tensile stress is close to 1/3 of the material yield strength, the rail is in dangerous condition. If the rail tensile stress is more than half of the yield strength, it is a very dangerous state. Thus, we need continuous monitoring when it works. If tensile stress is more than 2/3 or 3/4 of its yield strength. The rail is in a critical state of crack, we must stop its work.

Conclusion

(1) Based on acoustoelasticity theory, the principle of residual stress ultrasonic testing is analyzed. The stress sensitivity of different types of ultrasonic are compared, and it is obtained that longitudinal wave propagation along the stress direction is most sensitivity.

(2) According to theory of LCR wave, residual stress ultrasonic testing system for pipe and rail component is built.

(3) The residual stress in pipeline welding line, inner surface of steel pipe and rail are tested. Through hydrostatic experimental, the accuracy and practicability and universally application fields of the ultrasonic method can be verified.

Acknowledgements:
This work was sponsored by Natural Science Foundation of China (No. 51275042).

References

[1] Totten, G. E., Howes, M. & Inoue, T. Handbook of Residual Stress and Deformation of Steel. ASM International Publishers, USA, 2002 vol. 1, 417-444.

[2] Rossini, N. S., Dassisti, M., Benyounis, K. Y. & Olabi, A. G.. Methods of measuring residual stresses in components. Mater. and Des. 2012, 35, 572–588. https://doi.org/10.1016/j.matdes.2011.08.022

[3] Gnaupel-Herold, T., Prask, H. J., Clark, A. V., Hehman, C. S. & Nguyen, T. N. A comparison of neutron and ultrasonic determinations of residual stress. Meas. Sci. Technol. 2000, 11, 436–444. https://doi.org/10.1088/0957-0233/11/4/315

[4] Schajer, G. S. Relaxation methods for measuring residual stresses: techniques and opportunities. Experimental Mechanics. 2010, 50, 1117–1127. https://doi.org/10.1007/s11340-010-9386-7

[5] Bray, D. E., Junghans, P., Application of the Lcr Ultrasonic Technique for Evaluation of Post-weld Heat Treatment in Steel Plates. NDT&E International. 1995, 28, 235-242. https://doi.org/10.1016/0963-8695(95)00020-X

[6] Rose, J. L. Ultrasonic Waves in Solid Media. Cambridge: Cambridge University Press. 1999.

[7] Viktor, H. Structural and Residual Stress Analysis by Nondestructive Methods. Netherlands: Elsevier Press, 1997.

[8] Wentao Song, Chunguang Xu, Qinxue Pan, et al. Regulation of Residual Stress in Elastic Solid Component with High-energy Acoustic Field. AIP Conference Proceedings, 2013, 1511, 196-203. https://doi.org/10.1063/1.4789179

Residual Stresses 2016: ICRS-10 Materials Research Forum LLC
Materials Research Proceedings **2** (2016) 247-252 doi: http://dx.doi.org/10.21741/9781945291173-42

Residual Stress and Contact Force Study for Deep Cold Rolling of Aero-engine Material

Arvind Prithiviraj[a]*, Wang Wei[b]

Advanced Remanufacturing and Technology Centre (ARTC), Agency for Science, Technology and Research (A*STAR), 3 Cleantech Loop, #01-01, CleanTech Two, Singapore 637143

[a]arvindp@artc.a-star.edu.sg

Keywords: Residual Stress, XRD, Deep Rolling, Burnishing, Mechanical Surface Treatment, IN718

Abstract. Deep Cold Rolling (DCR) process is used in various industries to improve the fatigue life of metallic parts by introducing work hardening, deep layer of compressive residual stresses and polished surface finish. In this paper, the influences of the angled design for a hydrostatic tool and its indentation depth to impart compressive residual stresses are studied and compared to its straight tool counterpart for treatment of IN718 material. Residual stress depth profile measurements, using the XRD technique, were employed to determine differences caused by using the angled and straight tool design. Higher rolling forces are measured in an angled tool design with a high indentation depth as compared to a straight tool design caused by the slip stick effect on the internal parts of the tool. This leads to high plastic deformation in the test material significantly affecting the compressive residual stress depth profile depending on its exisiting state.

Introduction

Aero engine metallic parts experience high levels of mechanical and thermal loading, high and low cycle fatigue and foreign object damage (FOD)[1] in service that can lead to early retirement. Suitable mechanical surface treatments applied onto aero engine metallic parts can prolong fatigue life, improve wear resistance and avoid stress corrosion by introducing work hardening, deep compressive residual stresses and polished surface finish[1,2].

Commonly used mechanical surface treatments in the aero engine manufacturing industry are shot peening (SP) and laser shock peening (LSP). Deep cold rolling (DCR), another mechanical surface treatment, uses a rolling ball element that is pressed at high pressure against a metallic part to impart a deep layer of compressive residual stress. Its process parameters that significantly affect the mechanical properties are hydrostatic pressure, rolling tool, contact angle of the tool, stepover, feed rate, yield strength and geometry of the material[6,7]. DCR is a cost effective localised treatment process that can be easily integrated to a robot arm or CNC platform. SP and DCR, are capable of generating high dislocation densities in the near surface regions that inhibit crack initiation although the residual stresses tend to relax faster when the part experiences high temperatures[1,3,4,5]. Deep compressive residual stresses are achieved through LSP and DCR, which help to arrest the crack during the crack propagation stage after FOD impacts and corrosion damages[1,2]. The ability of DCR to generate high cold work at the near surface as well deep compressive residual stresses makes it a good choice for mechanical surface treatment. DCR creates a nanocrystalline surface layer in the range of 3μm[4]. The nanocrystalline structure acts as a good resistance to fatigue crack initiation up to a moderate temperature but deteriorates the creep resistance at elevated temperature[4].

The magnitude of the rolling force applied on to the part is a combined effect of the hydrostatic pressure, ball diameter and contact angle of the tool and the part geometry. Higher rolling force imparts the maximum and deeper layer of compressive residual stresses. When the part has reached its work hardened state, the high rolling forces will shift the maximum compressive stresses deeper in the material's depth thereby reducing the compressive stress state at the surface[1,2,7,8,9]. Severe rolling on a work hardened material will cause tensile stress state at the surface, which is called near surface work softening[2] or cycle softening[1], resulting in a negative effect of decline in hardness and fatigue life and poor surface finish. The factors affecting severe rolling are high rolling force, lower step over value and number of treatment passes[7,8].

The angled tool design is suitable for a curved geometry if the available platform is 3-axis CNC and a complex internal geometry using multi axis CNC and robot. However, during the process application development stage, when the component geometry is unknown or due to the platform limitations, DCR tool design is not considered. In this paper, the influences of the angled design for a hydrostatic tool and its indentation depth to impart compressive residual stresses are studied and compared to its straight tool counterpart for treatment of IN718 material. Residual stress depth profiles are measured by XRD technique.

Experiment and Measurement Procedures

Process Platform Selection and Experiment Plan. The DCR process is integrated with a robot (ABB model no. 6660), which is robust and suitable to handle high process forces. The DCR tool used in the experiment is a hydrostatic single ball point tool with a burnishing ball diameter Ø6.3mm from Ecoroll AG Werkzeugtechnik. Two variants of the hydrostatic tool were used, straight and angled design as shown in Fig. 1. The angled design is indexed to 15deg so that its burnishing ball element is normal to surface when used in the experiments. A hydraulic pump unit supplies pressurised coolant in the range of 50 to 400 bar to the tool, thereby activating its constant stroke length of 6mm. The burnishing ball element is made of Tungsten Carbide, which has a higher bulk modulus than the IN718 specimens. The IN718 test material is machined into flat specimens of dimensions 80mm (L) X 20mm (W) X 10mm (H). The test specimens are annealed first to relieve the stress.

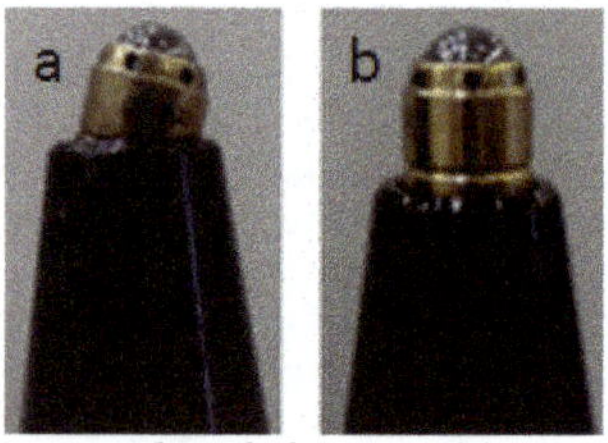

Fig. 1 Single point tool angled variant (a) straight variant (b)

The first experiment set is the track width trials in which track lines caused by plastic deformation for various pressure settings, tool indentation depth and tool design are drawn on the IN718 test piece. The step over, the distance between the subsequent track line in a surface area treatment for a required overlap, is calculated from the track width data.

Table 1 DCR process factors and the selected levels for the 1st and 2nd experiment sets.

Process Factors	1st Experiment Set Levels	2nd Experiment Set Levels
Pressure [bar]	100, 200, 300, 400	200, 400
Indentation Depth [mm]	2, 3, 4	2, 4
Tool Design	Straight, Angled	Straight, Angled
Feed Rate [mm/min]	1000	1000

The second experiment set is for the surface area treatment on the IN718 test pieces for various pressure settings, tool indentation depth and tool type. A zig zag tool path was adopted for the trials.

Materials Research Proceedings **2** (2016) 247-252 doi: http://dx.doi.org/10.21741/9781945291173-42

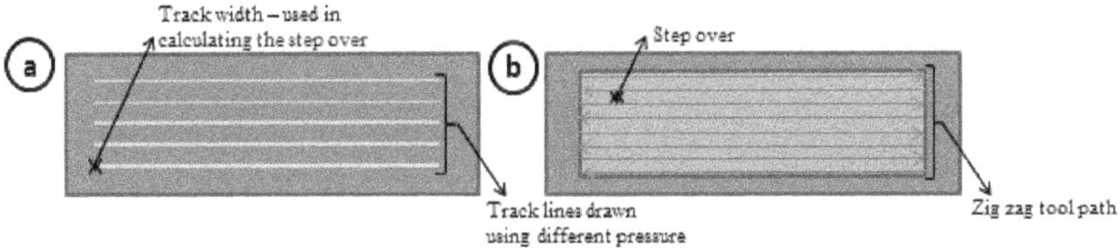

Fig. 2 Illustration of track width in set 1(a) and surface area treatment using zig zap tool path in set 2(b)

Track Width Measurement. The track width is measured in microscopic images of treated test pieces captured using Zeiss Stemi 2000 and is used to determine the step over value for a required overlap of 90%.

Rolling Force Measurement. The rolling forces in three directions are measured online during the track width trials with a Kistler dynamometer type 9129AA, which is fixed underneath the processed IN718 test specimen.

Surface Roughness Measurement. The surface roughness is measured using the Taylor Hobson Talysurf stylus profilometer and the measurements are taken in longitudinal (along the rolling) and transverse (perpendicular the rolling) directions.

Residual Stress Measurement. The residual stresses are measured using Stresstech's Xstress 3000 G3R X-ray Diffraction (XRD) in longitudinal and transverse directions. The measurement mode is modified-χ using two detectors with the 2Θ angle set at 156° and the residual stresses are calculated using the $\sin^2\psi$ method. The elastic constants for IN718 were Young's modulus = 207.6 GPa and Poisson's ratio = 0.29

Layer removal at the measurement location for the subsurface residual stress measurement is by electrochemical polishing, using a mixture of perchoric acid and methanol. There is no additional stress induced due to the layer removal by electrochemical means, but stress relaxation occurs resulting in redistribution and are corrected using a standard Moore & Evans method[10]. The layer removal and the subsequent residual stress measurements are carried out up to the depth of 200μm.

Results and Discussions

The averaged normal direction rolling force is the force acting downward on the test material and it is averaged for the distance of the track width line and the measurements are shown in Fig. 3a. The pressure has an obvious dominance on the rolling force because it increases the contact area of the burnishing ball on the test material. The increase in rolling force remains constant for the straight tool design with varied indentation depth while higher force is measured in the angled tool design as a result of slip stick effect due to the inner sealing and other internal parts that are sliding over each other in the tool.

In the track width comparison plot shown in Fig. 3b, the track width is increased by the pressure causing the burnishing ball to indent deeper in the test material. The track width of the angled tool design is higher compared to the straight tool design; this is because of the higher force in the angled tool. The stability of the angled tool is increased by the indentation depth resulting is higher track width while the straight tool design is stable throughout.

Materials Research Proceedings **2** (2016) 247-252 doi: http://dx.doi.org/10.21741/9781945291173-42

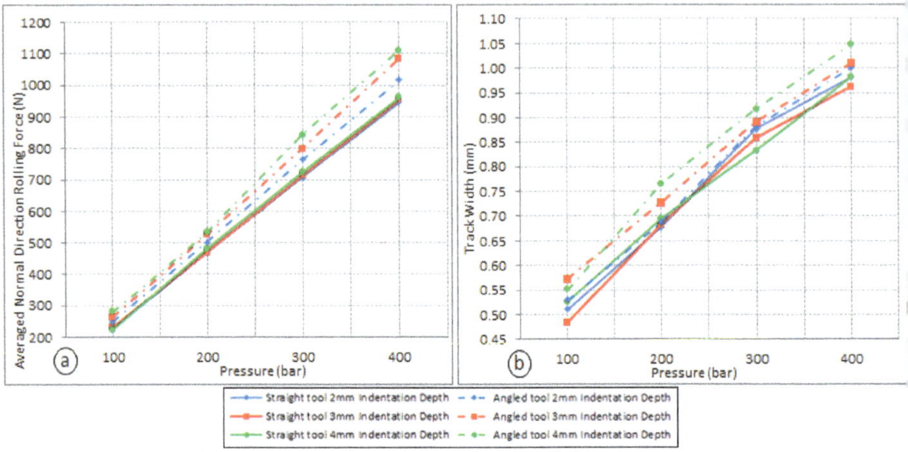

Fig. 3 Averaged normal direction rolling force (a) and track width (b)

The initial surface roughness Ra was 0.06μm in the longitudinal and 0.47μm in the transverse directions. The Ra measured after DCR is 0.02μm and it is identical in the longitudinal direction for varied pressure, indentation depth and tool type with a fixed overlap of 90%. For transverse direction, the Ra for 400 and 200 bar pressures are 0.17μm and 0.09μm respectively with negligible effects from indentation depth and tool type.

The residual stress depth profile measured after DCR treatment in Fig. 4 and 5 shows transverse direction stresses were higher than the longitudinal direction. In Fig 5a, the maximum residual stresses for 200bar pressure are -780MPa in longitudinal and -1060MPa in transverse directions and were measured at the depth of 100μm and 50μm respectively. In Fig. 5b, the maximum residual stresses for 400bar pressure are -750MPa in longitudinal and -1075MPa in transverse directions and were measured at the depth of 110μm and 100μm respectively. In Fig. 4 and 5, there is an increase in maximum compressive residual stress in transverse direction for the 200bar and 400bar pressure in an angled tool with high indentation depth compared to straight tool. In Fig 5b, there is an increase in the surface compressive residual stress in the transverse direction for 400bar pressure in an angled tool with high indentation depth compared to straight tool. It is observed that the maximum compressive residual stress profile for angled tool was shifted deeper into material depth due to the high rolling force as seen in Fig. 3a.

The increase in the surface and sub surface compressive residual stresses and the consistent surface roughness Ra are required to be further investigated with fatigue analysis, to evaluate the significance of high rolling force from the angled tool type with varied indentation depth. The influence of the angled tool design should be carefully studied using residual stress and force measurements for pressure greater than 400 bar or different ball diameter and the angle of the tool body, to investigate if high rolling force can cause cycle softening at the surface, leading to reduction in compressive residual stress which is detrimental to fatigue life.

Materials Research Proceedings **2** (2016) 247-252 doi: http://dx.doi.org/10.21741/9781945291173-42

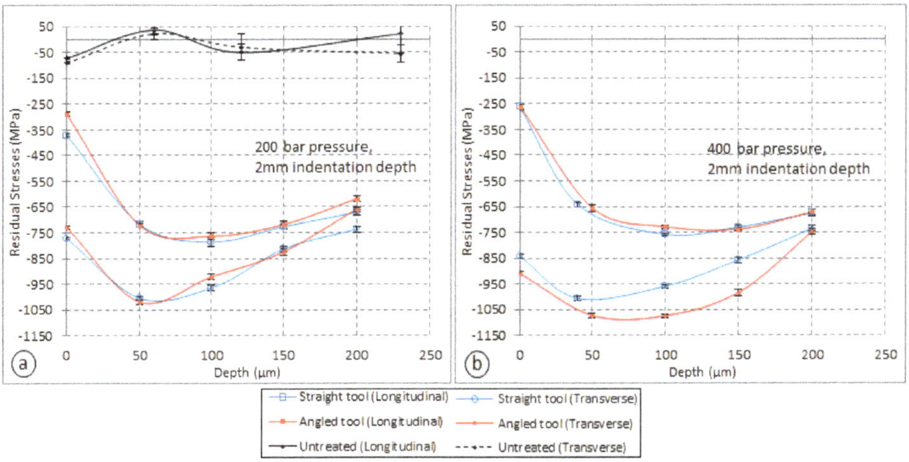

Fig. 4 Residual stress depth profiles for 200bar pressure (a) and 400bar pressure (b) with 2mm indentation depth

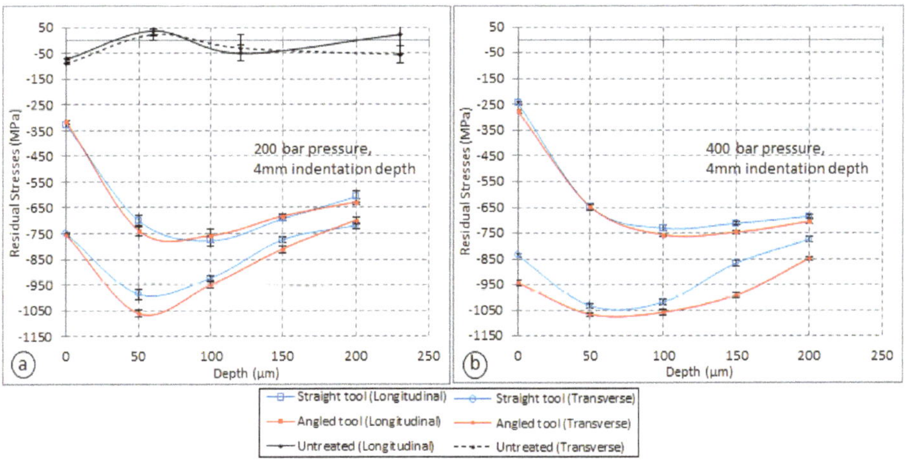

Fig. 5 Residual stress depth profiles for 200bar pressure (a) and 400bar pressure (b) with 4mm indentation depth

Summary and conclusion

DCR experimental trials were conducted using Ø6.3mm ball diameter tool that has two variants, a straight and an angled tool design, to study the effects on the averaged normal direction rolling force, and residual stress depth profiles caused by the tool design and its indentation depth. In the track width and average rolling forces analysis, the angled tool caused higher track width for increasing indentation depth and rolling pressure corresponding with the high rolling force measured. Its effect on the residual stress depth profile is an increase in the surface compressive stress and maximum compressive stress for transverse direction using 400 bar pressure. The surface roughness Ra is not affected by the high rolling force. Fatigue analysis is proposed to study the influence of the increase in compressive residual stress profile using the angled tool design as a result of high rolling force with varied indentation depth.

Acknowledgements
The author would like to thank engineers Mr. Alex Goh Qi Ming, Mr Buenaventura Rizaldy Jr Tubera and Mr Jermin Soh Ming Jie who helped in the experiments and measurements, and show sincere appreciation to Mr. Stefan Zenk from Ecoroll AG Werkzeugtechnik for his technical guidance.

References
[1] I. Altenberger, U. Noster, B. L. Boyce, J. O. Peters, B. Scholtes, and R. O. Ritchie, "Effects of Mechanical Surface Treatment on Fatigue Failure in Ti-6Al-4V: Role of Residual Stresses and Foreign-Object Damage," *Materials Science Forum*, vol. 404–407. pp. 457–462, 2002. http://dx.doi.org/10.4028/www.scientific.net/MSF.404-407.457

[2] I. Altenberger, "Deep rolling–the past, the present and the future," in *Proceedings of 9th international conference on shot peening*, 2005, pp. 6–9.

[3] P. Juijerm and I. Altenberger, "Effective boundary of deep-rolling treatment and its correlation with residual stress stability of Al–Mg–Mn and Al–Mg–Si–Cu alloys," *Scripta Materialia*, vol. 56, no. 9, pp. 745–748, 2007. http://dx.doi.org/10.1016/j.scriptamat.2007.01.021

[4] I. Altenberger, R. K. Nalla, Y. Sano, L. Wagner, and R. O. Ritchie, "On the effect of deep-rolling and laser-peening on the stress-controlled low- and high-cycle fatigue behavior of Ti-6Al-4V at elevated temperatures up to 550°c," *International Journal of Fatigue*, vol. 44. pp. 292–302, 2012. http://dx.doi.org/10.1016/j.ijfatigue.2012.03.008

[5] C. M. Gill, N. Fox, and P. J. Withers, "Shakedown of deep cold rolling residual stresses in titanium alloys," *Journal of Physics D: Applied Physics*, vol. 41, no. 17, p. 174005, 2008. http://dx.doi.org/10.1088/0022-3727/41/17/174005

[6] A. Rodríguez, L. N. López de Lacalle, A. Celaya, A. Lamikiz, and J. Albizuri, "Surface improvement of shafts by the deep ball-burnishing technique," *Surface and Coatings Technology*, vol. 206, no. 11–12. pp. 2817–2824, 2012. http://dx.doi.org/10.1016/j.surfcoat.2011.11.045

[7] A. M. Abrão, B. Denkena, B. Breidenstein, and T. Mörke, "Surface and subsurface alterations induced by deep rolling of hardened AISI 1060 steel," *Production Engineering*, vol. 8, no. 5, pp. 551–558, 2014. http://dx.doi.org/10.1007/s11740-014-0539-x

[8] M. Beghini, L. Bertini, B. D. Monelli, C. Santus, and M. Bandini, "Experimental parameter sensitivity analysis of residual stresses induced by deep rolling on 7075-T6 aluminium alloy," *Surface and Coatings Technology*, vol. 254. pp. 175–186, 2014. http://dx.doi.org/10.1016/j.surfcoat.2014.06.008

[9] F. Klocke, V. Bäcker, H. Wegner, B. Feldhaus, H. U. Baron, and R. Hessert, "Influence of process and geometry parameters on the surface layer state after roller burnishing of IN718," *Production Engineering*, vol. 3, no. 4–5. pp. 391–399, 2009. http://dx.doi.org/10.1007/s11740-009-0182-0

[10] M. G. Moore and W. P. Evans, "Mathematical correction for stress in removed layers in X-ray diffraction residual stress analysis," SAE Technical Paper, 1958.

Residual Stresses 2016: ICRS-10 Materials Research Forum LLC
Materials Research Proceedings **2** (2016) 253-258 doi: http://dx.doi.org/10.21741/9781945291173-43

Influence of the Pre-Stressing on the Residual Stresses Induced by Deep Rolling

Nataliya Lyubenova[1,a*], Maxime Jacquemin[1,b] and Dirk Bähre[1,c]

[1]Institute of Production Engineering, Saarland University, Campus A 4.2, 1st floor, D-66123, Saarbruecken, Germany

[a]nataliya.lyubenova@uni-saarland.de, [b]maxime.jacquemin7@etu.univ-lorraine.fr, [c]d.baehre@mx.uni-saarland.de

Keywords: Residual Stresses, Mechanical Surface Treatment, Deep Rolling, Pre-stressing, Finite Element Simulation

Abstract. Deep rolling is a mechanical surface treatment, which main aim is to increase the fatigue life of components by reducing their roughness, increasing the surface hardening and inducing compressive residual stresses. The increase of certain process input parameters such as the applied pressure or the number of overturns leads directly to the raising of the induced compressive residual stresses. Nevertheless, a saturation point is always achieved, where the further increase of the parameters' levels does not change the induced residual stresses. For other mechanical surface treatments, like shot peening, several pre-stress techniques were employed in order to further increase the induced residual stresses without raising the shot intensity or the coverage percentage. Pre-stressed shot peening with the means of bending or torsion is an established processing. Up to now, a few investigations are available regarding the pre-stressing techniques applied to deep rolling. Therefore, this paper offers a newly designed finite element model, built to calculate the induced residual stresses by bending, consequent deep rolling and springback. A four point bending setup with different pre-stress levels was employed and the influence of pre-stress levels on the induced residual stresses was investigated. Additionally, the applied deep rolling pressure was also varied in order to optimize this hybrid processing. At the end, the anisotropy of the induced longitudinal and transverse residual stresses due to the bending and deep rolling was analyzed.

Introduction

Manufacturing processes that selectively induce compressive residual stresses (CRS) in the critical areas of highly loaded components are gaining increasing importance as design tool as they are able to increase components' fatigue strength. Several processes like shot peening, hammering and autofrettage are well established and have the ability to introduce CRS from several hundred micrometers to few millimeters in depth. Among these processes stays the deep rolling (DR), which is a recognized surface mechanical treatment that attracted the interest of the scientific community in the thirties of the last century. It has the advantage that along with the induced CRS, it reduces also the roughness, thus preventing new crack formations. All mechanical surface treatments mentioned above have in common that the raise of certain input parameters leads directly to higher and deeper induced CRS. Nevertheless, there is always a saturation level where the further increase of the input parameters does not lead to raising of the CRS and can even result in tensile residual stresses at the surface [1]. Some changes in the initial boundary conditions give the possibility to further exploit the capabilities of the mechanical surface treatments. For example, processing at elevated workpiece's temperatures leads to an increase of the induced CRS [2, 3] and longer fatigue life [2], while the decrease of the workpiece's temperature below room temperature results in higher hardness and hardness penetration depth of the treated components [4].

Another change in the initial boundary conditions can be a mechanical pre-stressing (PS) of the workpiece. The mechanism of the PS (in case of positive pre-loading) is the following: an external stress that causes elastic deformation is applied to the workpiece while the mechanical surface treatment is performed. When the PS is released, it causes a deformation in the direction opposite to the initial PS. This means that the PS adds CRS, increasing linearly from the center of the specimen towards the treated surface. The result is shifting in depth of the point, where the transition between compressive and tensile residual stresses is and thus the area loaded with CRS expands. PS mechanical surface treatments can be applied to any highly stressed parts which are exposed to operational bending or torsion, like leaf springs, Belleville springs, coil springs, torsion bars, propeller shafts, etc. The applied PS can be tension, bending or torsion.

The interest for the development of mechanical surface treatments in order to further increase their effectiveness rose in the forties of the last century. Staub and May [5] introduced an innovative process called stress peening, where the specimen is statically stressed in the direction of the operational loading during the conventional shot peening. They performed fatigue tests on not-shot peened, conventionally shot peened and PS shot peened specimens and observed an increase of the minimum fatigue life of the shot peened specimens by an average of 350 % while the PS shot peened specimens showed a minimum fatigue life 740 % longer than the one of not-shot peened specimens. Barrett et al. [6] investigated the effect of the elastic PS on the magnitude of CRS induced by PS shot peening of aluminum plates. They bent their specimens at 87 % of the yield strength of the material and found out that the peak of CRS in the direction longitudinal to the bending curvature was enhanced from 405 MPa for the not-stressed plate to -594 MPa for the PS plate (app. 47 % improvement). Nevertheless, the depth of the achieved CRS did not change with the application of a PS. In the perpendicular direction, the measured residual stresses were lower than the ones in longitudinal direction and even lower than those measured in not-stressed treated specimen. Xu et al. [7] studied the influence of the direction (positive or negative) of the PS on shot peening. They bent their specimens at +38 %, +57 %, +76 % and -75 % of the material's yield strength and found out that the positive bending increased the endurance limit while the negative bending significantly decreased the endurance limit, even compared with not-shot peened specimens. The measurements of the surface and depth distribution of residual stresses showed that enhancing the positive bending results in higher surface-, higher magnitude- and greater depth of the induced CRS. On the contrary, the negatively bent specimen showed no presence of CRS but rather low tensile residual stresses. Some experiments were also made about the PS of a workpiece for the deep rolling process. Müller [8] applied bending using a four point bending setup and performed consecutive DR treatment in directions longitudinal and transverse to the bending. In the longitudinal variant, the induced surface CRS did not change with the increasing PS level. Still, the enhancing of the PS led to deeper distributed CRS, which for PS at 80 % were still negative at 1 mm depth. The transverse variant resulted in a quite different residual stress distribution. The surface stresses raised significantly (more than 100 % for 80 % bending), while in depth the zero stress plateau remained unchanged at 0.80 mm depth.

Despite the available experimental data about the PS mechanical surface treatments, the prediction of the residual stress distribution still is a challenging task, when varying the numerous input parameters. The residual stress investigation is always supported by time-consuming and expensive experimental measurements. It is well known that finite element modeling (FEM) is a powerful supplement to the experimental work as it offers the opportunity to solve complex engineering problems and to simulate different types of processing. Several attempts were made to simulate the DR process [9-11] and the results qualitatively and partly quantitatively fitted the experimental verifications. Therefore, this paper offers a newly designed finite element modeling of a hybrid processing, namely elastic bending using four point bending set up, consecutive DR treatment and a springback. The model is able to calculate the resulting residual stresses for different values of bending magnitude and applied DR pressure.

Materials Research Proceedings **2** (2016) 253-256 doi: http://dx.doi.org/10.21741/9781945291173-43

Finite Element Modeling Setup

The FEM used in this paper consists of three parts. First, an elastic four-point bending at levels 30 %, 60 % and 70 % of the material's yield strength was performed using the standard module of ABAQUS CAE 6.14. Then, with the means of pre-defined field, the PS was determined as an initial stress state for the DR treatment (DR pressure applied on the DR tool was at levels 20 MPa, 30 MPa and 40 MPa), performed in ABAQUS CAE 6.14 explicit. The final springback operation was made in ABAQUS standard, using a pre-defined field from the DR operation. The material assigned for the workpiece was AISI 4140 steel with Young's modulus of 210 GPa, Poisson's ratio 0.28, Yield strength of 997 MPa and ultimate strength of 1144 MPa. The applied material model was elastic-plastic with bi-linear kinematic hardening, described in details in [10, 11]. The geometry of the workpiece was a plate with dimensions: length = 18 mm, width = 3.75 mm and thickness = 2 mm. The DR tool consisted of a sphere modeled as a rigid body of diameter = 3 mm. The DR process was applied along the bending curvature on the tensile side of the bending set up. Concerning the meshing of the workpiece, structured C3D8R hexahedral elements were used. A convergence study was performed, where the results from the FEM of the bending step were compared with the analytical solution of the bending. A progressive meshing strategy was employed to obtain more accurate results in the areas of interest, using a smallest element size of 0.04 mm and a biggest size of 0.1 mm. The FEM results differed from the analytical solution with 2.25 %, which is an acceptable error. Due to the complexity of the FEM, numerous boundary conditions were assigned. During the bending and the springback, the workpiece was restrained at the two supports of the four-point set up. The face opposite to the one on which the DR is applied was encastred during the DR process. The DR tool was only allowed to move in the vertical direction during the application and the release of the DR pressure but was free to rotate along the DR path during its movement. The DR treatment was performed with a constant velocity of 1 mm/s and the friction coefficient between the workpiece and the DR sphere was assigned to 0.1.

Results and Discussion

Fig. 1 shows the residual stress vs. depth profiles w/o PS and with 30 % PS after DR pressure variation.

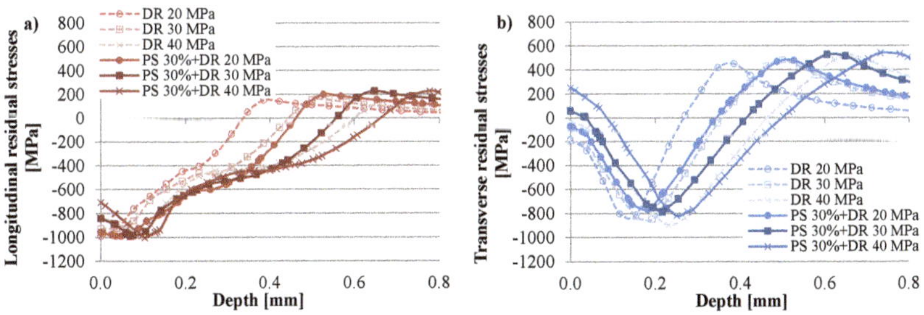

Fig. 1 – Residual stress vs. depth profiles for 30 % PS, DR pressure variation, a) longitudinal direction and b) transverse direction

Fig. 1 a) illustrates the longitudinal (along the DR trace) residual stresses, while Fig. 1 b) the transverse (transverse to the DR trace) residual stresses. The typical DR anisotropy of the stresses in both directions is clearly visible here. The anisotropy is noticeable with- as well as w/o PS. It can be noticed that in both directions, increasing the PS does not enhance the maximum of CRS. However, there is a significant raise in the depth of CRS in the longitudinal direction which reaches 650 µm for a PS of 30 % and DR with a pressure of 40 MPa. In the transverse direction, the depth of CRS raises with increasing PS levels but the surface residual stresses deteriorate with the application of PS.

Materials Research Proceedings **2** (2016) 253-256 doi: http://dx.doi.org/10.21741/9781945291173-43

Fig. 2 displays the residual stress vs. depth profiles after DR pressure variation with and w/o PS of 70 %. Here, it is also noticeable that the improvement in the CRS field is higher in the longitudinal direction. Even the surface stresses aggravate slightly, the CRS depth increases to 0.8 mm when raising the PS level. For the transverse direction, the PS reduces the surface- and the maximum of the residual stresses but still has a positive effect on induced CRS in depth.

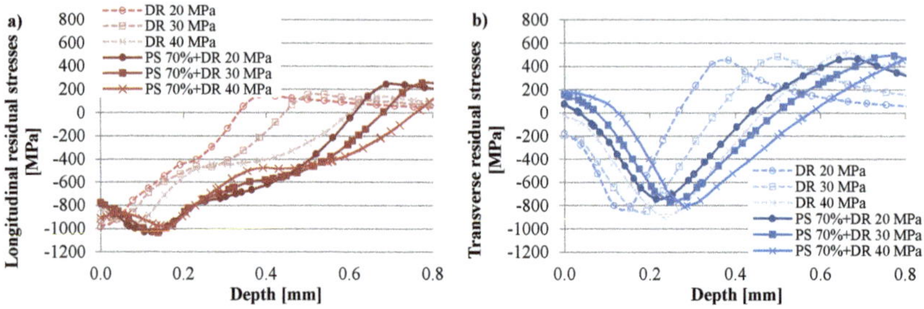

Fig. 2 – Residual stress vs. depth profiles for 70 % PS, DR pressure variation, a) longitudinal direction and b) transverse direction

In Fig. 3 are plotted the longitudinal residual stresses after variation of the PS level, as the observation was made that the CRS in the direction along the bending curvature/DR trace augment more than these in the transverse direction.

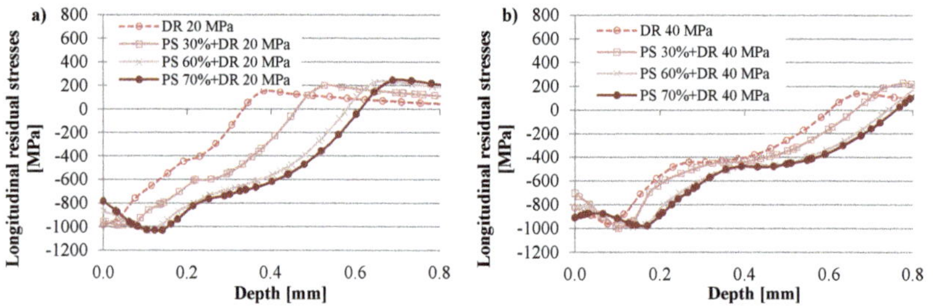

Fig. 3 – Longitudinal residual stress vs. depth profiles for different PS levels, a) DR pressure 20 MPa and b) DR pressure 40 MPa

At DR pressure of 20 MPa, see Fig. 3 a), the depth of CRS increases significantly for PS of 30 % and 60 %, and reaches a saturation point at 70 %, where almost no further improvement is observed. For the DR pressure of 40 MPa, shown on Fig. 3 b), the improvement when applying PS is not so strong but it must be taken into account that the CRS level w/o PS is higher than that at 20 MPa.

In Fig. 4 are represented the calculated areas under the curve for the tensile and compressive residual stresses for the longitudinal and transverse residual stresses profiles. In the longitudinal direction (Fig. 4 a)), the variation of the DR pressure and the PS level has no influence on the tensile residual stresses. However, by looking at the CRS, two observations can be made: first, higher DR pressure results in a higher amount of CRS and second, raising the PS level also enhances the CRS. Still, a saturation point is visible at 40 MPa, where the change in the PS has almost no impact on the CRS. The transverse residual stresses areas are plotted in Fig. 4 b). Here, it is obvious that increasing the DR pressure as well as the PS leads to higher tensile stresses while the CRS remain almost intact.

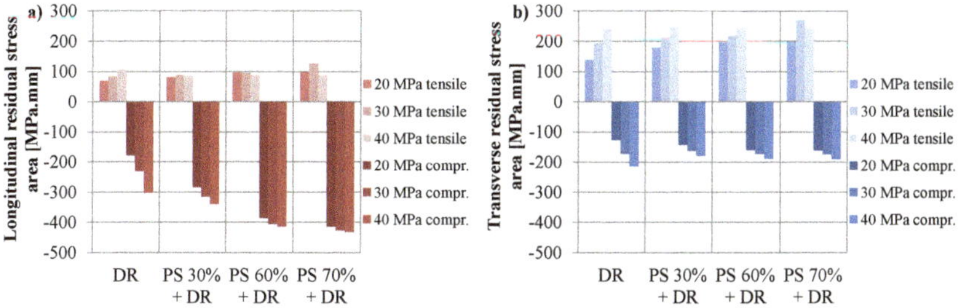

Fig. 4 – Calculated residual stress areas, variation of the PS, variation of the DR pressure, a) longitudinal residual stress areas and b) transverse residual stress areas

The results shown in this paper are expected and can be explained by the following realizations. To begin with, the observed anisotropy of the residual stresses induced in longitudinal and transverse directions by the DR process w/o PS is well described in many publications [12, 13]. The plastic deformation mechanism and shrinkage of the material after releasing of the DR pressure is always different in the both directions. Second, when applying a PS and DR in the same direction, the springback superimposes with the shrinkage of the material, thus shifting the CRS to greater depths. Therefore, the enhancement due to the PS is greater in the longitudinal direction (along the DR path/bending curvature). Müller [8] measured the stresses in both directions and observed the same trends. In the presented investigations was also modelled a pre-stressed DR in a direction which is transverse to the bending curvature. The results showed almost no enhancement of the CRS field and at higher PS levels high sub-surface tensile residual stresses appeared.

Summary and Conclusion

This paper introduced a newly designed finite element model of a hybrid mechanical surface treatment consisting of elastic PS using a four point bending setup, consecutive deep rolling treatment and a springback. The model offered the possibility to vary the applied PS and the DR pressure and to investigate the resulting surface and depth residual stresses. The DR treatment was applied along the bending curvature on the positively (tensile) loaded side of the workpiece. The resulting residual stresses in the directions longitudinal and transverse to the DR treatment were analyzed. Some important findings were concluded: it was observed a strong anisotropy in the stresses induced in longitudinal and transverse directions even w/o PS. This can be explained by the different portions of the plastic stretching in both directions. An enhancing of the CRS in the longitudinal direction was found by both increasing the DR pressure and PS level. Still, at higher DR pressures the PS had less impact than at lower DR pressures. In the transverse direction, the PS had minor effect on the CRS and resulted in slightly higher tensile residual stresses at the surface. The calculated areas under the curve of the tensile- and compressive stress vs. depth profiles confirmed these findings and gave a general overview of the interactions between the applied input parameters (DR and PS) and the resulting stresses. It can be concluded that the tensile PS has an enhancing effect on the CRS layer in longitudinal direction induced by DR. The surface tensile residual stresses in transverse direction have deteriorating influence regarding the surface crack initiation. Still, the presence of thicker CRS layer can lead to the so-called "crack-arrest" and to retard the crack propagation. The maximum of the CRS is still limited by the yield strength of the material but it can be achieved at lower DR pressures, which will result in lower deformations induced during the processing.

Materials Research Proceedings **2** (2016) 253-256 doi: http://dx.doi.org/10.21741/9781945291173-43

References

[1] G. Bernstein, B. Fuchsbauer, Festwalzen und Schwingfestigkeit, Z.f. Werkstofftechnik 13 (1982), 103 – 109. http://dx.doi.org/10.1002/mawe.19820130309

[2] I. Nikitin, I. Altenberger, B. Scholtes, Effect of deep rolling at elevated and low temperatures on the isothermal fatigue behavior of AISI 304, Proceedings of 9th International Conference on Shot Peening, Paris (2005), 185-190.

[3] P. Juijerm, Fatigue behavior and residual stress stability of deep-rolled aluminium alloys AA5083 and AA6110 at elevated temperatures, dissertation, Universität Kassel, Institut für Werkstofftechnik Metalische Werkstoffe, (2006).

[4] D. Meyer, E. Brinksmeier, F. Hoffmann, Surface hardening by cryogenic deep rolling, Proceedings of 1st CIRP Conference on Surface Integrity (CSI), Procedia Engineering 19 (2011), 258 – 263. http://dx.doi.org/10.1016/j.proeng.2011.11.109

[5] C. Straub, D. May, Stress Peening, The Iron Age (1949), 66–70.

[6] C. F. Barrett, and R. Todd, Investigation of the effects of elastic pre-stressing technique on magnitude of compressive residual stress induced by shot peen forming of thick aluminum plates, Proceedings of the 2nd International Conference on Shot Peening (1984), 15-21.

[7] J. Xu, D. Zhang, B. Shen, The Fatigue Strength And Fracture Morphology Of Leaf Spring Steel After Prestressed Shot Peening, Proceedings of the 1st International Conference on Shot Peening (1981), 367-374.

[8] E. Müller, Evolution of the residual stresses by stress rolling, Proceedings of the 9th International Conference on Shot Peening (2005), 436-441.

[9] D. Trauth, F. Klocke, P. Mattfeld, A. Klink, Time-Efficient Prediction of the Surface Layer State after Deep Rolling using Similarity Mechanics Approach, Procedia CIRP 9 (2013), 29 – 34. http://dx.doi.org/10.1016/j.procir.2013.06.163

[10] N. Lyubenova, D. Baehre, Finite Element Modelling and Investigation of the Process Parameters in Deep Rolling of AISI 4140 Steel, Journal of Materials Science and Engineering B 5 (7-8) (2015), 277-287.

[11] N. Lyubenova, L. Thuillier, D. Baehre, Modeling of the Input Parameters and Investigation of the Surface Residual Stresses and Deformations in Deep Rolling, Proceedings of the 29th International Conference on Surface Modification Technologies (2015), 86-93.

[12] E. Müller, Comparison of the Residual Stresses Induced by Shot (Stress) Peening and Rolling in Spring Steel, in Shot Peening (ed L. Wagner), & Co. KGaA, Weinheim, FRG (2002).

[13] Schulze, Surface Layer States after Mechanical Surface Treatments, in Modern Mechanical Surface Treatment: States, Stability, Effects, Wiley-VCH Verlag GmbH (2005).

Residual Stresses 2016: ICRS-10 Materials Research Forum LLC
Materials Research Proceedings **2** (2016) 259-264 doi: http://dx.doi.org/10.21741/9781945291173-44

Measurements of Surface and Near-surface Residual Stress in 4330 Low Alloy Carbon Steel Weld Clad Components

G. Benghalia[1,a *], S. Rahimi[1,b] and J. Wood[2,c]

[1]Advanced Forming Research Centre (AFRC), University of Strathclyde, Glasgow, UK

[2]Deparment of Mechanical and Aerospace Engineering, University of Strathclyde, Glasgow, UK

[a]gladys.benghalia@strath.ac.uk, [b]salah.rahimi@strath.ac.uk, [c]j.wood@strath.ac.uk

Keywords: Residual Stress, X-Ray Diffraction (XRD), Electronic Speckle Pattern Interferometry (ESPI), Incremental Centre Hole-drilling, Finite Element Modelling, Weld Cladding

Abstract. Weld cladding of low alloy carbon steel generates compressive residual stress in the clad layer, in turn potentially improving resistance to fatigue failure, depending on the material used for cladding. The current paper summarises the results of investigations on the magnitude and distribution of residual stress in these weld clad components, undertaken using different techniques including X-ray diffraction, and incremental centre hole drilling based on both strain gauge rosettes and electronic speckle pattern interferometry. Results confirm the presence of tensile residual stress when cladding with Inconel 625 beyond the initial clad profile and compressive residual stress when cladding with 17-4 PH steel. The complementary nature of XRD and hole drilling techniques is highlighted with considerations regarding the weld clad profile and stress distribution with depth. Modelling of residual stress induced by weld cladding using a thermal transient analysis is presented. Simplification of the weld cladding process is shown to provide good correlation with experimentally measured residual stress. Complexities in modelling material behaviour and hence accurate prediction of residual stress are discussed. Chemical composition of the weld into the heat-affected zone and substrate is presented for both weld clad materials, highlighting the effects of alloying and diffusion on chemical composition. Given the complexities in obtaining accurate thermo-mechanical material properties required for modelling, and that residual stress profiles are measured to a limited depth into the clad layer, recommendations are made for the continuation of both experimental and simulation studies.

Introduction

Residual stresses present in a component post-manufacture prior to loading, are of crucial importance to ensure accurate determination of the performance of a component in service. Such stresses are described as arising due to misfits between regions, different parts or different phases [1]. Unprocessed raw materials will invariably contain residual stresses, with manufacturing processes acting as additional sources of residual stress resulting in alteration of the distribution and status of residual stress. The understanding of these elastic, self-equilibrating stresses requires knowledge of the processes experienced by the component (e.g. heat treatment, mechanical work or joining), and the extent to which these processes alter the residual stress magnitude and distribution.

Residual stresses can be categorised as macro and micro residual stresses. Macro stresses are those that are present at scales comparable to the size of the component, also known as Type I residual stress, whilst micro stresses are in the scale of several grains (Type II) and atomic scales (Type III) [2]. These stresses arising due to elastic strain energy in the lattice structure of a material can act as a driving force for crack initiation and accelerate crack propagation. It is commonly known that the presence of tensile residual stresses negatively impacts fatigue life, whilst the presence of compressive residual stresses are beneficial for fatigue life. The presence of a compressive residual stress is therefore particularly beneficial at the surface of a component where operational stresses are often highest. Upon superpositioning of the residual stresses with cyclic operational stresses, the entire stress cycle would ideally remain in the compressive region, resulting in retardation of crack propagation. Crack initiation is however thought to be uninfluenced by the presence of compressive residual stresses [3]. Compressive residual stresses can improve corrosion performance and resistance to stress corrosion cracking in critical environments [4]–[6]. In this study, residual stresses induced by weld cladding, using different clad materials, are investigated. A cylinder of 4330 low alloy carbon steel is weld clad with nickel-chromium-based superalloy Inconel 625 and 17-4 precipitation hardening martensitic stainless steel. The weld cladding process, due to the joining of dissimilar materials, involves thermo-mechanical and chemical mechanisms that give rise to the generation of residual stresses with complex distribution. A shrinkfit effect occurs due to the effective shrinking of the cladding relative to the substrate. Constraint on thermal expansion and contraction is produced through the heating and cooling cycle during welding. Volume changes due to phase transformation, precipitation or chemical reactions can result in chemically induced residual stresses. The deposition process and subsequent cooling of the clad component can be controlled to modify residual stresses depending on the spatial and temporal variation of the process [7], [8]. Furthermore the original material properties and microstructures are altered during the joining of materials through the effects of alloying and diffusion [9]. On a microscopic level, varying values of thermal properties in constituent phases can also generate thermally induced residual stresses [10].

Herein the residual stress distribution in near-surface regions of the weld clad components are investigated using several measurement techniques. The determination of residual stresses can be achieved through means of relaxation or diffraction [11]. In this study residual stress measurements have been carried out based on both principals using X-Ray diffraction (XRD) and hole drilling. The generation of residual stress due to weld cladding has also been modelled using a thermal transient analysis.

Weld Cladding Process

Weld overlay cladding of hollow cylinders, made from 4330 low alloy carbon, was achieved using hot-wire Gas Tungsten Arc Welding (GTAW) to deposit a layer of clad material onto the outer diameter of the cylinders to create a bond between the two materials. Two cladding materials, Inconel 625 and 17-4 PH, were used on two different cylinders. For each sample, a two-pass deposition process was utilised resulting in a total clad layer thickness of 6 mm, with cylinder dimensions noted in Fig. 1(a).

Residual Stress Measurements

X-Ray Diffraction. Surface residual stress measurements were obtained by XRD using a PROTO-LXRD diffractometer and $\sin^2\psi$ method according to the NPL good practice guide [12], [13]. The stresses were calculated from the strains of the {311} and {211} Bragg reflections at 155.2° and 155.1° Bragg angles, considering elastic Young's modulus of 168 GPa and 184 GPa and Poisson's ratios of 0.29 and 0.272 for Inconel 625 and 17-4 steel, respectively. Weld bead peaks were utilized as measurement points, with a path assigned from the reference face spanning 20 peaks. Measurements were obtained in perpendicular axial (φ = 0°) and hoop (φ = 90°) directions as illustrated in Fig. 1(a).

Hole-drilling using optical technique. The StressTech Prism system was used in the measurement of residual stresses using hole drilling based on electronic speckle pattern interferometry (ESPI) [14]. The sole requirement regarding surface preparation is a clean surface with low reflectivity, the clad cylinder therefore ideal. Drill diameters determine the depth to which measurements can be obtained, residual stresses obtained to a depth of 1.2 mm.

Hole-drilling using strain gauge rosettes. The incremental centre hole drilling (ICHD) method, accurately obtaining measurements to a depth of approximately half of the drill diameter [15], was also undertaken using the SINT Technology MTS3000 system to obtain residual stress distributions. Strain gauge installation in diametrically opposing locations required machining of an average of 306 μm and 586 μm for Inconel 625 and 17-4 PH clad profiles respectively.

Materials Research Proceedings **2** (2016) 259-264 doi: http://dx.doi.org/10.21741/9781945291173-44

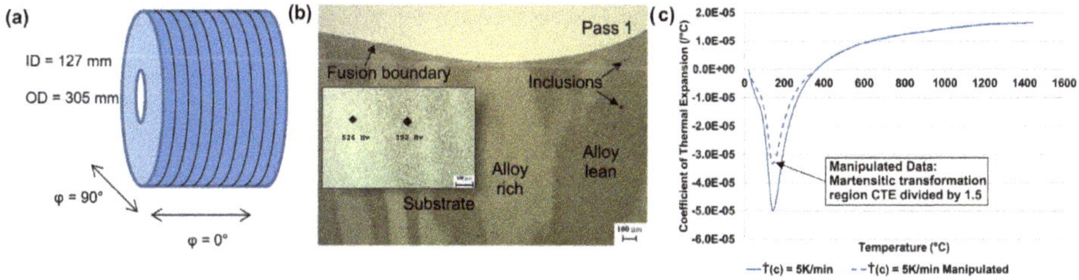

Fig. 1. (a) Orientation of the sample for XRD stress measurements (i.e. φ-angles) (b) fusion boundary of Inconel 625 clad - magnification factor x50 (c) manipulation of 17-4 PH coefficient of thermal expansion (CTE).

Simulation of the Weld Cladding Process

A 2D axisymmetric model was utilised for the simulation of the weld cladding process, assuming weld passes to be deposited on the entire outer diameter at one time step. The thermal transient analysis consisted of heating the pipe to a temperature of 300°C prior to application of the weld material at melt temperature. The entire component was then slowly cooled using a convective heat transfer coefficient of h = 10 W/m^{2}K, with latent heat effects neglected and radial edges insulated to simulate no axial heat transfer. An elastic-perfectly plastic material was assumed utilizing experimentally obtained thermal and mechanical properties for clad and heat-affected zone (HAZ) regions. Fig. 1(b) shows the fusion boundary and heavy micro-segregation in the HAZ of the Inconel 625 clad.

Results

X-Ray Diffraction. Fig. 2 indicates the presence of primarily compressive residual stress on the Inconel 625 weld clad profile in both axial and hoop directions, with a maximum generally around -300 MPa. It can also be seen that a number of points illustrate the presence of tensile residual stress, although those close to a zero value possess error bars in the region that would suggest these may also be low compressive stresses. Hoop stresses shown in Fig. 2(b) appear to be tensile towards the reference edge of the cylinder, however this is not observed for axial stresses. Given the nature of the weld clad profile, insufficient diffraction may arise should the beam extend across more than one weld bead and this will be observed as a variation in stress values. Furthermore geometrical variations, surface irregularities and sampling population can give rise to variations. It should also be noted that strain values are averaged in the measurement region and this is reflected ultimately in stress values.

Residual stress measured in the 17-4 PH clad surface showed compressive residual stress at the majority of measurement locations, slightly less compressive than in the case of Inconel 625. In the axial stress component a small number of points indicated low tensile residual stress, Fig. 3(a), however the hoop stress component highlighted that all residual stresses measured were compressive, Fig. 3(b).

Calculated error values are also generally smaller in the case of the 17-4 PH clad cylinder, likely due to the increased sampling population in the martensitic microstructure.

Hole drilling techniques and comparisons with FE model. The hole drilling methods utilized obtain residual stresses in the near surface region, serving as a means for comparison with the simulation model. Fig. 4 and Fig. 5 compare residual stress components obtained through simulation and experimental methods. In the case of the Inconel 625 clad layer good correlation can be seen between the simulation and ICHD residual stress for both stress components. In the case of ESPI, compressive residual stress is observed in both stress components at the surface, to a similar level as was observed in XRD measurements. ICHD A and B represent two measurements at diametrically

Materials Research Proceedings **2** (2016) 259-264 doi: http://dx.doi.org/10.21741/9781945291173-44

opposed locations on the same component, undertaken for the purposes of repeatability, while average XRD values are plotted.

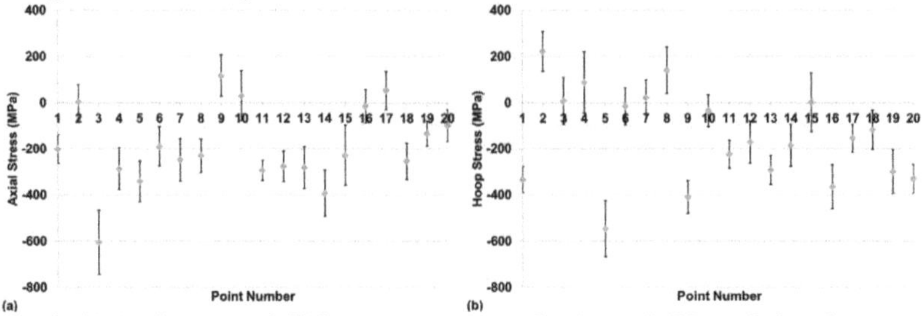

Fig. 2. (a) Axial stress and (b) hoop stress measured in Inconel 625 as-clad surface - measurements obtained on 20 consecutive weld bead peaks using XRD.

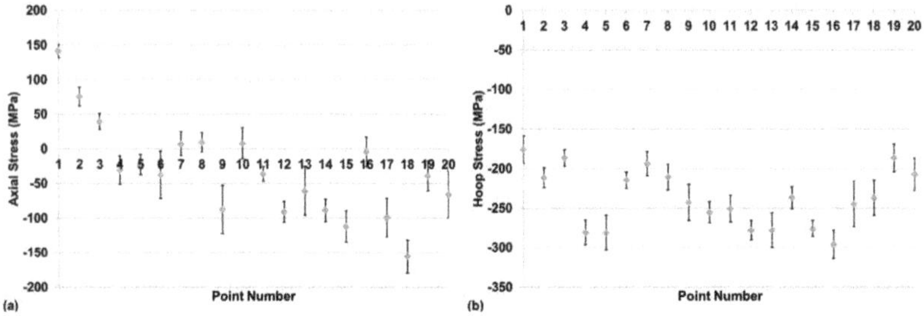

Fig. 3. (a) Axial stress and (b) hoop stress measured in 17-4 PH as-clad surface - measurements obtained on 20 consecutive weld bead peaks using XRD.

Towards the greatest depth (≈1.2 mm) of the ESPI measurement it can be seen that the residual stress levels matches more closely with those obtained in the simulation.

Comparing experiment and simulation for a 17-4 PH clad, shows that in the case of axial stress good correlation is achieved for hole drilling methods with the simulation (Fig. 5(a)). Hoop stresses show that ESPI values are almost zero at the surface, increasing in compression towards a depth of 1.2 mm, while ICHD values match closely with simulation.

Discussion

As highlighted, correlation between XRD and ESPI methods was seen to be good at the component surface. The complementary nature of XRD and ESPI hole drilling has therefore been demonstrated herein, along with the inability of the simulation model to capture stress variation at the surface due to the weld clad profile. In terms of residual stress measurement, techniques providing bulk residual stress, such as the contour method [16] and deep-hole drilling [17], would allow increased understanding of residual stresses throughout the clad cylinder, as opposed to near-surface stresses presented in this study.

Considerations on weld geometry. In the Inconel 625 clad cylinder both XRD and ESPI hole drilling measurements indicated compressive surface residual stresses, however residual stress in the clad layer measured by ICHD is tensile and in agreement with simulation results. This may be due to (i) the weld bead geometry influencing the residual stress measured by both XRD and ESPI hole drilling, or (ii) the FE simulation not being able to predict the real process and in turn levels of residual stress. It is expected that results obtained by ICHD do not match surface results of ESPI hole drilling and XRD, as the surface residual stress profile is changed by surface preparation for strain gauge installation.

Materials Research Proceedings **2** (2016) 259-264 doi: http://dx.doi.org/10.21741/9781945291173-44

Material behaviour. In Fig. 5, simulation results are presented for two cases: original and manipulated. In the manipulated case (simulation B), experimental data obtained for coefficient of thermal expansion (CTE) was manipulated as shown in Fig. 1(c) to account for the rapid cooling rate during the welding process. Studies have been conducted into dilatation during a cooling rate of 234 K/s [18], however this level of cooling could not be obtained experimentally. Increased accuracy in experimental data is required to account for CTE variation, particularly during the martensitic transformation to ensure that accurate material behavior is modelled.

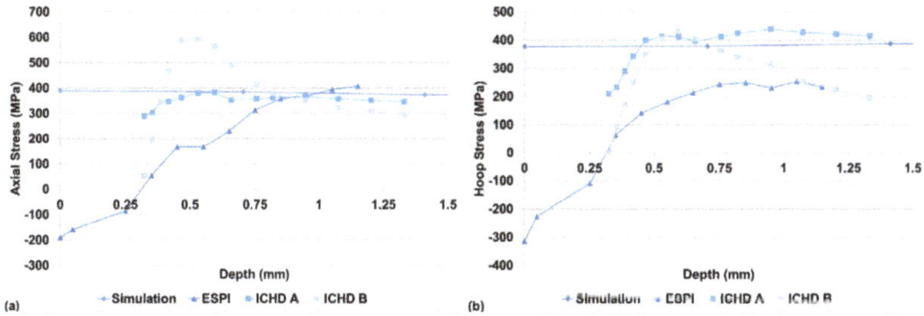

Fig. 4. Comparing simulation and experimental (a) axial and (b) hoop residual stress in the Inconel 625 clad layer.

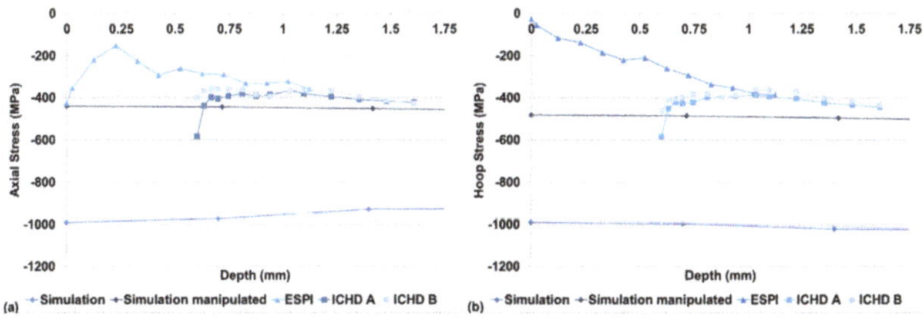

Fig. 5. Comparing simulation and experimental (a) axial and (b) hoop residual stress in the 17-4 PH clad layer.

Alloying and diffusion. Fig. 1(b) highlighted heavy HAZ micro-segregation, chemical analysis of these regions indicating diffusion from the clad layer into the substrate. Higher chromium and nickel levels were present in alloy rich regions increasing hardness values as shown. Furthermore chemical analysis of clad passes illustrated higher iron levels in the first Inconel 625 pass, diffused from the 4330 substrate. Similarly substrate diffusion increased iron levels in the first 17-4 PH clad pass. These factors will impact material properties and resulting residual stresses. Care must be taken to ensure that crack susceptibility and tensile residual stresses do not result from the weld cladding process. Microstructural modelling would enable greater control of residual stress generation during weld cladding.

Conclusions

Weld cladding of a 4330 cylinder was achieved using a hot-wire GTAW process using Inconel 625 and 17-4 PH as clad materials. The major observations of this work are concluded as follows:

- The residual stresses throughout the component were measured using XRD and two hole drilling techniques, one based on electronic speckle pattern interferometry (ESPI) and the other based on strain gauge rosettes. These methods are shown to be complementary.
- Good correlation was obtained between experiment and simulation upon manipulation of CTE values for 17-4 PH to account for high cooling rates during welding.

- XRD and ESPI hole drilling results indicated that weld bead geometry influenced surface measurements, with stresses near-surface transforming from compressive to tensile for an Inconel 625 clad.
- Discrepancies in the results of different residual stress measurements techniques have been highlighted. The study indicates the need for a range of residual stress measurement techniques to gain knowledge of the residual stress distribution throughout the component.
- Weld cladding with Inconel 625 resulted in the presence of tensile residual stresses in the clad layer beyond the initial weld bead profile, while compressive residual stresses were observed in the 17-4 PH clad layer.

References

[1] P. J. Withers and H. K. D. H. Bhadeshia, "Residual Stress: Part 2 - Nature and Origins," *Mater. Sci. Technol.*, vol. 17, no. April, pp. 366–375, 2001. http://dx.doi.org/10.1179/026708301101510087

[2] F. A. Kandil, J. D. Lord, A. T. Fry, and P. V. Grant, "A Review of Residual Stress Measurement Methods - A Guide to Technique Selection," Teddington, 2001.

[3] L. Wagner, "Mechanical Surface Treatments on Titanium, Aluminum and Magnesium Alloys," *Mater. Sci. Eng. A*, vol. 263, pp. 210–216, 1999. http://dx.doi.org/10.1016/S0921-5093(98)01168-X

[4] A. F. Liu, *Mechanics and Mechanisms of Fracture: An Introduction*. Ohio, USA: ASM International, 2005.

[5] S. Rahimi, K. Mehrez, and T. J. Marrow, "Effect of surface machining on intergranular stress corrosion cracking (IGSCC) in sensitised type 304 austenitic stainless steel," *Corros. Eng. Sci. Technol.*, 2016. http://dx.doi.org/10.1080/1478422X.2015.1122295

[6] S. Rahimi and T. J. Marrow, "Effects of orientation, stress and exposure time on short intergranular stress corrosion crack behaviour in sensitised type 304 austenitic stainless steel," *Fatigue Fract. Eng. Mater. Struct.*, vol. 35, pp. 359–373, 2011. http://dx.doi.org/10.1111/j.1460-2695.2011.01627.x

[7] G. Schnier, J. Wood, and A. Galloway, "An Experimental Validation of Residual Stresses in Weld Clad Pipelines," in *Research and Applications in Structural Engineering, Mechanics & Computation: Proceedings of the Fifth International Conference on Structural Engineering, Mechanics & Computation*, 2013, pp. 613–617. http://dx.doi.org/10.1201/b15963-113

[8] G. Schnier, J. Wood, and A. Galloway, "Investigating the Effects of Process Variables on the Residual Stresses of Weld and Laser Cladding," *Adv. Mater. Res.*, vol. 996, pp. 481–487, Aug. 2014. http://dx.doi.org/10.4028/www.scientific.net/amr.996.481

[9] G. Benghalia and J. Wood, "Autofrettage of Weld Clad Components," *Procedia Eng.*, vol. 130, pp. 453–465, 2015. http://dx.doi.org/10.1016/j.proeng.2015.12.239

[10] G. Benghalia and J. Wood, "Material and residual stress considerations associated with the autofrettage of weld clad components," *Int. J. Press. Vessel. Pip.*, vol. 139–140, pp. 146–158, 2016. http://dx.doi.org/10.1016/j.ijpvp.2016.02.003

[11] P. J. Withers and H. K. D. H. Bhadeshia, "Residual Stress: Part 1 – Measurement Techniques," *Mater. Sci. Technol.*, vol. 17, no. April, pp. 355–365, 2001. http://dx.doi.org/10.1179/026708301101509980

[12] M. E. Fitzpatrick, A. T. Fry, P. Holdway, F. A. Kandil, J. Shackleton, and L. Suominen, "Measurement Good Practice Guide No. 52: Determination of Residual Stresses by X-Ray Diffraction - Issue 2," Teddington, UK, 2005.

[13] ASTM International, *Standard Test Method for Residual Stress Measurement by X-Ray Diffraction for Bearing Steels*. West Conshohocken, PA, USA, 2012.

[14] R. Jones and J. A. Leendertz, "Elastic constant and strain measurements using a three beam speckle pattern interferometer," *J. Phys. E Specif. Instruments*, vol. 7, pp. 653–657, 1974.

[15] P. V. Grant, J. D. Lord, and P. S. Whitehead, "Measurement Good Practice Guide No. 53 - Issue 2: The Measurement of Residual Stresses by the Incremental Hole Drilling Technique," Teddington, UK, 2006.

[16] M. B. Prime, "Cross-sectional mapping of residual stresses by measuring the surface contour after a cut," *J. Eng. Mater. Technol.*, vol. 123, pp. 162–168, 2001. http://dx.doi.org/10.1115/1.1345526

[17] R. H. Leggatt, D. J. Smith, S. D. Smith, and F. Faure, "Development and experimental validation of the deep hole method for residual stress measurement," *J. Strain Anal.*, vol. 31, no. 3, 1996. http://dx.doi.org/10.1243/03093247V313177

[18] L.-E. Lindgren, *Computational Welding Mechanics: Thermomechanical and Microstructural Simulations*. Cambridge, UK: Woodhead Publishing Limited, 2007.

Residual Stresses 2016: ICRS-10 Materials Research Forum LLC
Materials Research Proceedings **2** (2016) 265-270 doi: http://dx.doi.org/10.21741/9781945291173-45

Accuracy and Stability of 2D-XRD for Residual Stress Measurement

Bob B. He*

5465 East Cheryl Parkway, Madison, WI 53711, USA

bob.he@bruker.com

Keywords: Residual Stress, 2D Detector, Accuracy, Stability

Abstract. Stress measurement with two-dimensional X-ray diffraction is based on the direct relationship between the stress tensor and diffraction cone distortion. Since more data points at more orientations are used for stress calculation, 2D-XRD can measure stress with high sensitivity, high speed and high accuracy. It is especially suitable for samples with large crystals, textures and curved surface. Ten almen strips were used for Gage R&R test of a 2D-XRD system for stress measurement in 1998. Since then there have been many advances in two-dimensional X-ray diffractometer, including detector technology, X-ray source and optics, goniometer, and data evaluation software. This paper covers accuracy and stability of stress measurement on the same almen strips with a 2D-XRD system.

Introduction

Two-dimensional x-ray diffraction (abbreviated as XRD^2) pattern contains information in a large solid angle which can be described by the diffraction intensity distribution in both 2θ and γ directions [1]. The 2θ peak position can be determined from each section of the diffraction ring within a γ range. The d-spacing variation at different orientations due to stress results in 2θ peak position variation at different γ angle and sample orientation. For a particular sample orientation, the variation of 2θ along γ angle is observed as the distortion of the diffraction ring on 2D diffraction pattern. A set of 2D diffraction patterns collected at various sample orientations can be used to evaluate the stress tensor in the sample. Since each 2D diffraction pattern covers one or several diffraction rings with large γ range, the XRD^2 method can measure stress with high accuracy and high speed, especially when dealing with texture, large grain size, small sample areas and curved sample surface [2-5].

The accuracy of stress measurement by XRD^2 system depends on many factors, such as, the spherical error of the goniometer, sample position, the detector, profile fitting algorithms and data collection strategy. Gage repeatability and reproducibility (Gage R&R) test on residual stress measurement with an XRD^2 system was reported in 1998 [6]. The test was performed by three operators with three measurements on 10 almen strips. The test shows that the overall measurement error is 6% or less. The repeatability error of the system is 5.5% or less, and the reproducibility error is 2.3% or less. The average stress for all 90 measurements is -628 ±19 MPa (3.1%). The sample from the same set of almen strips has been used to evaluate many XRD^2 systems in terms of the accuracy and system stability in the past 18 years. This paper covers accuracy and stability of stress measurement with an XRD^2 system containing recent advances in X-ray source, goniometer and detector.

Fundamental Equation for Stress Measurement

Fig. 1 illustrates a diffraction cone for backward diffraction. The regular diffraction cones (broken lines) are from a sample with no stress, so the 2θ angles are constant at all γ angles. The solid line is the cross sections of the distorted diffraction cone due to stresses. For a stressed sample, 2θ becomes a function of γ and the sample orientation (ω,ψ,ϕ). The fundamental equation for stress measurement with XRD^2 can be given as [3]:

$$\boldsymbol{S}_1(\sigma_{11}+\sigma_{22}+\sigma_{33})+\tfrac{1}{2}\boldsymbol{S}_2(\sigma_{11}h_1^2+\sigma_{22}h_2^2+\sigma_{33}h_3^2+2\sigma_{12}h_1h_2+2\sigma_{13}h_1h_3+2\sigma_{23}h_2h_3)=\ln\left(\frac{\sin\theta_0}{\sin\theta}\right) \quad (1)$$

where S_1 and $\frac{1}{2}S_2$ are the macroscopic elastic constants, and $\boldsymbol{h}_1$, $\boldsymbol{h}_2$ and $\boldsymbol{h}_3$ are the three components of the unit diffraction vector expressed in sample coordinates. For Eulerian geometry:

$$\begin{aligned}
h_1 &= \sin\theta(\sin\phi\sin\psi\sin\omega+\cos\phi\cos\omega)+\cos\theta\cos\gamma\sin\phi\cos\psi-\cos\theta\sin\gamma(\sin\phi\sin\psi\cos\omega-\cos\phi\sin\omega)\\
h_2 &= -\sin\theta(\cos\phi\sin\psi\sin\omega-\sin\phi\cos\omega)-\cos\theta\cos\gamma\cos\phi\cos\psi+\cos\theta\sin\gamma(\cos\phi\sin\psi\cos\omega+\sin\phi\sin\omega)\\
h_3 &= \sin\theta\cos\psi\sin\omega-\cos\theta\sin\gamma\cos\psi\cos\omega-\cos\theta\cos\gamma\sin\psi
\end{aligned} \quad (2)$$

Eq. 1 can be further reduced for various stress states and instrument conditions. For example, if only the data points at γ=90° or γ=-90° are considered, the equation can be reduced to the fundamental equation for stress measurement with point detector [1]. For most materials, X-ray diffraction can measure only a very thin layer on the surface, it is reasonable to assume that the average normal stress in the surface normal direction is zero within such a thin layer, i.e. $\sigma_{33}=0$. In this case, the other five components of the stress tensor can be measured without accurate stress-free d-spacing.

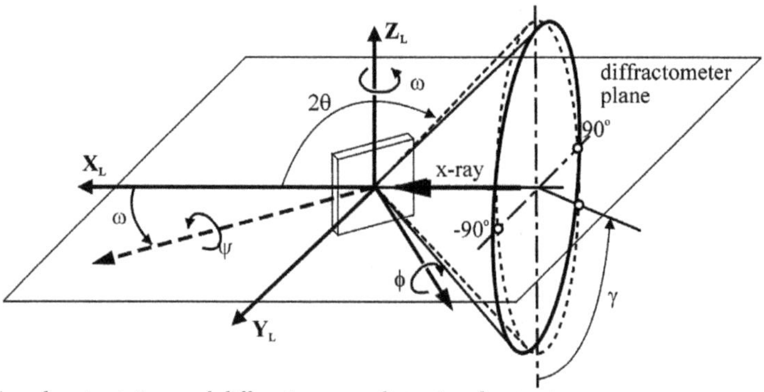

Fig. 1. Sample orientation and diffraction cone distortion due to stress.

Instrumentation and Data Collection Strategy

The XRD^2 system used for the experiment is Bruker D8 DISCOVER diffractometer. As shown in Fig. 2, it contains IμS microsource X-ray generator with Cr-Kα radiation, VÅNTEC-500 2D detector, Eulerian cradle, dual lasers and video microscope. The 2θ of Fe (211) peak with Cr Kα radiation is approximately 156°, so the incident angle is set at θ_1 (ω) = 78°. The detector is set at θ_2=71° (swing angle α=149°) at sample-to-detector distance 234 mm. The diffraction vector direction is determined by the incident angle and the 2D detector covers about 30° 2θ range, which is sufficient to cover the (211) ring with complete background. The collimator size is 0.5 mm. The cross point of the dual laser beam indicates the instrument center which can be observed with the video microscope. The motorized X-Y-Z stage on the Eulerian cradle allows precise position of the sample. The generator power setting in this experiment is 15W (40kV/0.375mA).

Materials Research Proceedings **2** (2016) 265-270 doi: http://dx.doi.org/10.21741/9781945291173-45

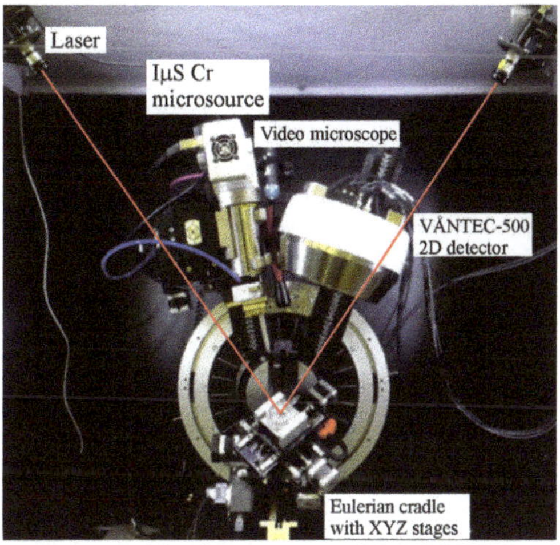

Fig. 2. The XRD² diffractometer used for the experiment: D8 DISCOVER with Cr IμS microsource, VÅNTEC-500, and Eulerian cradle.

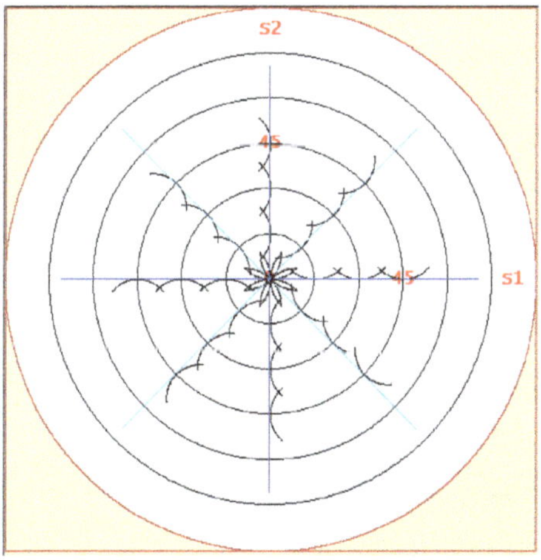

Fig. 3. Data collection strategy schemes with 32 frames at ψ = 0, 15°, 30°, 45° with complete ϕ rotation of 45° steps.

The diffraction vector is in the normal direction of the measured crystalline planes. It is not always possible to have the diffraction vector in the desired measurement direction. The stress components within the surface plane are calculated by elasticity theory from the measured strain in other directions. The final stress measurement results can be considered as an extrapolation from the measured values. For example, in the conventional $\sin^2\psi$ method, the stress in the sample surface

Materials Research Proceedings **2** (2016) 265-270 doi: http://dx.doi.org/10.21741/9781945291173-45

(ψ=90°) is calculated from the measured strains at several ψ-tilt angles. The same is true with an XRD^2 method. Similar to a pole-figure, the diffraction vectors corresponding to a data set can be mapped in a data collection strategy scheme. By evaluating the scheme, one can collect a data set optimized for the intended stress components. Fig. 3 illustrates the scheme used for data collection with this experiment. The small arcs represent the trace of the diffraction vector corresponding to the data set. S1 and S2 are two sample orientations. In this scheme, 32 frames are collected at ψ=0°, 15°, 30° and 45° at eight ϕ angles with 45° intervals. This scheme produces comprehensive coverage on the scheme chart in a symmetric distribution. The data set collected with this strategy can be used to calculate the complete biaxial stress tensor components and shear stress ($\sigma_{11}, \sigma_{12}, \sigma_{22}, \sigma_{13}, \sigma_{23}$). A proper scheme is selected based on the interested stress components, the goniometer, desired measurement accuracy and data collection time.

Stress Analysis

The stress calculation is done with Bruker DIFFRAC.LEPTOS software version 7.9. Fig. 4 shows the data evaluation setting. The data integration region is defined by 2θ range of 150° to 160° and γ range of -70° to -110°. The 40° γ range is divided into 8 subregions, 5° for each subregion. The counts within each subregion are integrated into a diffraction profile and the 2θ peak position is determined by one of the five peak evaluation algorithms. In this experiment, Pearson VII function is used to fit the profile and evaluate the 2θ peak position. Fig. 5 shows the stress results from one of the data set. The charts above "A" are the fitted data points on 2D frames. The charts above "B" are fitted data points in γ-2θ rectangular coordinates with magnified 2θ scale, in which, black line indicates $2\theta_0$, blue cross and line indicates the data points from the profile fitting of each subregion, and red line represents the calculated diffraction rings from the stress results. The scattering of the crosses about the red line represents the quality of the data, affecting the standard deviation of the stress results. By click on any data point, the integrated profile displays above "C".

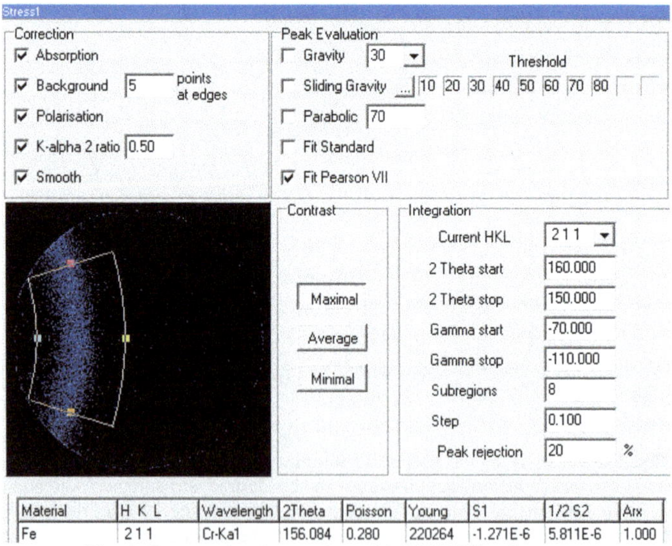

Fig. 4. Data evaluation setting with LEPTOS software.

Materials Research Proceedings **2** (2016) 265-270 doi: http://dx.doi.org/10.21741/9781945291173-45

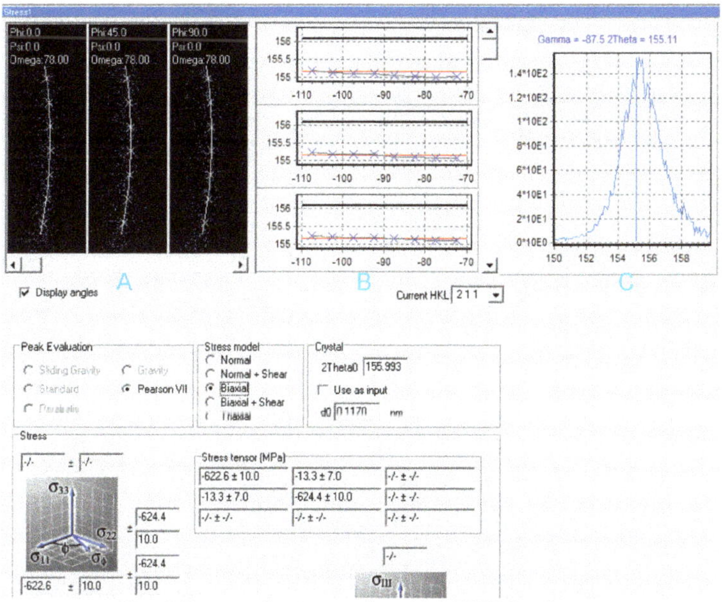

Fig. 5. Data evaluation results with LEPTOS software.

Accuracy and Stability

The stress-free Fe powder sample is used to check the instrument accuracy. Any spherical confusion of the goniometer and sample position error produces a fictitious stress value. With 60 second frames, the measured stress values are: σ_{11}=13 MPa, σ_{22}=9 MPa and standard deviation 12 MPa. With 180 second frames, the measured stress values are: σ_{11}=8 MPa, σ_{22}=10 MPa and standard deviation 9 MPa. The results satisfy the accuracy specification for the D8 Discover diffractometer.

Stability of the instrument is tested with one of the almen strips over 65 hours, with data collection of 120 second and 600 second per frame data set alternatively for 10 times. Fig. 6 shows the measurement results over the data collection time. Stress values from 600s frames are more stable compared with the results from 120s frames. The average values with 120s frames are: σ_{11} = -682 MPa, σ_{22} = -664 MPa with standard deviation 20 MPa. With 600 second frames, the average stress values are: σ_{11} = -625 MPa, σ_{22} = -616 MPa with standard deviation 10 MPa (1.6%). The 600s results seem to be more consistent with the average value of 628 MPa measured 18 years ago, assuming there is no stress relaxation in the almen strip. The results from the 120s frames have twice the standard deviation and more systematic error compared with the results from 600s frames.

Since the 120s data set and 600s data set are collected alternatively, the discrepancy is not likely due to the motion accuracy of the instrument, but most likely from the counting statistics. One of each integrated profiles are displayed in Fig. 6. It can be seen that the 600s profile is much smoother than the 120s profile. In addition to the data collection time, there are many ways to improve the counting statistics, for instance, more powerful X-ray source or generator setting, large collimator size, and short source to sample distance. For instance, the 100 mm gap between the collimator and sample can be significantly reduced. As previously reported with a different system, the results from a data set of 5 second frames have less than 2% standard error [3].

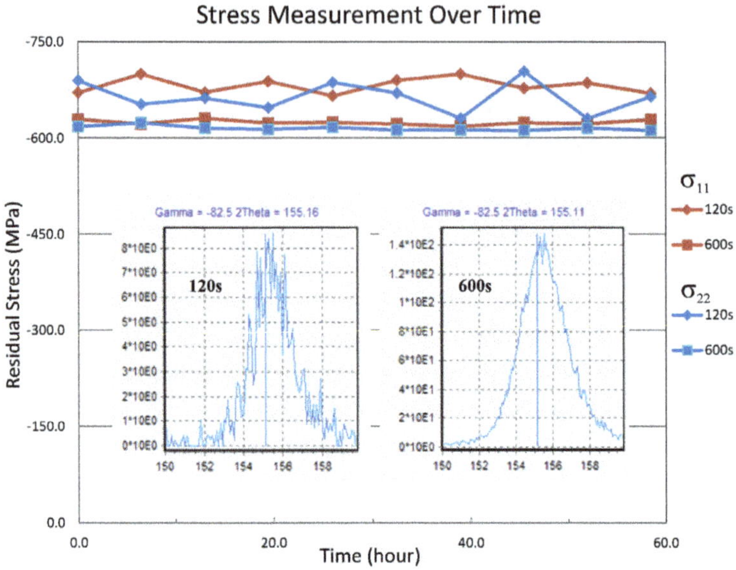

Fig. 6. Stress measurement results over time.

Summary

The XRD^2 system can measure residual stress with high accuracy and stability. The sample position is maintained during repeatable data collection with high accuracy and stability. The stress measurement accuracy in terms of standard deviation and systematic error is mostly due to counting statistics. The residual stresses in the almen strips did not show a detectable change over 18 years.

References

[1] B. B. He, *Two-dimensional X-ray Diffraction*, John Wiley & Sons, Hoboken, New Jersey 2009. http://dx.doi.org/10.1002/9780470502648

[2] B. B. He, U. Preckwinkel and K. L. Smith, Advantages of Using 2D Detectors for Residual Stress Measurement, *Advances in X-ray Analysis,* Vol. 42 (1998) pp. 429-438.

[3] B. B. He, MicroGap area detector for stress and texture analysis, *Materials Science Forum*, Vol. 681 (2011), pp. 19-24. http://dx.doi.org/10.4028/www.scientific.net/MSF.681.19

[4] B. B. He, XRD^2 stress measurement for samples with texture and large grains, *Materials Science Forum*, Vols. 768-769 (2014) pp. 227-234. http://dx.doi.org/10.4028/www.scientific.net/MSF.768-769.227

[5] T. Miyazaki and T. Sasaki, A comparison of X-ray stress measurement methods based on the fundamental equation, *J. Appl. Cryst.* Vol. 49 (2016) pp. 426–432. http://dx.doi.org/10.1107/S1600576716000492

[6] B. B. He, K. L. Smith, U. Preckwinkel and W. Schultz, Gage R&R study on residual stress measurement system with area detector, *Materials Science Forum* Vols. 347-349 (2000) pp. 101-106. http://dx.doi.org/10.4028/www.scientific.net/MSF.347-349.101

Materials Research Proceedings 2 (2016) 271-276 doi: http://dx.doi.org/10.21741/9781945291173-46

Turning Residual Stresses in Functionally Graded Steel Components

Wolfgang Zinn [1, a *], Marcel Tiffe [2,b], Dirk Biermann [2,c] and Berthold Scholtes [1,d]
[1]Institute of Materials Engineering, University of Kassel, Germany
[2]Institute of Machining Technology, Technical University of Dortmund, Germany
[a]zinn@uni-kassel.de, [b]tiffe@isf.de, [c]biermann@isf.de, [d]scholtes@uni-kassel.de

Keywords: Quenched and Tempered AISI 6150, Dry Turning, Residual Stresses

Introduction

Metallic components with graded materials properties are of particular importance for the realization of lightweight design. The scientific basis of such concepts was provided by the Collaborative Research Centre TRR 30, funded by the German Research Foundation DFG between 2006 and 2015. As part of the project, thermo-mechanically graded components were manufactured by hot metal forming processes [1, 2]. Their characteristic properties are to a large extent determined by final machining operations. In this context, residual stresses play an important role for strength and lifetime of the produced parts. A survey about near surface residual stress states after hard turning of differently heat treated quenched and tempered steel AISI 6150 (51CrV4) is given in [3]. In the present paper, results of similar investigations carried out of Jominy end quench samples are reported and compared with results of samples with homogeneous microstructures.

Experimental Setup

The Jominy end quench test samples had a diameter of 25 mm and a length of 100 mm. After austenitizing and quenching the shafts were machined using uncoated polycrystalline boron nitride (PCBN) inserts with the ISO-Code CNGA120408. The tool had a chamfer with a width of $b_\gamma = 0.15$ mm and an angle of $\gamma_f = 25°$. The tool holder was of the type DCLNL2525 and had a nominal rake angle of $\gamma = -6°$ and an effective rake angle of $\gamma = -31°$ in the chamfer area. The turning processes were carried out in two passes, part left and right, starting in each case from both ends of the six specimens (See Fig. 1).

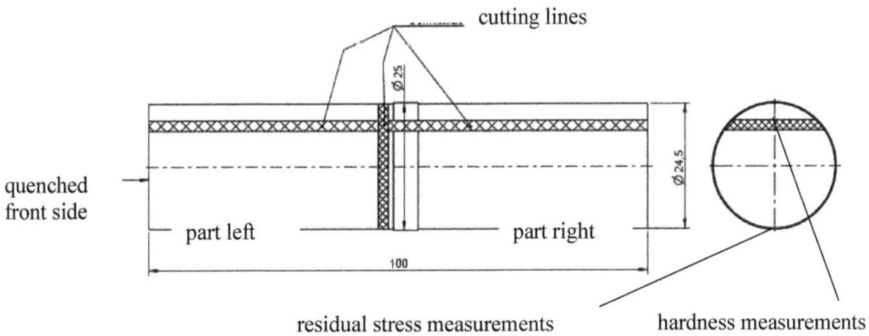

Fig. 1: Jominy end quench samples with cutting planes and measuring areas

Materials Research Proceedings **2** (2016) 271-276 doi: http://dx.doi.org/10.21741/9781945291173-46

In Tab. 1 the process parameters used for the turning operations are listed, which are identical to those of experiments in [3]. From the hardened and turned Jominy-samples 5 mm thick segments were prepared by longitudinal cutting for hardness testing and microstructural investigations. Hardness measurements HV2 were carried out along the polished surface of both specimen parts and subsequently, the sample surfaces were etched for microstructural analyses. At both specimen parts (left and right part, see Fig. 1) residual stress depth distributions in axial and tangential direction were determined by stepwise electrolytical polishing and applying standard X-ray diffraction technique. A laboratory Ψ-diffractometer was used. Lattice strains of {211}-lattice planes in 11 Ψ-directions in the range $-45° \leq \Psi \leq +45°$ and $148° \leq 2\Theta \leq 164°$ were determined using CrKα-radiation. The measured area was circular with 1 mm diameter.

Table 1: Turning parameter

Sample No.	feed [mm]	cutting speed [m/min]	depth of cut [mm]
1	0.050	100	0.25
2	0.050	250	0.25
3	0.125	175	0.25
4	0.200	100	0.25
5	0.200	250	0.25
6	0.300	250	0.25

Experimental Results

In Fig. 2 results of residual stress analyses in axial direction after hard turning applying different process parameters of identical materials states taken from [3] are summarized. In this case cylindrical specimen with homogeneous microstructures of differently annealed martensite, characterized by the hardness values indicated, were machined. Characteristic residual stress depth distributions result which are mainly influenced by the feed applied. For the softer materials states with a hardness of 322 HV tensile residual stresses were detected at the surface. With increasing materials hardness, these values are shifted in the direction of compression. Below the surface compressive residual stress maxima occur. Their amount and surface distance increases with increasing feed. Similar observations were made for the residual stress distributions in tangential direction.

The scientific issue of the present work was to clarify whether results of homogeneous materials states can be transferred to components with graded and locally inhomogeneous microstructures. For this purpose, Jominy end quench samples were investigated which before heat treatment had a line-shaped microstructure of ferrite and pearlite. After the quenching process at the quenched end a martensitic structure was determined which continuously changed to a ferritic-pearlitic structure with increasing distance from the end face. This can also be seen from the hardness distributions shown in Fig. 3. The pronounced hardness fluctuations especially in regions of lower hardness can be attributed to the inhomogeneous microstructure of the starting condition of the material. Note that maximum hardness values close to the end face of the material are quite higher than the highest hardness of the materials investigated in [3] (see Fig. 2).

In Fig. 4 axial residual stress depth distributions measured at Jominy end quench specimen are presented, measured at different distances from the quenched end face. In addition, hardness values measured at these positions as well as the characteristic parameters of the turning processes applied are indicated. In all cases near the surface small tensile or compressive residual stresses are measured followed by local compressive residual stress maxima. Their distance from the surface increases with increasing feed values applied. Accordingly, the thickness of the surface layer with compressive residual stresses increases.

Materials Research Proceedings **2** (2016) 271-276 doi: http://dx.doi.org/10.21741/9781945291173-46

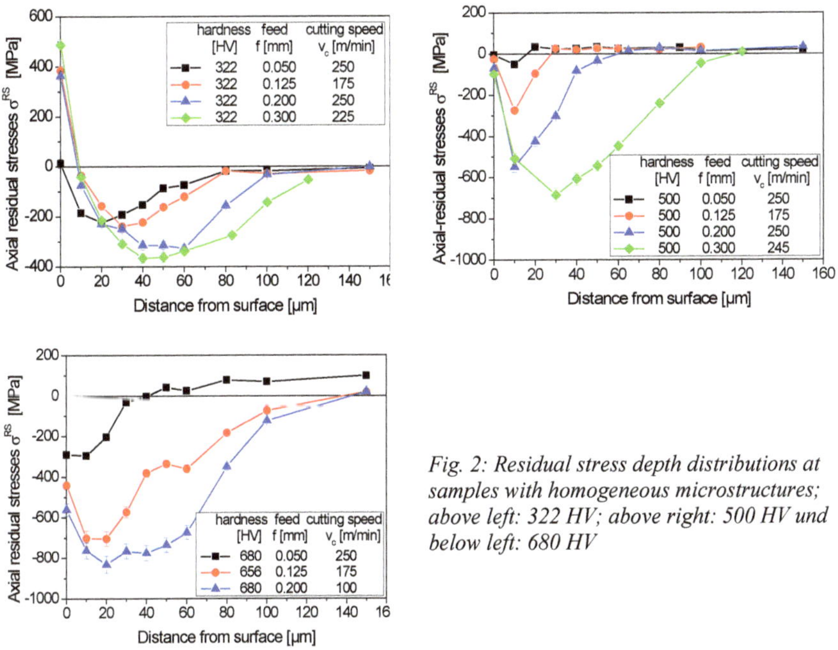

Fig. 2: Residual stress depth distributions at samples with homogeneous microstructures; above left: 322 HV; above right: 500 HV und below left: 680 HV

Similar observations were made for the residual stress component measured in tangential direction (see Fig. 5). For lower feed values, at the surface tensile residual stresses exist, which become compressive for higher feed values. Compressive residual stress maxima occur at surface distances between 10 µm and 30 µm. Amounts of compressive residual stresses are higher as in axial direction and have values between -500 MPa and -1000 MPa, depending on feed values applied.

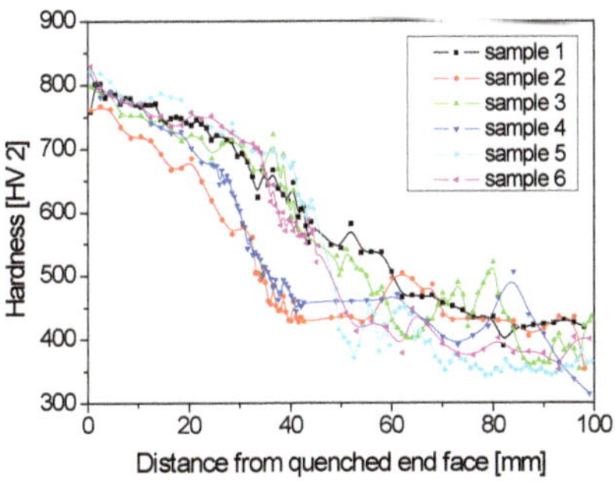

Fig. 3: Hardness distributions as function of the distance from end face

Materials Research Proceedings **2** (2016) 271-276 doi: http://dx.doi.org/10.21741/9781945291173-46

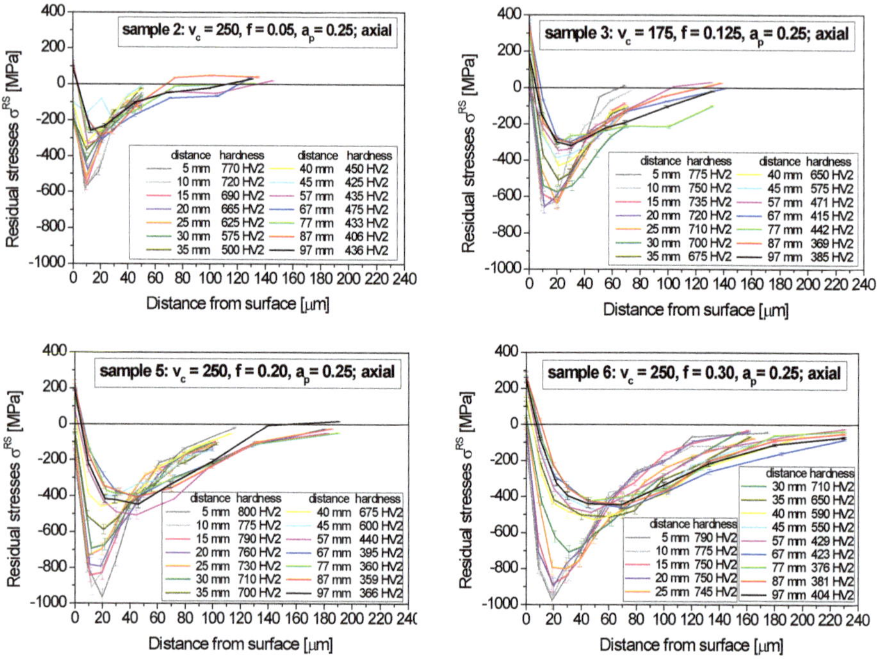

Fig. 4: Axial residual stress depth distributions of Jominy end quench samples after hard turning with different feed f (v_c: cutting speed; a_p: depth of cut)

While in axial direction shear components are small and reach maximum values of -50 MPa, in tangential direction distinct shear stress values are observed (see Fig. 6). Their depth distributions depends from applied feed values in a characteristic way. For smaller feed values higher shear stress amounts, but restricted to thinner surface layers are found compared to turning processes with higher feed values.

The thickness of the surface layer affected by the turning process can be assessed using the depth distributions of integral widths (see Fig. 7). For the cases presented in Fig. 7 (left) at a surface distance of roughly 50 µm integral width values comparable with those measured for the not machined state are found. For the harder materials states, turning leads to reduced integral width values while for softer materials states an increase can be found. This is a common observation in the case of plastic deformation of steels with different hardness and related to the formation and annihilation of dislocations as well as to different scattering domain sizes. For higher feed values (see Fig. 7, right) the thickness of the affected surface layer is increased and even larger than 150 µm.

Materials Research Proceedings **2** (2016) 271-276 doi: http://dx.doi.org/10.21741/9781945291173-46

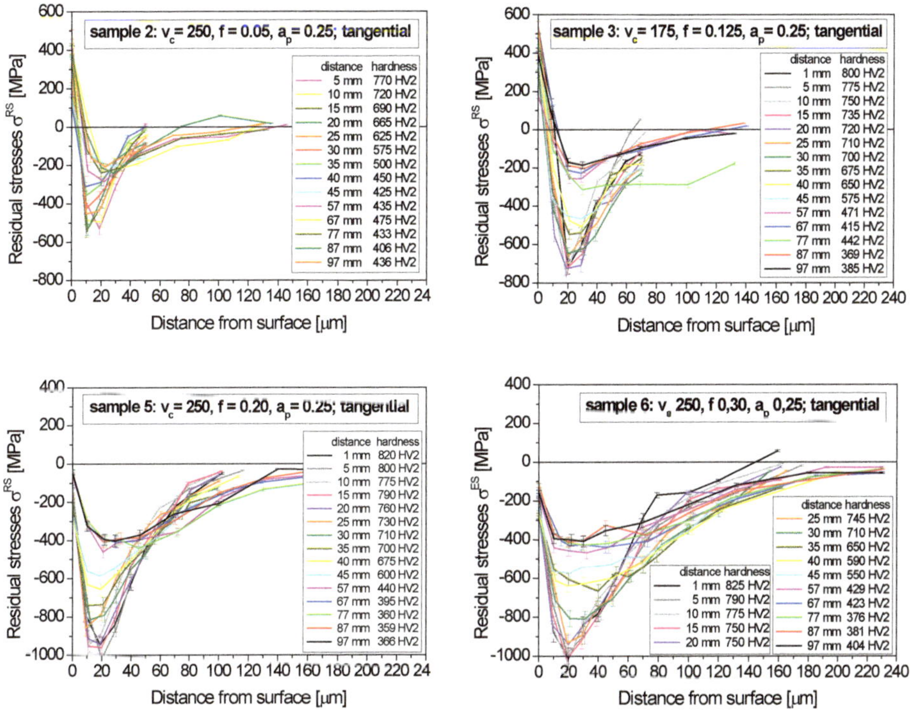

Fig. 5: Tangential residual stress depth distributions of Jominy end quench samples after hard turning with different feed f (v_c: cutting speed; a_p: depth of cut)

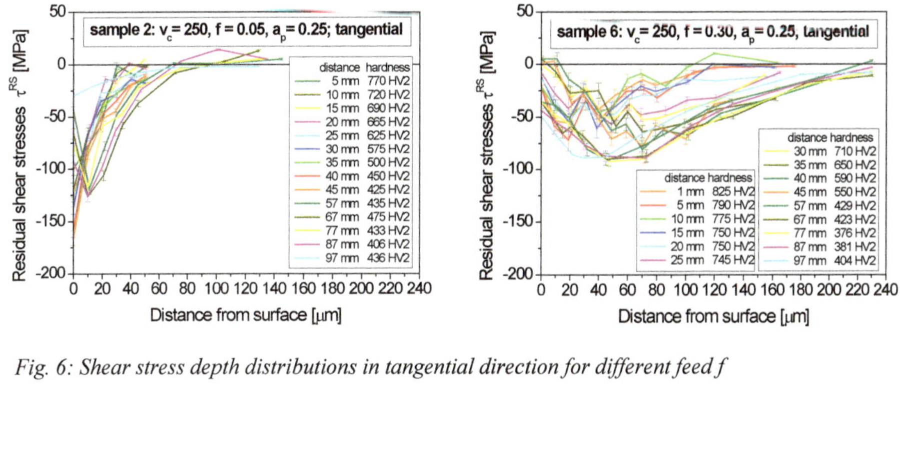

Fig. 6: Shear stress depth distributions in tangential direction for different feed f

Materials Research Proceedings **2** (2016) 271-276 doi: http://dx.doi.org/10.21741/9781945291173-46

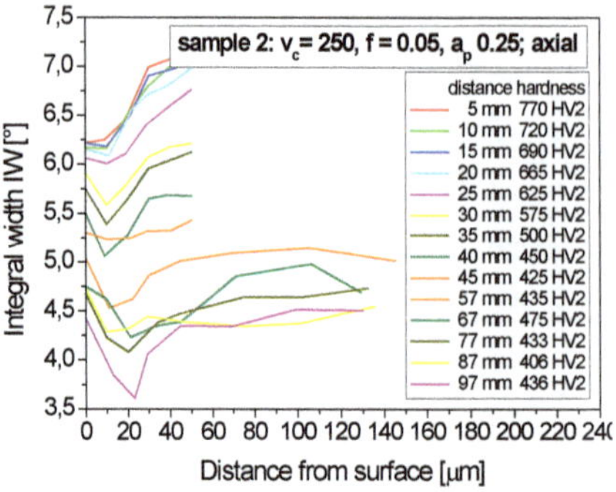

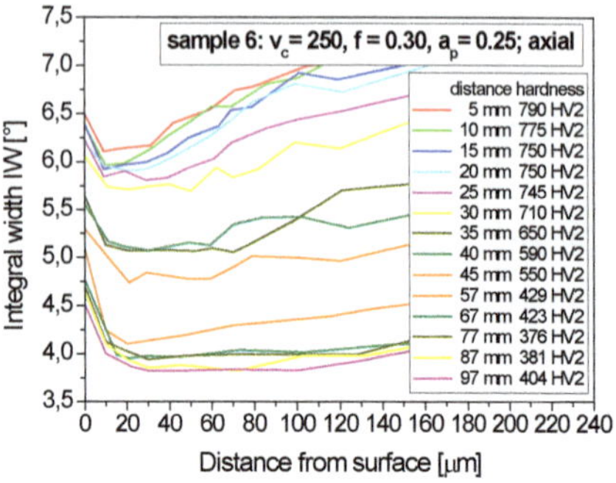

Fig. 7: Integral width depth distributions for different feed f

Summery and Conclusion

For given process parameters turning operations at steels with graded microstructures result in locally different residual stress depth distributions depending on local materials hardness. From the results presented it can be concluded that residual stress data gained from hard turning experiments with uncoated polycrystalline boron nitride (PCBN) inserts at samples with homogeneous microstructures can be transferred to graded inhomogeneous parts, if comparable materials states are considered. In all cases, in axial as well as in tangential direction, a pronounced residual stress maximum below the surface is formed, the amount and the surface distance of which increases with materials hardness and turning feed.

Acknowledgement

The authors gratefully acknowledge the support of this work by the German Science Foundation DFG in the frame of SFB/TR TRR 30 "Process-integrated production of functionally graded structures on the basis of thermo-mechanically coupled phenomena". Special thank goes to Julia Richter and Jannik Mohr for execution of the residual stress measurements.

References

[1] H.-P. Heim, D. Biermann, H. J. Müller: Proceedings 1st International Conference on Thermo-Mechanically Graded Materials, Verlag Wissenschaftliche Scripten, Auerbach, Germany, 2012, ISBN: 978-3-942267-58-8

[2] H.-P. Heim, D. Biermann, W. Homberg: Functionaly Graded Materials in Industrial Mass Production, Vol. 2, Verlag Wissenschaftliche Scripten, Auerbach, Germany, 2013, ISBN: 978-3-942267-91-5

[3] Lebsanft, M.; Tiffe, M.; Zabel, A.; Zinn, W.; Biermann, D.; Scholtes, B.: Residual Stresses in Different Heat Treated Workpieces after Turning; Proceedings of the 9th European Conference on Residual Stresses, Troyes, 07. - 09. 07.2014, Advanced Materials Research Vol. 996 (2014), p. 652 – 657

Residual Stresses 2016: ICRS-10 Materials Research Forum LLC
Materials Research Proceedings **2** (2016) 277-282 doi: http://dx.doi.org/10.21741/9781945291173-47

Residual Stress Condition of Tubular Laser Welds of an AZ31 Magnesium Alloy

Thomas Nitschke-Pagel[1], Klaus Dilger[2]

Institut für Füge- und Schweißtechnik, Technische Universität Braunschweig

Langer Kamp 8, D-38106 Braunschweig, Germany

[1]t.pagel@tu-braunschweig.de, [2]k.dilger@tu-braunschweig.de

Keywords: Magnesium Alloy, Laser Welding, Welding Residual Stresses, Numerical Model

Abstract. Laser welded overlap tubular joints of an AZ31 magnesium alloy show an unexpected failure behaviour under axial and tensional cyclic loading. In the as-welded condition the fatigue cracks propagate not from the locations with the highest load stress concentration but from other sites. After a stress relief annealing at relatively low temperatures the crack initiation sites change to the expected locations This specific behaviour can be found on overlap connections of tubular joints as well as in aluminium as in magnesium alloy joints. Combined FE-calculations and residual stress measurements by means of XRD and neutron diffraction reveal that the particular constraint conditions of tubular joints lead to residual stress distributions which are distinguished remarkably from those of flat plates with linear welds due to the self-constraining geometry. The generation of a non-uniform circumferential residual stress condition with particular locations of high tensile and compressive residual stresses may explain the observed failure behaviour. After an annealing process with strong residual stress relaxation the fatigue cracks start at the expected locations. A shrinkage model for the description of the residual stress generation in tubular joints explains the local residual stresses over the entire thickness in the weld zones and in the adjacent material.

Introduction

Laser welding is a commonly used joining technique in the automotive industry and also in other technical fields especially for applications with materials which are difficult to weld with traditional techniques like spot welding or arc welding procedures [5]. Examples are heat treated or cold formed steels, precipitation hardened aluminium alloys and magnesium alloys, connections with plates of quite different thicknesses or mixed material connections. Here the strongly limited heat input and the small width of the laser welds enables to avoid greater loss of the material strength, the generation of smaller diffusion zones with detrimental intermetallic phases and finally the limitation of distortion problems. An important aspect is that due to the concentrated heat input high welding speed is possible which furthers also the productivity of the process but, however in combination with expected high tensile residual stresses.

A general problem of laser welding is that usually high quality requirements for the preparation of the edges of parts which shall be connected must be fullfilled because the process cannot compensate varying welding gaps. This is the reason, why laser welding for joining processes of thin walled structures is usually applied for overlap joints. Here the quality of the process can be guaranteed much easier without special preparation requests. The problem of such connections is that in overlap laser welded joints a sharp notch between the welded plates is necessarily created. Under cyclic load conditions this crack like notch is responsible for a very poor fatigue strength of these joints.

The load conditions in the surrounding of the gap can be described with local stress approaches which try to describe the local stress condition at the gap. This could be shown for steels and for aluminium alloys. A summary of local approaches which are based mainly on Neuber's rule can be

Materials Research Proceedings **2** (2016) 277-282 doi: http://dx.doi.org/10.21741/9781945291173-47

used for the fatigue assessment is given in [1]. Two general problems must be taken into account. The notch radius is one of the most important parameters because it determines the amount of the effective stresses in the notch. Since neither the radius nor the local stresses can be exactly measured precisely the radius is set in practice conservative to zero which leads to a so called fictitious notch radius of 1mm. The second difficulty is that in the local approaches residual stresses due to welding are used like mean stresses. Since a RS measurement in the gap region is usually impossible the residual stresses used in these calculations are assumptive and usually set to the yield strength. Nevertheless recent investigations [2] on overlapping tubular joints of aluminium and magnesium alloys reveal that obviously residual stresses are of great importance for the failure behavior under cyclic loading. Fatigue test on tubular joints as shown in Fig.1 have shown that under axial loading the fatigue cracks start at the gap between the outer and the inner tube as expected. However in the as-welded state the failure occurs on the side of the outer tube, where the load stresses are necessarily lower due to the larger diameter. After a stress relief treatment the crack initiation site moves to the inner tube with the lower diameter. It is assumed that particular high tensile residual stresses at the gap on the outer tube side are responsible for this specific failure behavior [2].

Investigations on laser welded tubular joints

Overlapping tubular joints of a wrought AZ31 Magnesium-Aluminium-Zinc-alloy as shown in Fig. 1 were laser-welded using a 3kW Nd:YAG-Laser. The welding speed was 47 mm/sec, the heat input 23.2 J/mm. The shape of the weld seam is shown also in the micrograph Fig.1. The distribution of the microhardness (Martens hardness in N/mm²) in the weld seam and in the adjacent zones show that obviously no significant changes of the mechanical properties are generated by the local heat input. The higher and lower hardness values in the center of the joint on both sides of the weld seam are related to measurement errors because the measurements were carried out by a surface scan where some points were too close to the gap.

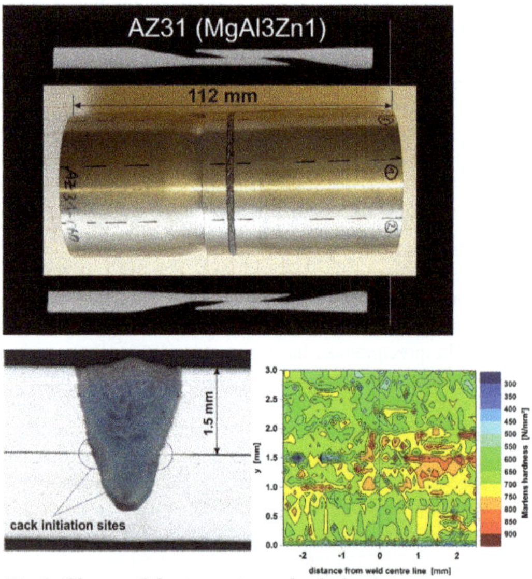

Fig.1: Shape of the investigated tubular joints, micrograph with overview of the weld seam and through-thickness martens-hardness distribution.

The aim of the investigations was to determine the residual stresses in the zone of greatest interest, that is to say, at the gap between the welded tube shells. Therefore residual stress measurements at the Stress-Spec Instrument of the Heinz-Meier-Leibnitz-Centre, Munich, were carried out by means of neutron diffraction. The setup of a measurement is presented in Fig.2. The residual stresses were measured in different layers in as-welded samples and after different static loads with regard to a load induced residual stress relaxation. The spatial resolution was 1x5x1 mm³. Additional surface measurements were carried out by means of X-ray diffraction in order to determine the near surface RS and texture pole figures. The measurements were performed with a standard ψ-diffractometer with an intersectioned Eulerian cradle using CuK_{α}-radiation. With regard doi: http://dx.doi.org/10.21741/9781945291173-47

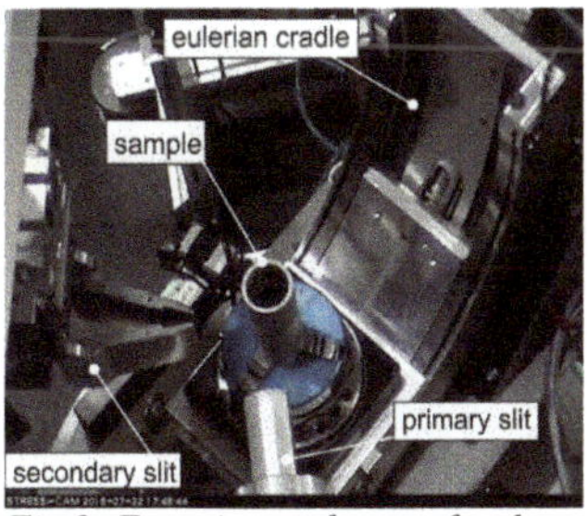

Fig.2: Experimental setup for the neutron measurements (Stress-Spec, MLZ).

to the RS calculation the X-ray elastic constant for the investigated material was experimentally determined to be $1/2s_2 = 20.65\ 10^{-6}$ mm²/N.

Additionally the residual stresses in the tubular joints were calculated using a finite element model (Sysweld). The welding process was simulated with a gaussian heat source which was calibrated with measured temperature profiles. Finally the temperature profile was adjusted by fitting the fusion line to fusion line which can be identified in the micrograph (Fig.3). The required temperature dependent material properties were determined up to a temperature of 500°C. The temperature dependency of the yield strength and the E-modulus is shown in Fig.4. Since the RS condition of an AZ31 alloy is generated by a hindered shrinkage of the weld seam the yield strength at the balancing temperature limits the maximum size of the expected RS. Therefore the RS in the investigated samples should not exceed an amount of 160 MPa (T=80°C).

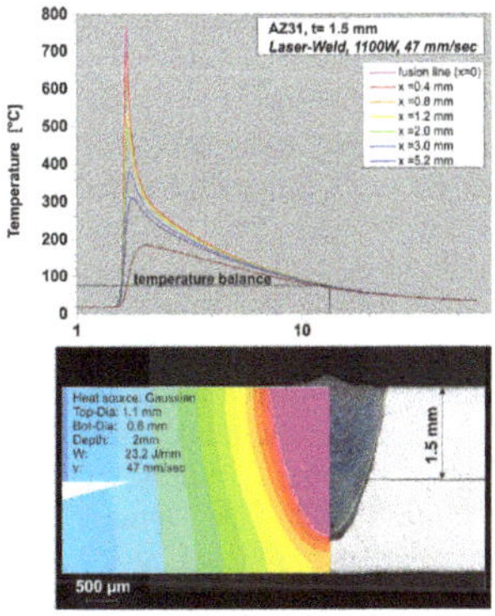

Fig.3: calculated cooling curves and temperature field with adjusted fusion line.

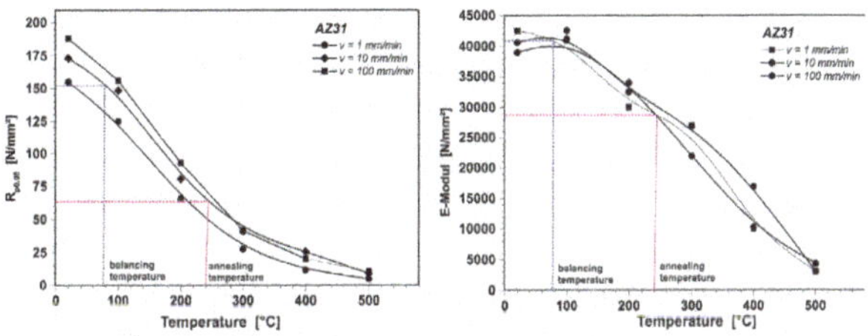

Fig.4: Temperature depending mechanical properties of the magnesium alloy AZ31

Results of the investigation

Fig.5 shows the results of residual stress calculations based upon the described FE-model. The distributions of the axial and hoop RS in the cross section reveal that the particular RS condition of the tubular joints shows a strong discrepancy to distributions which are expected in welded plates of similar thickness. Both RS components are not symmetrical in relation to the weld centre line and the highest tensile residual stresses in the hoop direction can be found on the root side and near the inner surface of the joints. Additionally a tensile RS peak is also present at the gap in the outer tube shell. In axial direction also nonsymmetrical RS can be found around the weld seam with a compressive peak at the outer surface and tensile RS peaks at the gap and at the inner surface. The high tensile peak of the axial component is located directly at the gap of the load-carrying part of the outer tube and this is an important sign which can be used to explain the unexpected failure behavior under axial loading in the as-welded condition.

Materials Research Proceedings **2** (2016) 277-282 doi: http://dx.doi.org/10.21741/9781945291173-47

In Fig.6 the results of the neutron diffraction experiments are summarized. Here the axial direction (e.g. parallel to the axial load) RS in the described layers are presented in the as-welded state and after different loads. The load of 10 kN correlates to a nominal shear stress of approximately 58 N/mm² while the fatigue limit at 10^7 was found at 1.7 kN (10 N/mm²). The measured RS distributions do not really match with the calculated ones. At certain locations (inner tube, inner layer) an agreement of the tendency of the residual stresses seems to be present but the majority of the calculated characteristic RS peaks could not be found by the measurements.

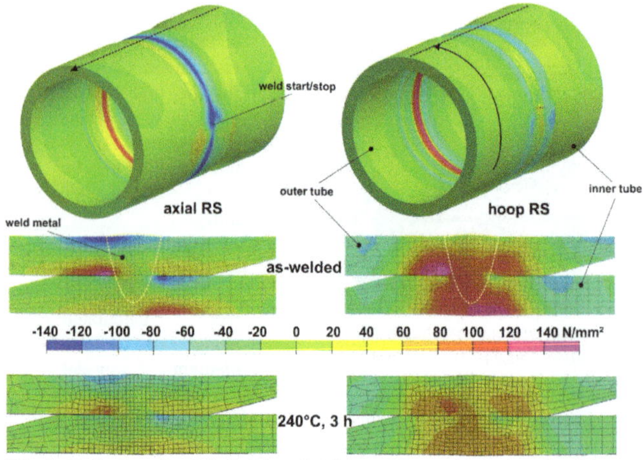

Fig.5: Calculated axial (left hand side) and hoop (right hand side) residual stresses in the as welded condition and after stress relief annealing at 240°C.

Two aspects must be taken into account to explain the disagreement between the measured and the calculated RS. The average information depth of the neutron experiments covers a depth of 1 mm and that is to say that the residual stresses are representative for a distance of approximately 0.5…0.6 mm from the related surface. With regard to a high local resolution in the weld gap region the FE-mesh size in the weld zone was chosen to 0.15 mm in depth. That means that the presented informations do not represent exactly the same layers. However it must be confessed that the neutron experiments are not really applicable to obtain the required local residual stresses with the required resolution. Higher local resolution was not applicable due to the poor diffraction conditions given by the material. Furthermore the experiments do not reveal a significant load induced relaxation of the residual

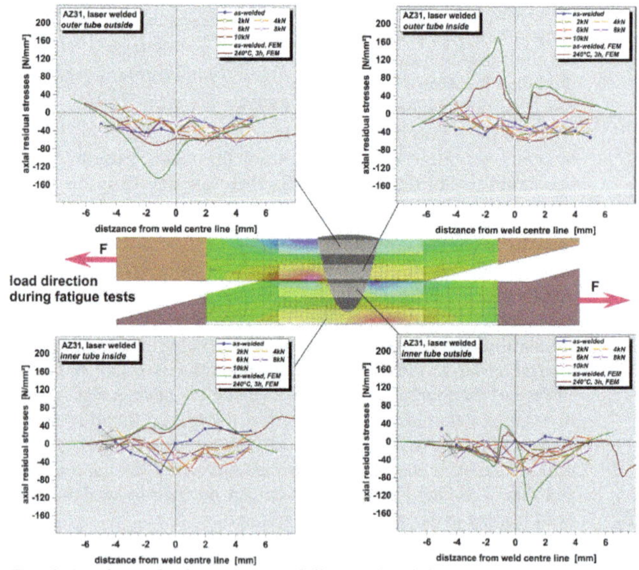

Fig.6:Axial residual stresses in different depth layers measured with neutron diffraction after different axial loads.

Materials Research Proceedings **2** (2016) 277-282 doi: http://dx.doi.org/10.21741/9781945291173-47

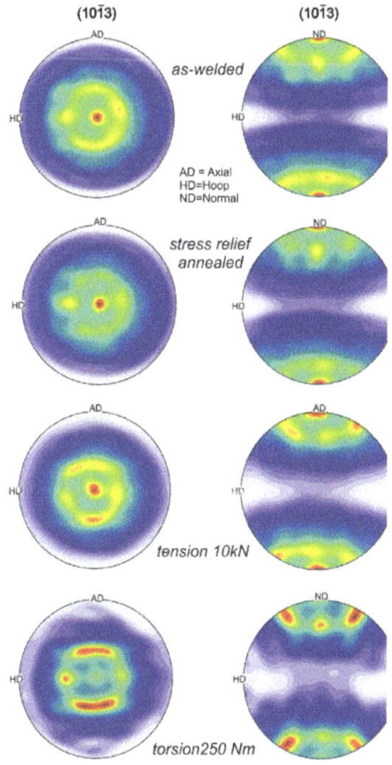

Fig.7: (10$\bar{1}$3)- pole figures determined in the weld seam.

stresses which may also be caused by the low initial residual stresses in the investigated areas.

The quality of the measurement results are additionally strongly influenced by a distinctive texture which is present in the investigated material independent from the location in the weld zone or in the base material. Fig.7 shows the pole figures measured in the weld seam with XRD using CuK_α-radiation at the (10$\bar{1}$3)-plane, which was used also for the residual stress measurements. The pole density distribution in the as-welded state and after stress relief annealing shows the fibre texture which is typical for the base material of cold rolled magnesium alloys [3,4]. It can be found in the weld material as well as in the base material. An annealing at 240°/3h does not change the observed structure. As the figures reveal after tensional and torsional loading the pole density distributions change due to the plastic deformations. Here the torsion leads to stronger changes in the crystal orientation. The consequence is that after different loads the quality of the results increases. In axial direction the measured residual stresses at the outer surface are more or less constant under tensional loading. This is not really surprising due to the initial compressive RS. However in the weld material the hoop RS decrease continuously with increasing shear stresses and that is an indicator of strong plastic deformations although the nominal load amount is relatively low in relation to the yield strength of the material (Fig.8). This behaviour is also represented by the width of the diffraction lines which increase after loading.

Discussion and conclusions

The results of the residual stress measurements have shown that in the investigated tubular samples the determination of residual stresses is rather difficult due to the influence of bad diffraction conditions. Comparing XRD-measurements which are published in literature the chosen (10$\bar{1}$3)-lattice using CuK_α-radiation is not the recommended one [4] but in fact the experiments showed that other combinations did not allow measurements with higher reliability. Furthermore the measurements with XRD as well as with neutrons could performed at the same lattice plane. The most important problem was the influence of texture which complicated strongly the residual stress determination. Therefore a good agreement between the experimental results and the RS calculation could not be achieved. Nevertheless the calculated residual stress distributions are very feasible and can be used to evaluate the failure behavior of the fatigue loaded samples as described in the introduction. The particular residual stress distribution with

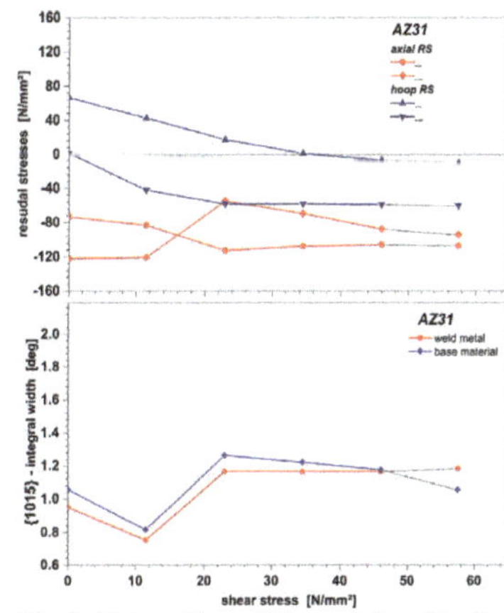

Fig.8: Relaxation of the axial residual stresses under axial loading.

Materials Research Proceedings **2** (2016) 277-282 doi: http://dx.doi.org/10.21741/9781945291173-47

nonsymmetric tensile and compressive RS peaks as shown in Fig.5 can be explained by a shrinkage model as shown in Fig.7. The locally heated zone in the weld seam and its neighbourhood is acting like a "belt" in the cooling phase. This leads to an elastic reduction of the tube radius. The missing symmetry is due to the geometry of the overlapping tubes which furthers a heat accumulation at the free end of each tube shell. Therefore the free side of each tube is more or less RS free while the high RS peaks are concentrated on the opposite side with a higher constraint. The radius reduction generates a bending moment on the constraint side which generates a tensile stress peak in the outer tube and a tensile RS peak at the inner tube. Annealing at 240° has shown a significant release of the RS peaks which makes the observed failure behaviour plausible. Therefore the calculated results can be evaluated as the more reliable ones than the measured ones. The change of the crack initiation sites due to the local amount of tensile residual stresses in the different stadiums is a definite indicator for the relevance of residual stresses for the fatigue behavior of the investigated joins.

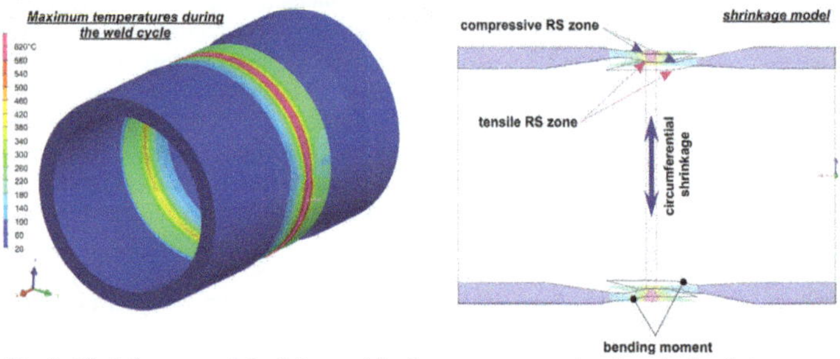

Fig.9: Shrinkage model of the residual stress generation in tubular joints.

Acknowledgement

The presented investigations were carried out in the research project Ni508/11-2 which was generously supported by the Deutsche Forschungsgemeinschaft (DFG). The authors would like to express their thanks for the support.

References

[1] Sonsino,C.M.; Fricke,W.; Radaj,D.: Fatigue assessment of welded joints by local approaches.2nd edition. Woodhead publishing Ltd., Cambridge (UK), 2006

[2] Exel,N.: PhD-Thesis, TH Darmstadt 2014, Shaker Verlag Aachen

[3] Noster,U.:PhD-Thesis, Universität Kassel 2003, Forschungsberichte aus dem Institut für Werkstofftechnik - Metallische Werkstoffe, Band 5.

[4] Krauß,M.: PhD-Thesis, Universität Kassel 2006, Forschungsberichte aus dem Institut für Werkstofftechnik - Metallische Werkstoffe, Band 8.

[5] Masny,H.; Ouaissa,B.: Bundesministerium für BMWi, Bonn * (2003) Seite 1-112

Materials Research Proceedings **2** (2016) 283-288 doi: http://dx.doi.org/10.21741/9781945291173-48

Effects of Hydrostatic 2nd Kind Residual Stresses and of Carbon Partitioning During Martensitic Quenching of Low Alloy Steel

Jeremy Epp

Stiftung Institut fuer Werkstofftechnik, University of Bremen, Badgasteiner Str. 3, Bremen, Germany

epp@iwt-bremen.de

Keywords: In-situ XRD, Retained Austenite, Carbon Partitioning, Hydrostatic Stress, Triaxial Stress State

Abstract. In situ X-ray diffraction measurements were performed at the ESRF in Grenoble, France during quenching of two steel grades: the ball bearing steel AISI52100 and the case hardening steel AISI5120. Diffraction frames were recorded during the complete heat treatment cycles and analyzed in order to determine the temperature- and time-dependent evolution of phase contents and lattice parameters. In the case of the AISI52100 with austenitizing at high temperature, the generation of high compressive hydrostatic stresses of 2nd kind was determined. In the case hardening steel, the dominating effect is a carbon enrichment of the austenite accompanied by the generation of compressive stresses. For the ball bearing steel austenitized at low temperature, both effects take place.

Introduction

Martensitic transformation in steels has now been investigated for more than 100 years [1] The interest in martensitic transformations is still very high as numerous industrial applications use this transformation to improve wear, mechanical and fatigue properties of parts in engineering components [2]. Moreover, new interest on fundamentals of martensitic transformations appeared in the last decades with the development of computer simulation, where kinetics, distortions and other phenomena have to be well described to reach reliable simulation results.

In situ X-ray diffraction analysis has become a powerful method of materials characterization stimulated by constant advances in instrumentation and data processing. This method allows obtaining time-resolved quantitative information about every single phase present in the investigated material.

In the present study, in situ X-ray diffraction experiments were performed with synchrotron X-ray radiation at ESRF on beamline ID11 during heat treatment of two low alloy steels with varying parameters. Different carbon contents in solution were set in hypereutectoid steel, leading to changing behaviors during quenching. Diffraction frames were recorded during the complete heat treatment cycles and analyzed to determine the temperature- and time-dependent evolution of phase contents and lattice parameters. Based on the data, different effects were investigated during quenching.

Experimental methods

Materials. A typical ball bearing steel 100Cr6 (AISI 52100) and a carburizing steel grade 20MnCr5 (AISI 5120) were investigated. The chemical composition of the 100Cr6 steel is Fe-0.95C-1.45Cr-0.44Mn-0.21Si-0.11Ni-0.10Cu-0.05Mo-0.01P-0.005S-0.004Al Mass-% while the chemical composition of the 20MnCr5 steel is as follows: Fe-0.204C-1.35Mn-1.02Cr-0.23Si-0.011S-0.03Mo-0.1Ni-0.03Al-0.1Cu Mass.-%. The initial microstructure of the 100Cr6 steel exhibits a ferritic matrix with globular carbides while the steel grade 20MnCr5 presents a ferrite/pearlite microstructure with a banded morphology. The width of the ferrite/pearlite bands is about 25 μm.

In-situ X-ray diffraction experiments. In situ X-ray diffraction experiments were performed at the European Synchrotron Radiation Facility (ESRF) in Grenoble, France, on beamline ID11. The experiments were executed with a heating device (ETMT, Instron) allowing a controlled heating of samples with different section of 1.5 mm with a length of 40 mm by resistivity. Following heat treatment cycle was used for both steel grades: heating with a rate of 5 K/s up to different austenitizing temperature (Ta) between 800 °C and 940 °C followed by a soaking time of 15 min and quenching to room temperature (RT). The cooling rates led to a $t_{8/5}$ time around 2.5 s. Constant argon purging was used to avoid oxidation of the surface. The temperature was controlled by a type K thermocouple welded on the surface of the samples. During the entire heat treatment cycle, diffraction frames were recorded in transmission mode with a FRELON camera using an exposition time of 0.7 seconds for each frame during quenching with additional time for read-out of 0.1 s. The beam energy was 71 keV and the beam size was set at maximum (100µm high and 300 µm width). The recorded frames were integrated with the program Fit2d and then analyzed with the Rietveld refinement software TOPAS© (Bruker-AXS). Detailed description of the experimental method and data analysis can be found in [3].

Results and discussion

Phase transformations during quenching. After integration of the 2d-diffraction frames and analysis of the standard intensity-vs 2Theta diffraction patterns by the Rietveld method, the evolution of phase content could be precisely described. Fig. 1 shows the evolution of austenite content during quenching for different initial austenitizing temperatures of the 100Cr6 material and for one 20MnCr5 sample austenitized at 900 °C. In the case of the 100Cr6, changing amount of undissolved carbides (Fe_3C) remains in the microstructure after austenitizing depending on the temperature, as expected. The kinetics of the martensitic transformation is well described. After quenching, the amount of retained austenite present in the 100Cr6 samples is 9, 14, 21 and 30 Mass-% for the austenitizing temperatures 825, 870, 905 and 940 °C respectively, while the 20MnCr5 steel exhibits about 5 Mass % retained austenite. In the case of the 20MnCr5, small quantity of bainite formed above Ms (about 10 Mass-%).

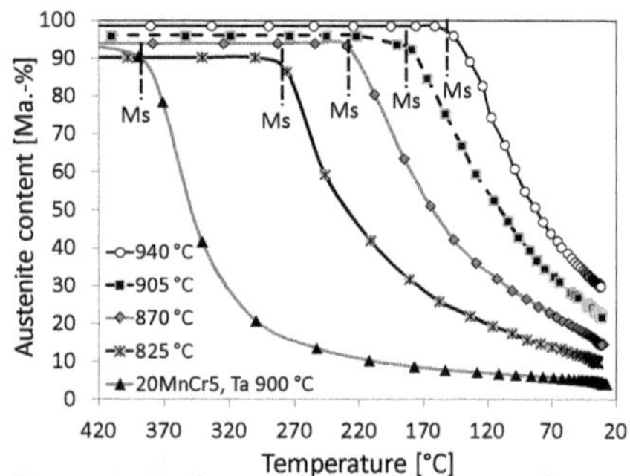

Figure 1: Evolution of austenite content during quenching of 100Cr6 samples with 4 austenitizing temperatures and a 20MnCr5 sample

Evolution of lattice parameters. From the Rietveld refinements, the lattice parameters of all present phases could be extracted. For martensite a nearly linear contraction with decreasing temperature could be observed. On the other hand, the evolution of austenite lattice parameter during cooling exhibits non-linear evolution for all experiments. The austenite lattice parameters are plotted in Fig. 2a as a function of the temperature during quenching of the considered four 100Cr6 samples and one 20MnCr5 sample. With increasing austenitizing temperature for 100Cr6 (corresponding to an increasing carbon content in solution), the lattice parameters are shifted to higher values, as expected. The thermal contraction is linear with decreasing temperature and almost parallel for all experiments until Ms is reached. During further cooling, it can be observed that for the 100Cr6 samples, a change of slope takes place with an accelerated decrease towards room temperature while the 20MnCr5 sample exhibits the opposite effect: a non-linear decrease with continuously decreasing slope takes place below Ms.

In order to observe these changes of slope during cooling, the presented data were corrected by the respective linear contraction taking place above Ms ($\Delta a = a^{measured} - a^{thermal}$) and plotted again as a function of the temperature (Fig. 2b). All values are close to 0 with small scattering

Materials Research Proceedings **2** (2016) 283-288 doi: http://dx.doi.org/10.21741/9781945291173-48

until Ms is reached. During further cooling, the continuous decrease previously observed for the 100Cr6 samples can be clearly seen, while the 20MnCr5 sample exhibits a strong increase.

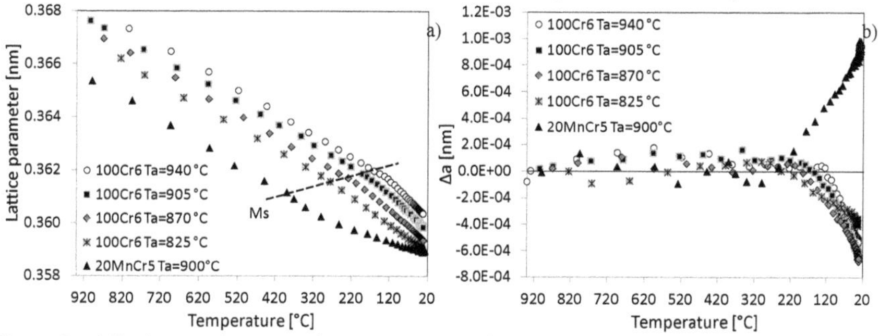

Figure 2: a) Evolution of austenite lattice parameter during quenching of 100Cr6 and 20MnCr5 sample; b) Deviation of the measured austenite lattice parameter from the extrapolated linear thermal contraction

Generation of hydrostatic stresses in austenite. Few studies can be found in the literature about residual stresses in retained austenite during or after quenching of steel. Several authors describe the residual stress state within retained austenite existing at room temperature as a hydrostatic residual stress state under high compressive stresses [4- 6]. The reason for this is a very large volume expansion associated with the martensitic transformation (close to 3%). Of course, due to shear-processes during the martensitic transformation, local residual stresses will not be purely hydrostatic in a single austenite region. However, XRD techniques give average information of thousands of "grains" and therefore the average information contained in the measured volume could be predominantly hydrostatic. In the literature, two different sets of data concerning the evolution of austenite lattice parameter at room temperature as a function of the carbon content can be found: the one based on room temperature measurements after quenching (retained austenite) and the one based either on high temperature measurements of austenite with thermal expansion correction or on Mn-/ Ni-stabilized austenite at RT. From the literature data, equations were developed based on the room temperature measurements ($a^{RT}=0.3556 + 0.00443 \times \%$ C) and on the high temperature measurements ($a^{stress\text{-}free}=0.3573 + 0.00327 \times \%$ C) with %C in Mass.-% [4]. It can be assumed that difference between the high temperature data and the retained austenite at room temperature should be due to generation of hydrostatic residual stresses caused by the large transformation strain.

Both equations from the literature can be used to quantify the possible existing hydrostatic residual stress state in retained austenite depending on the carbon content in solution. If it is assumed that the high temperature lattice parameters describe the stress-free state while the equation based on room temperature measurements represents the retained austenite under stress after quenching, Eq. 1 can be used to calculate the corresponding strain (ε^{RA}) in retained austenite. The resulting hydrostatic residual stresses at room temperature can be calculated by Eq. 2 with E^{RA}=207 GPa and ν^{RA}=0.28.

$$\varepsilon^{RA} = \frac{\Delta a}{a^{stress-free}} = \frac{a^{RT} - a^{stress-free}}{a^{stress-free}} \qquad (1) \qquad\qquad \sigma^{RA} = \varepsilon^{RA} \times \frac{E^{RA}}{\left(1 - 2\nu^{RA}\right)} \qquad (2)$$

The evolution of the literature data of the stress-free lattice parameter of retained austenite ($a^{stress\text{-}free}$) and of the lattice parameter of retained austenite after quenching (a^{RT}) depending on the carbon content in solution is presented in Fig. 3a. From these theoretical evolutions, it can be observed that at values close to 1.4 Mass-% C both lattice parameters are almost similar and with decreasing carbon content, the gap is increasing continuously. From these data and together with equations 1 and 2, resulting theoretical hydrostatic residual stresses were calculated and plotted in

Materials Research Proceedings **2** (2016) 283-288 doi: http://dx.doi.org/10.21741/9781945291173-48

Fig. 3b. At 1.4 Mass-% C, the residual stresses are close to 0. With decreasing carbon content, the theoretical values go to always increasing compressive residual stresses. For 0 % C a theoretical value of -2200 MPa is obtained.

The experimental data collected at RT within the frame of the present study were then used to calculate hydrostatic residual stresses in retained austenite based on equations 1 and 2 using the stress-free lattice parameter from the literature and plotted in Fig. 3b. It can be observed that the experimental values can be divided into two zones. For %C > 0.8 Mass.-% the values are close to the theoretical line with values between -750 MPa to -350 MPa, but with a slight shift. In previous papers, these values have been confirmed and it has been demonstrated that the strong decreasing austenite lattice parameter below Ms (Fig. 2) is due to the generation of strong hydrostatic compressive stresses due to the martensitic transformation with a continuous evolution during cooling [3, 7]. Moreover, the hydrostatic nature of the stresses could also be demonstrated by tensor measurements [3, 7].

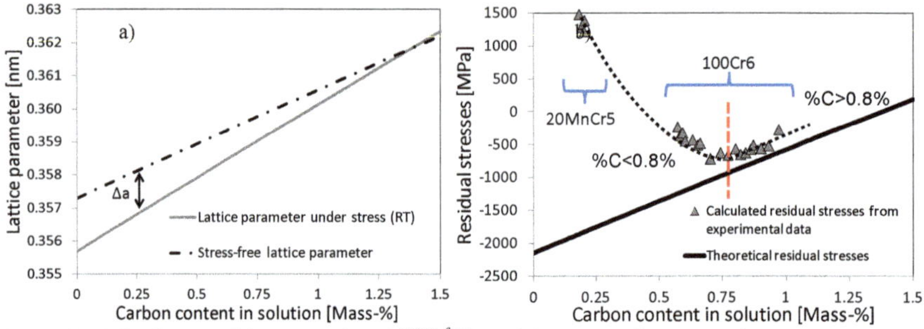

Figure 3: a) Evolution of the stress-free ($a^{stress\text{-}free}$) and the retained austenite lattice parameters (a^{RT}) as a function of the carbon content in solution (data from the literature); b) theoretical hydrostatic residual stresses in austenite (black line) and values calculated from present experimental data (triangles)

When the carbon content in is below 0.8 Mass-% a continuous deviation of the experimental values from the theoretical line can be observed. In previous publications, it has been demonstrated that in the 20MnCr5 steel a carbon enrichment of the austenite takes place during martensitic transformation leading to a carbon content of more than 0.5 Mass.-% in austenite at room temperature [3, 8]. Similar effects of carbon enrichment of austenite during quenching were already reported in the literature for low to medium carbon steels [1, 9]. Therefore, the value of hydrostatic tensile residual stresses in retained austenite for the 20MnCr5 steel in Fig. 3b is erroneous as the stress-free lattice spacing has to take into account the modified carbon content in solution. With decreasing carbon content in solution in the 100Cr6 steel, Ms increases and therefore carbon diffusion is accelerated. In the case of the 100Cr6 with low carbon content in solution, self-tempering effects will also increasingly take place with decrease C-content due to increasing Ms temperature and accelerated diffusion. As shown in [3] for the present experiments, inhomogeneous carbon distribution at nm scale could be observed by atome probe tomography and pronounced carbon depletion in martensite could be demonstrated. Therefore, a certain carbon enrichment of the austenite will also take place in the 100Cr6 steel for high Ms temperatures.

If the theoretically calculated residual stress distribution (Fig. 3b) is assumed to be the real residual stress state in retained austenite, the gap between these values and the experimentally determined values might be attributed to the effect of carbon enrichment of austenite during quenching. In order to determine the C-content in solution resulting from these assumptions, a least squares fit of the experimental data with the calculated distribution was done by implementing a dependence of the C-content in solution on the nominal C-content determined

Materials Research Proceedings **2** (2016) 283-288 doi: http://dx.doi.org/10.21741/9781945291173-48

above Ms ($\%C^{above\ Ms}$). Following equation giving the evolution of C-content in solution in austenite after quenching ($\%C^{RT}$) was obtained:

$$\%C^{RT} = -0.68 \times \%C^{above\,Ms} + 0.67 \times \left(\%C^{above\,Ms}\right)^3 + 1.02 \qquad (3)$$

The new resulting stress distribution compared to the theoretical line is shown in Fig 4a while the evolution of the resulting carbon content in retained austenite as a function of the carbon content in solution above M_s is given in Fig. 4b. Below 0.8 Mass-% C, a carbon enrichment of retained austenite after quenching is resulting. The gap between the theoretical line and the obtained behavior is continuously increasing with decreasing carbon contents. Carbon contents in retained austenite up to 1.0 Mass-% C are obtained after quenching for 0.1 Mass-% C in solution before quenching. This is in a similar range as values that have been reported in the literature [9- 11]. However, it has to be kept in mind that the determined carbon enrichment is a first estimation based on several assumptions. Moreover, according to the considered steel grade and heat treatment, variations are expected to occur.

Based on these results, it seems that carbon partitioning from martensite to retained austenite during quenching can occur for carbon contents in solution above Ms below 0.8 Mass-% C for the given cooling condition. The resulting carbon enrichment of the austenite increases for decreasing carbon contents in solution. Superimposed to this, high hydrostatic compressive residual stresses are present in retained austenite after quenching. For carbon contents in solution between 0.1 and 1 Mass-% C, compressive residual stresses in the range of -500 and -2000 MPa are resulting.

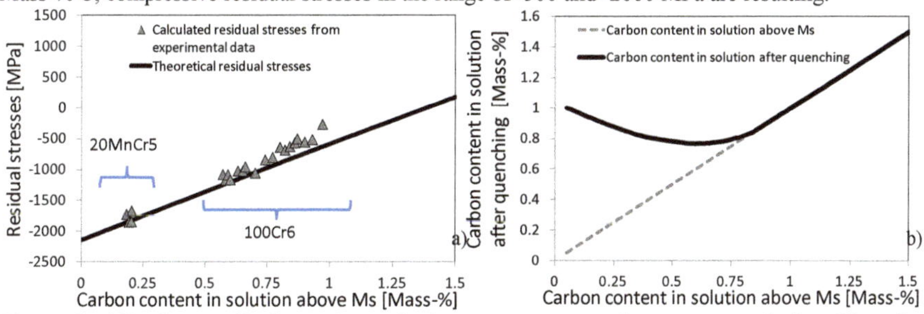

Figure 4: a) Evolution of hydrostatic residual stresses in retained austenite recalculated based on the increased carbon content in solution after quenching given in b; b) Evolution of fitted carbon content in solution in retained austenite after quenching depending on the original carbon content in solution above Ms

Based on the present results and on the in situ experiments, it is possible to determine the development of residual stresses in austenite during cooling. For experiments with carbon content in solution above 0.8 Mass-%, no carbon enrichment in austenite occurs. Therefore, a linear extrapolation of the austenite lattice parameter measured at temperatures above Ms, down to room temperature can be used to obtain the stress free lattice parameter. For the experiments with carbon content in solution below 0.8 Mass-%, the carbon contents at RT obtained from Fig. 4a can be used as end value. A specific evolution between the nominal carbon content at Ms and the final increase C-content at RT has been taken into account and is described in [3]. The stress calculations were performed by using Eq. 1 and 2 during cooling, taking into account the temperature-dependent changes of elastic properties as given in [3]. The evolutions of hydrostatic stresses in austenite during cooling are presented in Fig. 5. It can be observed that for the experiments made with the 100Cr6 steel, the increase of compressive stresses is very moderate at the beginning of the transformation, in particular for the experiments with austenitizing above 900 °C. After certain undercooling the stresses increase continuously with decreasing temperature until room temperature is reached. Towards room temperature it can be observed that small cooling steps lead to a large increase of the compressive stresses. This is also pronounced for the 20MnCr5 steel. However, it should be

Materials Research Proceedings **2** (2016) 283-288 doi: http://dx.doi.org/10.21741/9781945291173-48

remarked that especially for the experiments conducted with 20MnCr5 samples, the assumed evolution of the stress-free lattice parameters is subjected to uncertainties as the changes due to carbon enrichment might deviate from the real case.

The residual stress values reached at RT after cooling are in the range of -1800 and -750 MPa. For the experiments with austenitizing at 905 and 940 °C no carbon enrichment of the austenite is expected to occur and therefore the lattice parameter evolution above M_s was directly extrapolated to RT to determine the stress-free state. The end values reached correspond to the residual stress values obtained by using the literature data as presented in Fig. 3b what confirms that the presented results are reliable.

It should be kept in mind that a part of the results is based on assumptions concerning the carbon enrichment of retained austenite. Indeed, diffraction methods are sensitive to stresses as well as to changes of chemical composition. A proper separation of both effects is therefore almost impossible without assumptions. However, as observed with the help of the in situ experiments, as well as with more advanced techniques, clear evidences of carbon inhomogeneities were obtained [3]. Moreover, for C > 0.8 Mass-% C, the experimentally determined residual stresses are close to the literature values.

Summary

In situ XRD measurements were performed at the ESRF during quenching of two steel grades: 100Cr6 (AISI52100) and 20MnCr5 (AISI5120). 2D diffraction frames were recorded during the heat treatment cycles. The integrated diffraction frames were analyzed with the Rietveld-method.

By investigating the changes of austenite lattice parameter during the quenching process, it could be observed that both steel grades exhibit a non-linear evolution below Ms but in opposite direction: a slowed decrease in the case of the 20MnCr5 and an accelerated decrease for the 100Cr6 steel. For %C > 0.8 Mass-%, no pronounced carbon diffusion is expected to take place and the generation of high hydrostatic compressive stresses in the austenite were determined and validated by different methods. For carbon contents below 0.8 Mass.-%, carbon partitioning from martensite to austenite takes place leading to a shift of the stress-free lattice parameter. Considering a carbon enrichment which is dependent of the initial carbon content in solution, and using a least square fit based on theoretical and experimental values, a possible carbon enrichment function was determined as well as expected hydrostatic compressive residual stresses in austenite. According to these calculations, carbon enrichment takes place and reaches a maximum for initial carbon content in solution of 0.1 Mass-% with values up to 1 Mass-%. Further investigations are ongoing to verify these results.

Acknowledgments

The authors gratefully acknowledge the Deutsche Forschungsgemeinschaft (DFG) for financial support of project C01 in the Transregional Collaborative Research Centre 136 „Process Signature" as well as the ESRF for provision of synchrotron radiation facilities and financial support.

References

[1] Olsen, G.B., Owen, W.S., Martensite, first ed., ASM International, 1992.

[2] Metals Handbook, Vol. 1: Properties and Selection: Irons, Steels, and High-Performance Alloys, tenth ed., ASM International, 1990.

[3] J. Epp, PhD Thesis, University of Bremen, Shaker Verlag, Aachen (2016).

[4] L. Cheng, A. Böttger,Th.H. de Keijser, E. J. Mittemeijer, Scripta metal mat. 24 (1990) 509-514. http://dx.doi.org/10.1016/0956-716X(90)90192-J

[5] K. Y. Golovchiner, Physics of Metals and Metallography 37-2 (1974) 126-130.

[6] M. Villa, K. Pantleon, M. A. J. Somers, Acta Mat.65 (2014) 383–392. http://dx.doi.org/10.1016/j.actamat.2013.11.007

[7] J. Epp, Adv. Mat. Res. 996 (2014) 525-531.

[8] J. Epp, T. Hirsch, C. Curfs, Metal. Mater. Trans. 43A (2012) 2210 – 2217. http://dx.doi.org/10.1007/s11661-012-1087-7

[9] T.Y. Hsu, Journal de physique IV 5-8 (1995) 351-354.

Residual Stresses 2016: ICRS-10 Materials Research Forum LLC
Materials Research Proceedings **2** (2016) 289-294 doi: http://dx.doi.org/10.21741/9781945291173-49

Residual Stress in Stainless Steels after Surface Grinding and its Effect on Chloride Induced SCC

Nian Zhou[1, a *], Ru Lin Peng[2, b], Rachel Pettersson[3, c] and Mikael Schönning[4, d]

[1]Department of Material Science, Dalarna University, SE-79188 Falun, and KTH, SE-10044 Stockholm, Sweden

[2]Department of Management and Engineering, Linköping University, SE-58183 Linköping, Sweden

[3]Jernkontoret, SE-11187 Stockholm, and KTH, SE-10044 Stockholm, Sweden

[4]Corrosion Department, Avesta Research Center - Outokumpu Stainless AB, SE-77422 Avesta, Sweden

[a] nzh@du.se, [b] ru.peng@liu.se, [c] rachel.pettersson@jernkontoret.se, [d] mikael.schonning@outokumpu.com

Keywords: Grinding, Stainless Steel, Residual Stress, Stress Corrosion Cracking

Abstract. The induced residual stresses in stainless steels as a consequence of surface grinding as well as their influence on the chloride induced stress corrosion cracking (SCC) susceptibility have been investigated. Three types of materials were studied: 304L austenitic stainless steel, 4509 ferritic stainless steel and 2304 duplex stainless steel. Surface grinding using 60# and 180# grit size abrasives was performed for each material. Residual stress depth profiles were measured using X-ray diffraction. The susceptibility to stress corrosion cracking was evaluated in boiling $MgCl_2$ according to ASTM G36. Specimens were exposed without applying any external loading to evaluate the risk for SCC caused solely by residual stresses. Induced residual stresses and corrosion behavior were compared between the austenitic, ferritic and duplex stainless steels to elucidate the role of the duplex structure. For all materials, the grinding operation generated tensile residual stresses in the surface along the grinding direction but compressive residual stresses perpendicular to the grinding direction. In the subsurface region, compressive stresses in both directions were present. Micro-cracks initiated due to high grinding-induced tensile residual stresses in the surface layer were observed in austenitic 304L and duplex 2304, but not in the ferritic 4509. The surface residual stresses decreased significantly after exposure for all specimens.

Introduction

Stainless steels are widely used in a variety of applications due to their combination of good mechanical properties and high corrosion resistance. There are a large number of stainless steel grades with different chemical compositions and microstructures. Since the microstructure has a decisive effect on the properties, stainless steels are often categorized by the microstructure, for example austenitic and ferritic stainless steels; different categories are suited for different applications [1].

Stress corrosion cracking (SCC) occurs under the simultaneous interaction of three factors: a corrosive environment, a susceptible material and the presence of tensile stresses [2]. For stainless steels, the chloride ion, which exists in many environments, is unfortunately found to make them prone to stress corrosion cracking. The stress corrosion cracking behavior of stainless steels has been widely investigated during the last decades. Experimental results show austenitic stainless steels are susceptible to SCC while ferritic grades are quite resistant [3]. Due to the combination of austenitic and ferritic structures, duplex stainless steels generally have higher Cl-SCC resistance compared to the austenitic grades, although this may depend on the actual testing conditions [4]. In addition to

Materials Research Proceedings **2** (2016) 289-294 doi: http://dx.doi.org/10.21741/9781945291173-49

microstructural effects, residual stresses also have a significant impact. For example, laser peening of ANSI 304 austenitic stainless steel has been demonstrated to improve the SCC resistance; the introduced high level compressive residual stress and grain refinement were proposed as the two main factors [5]. High surface tensile residual stresses introduced by milling as well as a detrimental surface profile have been reported to generate branched cracks on 316Ti austenitic stainless steel surface after exposure in boiling $MgCl_2$ solution [6].

Grinding is an important and widely used surface finishing process when fabricating stainless steels. During grinding, both plastic deformation and heating occur in the surface layer, which in turn change the residual stress state in the material. However, little research has been performed to evaluate the role of pure residual stress in the risk for SCC. The aim of this study is to increase the understanding of residual stresses induced by surface grinding of stainless steels as well as their role in crack initiation and propagation. In addition, the role of the ferritic and austenitic microstructure in a duplex stainless steel has also been investigated.

Experimental work

Three grades of stainless steels, austenitic 304L, ferritic 4509 and duplex 2304, were investigated in this study. The as-delivered materials had been solution annealed at 1100°C, quenched, pickled and roll leveled. The materials were supplied with 2B surface finish and as test coupons of 400mm×150mm×2mm in dimensions. The chemical compositions and measured mechanical properties perpendicular to the rolling direction at room temperature are given in Table 1.

Table 1 Chemical compositions (wt.%) and mechanical properties, measured perpendicular to the rolling direction at room temperature, of stainless steels investigated

Grade	C	Si	Mn	P	S	Cr	Ni	Mo	Nb	N	Cu	Co	Ti	Rp0.2(MPa)	Rm (MPa)	Elongation (%)	Hardness (HB)
304L	0.02	0.32	1.55	0.03	0	18.2	8.11	-	0.01	0.07	0.31	0.16	-	230	642	54	170
4509	0.01	0.48	0.46	0.03	0	17.9	0.2	-	0.37	-	-	-	0.13	345	462	32	175
2304	0.02	0.39	1.48	0.03	0	23.4	4.84	0.36	-	0.13	0.22	-	0.01	590	739	30	228

The grinding operations were conducted on a Chevalier FSG-2A618 grinding machine using grinding belts with conventional aluminum oxide grit. The detailed grinding set-up is described in [1]. All the grinding operations were performed along the rolling direction of the material. A fixed grinding speed v_s=23m/s, a fixed feed rate v_w=8m/min and a fixed motor power of 600W were used. Grinding was first performed using 60# grit size abrasive to remove the as-delivered material surface, and then followed by grinding with new 60# and 180# grit size abrasives respectively to get the final surface finish for each material. No grinding lubricant was used during the operation.

The influence of residual stress from surface grinding on chloride induced stress corrosion cracking susceptibility was investigated without applying external loading. Two 30mm by 30mm square-sized specimens were cut out from each test coupon with different grinding conditions. All specimens were exposed for 20h in boiling $MgCl_2$ solution at 155°C±1°C according to ASTM G36. Detailed corrosion procedures are described in [2].

The in-depth residual stress profiles were measured by X-ray diffraction for all the ground specimens. Cr-K_α radiation was used; residual stresses were determined from the measured diffraction peak at 2θ~128° for the {220} lattice planes of the austenitic phase and at 2θ~154° for the {211} lattice planes of the ferritic phase respectively. Residual stresses were calculated based on the $\sin^2\psi$ method [3]. Controlled electrolytic polishing was used to remove the material and measure the in-depth profiles; no correction was made for possible stress relaxation due to polishing. Surface stresses of all types of specimens were also measured after corrosion tests using the same method. The surface morphology after exposure was investigated using a FEG-SEM Zeiss Ultra 55 scanning electron microscopy (SEM). Both SEM and ECCI (electron channeling contrast imaging, Hitachi FEG-SEM SU-70) were used for examination of cross-sections after corrosion testing.

Materials Research Proceedings **2** (2016) 289-294 doi: http://dx.doi.org/10.21741/9781945291173-49

Results and Discussion

The in-depth residual stress profiles parallel ($\sigma_{\parallel}$) and perpendicular ($\sigma_{\perp}$) to the grinding directions of all three types of materials ground by both 60# and 180# abrasive grit size as final surface finish are presented in Figure 1. For the duplex 2304, the results are presented as phase stresses and macro-stresses. Phase stresses are residual stresses measured in the austenitic phase (FCC), σ^{γ}, and ferritic phase (BCC), σ^{α}, respectively. The macro-stress (σ^{M}) is the homogeneous residual stress on a macroscopic scale along at least one direction [3] and is calculated according to the equation: $\sigma^{M} = V^{\gamma}\sigma^{\gamma} + (1-V^{\gamma})\ \sigma^{\alpha}$ (where V^{γ} is the volume fraction of the austenitic phase). As shown in Figure 1(a) and (b), the grinding operations generated tensile $\sigma_{\parallel}$ but compressive $\sigma_{\perp}$ in the surface layer for both austenitic 304L and ferritic 4509. The tensile $\sigma_{\parallel}$ was highest in the ground surface and dropped rapidly to compression in the subsurface. The compressive $\sigma_{\perp}$ showed a relative low value in the surface and increased rapidly to reach a peak value in the subsurface. The tensile surface residual stresses were higher for the austenite 304L than ferrite 4509. However, grinding induced residual stresses showed different features in the austenitic and ferritic phases in the duplex 2304. In the austenitic phase (Figure 1(c1)), tensile $\sigma_{\parallel}$ and compressive $\sigma_{\perp}$ were observed; the measured tensile $\sigma_{\parallel}$ dropped gradually from the surface to the subsurface while the compressive stress showed a peak value in $\sigma_{\perp}$ the subsurface region. This is very similar to the trend seen for the austenitic 304L steel. In the ferritic phase (Figure 1(c2)), except for the $\sigma_{\parallel}$ for grinding using 60# abrasive in the surface layer, both $\sigma_{\parallel}$ and $\sigma_{\perp}$ were compressive with a subsurface peak value and $\sigma_{\perp}$ was more compressive than $\sigma_{\parallel}$. On the macroscopic scale (Figure 1(c3)), similar trends were observed to the single phase materials. For all the results shown in Figure 1, grinding with 60# grit size abrasive generated higher residual stresses in both surface and subsurface region and a higher penetration depth compared with grinding using 180# as final surface finish.

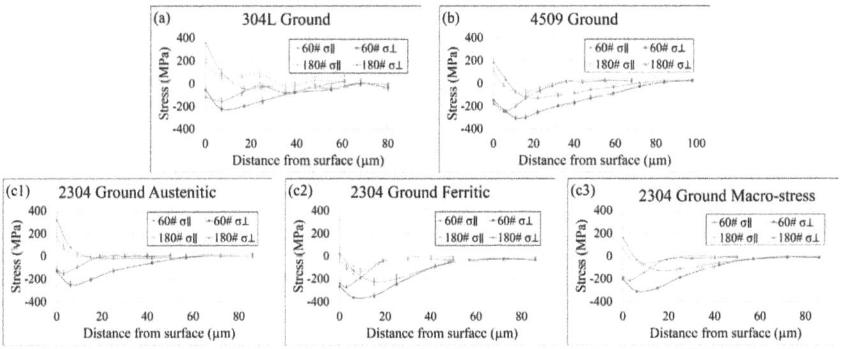

Figure 1 In-depth residual stress profiles of (a) 304L ground specimens, (b) 4509 ground specimens, and (c) 2304 ground specimens

Residual stress in grinding is caused by the combination of thermal and mechanical effects. During grinding, heat is generated in the surface layer of the workpiece material, thus a temperature gradient is formed from surface to the bulk. Since the bulk material hinders contraction of the surface layer during the cooling period, surface tension can be generated both along and transverse to the grinding direction [5]. However, an anisotropic surface residual stress field with tensile $\sigma_{\parallel}$ and compressive $\sigma_{\perp}$ was observed in 304L, 4509 and the macro-stresses of 2304, indicating that the mechanical effect, i.e. anisotropic plastic deformation, is more dominant than the thermal effect in this study. During grinding, the surface layer experienced overall compressive plastic deformation in the grinding direction and tensile deformation in the transverse direction, thus the constraint by the material beneath the surface layer resulted in surface tensile $\sigma_{\parallel}$ but compressive $\sigma_{\perp}$ after the grinding zone moved away [6]. Higher tensile residual stresses were measured from surface to subsurface in 304L than in 4509 in both directions, even though same grinding parameters were used. The lower thermal conductivity of the austenitic stainless steel is probably the main contributory factor. In addition, the in-depth residual stress profiles showed different trends for the ferritic 4509 and the ferritic phase of the duplex 2304. This is attributable to the complicated interaction between the austenitic and ferritic phases in the duplex material during grinding due to their difference in both thermal and mechanical properties. For all three materials, using smaller grit size abrasive (180#) introduced lower deformation in both surface and subsurface layers, thus lower residual stress with lower penetration depth were observed in both directions.

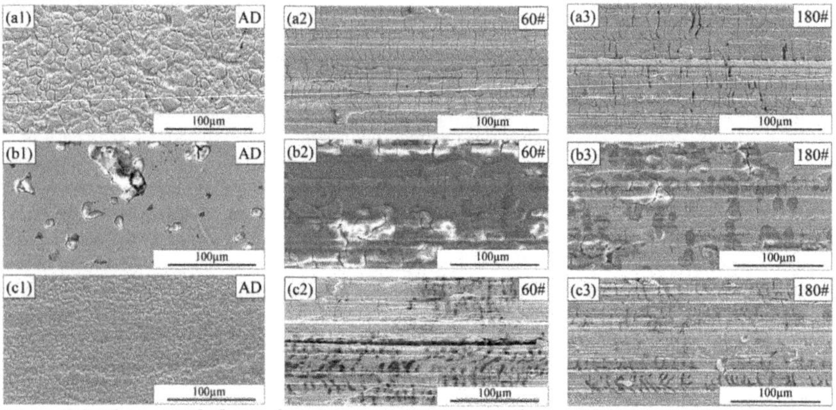

Figure 2 Surface morphology of as-delivered (AD) material, ground 60# and 180# specimens after exposure without external loading: (a) austenite 304L, (b) ferrite 4509, (c) duplex 2304

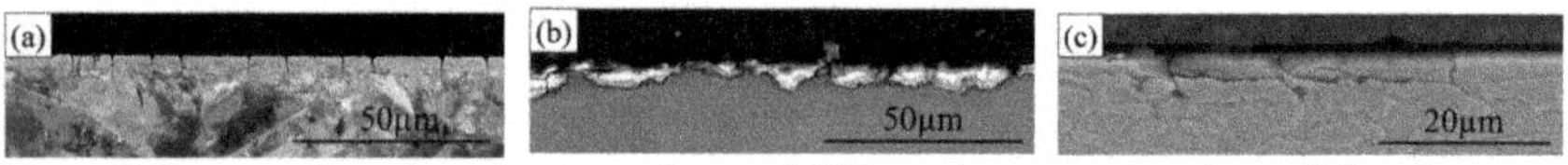

Figure 3 Cross-section microstructure of ground 60# specimens sectioned parallel to the grinding direction after exposure without external loading: (a) austenite 304L, (b) ferrite 4509, (c) duplex 2304

Surface morphology and cross-section microstructure were investigated after corrosion testing for all ground specimens, selected images of the three types of materials are shown in Figure 2 and Figure 3. Extensive branched micro-cracks were present in the ground surfaces of both the austenite 304L and the duplex 2304 in spite of the absence of external loading. It should be noted that no such cracking was observed on the as-delivered surface. Both the surface and cross-section investigations showed the micro-cracks were primarily oriented perpendicular to the grinding marks; while in the parallel direction, much fewer micro-cracks with lower penetration depth were observed. This correlates to the high levels of tensile stresses measured parallel to the grinding direction in the surface layer. The results strongly indicate that tensile residual stress in the ground surface makes both materials susceptible to SCC in chloride existing environment even in the absence of any external loading, while compressive residual stress transverse to the grinding direction can retard cracking. For the austenitic 304L, the majority of the micro-cracks arrested at the depth where the

Materials Research Proceedings **2** (2016) 289-294 doi: http://dx.doi.org/10.21741/9781945291173-49

residual stress profiles shifted from high tensile level to low or no tensile residual stresses, indicating that a threshold tensile stress is required to initiate cracks and sustain their propagation in the boiling $MgCl_2$ solution. Tensile $\sigma_{\parallel}$ in austenitic phase and compressive $\sigma_{\parallel}$ in ferritic phase were measured for ground surfaces of 2304, however, micro-cracks were observed to initiate and propagate in both phases. The macroscopic residual tensile stress thus seems to play a more dominant role than the single phase stresses. The penetration depth of the micro-cracks correlated to the position where the macro-stress shifted from tension to compression, and cracks also tended to be arrested at phase boundaries. For both materials, less micro-cracks with lower penetration appeared by using smaller grit size (180#) abrasives after exposure. The penetration depth of micro-cracks in duplex 2304 was smaller than that for the austenitic grade, even though the same grinding and exposure conditions were used. In the case of the ferritic 4509 a large number of pits, with varied size and filled with corrosion products, were observed after exposure. The extensive pitting may reflect the lower alloying content of this steel. Many more pits were observed on the ground surface than the as-delivered material; in addition, pitting was more severe for the ground 60# than 180# specimens, suggesting that the roughness, topography, deformation and residual stress of the surface layer contributes to an increasing pitting susceptibility. Cross-sectional investigations showed the pits were largely limited to the deformed surface and subsurface layers, and that some cracks appeared to have initiated from the pits. It is suggested that the extensive pitting on 4509 has to a large extent suppressed stress corrosion cracking, even though some cracks can emanate from pits.

Measured surface stresses along the grinding direction of all ground specimens before and after exposure without external loading are plotted and compared in Figure 4. As shown in the figure, for all the specimens, the grinding induced surface tensile $\sigma_{\parallel}$ were observed to be reduced after exposure. For all the three materials, grinding by 60# abrasive grit size generated much higher tensile $\sigma_{\parallel}$ than 180# in the ground surface; however, measured surface stresses showed similar values for both grinding conditions after exposure. The micro-cracks in austenitic 304L and duplex 2304 and the pits in 4509 are proposed as the main factors that caused surface stress release after exposure.

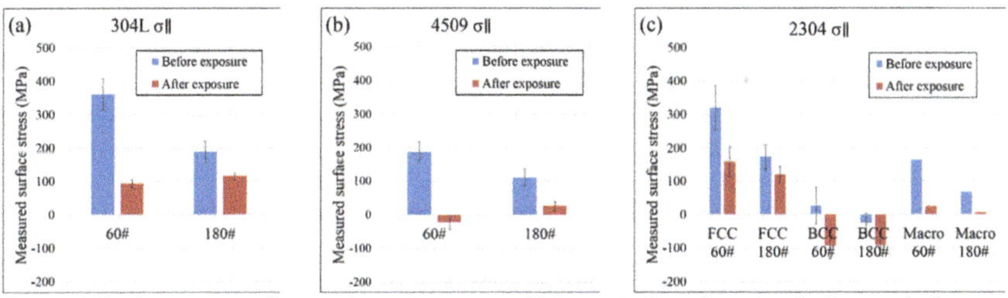

Figure 4 Surface stress of ground specimens parallel to grinding direction before and after testing in $MgCl_2$ solution: (a) austenite 304L, (b) ferrite 4509, (c) duplex 2304

Conclusions

Grinding generated surface macroscopic tensile residual stresses along the grinding direction and compressive residual stresses perpendicular to the grinding direction for the austenitic 304L, ferritic 4509 and duplex 2304 stainless steels. In the subsurface region, compressive stresses were present in both directions. Residual stresses were more tensile for the austenitic phase in the duplex steel than the ferritic phase. Results indicate that anisotropic mechanical effects dominant over isotropic thermal effects for the grinding operation in this study. Abrasive grit size had a significant influence on both the level and the penetration depth of residual stress.

Grinding induced surface tensile residual stresses caused micro-cracks to initiate from the surface during exposure to boiling magnesium chloride solution even in the absence of external loading for both 304L and 2304. The cracks arrested when reaching the depth with low or no tensile stresses. For the duplex 2304, cracks were present in both phases but tended to stop at phase boundaries. Less cracking was seen in the ferritic 4509. This was attributed to the extensive pitting which is a consequence of the lower alloying level of this steel.

The formation of micro-cracks or pits was proposed as the main factor contributing the reduction of surface stresses after exposure in the magnesium chloride solution.

Acknowledgement

This work is financially supported by Outokumpu Stainless Research Foundation, Region Dalarna, Region Gävleborg, Länsstyrelsen Gävleborg, Jernkontoret, Sandviken kommun and Högskolan Dalarna within the Steel Industry Graduate School.

References

[1] Outokumpu, Handbook of Stainless Steels, Outokumpu Oyj, Finland, 2013.

[2] D. Jones, Principles and prevention of corrosion, Pearson Education (US), 1996.

[3] Outokumpu, Corrosion Handbook, Outokumpu Oyj, Finland, 2015.

[4] E. Johansson and T. Prosek, "Stress corrosion cracking properties of UNS S32101 - a new duplex stainless steel with low nickel content," in *Corrosion 2007 Conference & Expo*, 2007.

[5] J. Lu, K. Luo, D. Yang, X. Cheng, J. Hu, F. Dai, H. Qi, L. Zhang, J. Zhong, Q. Wang and Y. Zhang, "Effects of laser peening on stress corrosion cracking (SCC) of ANSI 304 austenitic stainless steel," *Corrosion Science,* vol. 60, pp. 145-152, 2012. http://dx.doi.org/10.1016/j.corsci.2012.03.044

[6] K. Lyon, T. Marrow and S. Lyon, "Influence of milling on the development of stress corrosion cracks in austenitic stainless steel," *Journal of Materials Processing Technology,* vol. 218, pp. 32-37, 2015. http://dx.doi.org/10.1016/j.jmatprotec.2014.11.038

[7] N. Zhou, R. L. Peng and R. Pettersson, "Surface Integrity of 2304 Duplex Stainless Steel After Different Grinding Operations," *Journal of Materials Processing Technology,* vol. 229, pp. 294-304, 2016. http://dx.doi.org/10.1016/j.jmatprotec.2015.09.031

[8] N. Zhou, R. Pettersson, R. L. Peng and M. Schönning, "Effect of Surface Grinding on Chloride Induced SCC of 304L," *Materials Science and Engineering: A,* vol. 658, pp. 50-59, 2016. http://dx.doi.org/10.1016/j.msea.2016.01.078

[9] I. Noyan and J. Cohen, Residual Stress Measurement by Diffraction and Interpretation, Springer, 1987.

[10] J. Davim, Surface intergrity in machining, Springer London Ltd, 2010. http://dx.doi.org/10.1007/978-1-84882-874-2

[11] G. Q. Guo, Z. Q. Liu, X. J. Cai, Q. L. An and M. Chen, "Investigation of Surface Integrity in Conventional Grinding of Ti-6Al-4V," *Advanced Materials Research,* Vols. 126-128, pp. 899-904, 2010. http://dx.doi.org/10.4028/www.scientific.net/AMR.126-128.899

[12] V. Hauk, Structural and Residual Stress Analysis by Nondestructive Methods, Elsevier Science, 1997.

Residual Stresses 2016: ICRS-10 Materials Research Forum LLC
Materials Research Proceedings **2** (2016) 295-298 doi: http://dx.doi.org/10.21741/9781945291173-50

How Precise can be the Residual Stress Determined by X-Ray Diffraction? A summary of the Possibilities and Limits

Eckehard Mueller

University of Applied Sciences Bochum, Lennershofstr. 140, D- 44801 Bochum, Germany

eckehard.mueller@hs-bochum.de

Keywords: Calibration, Residual Stress Measurements, Round Robin Tests

Abstract. Many springs are shot peened and the quality of shot peening is essential for the fatigue life. Today the determination is often done via x-ray diffraction. The lattice distance is measured and out of this information the residual stress is determined (and not directly measured). For this kind of measurement an absolute measurement is not available. The only way is to calibrate it in some way. It is shown how precise measurements today are in relation to different x-ray diffractometers and a specimen must be designed to get something like a usable calibration sample. The difference between statistical and systematic errors is shown and the consequences of these errors are discussed.

Introduction

Today the determination of residual stresses for many products is a common procedure, e.g. to prove the efficiency of the shot peening process or other hardening processes. Mostly it is done with the help of the x-ray diffraction method, because it is fast and not so expensive. The demands of the automotive industry concerning the accuracy and the number of measurements are still increasing. The question is whether precise measurements can be even performed. Here, round robin tests are reported designed to calibrate such a x-ray-diffractometer.

Basics of stress determination by x-rays

One popular method to determine the residual stresses in springs is the x-ray method. The idea is the measurement of the lattice distance within a solid or spring steel. The basic method is called Bragg reflection. A detailed description of the method will be found in the literature: [1;2]. A very brief summary is given here. X-rays with the wavelength λ are sent under certain angles Ψ to the surface normal and the diffraction angle 2θ with the maximum intensity is determined. The following equation can then be used:

$$\varepsilon = (1 + \nu) / E * \sigma \sin^2 \psi - \nu / E * (\sigma_{11} + \sigma_{22}) \tag{1}$$

From the measured reflection angle a lattice spacing $D = \lambda / (2 * \sin \theta)$ is determined and is compared with the lattice spacing D_0 without any stress ($\varepsilon = (D - D_0)/D$). (E is the Young's modulus, ν is the Poisson's ration, $\sigma_{11} + \sigma_{22}$ are stresses in the main direction on the surface)

The main aspect is that the stress is not measured directly. The lattice parameter is measured at differentangles Ψ and a slope m is calculated that depends on material constants and the stress thus:

$$\sigma \ (m = (1 + \nu) / E * \sigma) \tag{2}$$

Out of this equation the stress σ can be calculated or determined. These considerations show that it is better to speak about determination of residual stress instead of measurement.

Errors of measurements

Introduction

Every physical measurement has at least two errors: a statistical (or reading) error and a systematical error, which depends on of the equipment. Both errors will be discussed in the following section.

Statistical error

By measuring with more Ψ angles or by measuring each angle several times the statistical error can be reduced, if there is a linear dependence. Today this kind of error can be reduced in the range of 1 % with some efforts.

Systematic error

Surface error

This x-ray measurement is influenced of the roughness of the surface. For instance, for shot peened surfaces a lower compressive residual stress is usually determined, because there is a relaxation of the residual stresses at the tops of the "surface mountain". It can be more than 25 % less than the compressive residual stress in the surface [3;4]. The consequence is that in the "*guideline of measuring residual stresses of shot-peened springs by x-ray diffraction*" [5] it was suggested that a measurement at 100 μm depth or more gives a more reliable value.

Machine error

The main systematical error is in the machine error. The whole x-ray diffractometer must be aligned. That means the path of the x-rays through the apertures to the detector is in a straight line and must correspond with the detection equipment. Minimal deviations can give results, which show more and less residual stresses. The author has see this several times, because he has access to five diffractometers of the same type. No equipment is available to gauge such a machine like the International Prototype Metre.

The only possibility is to organize round robin tests to optimize the accuracy, which is described in the next section.

Calibration of an x-ray-diffractometer

Basics

The long-term stability of an x-ray diffractometer can be monitored using a sample, which has (high) residual stresses. If the sample is stored at normal constant room temperature and no surface corrosion is possible, it can be used many years. This situation is unsatisfactory, because you do not know the systematical deviation of the machine.

Round robin tests

General demands

Round robin tests are useful if you have lots of members to get a smaller error of the mean value, which is calculated. The other demand is to have a long-term stability of the sample if it is handled under adequate conditions. The shorter the experiment last the better it is.

General overview of the round robin tests

In the last 10 years several round robin tests have been made on spring steels. The following table shows the different tests in an overview.

The round robin tests with round material (wire) show dependence between the value of the residual stress and the measuring spot diameter. These experiments are useful in another way, but not for the considerations to get an exact value for calibration. The further the value is away from zero MPa the more meaningful is of the result. The last round robin test fulfils all of the demands, which gives high confidence to the results. There were about 30 participants within a short time. Many measurements at different laboratories in Europe were done on a compressive residual stress sample, which has a high stability.

Materials Research Proceedings **2** (2016) 394-298 doi: http://dx.doi.org/10.21741/9781945291173-50

Table 1: different round robin tests

year	sample(s)	organizer	mean value	reference
2006 – 2009	flat	GKN, Mr. Lietzau	-350 MPa	[6[
2006 – 2011	flat/round	Rigaku, Mr. Yokohama	-800 MP	[7]
2013	round	VDFI, Prof. Mueller	-650 MPa	[8]
2012 – 2014	flat	GKN, Mr. Lietzau	+550/-1120 MPa	[9[

Demands on the sample

The sample should not have curvatures at the surface and the roughness should be as low as possible (see section below). On the other hand, the sample should have a high residual stress to be far away from the zero point. The consequence is that the tensile strength of the material must be high. If one takes all aspects into consideration spring steel, which is deep rolled, fulfills the demands. It is also easy to produce.

Materials for normal leaf springs were taken and after hardening the material was quenched to get a tensile strength R_m = 1500 MPa. The surface was ground to remove the decarburization, which is always on the surface. To be sure, at least 1 mm was removed. Afterwards an area of 55 mm * 55 mm was deep rolled in a meandering pattern like shown in fig. 1. Deep rolling at this high tensile strength gives low roughness combined with high compressive residual stresses at the surface.

Practical realization

A practical realization is shown in fig. 2. Two samples were deep rolled with an HG6-tool (from Ecoroll). The track distance Δx was optimized to Δx = 0.15 mm [10]. The diameter of the ball was 6 mm and the pressure was 100 bar. The roughness R_z is between 5 and 12 µm. Perpendicular to the rolling track you produce compressive residual stresses σ of nearly $\sigma = 2/3 * R_m$, which means around 900 MPa.

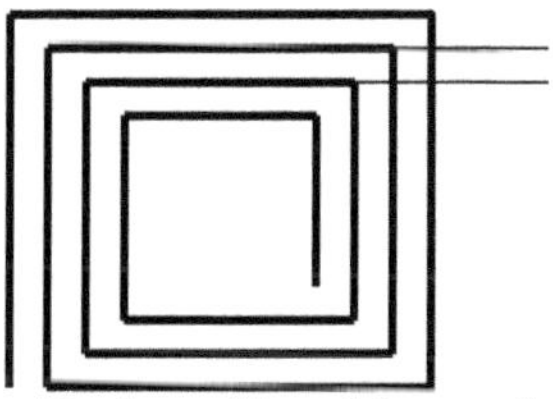

Fig. 1: tracking of the sample

Fig. 2: realization of the samples

Mr. Lietzau of the company GKN (Germany) organized several times a great round robin test in Europe. He uses a sample of the same structure, which was also made by the author. About 30 different laboratories measured the sample within less than two years. The result is that a sample is available with a known compressive residual stress σ = -1120 MPa +/- 43 MPa [9]. A deviation of more than 100 MPa from the average was seen at some labs.

The samples shown in fig. 2 were produced from this "mother sample". They have a residual stress of about σ = -925 MPa with nearly the same absolute error. These samples can be used to calibrate x- ray measuring equipment [11].

Conclusions

When comparing residual stresses from different labs, one must keep in mind that there may be great differences measuring the same objects (e.g. springs). To minimize the variation a calibration sample that has been used in round robin tests is useful. Laboratories have to make their own samples to monitor the long- term stability. Today an absolute calibration of an x-ray diffractometer is not possible and the measurements have a systematic uncertainty of at least 5 %.

In many (delivery) specifications very small errors are claimed, which are in no relation to the systematic uncertainty of 50 MPa respectively 5 % . One way to solve the problem is to organize round robin tests with a huge number of participants.

Acknowledgements

I thank Mr. Lietzau of GKN for his great efforts to organize and evaluate this big round robin test. He is permanently working to improve the uncertainties of the calibration x-ray measuring equipment.

References

[1] Noyan I. C. u. J. B. Cohen: Residual Stress, Springer Verlag, New York, 1987. http://dx.doi.org/10.1007/978-1-4613-9570-6

[2] SAE-Paper: Residual Stress Measurements by X-Ray Diffraction, No.: SAE J784a, Warrendale, 1971.

[3] Müller, E.: Messung von Lastspannungen bei verschiedenen Oberflächenrauhigkeiten mit Hilfe eines Röntgendiffraktometers, HTM 48 (1993) 1/93, pp. 50 – 52.

[4] Müller, E.: The Difficulty of Measuring the Residual Stress at Shot Peened Surfaces, MFN (Metal Finishing News), Vol.11, Nov. 2010, pp. 60 – 61.

[5] VDFI (German Spring Association): Guideline 8501: Messung von Eigenspannungen durch Röntgenografie an kugelgestrahlten Federn, Verband der deutschen Federnindustrie, Hagen (Germany), Febr. 2012.

[6] Lietzau, J. Ringversuch Eigenspannungen, talk given at the Sitzung des AWT FA13 „Eigenspannungen", Freiburg, (Germany) 29.04.2008

[7] Yokohama, R.: X-ray Residual Stress Evaluation of Springs, Rigaku Cooperation, talk given in Donnguan (China), June 2013.

[8] Müller, E.: Ringversuch – Eigenspannungen, Verband der deutschen Federnindustrie, talk given at the Jahreshauptversammlung, Leipzig (Germany) , May 2014.

[9] Lietzau, J.: Materials Engineering Report No. 18_L13167, released 22 Oct. 2014, GKN Driveline Int. GmbH, Lohmar, 2014.

[10] Müller, E.: Evolution of the residual stresses by stress rolling, in V. Schulze u. A. Nikul-Lari (Editor): Shot Peening and other Mechanical Surface Treatments, IITT-International, Noisy-le-Grand, 2005, pp. 436– 441.

[11] Müller, E.: Die Messung von Eigenspannungen mittels Röntgendiffraktion und die Sinnhaftigkeit der Fehlernangabe, in Borsutzki, B und Moninger, G (Editor): Fortschritte in der Werkstoffprüfung für Forschung und Praxis, Verlag Stahleisen, Düsseldorf, 2015, pp. 215 – 219.

Residual Stresses 2016: ICRS-10 Materials Research Forum LLC
Materials Research Proceedings **2** (2016) 299-304 doi: http://dx.doi.org/10.21741/9781945291173-51

Challenges in the Calculation of Residual Stresses in Thick-walled Components

Jakob Klassen[1, a*], Thomas Nitschke-Pagel[1,b] and Klaus Dilger[1,c]

[1]Langer Kamp 8, 38106 Braunschweig, Germany

[a]j.klassen@tu-braunschweig.de, [b]t.pagel@tu-braunschweig.de, [c]k.dilger@tu-braunschweig.de

Keywords: Residual Stresses, Numerical Welding Simulation, Multi-Layer Welds

Abstract. Numerical simulations of welding induced residual stresses and distortion is still a challenging task, especially with regard to thick-walled structures. These structures are mostly welded with high numbers of layers. Usually adjacent welds influence each other. Therefore, simple 2D calculations become unsuitable, thus, leading to the necessity of 3D transient calculations. In turn 3D multi-layer welding simulations require large finite element models. To account for steep temperature gradients in the weld and the heat affected zone each layer has to be modeled using rather fine elements. Thus, numerical models become very large regarding the degrees of freedom. Significant simplifications are common practice, although it is known that these might cause inaccurate or even wrong results. The current paper presents 3D transient finite element simulations of welding residual stresses and distortions in multi-layer welds of 29mm thick-walled components of a construction steel grade S355N. Furthermore, several modelling simplification approaches were applied. During welding temperature profiles were recorded to validate numerical models. After each layer residual stress measurements were carried out by means of X-ray diffraction (XRD). Furthermore, some weld fill levels were investigated by neutron diffraction (ND). This allowed the comparison of real through-thickness residual stress distributions with numerical simulations.

Introduction

Residual stresses are known to influence fatigue strength of welded components. Especially welding residual stresses in site joints of larger structures are of interest already in design state. However, these stresses are hard to determine. One of the reasons can be seen in difficulties of accessibility and another one in the lack of measuring equipment availability. And even if one is able to measure residual stresses in the areas of interest, which are in the weld itself or in the heat affected zone, those measurements are limited to the near surface residual stresses. Through thickness residual stress states remain unknown. As numerical welding simulations are not limited to those kinds of restrictions, designers of welded structures are interested in using FEM for residual stress calculations since the beginning of the last century, e.g. [1]. However, numerical simulations of welding induced residual stresses and distortion is still a challenging task, especially with regard to thick-walled structures. These structures are mostly welded with high numbers of layers. Furthermore, adjacent welds influence each other and make a simple 2D calculation unsuitable by means of calculated result quality, thus leading to the necessity of 3D transient calculations. However, 3D multi-layer welding simulations require large finite element models with rather fine geometric discretization to account for steep temperature gradients in the weld and the heat affected zones. Thus, numerical models become very large by means of degrees of freedom. These large numerical models again result in high computational times. To reduce calculation times simplifications are common practice. Examples of which are thermal cycles (TC) instead of transient calculations or lumping several weld beads to larger layers. However, these simplifications cannot be verified through experimental testing. Nevertheless, these approaches are often used although it is known that simplified calculations of residual stresses might cause inaccurate or even wrong results.

Materials Research Proceedings **2** (2016) 299-304 doi: http://dx.doi.org/10.21741/9781945291173-51

As stated above, one example for simplifications in numerical simulations of multi-layer welding is the lumping technique. This technique has been often applied to 2D modelling [2], [3]. Although it is often enough stated that its application on 3D would not be easy [4], lumping still is sometimes applied to 3D multi-layer welding. It is noted here that numerical modelling is somewhat different for distortion and residual stress calculations. Some approaches lead to accurate distortions while residual stresses do not agree with measured results.

The presented paper intends to show some results of an ongoing research project. The focus of this research is to investigate feasibility of lumping technique on transient 3D numerical residual stress simulations. A 29mm thick-walled specimen of a general construction steel (German steel grade S355) with a V-groove was welded in experimental investigations. Welding experiments were accompanied by temperature measurements and layer-wise residual stress determinations by means of X-ray diffraction. Furthermore, measurements were carried out at the E3 instrument at Helmholtz-Zentrum Berlin using neutron diffraction for through-thickness residual stress state determination. These measurements were needed for validating numerical models and to verify calculated results.

The comparison between the full numerical models and the experiment reveals close agreement. However, comparison of the full numerical model with simplified approaches showed partially rather large discrepancies, although temperature profiles used for validation were comparable. At this state of the work it can be concluded that lumping is applicable to multi-layer welds by means of calculated distortions. However, caution must be used when applying this technique on numerical residual stress simulations.

Experimental set-up

Welding experiments were carried out on several butt welded 29mm thick steel plates with a V-groove (see Fig. 1). The overall specimen's dimensions were 350mm x 350mm x 29mm. Welding parameters (current, voltage, wire feed and travel speed) were set equal for each weld bead. This simply allows variations of weld build-up in numerical modelling and keeping energy input in balance.

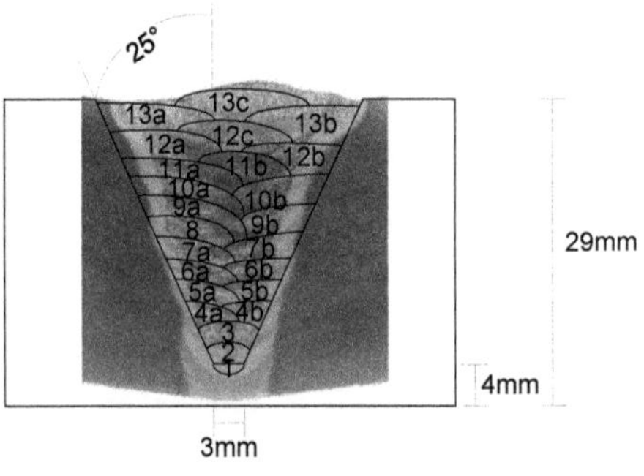

Fig. 1: Macrosection of the specimen after final passes and weld geometry from the welding procedure specification including weld bead numbering

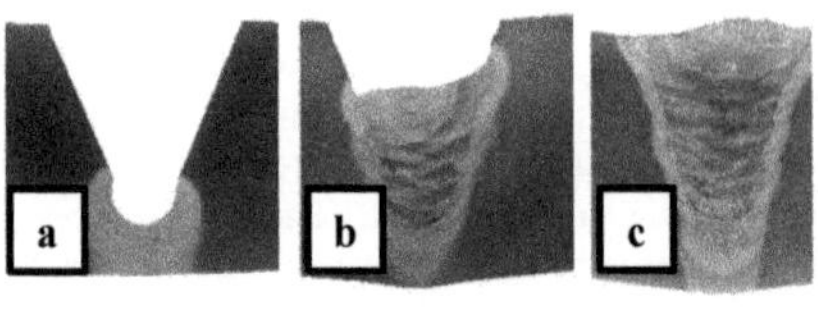

Fig. 2: Macrosections of different weld fill levels. After root pass (a), half weld fill level (b), after final capping passes (c)

Macrosections were taken after each pass for most precise in layers modelling (Fig. 2). Furthermore, each weld pass was followed by displacement measurements (compare Fig. 4) as well as near surface residual stress determinations by means of X-ray diffraction (XRD) (compare Fig. 3). Diffraction lines ({211}-patterns) were obtained with CrKα-radiation using a 2mm collimator.

Materials Research Proceedings **2** (2016) 299-304 doi: http://dx.doi.org/10.21741/9781945291173-51

Afterwards residual stresses were calculated with the $\sin^2\psi$-method as described in [5]. Through thickness residual stress measurements by means of neutron diffraction (ND) were carried out at the E3 instrument at Helmholtz-Zentrum Berlin. Since beam time is limited and specimen's thickness require high measuring times, only two different weld fill levels were measured so far (b and c in Fig. 2).

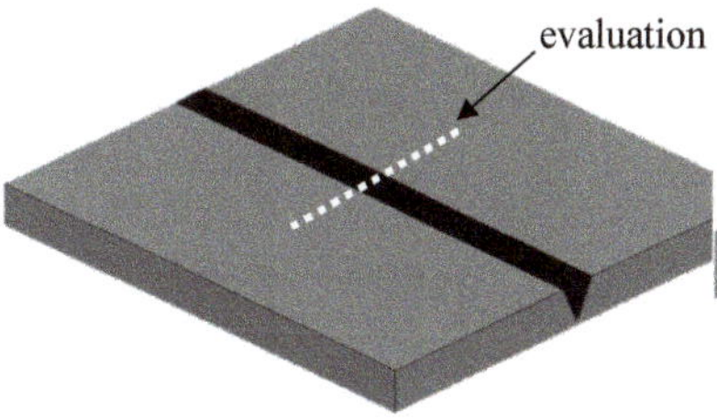

Fig. 3: Measuring and evaluation path for XRD and FEM

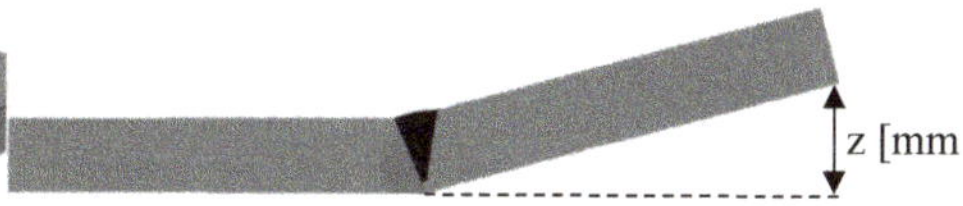

Fig. 4: Measuring location for out of plane distortion

Numerical welding simulations

Finite element analyses (FEA) were carried out using the commercially available welding simulation software SYSWELD. Phase transformation as well as temperature-dependent plasticity was taken into account. Material data was taken from software's data base. This data have been used before and have shown satisfactory results. However, the aim of this work was to investigate relative differences in calculated results through modelling variations. Therefore, not much effort was put into generating steel batch dependent material data.

In addition to the fully transient 3D simulations based on experimentally obtained data two simplifications were carried out, which are lumping several weld beads to larger layers as well as applied thermal cycle technique (TC). In the latter transient and therefore time consuming calculations are avoided. Instead, a temperature profile is imposed on all elements of one bead at the same time. Thus, influences of transient effects (e.g. heat transfer or gap opening in front of flux) remain unconsidered. Each numerical model and its FE mesh are shown in Fig. 5. This part of the investigation has been carried out on a numerically basis only, using a model of half the length to save calculation times.

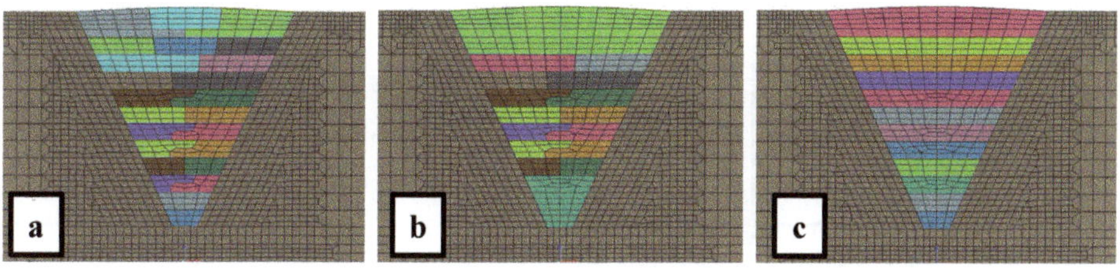

Fig. 5: Weld build-up and FE mesh in numerical models. Fully transient model with moving heat source and thermal cycle approach model based on macrosections from experiments (a); simplified by means of lumping the first and the last three layers (b); simplified by means of lumping in layers (c)

Materials Research Proceedings **2** (2016) 299-304 doi: http://dx.doi.org/10.21741/9781945291173-51

Results

Fully transient calculations with moving heat sources are time consuming. Still, these calculations were done first to validate the numerical model by means of defined boundary conditions and used material data etc.. As can be seen in the following (Fig. 6 through Fig. 9) results of numerical welding simulation and experiments correspond well. Slight deviations as in Fig. 7 can be explained by geometrical imperfections and discrepancies in some material data. Latter was set equal for weld and parent material for simplification reasons. Even through-thickness residual stresses and out of plane distortions correspond well. Standard deviation of measured results was ±30 MPa at maximum. However, standard deviations are not shown in diagrams for clarity reasons.

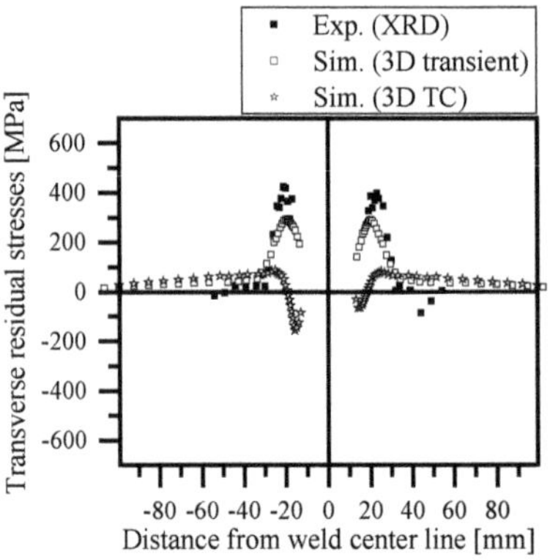

Fig. 6: Comparison of transverse residual stresses on top surface perpendicular to the weld in the middle of the specimen

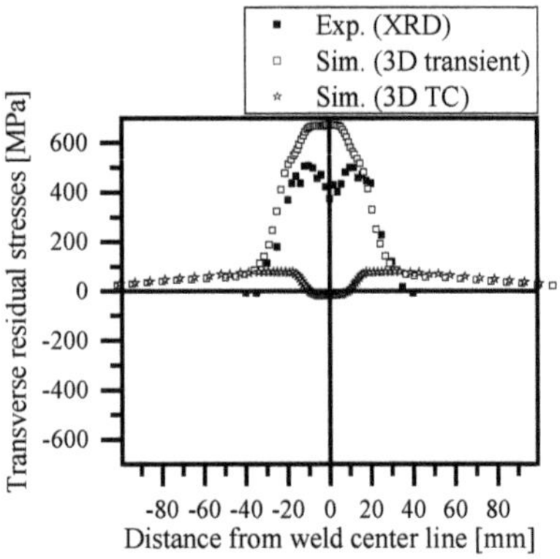

Fig. 7: Comparison of transverse residual stresses on bottom surface perpendicular to the weld in the middle of the specimen

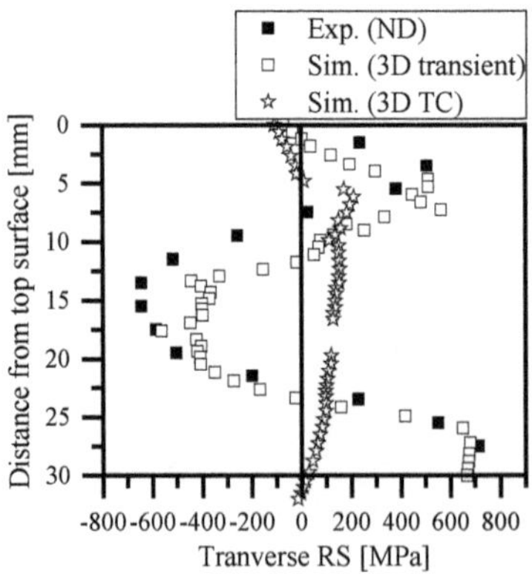

Fig. 8: Comparison of calculated and measured through-thickness residual stresses in transverse direction in the middle of the weld.

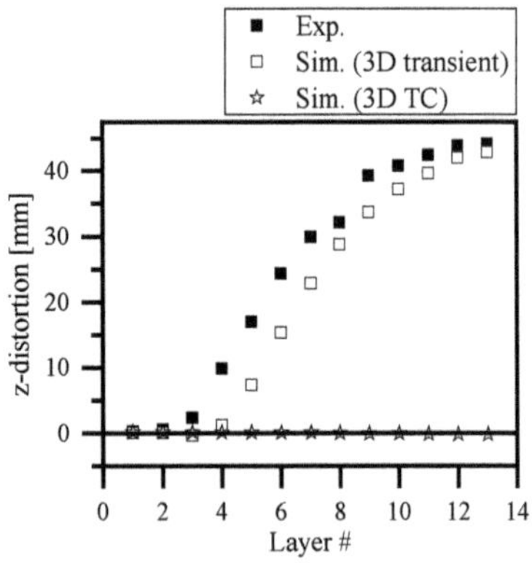

Fig. 9: Comparison of calculated and measured out of plane deformation

Materials Research Proceedings **2** (2016) 299-304 doi: http://dx.doi.org/10.21741/9781945291173-51

The applied thermal cycle approach (TC) is not suitable for simplifying this numerical model. Although there is a significant reduction in CPU hours, the results are not trustworthy. Neither by means of residual stresses nor calculated displacement, not even in direction. Similar findings were published by [6] and explained by significant differences in heat input and accumulated shrinking.

Further simplification approaches were investigated with smaller models and on numerical basis only. One reason for this is the limitation in actual welding processes regarding energy input and therefore a lack in experimentally obtainable data for validation. Another reason is calculation times. Since it should be sufficient to compare qualitative differences in applied simplifications, smaller models were set up for this purpose.

Results of applied lumping compared with fully transient calculations are shown in the following figures. As can be seen in Fig. 10 the out of plane deformation is comparable for lumping the first and last layers. Increasing lumping for all beads of each layer led to strong deviations. However, through-thickness residual stresses in transverse direction to the weld show significant discrepancy (Fig. 11) in sign and quantitative values for each model. The same applies to residual stresses in other directions as well as the calculated phase proportions (not shown here).

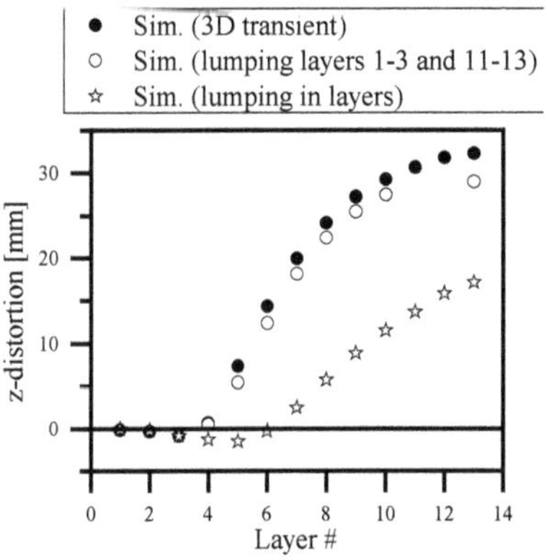

Fig. 10: Comparison of calculated out of plane deformation in fully transient model and models with applied lumping

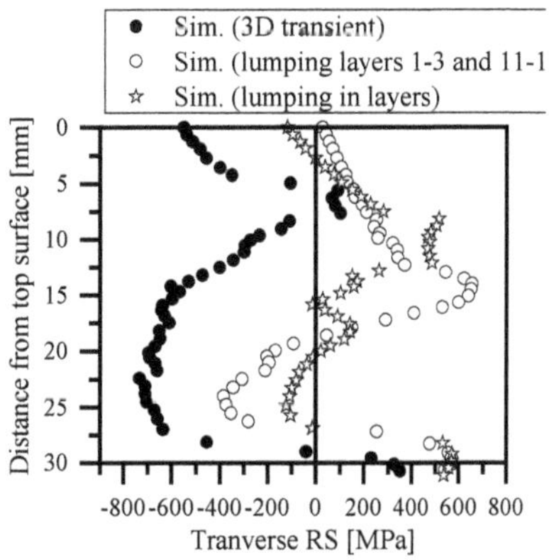

Fig. 11: Comparison of calculated through-thickness transverse residual stresses in the middle of the weld in fully transient model and models with applied lumping

Discussion

It was shown that residual stresses may be calculated correctly even for multi-pass welds. However, it is noted that fully transient welding simulations with moving heat sources are needed for accurate prediction of residual stress distributions. Similar findings based on other numerical simplification approaches were published by [7]. Furthermore, it can be found in literature that simplifications by means of lumping or applying thermal cycles is sufficient for predicting welding induced distortion. This, however, has to be strictly limited to validated models and never should be expanded on or mistaken for residual stress calculations.

Although reasons for false results through different simplification approaches are evident, they still are tempting and used rather often. On the one hand, experimental data for validation is hard and costly to obtain. On the other hand, these data is not even needed if simplifications are applied, since it is just not possible to weld in real life as it is in finite element analyses.

Materials Research Proceedings **2** (2016) 299-304 doi: http://dx.doi.org/10.21741/9781945291173-51

It is noted that presented approaches were limited to the same energy input over all states. Energy balance was to keep as a premise. It may be possible to adjust energy input or weld pool geometry in simplified numerical modeling to reach higher precision levels. However, this would not only mean a large number of iteration steps, but also would be a trial and error approach. This, however, is not an expedient task. If welding simulation is to be used in industrial practice, numerical engineers will have to deal with data from their shops or welders on site. And these data will mostly be limited to current, voltage and torch travel speed. Therefore, the overall aim of numerical welding simulation should be providing solutions not only fast but also based on very little data.

Acknowledgement
We thank HZB for the allocation of neutron radiation beam time. Special thanks to Dr. Boin for his support during beam time at HZB. We also thank Ms. Lepper for her contribution to numerical modelling.

References

[1] G. Bierett, „Versuche zur Ermittlung der Schrumpfspannungen in geschweißten Stumpfnahtverbindungen,“ *VDI Zeitschrift des Vereins Deutscher Ingenieure,* Bd. 78, pp. 709-715, 1934.

[2] Y. Ueda, K. Iida, M. Saito und A. Okamoto, „Finite Element Modeling and Residual Stress Calculation for Multi-pass Single Welded Joint between a Plate and Penetrating Pipe,“ *Modeling of Casting, Welding and Advanced Soldification Processes V; The Minerals, Metals and Materials Society,* pp. 219-227, 1991.

[3] Y. Ueda und K. Nakacho, „Simplifying Methods for Analysis of Transient and Residual Stresses and Deformations due to Multi-pass Welding,“ *Transactions of Joining and Welding Research Institute, Osaka University,* pp. 95-103, 31 März 1982.

[4] J. Hong, C.-L. Tsai und P. Dong, „Assessment of numerical procedures for residual stress analysis of multipass welds,“ *Welding journal,* Bd. 77, Nr. 9, pp. 372-382, 1998.

[5] M. P. Macherauch E., „Das $\sin^2\psi$-Verfahren der röntgenografischen Spannungsmessung,“ *Zeitschrift für angewandte Physik,* Bd. 13, 1961.

[6] J. Dike, C. Cadden, R. Corderman, C. Schultz und M. McAninch, „Finite Element Modeling of Multipass GMA Welds in Steel Plates,“ in *Proceedings of the 4th International Conference on Trends in Welding Research*, Tennessee, 1995.

[7] H. Serizawa, S. Nakamura, K. Kanbe, Y. Fujita, S. Asai und H. Murakawa, „Numerical analysis of deformation in multi-pass circumferential TIG welding with narrow gap,“ *Welding in the World,* Nr. 57, pp. 615-623, 10 Mai 2013.

Residual Stresses 2016: ICRS-10 Materials Research Forum LLC
Materials Research Proceedings **2** (2016) 305-310 doi: http://dx.doi.org/10.21741/9781945291173-52

Residual Stress Relaxation in Welded Steel Joints – an Experimentally-based Model

Jonas Hensel[1,a *], Thomas Nitschke-Pagel[1,b] and Klaus Dilger[1,c]

[1]Institute of Joining and Welding, TU Braunschweig, Langer Kamp 8, 38106 Braunschweig, Germany

[a]j.hensel@tu-braunschweig.de, [b]t.pagel@tu-braunschweig.de, [c]k.dilger@tu-braunschweig.de

* corresponding author

Keywords: Residual Stresses, Residual Stress Relaxation, Welding, Fatigue, Steel

Abstract. Residual stresses may affect the fatigue strength of welded components significantly. Structural design concepts for fatigue loaded welds do not account for real residual stress conditions but rather generally estimate high tensile residual stresses. The assumption of high tensile residual stresses in current engineering practice is resulting in over-conservative designs. The consideration of real residual stress conditions in the design process is one of the major objectives in current research on structural engineering. In order to achieve this objective, one must be able to describe the residual stress generation due to manufacturing and the relaxation of residual stresses during component life time. However, nowadays it is not practical to describe the relaxation process by means of numerical or analytical methods. This work describes an experimentally-based model for the estimation of the stabilized residual stresses in welded steels. The model is capable of describing residual stress relaxation depending on the initial residual stresses, the load magnitudes and the material strength. The model is based on XRD residual stress measurements during the fatigue life of typical welded joints. The samples used here are longitudinal fillet welded gussets made from low-carbon high-strength construction steels S355NL (yield strength 360 MPa) and S960QL (yield strength 960 MPa). Finally, the model is extended to butt welded joints using experimental data from the literature.

Introduction

Residual stress effects in fatigue design of welded steels. The fatigue strength of welded components is of major concern for component safety and durability. In engineering practice fatigue strength is mainly examined by means of the nominal stress approach. Speaking generally this method compares expected fatigue loads in the net section of the weld details to a reference design S-N curve [1, 2]. The reference S-N curves are provided by technical standards or design recommendations and are specific to certain weld details such as un-welded base metal, butt welds, cruciform joints or longitudinal fillet welded gussets. Each design S-N curves relates the expected fatigue load to a specific number of load cycles until this load can be born by the component without

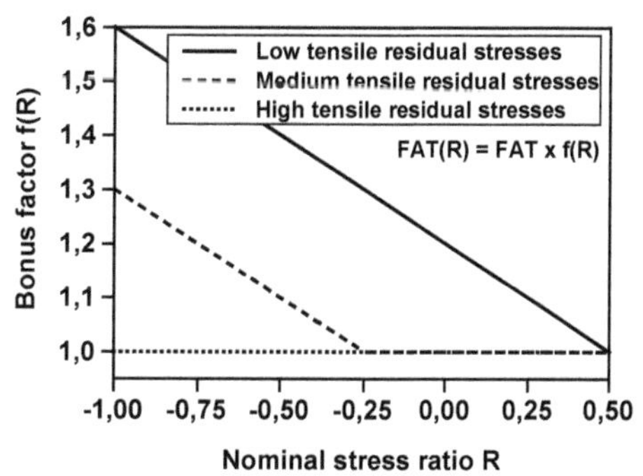

Fig. 1 Effect of tensile mean stresses on the design fatigue strength for welded steels in dependence of the residual stress conditions [1]

failure. Codes for structural design assume conservatively high tensile residual stresses resulting from the entire manufacturing process including welding and mounting are present [3]. This results in mean stress independency of the design S-N curves if real residual stress conditions are unknown.

However, post-weld treatment methods such as thermal stress relieve are capable of reducing residual stresses leading to an increase of fatigue strength. Thus the IIW-recommendations for fatigue design of welded structures allow for the consideration of the mean stress effect in case of proven low or medium tensile residual stresses. This can be executed by means of a bonus factor manipulating the design fatigue strength of a weld detail based on the applied stress ratio, Fig. 1. One of the major uncertainties is how users of the design code can judge whether the proven residual stresses in the component of interest are "high" or "low". Consequently residual stresses and their effects are often conservatively over-estimated leading to un-economical structural design.

The reference fatigue strengths for different weld details given by above mentioned design codes are usually based on experimental fatigue testing of small scale specimens with typically low residual stresses. The lack of residual stresses on the laboratory scale is compensated by testing at high tensile mean stresses respectively positive stress ratios $R = \sigma_{min}/\sigma_{max} = 0.5$ or $R = 0$ [1, 2]. The applied mean stresses are supposed to reflect tensile residual stresses in large scale components conservatively.

In terms of fatigue design these codes equate residual stresses and load mean stresses, although residual stresses may be subject to change during component lifetime. Thus residual stresses may be assumed too high leading to the fatigue resistance being under-estimated accordingly. This hidden potential of fatigue strength could be used in the future in order to accomplish a more economical design of welded structures.

Residual stress relaxation in cyclically loaded welded steels. Speaking generally, residual stresses in steels are limited by the yield strength of the material. Residual stresses are hence degraded under mechanical loading as far as the sum of residual stresses and load stresses theoretically exceeds the yield strength. In case of fatigue loaded welds with high tensile residual stresses the material's yield strength is equal to the maximum stress of the fatigue loading [3]. It was pointed out that residual stresses far below the yield strength can have severe influence on the fatigue strength especially in case of high strength steels because of their stability under service loading [4].

The general mechanisms of residual stress relaxation were described earlier and classified in four groups [5]. These four cases illustrate residual stress relaxation according to the static yield strength, the cyclic yield strength and according to combinations of both, Fig. 2. Case one describes the situation of stable residual stresses which can be observed if both residual stresses and maximum load stresses are low. Case two describes continuous residual stress relaxation according to the cyclic yield strength. Here residual stress relaxation occurs due to cyclic softening of the material. Cases three and four describe residual stress relaxation according to the static yield strength during initial loading. Additionally case four includes cyclic softening which leads to continuous cyclic residual stress degradation as well. However, residual stress degradation in welded construction steels is usually governed by case three. Case four applies to welded and heat treated high strength steels with martensite and bainite microstructure (respectively the heat affected zone of ferritic steels) but is usually of second order compared to case three [6, 7]. Further it was shown that the combined static and cyclic residual stress relaxation is approximately finished after N = 10,000 load cycles [6]. It has been shown to be practical to consider the maximum respectively minimum

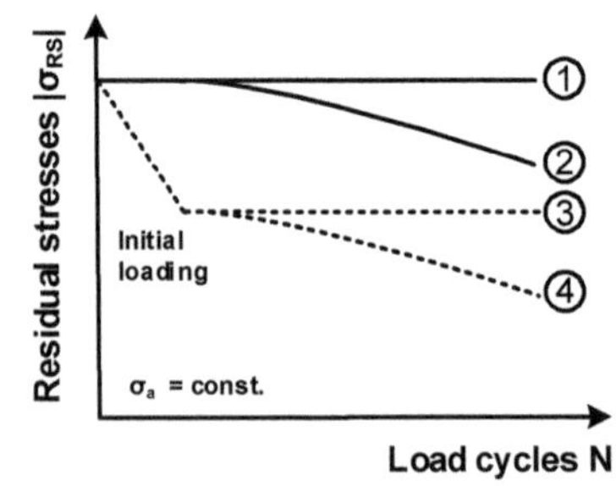

Fig. 2 Residual stress degradation due to cyclic loading according to Vöhringer [5]

load stress during a certain load cycle for the consideration of tensile and compressive residual stresses respectively. Additionally residual stresses are mostly considered in the loading direction perpendicular to the fatigue crack growth direction. All results from residual stress measurements shown here refer to the residual stress component in the loading direction.

Residual stress relaxation in welds can be observed due to stress concentration at weld notches or simply due to high maximum stresses. One example for such behavior of residual stresses in butt-welded joints under static and cyclic loading is given in Fig. 3 [8]. Samples were loaded incrementally statically and cyclically and residual stresses were determined after each load step. Static loading led to residual stress degradation in both steel grades according to case three described by *Vöhringer*.

Cyclic constant amplitude loading until N = 10,000 load cycles led to higher residual stress degradation. Residual stresses after N = 10,000 load cycles were degraded depending on the magnitude of the initial residual stresses and the maximum load stresses. Both materials S355J2 and S690Q showed similar behavior in terms of cyclic residual stress relaxation. Not shown here is that residual stresses had stabilized at this stage of cyclic loading.

Vöhringer's case three applies only partially to the cyclically loaded butt-welds due to cyclic plasticity effects at the weld toe. The increase of residual stress relaxation under cyclic loading is explained by the application of full load cycles (compared to single tension loading in the static load case) and due to cyclic softening of the material. However, in engineering practice it is normally not practical to establish the cyclic mechanical properties of a certain material. Consequently the designers demand for a model that is capable of estimating the stabilized residual stresses as a function of the base material strength and the applied load stresses. Although residual stress relaxation in general is quite well understood, more experimental data is needed to derive such an engineering model.

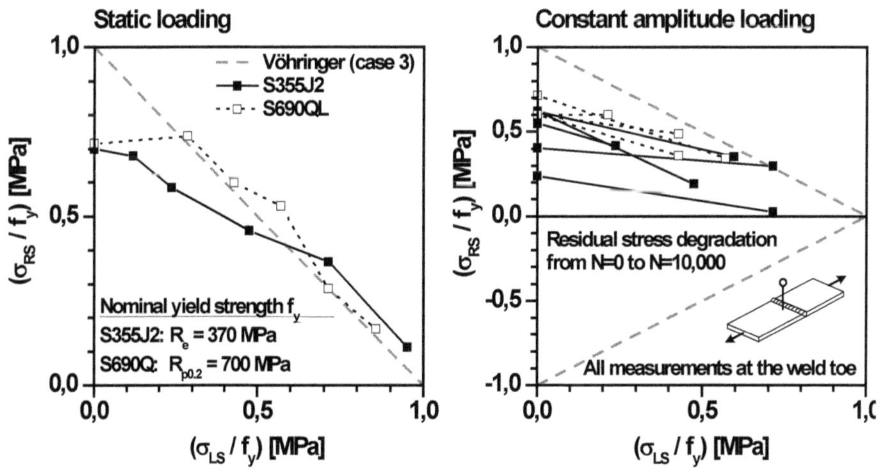

Fig. 3 Static and cyclic residual stress relaxation at the weld toe of butt-welded steel joints [8]. Nominal load stresses σ_{LS} reflect the maximum applied stress in the net section (in case of cyclic loading: Mean stress plus stress amplitude)

Additional experimental investigations on residual stress relaxation in welded steels

Linear welds such as butt-welds show relatively low stress concentration while another group of welded joints show rather high stress concentration. Accordingly these weld types are characterized by higher plasticity effects at the weld notch due to local load stress increase. One of such weld types with higher stress concentration (kt ≈ 3 at a notch radius ρ_k = 1 mm and sheet thickness t = 10 mm) is the longitudinal fillet welded gusset with relatively high tensile residual stresses in the as-welded condition, Fig. 4. This sample type was used here for further investigations on residual stress relaxation.

Samples were made from low-alloyed construction steels with yield strengths of f_y = 360 MPa (S355NL) and f_y = 1000 MPa (S960QL). The one-layered fillet welds were produced using general metal arc (GMA) welding with solid wire electrodes of matching strengths. The welding parameters were chosen in order to ensure high safety against cold cracking. The high strength material S960QL was pre-heated to 100 °C accordingly.

Residual stresses were determined by means of X-ray diffraction (XRD) and the $\sin^2\Psi$-method. Diffraction lines of the {211}-patterns were obtained with Cr-radiation. The irradiated area was controlled using a collimator with a diameter of 2 mm. Residual stresses were measured at the location of crack initiation under fatigue loading and determined initially "as-welded" as well as after repeated cyclic loading. As discussed above residual stresses were measured after 10,000 load cycles to establish the stabilized residual stresses. The initial residual stresses were controlled by heat input and post-weld treatment methods and samples with both compressive and tensile residual stresses at the weld notch were chosen for these investigations.

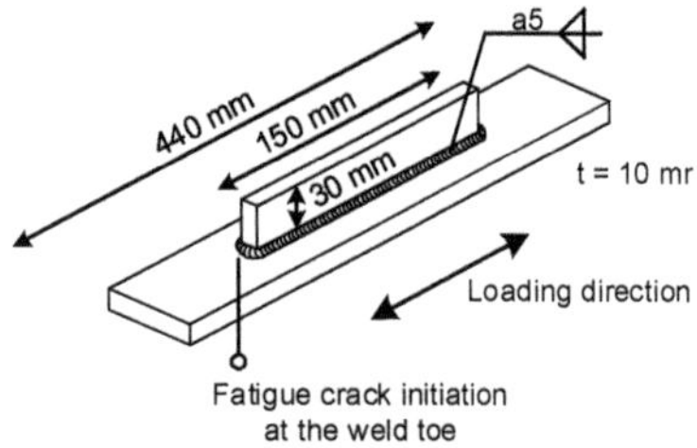

Fig. 4 Specimen type used: Longitudinal fillet welded gusset

Residual stress relaxation

Samples of both steel grades were loaded statically in tension. The load was increased incrementally and residual stresses were measured after each load step on the unclamped samples. It can be seen from Fig. 5 that residual stresses in S355N were degraded continously although the load stresses were comparably small. Specimens made from S960Q showed a similar tendency up to 70 % of the yield strength. At the highest load level considerably high compressive residual were generated in S960Q whereas this was not observed in S355N. Case three by Vöhringer applies not to this test series which can be explained with the high stress concentration at the location of residual stress measurements. Stress increase at the weld toe led to higher plasticity effects than in case of butt-welded samples.

Cyclic loading led to residual stress relaxation both of initial tensile and compressive residual stresses. The residual stress relaxation at comparable load levels was higher than in case of static loading. This is explained by the same effects as mentioned for butt-welded joints. Samples of S960Q generally showed higher residual stress stability which is caused by the very low level of initial residual stresses. Post-weld treated samples did not show different behavior in terms of residual stress relaxation than as-welded samples. Plasticity effects leading to residual stress relaxation are higher at higher levels of residual and load stresses.

Materials Research Proceedings **2** (2016) 305-310 doi: http://dx.doi.org/10.21741/9781945291173-52

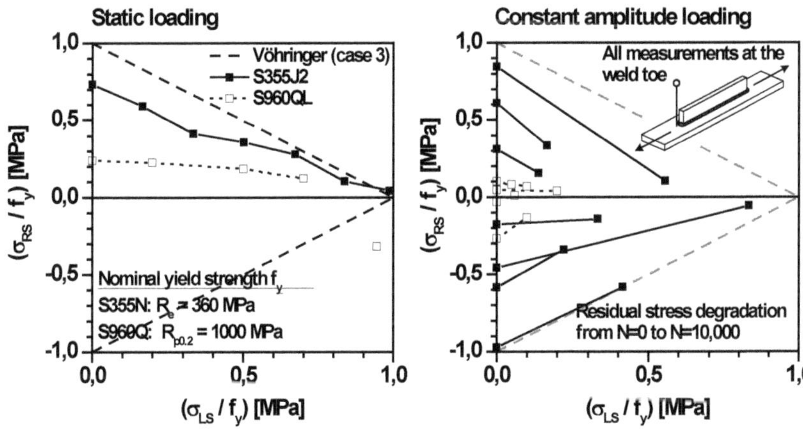

Fig. 5 Static and cyclic residual stress relaxation at the weld toe of fillet-welded longitudinal stiffeners. Nominal load stresses σ_{LS} reflect the maximum applied stress of the same sign as the initial residual stresses in the net section (in case of cyclic loading: Mean stress plus stress amplitude (tensile initial residual stresses) respectively mean stress minus stress amplitude (compressive initial residual stresses))

Engineering model

The authors propose an engineering model for the estimation of residual stress relaxation based on the experimental data, Fig. 6. This model reflects the demonstrated effects of cyclic loading on the residual stresses at the weld notch for butt-welded and fillet welded steels. The stabilized residual stresses $\sigma_{RS,N=10,000}$ can be estimated based on the yield strength of the material f_y, the initial residual stresses $\sigma_{RS,N=0}$ and the highest load stresses σ_{LS}. The highest load stresses σ_{LS} reflect the maximum and minimum stress during fatigue loading depending on the sign of the initial residual stresses (tensile residual stresses: $\sigma_{LS}=\sigma_{max}$; compressive residual stresses: $\sigma_{LS}=\sigma_{min}$), Eq. 1.

$$\frac{\sigma_{RS,N=10,000}}{f_y} = -\frac{\sigma_{RS,N=0}}{f_y} \cdot \left|\frac{\sigma_{LS}}{f_y}\right| + \frac{\sigma_{RS,N=0}}{f_y}. \qquad (1)$$

This approach is based on *Vöhringer*'s case four. It is assumed that cyclic softening leads to residual stress degradation depending on the magnitude of the initial residual stresses. Residual stresses are degraded fully at load levels as high as the static yield strength of the metal. Low residual stresses are predicted to be more stable under mechanical loading.

This model is designed to over-predict stabilized residual stresses on purpose. The reason for this is that mainly tensile residual stresses are of interest for structural engineers. In this matter, over-predicted tensile residual stresses are conservative for the fatigue design. However, this model is not capable of predicting the generation of residual stresses due to local high strains as could be seen in case of longitudinal stiffeners made of S960Q. Further it should not be used without proof of correctness in general and more specific for the prediction of compressive residual stress relaxation.

Materials Research Proceedings **2** (2016) 305-310 doi: http://dx.doi.org/10.21741/9781945291173 52

Outlook

This model was derived and tested for the fatigue life prediction of welded longitudinal stiffeners (more on this can be found here [9]). It has shown that the prediction of stabilized residual stresses can be used to correct for the effective stress ratio under consideration of residual stresses and load mean stresses The application of the bonus factor concept according to IIW-recommendations and thus the entire fatigue design becomes now more effective since the bonus factor is no longer depending on the subjective judgment of the residual stress state. It becomes rather possible to account quantitatively for the stabilized residual stresses.

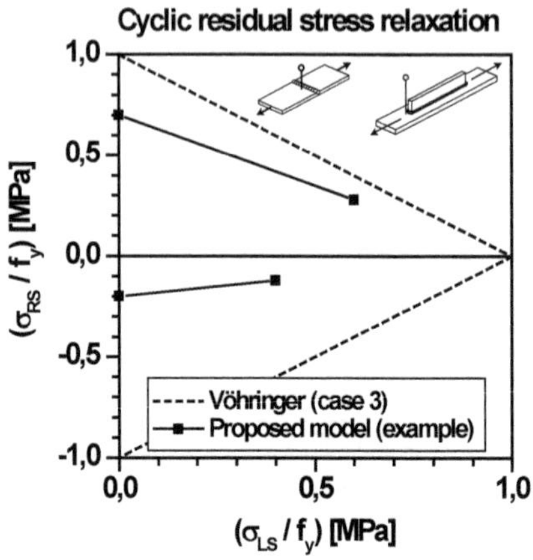

Fig. 6 Proposed model for estimation of stabilized residual stresses in welds

Acknowledgements

The authors thank the German Research Foundation (DFG) for funding. This work was part of the TP-G2 of the DFG-AIF research cluster IBESS (Integrale Bruchmechanische Ermittlung der Schwingfestigkeit von Schweißverbindungen).

References

[1] A. Hobbacher, *Recommendations for Fatigue Design of Welded Joints and Components,* New York: Welding Research Council, 2009.

[2] *DIN EN 1993-1 Eurocode 3: Bemessung und Konstruktion von Stahlbauten,* Berlin: Beuth Verlag, 2010.

[3] T. Gurney, Fatigue of Welded Structures, London: Cambridge University Press, 1968.

[4] H. Wohlfahrt, "Einfluss von Eigenspannungen und Mittelspannungen auf die Dauerschwingfestigkeit," in *Dauerfestigkeit und Zeitfestigkeit, VDI-Bericht 661*, Düsseldorf, VDI-Verlag, 1988, pp. 99-127.

[5] O. Vöhringer and H. Wohlfahrt, "Abbau von Eigenspannungen," in *Eigenspannungen - Entstehung-Messung-Bewertung*, Oberursel, DGM-Informationsgesellschaft, 1983, pp. 144-156.

[6] T. Nitschke-Pagel, Eigenspannungen und Schwingfestigkeitsverhalten geschweißter Feinkornbaustähle. Dissertation TU Braunschweig, Clausthal-Zellerfeld: Papierflieger, 1995.

[7] H. Wohlfahrt, "Zum Eigenspannungsabbau bei der Schwingbeanspruchung von Stählen," *Härtereitechnische Mitteilungen,* no. 28, pp. 288-293, 1973.

[8] M. Farajian, Stability and Relaxation of Welding Residual Stresses. Dissertation TU Braunschweig, Aachen: Shaker Verlag, 2011.

[9] J. Hensel, T. Nitschke-Pagel and K. Dilger, "XIII-2642-16: Engineering model for the quantitative consideration of residual stresses in fatigue design of welded components," in *IIW-Annual Assambly*, Melbourne, 2016.

Residual Stresses 2016: ICRS-10 Materials Research Forum LLC
Materials Research Proceedings **2** (2016) 311-316 doi: http://dx.doi.org/10.21741/9781945291173-53

Investigations of the Residual Stresses and Surface Integrity Generated by a Novel Mechanical Surface Strengthening

Dennise Tanoko Ardi[1, a*], Wang Wei[1,b], Iain Parr[2,c], Goetz Feldmann[3,d], Ampara Aramcharoen[4,e], Chow Cher Wong[4,f]

[1]Advanced Remanufacturing and Technology Centre, 3 Cleantech Loop #01-01, Singapore 637143, Singapore

[2]Rolls-Royce Plc, ELT-10, PO Box 31, Derby DE24 8BJ, United Kingdom

[3]Rolls-Royce Deutschland Ltd & Co KG, Hohemarkstraße 60-70, 61440 Oberursel, Germany

[4]Rolls-Royce Singapore, Advanced Technology Centre, 1 Seletar Aerospace Crescent, Singapore 797565, Singapore

[a*]denniseta@artc.a-star.edu.sg, [b] wangwei@artc.a-star.edu.sg, [c] iain.parr@rolls-royce.com, [d] goetz.feldmann@rolls-royce.com, [e] Ampara.aramcharoen@rolls-royce.com, [f] chow.wong@rolls-royce.com

Keywords: Cold Work, Electron Back-Scatter Diffraction, Residual Stress, RR1000, Surface Strengthening, X-Ray Diffraction

Abstract. A novel mechanical surface treatment has been investigated for its ability to introduce compressive residual stresses as well as low cold work and surface roughness to metallic components, all of which are known to contribute to fatigue performance enhancement. Comprehensive evaluation of the surface integrity is therefore crucial for surfaces of load bearing components where fatigue life is a concern. The novel treatment involves submerging a work piece within a vibratory chamber filled with hardened stainless steel media, analogous to the mass finishing process. During the treatment, the work piece's surface is peened and polished simultaneously through the normal and shear stresses generated by impacts between the work piece and steel media. The surface integrity generated by this treatment is intimately related to the processing parameters. This work focuses on measurements of residual stresses and cold work distribution in the near-surface layers as well as surface topography generated at different stages of processing. Such measurements allow for process optimisation as well as a better understanding on the contribution of the different aspects of surface integrity to mechanical performance, particularly fatigue.

Introduction

Residual stresses and surface integrity are known to influence fatigue performance of metallic alloys [1,2]. Tensile residual stresses are known to be undesirable as they promote the propagation of fatigue crack resulting in a fatigue debit (i.e. reduction in life before failure) [1]. Compressive residual stresses on the other hand are desirable for fatigue life enhancement [3,4]. Various surface treatments have therefore been developed with the aim of delivering desirable compressive residual stresses near the free surface where fatigue cracks are typically initiated [5].

The novel surface strengthening being investigated in the current work aims to introduce desirable compressive residual stresses while also reducing surface roughness and the amount of cold work in sub-surface layers. All of these are postulated to generate fatigue performance improvements.

Experimental methods

Materials. The material being investigated in the current work is a nickel based superalloy RR1000 manufactured through a powder metallurgy route (see Table 1 for chemical composition). The alloy is typically utilised in the hot section of aero-engines for its excellent mechanical properties at elevated temperatures.

Materials Research Proceedings **2** (2016) 311-316 doi: http://dx.doi.org/10.21741/9781945291173-53

Table 1: Chemical composition of RR1000 (weight %, balance nickel) [6]

Co	Cr	Mo	Al	Ti	Ta	Hf	C	B	Zr
18.5	15	5	3	3.6	2	0.5	0.027	0.015	0.055

Plain rectangular specimens of RR1000 (80 x 20 x 10 mm) were machined from a forged disc after solution heat treatment and age homogenisation. Heat treatments were applied to the alloy as a means to control the grain size. Coarse (approximately 25-60 μm) grain size microstructures were developed by employing a controlled heat treatment process. Different grain sizes can be developed to optimise the alloy's performance according to its operating conditions. The coarse grain microstructure being investigated here is optimised for its high temperature performance as it is less susceptible to creep deformations. Fly-cutting was employed as the final machining process to generate a surface integrity typically seen on the nickel disc components manufacturing. Three distinct processing routes were applied to the specimens, as summarised in Fig. 1. Brief descriptions of the different processes involved are shown in the following sections.

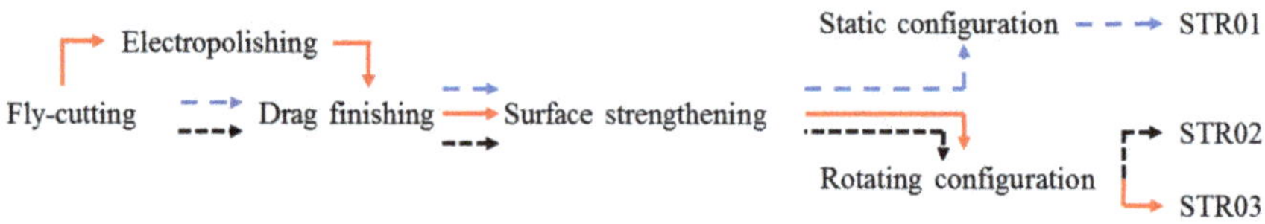

Figure 1: Processing routes applied to the specimens being investigated in this work

Drag finishing. Drag finishing aims to remove burrs generated during fly-cutting. All specimens were drag finished where they were rotated at a controlled speed through a wet mixture of SiC cutting chips/media while being held stationary with respect to the rotary spindle. MultiFinish MFD500 was utilised to complete the drag finishing process, with an immersion depth and rotation speed of 220 mm and 100 RPM respectively. The spindle was programmed to change its rotational direction (clockwise/anti-clockwise) every 5 minutes during the entire processing time of 30 minutes.

Surface strengthening. A novel surface strengthening process that generates desirable surface finish as well as compressive residual stresses has been jointly developed with Rolls-Royce. During the process, the specimens were rigidly fixed to a structure connected to the vibratory trough filled with stainless steel media with either no movement (static configuration) or rotational movement (rotating configuration) allowed during the process. The vibration generated by the trough produced kinetic agitation of the steel media. When these agitated media come into contact with the surface of the specimen, some of the kinetic energy from the media is used to introduce plastic deformation on the specimens' surface. Several possible configurations of the process have been studied previously [7-10] and were found to generate promising results. The surface strengthening process covered in this work was completed using a Walther Trowal TMV175/85VP vibratory trough. Details of the media geometry and surface treatment parameters are Rolls-Royce's proprietary information and are thus omitted from this publication.

Characterisation.

Residual stresses. In all stress measurements, manganese radiation ($\lambda = 2.1031$Å) at 30 kV and 6.7 mA was used to acquire the {311} diffraction peak at 2θ angle of around 156°. This particular lattice plane was found to be representative of the bulk property of the γ/γ' structure. The Young's modulus and Poisson's ratio of the alloy used for the {311} reflection are 220 GPa and 0.300 respectively. All measurements were carried out using Stresstech Xstress G3 X-Ray Diffraction (XRD) equipment with an averaged depth of penetration of approximately 5 μm.

Materials Research Proceedings **2** (2016) 311-316 doi: http://dx.doi.org/10.21741/9781945291173-53

$Sin^2\psi$ measurements were performed at ψ offsets between -36° and 39°. Linear regressions were performed on the acquired d vs $Sin^2\psi$ data to determine the residual stress. Readers are directed to EN15305, which all residual stress measurements performed in this work follow for further details on the methodology. XRD Measurements were carried out in the transverse direction, along the width of the sample.

Layer removal technique is necessary to generate the residual stress profile over the depth of the sample due to limited X-ray penetration. The surface layer was successively removed through electro-polishing. A stylus profilometer was employed to measure the depth of the layer removed through electro-polishing using a step-height measurement. A number of measurements at various layer depths were completed for each specimen in order to observe the variation of residual stress with depth under the free surface. No correction factors were applied for residual stresses measurements in the sub-surface as the layers removed were small relative to the thickness of the sample.

Areal topography. Areal topography measurements were completed using an optical system, Taylor Hobson CCI white light interferometry (WLI), with 20× objective lens and field of view of approximately 900 by 900 µm. The raw data was then filtered using a Robust Gaussian filter with a prescribed cut-off wavelength of 800µm to eliminate waviness from the parameters computation. All parameters were then computed using TalyMap Gold software while ensuring compliance to the prevailing standards for areal topography characterisation, ISO25178.

Cold work. Electron Backscatter Diffraction (EBSD) technique has been used as a tool to estimate cumulative plastic strain using different available metrics based on local misorientations [11] and its usefulness has been demonstrated for evaluation of mechanically treated surfaces using grain orientation spread (GOS) [12]. GOS describes the average deviation in orientation between individual point in a grain and that of the grain to evaluate plastic strain within the material. High GOS values are related to high degree of plastic deformation (i.e. cold worked) and vice versa. Using a statistical analysis, an estimate of the depth of the cold worked or deformed zone can be obtained. The data can also be displayed visually using orientation imaging microscopy (OIM) to show maps of scanned areas.

EBSD is particularly sensitive to the surface topography as the diffraction signal comes from the top few nanometres of the crystal lattice. The cross-sectioned surface was first of all polished using 1200 grit SiC paper before ion-milled to remove 25 µm of material layer using JEOL cross sectional polisher followed by a short surface ion-milling to ensure a good surface topography for EBSD mapping. EBSD mapping was completed over an area of 250 by 250 µm with a step size of 0.3 µm using an Oxford EBSD detector – NordlysMax. The acceleration voltage and tilt angle used during the EBSD mapping were 20 kV and 70° respectively.

Ignoring the effects of low-angle grain boundaries and other microstructural features, full-width at half maximum (FWHM) obtained from the peak broadening can also be utilised to estimate the amount of cold work using an appropriate calibration curve [13].

Results

Surface topography is commonly characterised using the arithmetic roughness average (S_a) parameter. From Fig. 2, it can be observed that the averaged roughness is reduced from 0.63 µm after fly-cutting to 0.43 µm after drag finishing and finally to 0.29 µm after surface strengthening. The regular structure of the topography was also observed to diminish with surface treatment particularly the surface strengthening process. This reflects the random nature of the media collision with the specimen's surface. The final topography is expected to be favourable for the fatigue performance.

Residual stress depth profiles obtained from XRD measurements are shown in Fig. 3. The measurements were completed on specimens at different stages of surface treatment: as-machined (AM), as-machined & drag finished (DF), electropolished (EP), electropolished & drag finished (EPDF) and surface strengthened (STR01-03, refer to Fig. 1).

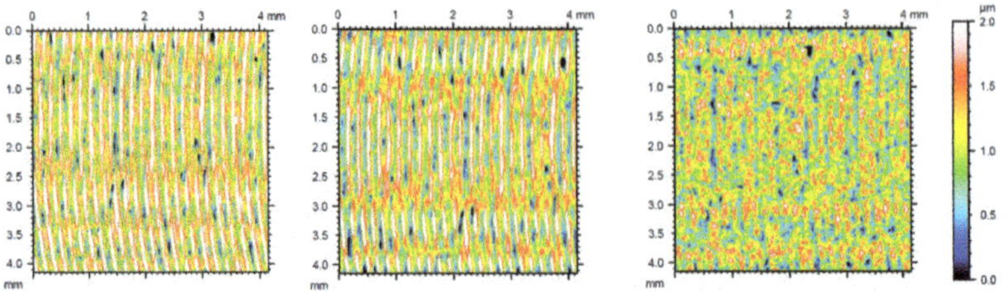

Figure 2: Surface topography changes from as-machined (left) to drag finished (middle) and surface strengthened, STR01 (right)

Fig. 4 shows the grain orientation spread (GOS) maps obtained from the EBSD analysis on coarse grain samples. Regions with low cold work are shown in blue where GOS value is low. It can be seen that the as-machined surface (Fig. 4a) has a certain degree of cold work and further processes (drag finishing (Fig. 4b) and mechanical surface strengthening (Fig. 4d-e)) were not found to result in observable reduction in cold work. Electropolishing (approx. 160 µm of material removal) on the other hand was found to be capable of removing virtually all the cold worked layer. Drag finishing applied to the electropolished surface was not found to introduce appreciable cold work (Fig. 4c). Comparing the GOS maps obtained from surface strengthened specimens, STR03 was found with the least amount of cold work (Fig. 4f). The results shown in the GOS maps are consistent with the full-width at half maximum values obtained from the XRD measurements at the surface (see Table 2).

Table 2: Full-width at half maximum (FWHM) readings and the corresponding estimated cold work obtained from different samples at the surface taken from the transverse direction.

Sample	***AM***	***DF***	***EP***	***EPDF***	***STR01***	***STR02***	***STR03***
FWHM (°)	***6.55± 0.2***	***5.77± 0.1***	***2.72± 0.3***	***4.45± 0.1***	***4.91± 0.2***	***5.77± 0.2***	***4.43± 0.1***
Cold work (%)	***48.5***	***37.7***	***5.6***	***21.8***	***27.0***	***37.7***	***21.6***

Discussions

High tensile stresses observed at the as-machined surface are not unusual for nickel based superalloys as large amount of heat can be accumulated at the surface during the cutting operation due to the alloy's low thermal conductivity. The tensile residual stress layers are compensated by the compressive residual stress layers, extending up to 180 µm deep in the coarse grain example.

Drag finishing only influenced residual stresses at the surface without generating significant impact on the sub-surface residual stresses, perhaps due to the insignificant material removal by the process (< 1 µm). The surface strengthening process was found to induce desirable compressive residual stresses to the material as evident from the residual stress profiles being dominated by compressive stresses after the surface strengthening process.

Electropolishing was found to be capable of removing the tensile stressed layers generated from the fly-cutting. Drag finishing after electropolishing was observed to imbue compressive stresses in the thin layer near the surface (approximately < 20 µm).

Interestingly, the final residual stress profiles after surface strengthening (i.e. STR01-03) are largely similar despite the observed variations in the residual stresses prior to the surface strengthening treatment. The final cold work at the surface on the other hand was observed to be sensitive to the processes prior to the strengthening treatment. When the degree of cold working is high prior to the strengthening treatment, the final cold work after strengthening tend to also be high

Materials Research Proceedings **2** (2016) 311-316 doi: http://dx.doi.org/10.21741/9781945291173-53

and vice versa. Minimal degree of cold work is desired for improved high temperature fatigue performance as high degree of cold work has been linked to increase in the relaxation of compressive residual stresses [14]. Therefore it is important to optimise the processes prior to the strengthening treatment in order to minimise the degree of cold working on the surface.

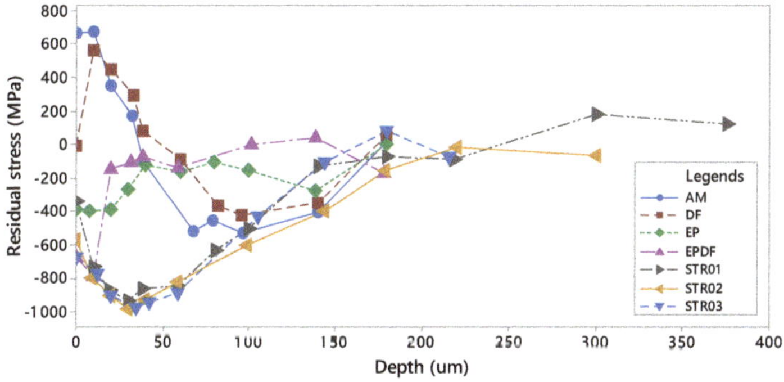

Figure 3: Residual stress depth profiles for coarse grain specimens.

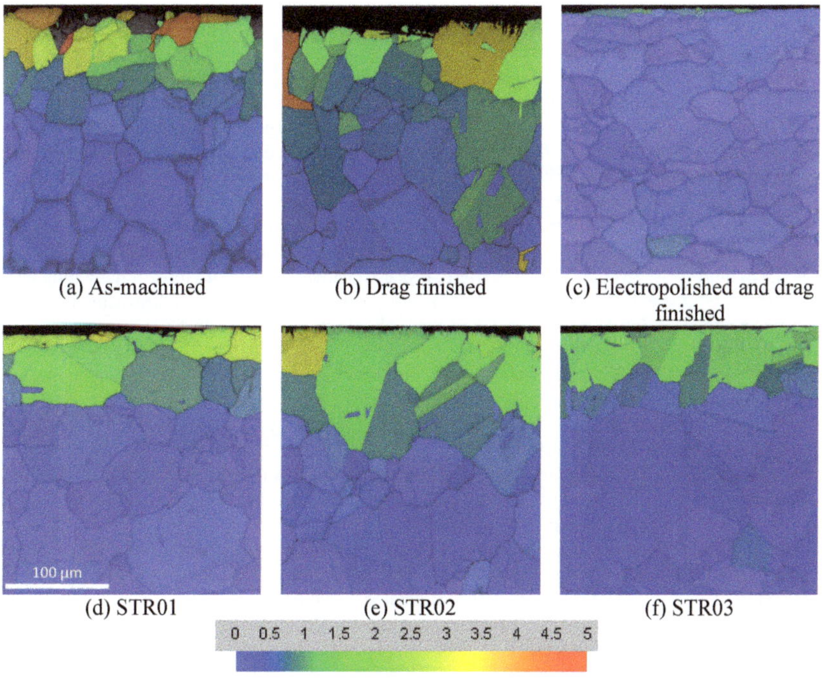

Figure 4: Grain orientation spread (GOS) maps for coarse grain RR1000 specimens. All images were obtained at the same magnification thus the scale bar on (d) is applicable to all maps.

Conclusions

The novel mechanical surface strengthening process was found to be capable of introducing desirable compressive stresses, but it has only a limited influence on the final amount of cold work (based on FWHM). It is therefore crucial to ensure that the cold work is removed prior to the strengthening treatment. A gentle finishing process such as electropolishing was found to be effective for cold worked layers removal. Such processes can potentially be employed before the mechanical strengthening process to generate a final surface with desirable compressive residual stresses whilst minimising the amount of cold work being introduced into the material to optimise its fatigue performance.

References

[1] D. Novovic, R. C. Dewes, D. K. Aspinwall, W. Voice and P. Bowen, The effect of machined topography and integrity on fatigue life, Int. J. Machine Tools & Manufacture 44 (2004) 125-134. http://dx.doi.org/10.1016/j.ijmachtools.2003.10.018
[2] D. T. Ardi, Y. G. Li, K. H. K. Chan, L. Blunt and M. R. Bache, Surface topography and the impact on fatigue performance, Surface Topography Metrology and Properties 3 (2015) no. 1.
[3] L. Wagner, Mechanical surface treatments on titanium, aluminium and magnesium alloys, Materials Science and Engineering A 263 (1999) 210-216. http://dx.doi.org/10.1016/S0921-5093(98)01168-X
[4] K. A. Soady, Life assessment methodologies incorporating shot peening process effects: mechanistics consideration of residual stresses and strain hardening part 1, Materials Science and Technology 29 (2013) no. 6.
[5] V. Schulze, Modern mechanical surface treatment: States, Stability, Effects, Wiley, 2005. http://dx.doi.org/10.1002/3527607811
[6] M. C. Hardy, C. Herbert, J. Kwong, W. Li, D. A. Axinte, A. R. C. Sharman, A. Encinas-Oropesa and P. J. Withers, Characterising the integrity of machined surfaces in a powder nickel alloy used in aircraft engines, Procedia CIRP 13 (2014) 411-416. http://dx.doi.org/10.1016/j.procir.2014.04.070
[7] G. G. Feldmann and T. Haubold, Mechanical surface treatment technologies for improving HCF strength and surface roughness of blisk rotors, Materials Science Forum 768-769 (2013) 510-518. http://dx.doi.org/10.4028/www.scientific.net/MSF.768-769.510
[8] G. Feldmann, W. Wei, T. Haubold and C. C. Wong, Application of vibropeening on aero-engine component, Procedia CIRP 13 (2014) 423-428. http://dx.doi.org/10.1016/j.procir.2014.04.072
[9] M. D. Sangid, J. A. Stori and P. M. Ferriera, Process characterization of vibrostrengthening and application to fatigue enhancement of aluminum aerospace components - Part I. Experimental study of process parameters, Int. J. Adv. Manufacturing Technology 53 (2010) 545-560. http://dx.doi.org/10.1007/s00170-010-2857-2
[10] M. D. Sangid, J. A. Stori and P. M. Ferriera, Process characterization of vibrostrengthening and application to fatigue enhancement of aluminum aerospace components - Part II: Process visualisation and modelling, Int. J. Adv. Manufacturing Technology 53 (2011) 461-575. http://dx.doi.org/10.1007/s00170-010-2858-1
[11] M. Kamaya, A. J. Wilkinson and J. M. Titchmarsh, Quantification of plastic strain of stainless steel and nickel alloy by electron backscatter diffraction, Acta materialia 54 (2006) 539-548. http://dx.doi.org/10.1016/j.actamat.2005.08.046
[12] D.J.Child, G.D.West and R.C.Thomson, Assessment of surface hardening effects from shot peening on a Ni-based alloy using electron backscatter diffraction techniques, Acta Materialia 59 (2011) 4825-4834. http://dx.doi.org/10.1016/j.actamat.2011.04.025
[13] W.Li, P.J.Withers, D.Axinte, M.Preuss and P.Andrews, Residual stresses in face finish turning of high strength nickel-based superalloy, J. Mat. Processing Technology 209 (2009) 4896-4902. http://dx.doi.org/10.1016/j.jmatprotec.2009.01.012
[14] P.S.Prevey, The effect of cold work on thermal stability of residual compression in surface enhanced IN718, Metal Finishing News 6 (2005).

Residual Stresses 2016: ICRS-10 Materials Research Forum LLC
Materials Research Proceedings **2** (2016) 317-322 doi: http://dx.doi.org/10.21741/9781945291173-54

Evaluation of Thickness and Residual Stress of Shallow Surface Regions from Diffraction Profiles

Wulf Pfeiffer[a*] and Eduard Reisacher[b]

Fraunhofer Institute for Mechanics of Materials IWM, Wöhlerstr. 11, 79108 Freiburg, Germany

[a]wulf.pfeiffer@iwm.fraunhofer.de*, [b]eduard.reisacher@iwm.fraunhofer.de

Keywords: Residual Stresses, Stress Gradient, Diffraction Profile, Shot Peening, Ceramic

Abstract. Residual stresses resulting from surface treatment methods like shot peening generally show a more or less steep stress gradient versus depth. X-ray diffraction is the most used method for depth profiling in case of near-surface stress gradients due to the small penetration depth of X-rays resulting in a high depth resolution. When the stress gradient is very steep and the stressed surface layer is in the range of the penetration depth of the X-rays, asymmetric diffraction lines may occur and determination of the diffraction line position has to be done with great care. In case of the stressed surface layer being thinner than the penetration depth of the X-rays a (partial) splitting of the diffraction line may occur which may allow to separately calculate both the mean stresses of the layer and the base material and additionally to determine the thickness of the stressed surface layer. For a shot peened silicon nitride ceramic the evaluation of the mean stress in the thin surface layer, the stress in the underlying base material and the thickness of the stressed surface layer on basis of profile fitting methods is demonstrated with special emphasis on the influence of the method used for determining the diffraction line position.

Introduction

Residual stresses resulting from surface treatment methods like shot peening generally show a more or less steep stress gradient versus depth. X-ray diffraction is the most used method for depth profiling in case of near-surface stress gradients due to the small penetration depth of X-rays resulting in a high depth resolution. In case of steep stress gradients being in the range of the penetration depth of X-rays various methods for calculation of the stress-depth profiles from stress-penetration depth profiles have been presented [1, 2, 3]. Although it is now well known that such stress gradients may result in asymmetric diffraction lines [4, 5, 6] little attention has been applied to the method to be used for the determination of the relevant diffraction angle from the intensity distribution of the diffraction line. Only the center of gravity of the complete diffraction line can be related to the depth inspected by the X-rays with the information being weighted by the exponential decrease of intensity versus depth [4, 7]. In case of using profile fitting methods asymmetric functions and the center of gravity of the fitted diffraction line shall be used.

X-ray investigations on ceramics, shot peened with small shot sizes, show additional effects on the diffraction profiles. When the surface layer influenced by shot peening is significantly smaller than the penetration depth of the X-rays, a significant part of the diffraction profile is built up by the unstressed base material. Due to the generally small widths of diffraction lines of ceramics, a more or less clear separation of the contribution of the surface layer and the base material may be obtained. Using profile fitting techniques may allow to separately calculating the stress states of both the surface layer and the base material as well as the thickness of the surface layer.

Effect of the stress gradients – penetration depths ratios on diffraction profiles

The effect of the stress gradient with respect to the penetration depth is sketched in Fig. 1. If the penetration depth is less than the stressed surface layer (1), an asymmetric diffraction profile may

Materials Research Proceedings **2** (2016) 317-322 doi: http://dx.doi.org/10.21741/9781945291173-54

occur. The location of the broadening of one flank of the diffraction line is depending on whether the near surface stress is compressive or tensile and is more or less continuous. In case of very steep stress gradients with the reversal point of the stress profile being close to the surface (2) and the thickness of the stressed surface layer being significantly smaller than the penetration depth of X-rays, the diffraction profile may show a splitting at one side of the profile. This is the situation handled in this paper.

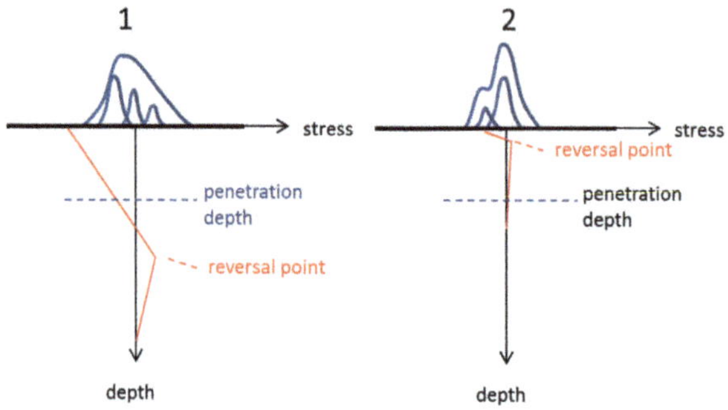

Figure 1: Sketch of diffraction lines resulting from different stress gradients – penetration depths ratios

Experimental details

Material. The material investigated was a commercially-available hot pressed silicon nitride ceramic ("HPSN-Black") fabricated by FCT Ingenieurkeramik GmbH, Germany. The most important material characteristic are a Young's modulus of 326 GPa, Poisson's ratio of 0.268, 4-point bending strength of 900 MPa and fracture toughness of 8 MPa·m$^{1/2}$. Stripes with a length 76 mm, width of 19 mm and thicknesses of 0.9 mm were used.

Shot Peening. The shot peening was carried out with an injection system with gravimetric feeding. The pressurized air and the shot are applied to the jet nozzle in two different tubes. The shot is accelerated in the nozzle. The shots used were tungsten carbide beads with a diameter of 45 µm – 135 µm (mean 90 µm). The peening pressure was 0.3 MPa.

Determination of residual stresses. The diffraction profiles of {411} –lattice planes of β-silicon nitride occurring at 2θ ~ 125° using CrKα-radiation were recorded for 15 ψ-angles between -56.8° and +56.8° using a Bruker D8 in side inclination mode and a Braun position sensitive detector. A X-ray elastic constant $1/2s_2$ of 3.89 GPa^{-1} was applied. The penetration depth from which 64% of the diffracted X-rays arose was 11 µm. A biaxial surface stress state with $\sigma_{33} \equiv 0$ MPa was assumed for fitting the resulting $\sin^2\psi$ distributions of the lattice strains. Fitting and separation of the diffraction profiles, determination of the peak positions and calculation of stresses was done using an enhanced version of the Bruker software "STRESS", originally designed by Fraunhofer IWM.

Determination of layer thickness. From the integral intensities of separated diffraction lines the thickness S_{layer} of the stressed surface layer can be calculated by Eq. 1:

$$S_{layer} = \frac{\ln \frac{I_{layer}(\theta,\psi)}{I_{layer}(\theta,\psi)+I_{substrate}(\theta,\psi)}}{k(\theta,\psi)\cdot\mu} \quad (1)$$

with I_{layer} = integral intensity of the diffraction line belonging to the stressed surface layer, $I_{substrate}$ = integral intensity of the diffraction line belonging to the non-stressed base material (substrate), k = factor counting for the penetration depth and µ = linear absorption coefficient. All these quantities

Materials Research Proceedings **2** (2016) 317-322 doi: http://dx.doi.org/10.21741/9781945291173-54

are dependent on the tilt angle ψ (and Bragg angle θ). Thus, calculations for different tilt angles ψ allow evaluating the scatter of results.

Results

Diffraction profiles. Fig. 2, left shows the {411} – $K\alpha_{1+2}$ diffraction lines of the shot peened silicon nitride recorded at 15 ψ-angles between -56.8° and +56.8°. The asymmetry of the diffraction profiles and an indication of peak splitting in some cases can be seen. In Fig. 2, right, the diffraction profiles recorded at ψ = -56.8° and ψ = 0° are shown. At ψ = -56.8° the slope of the right hand flank is less than the slope of the left hand flank. Together with the observation of a more or less continuous intensity course it can be concluded that the thickness of the stressed surface layer is in the range of the penetration depth (see clause 1, Fig. 1). At ψ = 0° a significant shoulder (peak splitting) at the left hand flank of the diffraction profile can be obtained. This indicates the thickness of the stressed surface layer being less than the penetration depth (see clause 2, Fig. 1).

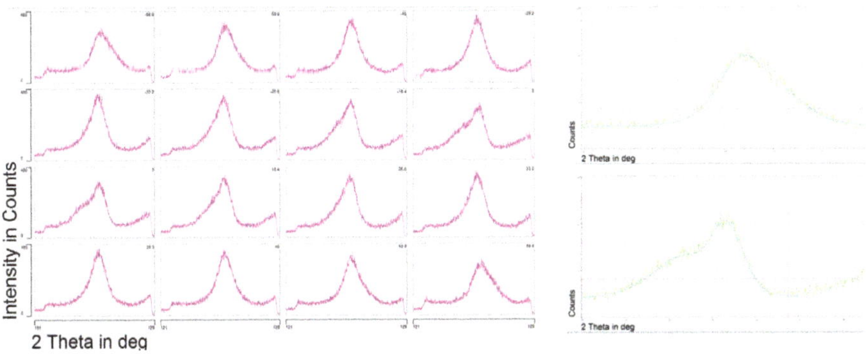

Figure 2: Left: {411} - $K\alpha_{1+2}$ diffraction lines of the shot peened silicon nitride recorded at 15 ψ angles between -56.8° and +56.8°.

Right: Diffraction profile recorded at ψ = -56.8° (top) and ψ = 0° (bottom), respectively, fitted by a Pseudo Voigt function doublet; displayed 2θ-range is 122° - 128°, the vertical lines indicate the range of fitting.

Calculation of residual stresses. Calculation of residual stresses was performed using 3 different methods for the determination of peak positions: (i) the (sliding) center of gravity, (ii) peak fitting using the split Pearson Seven function, (iii) peak splitting using Pseudo Voigt functions.

(i) Center of gravity. A rough check whether the diffraction profiles are influenced by no-ideal diffraction situations, stress gradients and other physically caused circumstances is possible on basis of the sliding-center-of gravity method [1]. In this method the center of gravity is calculated using a gradually increasing intensity threshold. The result of this peak treatment (including background and Rachinger-$K\alpha_2$ subtraction) is shown in Fig. 3.

Remembering that the intensity threshold determines both the location and amount of the diffracting volume this quick check already indicates that at different depths different stress states are present. But the contribution of different depths to the diffraction profile cannot be concluded exclusively on the threshold as the peak intensities may be dominated by highly stressed surface layers or by a large volume with homogeneous stress below the surface. Nevertheless, from the slope of the $\sin^2\psi$-distribution (see Fig. 4), increasing with higher $\sin^2\psi$-values (reduced penetration depths, respectively), a higher amount of stress acting near the surface can be concluded.

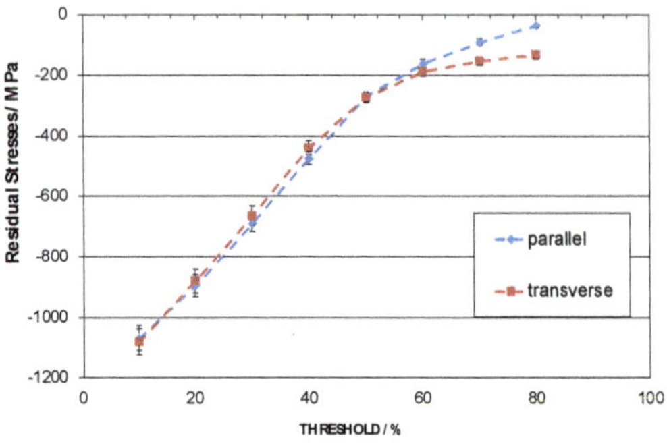

Figure 3: Dependency of the residual stress (parallel and transverse components) on the intensity threshold used for the calculation of the center of gravity of the diffraction line.

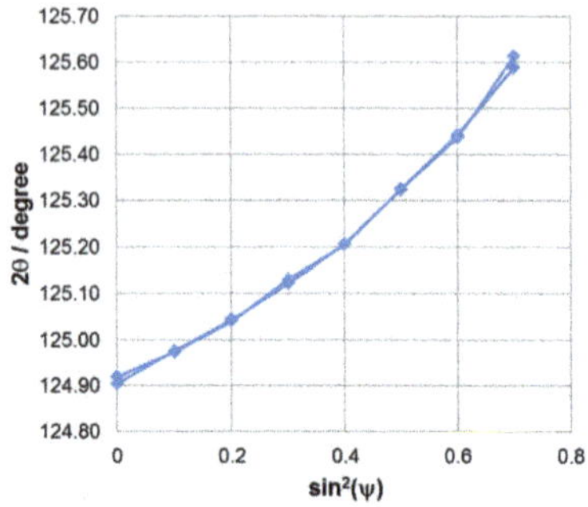

Figure 4: Diffraction angle - $sin^2\psi$-distribution of the shot peened specimen using the center of gravity and an intensity threshold of 10%.

(ii) Split Pearson Seven. Peak fitting using a reasonable mathematical function is one of the most applied methods for determining the diffraction line position. Using more or less of the complete peak a statistically well covered result can be expected. Fitting asymmetric diffraction profiles needs asymmetric functions like the split Pearson Seven function consisting of two slightly shifted Pearson Seven functions applied to the flanks of the diffraction line. This may result in a good peak fit but the physical meaning behind that fit is somewhat questionable when using the 2θ- position at maximum intensity (which is the common procedure of commercial software). The slope of the resulting $\sin^2\psi$-distribution would indicate a tensile stress which would obviously not be expected for a shot peened surface layer. But, this agrees somewhat with the result of the center of gravity method (see Fig. 3) when using a high intensity threshold. In addition, the slope of the $\sin^2\psi$-distribution, decreasing with increasing $\sin^2\psi$ (decreasing penetration depth) indicates a tendency to smaller tensile stresses near the surface. Nevertheless, a stress calculation based on the Split Pearson Seven peak fit would lead to wrong results at all.

(iii) Peak splitting. Basis of this peak treatment was the assumption of shot peening using beads of only 90 µm diameter would affect only a thin surface layer with a more or less stress free base material beneath. This assumption was confirmed qualitatively by the stress calculation using the sliding center of gravity method and the curvature of the $\sin^2\psi$-distribution. The starting parameters for the separation of the diffraction profiles using two independent Pseudo Voigt functions -including the $K\alpha_1$ and $K\alpha_2$ diffraction lines were chosen from the peak with the most significant shoulder in one flank (see Fig. 2).

Materials Research Proceedings **2** (2016) 317-322 doi: http://dx.doi.org/10.21741/9781945291173-54

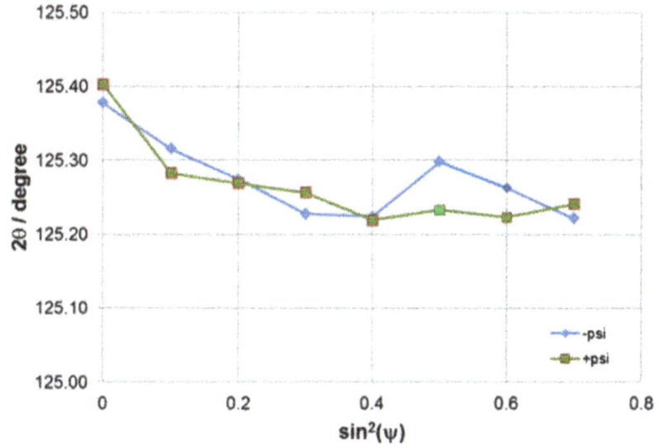

Figure 5: Diffraction angle - sin^2y-distribution of the shot peened specimen using a split Pearson Seven fit. A linear regression of the distribution would indicate a tensile stress of 198 ± 50 MPa.

In Fig. 6 the $\sin^2\psi$-distributions resulting from the peak splitting procedure are shown. Both distributions show a more or less linear behavior indicating that no significant stress gradient could be resolved in the near surface and the base material. From the slopes a high compressive residual stress of 2936 MPa ± 118 MPa can be calculated for the surface layer affected by shot peening; for the base material, not affected by shot peening, small tensile residual stresses of 125 MPa ± 8 MPa are derived. Tensile residual stresses in the base material are needed to compensate compressive residual stresses in the surface layer.

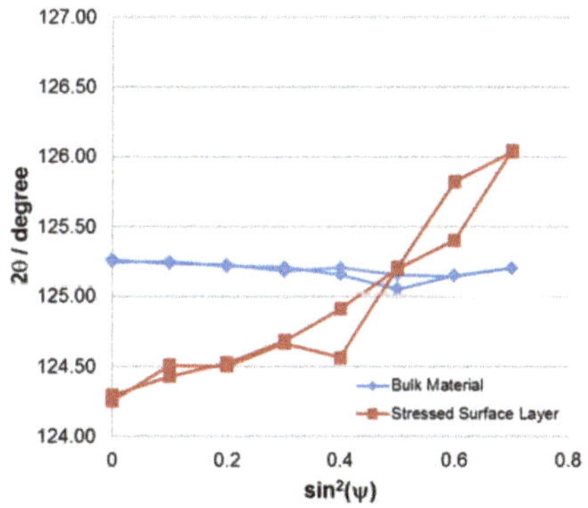

Fig. 6: Diffraction angle - sin^2y-distributions belonging to the stressed surface layer and the bulk material, respectively. The two distributions are derived from the peak splitting procedure.

Thickness of the stressed surface layer

Based on Eq. 1 the thickness of the stressed surface layer was calculated from two measurement directions phi = 0° and 90° (longitudinal and transverse). Due to the symmetry of the stress state resulting from shot peening, both measurements should give comparable values. This is confirmed by the results:

Phi = 0°: Thickness of the stress surface layer = 4.95 ± 1.05 μm
Phi = 90°: Thickness of the stress surface layer = 4.94 ± 0.7 μm

Materials Research Proceedings **2** (2016) 317-322 doi: http://dx.doi.org/10.21741/9781945291173-54

Conclusion

X-ray diffraction measurements were performed on silicon nitride ceramics shot peened with 90 µm shot. The diffraction profiles show some asymmetry and in some cases evidence of peak splitting. The result of residual stress calculation is dramatically dependent on the method used to determine the diffraction line position. Fitting of distinct peaks to the intensity distributions leads to reasonable residual stress results for the surface layer influenced by the shot peening and the underlying base material. In addition, the thickness of the stressed surface layer can be calculated on basis of the intensities of the separated diffraction peaks. The results show, that the shot peening process led to extremely high compressive stresses in a very thin surface layer. The presented method may be applicable to all near surface stress states where the penetration depth is significantly larger than the thickness of the stressed surface layer.

References

[1] C. Genzel, Non-Destructive Analysis of Residual Stress Gradients $\sigma_{ij}(z)$ in the Near-Surface Region by Diffraction Methods – Problems and Attempts at Their Solution, Proc. Of the 5th Intern. Conf. On Residual Stresses (1997) 514-521.

[2] B. Eigenmann, Non-Destructive and Partially Destructive Determination of Residual Stress State with Steep Stress Gradients, Advances in X-Ray Analysis 40 (1997).

[3] M. Härting, A Seminumerical Method to Determine the Depth Profile of the Three Dimensional Residual Stress State with X-Ray Diffraction, Acta Mater., Vol 40, No 4 (1998) 1427-1430. http://dx.doi.org/10.1016/S1359-6454(97)00284-X

[4] W. Pfeiffer, The role of the peak location method in X-ray stress measurement, Proc. of 4th International Conference on Residual Stresses; James, M.R. (Ed.); Society for Experimental Mechanics, Bethel, USA (1994) 148-155.

[5] M. Klaus, Röntgendiffraktometrische Ermittlung tiefenabhängiger Eigenspannungsverteilungen in Dünnschichtsystemen mit komplexem Aufbau. Diss. Tu Berlin (2009).

[6] M. Klaus, et al., Application of energy-dispersive diffraction to the analysis of highly Inhomogeneous residual stress fields in thin film structures, Powder Diffraction 24 [Suppl.1] (2009), S82-S86. http://dx.doi.org/10.1154/1.3134362

[7] P.D. Evenschor, Zur röntgenographischen Ermittlung von Dehnungen beim Vorliegen von Dehnungsgradienten, Z. Metallkde. 74 (1983) 119.

Residual Stresses 2016: ICRS-10 Materials Research Forum LLC
Materials Research Proceedings **2** (2016) 323-328 doi: http://dx.doi.org/10.21741/9781945291173-55

Evaluation of Residual Stress Determinations Conducted with Laser Ablation and Optical Displacement Measurement

Peter Weidmann[1, a *], Giancarlo Pedrini[2, b], Venancio Martínez-García[3, c], Martin Wenzelburger[3], Andreas Killinger[3], Siegfried Schmauder[1], Rainer Gadow[3], Wolfgang Osten[2]

[1]University of Stuttgart, IMWF, Pfaffenwaldring 32, 70569 Stuttgart, Germany

[2]University of Stuttgart, ITO, Pfaffenwaldring 9, 70569 Stuttgart, Germany

[3]University of Stuttgart, IFKB, Allmandring 7b, 70569 Stuttgart, Germany

[a]peter.weidmann@imwf.uni-stuttgart.de, [b]giancarlo.pedrini@ito.uni-stuttgart.de, [c]venancio.martinez@gsame.uni-stuttgart.de

Keywords: Residual Stress, Optical Determination, Laser Ablation

Abstract. The presented work is related to the determination of residual stresses in ceramic coatings, where the material removal is done by laser ablation and the occurring surface deformation is measured with digital holography. Insights into the underlying assumptions and calculations to determine the depth dependent stress values from extensive displacement measurements shall be given. Moreover, the inhomogeneous elastic material properties and deviations from idealized removal geometries (in this case holes and notches) have to be considered. To cover all involved fields, i.e. thermodynamics and continuum mechanics, numerical studies for conventional and laser hole drillings were conducted and the ablation process was simulated. As the laser cannot produce ideal cylindrical or rectangular geometries, appropriate mathematical approximations are applied to choose a representative profile of the removed material. Furthermore, the calculation of stresses based on full field displacement measurements is done with an own program in several depth increments. These calculations are based on numerical calibration data for the elastically inhomogeneous compound of coating and substrate. Especially notch-like forms of material removal, where case sensitive weighting functions are applied for evaluation, can lead to reliable results. The physical experiments were done on aluminium plates with applied alumina layers.

Introduction

The reduction of manufacturing costs and enhancement of product lifetime are central goals for many modern research areas. Especially components which are supposed to withstand aggressive thermal, mechanical or chemical environments can be challenging in their production and maintenance. A reliable way to protect metals from high thermal and abrasive loads, i.e. turbine blades, can be achieved by coating them with suitable ceramics by atmospheric plasma spraying (APS). Hereby, a ceramic powder is injected into a plasma jet, molten and accelerated towards the specimen. During the impact of the liquid particle it is highly deformed and will quench almost immediately due to the comparably low heat capacity which leads to high tensile stresses in the solidified ceramic. When the coating is finished the whole component will bear a non-homogeneous temperature distribution, but even for a uniform temperature the differences in thermal expansion coefficients of substrate and coating will lead to further residual stresses during the cooling-down. While this residual stress cannot be avoided, it is possible to conduct the process in such ways, that a sometimes even beneficial state can be achieved, i.e. a compressive residual stress within the brittle coating. Nevertheless it is of great importance to know the existing stress state of the investigated part in order to ensure high quality products. A profound determination technique for this case is the incremental hole drilling (IHD). Hereby, a mechanical drill removes incrementally material out of the

Materials Research Proceedings **2** (2016) 323-328 doi: http://dx.doi.org/10.21741/9781945291173-55

stress afflicted component. As a new equilibrium is achieved, the area around the hole deforms. This deformation is mostly measured by a strain gauge and with a proper case sensitive calibration the underlying residual stress can be derived from it [1], [2].

The drawbacks of this method are the necessity of mechanical contact with the specimen, the specimen preparation with strain gauges and the overall time-consuming drilling operating. Therefore, a new approach shall be investigated, where the material removal is conducted via laser ablation and the occurring surface displacements will be measured by digital holography (DH).

Boundary conditions and set-up

The majority of investigations were conducted on an aluminium sample (E = 68 GPa, μ = 0.34) with a thickness of 6 mm on which a 200 μm Al_2O_3 coating (E = 166 GPa, μ = 0.23) was applied by APS. The specimens are initially grit blasted in order to enhance the adhesion between substrate and coating, which leads to a preliminary surface near hydrostatic residual compressive stress. Afterwards, the coating is applied and the overall residual stress distribution can be determined after the specimen has cooled down to room temperature (Fig. 1). For this determination a mechanical IHD operation was conducted, where an integral method (IM) evaluation scheme with a case sensitive calibration was applied [1]. Further numerical studies showed, that the high values after 200 μm depth might be mainly related to a plastification process due to the stress intensification around the hole. The relaxation of the coating tends to stress the aluminium 50% over its yield strength, which leads to an overestimation of the residual stress determination.

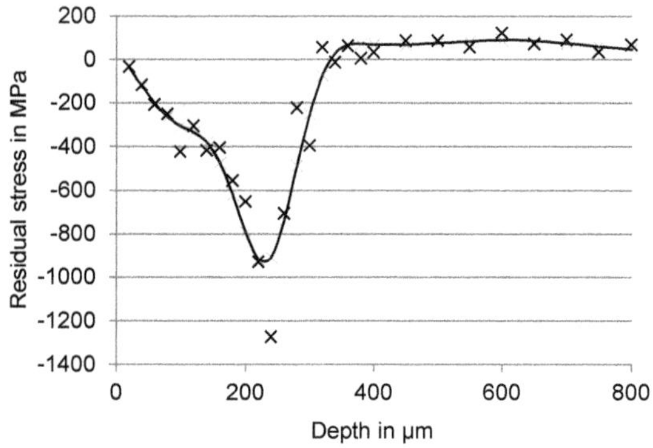

Fig. 1: Residual stress distribution, determined with classical IHD and case sensitive calibration

As the next step, the drilling and displacement measurement shall be done by optical manners. To measure the surface displacement, digital holography was applied. Hereby, green laser light (wavelength 532 nm) is used to illuminate the specimen under at least one direction. The reflected light is overlaid with a reference beam, this leads to an interference pattern which is recorded with a CCD sensor. When the surface changes due to residual stress relaxation, the wavelength of reflected light also changes, which leads to another interference pattern. By overlaying reference- and final-pattern the underlying displacement can be calculated [3].

For the material ablation a SpitLight 600 laser from InnoLas© with a wavelength of 1064 nm, a pulse length of 7 ns, a repetition rate of 20 Hz and an energy of 40 mJ was used. Due to the high energy densities (> 10^6 W/cm²) the subjected material sublimates almost suddenly and a high back pressure builds up which accelerates the ablated material out of the hole. As this thermodynamic process imposes heat, a numerical simulation was conducted in order to investigate possible heating effects [4]. It was shown, that the majority of heat is transported out of the specimen by the

Materials Research Proceedings **2** (2016) 323-328 doi: http://dx.doi.org/10.21741/9781945291173-55

sublimation of material. The resulting heat affected zone is smaller than 10 µm, which is also observed in corresponding micrographs of the holes cross-section (Fig. 2). Therefore, only a neglectable amount of heat is really imposed into the remaining material, which means that further surface strains are hereby not achieved.

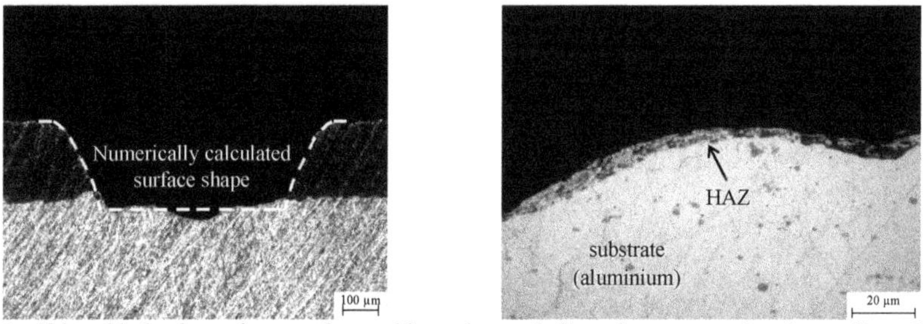

Fig. 2: Ablated hole shape for spot focused laser beam (left) and corresponding heat affected zone (HAZ) [5]

Residual stress evaluation for hole-like structures

The first residual stress determinations were conducted on simple hole-like structures like in Fig. 2. Here, an unshaped laser beam was used for the ablation process and led to a non-cylindrical structure. A well-known procedure to calculate the residual stress field of extensive displacement measurement was introduced by Schajer and Steinzig in 2005 [5], [6]. It relies on the integral method, where the radial profiles of pure shear and hydrostatic stress induced surface displacements are used for calibration. Hence, several finite-element-method (FEM) models have been made to calculate these profiles for the laser hole geometries in several depth steps. Therefore, white-light confocal microscopy images of the surface (*x*-*y*-plane) have been made, in order to evaluate the hole geometry for subsequent increments. Those images were then approximated with a best-fit-algorithm by frustums of a cone aka the equation

$$(x - x_m)^2 + (y - y_m)^2 = [a/h(h - z)]^2, \quad (1)$$

where x_m and y_m are referring to the middle, *a* to the radius and *h* to the frustums height. Therefore, a convenient description of the material removal was achieved and the transfer into a FEM program could be eased. A more detailed description can be found in [7]. Despite the efforts, the resolution and deviation are not as good as the ones for classical IHD (Fig. 3). Yet, the qualitative trend is comparable to the IHD and, therefore, the feasibility of this method could be shown. But to enhance the method, especially the shape of the removed material must be improved.

Residual stress evaluation for improved ablation of notch structures

While the ablation of a simple hole structure doesn't need further equipment, it lacks necessary geometrical accuracy. Therefore, the system was enhanced with a spatial light modulator (SLM), in order to achieve material removal in more sophisticated forms. For hydrostatic residual stress states, like in most APS coatings, a stress determination in one direction is enough. Therefore, a notch was produced by laser ablation with considerable geometrical accuracy in several increments. One increment is achieved by applying approximately 200 pulses over the notch-length within one sequence. The displacement is then measured after every depth increment via digital holography perpendicular to the notch. Rigid body movements are simply filtered by adjusting the outer values regarding their angle and absolute displacements to zero. The basic evaluation scheme relies on the integral method [8]. That means a calibration for the given material and removal geometry has to be

Materials Research Proceedings **2** (2016) 323-328 doi: http://dx.doi.org/10.21741/9781945291173-55

calculated step- and load-wise with proper FEM models in advance. For this reason several confocal microscopy images have been taken after different sequences to find the real notch depths. The achieved notch length was 1.1 mm for a width of 0.15 mm, while the ablation depth per sequence was approximately 6.03 µm. The FEM calibration was done for increments with a thickness of 10 µm.

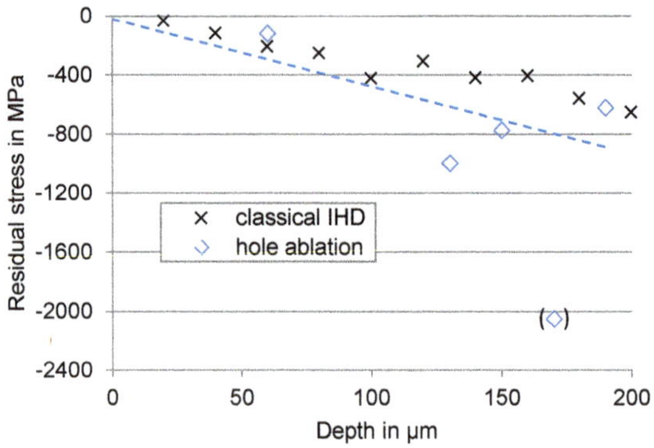

Fig. 3: Residual stress determination with incremental laser ablated hole-like structure

Compared to hole-like ablations, the relieved surface displacements are no longer radial symmetrical, why the formerly described evaluation cannot be applied. Of course, the values close to the notch tend to be the highest, while they gradually decrease towards the outside (Fig. 4). Hence, it is self-evident to weight the displacements according to their height, as for normal distributed absolute errors higher values are more reliable. To find a suitable weighting function the displacement field (along the x- and y-direction) is represented by two functions: one along the parallel side of the notch as a power function

$$f(y) = p_1 \cdot y^{p_2} + p_3 \tag{2}$$

and one in the middle of the notch, parallel to the x-axis as an arc tangent function

$$f(x) = -p_4 \cdot \mathrm{atan}\big(p_5(x - p_6)\big) + p_7. \tag{3}$$

The parameters p_1 to p_7 were identified with the commercial tool Matlab©. With this regression, a weighting distribution can simply be calculated by writing the values of *f(y)* in columns of a matrix which has the size of the measurement field. The same is done for *f(x)* in an equal matrix, but line wise. The product of both matrices leads to a measurement data based weighting distribution *F*, which should be normalized to its maximum value. Its shape is overall similar to the one of Fig. 4, but it is mirrored symmetrical to the x-axis. A weighted mean displacement for all measurement points *i* can then easily be calculated by

$$\overline{U} = \textstyle\sum_i^n [F_i \cdot U_i] \,/\, \sum_i^n [F_i]. \tag{4}$$

The same procedure has to be applied on the calibration fields *K* in order to be able to calculate the residual stress distribution σ with the matrix operation $\overline{K} \cdot \sigma = \overline{U}$. Hence, the displacement- and calibration-fields are described by their mean value over the same area. As it is difficult to ablate an

Materials Research Proceedings **2** (2016) 323-328 doi: http://dx.doi.org/10.21741/9781945291173-55

exactly defined depth, it is reasonable to calculate the calibration coefficients $\overline{K}$ for a given depth by a bivariate interpolation scheme [8]. After the primary residual stress distribution was determined, the results can be regularized with the Tikhonov regularization [9]. To investigate the effects of weighting, a numerical study for an incremental laser notch ablation was done. Therefore, a corresponding FEM model was used, where a residual stress of 200 MPa was applied. The resulting surface displacements were then artificially spread by means of a Gaussian distribution with a standard deviation of 8 nm. The resulting deviation of residual stress was 18 MPa. By using the proposed weighing, this deviation could be lowered to 12 MPa.

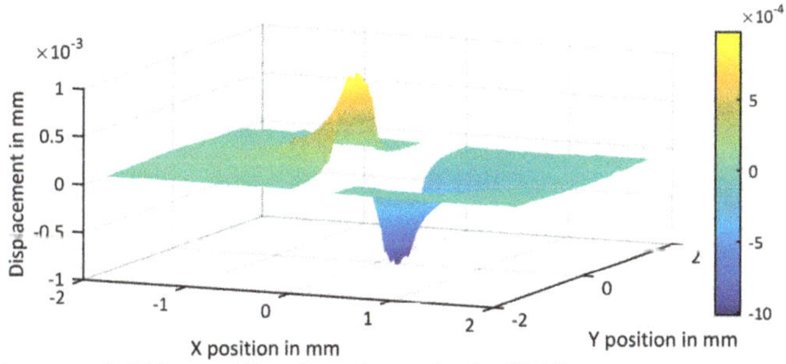

Fig. 4: Displacement field for an ablated notch at a depth of 0.12 mm

In Fig. 5 the results of the residual stress determination with laser ablation material removal are plotted together with the results from the classical incremental hole drilling. The resolution is significantly better than for simple hole ablation (Fig. 3), while the deviation compared to the regularized data is acceptable.

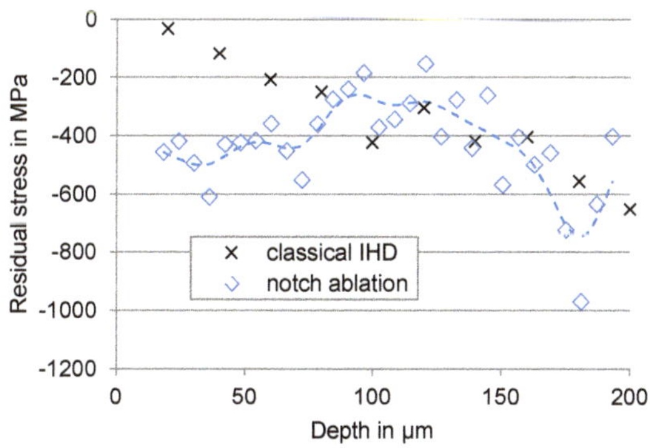

Fig. 5: Residual stress determination with laser notch ablation

Compared to the IHD, the deviation seems higher, which might be related to geometrical inaccuracies and slight depth variations. But solely because of the finer depth steps, small errors lead to bigger notable spreading of results. Noteworthy is also the difference of values up to the depth of 100 µm. While the IHD shows a linearly rising stress, the notch ablation method indicates surface stresses of roughly 500 MPa. A comparative XRD measurement revealed a surface stress of -596 ± 44 MPa. This might indicate a better sensitivity to surface near stresses for the laser notch method.

Conclusions and outlook

First results proof the applicability of laser ablation to relax residual stress, where the resulting surface displacements are measured by digital holography and evaluated according to the integral method. The laser ablation seems to imply only neglectable amounts of heat, while the resulting hole geometry can be approximated by simple geometrical forms and, therefore, calibration values can be obtained numerically. A spot focused laser beam leads to some variations in the holes geometry, but overall axial symmetrical removals can be evaluated by using the approach of [5], [6], even if not ideal cylindrical holes and inhomogeneous specimen are present. Hereby, the results of residual stress profiles show the same qualitative behaviour as IHD determinations, but tend to a higher deviation. The main reason is related to uncertainties of the ablated hole structure.

By using SLM to ablate material in notch-form, the structure of removal is significantly more reliable regarding its dimensions. The extensive displacement measurements can be evaluated with a special integral method based scheme and data dependent weighting functions to reduce the corresponding deviation considerably. Hence, especially surface near stresses could be determined more accurate compared to classical IHD determinations. Now continuing analyses should be performed in order to validate the results further.

Acknowledgement

This work was supported by the German Research Foundation (DFG) under Grant Nos. GA746/10, OS111/37 and SCHM746/120.

References

[1] P. Weidmann, U. Weber, S. Schmauder, and V. Martinez Garcia, "Investigation of Influence Factors on Residual Stress Determination within Coated Surfaces in Consideration of the Differential and Integral Method," *Advanced Materials Research*, vol. 996, pp. 307–312, 2014. http://dx.doi.org/10.4028/www.scientific.net/amr.996.307

[2] M. Wenzelburger, D. Lopez, and R. Gadow, "Methods and application of residual stress analysis on thermally sprayed coatings and layer composites," *Surface and Coatings Technology*, vol. 201, no. 5, pp. 1995–2001, 2006. http://dx.doi.org/10.1016/j.surfcoat.2006.04.040

[3] T. Kreis, *Handbook of Holographic Interferometry*. WILEY-VCH Verlag, 2004. http://dx.doi.org/10.1002/3527604154

[4] P. Weidmann, U. Weber, S. Schmauder, G. Pedrini, and W. Osten, "Numerical calculation of temperature and surface topology during a laser ablation process for ceramic coatings," *Meccanica*, vol. 51, no. 2, pp. 279–289, 2016. http://dx.doi.org/10.1007/s11012-015-0220-2

[5] G. Schajer and T. Rickert, "Incremental Computation Technique for Residual Stress Calculations Using the Integral Method," *Experimental Mechanics*, vol. 51, pp. 1217–1222, 2011. http://dx.doi.org/10.1007/s11340-010-9408-5

[6] G. S. Schajer and M. Steinzig, "Full-field calculation of hole drilling residual stresses from electronic speckle pattern interferometry data," *Experimental Mechanics*, vol. 45, no. 6, pp. 526–532, 2005. http://dx.doi.org/10.1007/BF02427906

[7] G. Pedrini, V. Martinez-Garcia, P. Weidmann, M. Wenzelburger, A. Killinger, U. Weber, S. Schmauder, R. Gadow, and W. Osten, "Residual Stress Analysis of Ceramic Coating by Laser Ablation and Digital Holography," *Experimental Mechanics*, Dec. 2015.

[8] ASTM837-13, *Standard Test Method for Determining Residual Stresses by the Hole-Drilling Strain-Gage Method*, no. E 837–13. ASTM, 2013.

[9] G. S. Schajer, "Measurement of Non-uniform Residual Stresses Using the Hole-Drilling Method. Part II - Practical Application of the Integral Method," *Transaction of the ASME*, vol. 110, pp. 344–349, 1988. http://dx.doi.org/10.1115/1.3226060

Materials Research Proceedings **2** (2016) 329-334 doi: http://dx.doi.org/10.21741/9781945291173-56

Evaluation of the Three-dimensional Welding Residual Stresses Based on the Eigenstran Methodology via X-Ray Measurements

Masaru Ogawa[1, a *] and Takehiro Ishii[2, b]

[1]Faculty of Engineering, Yokohama National University, 79-1, Tokiwadai, Hodogaya-ku, Yokohama 240-8501, Japan

[2]Domestic Preventive Maintenance Group, Mitsubishi Hitachi Power Systems, 1-1, Saiwai-cho 3-chome, Hitachi-shi, Ibaraki 317-8585, Japan

[a]ogawa-masaru-jv@ynu.ac.jp, [b]takehiro_ishii@mhps.com

Keywords: X-Ray Diffraction, Three-dimensional Residual Stresses, Non-destructive, Inverse Problem, Eigenstrain, Weld

Abstract. In order to predict the lifetime of welded structures, it is important to evaluate three-dimensional residual stresses nondestructively. However, X-ray diffraction is useful for measuring residual stresses only on the surface. Neutron diffraction is known to be an effective nondestructive method for measuring residual stresses till a depth of several tens of millimeters. Nevertheless, it is not a technique useful for on-site measurements. This is because neutron diffraction is achievable only within special irradiation facilities. In this paper, the authors have proposed a non-destructive method to estimate three-dimensional residual stresses by using X-ray measurements. In this method, the residual stresses for an entire structure are calculated from eigenstrains by using an elastic FEM (Finite Element Method) analysis. Eigenstrains can be estimated from elastic strains measured through X-ray diffraction using inverse analysis. The relationship between the three-dimensional eigenstrains and the surface elastic strains can be obtained if the Young's modulus, Poisson's ratio, and dimensions of the structure are known. This study aims to demonstrate the effectiveness of the proposed method on an actual butt-welded plate made of stainless steel SUS430. To evaluate the accuracy of this method, the residual stresses estimated using the method was compared with that measured using X-ray diffraction. In order to maximize the estimation accuracy of this inverse analysis, we considered the eigenstrain distributions along both the welding and thickness directions, as well elastic strains measured on the weld metal were used to estimate.

Introduction

It is important to predict crack propagation for observed cracks to prevent fatigue fractures and stress corrosion cracking (SCC) in welded structures. However, it is difficult to estimate crack growth rate without information on welding residual stresses. Three-dimensional welding residual stresses can be simulated qualitatively by using thermal elastic-plastic FEM analysis [1,2]. Nevertheless, it is recommended that each welded structure be quantitatively measured because individual differences in welding residual stress between them can be high. Currently, diffraction methods, including X-ray diffraction and neutron diffraction [3,4], are used as nondestructive measurement methods. However, it is impossible to measure three-dimensional stress distribution on-site by using these methods. X-ray diffraction is useful to measure only surface residual stresses. Higher energy diffraction methods, including synchrotron and neutron diffractions, are available only within special irradiation facilities. Therefore, the targeted welded parts have to be destructively cut from an actual welded structure for measuring stress distribution. The cutting process releases some of the residual stresses, which affects the accuracy of the entire measurement process.

As an alternative, the Bead Flush method, which is based on the eigenstrain methodology [5], has been proposed [6]. In this method, three-dimensional residual stresses are calculated by conducting

Materials Research Proceedings **2** (2016) 329-334 doi: http://dx.doi.org/10.21741/9781945291173-56

elastic FEM analysis on eigenstrains, which in turn are estimated through inverse analysis [7] on released strains measured using strain gauges during the removal of the weld reinforcement. Eigenstrains are defined as a sum of inelastic strains [5], but they are different from the sum of physical inelastic strains such as plastic, thermal, and transformation strains [8]. Here, the smoothening of excess weld metal can be regarded as a non-destructive process essentially because it can be eliminate a toe of weld that has the potential to become a crack starter. However, processing strains that arise on the machined surface owing to this method can worsen the accuracy of strain gauges that measure released strains.

In order to overcome the disadvantages of the Bead Flush method, the first author has proposed the use of X-ray diffraction instead of strain gauges to estimate both the welding eigenstrains and processing strains [9]. The effectiveness of this method has been proved via numerical simulation. Besides, experimental demonstrations of this method have been conducted on an actual butt-welded plate made of 2Cr-1Mo ferritic steel [10]. However, the accuracy of estimating residual stresses along the perpendicular to the welding direction is not sufficient. Numerical simulation has confirmed the higher accuracy of this method when residual strains on machined surface measured by X-ray diffraction are used for estimation [11].

In the present study, the authors have estimated residual stresses of welded plates made of SUS430 by using this method on elastic strains measured over the weld metal [12]. For this estimation, it is assumed that welding eigenstrains are constant along the thickness direction even though the butt-welded joint has single U-groove before welding. The eigenstrain distribution along the welding direction is also neglected during inverse analysis. The residual stresses of welded joints made of SUS430 are estimated using this method after considering eigenstrain distributions along both the welding and thickness directions. Parameters to express eigenstrain distributions are determined following the Response Surface methodology. To evaluate the estimation accuracy, the residual stresses estimated using this method are compared with those measured by X-ray diffraction.

Analytical Procedure

The elastic strains $\{\varepsilon_e\}$ of the concerned elements and the welding eigenstrains $\{\varepsilon^*\}$ for the entire structure can be related through an intermediary - an elastic response matrix $[\boldsymbol{R}_e]$ as:

$$\{\varepsilon_e\} = [\boldsymbol{R}_e]\,\{\varepsilon^*\} \tag{1}$$

The i-th column of the response matrix can be determined by imposing a unit eigenstrain on the i-th component of $\{\varepsilon_e^*\}$ during elastic FEM analysis. Based on Eq. (1), the surface elastic strains before and after the removal of the weld reinforcement - $\{\varepsilon_{eb}\}$ and $\{\varepsilon_{ea}\}$ respectively - can be expressed, as follows:

$$\{\varepsilon_{eb}\} = [\boldsymbol{R}_b]\,\{\varepsilon^*\} \tag{2}$$

$$\{\varepsilon_{ea}\} = [\boldsymbol{R}_a]\,\{\varepsilon^*\} + [\boldsymbol{R}_a]\,\{\varepsilon_{pa}^*\} \tag{3}$$

where $[\boldsymbol{R}_b]$ and $[\boldsymbol{R}_a]$ respectively denote the response matrix before and after the removal. $\{\varepsilon_{pa}^*\}$ is the processing strain that remains after the smoothening. By jointing Eq. (2) and Eq. (3), we obtain the following equations [18]:

$$\{\varepsilon_{eb}\ \varepsilon_{ea}\}^T = [\boldsymbol{R}]\,\{\varepsilon^*\ \varepsilon_{pa}^*\}^T \tag{4}$$

$$[\boldsymbol{R}] = \begin{bmatrix} \boldsymbol{R}_b & \boldsymbol{0} \\ \boldsymbol{R}_a & \boldsymbol{R}_a \end{bmatrix} \tag{5}$$

Actual nondestructive measurements of elastic strains contain observational errors $\{\varepsilon_{err}\}$, as given below:

$$\{\varepsilon_{ebm}\ \varepsilon_{eam}\}^T = [\boldsymbol{R}]\ \{\varepsilon^*\ \varepsilon_{pa}^*\}^T + \{\varepsilon_{err}\} \tag{6}$$

where $\{\varepsilon_{ebm}\}$ and $\{\varepsilon_{eam}\}$ are both elastic strains measured using nondestructive methods. The estimated values of welding eigenstrains $\{\varepsilon_{est}^*\}$ and processing strains $\{\varepsilon_{pa\_est}^*\}$ can be obtained from elastic strains measured by X-ray diffraction, through inverse analysis, as shown below:

$$\{\varepsilon_{est}^*\ \varepsilon_{pa\_est}^*\}^T = [\boldsymbol{R}]^+\ \{\varepsilon_{ebm}\ \varepsilon_{eam}\}^T. \tag{7}$$

where $[\boldsymbol{R}]^+$ is the Moore and Penrose generalized inverse matrix [7] of $[\boldsymbol{R}]$, which is given by:

$$[\boldsymbol{R}]^+ = [\boldsymbol{R}]^T\ [\boldsymbol{R}]\ ([\boldsymbol{R}]^T\ [\boldsymbol{R}]\ \ [\boldsymbol{R}]^T\ [\boldsymbol{R}]\ )^-[\boldsymbol{R}]^T. \tag{8}$$

Target Joint and Measurements by X-ray Diffraction

The dimensions of the target plate (Fig. 1) were $80\times100\times10$ mm^3. The widths of the weld bead on the top and bottom surfaces were 16 mm and 8 mm, respectively. The welding conditions are listed in Table 1. Ferritic weld wires were used to join stainless steel SUS430 to enable easier measurements over the weld metal by using X-ray diffraction. It took nine weld passes to fill single U-groove. Except during the first weld pass, the heat source was oscillated to avoid poor penetration along the boundaries between the base and weld metals. The excess metal on the top surface was removed using a flat-surface grinding machine.

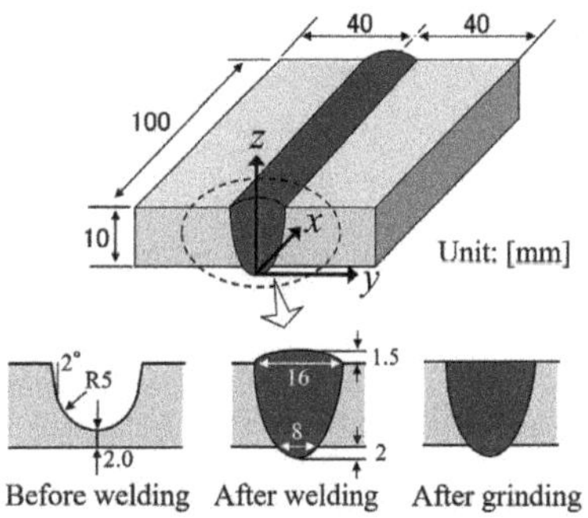

Fig. 1 Butt-welded

Table 1 Welding conditions.

Welding method	TIG
Welding current	200 A
Welding voltage	10 V
Welding speed	10 cm/min
Welding electrodes	YT-160
Diameter of electrodes	2.4 mm
Shield gas	Ar
Preheating	No
Inter-pass temperature	Under 150°C
Welding oscillation	Yes

Elastic strains along the welding direction and the perpendicular to the welding direction on the top surface were measured using X-ray diffraction to estimate welding eigenstrains. Fig. 2 shows the measurement points before and after the removal of weld reinforcement. Measurement conditions of X-ray diffraction were typical, as shown in Table 2. This X-ray diffraction is based on the $\sin^2\psi$ method. Irradiation area of X-ray diffraction was set at 2×2 mm^2, as normal range. Note that x, y, and z are the welding direction, the perpendicular to the welding direction from the center of the weld, and the thickness direction, respectively, as shown in Fig. 1.

Stabilizing Solution in Inverse Analysis

FEM Model. An FEM model used in this estimation is shown in Fig. 3. This model is symmetric about the weld line. The numbers of nodes and elements before the removal of the excess metal are 1397 and 1020, respectively. The Young's modulus and Poisson's ratio were set at 200 GPa and 0.3, respectively. For FEM analysis, the commercial software ANSYS (Cybernet Systems Company, Limited, Japan), was used.

Stabilizing Solutions. Solutions obtained from this inverse analysis are relatively sensitive because the three-dimensional eigenstrain distributions are being estimated from measurements of two-dimensional surface strains that include observational errors. During this study, the method of functional approximation of welding eigenstrains and the artificial noise method [7] were introduced. These are described below.

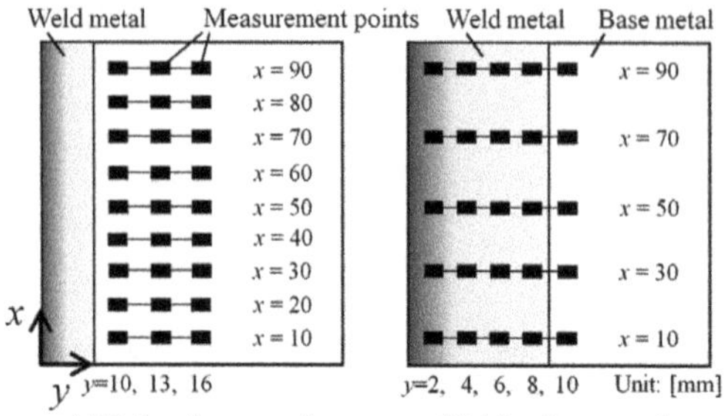

(a) Before the removal. (b) After the removal.

Fig. 2 Measurement points before and after the removal.

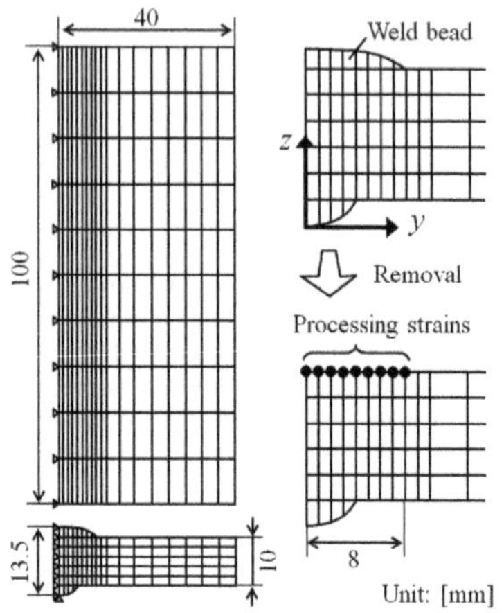

Fig. 3 An FEM model before the removal.

Table 2 Measurement conditions.

Target	Cr
Tube voltage and current	40 kV, 200 mA
Bragg's angle ($2\theta_0$)	156 °
Incident angles (Ψ)	0, 18.4, 26.7, 33.2, 39.2, 45.0°

First, logistic functions were used to express welding eigenstrains [13], as shown below:

$$\{\varepsilon^*_s\}(y) = \sum_{i=1}^{4} \frac{\{a_{si}\}}{1+\exp(p+q_i y)}, \quad p = -5.0,\ q_1 = 0.60,\ q_2 = 0.40,\ q_3 = 0.30,\ q_4 = 0.25 \tag{9}$$

where the subscript s denotes any of the three directions - welding (x), perpendicular to the welding (y) and thickness (z) directions. The constants p and q_i are determined such that eigenstrains are distributed within 40 mm in the y direction [14]. Therefore, the number of elements in the unknown vector $\{a_{si}\}$ becomes 12 (four functions in each direction). In the previous study, it was assumed that eigenstrains were not distributed along the welding and thickness directions [12]. In order to improve the estimation accuracy in this study, eigenstrain distributions in these directions were considered using the equation below instead of Eq. (9) [10].

$$\{\varepsilon^*_s\}(x, y, z) = \sum_{i=1}^{4} \frac{\{a_{si}\}(\beta+1)}{1+\exp(p+q_i\sqrt{y^2+(t-z)^2/\alpha})} \tag{10}$$

where it is better to set constant t as the gross thickness that includes bead heights on the top and bottom. In this study, the value of t was set at 14 mm. The most probable values of α and β are determined based on the Response Surface methodology. Additionally, it was assumed that processing strains formed on the machined surface were constant along the welding direction, and that their values from y=0 mm to 2 mm and from 3 mm to 8 mm are equal. Consequently, the total number of unknown parameters of the processing strains was 6 (2 surfaces in each direction).

The other approach to improve well-posedness of this inverse analysis is the used of the artificial noise method [7]. When the response matrix $[\boldsymbol{R}]$ with rank n is decomposed, the following equations can be obtained.

Materials Research Proceedings **2** (2016) 329-334 doi: http://dx.doi.org/10.21741/9781945291173-56

$$[\boldsymbol{R}]^{+} = [\boldsymbol{U}]\,[\boldsymbol{B}]^{-}\,[\boldsymbol{V}]^{\mathrm{T}} \tag{11}$$

$$[\boldsymbol{B}]^{-} = \begin{bmatrix} \boldsymbol{B}_n^{-} & \boldsymbol{0} \\ \boldsymbol{0} & \boldsymbol{0} \end{bmatrix},\quad [\boldsymbol{B}_n]^{-} = \begin{bmatrix} 1/\mu_1 & & & 0 \\ & 1/\mu_2 & & \\ & & \ddots & \\ 0 & & & 1/\mu_n \end{bmatrix},\quad \mu_1 \geqq \mu_2 \geqq \dots \geqq \mu_n \geqq 0. \tag{12}$$

If the absolute value of the singular value μ_j ($1 \leqq j \leqq n$) reduces, measurement errors exert a larger effect on solutions. To eliminate this influence in the artificial noise method, $[\boldsymbol{B}_{n\_\gamma}]^{-}$ is used instead of $[\boldsymbol{B}_n]^{-}$, as shown below:

$$[\boldsymbol{B}_{n\_\gamma}]^{-} = ([\boldsymbol{B}]^2 + \gamma[\boldsymbol{I}])^{-}\,[\boldsymbol{B}]^{\mathrm{T}} \tag{17}$$

where $[\boldsymbol{I}]$ denotes a unit matrix. Solutions can be stabilized by increasing the value of the real number γ. In this inverse analysis, the L-curve method [15] was applied to determine the most probable value of γ.

Estimated Results

The most probable values of α and β were determined from a response surface (Fig. 4) to be 0.7 and 0.010, respectively. The vertical axis in this graph is a residual norm that is a difference between the estimated and measured residual strains on the measurement points. Therefore, the most probable values can be found at the minimum value.

The residual stresses at x=50 mm on the bottom surface estimated using this method are compared with residual stresses directly measured using X-ray diffraction (Fig. 5). The dashed green lines are the estimated values without considering eigenstrain distributions in the x and z directions. The residual stresses, estimated using α and β, are drawn by using solid blue lines. An almost equivalent estimation accuracy can be observed at x=50 mm on the bottom surface (Fig. 5(a)). On the other hand, estimation accuracy can be improved at y=10 mm by considering eigenstrain distributions (Fig. 5(b)).

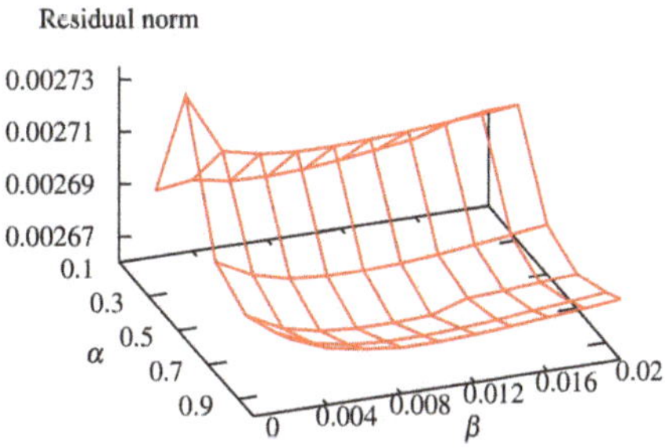

Fig. 4 A response surface to determine the most probable values of α and β.

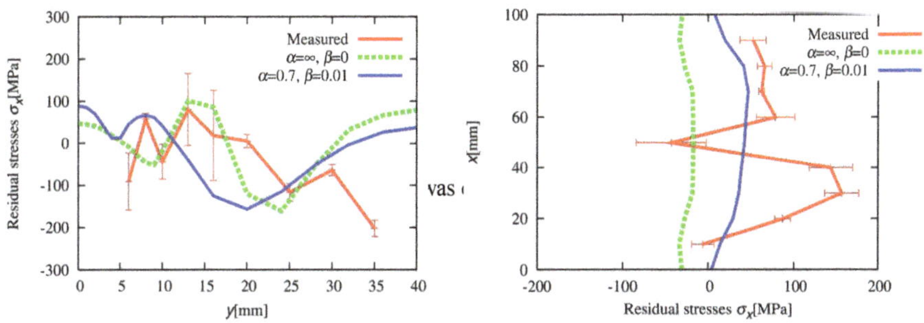

(a) At x=50 mm. (b) At y=10 mm.

Fig. 5 Exact and estimated residual stresses along the x direction on the bottom surface. The red lines are measured values using X-ray diffraction. Estimated residual stresses with and without considering eigenstrain distributions are drawn by the solid blue line and the broken green line, respectivley.

However, it might be better to use higher-degree equation for the approximation because residual stresses measured by X-ray diffraction were largely oscillated along the welding direction, as Fig. 5(b). Eigenstrain distributions might be complicated by the oscillation of the heat source during welding.

Summary

The method that is based on the eigenstrain methodology and X-ray diffraction was applied on an actual butt-welded joint made of SUS430 to evaluate three-dimensional residual stresses. The estimation accuracy of this method was improved by considering the eigenstrain distributions along the welding and thickness directions.

Acknowledgements

The authors acknowledge Tonen General Sekiyu Research/Development Encouragement & Scholarship Foundation and Mazak Foundation for their generous financial assistance. As well, this work was supported by JSPS KAKENHI Grant Number 26820005. We would also like to thank Editage for English language editing.

References

[1] Z. Saternus, W. Piekarska, M. Kubiak, T. Domanski and L. Sowa, Procedia Engineering 136 (2016) 95-100. http://dx.doi.org/10.1016/j.proeng.2016.01.180

[2] Y. Ye, J. Cai, X. Jiang, D. Dai and D. Deng, Advances in Engineering Software 86 (2015) 39-48. http://dx.doi.org/10.1016/j.advengsoft.2015.04.001

[3] A. Skouras, A. Paradowska, M.J. Peel and M.J. Pavier, International Journal of Pressure Vessels and Piping 101 (2013) 143-153. http://dx.doi.org/10.1016/j.ijpvp.2012.11.002

[4] H.J. Kirkwood, S.Y. Zhang, A.S. Tremsin, A.M. Korsunsky, N. Baimpas and B. Abbey, Materials Today: Proceedings 2S (2015) S414-S423.

[5] T. Mura, Micromechanics of Defects in Solids, Martinus Nijhoff Publishers, 1987. http://dx.doi.org/10.1007/978-94-009-3489-4

[6] H. Nakamura, Y. Naka W. Park and H. Kobayashi, Current Topics in Computational Mechanics, ASME PVP, 305 (1995) 49-56.

[7] S. Kubo, Inverse Problems, Baifukan, 1992 (in Japanese).

[8] K. Masuda and H. Nakamura, The Japan Society of Mechanical Engineers, Series-A, 76 (2010) 884-892 (in Japanese).

[9] M. Ogawa, The Japan Society of Mechanical Engineers, Series-A, 79 (2013) 1266-1277 (in Japanese).

[10] M. Ogawa, T. Ishii and S. Furusako, Journal of the Society of Materials Science, Japan, 64 (2015) 932-939 (in Japanese). http://dx.doi.org/10.2472/jsms.64.932

[11] M. Ogawa, International Journal of Computational Materials Science and Engineering, 3 (2014) 1450023. http://dx.doi.org/10.1142/S2047684114500237

[12] M. Ogawa and T. Ishii, Proceedings of the International Conference on Advanced Technology in Experimental Mechanics 2015, (2015) 36.

[13] K. Kumagai, H. Nakamura and H. Kobayashi, The Japan Society of Mechanical Engineers, Series-A, 65 (1999) 133-140 (in Japanese).

[14] Y. Ueda and N. X. Ma, Japan Welding Society, 11 (1993) 189-195 (in Japanese).

[15] P. C. Hansen, SIAM Review, 34 (1992) 561-580.

Residual Stresses 2016: ICRS-10 Materials Research Forum LLC
Materials Research Proceedings **2** (2016) 335-340 doi: http://dx.doi.org/10.21741/9781945291173-57

X-Ray Diffraction Measurements and Investigation of the Stress Relaxation in Autofrettaged AISI 4140 Steel Thick Walled Cylinders

N. Lyubenova[1,a*], J. Pineault[2,b], H. Brünnet[1,c], D. Bähre[1,d]

[1] Saarland University, Institute of Production Engineering, Campus A4.2, Saarbrücken, Germany

[2] Proto Mfg. Inc., 12350 Universal Drive, Taylor, Michigan, USA

[a]nataliya.lyubenova@uni-saarland.de, [b]xrdlab@protoxrd.com, [c]horstbruennet@gmx.de, [d]d.baehre@mx.uni-saarland.de

Keywords: Residual Stress, Autofrettage, Post-Machining, X-Ray Diffraction, Stress Correction

Abstract. Hydraulic autofrettage is a manufacturing process that induces favorable compressive residual stresses and is especially suitable for the treatment of internally pressurized components. If autofrettage is not the final treatment applied, the application of post-machining or other cold working processes can lead to a relaxation and redistribution of the stresses induced by the autofrettage process. In this paper, comprehensive X-ray diffraction residual stress measurements were performed and the influence of the applied autofrettage pressure and post-machining on the resultant residual stress vs. depth profiles was investigated.

Introduction

It is well known that compressive residual stresses are favourable because they act to close existing cracks in components and prevent the generation of new ones. When a component experiences in-service loading, applied tensile stresses will be shifted by the compressive residual stress (RS) field, if present. If the compressive RS field is of sufficient magnitude, the final stress state may still remain locally compressive despite the superimposed applied tensile stresses. There are several processes that are able to induce high magnitude compressive RS; e.g. shot peening, deep rolling and laser shock peening. The autofrettage (AF) process is especially suitable for treating internal geometries, e.g. components of the common rail diesel injection system. It leads to a beneficial and pronounced compressive RS vs. depth profile [1, 2] and several authors report an extension in fatigue life for components treated with this process [3, 4]. Its principle can be explained as follows [5]: when applying AF, a low-viscosity hydraulic medium is used to rapidly over-pressurize the treated component. If the resulting stresses exceed the yield strength of the material, then elasto-plastic deformation will result. Typically, the inner surface of the treated component deforms plastically while the outer surface of the component remains only elastically deformed. After releasing the AF pressure, the elastically deformed region of the component strives to return to its original state but is prevented from doing so by the inner plastically deformed region. This inhomogeneous deformation leads to the generation of compressive RS on the inner region of the component. This compressive region is compensated for with tensile stresses on the outer region of the component. The AF pressure is the most important processing input parameter and changing it leads to different RS vs. depth profiles in the part [6]. It has been shown that the AF process not only induces RS but also results in concomitant macroscopic shape deviations [7]. When high dimensional accuracy of the treated component is required, it may be necessary to perform a post-machining operation that could result in a redistribution and/or relaxation of the RS induced by the AF process. As such, the following paper presents investigations that include: pre-machining, autofrettage and post-machining.

Materials Research Proceedings **2** (2016) 335-340 doi: http://dx.doi.org/10.21741/9781945291173-57

X-ray diffraction (XRD) is an established, generally applicable, and time-proven method for near surface (up to 0.025 mm) RS measurements [8]. Accompanied with an electropolishing procedure, it offers the possibility to investigate RS at depth. In the presented paper, XRD techniques were employed to measure the RS vs. depth profiles in thick walled cylinders that were treated with different AF pressures and partially post-treated using a reaming operation. Due to the geometric constraints inherent to the specimens, they were sectioned in half to enable XRD based RS measurements on their inner diameter (ID). To account for the relaxation and redistribution of the RS induced by the AF process as a result of axial sectioning, strain gauges were applied to the inner and outer diameters (OD) of the specimens' surfaces prior to cutting. The strain gauges were then monitored during the sectioning process and the results obtained were used to correct the measured stresses for the effect of sectioning. Since the XRD technique samples a relatively thin surface layer, material removal techniques were applied so as to obtain the full through wall thickness RS profile. To this end, electropolishing techniques were employed and the resulting stress relaxation was corrected for using the analytical methods proposed by Moore and Evans [9].

Experimental setup

The samples used for the experiments are thick walled cylinders, manufactured from high strength AISI 4140 steel (hardened at 840 °C and tempered at 610 °C). Two measurement points (MP1 and MP2) were chosen to characterize the different treatment sequences on both sides of the thick walled cylinders. The process and measurement chains are shown in Table 1.

Table 1 – Process and measurement chains for the thick walled cylinders

MP1	MP2	Process and measurement chain		Process parameters	Residual Stresses
+	+		Raw part	Rolled Hardened at 840 °C Tempered at 610 °C	Present, unknown Randomly distributed
+	+		Pre- Machining (boring)	HSS twist drill Length: 125 mmm n = 200 1/min, v_f = 1 mm/s, ID = Ø 9H12	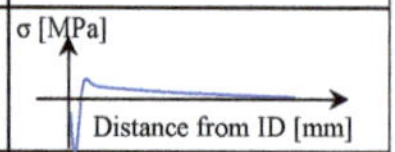
+	+		Autofrettage	Autofrettage pressure levels: P = 2000, 8000, 8500, 9000 and 9500 bar	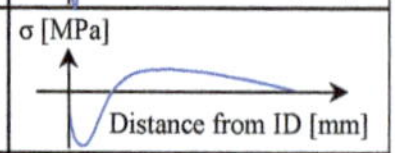
-	+		Post-Machining (reaming)	HSS – E 6 blades reamer n = 200 1/min vf = 0.6 mm/s OD (up to middle) = Ø 10H7	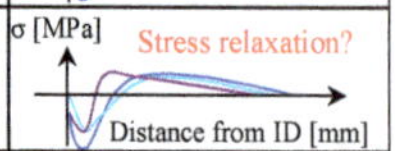
+	+		Strain gauges application	N.D.	Unchanged
+	+		Axial cutting/ strain gauges reading	N.D.	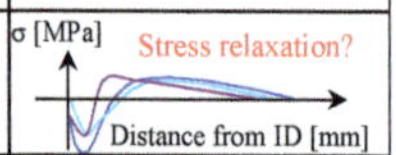
+	+	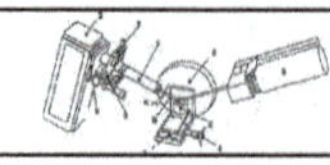	XRD Measurements/ Electropolishing	See Table 2	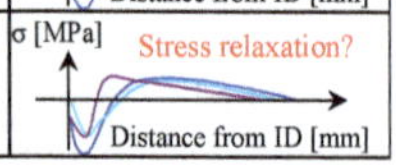

Samples were prepared by first boring (n = 200 1/min, v_f = 1 mm/s) an inner diameter of 9 mm, using a HSS twist drill. The cylinders were then treated with AF pressures ranging from 2000 to 9500 bar. The treatment with 2000 bar did not lead to a plastification of the cylinders' surface and RS was therefore not generated; the cylinders treated with an AF pressure of 2000 bar are thus considered to be untreated. A reaming operation (n = 200 1/min, v_f = 0.6 mm/s) to an inner diameter of 10 mm was then performed on the right side of the cylinder leaving the left half of each cylinder

Materials Research Proceedings **2** (2016) 335-340 doi: http://dx.doi.org/10.21741/9781945291173-57

(MP1) bored and autofrettaged, whereas the right half (MP2) was additionally reamed. The reason for this additional treatment was to investigate the RS relaxation and redistribution due to this post-AF treatment. Strain gauges were then applied to the inner and the outer surfaces of the specimens and an axial cut provided access for XRD RS measurements on the inner diameter of the specimens. After the axial cut was performed, the stress relaxation on both the inner and outer diameters of each specimen was calculated via the strain gauge readings. For the XRD RS measurements the setup parameters shown in Table 2 were employed.

Table 2 - X-ray diffraction setup Parameters

Instrument: Proto LXRD	Crystallographic Plane: {211}	Oscillation(s): Beta 4.00°
Software Version: XRDWIN 2.0 Build 86	Bragg Angle (2θ): 156.4°	Collection Time: 0.25 seconds x 20 exposures
Goniometer Configuration: Psi	Aperture: 2 mm	Peak Fit: Gaussian 85%
Target: Cr ($K\alpha_{avg\ 2.291}$ Angstroms)	Psi Tilts: 22	LPA Correction On: Yes
Target Power: 30 kV, 25 mA	Tilts: (0°, ±25.00°, ±20.60°, ±15.85°, ±11.80° , ±3.74°)	Background Subtraction: Linear

The RS vs. depth profile was generated by means of sequential XRD measurements and material removal using electropolishing techniques to mid–wall thickness from the ID. RS measurements were then performed from the OD to mid-wall thickness. Once the RS measurements were completed, the pertinent data corrections were applied. First, the gradient correction was applied as per SAE HS-784 [10]. This correction accounts and corrects for steep RS gradients and their effect on sampling depth variations with incident x-ray beam angle. Next, the material removal correction was applied to the gradient corrected data; this correction is based on the analytical methods proposed by Moore and Evans [9] for cylinders and plates and is still the most widely used correction in industry. Finally, the measurement results obtained were corrected for stress relaxation due to sectioning using the strain gauge data. The two data sets obtained on the OD to mid-wall thickness and ID to mid-wall thickness were combined into one contiguous through wall RS profile.

Results

XRD measurements were performed for the untreated (AF 2000 bar) as well as the specimens treated with 8000, 8500, 9000 and 9500 bar AF pressure. In order to characterize statistical variability, three different specimens treated with the same AF pressure (8000 bar) were characterized using XRD. In Fig. 1, the residual hoop stress vs. depth for these specimens is plotted after gradient, depth and strain corrections were applied. The results for the bored and AF side (MP1) of the AF 8000 bar specimens plotted in Fig. 1a), indicate that the maximum induced RS for each specimen is just below the surface and ranges from -450 to -550 MPa in compression. It can also be seen that the RS becomes slightly tensile at a depth of approximately 2 mm. On the outer diameter of the specimens, a thin layer of compressive RS can be also observed which is thought to be due to the turning operation that was used for their pre-machining. Fig. 1b) depicts the hoop RS for the specimens autofrettaged with 8000 bar on the bored, autofrettaged, and reamed side of the specimen (MP2). The maximum induced compressive RS is lower in magnitude and on the order of -400 MPa. This reduction in RS is attributed to the reaming operation after the AF which removes a part of the most compressive loaded material thus leading to a redistribution and relaxation of the balancing tensile RS towards the OD. Brünnet et al. [7] investigated the influence of the post-machining, namely the reaming operation after AF, and confirms that it leads to a relaxation and redistribution of RS.

Materials Research Proceedings **2** (2016) 335-340 doi: http://dx.doi.org/10.21741/9781945291173-57

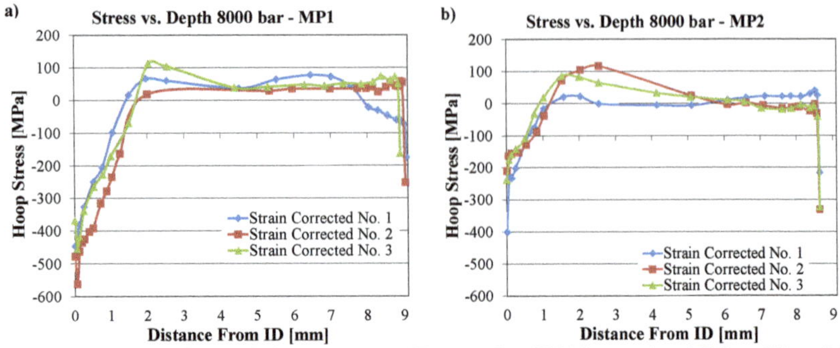

Fig. 1 – Corrected hoop RS for specimens AF treated at 8000 bar – a) at MP1 and b) at MP2.

The hoop RS vs. depth profile for the non-autofrettaged specimen can be seen in Fig. 2. On the bored side of the specimen (MP1), see Fig. 2a), a thin layer of compressive RS is visible near the ID and the OD. This can be attributed to the boring (ID) and turning (OD) operations. In Fig. 2b), the stress profile for the reamed side of the specimen (MP2) is plotted. It can be seen that the near surface measured RS was approximately 70 to 100 MPa less compressive as compared to MP1. This indicates that the reaming operation itself results in lower magnitude compressive RS as compared to the boring operation.

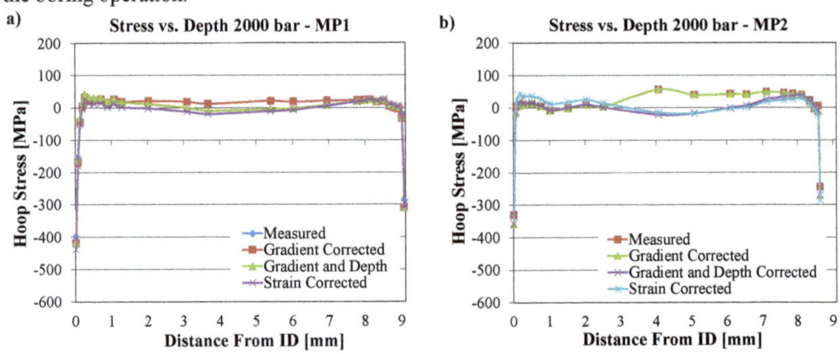

Fig. 2 – Hoop RS for specimens AF treated at 2000 bar – a) at MP1 and b) at MP2.

The hoop RS vs. depth profiles measured at MP1 for the specimens treated with different AF pressures are plotted in Fig. 3. Fig. 3a) depicts the "as measured" results. Here, it can be observed that the non-autofrettaged specimen (2000 bar) has a very thin layer (about 0.5 mm) of compressive RS on the ID that is produced during the boring operation. An unanticipated result was observed in the curves for the specimens that were autofrettaged at different pressures. It was expected that increasing the AF pressure would lead to a greater maximum induced compressive RS. Instead, the apparent maximum stresses remained approximately the same with increasing AF pressure and in some cases, it decreased. Nevertheless, the higher AF pressures produced a deeper compressive RS layer that was compensated for with higher subsurface tensile stresses. When examining the corrected results, see Fig. 3b), it is clear that even after the corrections are applied, the maximum compressive RS achieved does not increase with increasing AF pressure, and in some cases, it decreases slightly. If the difference in the RS vs. depth profiles for the three specimens treated with

Materials Research Proceedings **2** (2016) 335-340 doi: http://dx.doi.org/10.21741/9781945291173-57

an AF pressure of 8000 bar is considered (about 75-100 MPa) it can be concluded that, in the AF pressure range (8000 to 9500 bar), the measured RS is of similar magnitude.

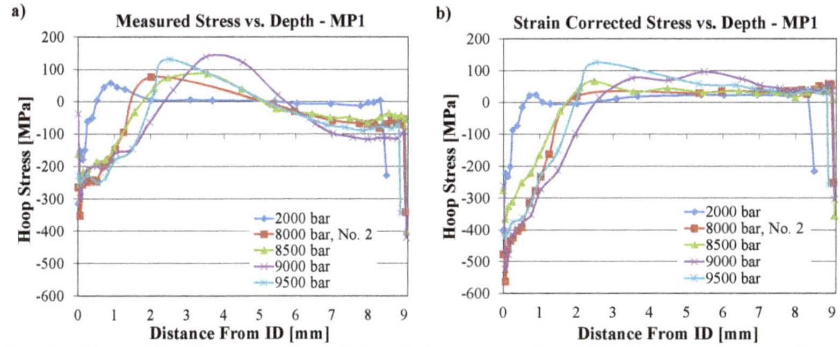

Fig. 3 – Hoop RS AF variation at MP1 – a) As measured and b) Gradient, depth and strain corrected.

The RS vs. depth profiles for the specimens treated with different AF pressures at MP2 (boring, AF and consecutive reaming) are plotted in Fig. 4. The "As measured" profiles can be seen in Fig. 4a). Here, it can be observed that material removed by the reaming operation leads to a significant relaxation in the compressive RS induced by the AF operation. When compared to results obtained at MP1 (where the AF process represents the final treatment applied), results obtained in the compressive RS region near the ID at MP2 are more than 100 MPa less compressive. It can also be observed that higher magnitude applied AF pressures result in higher magnitude stress relaxation. Notwithstanding, the apparent depth of the induced compressive RS remains unaffected. The "Gradient, depth and strain corrected" RS vs. depth profiles at MP2 for the different applied AF pressures can be seen in Fig. 4b). The corrections applied to the data lead to a shift, similar to the one observed at MP1, for all of the stress curves, most noticeably in the compressive region near the ID. The maximum stress relaxation induced by the reaming operation was about 200 MPa (see Fig. 3b) as a point of comparison to MP1). It can also be observed that the depth of the compressive RS layer increases by a few tenths of a millimeter as compared to the uncorrected data.

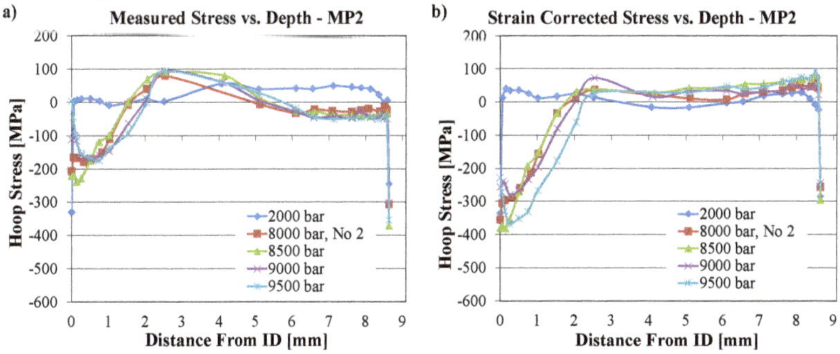

Fig. 4 – Hoop RS AF variation at MP2 – a) as measured and b) gradient, depth and strain corrected.

Materials Research Proceedings **2** (2016) 335-340 doi: http://dx.doi.org/10.21741/9781945291173-57

Conclusions

In this paper, extensive XRD based RS measurements on AF thick walled cylinders were presented. The effects of pre-machining were found to be localized to the ID and OD surfaces and tended to result in a thin compressive RS layer. With the application of the AF process, significant compressive RS was imparted to the ID of the samples with concomitant balancing tensile RS near the OD. The general trend observed was an increase in the compressive RS layer with increasing AF pressure. An unanticipated result was observed in the curves for the specimens treated at higher AF pressures. It was expected that increasing the AF pressure would lead to a greater maximum magnitude of induced compressive RS. Instead, the apparent maximum magnitude of RS remained approximately the same with increasing AF pressure and in some cases, decreased. Nevertheless, the higher AF pressures produced a deeper compressive RS layer that was compensated for with higher magnitude subsurface tensile stresses. Post-machining, i.e. the reaming operation applied after the AF process which removes a part of the most compressive loaded material, led to a redistribution and relaxation of the balancing tensile RS towards the OD.

References

[1] Davidson, T.E., Barton, C.S., Reiner, A.N., Kendall, D.P., "The Autofrettage Principle as Applied to High Strength Light Weight Gun Tubes", U.S. Army T. R. WVT-RI-5907, Watervliet Arsenal, Watervliet, USA, 1959.

[2] H. Brünnet, M. Hofmann, N. Lyubenova, D. Bähre, Effect of Different Manufacturing Process Steps on the Final Residual Stress Depth Profile along the Process Chain of Autofrettaged Thick-Walled-Cylinders, Proceedings of the 9th European Conference on Residual Stresses, Advanced Materials Research (Volume 996), 2014, pp. 676-681. http://dx.doi.org/10.4028/www.scientific.net/amr.996.676

[3] Lechmann, M., "Development of a Fatigue Fracture Design Concept for Internal Pressure Loaded Parts with Extensive Compressive Stress Fields", Dissertation, Stuttgart University, Germany, 2008.

[4] Herz, E., Thumser, R., Bergmann, J.W., Vormwald, M., "Endurance Limit of Autofrettaged Diesel-engine Injection Tubes with Defects", Engineering Fracture Mechanics 73, 2006, p. 3. http://dx.doi.org/10.1016/j.engfracmech.2005.06.006

[5] H. Brünnet, D. Bähre, T. Rickert, D. Dapprich, Modeling and Measurement of Residual Stresses Along the Process Chain of Autofrettaged Components by Using FEA and Hole-Drilling Method with ESPI, Proceedings of the 9th International Conference on Residual Stresses, Materials Science Forum 768-769, 2014, pp. 79-86, http://dx.doi.org/10.4028/www.scientific.net/MSF.768-769.79

[6] Happe, U. "Entwicklungstendenzen im Hochdrudcbau", Chemie Ingenieur Technik, 24. Jahrg. 1952/Nr. 7, 1952, pp. 411-413.

[7] Brünnet, H.; Yi, I.; Bähre, D., "Modeling of Residual Stresses and Shape Deviations along the Process Chain of Autofrettaged Components", Journal of Materials Science and Engineering A, Volume 1/7A, 2011, pp. 915-936.

[8] Totten, G., Howes, M., Inoue, T., "Handbook of Residual Stress and Deformation of Steel, Chapter Measurement and Prediction of Residual Stress and Distortion", ASM International, 2002, pp. 112.

[9] Moore, M.G., Evans, W.P., "Mathematical Correction for Stress in Removed Layers in X-Ray Diffraction Residual Stress Analysis", SAE Residual Stress Committee, 1956.

[10] Ahmad, A., et. al., "Residual Stress Measurement by X-ray Diffraction", SAE HS-784, SAE International, Warrendale, PA, 2003, p.85.

Residual Stresses 2016: ICRS-10 Materials Research Forum LLC
Materials Research Proceedings **2** (2016) 341-346 doi: http://dx.doi.org/10.21741/9781945291173-58

Contour, iDHD, and ICHD Residual Stress Measurements on a T-Section Plate

X. Ficquet[1a*], D. Douglas[1], K. Serasli[1], F. Bridier[2]

[1] VEQTER Ltd, 8 Unicorn Business Park, Whitby Road, Bristol BS4 4EX, UK

[2] DCNS Research, Technocampus Ocean, 44340 BOUGUENAIS, France

[a]xavier.ficquet@veqter.co.uk

Keywords: Residual Stress, Contour, iDHD, T-Section, Fillet Weld

Abstract. This paper presents a review of work carried out to ascertain the residual stresses present within a T- plate section. The contour method, the incremental deep hole drilling (iDHD) and the incremental centre hole drilling (ICHD) are coupled to determined longitudinal and transverse components of residual stress in the weld tope of the fillet weld in the as-welded condition. This paper presents first the measurement using the iDHD method and the measurement obtained using the contour method is then presented. The accuracy and resolution of the contour method results are directly linked to the quality of the electro- discharge machining (EDM) cut made. The challenges incurs by cutting complex geometry like this T- plate section are identified and their influence on residual stresses calculated by the contour method is quantified.

Introduction

Welding residual stresses play an important role when assessing the structural integrity of welded components. The ability to determine residual stresses at critical locations in components as well as throughout the component for structural integrity assessments is desirable. Current measurement techniques to measure the trough thickness residual stress includes the Incremental Deep Hole Drilling (DHD), Neutron Diffraction (ND) and the Synchrotron Xray (HEXRD). The Contour technique, more recent, is capable of measuring residual stresses in a single direction, in a two-dimensional plane through a specimen. A T-plate component is particularly challenging for the contour measurement due to the different change in section.In this paper, the residual stress of the T-section fillet weld has been determined using the iDHD, the contour and the Incremental Center-Hole Drilling (ICHD) techniques. A description of the t-section is first provided. A brief description of the iDHD technique is provided with a more details description of the contour technique. The contour measurement is described stage by stage, and highlight the problem when a wire breaks during the wire-EDM cutting process.

Component description and measurement locations

The component characterised in the present work comprises a main plate with a web stiffener plate fillet welded on the top surface of the main plate. A smaller flange plate is also fillet welded on the top of the web plate. All these components are made of high strength steel. The web plate was welded on both the main and flange plates using an automatic arc welding processes with a simultaneous welding on both sides of the web plate.

The iDHD measurement was carried out first, located at 207mm longitudinally away from the front of the specimen at weld toe 1 and carried out to a depth of $0.44t_0$ from the weld toe surface through the hull plate, where t_0 is the thickness of the hull plate, see Fig. 1. Then the contour measurement was performed

70mm longitudinally away from the iDHD measurement towards the front end of the specimen. The contour cut was made at 90° to the fillet welds in order to obtain a map of the longitudinal residual stresses.

Materials Research Proceedings **2** (2016) 341-346 doi: http://dx.doi.org/10.21741/9781945291173-58

Finally, a total of three ICHD measurements were carried out on the Contour cut surfaces after completion of the Contour measurement, see Fig. 2b. ICHD1 was located at 0.1t0 below the weld toe 2, ICHD2 was located at the same depth below weld toe1. ICHD3 was located at $0.18t_0$ below weld toe 1 on the opposite contour cut surface to ICHD 2.

(a)

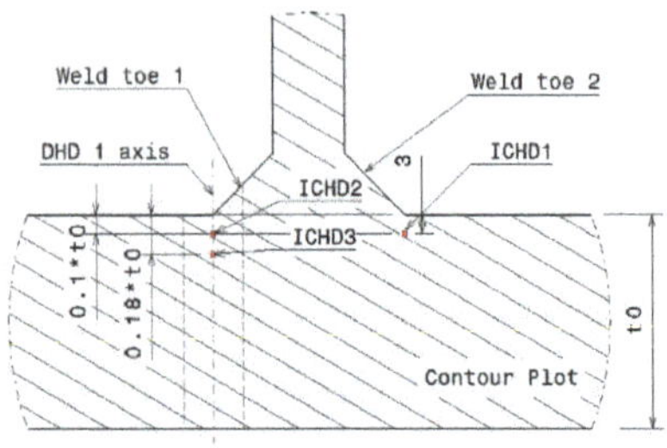

(b)

Fig. 1: (a) Distorted photograph of the specimen , (b) Drawing showing the location of the ICHD measurements on the Contour measurement surface.

Residual Stress Measurement description

In order to define the full residual stress distribution in the joints two techniques were employed. The well-established technique iDHD was combined with the contour method, a more recent technique. The iDHD technique can provide bi-axial residual stress through the thickness of the joint. This technique can measure steep gradient through the thickness, however the results are only provided along the measurement line. The contour method has the advantage of measuring the full 2D uni-axial residual stress normal to a cut. Therefore providing the location of the maximum and minimum residual stress peaks. The iDHD technique has been covered in numerous papers, therefore only a brief description of the technique will be provided. The contour measurement will be developed in more details.

The iDHD measurement. The DHD technique is a semi-invasive, mechanical strain relief technique (i.e. the strain of the component is measured during stress relief from the removal of a small amount of material). The analysis of measured distortion was developed by Leggatt et al [1] and later extensively refined, [2]. The iDHD technique can be applied in two ways depending on the residual stresses assumed to be present within the component. The standard technique (DHD) involves trepanning through the component in a single stage and then measuring the final diameter of the deformed reference hole. However, if the component contains, for example, high magnitude, tri-axial residual stresses then during the trepanning process the material will undergo plastic relaxation and the deformed shape of the reference hole will not be representative of the original residual stress field and so provide inaccurate results. Therefore if it is thought that plasticity will occur during the DHD process then the modified technique, called Incremental-DHD or iDHD, will be used in which the trepanning process is carried out in multiple stages and the diameter of the reference hole is measured after each trepanning stage. With the modified technique plasticity is no longer an issue.

In order to estimate where plastic relaxation may occur, finite element (FE) analysis predictions of the residual stresses generated in similarly designed and manufactured components and/or previous measurements are reviewed. It was decided that for this iDHD measurement, a single incremental EDM step would be performed at approximately $0.8t_0$ deep from the fillet weld toe. With iDHD, a single, biaxial measurement point is produced per EDM increment, usually within a distance of +2mm from the EDM increment depth. Unlike the standard DHD technique where measurement points at specific depths can be prescribed, the iDHD technique cannot be so exact. The incremental depth of the EDM processcan be prescribed, but the location of the measured point depends upon the magnitude of residual stresses present and the level of plastic relaxation.

The contour measurement. The contour residual stress measurement technique is a destructive mechanical strain relief technique. The technique is based on a variation of Bueckner's elastic principle of superposition [3]. The technique provides a full uniaxial map of the residual stress perpendicular to a cut plane through the specimen. The contour method can be divided into 5 stages.

Stage 1: Sacrificial bushes, highlighted in blue in Fig. 2a, are attached to the specimen with conductive adhesive in line with the plane of measurement. A small pilot hole is then drilled through the bushes and specimen along the measurement plane, at 90° to the cutting direction and away from the zone of most interest to permit "self-constraint" during the cut. The sacrificial material attached to the specimen creates a constant cut area throughout the specimen and aids to cutting accuracy and prevent wire breakages [3, 4].

Stage 2: The specimen is clamped into a wire-EDM (electric-discharge machining) machine, and cut into two sections in a single pass with a 250μm diameter brass wire on "skim cut" settings. The cut is initiated from the pilot hole which keeps the specimen constrained. This is of critical importance during the wire-EDM cutting process in order to minimise plastic deformations and bulge error [4]. The cut creates two stress-free surfaces and hence relaxes the out-of-plane residual stresses.

The complex geometry of the specimen made the fitting and attachment of the sacrificial material difficult. The EDM wire may have vibrated due to the various transitions from cutting through the bushes to cutting through the specimen. Cutting artefacts were therefore left apparent on the cut surface in line with these various transitions, shown in Fig. 2b.

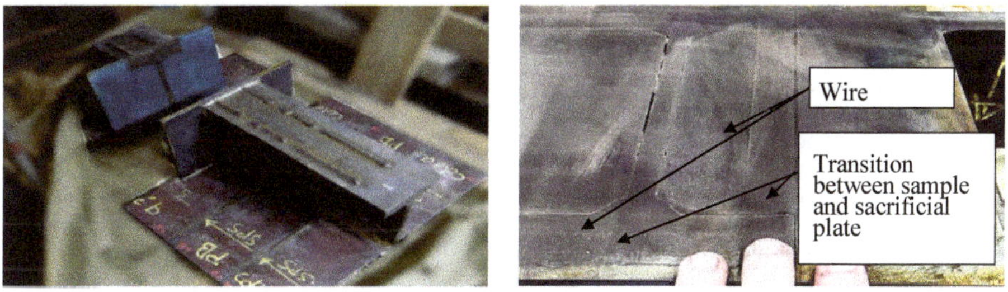

(a) (b)

Fig. 2: (a) Cross section showing sacrificial materials , (b) The wire breaks and artefacts on the cut surface

Stage 3: The two newly cut surfaces are measured with a coordinate measuring machine (CMM) to record their out-of-plane deformations resulting from the relaxation of the out-of-plane residual stresses. Therefore the longitudinal displacements of both contour cut surfaces due to longitudinal residual stress relaxation were measured using CMM with a node spacing of about 0.2mm in the weld regions and about

0.5mm elsewhere. The 6mm diameter, spherical probe tip of the CMM provides an uncertainty of about

±4μm.

A wire break near the weld toe during the EDM cutting meant that an important region of data was corrupted. As the residual stress profile was considered to be symmetrical about the stiffener, the wire break zone was repaired with mirrored data for the comparable region at the opposing weld toe, as seen in Fig. 3a.

Stage 4: The CMM surface data are combined, averaged and smoothed into one single 2D contour surface of deformations, Fig. 3b.

Fig. 3: (a) Raw CMM data (2D and 3D projections), (b) Smoothed CMM data (2D and 3D projections)

The CMM data smoothing was performed for the whole surface after data repairs. Aggressive smoothing leads to loss of details in the residual stress field which can lead to an apparent reduction in the peak magnitudes. Consequently a sensitivity trial with multiple smoothing coefficients was performed to find the minimum level of smoothing that removed excess noise but preserved the details of the residual stress field.

Stage 5: The new contour surface is inverted and introduced into a 3D finite element (FE) model as a boundary condition to reconstruct the original out-of-plane residual stress state of the specimen. Therefore a 3D FE model of the T-section specimen with dimensions limited to the 300mm wide region of interest was created. The model contained about 260,000 elements of type C3D8R. Boundary conditions were used to constrain the contour cut surface in the X and Y directions in order to have the model acting as a rigid body, Fig. 3a. Then a fine mesh with a node spacing of 0.5mm in the weld regions and 1mm elsewhere was generated. Finally the displacement data was introduced into the model at the FE nodes corresponding to the matrix elements. The longitudinal residual stresses were then obtained from the FE model after a general static step related to the introduced displacements and the requirement for the stresses across the cut surface to be in equilibrium. A sensitivity study of the mesh and the smoothing of the displacement data (from no smoothing to very aggressive smoothing) was carried out. Only results using optimum mesh and data smoothing are presented.

The ICHD measurement. The ICHD measurements represent the relaxed and redistributed residual stresses present after the contour cut. Therefore to obtain the initial residual stresses at the ICHD measurement locations before the contour cut, the residual stresses obtained from the contour measurements were then taken into account, more specifically the reconstructed in-plane residual stresses from the finite element analysis stage of the contour technique. A superimposition of the measured and reconstructed residual stresses at the measurement locations was therefore carried out for ICHD1, ICHD2 and ICHD3 to obtain the "corrected" residual stresses. The "corrected" residual stresses are meant to be the initial residual stresses at these locations before any material removal. It is important to note that the reconstructed residual stresses from the mirrored analysis of the contour measurement were used for this superimposition since they are thought to be the most accurate.

Results and discussion

The iDHD, Contour and ICHD measurements were converted into residual stresses using a Young's Modulus, E, and a Poisson's ratio for the parent material. The analyses used for the DHD/iDHD technique assumed isotropic, plane stress conditions; therefore the Poisson's ratio was not required for the iDHD analyses. Each of the measured stress components was normalised with the yield strength (σy) and is plotted as a function of the normalised depth from the plate.

Materials Research Proceedings **2** (2016) 341-346 doi: http://dx.doi.org/10.21741/9781945291173-58

Fig. 3b illustrates the map of the longitudinal residual stresses obtained from the mirrored contour analysis. The symmetric field of displacements generated a balanced residual stress field perfectly symmetric about the stiffener which is very similar to the residual stress field obtained from the original contour measurement for side 1 containing weld toe 1.

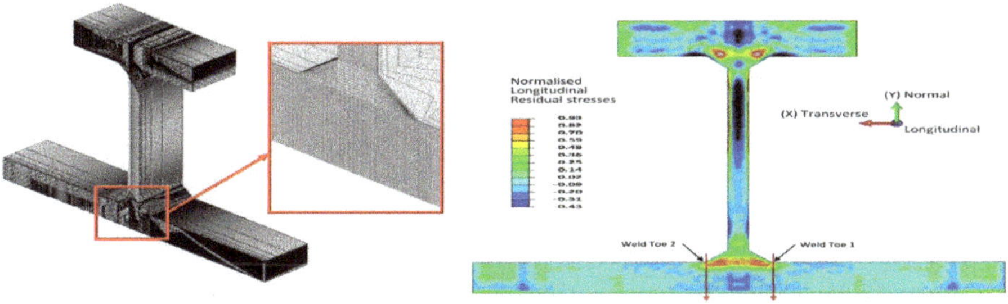

Fig. 3: (a) 3D FE model of the DCNS T-section specimen., (b) 2D Map of the longitudinal residual stresses obtained from the mirrored Contour analysis.

Fig.4 shows the comparison between the iDHD, the contour and the ICHD measurement in the transverse and longitudinal direction.

The DHD technique measured similar shaped longitudinal and transverse residual stress profiles with the longitudinal and transverse residual stresses being the most tensile throughout. The results showed relatively low tensile residual stresses around $0.15\sigma_y$ and $0.3\sigma_y$ a near to the weld toe surface in the transverse and longitudinal directions respectively. The longitudinal and transverse residual stresses both increased to initial peaks of $0.6\sigma_y$ and $0.35\sigma_y$ respectively at a depth of $0.07t_0$. The initial peaks were measured by the iDHD technique having found that the standard DHD technique experienced a small level of plasticity in this region as expected. The longitudinal and transverse residual stresses then decreased slightly before increasing again to maximum tensile peaks of about $0.7\sigma_y$ and $0.35\sigma_y$ respectively at a depth of $0.18t_0$. Finally, the longitudinal and transverse residual stresses decreased into compression at roughly $0.28t_0$ deep to reach low compressive values of roughly $-0.12\sigma_y$ and $-0.2\sigma_y$ respectively thereafter.

The ICHD measurements were carried out on the Contour cut surfaces after completion of the Contour measurement. The ICHD measurements represent the relaxed and redistributed residual stresses present after the contour cut. Therefore to obtain the initial residual stresses at the ICHD measurement locations before the Contour cut, the residual stresses obtained from the Contour measurements were then taken into account, more specifically the reconstructed in-plane residual stresses from the finite element analysis stage of the contour technique. A superimposition of the measured and reconstructed residual stresses at the measurement locations was therefore carried out for ICHD1, ICHD2 and ICHD3 to obtain the "corrected" residual stresses. It is important to note that the reconstructed residual stresses from the mirrored analysis of the contour measurement were used for this superimposition due to being the most accurate. The "corrected" residual stress results for ICHDs 2 and 3 are shown in Fig.4a.

Finally, the results from the mirrored contour analysis are compared with the longitudinal residual stresses obtained from the iDHD measurement at weld toe 1 in Fig.4b. Overall, the mirrored contour measurement did not find the two residual stress peaks at the surface, nor did it find the steep stress gradient measured using DHD. This was probably due to a loss in depth resolution of the Contour results due to too much smoothing of the CMM raw data within the analysis. However, the general trend and peak residual stresses of $0.6\sigma_y$ and $0.73\sigma_y$ for iDHD and $0.7\sigma_y$ for mirrored

Materials Research Proceedings **2** (2016) 341-346 doi: http://dx.doi.org/10.21741/9781945291173-58

contour were similar. Also both measurement techniques showed residual stresses around 0.3σy near to the weld toe surface and low compression at a depth around $0.4t_0$.

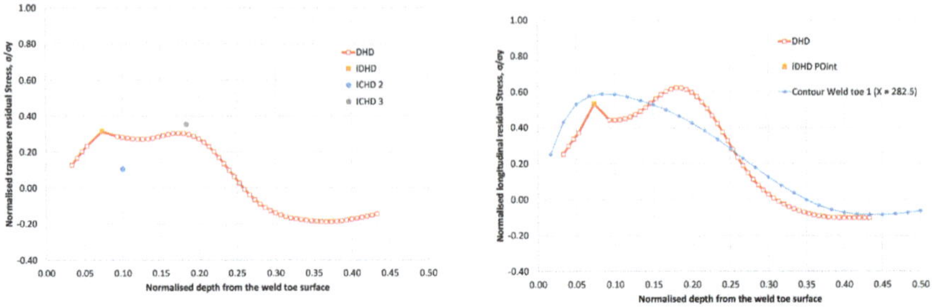

Fig.4: Comparison made between the transeverse (a) and longitudinal (b) residual stresses obtained from the mirrored Contour analysis and the iDHD measurements.

Conclusion

In the present work, DHD, ICHD and contour method have been coupled to measure the residual stress field within a complex welded part.

DHD and contour methods gave similar results in terms longitudinal stress under the fillet weld. ICDH and DHD results were also relatively close in terms of transverse stress but some scatter between the data was observed. Regarding the contour method, its application to such a complex part revealed all the interest of this method to fully describe the out-of-plane stress field. However, the involved contour cut induced artefacts and thereby scatter in the obtained results. Finally, if the occurrence of plastic stress relaxation is taken into account during DHD with local use of iDHD, the correction of plastic relaxation occurring during the contour method appears as a challenge to tackle.

Acknowledgements

This work is part of the research programm MUSICAS, which is partly financed by the French Single Interministerial Fund (FUI).

References

[1] R. H. Leggatt, D. J. Smith, S. Smith, and F. Faure, "Development and experimental validation of the deep hole method for residual stress measurement", Journal of Strain Analysis for Engineering Design, Vol 31, No 3, pp 177-186, (1996). https://doi.org/10.1243/03093247V313177

[2] Kingston, E. K., et al., "Novel applications of the deep-hole drilling technique for measuring through- thickness residual stress distributions", Journal of ASTM International, Vol 3, No 4, (2006). https://doi.org/10.1520/JAI12568

[3] Hosseinzadeh, F. et al. "Towards good practice guidelines for the contour method of residual stress measurement", Journal of Engineering (2014).

[4] Traore, Y. et al. "Measurement of the Residual Stress Tensor in a Compact Tension Weld Specimen", Experimental Mechanics, 53(4):605-618. https://doi.org/10.1007/s11340-012-9672-7

Materials Research Proceedings **2** (2016) 347-352 doi: http://dx.doi.org/10.21741/9781945291173-59

Characterisation of the Effect of Corrosion on the Residual Stresses in Girth Weld Pipe Using Ultrasonic Calibrated with Strain-relieving Measurement Techniques

R. Romac[1a*], D. Cave[1b], X. Ficquet[1c]

[1] VEQTER Ltd, 8 Unicorn Business Park, Whitby Road, Bristol BS4 4EX, UK

[a]remi.romac@veqter.co.uk, [b]douglas.cave@veqter.co.uk, [c]xavier.ficquet@veqter.co.uk

Keywords: Residual Stress, Girth Weld, Ultrasonic Measurement, Contour, iDHD, Corrosion

Abstract. The structural integrity of an oil and gas pipeline is increasingly factored into the design of new installations to ensure that operating risks are maintained low. In addition, the life extension of existing assets beyond their original anticipated design life, as a result of the current oil price environment and the need to optimize field development expenditure, is an ongoing challenge. Operators would like to extend pipeline service life, while many of the technologies required for the validation of their ongoing condition are not yet mature enough to provide confidence that this is a viable strategy. One of the issues considered as a key threat to pipeline integrity is corrosion. Therefore understanding the distribution and redistribution of residual stresses within a pipeline affected by corrosion can be of great benefit. A way to monitor in-situ the pipeline residual stress is to use the ultrasonic (UT) technique. The paper aimed to assess and calibrate the US technique on a pipeline mock-up in the presence of a typical local corrosion damage mechanism. Contour, iDHD, ICHD, XRD and ultrasonic measurements were carried out before machining a flaw to produce an accurate FE model of the pipe. The residual stress was then measured during the manufacture of the flaw and was compared with the FE prediction. Ultrasonic measurements were then carried out on the outer surface of the pipe and show a significant increase in the residual stress. The Ultrasonic technique can therefore, be used to monitor the changes in the residual stress which may be caused by corrosion.

Introduction

Across the globe, 5 metric tons of steel are degenerated every second with the offshore industry. The main fault on pipelines in the North Sea and the Gulf of Mexico is caused by internal corrosion. About 67% of the global pipeline are now older than 20 year which is the minimum design life [1]. Guideline provides guidance to estimate the life of a component in the presence of a flaw, analytically or with finite element analysis [2]. The finite element analysis can be less conservative than the analytical solution and, therefore, increase the in-service time of the pipeline. In order to create a representative FE model of the pipeline, an accurate description of the weld residual stress is important. In this study, the Contour, the Incremental Deep Hole Drilling (DHD), the Incremental Center-Hole Drilling (ICHD) and XRD residual stress measurement techniques were employed and compared on an as-welded section of pipeline.

In order to evaluate the life of a pipeline, the corrosion growth needs to be known. Ultrasonic technique is already used to measure the wall thickness of the pipes. This study, however, looks at the measure of the residual stress using ultrasound in an as-welded zone and in a zone with thinner wall thickness. To obtain this results the ultrasonic system needs to be first calibrated. The calibration can be done using a tensile test or using absolute residual stress values. In this study, the ultrasonic method was calibrated using strain relieving technique.

Materials Research Proceedings **2** (2016) 347-352 doi: http://dx.doi.org/10.21741/9781945291173-59

The Ultrasonic technique

The measurement of residual stress with Ultrasonic is mainly done via standard crack detection ultrasonic probes. The primary difference is about the way to use the probes. For each measurement, the pulse is sent to a transmitter probe via a Pulser-Receiver device. The voltage signal is sent to a Piezo-electric probe which vibrates and produces an ultrasonic wave sent into the Acrylic wedge. This wave travels through the acrylic wedge at a lower speed than the metallic material tested. As it comes out of the probes, the direction of propagation of the wave is parallel to the probe. Once it reaches the specimen, it starts to travel at an angle equal to the sinus of the wedge angle divided by the speed of the longitudinal ultrasound in the acrylic material, then multiplied by the speed of the longitudinal ultrasound in the specimen, as per the formula below:

$$\frac{\sin\theta_1}{V_{Lw}} = \frac{\sin\theta_2}{V_{Ls}}$$

Where θ_1 is the angle from the ultrasonic emitting probe to the normal to the specimen surface, θ_2 is the angle from the normal of the specimen surface to the refracted wave, and V_{Lw} and V_{Ls} are the material longitudinal ultrasonic velocity respectively for the wedge and the specimen.

If the angle of the probe is appropriately set-up on the wedge – which depends on the material tested – then the wave travels below the surface of the specimen at an angle parallel to the surface of the specimen. The wave is then picked up by a first receiver, distant from around 18.2mm and the second one around 34mm from the emitter probe. Both probes receive the longitudinal critically refracted (LCR) wave but at different times. The arrival time for each probe is measured and the difference of Time-of-Flight between the two probes are then calculated. Fig. 1a shows the cross section of the acrylic wedge.

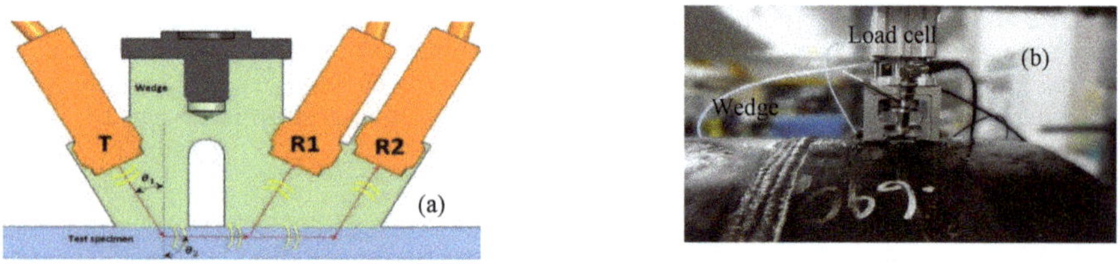

Fig. 1: (a) Diagram showing the Lcr technique principle, (b) Setup of the ultrasonic measurement

In order to calculate the residual stress, the following formula is used:

$$\Delta\sigma = \frac{E(t - t0)}{L11.t0}$$

L11 is called the acoustoelastic coefficient and translate of the slope of the difference of residual stress with a changing Time-of-Flight [3]. It is inherent to the material. E is the Young's modulus, t is the recorded Time-of-Flight (TOF) for the current measurement and t0 is the recorded Time-of-Flight at the initial state or stress-free state. For the Young's Modulus, a fixed value of 203GPa was used.

In order to determine the Time-of-Flight, several analyses are being used for the calculation. The common way of determining the Time-of-Flight between two ultrasonic signals is by picking up a specific event in the signal that could be comparable with different signals. The point must be an event that is equally happening for each echo and is only alterable because of the residual stress [4].

Materials Research Proceedings **2** (2016) 347-352 doi: http://dx.doi.org/10.21741/9781945291173-59

Absolute Measurement technique description

Absolute residual stress measurements were carried out on the pipe in order to determine the full residual stress field. The iDHD, Contour, ICHD and XRD techniques were used.

The iDHD technique can provide bi-axial residual stress through the thickness of the joint [5]. This technique can measure steep gradient through the thickness, however the results are only provided along the measurement line and documented in [6].

The Contour residual stress measurement technique is a destructive mechanical strain relief technique. The technique is based on a variation of Bueckner's elastic principle of superposition [7] and consists of three main stages. The technique provides a full uniaxial map of the residual stress perpendicular to a cut plane through the specimen. Therefore providing the location of the maximum and minimum residual stress peaks. The results are also documented in [6].

The XRD technique was used to provide non-destructive "surface" measurements, giving a single measurement data point which represents the average residual stresses acting within the first 10μm of the component surface.

The ICHD technique [6] provided residual stress profiles from 0.05mm – 1.00mm depth at each location. As mentioned previously, the ICHD measurements were carried out according to the ASTM standard assuming an isotropic linearly elastic material and with a plane stress condition.

Component description and measurement locations

The pipe specimen was made from ferritic steel API 5L X52. As illustrated in Fig. 3b, the specimen had a length of about 410mm and an outer diameter of 762mm. It contained a seam weld along the length of the pipe and a circumferential girth weld at mid-length of the pipe. A cut-out had been machined on the inner diameter at mid-length and has an axial length of about 150mm, reducing the thickness of the pipe from 25mm to about 19mm in the weld area. It was assumed that both the weld and parent metals had the same material properties with a Young's modulus of 203GPa, a Poisson's ratio of 0.3 and a yield strength of 450MPa.

A total of 75 XRD measurements were carried in the ID and OD across the weld at loc1 and Loc2. In this paper only the measurements carried on the weld centreline (WCL) and the heat affect zone (HAZ) at 20mm from the WCL in the OD and at 10mm from the weld root in the ID are presented. Three ICHDs were performed on the OD at LOC2, on the WCL, 20mm away from the WCL and 38mm away from the WCL for ICHDs 1, 2 and 3 respectively. The three other ICHDs were performed at LOC2 but on the inner diameter, on the WCL, 10mm and 30mm away from the WCL for ICHDs 4, 5 and 6 respectively.

Two iDHDs were performed at two specific locations: at the WCL and in the HAZ in order to determine the through thickness residual stresses induced by the welding process. Since ICHDs have been performed at LOC2, iDHD A, was performed at LOC1 so it could be drilled at a location where the weld hadn't been polished. It was performed through the WCL and drilled from the outer diameter of the pipe, with a hole diameter of 1.5mm and extracting a 5mm core. iDHD B was performed at LOC3, 90° from the seam weld and at 10mm axially from the WCL.

The Contour measurement location was an axial-radial plane through the pipe, opposite the initial pipe splitting saw cut location. The cut was arranged such that the contour surfaces would permit the hoop residual stresses to be measured.

Destructive residual stress measurements results and discussions

The results of the absolute measurements for the hoop direction can be seen in Fig. 2. Fig. 2a shows the through thickness residual stress at the WCL location. At the near OD surface the XRD results show a low compression value which rapidly increase to high tension as shown by the ICHD results. The ICHD results are clearly overestimating the hoop stress. The ICHD tend to overestimate the residual stress when its value are near the yield strength. The iDHD and contour show high tensile stress throughout the thickness of the weld. The iDHD gave value near yield stress while the contour

Materials Research Proceedings **2** (2016) 347-352 doi: http://dx.doi.org/10.21741/9781945291173-59

result was lower. The last ICHD measurement carried out at the ID agreed very well with the iDHD measurement.

Fig. 2b shows the through thickness residual stress at the HAZ location 10mm from the WCL. Again the XRD measurements show some low compression value near the OD and ID surfaces. Which raise rapidly to high tension value as shown by both ICHD measurement. At this location both Contour and iDHD provide the same results with higher tension value about 400MPa at the OD decrease slowly toward the ID to values around 250MPa.

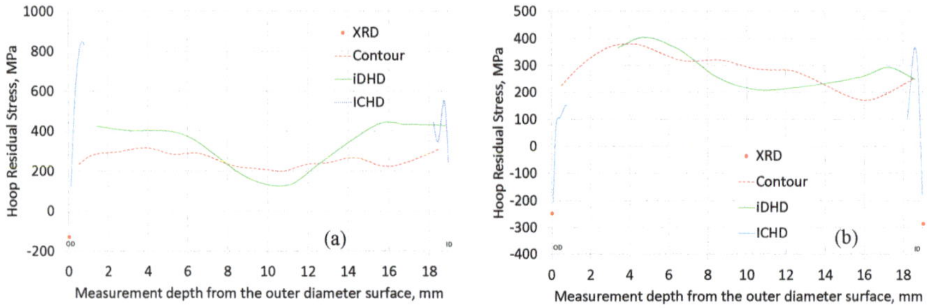

Fig. 3a shows the hoop residual stress from the contour measurement in a 2D map. High tensile stress is found at the girth weld location with a higher tensile value near the HAZ than in the WCL. The iDHD measurement however, has shown that at this location the tensile stress is in the same order of magnitude. Compression residual stress balancing the tensile stress can be seen from 25mm away from the WCL.

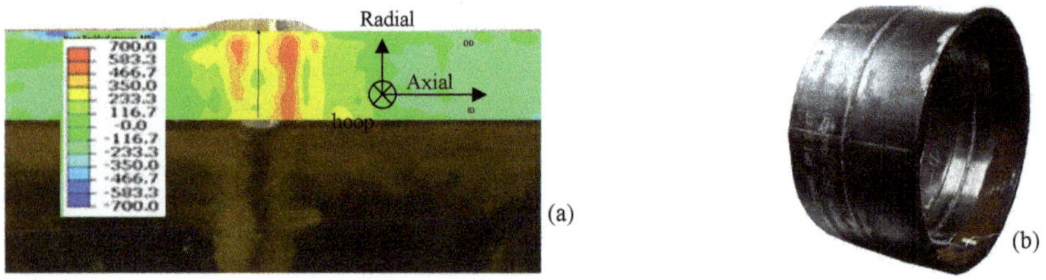

Fig. 3: (a) 2D map of the contour measurement, (b) Photograph of the specimen

Ultrasonic residual stress measurement results and discussions

There are several sources of uncertainty when measuring the residual stress using ultrasonic waves. One of the difficulty is the accuracy of the time-of-flight measurement. Indeed, although it was found that longitudinal waves are the most sensitive to residual stress [3], the resolution needed to detect a change in the acoustic speed is generally high. A material with an acoustoelastic coefficient of 1 would mean that 100MPa would represent a change of around 7ns. As a consequence, it is necessary that the signal acquisition device is accurate enough. The pulse is sent via an ultrasonic pulser at 15Hz into the transmitter probe. It is also sent towards a fast sampling oscilloscope to set-up the trigger reference point. The oscilloscope then displays the longitudinal critically refracted echoes picked-up by the two receivers at a resolution of 0.4ns. The uncertainty due to the sampling equipment was therefore found to be ±1.5MPa in the hoop direction which is very low.

Materials Research Proceedings 2 (2016) 347-352 doi: http://dx.doi.org/10.21741/9781945291173-59

Other sources of error when measuring can occur such as the temperature, the texture and the couplant thickness. It was seen [4] that a change of temperature can significantly affect the acoustic speed. Therefore, the temperature was monitored in lab condition for each test using a data logger and several thermocouples and was found to be constant around 21°C.

When using the contact method, ultrasonic gel needs to be applied at any interfaces between the wedge, the transducers and the specimen in order to allow the wave to be transmitted into the material to reach the receivers. Uncertainties can however occur if the thickness of this gel is not properly controlled. The slightest change in couplant thickness can induce hundreds of MPa differences. In order to minimize errors associated with couplant thickness, constant and consistent pressure needs to me maintained on the wedge. As it can be seen in Fig.1b this was controlled by using a pneumatic ram and a load cell to monitor the force to be 179.9±7.3N in the hoop direction. The probes used were screw-in Time-Of-Flight diffraction probes and therefore the gel thickness and contact between the probes and the acrylic wedge remained constant throughout the test. Finally, the two regions examined were compared against each other and so parent vs parent, HAZ vs HAZ and weld vs weld, therefore the textures should be comparable.

Calibration of the ultrasonic measurement

In order to determine the acoustoelastic constant L_{11} and t_0 a tensile test need to be carried out on a dog bone type stress-free sample extracted from the studied sample. The objective would be to measure the acoustic velocities for different applied load. A uni-axial strain gauges would be fitted on the tensile specimen to determine the stress due to the load. Hence, one could accurately determine the relation between the Time-of-flight measured and the stress in the material. However, this test can be avoided if the L_{11} constant is calibrated with absolute residual stress measurement which is the aim of this study. Therefore Incremental Centre Hole Drilling measurements averaged over their depths have been used to calibrate the acoustic velocities to the absolute residual stress magnitudes. Fig. 4a shows the calibrated ultrasonic measurements compared with the iDHD, ICHD and contour measurements. It can be seen that there is very good agreement with the DHD and Ultrasonic results. The acoustoelastic constant for this material was found to be 4.1 in the hoop direction, which is in the range of other materials tested. The calibration process also improved the determination of the stress free wave velocity in the hoop direction in this material and so providing a vertical datum, or stress free reference to adjust the overall levels by.

Once the ultrasonic calibration has been done, there is high interest in to measuring a different state of stress, i.e. weld corrosion, in order to characterise any residual stress changes which is the subject of the next section.

Corrosion simulation

Offshore pipelines are subject to internal corrosion reducing the structural integrity of the pipe. An, example corrosion mechanism to be considered is Preferential Weld Corrosion. Typically, the deepest PWC zones cover about 30% of the circumference of the pipe.

A study was carried out to evaluate the change in residual stress as a groove was incrementally machined out to about 5-6% of the pipe circumference. In order to measure the change in the residual stress field a technique similar to the DHD was used. A 1.5mm diameter hole was created and measured as the groove was machined as seen in Fig. 4b. The results of this work are not presented in this paper.

After the final groove increment at about 95% deep. Ultrasonic were carried out across the weld near the groove location (LOC4) and compared with the previous measurement without the groove, shows in Fig. 4b. The ultrasonic measurement used the same calibration found before the groove. The results show a much higher tensile peak of about 1500MPa than the as welded pipe location. The ultrasonic measurement after the groove seems to overestimate the residual stress. However, a clear difference in residual stress can be observed using a non-destructive technique of a pipe with a

Materials Research Proceedings **2** (2016) 347-352 doi: http://dx.doi.org/10.21741/9781945291173-59

groove i.e. with internal corrosion than an intact pipe. Further work could investigate the magnitude of the peak with different groove depth and geometries.

Conclusion

A residual stress study was carried out on a 762mm diameter and 19mm thick pipe representative of an offshore pipeline. Comprehensive measurements were carried out to determine the full residual stress field using the Contour, the iDHD, the ICHD and XRD. It was found that all four techniques were in good agreement. Using the ICHD technique, the ultrasonic method was calibrated to also provide absolute residual stress across the weld on the OD. A very good agreement was found between the ultrasonic measurement and the other destructive techniques. Offshore pipelines are subject to corrosion. A groove was machined internally to simulate some corrosion. The calibrated ultrasonic machine was then used on the groove location to measure the change of residual stress. The residual stress at the groove location was measured and found to be much higher than the measured stress at an intact location. The magnitude of the measured stress at the groove location was overestimated, however, the non-destructive technique clearly identifies a rise in the residual stress. Further investigation should be performed to characterise the overestimation.

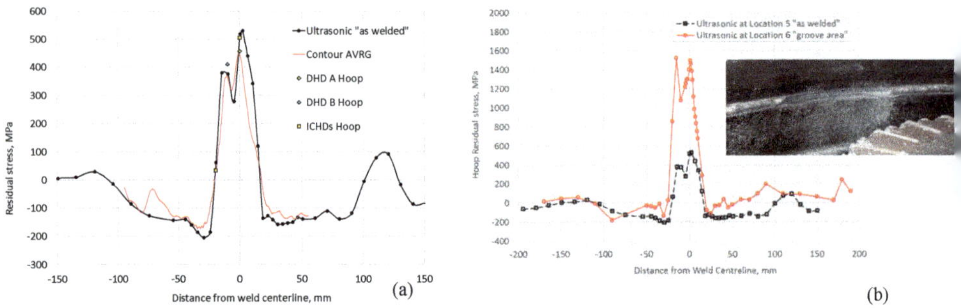

Fig. 4: Comparison of the ultrasonic measurement across the weld, (a) without grove compared with contour and iDHD, (b) with and without the presence of a groove

Acknowledgements

The authors would like to thank Dale McIntyre and Herman Rincon from ConocoPhillips (U.S.A) for providing the mock-up and supporting this research.

References

[1] DNV, recommended practice det norske veritas dnv-rp-f116, "Integrity management of submarine pipeline systems", (2009)
[2] BS7910, "Guide on methods for assessing the acceptability of flaws in metallic structures", British Standard Institution. (2005)
[3] A.A. Santos, D.E. Bray, "Ultrasonic Stress Measurements using PC based and commercial flaw detector", Review of Scientific Instruments, (2000). http://dx.doi.org/10.1063/1.1287339
[4] Y. Javadi, H. S. Pirzaman, M. H. Raeisi, M. A. Najafadabi, Ultrasonic inspection of a welded stainless steel pipe to evaluate residual stresses through thickness, (2013)
[5] Mahmoudi, A.H. et al., "A New Procedure to Measure near Yield Residual Stresses Using the Deep Hole Drilling Technique", Journal of Experimental Mechanics, , Vol. 49, No. 4, (2009)
[6] Serasli, K et al., "Measurements of residual stresses using mechanical strain relaxation methods and mapping in a thin-walled girth welded pipe mock-up, Pressure Vessels & Piping conference 2016
[7] Hosseinzadeh, F. et al. "Towards good practice guidelines for the contour method of residual stress measurement", Journal of Engineering (2014)

Residual Stresses 2016: ICRS-10 Materials Research Forum LLC
Materials Research Proceedings **2** (2016) 353-358 doi: http://dx.doi.org/10.21741/9781945291173-60

The Effects of Residual Stress on Elastic-Plastic Fracture: Two Diffraction Studies

Harry Coules[1,a*], Graeme Horne[1,b] and Matthew Peel[1,c]

[1] University of Bristol, Queens Building, University Walk, Bristol BS8 1TR, United Kingdom

[a]harry.coules@bristol.ac.uk, [b]graeme.horne@bristol.ac.uk, [c]matthew.peel@bristol.ac.uk

* Corresponding author

Keywords: Residual Stress, Fracture Mechanics, Neutron Diffraction, Synchrotron X-Ray Diffraction, Finite Element Analysis

Abstract. Brittle fracture in structural materials is strongly influenced by the presence of residual stresses. During more ductile fractures the influence of residual stress is reduced by plastic deformation prior to the initiation of fracture. In structural integrity assessments, it is therefore important to account for the interaction between applied loads and residual stresses. This interaction was studied in two experiments. In the first, residual stresses in aluminium alloy 7075-T6 three-point bend specimens were measured prior to loading using neutron diffraction. In the second, the stress field in compact tension specimens of ferritic pressure vessel steel was mapped using energy-dispersive synchrotron X-ray diffraction as the specimens were loaded. In both cases, finite element modelling of the elastic-plastic fracture was performed using residual stresses either reconstructed from, or validated using, experimental results. This method of analysis allowed us to compare the effect of residual stress on two different fracture mechanisms. The aluminum specimens fractured in brittle manner, and analysis of experimental load-displacement results showed that residual stresses were superimposed almost linearly with externally-applied stresses at the point of fracture initiation. In the steel specimens, fracture initiated by ductile tearing and it was shown using finite element analysis that residual stress had only a very small effect on the crack-driving force at higher levels of applied load. In fact, at the point of tearing initiation prior strain-hardening caused by the indentation process used to introduce residual stresses had a greater effect on the crack-driving force than the residual stress state itself.

Introduction

Residual stresses can strongly affect the initiation of fracture in brittle materials [1]. However, in more ductile fractures the effect is diminished by partial relaxation of the residual stress, which is enabled by plasticity prior to fracture initiation. As a consequence, some structural integrity assessment procedures such as the R6 procedure [2] and the British Standard BS7910:2013 [3] include methods to account for this effect. This is achieved using approximate methods. For example, in the Failure Assessment Diagram (FAD) approach used in R6, a plasticity correction factor V is introduced to represent the reduction in the effect of "secondary" stresses. Secondary stresses are defined as those stresses that can influence fracture initiation but do not contribute to plastic collapse when this is the dominant failure mechanism. Typically, residual and thermal stresses are classed as secondary. By contrast "primary" loads are those which affect both failure mechanisms and include, for example, pressurisation and gravity loads. Although it is relatively simple to apply, the plasticity correction method can give overly-conservative results for some geometries. In certain situations, disproportionately severe plastic deformation can occur close to a defect as a result of high residual stresses remote from it – a situation referred to as elastic follow-up. In cases where there is significant elastic follow-up, the V factor approach has the potential to give a non-conservative assessment result unless the elastic follow-up effect is considered [4].

In the past, the interaction between primary and secondary stresses during fracture has been widely studied using finite element analysis, often for the purpose of developing improved techniques for structural integrity assessment. However, there has been little experimental work on this topic due to the difficulty involved in measuring the complex stress fields which occur close to a crack tip. We performed two experiments to observe these stress fields and, using this data in conjunction with finite element analysis, investigated the effect of residual stress on the mechanism of elastic-plastic fracture [5], [6].

Method

Brittle fracture. Single Edge Notched Bend (SEN(B)) specimens of 7075-T6 aluminium alloy with dimensions 120 x 30 x 15 mm were prepared as shown on Fig 1. Half of the 16 SEN(B) specimens were compressed in the out-of-plane direction using a pair of circular indenters, before the "crack" (actually an Electrical Discharge Machined notch) was cut into them. The other half were notched in the as-received condition. The residual elastic strain field on the mid-thickness plane of the specimen in the region of the notch (see Fig 1) was measured using neutron diffraction. The measurements were performed using the SALSA diffractometer at the Institut Laue-Langevin (Grenoble, France). Measurements were taken at 2.5 mm intervals over most of the measurement plane and at 1 mm intervals in a small region surrounding the crack tip. Finally, both the indented and non-indented sets of specimens were loaded to failure in three-point bending and their apparent plane-strain fracture toughness was calculated.

Elastic-plastic finite element analysis was performed to estimate the elastic-plastic strain energy release rate at the point of fracture for both sets of specimens. This analysis used a residual stress field for the indented specimens that was reconstructed from the measured neutron diffraction data using an iterative technique [7]. Loading of the residually-stressed specimen was then simulated and the energy release rate was determined by calculating the J-integral as a function of through-thickness position on the crack front. To account for the effect of residual stress, the modified J-integral formulation due to Lei was used [8], [9].

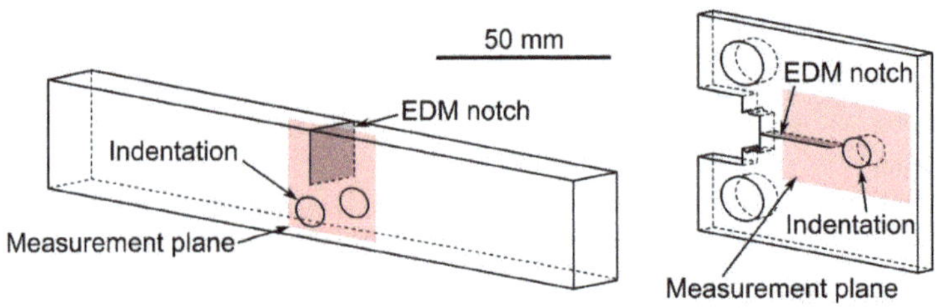

Figure 1: Indented Single Edge Notched (SEN(B)) and Compact Tension (C(T)) fracture specimens used in the experiments.

Ductile tearing. This experiment used modified Compact Tension (C(T)) specimens of a ferritic pressure vessel steel (Fig 1). As in the previous experiment, a residual stress field was introduced in some of the specimens by local compression ahead of the notch tip. However, the steel specimens exhibited a much more ductile fracture process and were loaded until the initiation of ductile tearing. Loading of the specimens was carried out on the I12 beamline at Diamond Light Source (Oxfordshire, UK). Energy-Dispersive X-ray Diffraction (EDXD) was used to measure the stress field surrounding the crack at incremental loading steps at a spatial resolution of 1 mm over most of the specimen, and 0.25 mm in a region surrounding the crack tip. X-ray radiography was used to detect the initiation of ductile tearing at the notch tip (Fig 2c).

Materials Research Proceedings **2** (2016) 353-358 doi: http://dx.doi.org/10.21741/9781945291173-60

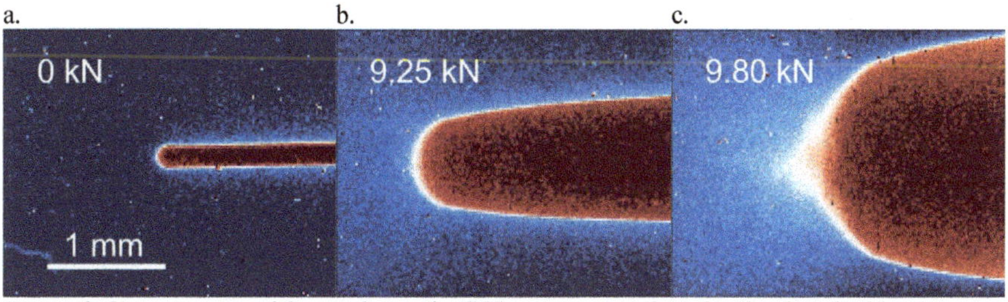

Figure 2: X-ray images of the notch tip of a C(T) specimen at three levels of applied load, showing the initiation of ductile tearing in c.).

As in the brittle fracture experiment, finite element analysis was performed to predict the specimens' elastic-plastic strain energy release rate. However, the initial residual stress field assumed to exist when modelling the steel C(T) specimens was derived from elastic-plastic modelling of the indentation process, validated using the EDXD measurements taken at incremental loading steps.

Results and discussion

In both experiments, we measured the residual stress field that resulted from out-of-plane compression followed by introduction of a crack. Results for each type of specimen under zero external load are shown in Fig 3. Significant opening-mode crack loading occurs under the action of residual stress alone in both the SEN(B) and the C(T) specimens.

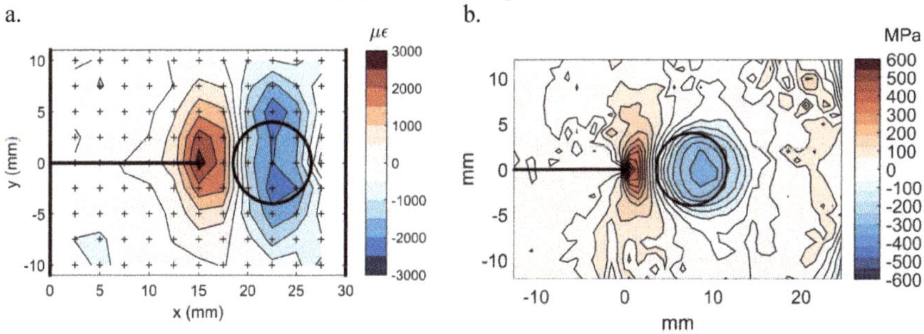

Figure 3: Measured residual elastic strain and residual stress fields around crack tips in two fracture specimens. In each case the crack-transverse component of strain/stress is shown. a.) Residual elastic strain in an aluminium alloy 7075-T6 three-point bend specimen, measured using neutron diffraction. Crosses indicate measurement gauge volume centres. b.) Residual stress in a modified compact tension specimen of pressure vessel steel, measured using EDXD.

The contribution of residual stress to the stress intensity factor for the SEN(B) specimens was calculated via the weight function method using neutron diffraction data from an indented but un-notched specimen. This calculation predicted a contribution of 11.3 MPa √m from residual stress. Under three-point bend loading, the SEN(B) specimens all failed in an almost completely brittle manner. The load at failure was used to calculate the contribution of externally-applied loading to the apparent Mode I stress intensity factor. The indented (i.e. residually-stressed) specimens fractured at a consistently lower load than the unindented ones. As shown in Fig 4, the difference in apparent fracture toughness was 13.2 MPa √m. Therefore the reduction in apparent Mode I plane strain

Materials Research Proceedings **2** (2016) 353-358 doi: http://dx.doi.org/10.21741/9781945291173-60

fracture toughness was similar to the contribution of residual stress to the crack-tip stress intensity, as would be expected for a brittle fracture.

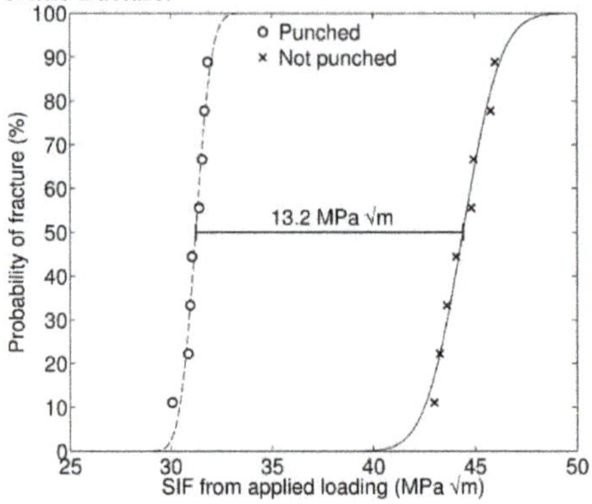

Figure 4: Curves indicating the cumulative probability of failure for indented and unindented SEN(B) specimens, based on the results of 16 fracture tests.

The stress field within the C(T) specimens was measured not only under zero external load (as shown in Figure 3b), but also at incremental loading steps up to the initiation of tearing. Figure 5 compares measured distributions of stress surrounding the crack in an indented specimen with those calculated using FEA. There is good agreement between the two methods and it was concluded that the FEA results were sufficiently accurate for reliable calculation of the elastic-plastic strain energy release rate.

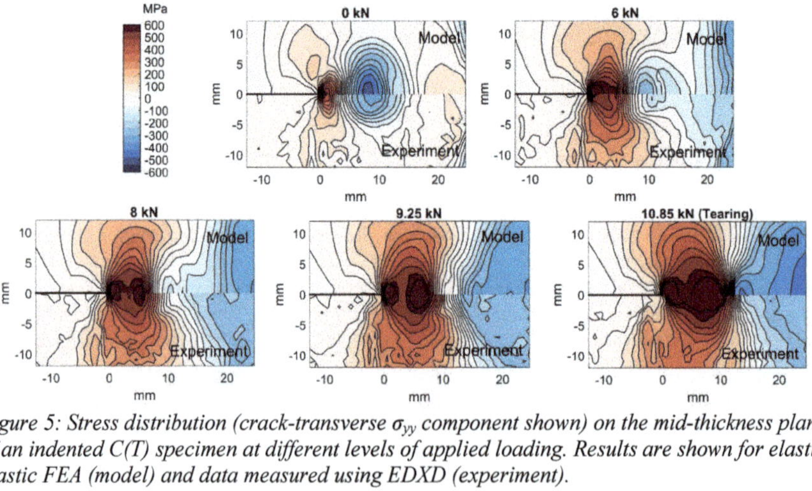

Figure 5: Stress distribution (crack-transverse σ_{yy} component shown) on the mid-thickness plane of an indented C(T) specimen at different levels of applied loading. Results are shown for elastic-plastic FEA (model) and data measured using EDXD (experiment).

Elastic-plastic equivalent stress intensity factor results calculated from finite element analysis of the SEN(B) and C(T) loading tests are shown in Fig 6. The crack-driving force is initially zero for the specimens containing no residual stress (solid blue lines), while there is a finite crack-driving force for the indented specimens (solid red lines). The SEN(B) specimens (Fig 6a) exhibited almost

Materials Research Proceedings **2** (2016) 353-358 doi: http://dx.doi.org/10.21741/9781945291173-60

perfectly brittle fracture; their K_J values deviate only a small amount from the result for a theoretical perfectly-elastic specimen (dashed lines). For the indented SEN(B) specimens, the model included an initial residual stress field reconstructed from neutron diffraction measurements. By comparing the through-thickness distribution of K_J at fracture to that for an unindented specimen, it was concluded that the reconstruction technique was a viable method for incorporating measured residual stress data into the elastic-plastic analysis of fracture.

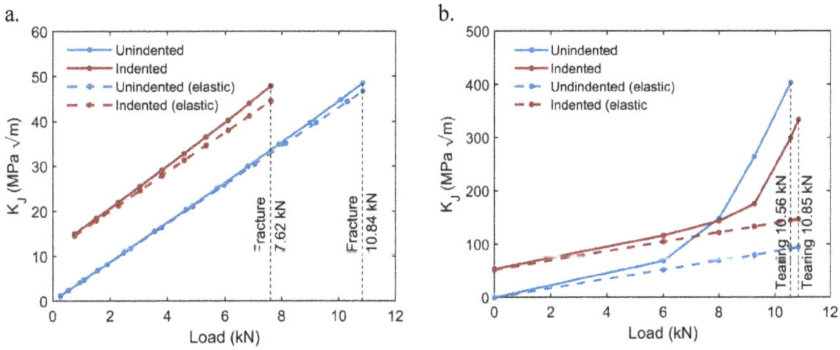

Figure 6: Elastic-plastic equivalent stress intensity factor at the mid-thickness of four specimens (two SEN(B) specimens and two C(T) specimens) as a function of applied load. Mode I SIF for a theoretical perfectly-elastic specimen is also shown in each case. a.) SEN(B) specimens, b.) C(T) specimens. Note different vertical scales.

The C(T) specimens (Fig 6b) show a much more ductile fracture than the SEN(B) specimens, and so deviate strongly away from the predictions of linear-elastic theory. Surprisingly, K_J for the unindented specimen rises above that for the indented specimen at higher levels of applied load. Using FEA, it was shown that this is due to strain-hardening of the material during indentation. This strain-hardening causes a spatial variation in initial hardness in the indented specimen which, like residual stress, affects the distribution of plastic deformation that occurs in the specimen as it is loaded to failure. Most structural integrity assessment procedures, including R6, contain relatively detailed treatment of residual stresses but put less emphasis on the effect of spatial variations in initial material hardness. These results suggest that hardness variations resulting from the processes which introduce residual stress may have a more significant effect on ductile fracture than is often considered. In future experiments, we plan to use the techniques developed here to study fracture occurring under elastic follow-up conditions, and the effect of primary/secondary stress interaction on elastic-plastic crack propagation.

Conclusions

1. The contribution of two well-characterised residual stress fields to the initiation of brittle fracture and ductile tearing has been analysed.
2. A method for incorporating measured residual stress data into elastic-plastic cracked-body finite element analysis of fracture via iterative reconstruction of the stress field was shown to be viable.
3. Variations in the strain-hardening state of material often occur in conjunction with residual stress. For ductile fractures, such variations can have a significant effect on the elastic-plastic strain energy release rate.

Materials Research Proceedings **2** (2016) 353-358 doi: http://dx.doi.org/10.21741/9781945291173-60

Acknowledgements
Access to the SALSA diffractometer was provided by the Institut Laue-Langevin (proposal no. 1-02-148) and access to the I12 beamline was provided by Diamond Light Source Ltd. (proposal no. EE11463). We are grateful to Mr Paulo Orrock, Dr Kiranmayi Abburi Venkata, Dr Thilo Pirling, Mr Sam Oliver, Dr Derreck Van Gelderen, Dr Thomas Connolley and the late Prof. David Smith who all assisted with the work described in this article.

References
[1] P. J. Withers, "Residual stress and its role in failure," *Reports on Progress in Physics*, vol. 70, no. 12, pp. 2211–2264, 2007. http://dx.doi.org/10.1088/0034-4885/70/12/R04

[2] *R6: Assessment of the Integrity of Structures Containing Defects, Revision 4, Amendment 11*. EDF Energy, Gloucester, 2015.

[3] *BS 7910:2013 - Guide to methods for assessing the acceptability of flaws in metallic structures*. BSi, 2013.

[4] R. A. Ainsworth, "Consideration of elastic follow-up in the treatment of combined primary and secondary stresses in fracture assessments," *Engineering Fracture Mechanics*, vol. 96, pp. 558–569, 2012. http://dx.doi.org/10.1016/j.engfracmech.2012.09.009

[5] H. E. Coules, G. C. M. Horne, M. J. Peel, S. J. Oliver, D. G. A. V. Gelderen, and T. Connolley, "Direct observation of elastic and plastic strain fields during ductile tearing of a ferritic steel," in *Proceedings of the ASME 2016 Pressure Vessels & Piping Conference*, 2016, no. 63345.

[6] H. E. Coules, D. J. Smith, P. J. Orrock, K. A. Venkata, and T. Pirling, "A combined experimental and modelling approach to elastic-plastic crack driving force calculation in the presence of residual stresses," *Experimental Mechanics (in press)*, 2016. http://dx.doi.org/10.1007/s11340-016-0171-0

[7] H. E. Coules, D. J. Smith, K. A. Venkata, and C. E. Truman, "A method for reconstruction of residual stress fields from measurements made in an incompatible region," *International Journal of Solids and Structures*, vol. 51, no. 10, pp. 1980–1990, 2014. http://dx.doi.org/10.1016/j.ijsolstr.2014.02.008

[8] Y. Lei, "J Calculation for a crack in a welding residual stress field following a FE welding simulation," in *Transactions of SMiRT 23 Manchester, United Kingdom - August 10-14, 2015*, 2015, p. 213.

[9] Y. Lei, "J-integral evaluation for cases involving non-proportional stressing," *Engineering Fracture Mechanics*, vol. 72, no. 4, pp. 577–596, 2005. http://dx.doi.org/10.1016/j.engfracmech.2004.04.003

Residual Stresses 2016: ICRS-10 Materials Research Forum LLC
Materials Research Proceedings **2** (2016) 359-364 doi: http://dx.doi.org/10.21741/9781945291173-61

Residual Stress Study of Al/Al Laminates Processed by Accumulative Roll Bonding

Lihong Su[1a*], Cheng Lu[1b], Huijun Li[1c], Vladimir Luzin[2d], Hui Wang[1e], Kiet Tieu[1f]

[1]School of Mechanical, Materials and Mechatronic Engineering, University of Wollongong, Australia

[2] Bragg Institute, Australian Nuclear Science & Technology Organisation, Australia

[a]lihongsu@uow.edu.au, [b]chenglu@uow.edu.au, [c]huijun@uow.edu.au, [d]vll@ansto.gov.au, [e]hw737@uowmail.edu.au, [f]ktieu@uow.edu.au

Keywords: Accumulative Roll Bonding, Residual Stress, Neutron Diffraction, Finite Element Model

Abstract. In this work accumulative roll bonding (ARB) was used to combine AA1050 and AA6061 sheets to produce AA1050/AA1050, AA6061/AA6061 and AA1050/AA6061 laminates with ultrafine grained (UFG) structure. Two sheets of starting materials were roll bonded with 200 °C preheating for 180 s before rolling. The through-thickness residual stress distribution of these laminates processed up to two cycles of ARB was determined by neutron diffraction with spatial resolution of 0.2 mm through 1.5 mm thickness. The measurements also required high accuracy of only few MPa since residual stresses in the laminates peaked at only about 15 MPa. The laminates composed of the same material (AA1050/AA1050 and AA6061/AA6061) showed symmetric residual stress profile with tensile stress at the centre of the sheets and compressive stress at the surfaces. The AA1050/AA6061 laminates showed asymmetric distribution with residual tensile stress in the AA1050 layer and compressive stress in the AA6061 layer. A finite element model (FEM) was used to simulate the residual stress distribution and the results were in agreement with the measured results qualitatively.

Introduction

Aluminium sheets are widely used as structural components in aircraft and automotive which has strict requirements of strength/weight ratio. Accumulative roll bonding (ARB) is an effective process to produce aluminium sheets with high strengths and it is also possible to bond two types of aluminium sheets in the hope that the properties of the original sheets would combine in the bonded laminates [1-3]. The materials processed by ARB generally experience larger strain magnitudes compared to traditional rolling. Furthermore, the ARB process includes rolling, cutting, surface grinding and stacking in different order, which makes the strain history complex in the bonded laminates. Microstructure, texture and mechanical properties distributions through the thickness of the laminates are non-uniform. This is especially true if different starting materials are used.

In addition, layers of different materials or materials at different thickness positions perform differently upon further deformation developing residual stress which is strongly dependent on the thermomechanical processing history of the materials. Residual stresses of traditional rolled sheets have been well reported [4-6]. Typical residual stress distribution of a rolled sheet material is tensile at the sheet centre and compressive at the sheet surfaces [6]. Residual stress of ARB processed sheets has not been evaluated up to now and it can be inferred that the residual stress distribution of the ARB processed sheets should be more complex than that of conventional rolling since the strain history of the ARB processed sheets is complex due to the cutting and stacking procedure.

For performance evaluation (e.g. fatigue), it is important to know the type and magnitude of the residual stress distribution in the laminates processed by ARB, especially the laminates composed of two different materials. The present work studied the through-thickness residual stress distribution of aluminium laminates produced by ARB using neutron diffraction techniques.

Experimental

Commercial aluminium alloy AA1050 and AA6061 sheets with initial thicknesses of 1.5 mm were used as starting materials. Three types of laminates AA1050/AA1050, AA6061/AA6061 and AA1050/AA6061 were produced by combining two sheets in a combination and rolling them together with 50% reduction so that this results in a laminate of the same 1.5 mm thickness after each ARB cycle. The as-received sheets were annealed to recrystallised condition and strips with size 400 × 50 × 1.5 mm^3 were cut from the annealed sheet for ARB processing. The rolling was performed on a rolling mill with 125 mm diameter rolls at a speed of 0.2 ms^{-1} under dry conditions. Pre-heating at 200 °C for 180 s was performed to the strips prior to each rolling cycle. To ensure good bonding, the surface for bonding of each strip was wire-brushed and cleaned before stacking. The details of the ARB process and the microstructure and mechanical properties of the ARB processed laminates can refer to Ref. [7].

Neutron diffraction residual stress measurements have been carried out at the ANSTO OPAL research reactor using the KOWARI strain scanner. The laminates used for residual stress measurements were 1-cycle roll bonded AA1050/AA1050, AA6061/AA6061 and AA1050/AA6061 laminates and 2-cycle roll bonded AA1050/AA6061 laminate. Residual stress distributions were measured through the entire thickness in the central part of all samples. A gauge volume of 0.4 × 0.4 × 25 mm^3 was used for measurements of 1-cycle roll bonded laminates with 7 through-thickness positions. The step size between each position was 0.2 mm. A gauge volume of 0.2 × 0.2 × 25 mm^3 was used for 2-cycle roll bonded AA1050/AA6061 laminate to resolve its 4-layer structure. The step size was 0.1 mm and the measurements were taken at 13 thickness positions. 5 measurements at different locations of the same depth were taken to achieve better averaging (schematically illustrated in Fig. 1). With the matchstick-like gauge volume a high spatial resolution was achieved in the thickness direction while providing sufficient scattering intensity. It should be noted that there was an overlap between each thickness position and there was an edge effect for all the measurements. The measurements were achieved with high accuracy of ~5 MPa for the 1-cycle laminates and ~10 MPa for the 2-cycle laminate.

A Si(400) monochromator with a take-off angle 79° was used to provide neutron wavelength of 1.73 Å. The Al (311) reflections at 2θ ~90.3° was used for the strain measurements.

At each position, d-spacings along three principle directions, rolling direction (RD), transverse direction (TD) and normal direction (ND) were measured. Residual stresses along two principle directions RD and TD and d_0 was calculated according to the method described in Ref. [8] by assuming that the normal stress is zero and there are no shear components.

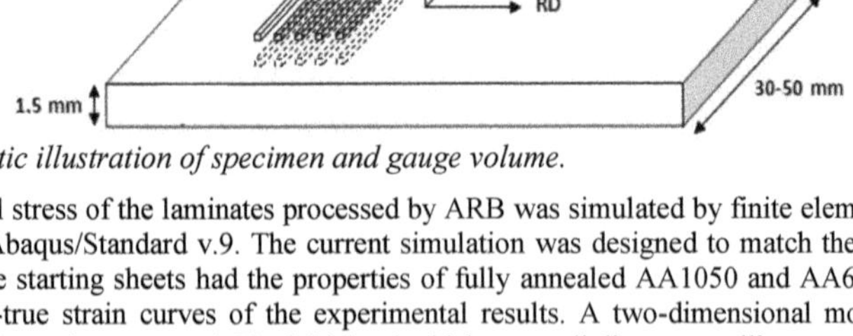

Fig. 1 Schematic illustration of specimen and gauge volume.

The residual stress of the laminates processed by ARB was simulated by finite element modelling (FEM) using Abaqus/Standard v.9. The current simulation was designed to match the experimental conditions. The starting sheets had the properties of fully annealed AA1050 and AA6061 by fitting the true stress-true strain curves of the experimental results. A two-dimensional model assuming plain strain deformation was used. The initial strip thickness, roll diameter, rolling speed and nominal reduction was set the same as in the experiment. The friction between the sheet and the roll surfaces was described by Coulomb's friction law with a friction coefficient of 0.25.

FEM of the real ARB process is difficult to achieve because the two sheets are separate prior to ARB and are joined together as one piece during roll bonding. However, for FEM, the two sheets will either be pre-joined as one piece or deform separately as two sheets during and even after ARB. In the current work, both cases were simulated. The case with two sheets deform separately was achieved by assuming a larger friction coefficient of 0.4 between the sheets.

Results and discussions

Fig. 2 shows the through-thickness residual stress distribution profiles of the ARB processed laminates. It can be seen from Fig. 2(a) and (b) that the through-thickness residual stress distribution for 1-cycle roll bonded AA1050/AA1050 and AA6061/AA6061 are similar to each other and are in accordance with the residual stress profile in rolled aluminium solid plate [6]. The profiles of the residual stresses in both RD and the TD are symmetric in general, with tensile residual stress zone at

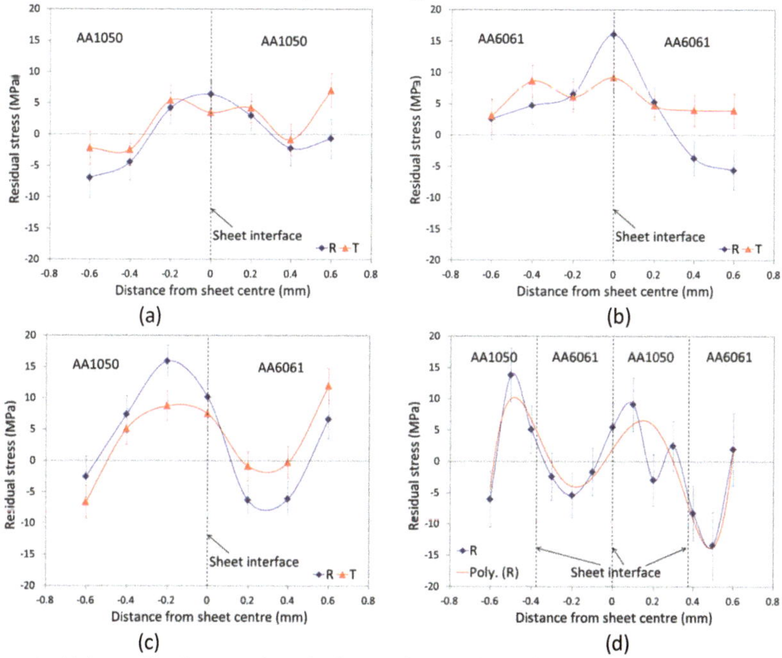

Fig. 2 Through-thickness distribution of residual stress for 1-cycle roll bonded (a) AA6061/AA6061, (b) AA1050/AA1050, (c) AA1050/AA6061, and (d) 2-cycle roll bonded AA1050/AA6061. (◆R Residual stresses along RD, ▲T Residual stresses along TD, —Poly. (R) polynomial curve fit of the residual stresses along RD of the 2-cycle roll bonded AA1050/AA6061)

the centre and compressive stress zones near the surfaces, except for a few locations which have residual stress values not in the general trend. Theoretically, the residual stresses close to the surfaces should be compressive to balance the central tensile zone so that overall residual stress is zero. However, residual stresses at one side of the surfaces of AA6061/AA6061 appear to be tensile instead of compressive and this is most probably due to a combination of unaccounted experimental errors and methodological biases. Since measurements in this experiment are at the limit of spatial

Materials Research Proceedings **2** (2016) 359-364 doi: http://dx.doi.org/10.21741/9781945291173-61

resolution and accuracy of the neutron diffraction technique and the absolute values of stresses are extremely small, even small experimental uncertainties ~5-10 MPa that were not accounted for, e.g. local stress variations, can lead to distortions affecting the real stress profile. In addition, the stress calculations were based on isotropic d_0 model, which in general is a good approximation, but it can also lead to a relatively significant and experimentally noticeable shift due to anisotropic microstresses because of the delicate stress distribution in our case. The distortion of stress profile must be more pronounced in AA6061/AA6061 due to its ability to acquire higher density of dislocations, thus the stress profile for this alloy tends to have more unbalanced stress distribution affected by biases.

The residual stress distribution of dissimilar AA1050/AA6061 laminates is quite different from that of the AA1050/AA1050 and AA6061/AA6061. It can be seen in Fig. 2(c) and (d) that tensile residual stress exists in the AA1050 layer and compressive residual stress exists in the AA6061 layer in general for both the 1-cycle and 2-cycle roll bonded sheets. Since the two alloys have different plastic properties (i.e. yield stress), residual stress arises as a result of the different deformation behaviours in the two alloys. After non-uniform deformation of the layers, a permanent curvature occurs (even though the curvature is very small) bringing stress momentum balance and finally shaping the stress profile. It is difficult to predict the residual stress distribution theoretically because of the complexity of the bonding process.

The measured residual stresses for all the laminates are in a range of ±15 MPa. This is also in accordance with the values reported in rolled aluminium [6, 9]. The residual stresses in rolling direction and transverse direction show similar characteristics but the absolute values are slightly smaller in transverse direction due to difference in constrain conditions. Since the laminates are only 1.5 mm thick and 0.2 mm spatial resolution of the neutron diffractometer is on its limit as well as accuracy of less than 5 MPa, any improvement of data in the neutron diffraction experiment is hardly possible.

Since FEM simulation cannot exactly reproduce the bonding process, two extreme cases were simulated: one with two sheets pre-joined as one piece and one with two sheets deform separately the entire process. Fig. 3 shows the FEM simulation results for the 1-cyle roll bonded AA1050/AA6061. The distorted FEM elements illustrate well the deformation behaviour and material flow during deformation. For the case with two sheets pre-joined as one piece, the interface deforms coherently. The mesh flows faster at the surfaces due to the friction between the sheet and the rolls. It can be seen that the mesh of AA1050 flows faster than AA6061. The thicknesses of the AA1050 layer and the AA6061 layer are the same and both have gone through 50% reduction from the initial state. The residual stress of the AA1050 layer is compressive and the AA6061 layer is tensile. For the case with two sheets deform separately, there is relative displacement between the mesh of the AA1050 layer and the AA6061 layer. The mesh also flows faster at the surfaces. The thickness of the AA1050 layer is slightly smaller than the AA6061 layer and the reduction ratios of the AA1050 and the AA6061 layer are 51.6% and 48.4%. The residual stresses of the two layers are different and are quite complex.

Fig. 4 shows the FEM simulated through-thickness distribution of residual stress along RD for 1-cycle roll bonded AA1050/AA1050 and AA1050/AA6061 in comparison. The simulated through-thickness residual stresses along TD (not shown here) have similar trend to those along RD and have slightly lower absolute values. It can be seen from Fig. 4(a) that the FEM predicted residual stress of AA1050/AA1050 along RD matches the experimental results in general and reproduce symmetric profile. This is not surprising, since for the laminates with the two starting sheets of the same alloy, the deformation is similar to conventional rolling. Although the two sheets also experienced bonding during the rolling, there was no relative movement between the two sheets since they have the same properties.

Materials Research Proceedings **2** (2016) 359-364 doi: http://dx.doi.org/10.21741/9781945291173-61

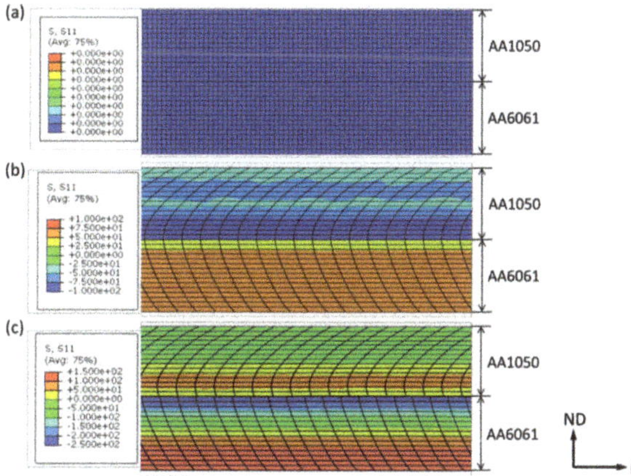

Fig. 3 FEM mesh and residual stress in rolling direction of 1-cyle roll bonded AA1050/AA6061, prior to roll bonding (a), and after roll bonding (b) two sheets pre-joined as one piece and (c) two sheets deform separately.

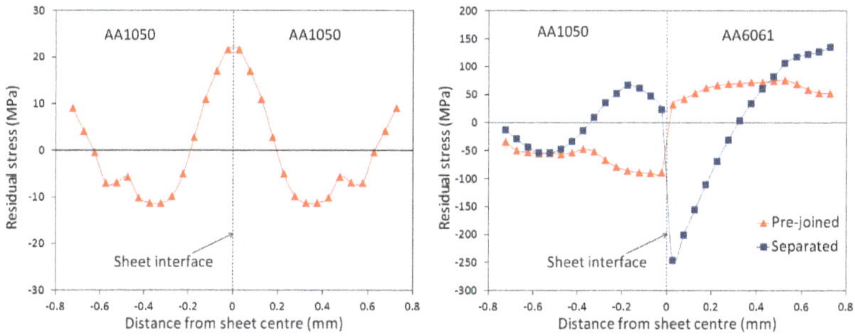

(a) (b)

Fig. 4 FEM simulated through-thickness distribution of residual stress along RD for 1-cycle roll bonded (a) AA1050/AA1050 and (b) AA1050/AA6061 with two sheets pre-joined as one piece and two sheets deform separately.

For the laminates with two starting sheets of different alloys, the two sheets are expected to deform differently because they have different plastic properties. In fact, they experience slightly different degree of reduction (larger reduction was observed for AA1050; it was slightly longer after the first cycle ARB). The FEM simulations are able to demonstrate this asymmetry in case of AA1050/AA6061 joint, However, the two extreme cases (pre-joint vs. separated) result in very different residual stress distributions, especially at the areas around the interface. This demonstrates that proper modelling of the contact phenomenon (e.g. amount of constraint versus amount of sliding) is the most critical element of the simulation. Qualitatively, the result of the two sheets deform separately (marked as Separated in Fig. 4(b)) agrees better with the experimental result. The absolute values of both cases are significantly larger than the values measured in experiment and also larger than the values of the simulated AA1050/AA1050.

The fact that the simulation results of the two sheets deform separately matches the experimental results better indicates that the two sheets deform more freely in the roll bonding process. Bonding

Materials Research Proceedings **2** (2016) 359-364 doi: http://dx.doi.org/10.21741/9781945291173-61

must occur in the very later stage of rolling deformation. The residual stress induced by the thermomechanical misfit of dissimilar materials is significantly larger in magnitude according to the simulation. The experimentally measured values, however, are very low. This may be because the simulation overestimates the difference in deformation resistivity of the two alloys and there may be a softening effect in the experiment due to the pre-heating before roll bonding.

Summary

Residual stress investigation was conducted on ARB processed AA1050/AA1050, AA6061/AA6061 and AA1050/AA6061 laminates of 1.5 mm thickness. Through-thickness stress profiles were determined with high spatial resolution (0.2 mm) and high measurement accuracy (<5 MPa). The results showed that the laminates composed of the same alloy (AA1050/AA1050 and AA6061/AA6061) showed symmetric residual stress profile with tensile stress at the centre and compressive stress under the surfaces. The AA1050/AA6061 laminates showed asymmetric distribution with residual tensile stress in the AA1050 layer and compressive stress in the AA6061 layer. FEM simulation of the laminates composed of the same alloy showed similar residual stress distribution as the experiment. The simulation results of the residual stress distribution in laminates composed of dissimilar alloys can only match the measured result qualitatively.

References

[1] T. Hausöl, H.W. Höppel, M. Göken, Tailoring materials properties of UFG aluminium alloys by accumulative roll bonded sandwich-like sheets, J. Mater. Sci. 45 (2010) 4733-4738. http://dx.doi.org/10.1007/s10853-010-4678-y

[2] M.Z. Quadir, O. Al-Buhamad, L. Bassman, M. Ferry, Development of a recovered/recrystallized multilayered microstructure in Al alloys by accumulative roll bonding, Acta Mater. 55 (2007) 5438-5448. http://dx.doi.org/10.1016/j.actamat.2007.06.021

[3] A. Weck, P. Bisaillon, L. Nong, T. Meunier, H. Jin, M. Gallerneault, Mechanical properties of the aluminum roll-bond laminate AA5005-AA5083-AA5005, Mater. Sci. Eng. A 528 (2011) 6186-6193. http://dx.doi.org/10.1016/j.msea.2011.04.037

[4] W. Razny, F.D. Fischer, G. Finstermann, W. Schwenzfeier, K. Zeman, The influence of some rolling parameters on the residual stresses after rolling, J. Mater. Process. Tech. 60 (1996) 81-86. http://dx.doi.org/10.1016/0924-0136(96)02311-4

[5] U.S. Dixit, P.M. Dixit, A study on residual stresses in rolling, Inter. J. Mach. Tool. Manu. 37 (1997) 837-853. http://dx.doi.org/10.1016/S0890-6955(96)00052-1

[6] M.B. Prime, M.R. Hill, Residual stress, stress relief, and inhomogeneity in aluminum plate, Scripta Mater. 46 (2002) 77-82. http://dx.doi.org/10.1016/S1359-6462(01)01201-5

[7] L. Su, C. Lu, A.K. Tieu, G. Deng, X. Sun, Ultrafine grained AA1050/AA6061 composite produced by accumulative roll bonding, Mater. Sci. Eng. A 559 (2013) 345-351. http://dx.doi.org/10.1016/j.msea.2012.08.109

[8] M.T. Hutchings, P.J. Withers, T.M. Holden, T. Lorentzen, Introduction to the Characterization of Residual Stress by Neutron Diffraction, Taylor & Francis Group, LLC, Broken Sound Parkway NW, 2005, pp. 201-204.

[9] B. Denkena, Y. Altintas, P. J. Arrazola, T. Aoyama, D. Axinte, S. Dreier, B. Denkena, Determination of Residual Stresses in Plate Material by Layer Removal with Machine-integrated Measurement, Procedia CIRP 24 (2014) 103-107. http://dx.doi.org/10.1016/j.procir.2014.07.137

Residual Stresses 2016: ICRS-10 Materials Research Forum LLC
Materials Research Proceedings **2** (2016) 365-370 doi: http://dx.doi.org/10.21741/9781945291173-62

Residual Stress Analysis on Oxide Layers Obtained by High Temperature Oxidation of Chromia-Forming Alloys

Ning LI[a], Ji XIAO[b], Nathalie PRUD'HOMME[c], Linwei LI[d] and Vincent JI[e*]

ICMMO/SP2M, Université Paris-Sud, bat. 410, 91405 Orsay Cedex, France

[a] ning.li@u-psud.fr , [b] ji.xiao@u-psud.fr , [c] nathalie.prudhomme@u-psud.fr,

[d] linwei.li2@u-psud.fr , [e] vincent.ji@u-psud.fr ,

Keywords: High Temperature Oxidation, Residual Stress, Water Vapor, X-Ray Diffraction Stress Analysis, Chromia-Forming Alloy

Abstract: The oxide layers formed during high temperature oxidation of metallic alloys depend on experimental conditions (oxidation gas composition, gas pressure, temperature, duration etc...) and often with complex structure or multilayer structure. Residual stress can be generated not only dues to oxide growth at high temperature (growth stress) but also during cooling of layer/metallic alloy system after oxidation (thermal stress). The determination of the level and the distribution of the residual stresses in oxide layers are very important to determine the influence of oxidation condition in one hand and to estimate the mechanical component's durability at high temperature in the other hand. Two Chromia-forming alloys have been studied: a nickel based Inconel 600 alloy (Ni-17%Cr-8%Fe-1%Mn) and a ferritic AISI 430 steel (Fe-17%Cr-1%Mn). Oxidation test has been carried out at different temperatures (from 600°C to 900°C) for various durations (from 2 h to 96 h) under different absolute humilities (from 0% to 19%). After oxidation of Inconel 600, the oxide layers are composed essentially by an external NiO layer and by an internal spinel $NiCr_2O_4$ layer. While the AISI 430 steel forms an external spinel $Mn_{1.5}Cr_{1.5}O_4$ layer and an internal Cr_2O_3 layer. The residual stresses (RS) have been analyzed by X-Ray Diffraction (XRD) method in each of oxide layers after oxidation tests. In oxide layers, the RS are compressive and the RS levels are more important in internal layer than those in external layer. Overall, the compressive RS in oxide layers increase with oxidation temperature, oxidation duration and absolute humidity.

Introduction

At high temperature, almost all metals are not thermodynamically stable in air or in atmosphere where oxygen exists, and a solid oxide scale will form on the surface. Oxidation of Chromia-forming alloys has drawn lots of attention in recent years, for their good oxidation resistance and low cost. It is true that lots of studies have been done to study the oxidation behavior of Chromia-forming alloy in dry air, but humidity exists in all stages of oxidation in reality. Therefore, studying the influence of water vapor on Chromia-forming alloys oxidation at high temperature has great importance. Some works have been done to study the influence of water vapor on oxidation [1-6], and several mechanisms have been proposed to explain the influence of water vapor on oxidation, such as Fuji and Meussner [2] proposed a dissociation mechanism, and suggested that hydrogen appears to play an important role in the oxidation process as an oxygen carrier between the separated inner and outer scale layers. Shen Jianian [3] suggested that breakaway oxidation happened in the water vapor containing atmosphere is the result of formation of microcracks and microchannels during oxidation. H. Asteman [4, 5] observed the evaporation of Chromia hydroxide. According to J. Ehlers the entry of molecular H_2O into the oxide scale was the main process leading to breakaway oxidation [6]. Obviously, the mechanism of Chromia-forming alloys oxidation with water vapor is still a contentious question.

It has been known that the residual stresses in the oxide scale are closely related to oxidation kinetic and the microstructure. Although some research works were already performed for Chromia-forming alloys to determine the residual stresses in the oxide scale after oxidation under dry air [7, 8], very few works were focused on residual stresses after oxidation in atmosphere with water vapor.

The purpose of this work was to investigate the influence of water vapor on Chromia-forming alloys at high temperature, and AISI 430 and Inconel 600 alloys were selected.

Experimental

The nominal chemical composition of the materials used in this study is listed in Table 1. Before oxidation, the samples were cut into a dimension of 10mm x 10mm x 1mm. All 6 surfaces of the samples were polished by SiC paper followed by silicon solution to make all the samples have the same surface state. Oxidation experiments were performed by thermal gravimetric analysis (TGA) (model SETARAM 92-16.18) under artificial air with different absolute humidity, and the precision of TGA analysis is more than 10^{-6} g. For AISI 430 alloy, the oxidation temperature is 700°C-800°C, and for Inconel 600 alloy the temperature is 600°C-900°C. The regulation of absolute humidity is realized by using a vapor generator (SETARAM D/WETSYS-2F) and the sensibility of regulation is about 1.5% in relative humidity. To ensure the reliability of the experiments, every oxidation conditions has been conducted at least 2 times and the results were reproducible.

After oxidation, the samples were investigated by using FEG-SEM (Field Emission Gun-Scanning Electronic Microscope) / EDX (Energy-dispersive X-ray spectroscopy) (ZEISS SUPRA 55VP) to study the surface morphology and the cross section, and by Grazing Incident X-ray diffraction (GIXRD) to identify the oxide phase formed during the oxidation.

In this study, European standard NF EN 15305 (version of April 2009) [9] has been applied to determine the residual stresses levels in different oxide layers. In our experiments, a high resolution configuration XRD system (Panalytical X'Pert MRD Pro) was used under Copper radiation (λ=0.154 nm), whatever the relative low diffraction angle 2θ, the determined peak position error is smaller than 0.005° in 2θ. For each selected plane family, more than 13 peaks are recorded. The calculation of the residual stress was performed via X'Pert Stress software. The precision for stress determination is about 30 MPa under our experimental conditions. The details of the operation parameters for determining residual stresses by XRD method is given in Table 2.

Table 1 Chemical composition of AISI 430 and Inconel 600 alloys (weight %)

	Fe	Cr	Mn	Si	S	C
AISI 430	Bal.	16-18	< 1.00	< 1.00	< 0.03	< 0.08
Inconel 600	6-10	14-17	< 1.00	< 0.5	< 0.015	< 0.15

Table 2 Operation parameters for determining residual stresses by XRD method

Analyzed zone	2mm x 2mm
ψ angle amplitude	[-60° ; 60°]
Number of ψ angle	13
diffraction angle 2θ	33.6° for $\{104\}_{Cr2O3}$, 35.1° for $\{311\}_{Mn1.5Cr1.5O4}$, 63° for $\{110\}_{NiO}$, 57° for $\{511\}_{NiCr2O4}$
2θ step	0.04°
Acquisition time	15s/step
Peak position error	0.005°
Young's modulus (GPa)	Cr_2O_3: 280 [10]; $Mn_{1.5}Cr_{1.5}O_4$: 250 [11]; NiO:190 [10]; $NiCr_2O_4$: 280 [10]
Poisson's ratio	Cr_2O_3: 0.29 [10]; $Mn_{1.5}Cr_{1.5}O_4$:0.27 [11]; NiO: 0,25 [10]; $NiCr_2O_4$: 0,29 [10]

Results and discussion

AISI 430 alloy. Figure 1 gives the 2° GIXRD patterns of AISI 430 alloy after oxidation in air with different humidity at 700°C, 800°C for 96h. It is clear that the oxide scales are mainly consisted of Cr_2O_3 and $Mn_{1.5}Cr_{1.5}O_4$ whatever the oxidation atmospheres, indicating that the water vapor doesn't affect the scale composition. The intensity of substrate peaks is larger in the atmosphere containing water vapor because the oxide scales formed during the oxidation are thinner than them formed in the dry air.

Observation of the cross section after oxidation in air with 5% absolute humidity at 800°C for 48h by FEG-SEM is given in Figure 2, and it shows clearly that the oxide scale consisted of two layers. With the FEG-SEM observation and XRD analysis, we can conclude that the outer layer is $Mn_{1.5}Cr_{1.5}O_4$ (with little Fe incorporation) and the inner layer is Cr_2O_3. Moreover, formation of thin and discontinuous SiO_2 was also observed at the scale/substrate interface. Isolated voids were found at the scale/substrate interface, which may be result from outward diffusion of ions.

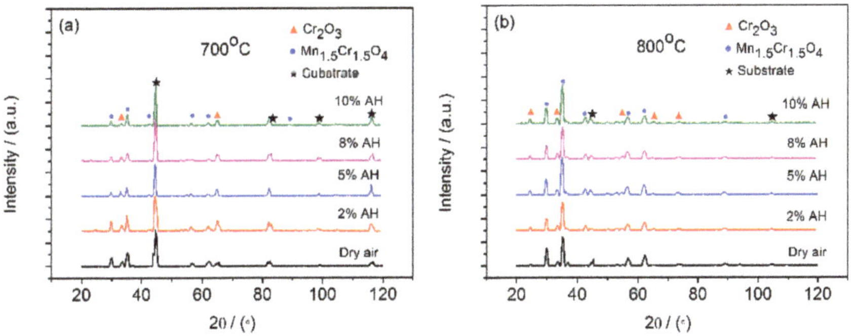

Figure 1 2° GIXRD pattern of AISI 430 after oxidation in air with different absolute humidity (AH) at 700°C for 96h (a), at 800 °C for 96h (b)

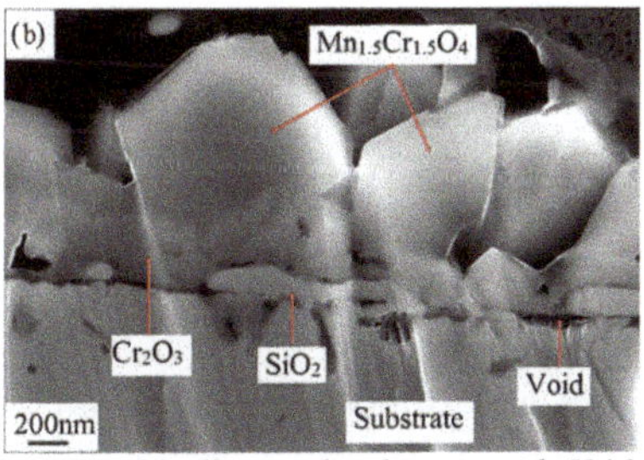

Figure 2 Cross section of the AISI 430 alloy oxidized in air with 5% humidity at 800°C for 48h

Residual stresses in protective Chromia layer after oxidation in air with different absolute humidity at different temperature are given in Figure 3. It is clear that the concentration of water vapor in the atmosphere can affect the residual stresses levels in the Chromia layer. At 700°C, the compressive residual stresses increase with the increase of water vapor. For 800°C, the higher the concentration of water vapor, the bigger the residual stresses at first 24h and smaller the residual stress after 48h. With the presence of water vapor, stresses relief was observed at 800°C.

Figure 4 shows the residual stresses in spinel layer after oxidation in air with different absolute humidity at different temperature, indicates that the residual stresses in this layer are increased with

Materials Research Proceedings **2** (2016) 365-370 doi: http://dx.doi.org/10.21741/9781945291173-62

the introduction of water vapor in the oxidation atmosphere. At all temperatures studied, the higher the concentration of water vapor, the bigger the level of residual stresses. The stresses relaxation is also observed in this layer, this might due to the effect of accommodation between the two oxide layers. The stresses in spinel layer decease with the deceasing of stress in the Chromia layer.

The residual stress values in Chromia layer is more important than those in $Mn_{1.5}Cr_{1.5}O_4$ spinel layer because the Chromia is in inner layer and the RS are generated not only by oxide formation but also by ion diffusion and the external layer formation. With the increase of absolute humidity (AH), the RS level increases in compression and it is probably related especially to oxide layer formation kinetics. Until now, there is no convincing argument to explain the role of the humidity for RS evolution. After 24h oxidation at 800°C, the residual stresses are relaxed and their level is reduced slowly.

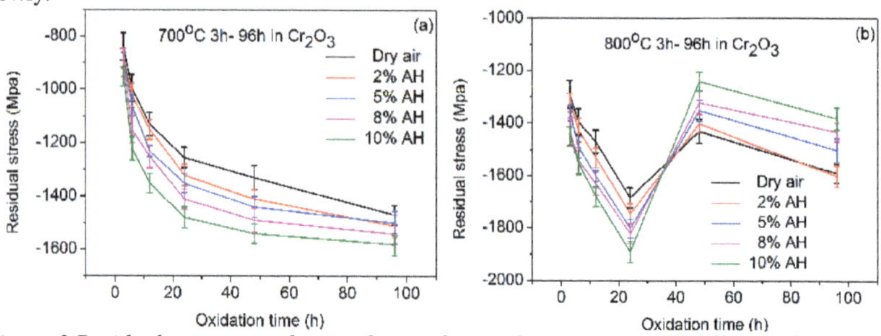

Figure 3 Residual stresses in chromia layer after oxidation in air with different absolute humidity (AH) at 700°C (a) and 800 °C (b)

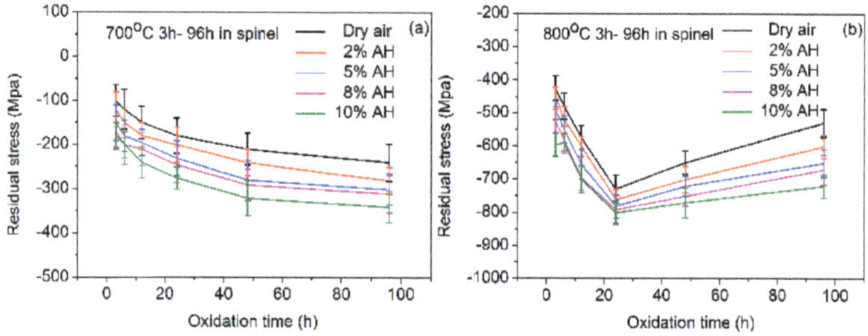

Figure 4 Residual stresses in spinel layer after oxidation in air with different absolute humidity (AH) at 700°C (a) and 800°C (b)

Inconel 600 alloy. Figure 5 gives the 2° GIXRD patterns of Inconel 600 alloy after oxidation in air with different humidity at 700°C, 800°C for 24h.The oxide layers obtained with or without water vapor contained the same mixture of phases (NiO, Cr_2O_3 and $NiCr_2O_4$), but the peak of NiO is more intense than that of Cr_2O_3 and $NiCr_2O_4$. The presence of water vapor has no effect on the nature of the phases in the oxide layer formed on the substrate.

Cross section of the Inconel 600 alloy oxidized at 800°C for 24h in dry air and in air with 5% absolute humidity is given in Figure 6. It is clear that there are two different oxide layers on the substrate, and the inner layer is thicker than the outer layer. The outer layer is mainly composed of Ni and O, but in the inner layer, the concentration of Cr increased obviously. With the XRD analysis, we can conclude that the outer layer is NiO and the inner layer is likely to be a mixed oxide of spinel type $NiCr_2O_4$ ($NiO + Cr_2O_3$) and Cr_2O_3.

Materials Research Proceedings **2** (2016) 365-370 doi: http://dx.doi.org/10.21741/9781945291173-62

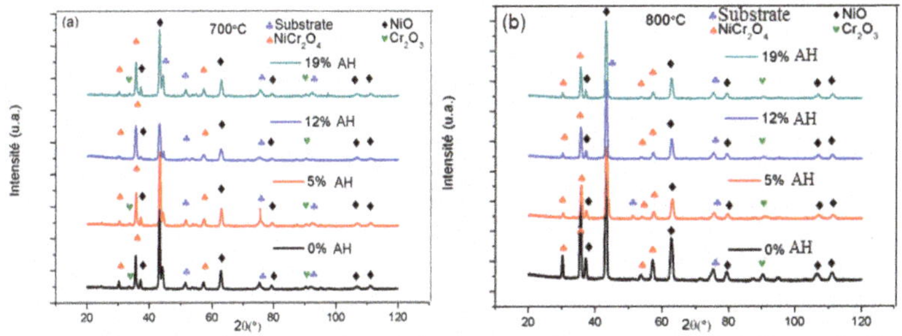

Figure 5 2° GIXRD pattern of Inconel 600 alloy after oxidation in air with different absolute humidity (AH) at 700°C for 24h (a), at 800 °C for 24h (b)

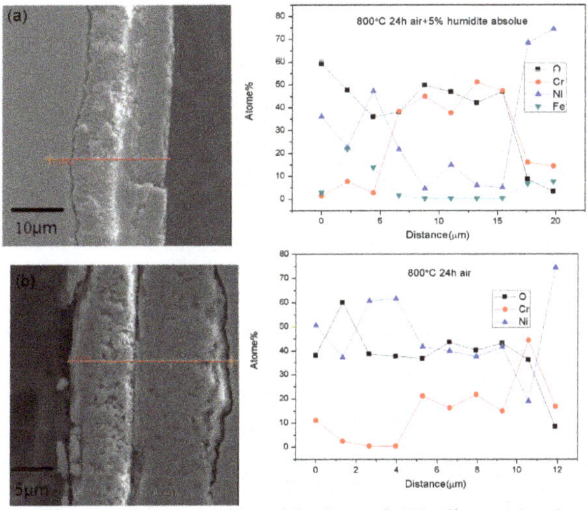

Figure 6 Cross section photos and EDX profiles of the Inconel 600 alloy oxidized at 800°C for 24h in air with 5% AH (a); in dry air (b)

The in-plane residual stresses levels in $NiCr_2O_4$ and NiO layers after oxidation in air with different absolute humidity at different temperatures are shown in Figure 7, indicating that the residual stresses in both layers are compressive. At 900°C, compared to other three temperatures, a significant increase of the stresses levels in the two oxide layers was observed. At 600°C, with or without water vapor, residual stresses in the layer of NiO are larger than those at 700°C and 800°C.

In addition, with the introduction of water vapor at 700°C and 800°C, the levels of residual stresses in the $NiCr_2O_4$ spinel oxide layer are quite similar to those in dry air; but at 900°C, the residual stresses in the $NiCr_2O_4$ spinel phase are much higher than those measured at 800°C.

The residual stress levels in $NiCr_2O_4$ spinel inner layer is more important than those in NiO external layer because the RS in inner layer are generated not only by oxide formation but also by ion diffusion and the external layer formation. With the increase of oxidation temperature, the RS values increase in compression and it is related directly to the increase of oxide layer thickness. There are less evolutions of residual stress when the absolute humidity is more than 5%; this is dues to perhaps

Materials Research Proceedings **2** (2016) 365-370 doi: http://dx.doi.org/10.21741/9781945291173-62

the relative short time of oxidation. With the increase of oxidation time, the RS levels can be relaxed and reduced.

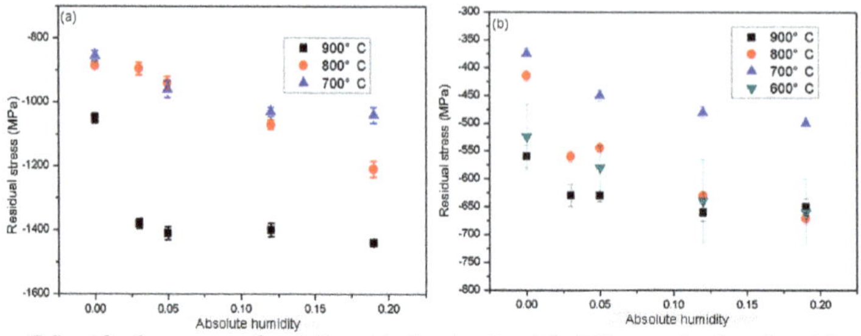

Figure 7 Residual stresses after 24h oxidation in air with different absolute humidity (AH) at different temperature in $NiCr_2O_4$ innner layer (a); in NiO external layer (b)

Conclusions

The oxidation of AISI 430 and Inconel 600 alloys have been carried out in air with different absolute humidity at different temperatures, the basic conclusions from this work can be given as follows:

(1) For AISI 430 alloy, the oxide scale is consisted of an inner Cr_2O_3 layer and an outer $Mn_{1.5}Cr_{1.5}O_4$ layer; and for Inconel 600 alloy, the oxide scale is composed of outer NiO layer and inner $NiCr_2O_4$ spinel layer;

(2) With the introduction of water vapor in the oxidation atmosphere, the composition of the oxide scale is not changed for these two alloys. But the water vapor can affect the surface morphology.

(3) For both alloys, with or without water vapor, the residual stresses in oxide scales are always compressive. The water vapor has obvious influence on residual stresses levels.

Reference

[1]. X. Cheng, Z. Jiang, D. Wei, J. Zhao, B.J. Monaghan, R.J. Longbottom, L. Jiang, Characteristics of oxide scale formed on ferritic stainless steels in simulated reheating atmosphere, Surf. Coat. Technol. 258 (2014) 257-267. http://dx.doi.org/10.1016/j.surfcoat.2014.09.019

[2]. C.T. Fujii, R.A. Meussner, The mechanism of the high-temperature oxidation of iron-chromium alloys in water vapor J. Electrochem. Soc. 111 (1964) 1215-1221. http://dx.doi.org/10.1149/1.2425963

[3]. S. Jianian, Z. Longjiang, L. Tiefan, High-temperature oxidation of Fe-Cr alloys in wet oxygen, Oxid. Met. 48 (1997) 347-356. http://dx.doi.org/10.1007/BF01670507

[4]. H. Asteman, J.E. Svensson, L.G. Johansson, M. Norell, Indication of chromium oxide hydroxide evaporation during oxidation of 304L at 873 K in the presence of 10% water vapor, Oxid. Met. 52 (1999) 95-111. http://dx.doi.org/10.1023/A:1018875024306

[5]. H. Asteman, J.E. Svensson, L.G. Johansson, Oxidation of 310 steel in H2O/O2 mixtures at 600°C: the effect of water-vapour-enhanced chromium evaporation, Corros. Sci. 44 (2002) 2635-2649. http://dx.doi.org/10.1016/S0010-938X(02)00056-2

[6]. J. Ehlers, D.J. Young, E.J. Smaardijk, A.K. Tyagi, H.J. Penkalla, L. Singheiser, W.J. Quadakkers, Enhanced oxidation of the 9%Cr steel P91 in water vapour containing environments, Corros. Sci. 48 (2006) 3428-3454. http://dx.doi.org/10.1016/j.corsci.2006.02.002

[7]. N. Li, J. Xiao, N. Prud'homme, Z. Chen, V. Ji, Residual stresses in oxide scale formed on Fe–17Cr stainless steel, Appl. Surf. Sci. 316 (2014) 108-113. http://dx.doi.org/10.1016/j.apsusc.2014.07.195

[8]. S. Chevalier, C. Valot, G. Bonnet, J.C. Colson, J.P. Larpin, The reactive element effect on thermally grown chromia scale residual stress, Mater. Sci. Eng. A 343 (2003) 257-264. http://dx.doi.org/10.1016/S0921-5093(02)00359-3

[9]. Test Method for Residual Stresses Analysis by X-ray diffraction, in: European Standard no NF15305, 2009.

[10]. A.M. Huntz, Stresses in NiO, Cr_2O_3 and Al_2O_3 oxide scales, Mater. Sci. Eng. A 201 (1995) 211-228. http://dx.doi.org/10.1016/0921-5093(94)09747-X

[11]. W.N. Liu, X. Sun, E. Stephens, M.A. Khaleel, Life prediction of coated and uncoated metallic interconnect for solid oxide fuel cell applications, J. Power Sources 189 (2009) 1044-1050. http://dx.doi.org/10.1016/j.jpowsour.2008.12.143

Residual Stresses 2016: ICRS-10 Materials Research Forum LLC
Materials Research Proceedings **2** (2016) 371-376 doi: http://dx.doi.org/10.21741/9781945291173-63

Neutron Optics Upgrades to the Residual Stress Diffractometer, KOWARI

Mark Reid[1,a*], Scott Olsen[1,b], Vladimir Luzin[1,c], Mark New[1,d], Norman Booth[1,e], Douglas Clowes[1,f], Tai Nguyen[1,g], Ferdi Franceschini[1,h], Adrian Ogrin[1,j], Steven Pangalis[1,k], Ania Paradowska[1,l], Nathan Larkin[2,m], Zengxi Pan[2,n], Nic Hoye[2,o] and Hiroshi Suzuki[3,p]

[1]Australian Centre for Neutron Scattering, ANSTO, Lucas Heights, 2234 NSW Australia

[2] University of Wollongong, Wollongong 2522 NSW Australia

[3] Japanese Atomic Energy Agency, 300-4352 Tsukuba, Japan

[a]markr@ansto.gov.au, [b]sol@ansto.gov.au, [c]vll@ansto.gov.au, [d]mln@ansto.gov.au, [e]normanb@ansto.gov.au, [f]dcl@ansto.gov.au, [g]tai@ansto.gov.au [h]ffr@ansto.gov.au, [j]aog@ansto.gov.au, [k]stevenp@ansto.gov.au, [l]anp@ansto.gov.au, [m]nlarkin@uow.edu.au, [n]zpan@uow.edu.au, [o]nhoye@uow.edu.au [p]suzuki-tokai@ss.em-net.ne.jp

Keywords: Residual Stress, Robotics, Texture, Neutron Optics & Alignment

Abstract. In the last 5 years a number of significant enhancements have been implemented on the neutron beam strain scanner Kowari at the OPAL reactor in Sydney Australia. These changes have resulted in reduced beam time losses when conducting experiments due to sample and stage alignment, and optics and sample changes. There have been 3 projects, starting in 2011 with a new manual slit system design and collision recovery system, in 2013 with a series of radial collimators and finally with the delivery of a 6 axis robot capable of texture measurements and of running up to 20 samples automatically.

Introduction

The thermal neutron residual strain scanner, Kowari, has been operational since 2007. On the residual strain stress scanning instrument KOWARI the spatial resolution ranges from 0.2 to 1000 mm^3. The resolution in lattice strain is of the order 0.01% giving a stress resolution in steels of ~ 20 MPa. The instrument has been operated accordingly to ISO/TS 21432-2005 to achieve reproducible and reliable stress measurements [1,2]. Initially sample alignment and changes to the neutron optics were both slow and ergonomically challenging.

A project commenced in 2011 to design and manufacture a new neutron optics exchange system, to improve the slit system through improved collision protection and appature control and to allow much faster change over between configurations of the neutron optics (currently, three options are available: slit system, radial collimator, open detector). The new system allows for rectangular gauge volumes of quite a large size range, and has extended toolkit for accurate alignment and built-in anti-collision system. After a successfully series of tests with collimators borrowed from the JRR-3M reactor a project commenced in 2013 to design and purchase a series of collimators.

Texture measurements have been undertaken for many years using a standard Euler cradle. Sample changes were thus manually made every 4-8 hours depending on sample and conditions. To increase productivity and lower exposures for users a 6 axis robotic sample changer began operating on KOWARI in 2015 which allows automated testing of 20 (in some cases more) samples.

Adjustable Slit system that is collision resistant and easy to align.

The first part of the project consisted of a rotation stage for the input and output optics system. This stage accurately and easily allowed the neutron optics (primary slits and secondary slits or

Materials Research Proceedings **2** (2016) 371-376 doi: http://dx.doi.org/10.21741/9781945291173-63

collimators) to be swung 90 or 180 degrees and then locked into position. The second part of the project consisted of a new kinematic mount design - a ball and socket design with tangential locating pins that, in the event of a collision would allow the slits to safely rotate away. The slits can then be quickly placed back into the correct position by hand. The slits had to have a very small profile so that they could be placed as close to the sample as possible, a limitation of 4 blade slit systems. The slits are manually adjusted in a vertical range from 1 to 30mm, blades of different sizes allow the horizontal opening to be varied, to any desired opening up to 10mm. The slits package can be adjusted in 2 directions and by 2 rotations (Fig1,2).

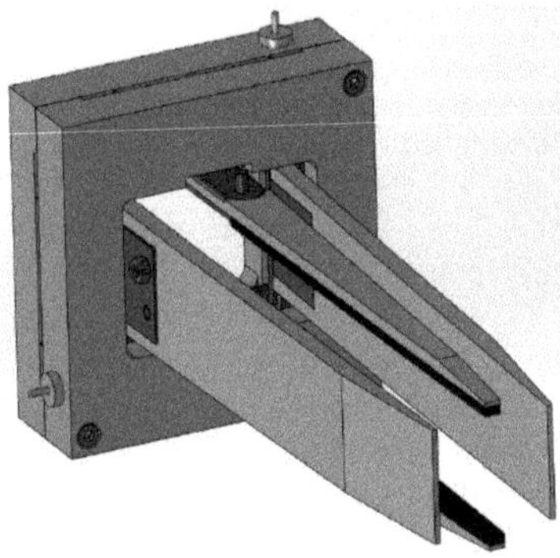

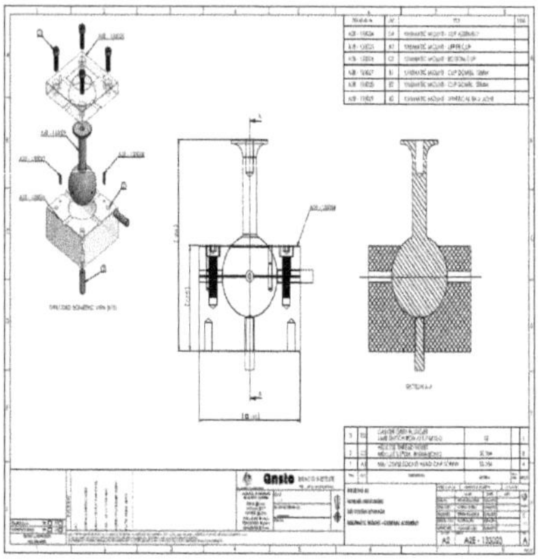

Figure 1. (Left) Solidworks model of the slit system.

Figure 2. (Right) Schematic of the ball and socket system installed to minimize damage to the slits in the event of a collision.

A Series of Radial Collimators.
A series of calculations was undertaken to (see Table 1) compare the performance of a collimator and the existing slit. The results indicated that 2mm and larger sized collimators would provide a similar neutron flux and a similar strain error, as long as there was full detector coverage. The collimators have a 14° degree vertical field of view and 5° horizontal (except for the 10mm collimator which has a 15° horizontal field of view).

Tests were conducted using collimators from JRR-3M, though they were not optimised for our detector size and solid angle. The collimators were built and optical tested in Japan by Hitachi Denki Kogyo and commissioning experiments ran both in Tsukuba and Sydney. On their arrival in 2014 their performance was characterised and they were found to perform as expected. Given in table 2 is a comparison of the performance of the new radial collimators to the existing slit system. It should be noted this is the optimal performance of the slit system and in reality the slits would be expected to not to perform this well as in real experiments they would have to be positioned further from the sample.

Materials Research Proceedings **2** (2016) 371-376 doi: http://dx.doi.org/10.21741/9781945291173-63

Counting statistics characterization

Strain error and intensity ratio

Size		0.5mm	3mm	5mm
Slit	Strain error	$2.1\text{x}10^{-5}$ (100%)	$2.2\text{x}10^{-5}$ (100%)	$2.2\text{x}10^{-5}$ (100%)
	Int. ratio	100%	100%	100%
RC*	Strain error	$4.3\text{x}10^{-5}$ (205%)	$3.0\text{x}10^{-5}$ (136%)	$3.1\text{x}10^{-5}$ (141%)
	Int. ratio	31%	60%	64%
RC**	Strain error	$3.0\text{x}10^{-5}$ (143%)	$2.1\text{x}10^{-5}$ (93%)	$2.2\text{x}10^{-5}$ (100%)
	Int. ratio	62%	110%	118%

* As-measured (~50% detector coverage)
** Assumption of 100% detector coverage

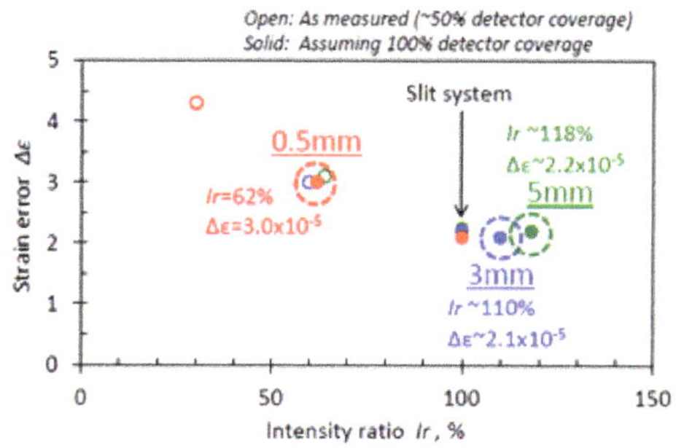

Table 1 (left) – Theoretical comparison of Collimator vs Slit performance for 0.5mm, 3 and 5mm.

Figure 3 (right) – Graphical comparison of the Intensity vs strain error

Size		2mm	3mm	5mm
Slit	Strain error	$4.33\text{x}10^{-5}$ (100%)	$4.66\text{x}10^{-5}$ (100%)	$4.93\text{x}10^{-5}$ (100%)
	Int. ratio	100%	100%	100%
RC	Strain error	$5.56\text{x}10^{-5}$ (129%)	$4.98\text{x}10^{-5}$ (107%)	$4.70\text{x}10^{-5}$ (95%)
	Int. ratio	80%	84%	103%

This highlights one of the major benefits of the RC's, they are positioned 500mm away from the measurement position allowing easy positioning of large samples. Additionally as shown in Fig. 4a & b, they were found to be insensitive to rotational orientation in both intensity and peak FWHM, making alignment trouble free. Since their introduction they have been heavily utilised.

Table 2 Comparison of the strain error and intensity ratio of the existing slit system and the new radial collimators.

Materials Research Proceedings **2** (2016) 371-376 doi: http://dx.doi.org/10.21741/9781945291173-63

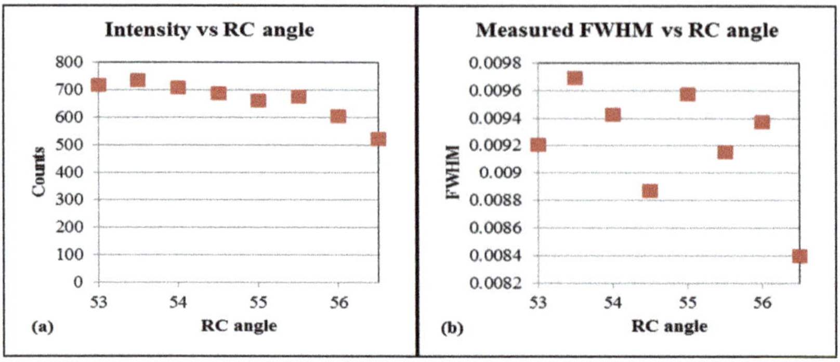

Figure 4 Intensity (a) and FWHM (b) of the measured peak vs. rotational (nominal) angle of the 2mm radial collimator.

A 6 axis robot for higher throughput and texture measurements.

Texture analysis on KOWARI has been undertaken using standard Eulerian cradles since 2007. Generally 3-5 sample changes were undertaken per day. Depending on the activation of the sample and the method of alignment between 10 and 30 minutes was lost per sample change. A 100 sample position 4 axis robot, an Epson E2S5553 had been used on the powder diffractometers at ANSTO since 2009 [3]. In 2011 it was decided that a 6 axis robot, capable of texture analysis would be purchased. At that time few if any facilities had this capability, we are currently aware of 2 facilities that can now do this, STRESS-SPEC at FRM II in Munich [4] and at the LANCSE facility in Los Alamos [5].

An Epson C3 6 axis robot was selected. The specifications are shown in Table 3. Software design commence at the University of Wollongong and an analysis of the robot's accuracy and repeatability with undertaken. The design requirement was <0.2mm, ideally 0.1mm. Tables 4 and 5 over show the results. A rotation in Chi is much simpler as it is a single joint movement only.

EPSON ROBOT MODEL		C3 6 Axis
Configuration		Articulated 6 Axis
Mounting Configurations		Table Top, Ceiling, Angled
		Wall (J1 motion restriction +/- 30 deg)
Payload		Rated 1 kg / Max. 3 kg (5 kg*)
Repeatability		+/-0.02 mm
Axis Rotation	J1 (Turning)	+/-170 deg
	J2 (Lower Arm)	-160 ~ +65 deg
	J3 (Upper Arm)	-51 ~ +225 deg
	J4 (Wrist Roll)	+/-200 deg
	J5 (Wrist Bend)	+/-135 deg
	J6 (Wrist Twist)	+/-360 deg
Horizontal Reach	to mounting face	665 mm
	to wrist center	600 mm
Vertical Reach	to mounting face	885 mm
	to wrist center	820 mm
Speed	J1	450 deg/sec
	J2	450 deg/sec
	J3	514 deg/sec
	J4	553 deg/sec
	J5	553 deg/sec
	J6	720 deg/sec
Cycle Time	1kg workload	0.37 sec

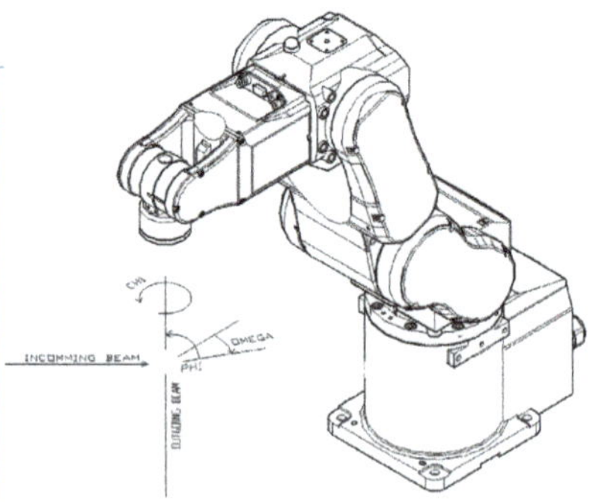

Table 3 (left) specifications for the 6 axis robot selected for texture analysis

Figure 4 – Schematic of the robot showing the principles degrees of rotation

Angle (Chi)	Error to Plane (μm)	Error to Circle (μm)	Total Error (μm)
0	15	27	31
10	6	-38	38
20	10	11	15
30	8	3	9
40	6	-21	22
50	2	1	2
60	11	2	12
70	29	26	39
80	12	18	22
90	11	-29	31

Table 4 – Test with Omega at 0 degrees, of rotations in Chi.

Angle (Phi)	Error to Plane (μm)	Error to Circle (μm)	Total Error (μm)
0	101	-5	101
10	36	4	36
20	82	12	83
30	65	-22	69
40	53	21	57
50	35	-6	36
60	97	0	97
70	105	-16	106
80	14	16	21
90	116	-3	116

Table 5 - Test with Omega at 30 degrees of rotations in Phi, a more complex operation..

After some final in house software modifications and a new 20 position sample changer were fabricated the first texture analysis experiments were undertaken on KOWARI and on the high intensity powder diffractometer WOMBAT in 2015. Comparisons with samples run on conventional Euler cradles have been undertaken (see Figures 7,8) – they show very close agreement. Aside from metallurgical samples geological samples and cultural heritage samples have been recently batch tested [6,7,8].

Figure 6. The 6 axis robot on KOWARI with the 20 position sample changer in the lower right.

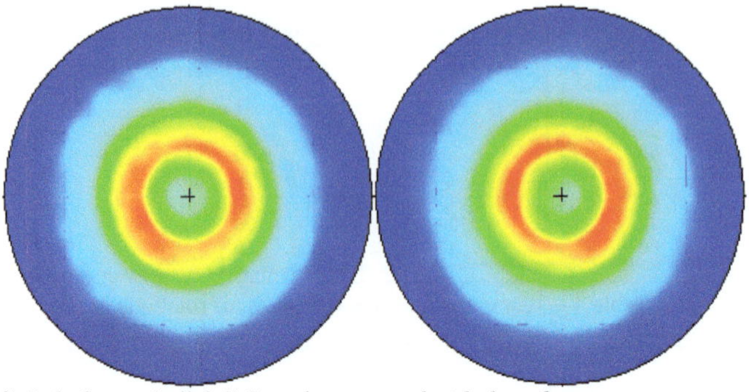

Figure 7 (Left) Pole figure Ag(1,1,1) Sample measured with the robot.

Figure 8 (Right) Pole figure Ag(1,1,1) Sample measured with the Eulerian Cradle.

Summary

Significant upgrades to the Neutron Residual Stress Scanner KOWARI at ANSTO have been undertaken in the last 5 years. There has been a significant reduction in beam time lost due to alignment and sample changing.

References

[1] A.Brule & O.Kirsten, Residual stress diffractometer KOWARI at the Australian Research Reactor OPAL: Status of the project 2006 Physics B 385-386, 1040-1042 (2006) 45

[2] O. Kirstein, V. Luzin & U. Garbe, The Strain-Scanning Diffractometer Kowari, Neutron News Vol. 2009 20(4)

[3] S.R.Olsen, S.A.Pullen, M.Avdeev, A 100 position robotic sample changer for powder diffraction with low background vacuum chamber, Journal of Applied Crystallography 43:2 (2010) 377-379. http://dx.doi.org/10.1107/S0021889810003614

[4] C.H.Randau, H. G. Brokmeier, W. M. Gan, M. Hofmann, M. Voeller, W. Tekouo, N. Al-hamdany, G. Seidl, and A. Schreyer. "Improved sample manipulation at the STRESS-SPEC neutron diffractometer using an industrial 6-axis robot for texture and strain analyses." Nuclear Instruments and Methods in Physics Research Section A: 794 (2015): 67-75. http://dx.doi.org/10.1016/j.nima.2015.05.014

[5] A.S.Losko, S. C. Vogel, H.M.Reiche, and H.Nakotte. A six-axis robotic sample changer for high-throughput neutron powder diffraction and texture measurements. Journal of Applied Crystallography 47, no. 6 (2014): 2109-2112. http://dx.doi.org/10.1107/S1600576714021797

[6] K.A. Sheedy, P. Munroe, F. FSalvemini, V. Luzin, U. Garbe and S.R Olsen, An incuse stater from the series 'Sirinos/Pyxoes', Journal of the Numismatic Association of Australia 26 (2015), 36-52.

[7] F.F.Salvemini, S.R.Olsen, V.Luzin, U.Garbe, J.Davis, T.Knowles, K.A.Sheedy, Neutron tomographic analysis: Material characterization of silver and electrum coins from the 6th and 5th centuries BCE, Mater. Charact. 118, 175-185 (2016). http://dx.doi.org/10.1016/j.matchar.2016.05.018

[8] S.R.Olsen, E.P.Gilbert, N.Booth, S.A. Pullen, P.Imperia, V.K. Peterson, U.Garbe et al. "Novel non-destructive sample analysis techniques using neutron scattering." 8th Australasian Congress on Applied Mechanics (2014): 723-730.

Residual Stresses 2016: ICRS-10 Materials Research Forum LLC
Materials Research Proceedings **2** (2016) 377-382 doi: http://dx.doi.org/10.21741/9781945291173-64

Extension of the Deep-Hole Drilling Method for the Measurement of in-Plane Residual Stresses in Composite Laminates

Carlos Garza[1,a*], Raj Das[2,b], Martyn J. Pavier[1,c], Anton Shterenlikht[1,d] and David J. Smith[1](†)

[1]Department of Mechanical Engineering, University of Bristol, Queen's Building, University Walk, Bristol, BS8 1TR, United Kingdom

[2]Department of Mechanical Engineering, University of Auckland, Auckland 1010, New Zealand

[a]carlos.garza@bristol.ac.uk, [b]r.das@auckland.ac.nz, [c]martyn.pavier@bristol.ac.uk, [d]mexas@bristol.ac.uk

(†) Deceased on November 13th, 2015

Keywords: Composite Laminates, Deep Hole Drilling, Finite Element Analysis, Residual Stress

Abstract. Deep-Hole Drilling (DHD) is a residual stress measurement method generally used in isotropic materials. This study provides an evaluation into applying DHD to determine residual and fit-up stress fields in anisotropic composite materials. In this method, a reference hole with a small diameter is firstly drilled through the thickness of the component. The diameter of the hole is measured accurately and then a cylindrical core of material around the hole is trepanned from the component, relaxing the residual stresses in the core. Finally, the diameter of the reference hole is re-measured and the change in diameter used to calculate the residual stress. For anisotropic materials, the calculation of residual stresses requires the evaluation of distortion coefficients which rely on the mechanical properties of the components. In this work, Lekhnitskii'solution for the distortion of a circular hole in an anisotropic plate and finite element analysis are used to determine these coefficients. Using this technique, the in-plane residual stresses in AS4/8552 composite laminates were experimentally measured and compared to finite element predictions as well as to classical lamination theory. The obtained results indicate that, when using DHD in laminated materials, an optimal ratio between the layers' thickness, the reference hole and trepan diameters is of significant importance since unrelaxed thermally induced residual stresses can lead to inaccurate measurements.

Introduction

Deep-hole drilling (DHD) is a technique that allows the measurement of residual stress fields through very thick components. This technique requires that a small diameter hole is drilled through a component that contains residual stresses. The hole diameter is accurately measured (generally using an air probe) as a function of both depth and angular position inside the hole. The residual stresses in the component are afterwards released by machining away a core of larger diameter from around the hole. The relaxation of residual stress distorts the shape of the reference hole. The reference hole diameter is subsequently re-measured at the same angular positions and depths. The reference hole distortion is lastly related to the residual stresses that existed in the component before drilling the hole.

The main assumption of this method is that the drilling of the reference hole has no effect on the residual stress state and that removing the core causes the residual stresses around the hole to be fully released in an elastic manner [1]. The core is additionally assumed to consist of independent lengths.

Materials Research Proceedings **2** (2016) 377-382 doi: http://dx.doi.org/10.21741/9781945291173-64

This indicates that a thick specimen is equivalent to a set of layers unconnected by through-thickness shear stresses [2]. The deep hole drilling technique has usually been used on metal parts such as in welds [3]–[5]and railway tracks [6]. In metallic components, EDM can be used to trepan a central core around the reference hole since the cutting stresses introduced by this method are not significant. An attempt to apply this technique to a laminated carbon-fibre composite was made by Bateman et al. [2]. In their investigation, EDM could not be used to machine the core around the reference hole because CFRP is not conductive and for this reason a diamond encrusted hole saw was selected instead, as it was for this study as well.

Analytical model

The DHD technique is based on the measured radial distortion to determine the components of the residual stress from the reference hole. The formulation to the case of far field biaxial stresses plus shear stress can be stated as

$$\bar{u} = \frac{1}{E}\left[f_{\theta_i}\sigma_x + g_{\theta_i}\sigma_y + h_{\theta_i}\tau_{xy} \right] \tag{1}$$

where for the case of isotropic materials (measurement angle θ)

$$f_{\theta_i} = 1 + 2\cos 2\theta_i$$
$$g_{\theta_i} = 1 - 2\cos 2\theta_i$$
$$h_{\theta_i} = 4\sin 2\theta_i$$

In the case of orthotropic materials these distortion coefficients depend on the mechanical properties of the component. Additionally, the relationship between the longitudinal direction of the material (fibre direction) and the measurement angle (air probe angular position during experiments) must be taken into consideration, see Figure 1.

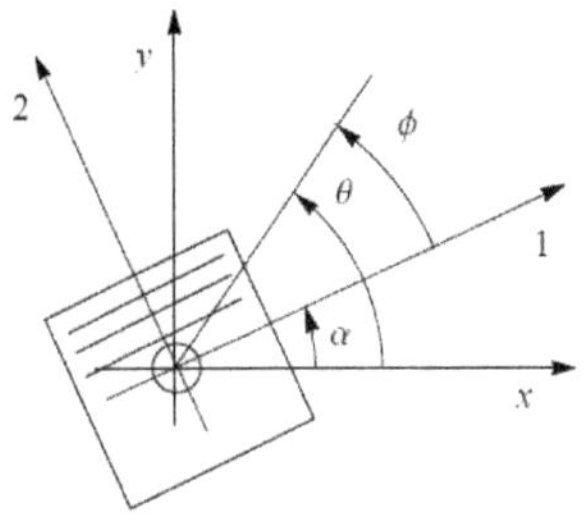

Fig. 1: Unidirectional laminate showing the relationship between the fibre direction (α) and measurement (θ) angles in reference to the global axes.

For this analysis, these coefficients for the studied material (AS4/8852) were calculated using the finite element software ABAQUS considering the mechanical properties given in Table 1 and the following values were determined

$$f_{\theta_i} = 1.49 + 5.21\cos 2\phi_i$$
$$g_{\theta_i} = 15.89 - 19.51\cos 2\phi_i$$
$$h_{\theta_i} = 30.04\sin 2\phi_i$$

Materials Research Proceedings **2** (2016) 377-382 doi: http://dx.doi.org/10.21741/9781945291173-64

Engineering constant		Value
E_1	[GPa]	135
E_2	[GPa]	9.6
E_3	[GPa]	9.6
σ_{12}	–	0.3
σ_{13}	–	0.3
σ_{23}	–	0.435
G_{12}	[GPa]	5.2
G_{13}	[GPa]	5.2
G_{23}	[GPa]	3.4

Table 1: Mechanical properties of AS4/8552.

The radial distortion is normally measured at nine different angles and at least three angles are required to obtain accurate results. The relation of the measured radial distortions and the residual stress is given by

$$\bar{\mathbf{u}} = -\frac{1}{E}\mathbf{M}\cdot\boldsymbol{\sigma} \tag{2}$$

The use of the minus sign is required because the radial distortions are measured after releasing the residual stresses. The matrices are expressed as

$$\bar{\mathbf{u}} = \begin{bmatrix} \bar{u}\big|_{\phi=\phi_1} \\ \vdots \\ \bar{u}\big|_{\phi=\phi_i} \\ \vdots \\ \bar{u}\big|_{\phi=\phi_N} \end{bmatrix}, \quad \mathbf{M} = \begin{bmatrix} f_{\phi_1} & g_{\phi_1} & h_{\phi_1} \\ \vdots & \vdots & \vdots \\ f_{\phi_i} & g_{\phi_i} & h_{\phi_i} \\ \vdots & \vdots & \vdots \\ f_{\phi_N} & g_{\phi_N} & h_{\phi_N} \end{bmatrix} \quad \text{and} \quad \boldsymbol{\sigma} = \begin{bmatrix} \sigma_x \\ \sigma_y \\ \tau_{xy} \end{bmatrix}$$

where N is the number of angular measurements. Lastly, the residual stresses are determined from the measurements of the radial distortions of the hole by

$$\boldsymbol{\sigma} = -E\mathbf{M}^{*}\cdot\bar{\mathbf{u}} \tag{3}$$

where $\mathbf{M}^{*} = \left(\mathbf{M}^{T}\cdot\mathbf{M}\right)^{-1}\cdot\mathbf{M}^{T}$ is the pseudo-inverse of $\mathbf{M}$.

Experimental procedures and results

An AS4/8552 composite laminate of 18 mm in thickness consisting of 10 layers with different orientations was used to perform the DHD measurement, see Figure 2. To allow the measurement of residual stress using DHD, front and rear bushes were bonded using high strength adhesive to the specimen, centred on the location of the measurement.

The use of bushes has several benefits provide a setup for the fixation of the essential components for gun drilling and air probing and contribute to guide the hole saw perpendicularly to the specimen. Additionally, they decrease the effects of the air probe entrance and exit during the measurements of the diameters and help to avoid fibres delamination when the drill enters and exits the specimen.

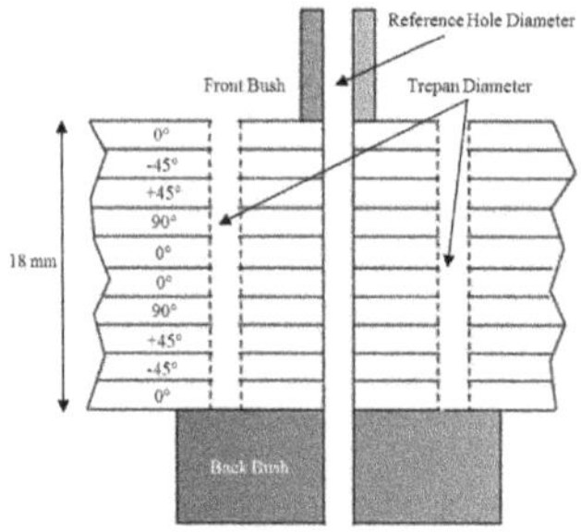

(a)

(b)

Fig. 2: (a) Illustration showing the layer arrangement specimens and the bushes required for a DHD experiment (b) composite laminate after conducting the DHD experiment

After drilling the reference hole, the air probe was introduced to measure precisely the hole diameter every 22.5° (i.e. 0°, 22.5°, 45°, 67.5°, 90°, 112.5°, 135°, 157.5° and 180°) around it and at every 0.1 mm through it, resulting in nine measurements of diameter in total for every axial position. The air probe is not able measure accurately near a surface, nevertheless the front and rear bushes allow measurements near the surface of the specimens. The air probe must be calibrated before and after the measurements of the hole diameters using 10 calibration rings, which allow to use calibration curves and interpret a diameter to each reading in millivolts. The calibration rings used have the following dimensions: 2.9796, 2.9893, 2.9997, 3.0096, 3.0204, 3.0306, 3.0397, 3.0496, 3.0595 and 3.0704 mm.

A hole saw was selected to perform the trepanning process which has the advantages of being faster than electrical discharge machining (EDM) and not requiring an electrolyte to be used. The hole saw used during the experiment has a diameter of 10 mm and is diamond tipped. Following the trepanning procedure, the diameter of the hole was re-measured using the air probe.

Lastly, the residual stresses were determined from the radial distortions of the reference hole to reveal the residual stress distributions shown in Figure 3. The principal material directions have been used in each layer. Therefore the stress component $\sigma_{11}$is longitudinal to the fibre direction while the stress component σ_{22} is in the transverse direction of the fibres. The experimental results are compared to Laminator [7] (software based on classical laminate theory) predictions and DHD simulations using the same reference hole and trepan diameters as in the experiments as well as different sizes of these two parameters additionally. The highest and lowest predicted stresses in the longitudinal direction are of the order of 20 and 65 MPa in compression, in the 0° and 90° plies respectively. On the other hand, the expected stresses in the transverse direction to the fibres remained approximately on 36-38 MPa on the specimen. The predicted magnitude of the shear stress is very low throughout the entire laminate.

A comparison between the DHD experiment and the DHD simulation using a reference hole of 3 mm of diameter and a trepan diameter of 10 mm shows a close agreement in the longitudinal, transversal direction as well as the in-plane shear component. However, these values significantly disagree with residual stress predictions using Laminator, which implies that a considerable amount of residual stresses were not relaxed after trepanning the core. For the proceeding of the DHD analysis, the core thickness is assumed to consist of a number of independent sections, each confined by two parallel planes normal to the reference hole axis. Each section is totally unaffected by the presence of other sections. As a consequence of the trepanning process, the stresses inside the core are totally released in a linear-elastic way.

Materials Research Proceedings **2** (2016) 377-382 doi: http://dx.doi.org/10.21741/9781945291173-64

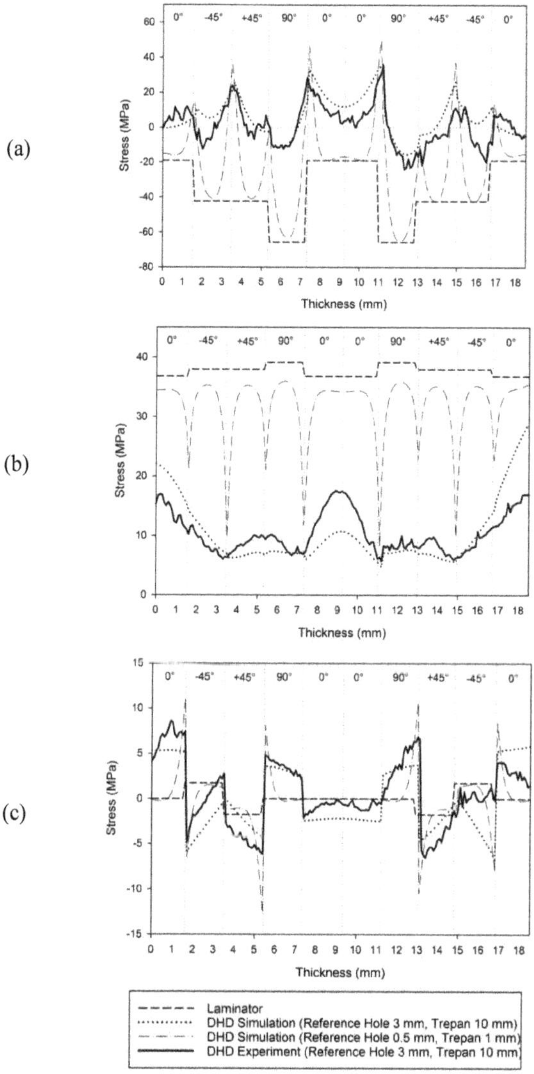

Fig. 3: Comparison of the residual stresses σ_{11}(a), σ_{22}(b) and in-plane shear τ_{12}(c) using Laminator, DHD experiment and simulations with different reference hole and trepan diameters.

In the case of composite laminates, these block lengths (layers) are influenced by the presence of adjacent layers since their stiffness and coefficient of thermal expansion variation may be sometimes higher than 15 times, thus affecting the area near their interface when expanding and contracting during the cure process. For this reason, the assumption that the trepanning process relaxes all residual stresses in the core cannot be valid. This issue can be nevertheless considerably resolved by using a smaller dimension for the reference hole and trepan diameters. After selecting 0.5 mm and 1 mm for the reference hole and trepan diameters respectively in a further DHD simulation the disagreement between the obtained results and Laminator predictions did not exceed a value of 5 MPa in all three cases in the central area of each layer.

Summary

The deep-hole drilling technique has become a standard method for the measurement of residual stress in isotropic materials, especially for thick components. The work described here is an extension to the method to allow the measurement of residual stress in orthotropic materials such as thick composite laminates.

After conducting a deep-hole drilling experiment on an 18 mm thick AS4/8552 composite plate, measurements showed a good agreement with a simulation using the same reference hole and trepan parameters, but a significant discrepancy was found when compared with correct predictions (Laminator). It was determined that, when using DHD in laminated materials, the reference hole and trepan diameters must be much smaller than the thickness of the layers in order to reduce the effect of remaining interlaminar shear stresses in trepanned cores which can significantly lead to inaccurate measurements.

Acknowledgements

The authors would like to acknowledge Mr. Steve Harding and VEQTER Ltd. for providing technical support for the experimental studies. The travel grant awarded by the Alumni Foundation of the University of Bristol as well as the financial support for this project provided by the National Council for Science and Technology (CONACyT), the Institute of Innovation and Technology Transfer (I^2T^2) and the Government of Nuevo León are likewise gratefully acknowledged.

References

[1] A. Mirzaee-Sisan, A. J. Fookes, C. E. Truman, D. J. Smith, T. B. Brown, and T. A. Dauda, "Residual stress measurement in a repair welded header in the as-welded condition and after advanced post weld treatment," *International Journal of Pressure Vessels and Piping*, vol. 84, no. 5, pp. 265–273, 2007. http://dx.doi.org/10.1016/j.ijpvp.2007.01.003

[2] M. G. Bateman, O. H. Miller, T. J. Palmer, C. E. P. Breen, E. J. Kingston, D. J. Smith, and M. J. Pavier, "Measurement of residual stress in thick section composite laminates using the deep-hole method," *International Journal of Mechanical Sciences*, vol. 47, no. 11, pp. 1718–1739, 2005. http://dx.doi.org/10.1016/j.ijmecsci.2005.06.011

[3] P. J. Bouchard, D. George, J. R. Santisteban, G. Bruno, M. Dutta, L. Edwards, E. Kingston, and D. J. Smith, "Measurement of the residual stresses in a stainless steel pipe girth weld containing long and short repairs," *International Journal of Pressure Vessels and Piping*, vol. 82, no. 4, pp. 299–310, Apr. 2005. http://dx.doi.org/10.1016/j.ijpvp.2004.08.008

[4] T. B. Brown, T. A. Dauda, C. E. Truman, D. J. Smith, D. Memhard, and W. Pfeiffer, "Predictions and measurements of residual stress in repair welds in plates," *International Journal of Pressure Vessels and Piping*, vol. 83, no. 11–12, pp. 809–818, 2006. http://dx.doi.org/10.1016/j.ijpvp.2006.08.012

[5] D. George and D. J. Smith, "Through thickness measurement of residual stresses in a stainless steel cylinder containing shallow and deep weld repairs," *International Journal of Pressure Vessels and Piping*, vol. 82, no. 4, pp. 279–287, 2005. http://dx.doi.org/10.1016/j.ijpvp.2004.08.006

[6] D. Stefanescu, P. A. Browne, C. E. Truman, and D. J. Smith, "Residual Stress Measurement within a European UIC60 Rail Using Integrated Drilling Techniques," *Materials Science Forum*, vol. 440–441, pp. 85–92, 2003. http://dx.doi.org/10.4028/www.scientific.net/MSF.440-441.85

[7] M. Lindell, "The Laminator: Classical Analysis of Composite Laminates (Version 3.7).

Materials Research Proceedings 2 (2016) 383-388 doi: http://dx.doi.org/10.21741/9781945291173-65

Residual Stresses in Selective Laser Melted Components of Different Geometries

Mark Reid[1,a*], Tim Sercombe[2,b], Anna Paradowska[1,c], Xiaopeng Li[2,d]

[1]ANSTO, Australian Centre for Neutron Scattering, Lucas Heights, NSW, Australia

[2]School of Mechanical and Chemical Engineering, The University of Western Australia, Perth, WA, Australia

[a]mark.reid@ansto.gov.au, [b]tim.sercombe@uwa.edu.au, [c]anp@ansto.gov.au, [d]xiaopeng.li@kuleuven.be

Keyword: Residual Stress, Neutron Scattering, Selective Laser Melting, Titanium

Abstract. As an emerging Additive Manufacturing (AM) technique, Selective Laser Melting (SLM) has found a promising application in biomedical field due to its advantages in fabricating Ti-6Al-4V components with specific and customised geometries. It has been reported that accumulated residual stress as a result of the high heating and cooling rates during SLM and the topological design of the components can largely influence the mechanical and functional properties of the fabricated components. In this study three notched samples were produced by SLM manufacturing. The samples were built using the same laser melting conditions on the same base size. The notches ranged in angle from 60° to 120°. Residual stresses in the three notched samples were analysed using the state-of-the-art neutron residual strain measurement instrument, Kowari, at the Australian Nuclear Science and Technology Organisation. A combination of gauge volumes were utilised to obtain high precision measurements at the notch tips and general measurements around the tips. This paper reports on the manufacture and measurement of differing residual stresses in the three SLM fabricated notches.

Introduction

As an emerging Additive Manufacturing (AM) technique, Selective Laser Melting (SLM) has found promising applications in a wide spectrum of industries, e.g. aircraft, automotive and biomedical sectors, due to its advantages in fabricating metal components with specific and customised geometries [1]. The process of SLM manufacture is highlighted in Fig 1. Initially an CAD model of the part to be generated is created (a), the model is then segmented into 50μm slices (b). The slice information is then fed to the SLM machine which melts the required pattern into a bed of powder (c). Once the pass is completed the bed is lowered by 50μm and a new layer of powder added (d) and the next melting pass conducted. Finally the completed part is removed from the powder bed (e) and the unused powder is recycled.

An important feature of SLM is that through optimisation of the fabrication parameters (e.g. laser energy, scanning speed, hatch spacing, etc.) and topological design of the components, these SLM fabricated metal components with desired mechanical and functional properties (e.g. ductility, elastic modulus, strength etc.) can be achieved [2]. It is reported that the accumulated residual stress as a result of the high heating and cooling rates during SLM and the topological design of the components can have a significant effect on the mechanical and functional properties of the fabricated components [3-5]. This leads us to deduce that understanding and tailoring the residual stress in SLM components is essential. As part of this understanding the effects of topological features, for a given fabrication method, on the residual stress within a component must be examined. Therefore, to investigate the effects of topography on the residual stress in SLM components, a state-of-art neutron scattering facility has been utilised for characterising the residual stress state SLM fabricated metal components of different topographies.

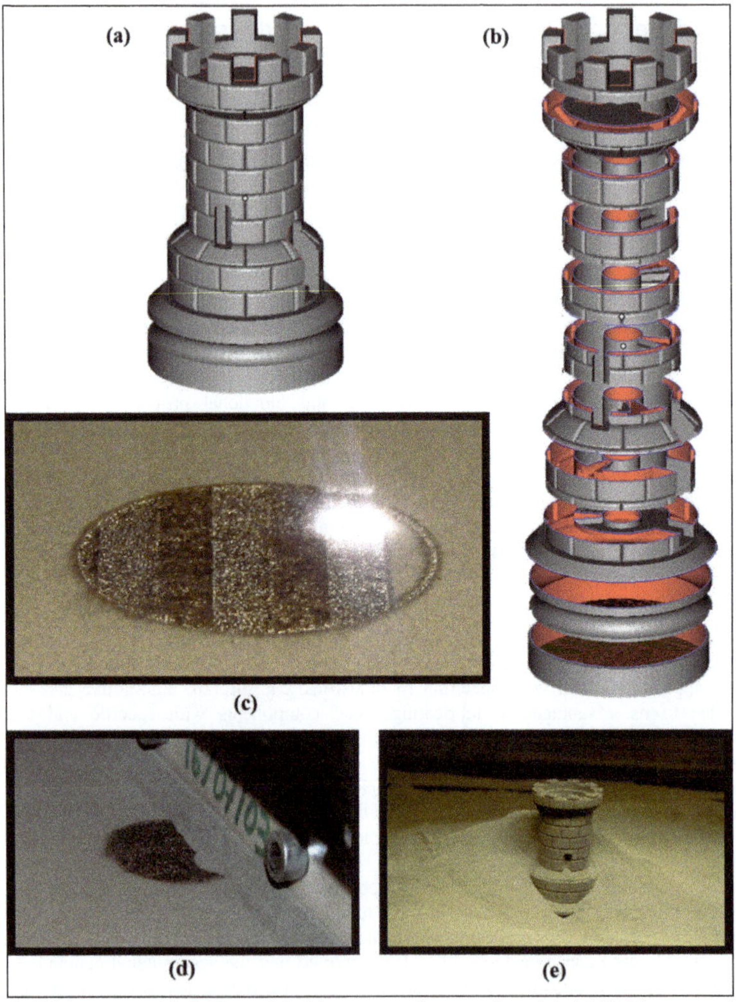

Figure 1. SLM process. (a) CAD model, (b) segmented autocad model, (c) SLM process, (d) new powder layer and (e) removal of final part.

Experimental

SLM: The Three Dimensional (3D) Printing group in The University of Western Australia produced three notched and one un-notched SLM samples using Ti-6Al-4V powder, size -53 µm, supplied by Falcon-tech (China). The geometries of the SLM samples are shown in Fig. 2a. The

Materials Research Proceedings **2** (2016) 383-388 doi: http://dx.doi.org/10.21741/9781945291173-65

samples were produced in a Realizer SLM-100 machine (ReaLizer GmbH, Germany). The process parameters utilised in manufacturing the components are given in Fig. 2b.

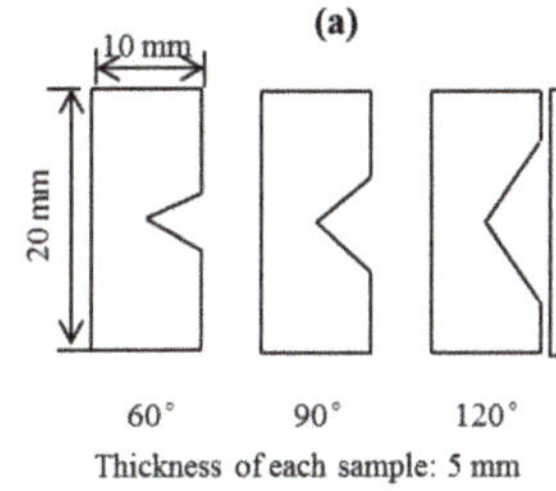

(b)

- Laser power: 200 W
- Laser spot size: ~40 μm
- Laser scan speed: 1000 mm/s
- Layer thickness: 50 μm
- Hatch spacing: 100 μm

Figure 2. (a) Cross-sections of the SLM fabricated notched Ti-64 components. (b) SLM process parameters

Residual Stress Measurements: The residual strain measurements were performed on the Kowari strain scanning instrument [6] at ANSTO. Because the basic principles of this technique are well known [7-9] only details specific to this measurement will be reported. Neutrons with a monochromatic wavelength are selected from a double focusing monochromator to enable single diffraction peaks to be sampled during strain scanning. Fig. 3b, gives the details of the instrument set-up for the current study and the material properties used in calculating the residual stresses. The sample gauge volume is defined by the primary beam vertical (pV) and horizontal (pH) slits and the receiving slits horizontal width (sH). The measurement locations and measurement direction definitions for the current study and given in Fig. 3a. Two different slit configurations were utilised in this study. The first, (pV,pH,sH) 1x1x1 mm was used to examine the bulk of the sample around the notch tips. The second which was used to obtain higher resolution measurement just below the notch tip used a (pV,pH,sH) 0.5x2x2 mm gauge volume. The second configuration was only used to measure the longitudinal and transverse directions. As data in the normal direction using the second setup could not be obtained as the gauge volume would have only been partially in the sample, the data for the normal direction from the first setup was extrapolated to provide normal data for the second setup to calculate stresses. Stresses in the longitudinal and transverse direction were calculated based on the assumption that the stress in the normal direction is zero. This allows deconvolution of a stress free reference lattice spacing.

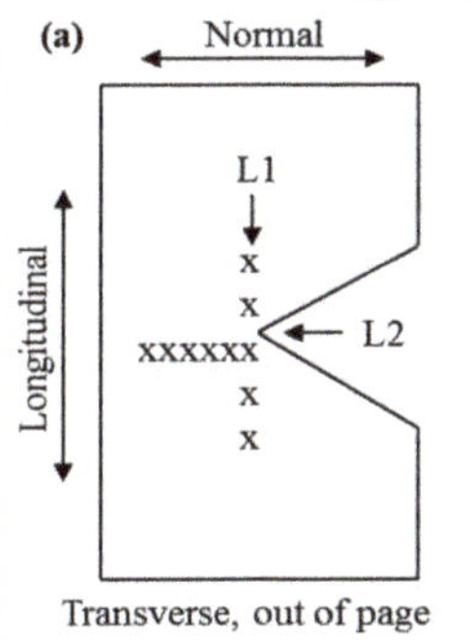

(b)

Neutron Wavelength	1.89 Å
Monochromator plane	Si(400)
Take off angle	88.2°
Measured plane	α-Ti (103)
E_{211}	120.6 GPa
ν_{211}	0.31

Figure 3. (a) Measurement locations (line1 and line 2) and direction definitions for the Kowari experiment. (b) Kowari experimental parameters

Materials Research Proceedings **2** (2016) 383-388 doi: http://dx.doi.org/10.21741/9781945291173-65

Results

The results of the residual stress measurements are given in Fig. 4, for the measurement line L1 (Fig 3a) and in Fig. 5, for the measurement line L2 (Fig 3a). The legends for the figures are as follows: Narrow, 60° notch; Med, 90° notch; Wide, 120° notch; Long, longitudinal stresses; Trans, transverse stresses.

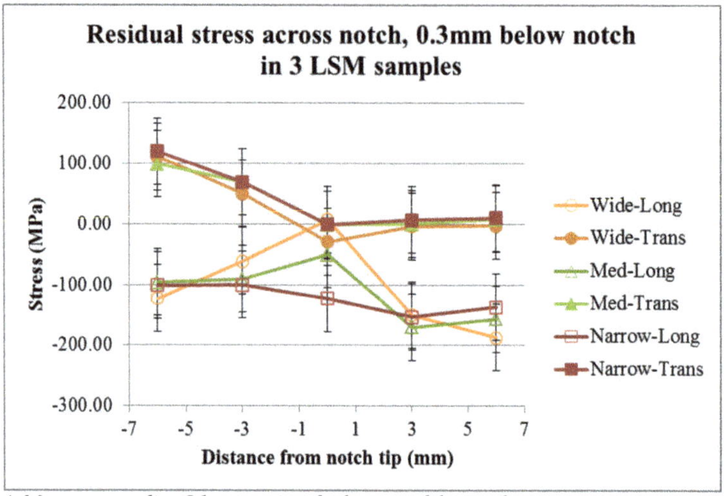

Figure 4. Measurement line L1 traversing the bottom of the notches

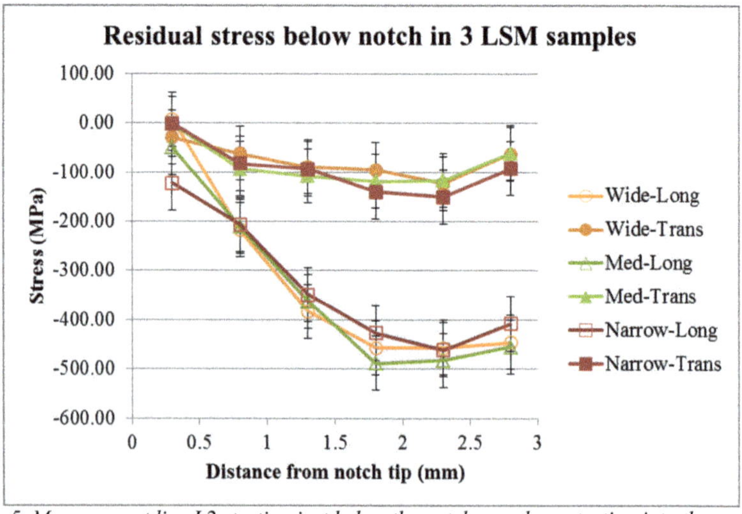

Figure 5. Measurement line L2 starting just below the notches and penetrating into the samples

Materials Research Proceedings **2** (2016) 383-388 doi: http://dx.doi.org/10.21741/9781945291173-65

Two features are of note in these results. In the case of the residual stresses observed across the base of the notches (Fig. 4) there is a strong asymmetry in the transverse residual stresses in all three of the notched samples. The maximum stress is observed at 6mm to one side of the notches, of ~125MPa in tension which then falls to ~0MPa below the notch and then remains at this stress out to 6mm on the other side of the notch. In the case of the longitudinal stresses there is reasonable symmetry observed. In the case of Fig. 5, where the stresses just below the notches and moving into the samples are shown there is very close agreement with the stresses in the transverse direction. This is generally the case for the longitudinal direction except at the position just below the notch tips. At this location (0.3mm below the notch) there is considerable spread in the observed longitudinal stresses. In the case of the narrow notch (60°) a compressive stress of ~150MPa is observed, while for the medium notch (90°) the compressive stress is ~50MPa and for the wide notch (120°) the stress is ~0MPa (also apparent in Fig. 4).

Discussion

The results of this study clearly demonstrate that the geometry of SLM manufactured components has an effect on the state of the residual stresses within the components. While this study was to an extent a proof of concept, and as such the authors consider the results somewhat qualitative, some intriguing results have been obtained. Looking at the case of line 2, below the notch and moving into the sample, the results clearly show that reducing the angle of the notch from 120° to 60° has resulted in an increase in the residual compressive stress at the notch. This is not a surprising result as the reduced notch angle will produce greater constraint at the notch tip and hence prevent the relaxation of any stresses developed during the manufacturing process. Different notches would also induce different thermal histories around the notches in different samples. This probably also contributes to the observed difference in residual stresses in samples with different notches. It should be noted that it would be expected the stresses will turn to tensile further into the samples (see [3,10] for example), however these positions were not measured in this experiment due to time constraints.

Turning to line 1, looking at the stresses across the base of the notches, these results are rather surprising. Given in Fig. 6, is the hatch pattern used in forming the components. As each layer is formed the pattern is rotated by 90°. At this stage the authors are unable to account for the observed asymmetry of the transverse residual stresses across the base of the notches. This effect will be the subject of future studies (i.e. samples of the same geometry but different hatch patterns will be examined).

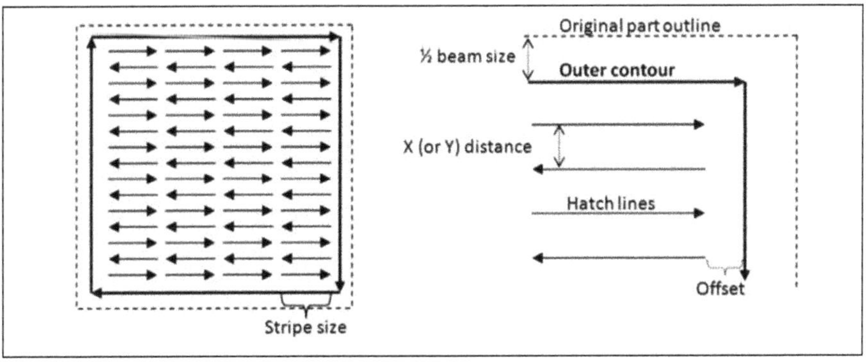

Figure 6. Hatch pattern used in SLM manufacture of the three notched components.

Materials Research Proceedings **2** (2016) 383-388 doi: http://dx.doi.org/10.21741/9781945291173-65

As noted at the start of the discussion this study is considered a feasibility / proof of concept study which the authors believe has been very successful. Several issues will be addressed in future studies in particular the choice of material for manufacturing the SLM components. Titanium is a strong coherent scatterer of neutrons (contributing to the diffraction peak), however it is also a strong incoherent scatter of neutrons (contributing to noise), thus producing a poor signal to noise ratio in neutron experiments and hence the relatively large error bars reported in Fig. 4 & 5. A second issue with titanium alloys is that during high cooling rates martensitic transformation can be induce (not addressed in this report) which can further complicate analysis of residual stresses. For these reasons future studies, which have already been submitted for consideration, will examine more neutron scattering friendly materials (for example austenitic or ferritic stainless steels). Finally the authors envision that the results of this and future studies will contribute real results to efforts to model and thus quantify the effects of geometry and other processing parameters on the residual stresses in SLM components.

Conclusion

Residual stresses have been measured in 3 SLM manufactured components of different geometries. Three notches, 60°, 90° and 120° were fabricated from Titanium-6%Aluminium-4%Vanadium powder. The residual compressive stresses at the notches were ~150, 50 and 0 MPa respectively. This is most likely due to the increased constraint with decreasing notch angle. An unexpected result was that across the notches an asymmetric tensile stress was observed with one side a stress (in all three samples) of 125MPa observed while the corresponding position on the other side of the notch showed virtually no stresses in the transverse direction. The longitudinal stresses either side on the notches were reasonably symmetric. An explanation for this unexpected stress distribution will be the subject of future studies.

References

[1] S.M. Giannitelli, D. Accoto, M. Trombetta, A. Rainer, Acta Biomaterialia 10 (2014) 580-594. http://dx.doi.org/10.1016/j.actbio.2013.10.024

[2] B. Vrancken, L. Thijs, J.P. Kruth, J. Van Humbeeck, Acta Materialia 68 (2014) 150-158. http://dx.doi.org/10.1016/j.actamat.2014.01.018

[3] X.P. Li, C.W. Kang, H. Huang, T.B. Sercombe, Materials & Design 63 (2014) 407-411. http://dx.doi.org/10.1016/j.matdes.2014.06.022

[4] X.P. Li, M. Roberts, Y.J. Liu, C.W. Kang, H. Huang, T.B. Sercombe, Materials & Design.

[5] X.P. Li, X.J. Wang, M. Saunders, A. Suvorova, L.C. Zhang, Y.J. Liu, M.H. Fang, Z.H. Huang, T.B. Sercombe, Acta Materialia 95 (2015) 74-82. http://dx.doi.org/10.1016/j.actamat.2015.05.017

[6] O. Kirstein,V. Luzin, Materials Australia, 41 (2008) 52-54.

[7] A. Allen, M.T. Hutchings, C.G. Windsor, C. Andreani, Adv. Phys. 34 (1985) 445-473. http://dx.doi.org/10.1080/00018738500101791

[8] M.T. Hutchings and A.D. Krawitz; NATO ASI Series 216E, Kluwer Academic Publishers, Dordrecht / Boston / London (1992).

[9] H.J Prask, P.C Brand, Neutrons in Research and Industry, Proc. of Int. Conf. on Neutrons in Research and Industry, ed: G. Vourvopoulos, SPIE Proceedings Series, 2867 (1997) 106-115. http://dx.doi.org/10.1117/12.267879

[10] X.P. Li, C. Kong, T. Becker, T. Sercombe, Adv. Eng. Mater (2016)

Residual Stresses 2016: ICRS-10 Materials Research Forum LLC
Materials Research Proceedings **2** (2016) 389-394 doi: http://dx.doi.org/10.21741/9781945291173-66

Residual Stresses Determination with Plasticity Effects by Electron Speckle-Interferometry Hole-Drilling Method

Leonid LOBANOV[1, a], Viktor SAVITSKY[1, b *]

[1]Paton Electric Welding Institute, the National Academy of Sciences of Ukraine, Kyiv, Ukraine

[a]holo@paton.kiev.ua, [b]viktor.savitsky@gmail.com

Keywords: Residual Stresses, Speckle-Interferometry, Hole Drilling, ESPI-HD Method, Plasticity Effect

Abstract. The electron speckle-interferometry hole-drilling (ESPI-HD) method for determination of residual stresses (RS) is based on drilling of a blind hole with a diameter of 0.5-1.0 mm and calculating stresses using displacements data measured by a speckle-interferometer. According to the standard procedure, calculations are made under the assumption of linear elastic behavior of the material. However, if the RS level is high, then plastic deformations caused by the stress concentration induced by the hole could lead to a stress calculation error. Unlike strain-gage hole-drilling (SG-HD) method which records strains only by three (six) strain gauges and elaborates in order to find the RS, the ESPI-HD method allows taking into account the displacements data at all points of the surface around the hole. Analysis of the displacements' variation across the circumference around the hole reveals local features related with plastic deformations. In this paper, a new approach for determination of RS by the hole drilling method that considers the plastic deformations has been proposed. This approach assumes determination of the plasticity effect by calculating stresses using displacements data at various sectors relative to the hole. Analysis of stresses variations calculated using values of displacements at different sectors allows retrieving the real stress state.

Introduction

One of the most widely used technique for determination of residual stresses (RS) is the strain-gage hole-drilling method (SG-HD method). There is an international standard [1] for determination of RS that is based on drilling of a small hole on the materials surface and measuring the surface strains using a special strain gauge rosette attached concentrically around the drilled hole. This method is applied in cases where material behavior is linear-elastic and plasticity effects are neglected. It is known that a hole is the stress concentrator that can result in significant plastic deformations around the drilled holes, if the nearby RS are high. It was shown that if the stressed state does not exceeds about 60% of the material yield stress σ_{02} the errors induced by yielding around the hole are less than 3% but the plasticity error increases with increasing RS [2, 3]. This is connected with the fact, that elastic deformations, which for stresses should be calculated, are added to unknown plastic deformation. It is known that deformation data at the three points measured around the hole are not enough to determine a presence or absence of plastic deformation. In the paper [3] the use of a four-element strain-gauge rosette has been considered to find "effect of plasticity" and real values of RS. Another extension of the hole drilling method is application of a non-contact speckle-interferometry technique to measure displacements around the hole [4-6]. The ESPI-HD method possesses such an important advantage as possibility of displacement data obtaining at a large amount of points near the hole. The analysis of the data allows determining of a presence of plastic deformation around the drilled hole, and to develop a new approach for RS measuring taking into account plastic deformations.

Materials Research Proceedings **2** (2016) 389-394 doi: http://dx.doi.org/10.21741/9781945291173-66

Determination of RS by the ESPI-HD method

The ESPI-HD method, similarly to the SG-HD method, assumes measure of displacements at the distance $2.5r_0$ (r_0 - is hole radius) of displacements u_x. The displacements data are connected with stresses by the following equation [2]:

$$u_x(\theta) = F(\theta)\sigma_{xx} + G(\theta)\sigma_{yy} + H(\theta)\tau_{xy} \quad (1)$$

where - $F(\theta)$, $G(\theta)$ and $H(\theta)$ – are the functions, which depend on geometric parameters of a hole and mechanical constants of the material, angle θ determines the coordinates ($x = 2.5r_0\cos(\theta)$; $y = 2.5r_0\sin(\theta)$) of the point, where the displacement u_x is measured.

Having the displacements data at the three points $u_x(2.5r_0, \theta_1)$, $u_x(2.5r_0, \theta_2)$ and $u_x(2.5r_0, \theta_3)$, located on a circle with a center at the drilling point, it is possible to compose the system of linear equations (1), and to find σ_{xx}, σ_{yy} and τ_{xy}. Using the displacements data u_x of more than hundred points, unknown values of stresses σ_{xx}, σ_{yy} and τ_{xy} are calculated by the least-squares method.

Mathematical simulation with the finite element method was conducted for estimation of plastic deformations influence on the accuracy of RS determination by the ESPI-HD method (Fig.1). As a result of simulation the values of displacements of material surface points depending on the level of stresses in the object were calculated. Then the displacement $u_x(2.5r_0, \theta)$ were selected at the points of surface, located at the distance $2.5r_0$ from the hole center. After that the values of the stresses σ_{xx}, σ_{yy} and τ_{xy} were found using equations (1). The finite element mesh used by FEM-simulation is presented at the Fig.1. For simulation the following data were used: diameter and depth of hole is 1 mm, object thickness is 4 mm. The area of plastic deformation is highlighted by red color in case when the predefined residual stresses $\sigma_{rs} = 0.85\sigma_{02}$.

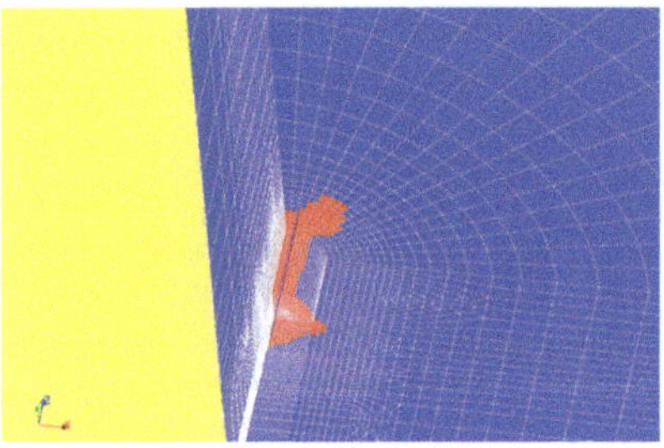

Fig. 1. Finite element mesh used for FE modeling of drilling processes

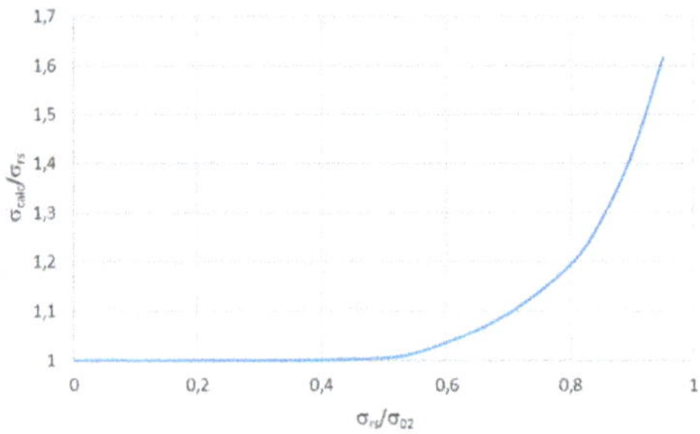

Fig. 2. Influence of the ratio σ_{rs}/σ_{02} on the accuracy of RS determination by the ESPI-HD method

To calculate the error in the determination of RS, depending on the level of stresses related to the material yield strength (σ_{rs}/σ_{02}), the following numeral experiments were held. The displacements $u_{x,calc}(2.5r_0, \theta)$ on the distance $2.5r_0$ for angles θ in the range $[0, 2\pi]$ were calculated for given ratio σ_{rs}/σ_{02}. Then the system of equations (1) was composed and the values $\sigma_{xx,calc}$ were calculated by the least-squares method. It was shown, that at the ratios $\sigma_{rs}/\sigma_{02} > 0.6$, the difference between σ_{rs} and $\sigma_{xx,calc}$ increases, and could reach 60% (Fig. 2). The area where plastic deformation occurs enlarges with increasing of RS. That results in higher error of RS determination by the ESPI-HD method, which assumes only elastic deformation (elastic stress relaxation) after drilling a hole. Thus, it is necessary to take into account influence of plastic deformation in the calculation of RS.

Materials Research Proceedings **2** (2016) 389-394 doi: http://dx.doi.org/10.21741/9781945291173-66

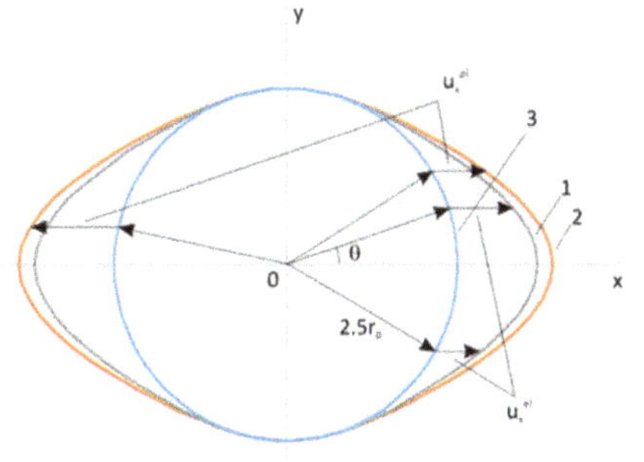

Fig. 3. Geometrical presentation of displacement u_x, calculated by FEM along the circumference of the radius $2.5r_0$ with the center at the point of drilling, from the angle θ: 1 – linear elastic model; 2 – elasto-plastic model; 3 – circle with radius $2.5r_0$. ($\sigma_{rs} = 0.85\sigma_{02}$).

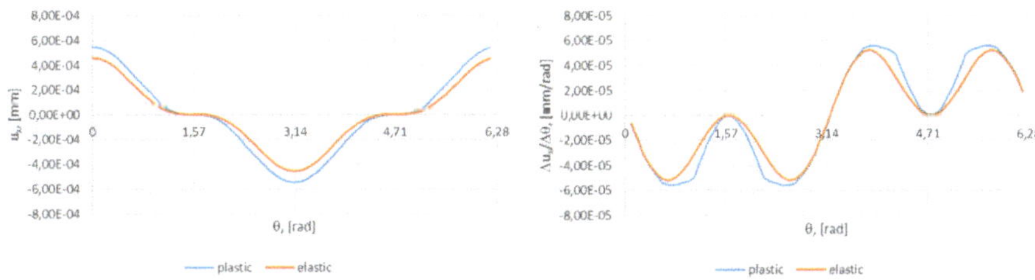

a) b)

Fig. 4. Dependency of displacements u_x (a) and their derivatives $\Delta u_x/\Delta\theta$ (b) from the angle θ, along the circumference with radius $2.5r_0$ centered at the point of drilling calculated using the FEM with linear-elastic and elasto-plastic models ($\sigma_{rs} = 0.85\sigma_{02}$).

Consider influence of plasticity effect on the results of RS calculation providing that $\sigma_{rs} = 0.85\sigma_{02}$. Geometrical presentation of displacement values u_x is shown in Fig. 3. The displacements u_x are set relatively from initial position of the point in a larger scale. The displacement curve $u_x^{pl}(\theta)$ calculated using elasto-plastic model is similar to the curve $u_x^{el}(\theta)$, that received at linear-elastic model (Fig.4, a). However, derivatives $\Delta u_x/\Delta\theta$ (Fig.4.b) calculated under the assumption of linear elastic and elasto-plastic behavior of the material, show that at the angles θ in the range $[0, \pi/6]$ the curves $\Delta u_x^{pl}/\Delta\theta$ and $\Delta u_x^{el}/\Delta\theta$ are practically identical, and the greatest deviations are observed at $\theta = \pi/3$. It can be concluded from this graph that the plastic deformations make the greatest influence on the displacements u_x of surface points around the hole for the angles θ in the range $[\pi/4, \pi/2]$. Consequently, plastic deformations near the drilled holes increase maximum values of measured displacements, but its contribution is not uniform along the circumference: on separate ranges of θ dependences of displacements $u_x^{el}(\theta)$ and $u_x^{pl}(\theta)$ fit within a constant summand, and on the other ranges they differ the more, than higher the ratio σ_{rs}/σ_{02}. Analysis of distribution of u_x along the entire circumference allows identifying presence of plastic deformations and their values.

The feature of the ESPI-HD method is an ability to calculate stresses using the displacement data u_x at points of the arc BC that could be chosen arbitrarily around the drilled hole (Fig. 5). Variation of size (defined by the angle ψ) and position (defined by the angle α) of the circular arc BC allows to develop a new technique for determining RS, which takes into account the plastic deformations. The arc is defined by the following way. Some point A with polar coordinates $(2.5r_0, \alpha)$ is marked on the circle. Regarding this point, the arc BC that includes all points with angular coordinates from $\alpha - \psi$

Materials Research Proceedings **2** (2016) 389-394 doi: http://dx.doi.org/10.21741/9781945291173-66

up to $\alpha + \psi$ is chosen. The displacement u_x on arc BC are used for calculation of stresses $\sigma_{xx,calc}(\alpha, \psi)$ from equations (1).

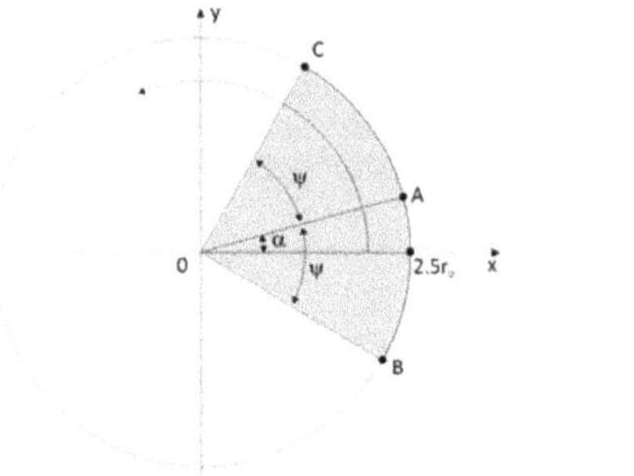

Fig. 5. Location of the arc BC relative to the drilled hole.

Dependences of stress values $\sigma_{xx,calc}(\alpha, \psi)$, that are calculated using the displacements data measured on the arc BC, on an angle α are presented in the polar (Fig.6, a) and Cartesian (Fig. 6, b) coordinate systems provided that $\sigma_{rs} = 0.85\sigma_{02}$. It is shown that decreasing the length of arc BC (lower values of ψ) leads to increasing $\Delta\sigma_{diff}(\psi)$ in the presence of plastic deformation, where

$$\Delta\sigma_{diff}(\psi) = \max_{0\leq\alpha<2\pi}\left|\sigma_{xx,calc}(\alpha, \psi)\right| + \min_{0\leq\alpha<2\pi}\left|\sigma_{xx,calc}(\alpha, \psi)\right| \quad (2)$$

These graphs show significant increase in the difference between the extreme and the averaged value of the stress $\sigma_{xx,calc}$ at decrease of the angle ψ, that is consequence of localization of areas with plastic deformation. One of the parameters that may characterize the plasticity effect is PE_{diff}:

$$PE_{diff}(\psi) = \sqrt{\Delta\sigma_{diff}(\psi)/\sigma_{xx,calc}(\pi)} \quad (3)$$

Dependency PE_{diff} on $1/\psi$ at different ratios σ_{rs}/σ_{02} is shown in Fig. 7. It is clear that PE_{diff} equal to zero in the absence of plastic deformation ($\sigma_{rs}/\sigma_{02} < 0.33$). A monotonic increase of PE_{diff} with increasing σ_{rs}/σ_{02} was observed. Thus, based on the analysis of the dependency $PE_{diff}(1/\psi)$, we can draw a conclusion about the absence or presence of plastic deformation after drilling a hole in the object. It should be noted that dependencies $PE_{diff}(1/\psi)$ obtained for different ratios σ_{rs}/σ_{02}, up to a factor coincide with each other. The figure shows that the ratio σ_{rs}/σ_{02} is uniquely determined by the slope k_ψ of the graph $PE_{diff}(1/\psi)$.

The developed ESPI-HD method for determination of RS taking into account the plasticity effects includes the following stages:

1. To calculate values of stresses $\sigma_{xx,calc}(\pi)$ by equations (1) using the displacements data along a circle with the radius $2.5r_0$ with the center in the point of a drilled hole.
2. For the set of values ψ in the range $[\pi/6, \pi/2]$ calculate stresses $\sigma_{xx,calc}(\alpha, \psi)$, $\alpha = [0,2\pi]$. For each value of ψ to calculate PE_{diff}. To build a curve $PE_{diff}(1/\psi)$. To determine the slope k_ψ of the curve.
3. To compute dependency of the slope k_ψ on σ_{rs}/σ_{02} for present experiment conditions (diameter and depth of the hole, thickness of the object, mechanical properties of the material) using FE simulation. To determine σ_{rs}/σ_{02} and σ_{rs}, substituting value k_ψ calculated in stage 2 into the dependency $k_\psi(\sigma_{rs}/\sigma_{02})$.

Materials Research Proceedings **2** (2016) 389-394 doi: http://dx.doi.org/10.21741/9781945291173-66

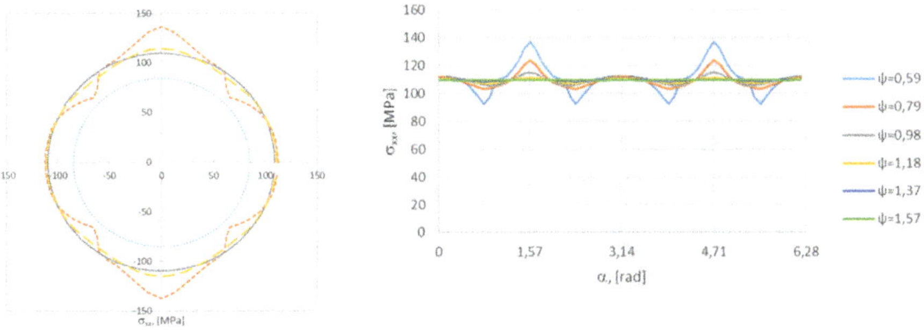

a) b)

Fig. 6. Dependencies $\sigma_{xx,calc}$ on α calculated by FEM for different values ψ providing that $\sigma_{rs} = 0.85\sigma_{02}$: a) – stress diagram in the polar coordinate system, where gray solid line is stresses $\sigma_{xx,calc}(\alpha, \psi = \pi)$, orange dashed line is stresses $\sigma_{xx,calc}(\alpha, \psi = \pi/2)$, red dashed line is stresses $\sigma_{xx,calc}(\alpha, \psi = \pi/4)$; b) – in the Cartesian coordinate system.

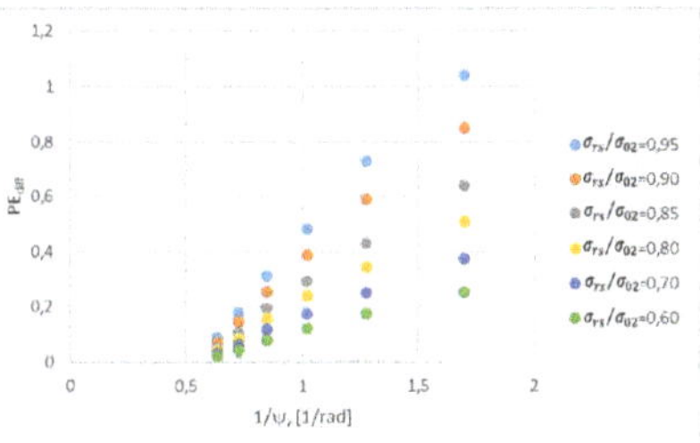

Fig. 7. Dependencies PE_{diff} on $1/\psi$ for different ratios σ_{rs}/σ_{02}, obtained by FE simulation.

Determination of RS by the ESPI-HD method considering the plasticity effect in welded samples

RS have been measured by the developed ESPI-HD method considering the plasticity effect in the welded joint made from aluminum alloy (Fig. 8). It is known that in heat-affected zone RS exceed $0.7\sigma_{02}$ of the material. A compact speckle-interferometer (Fig. 8, a) has been designed for determination of RS. The recorded interference fringe pattern is shown on Fig. 8, b. Values of $\sigma_{xx,calc}(\alpha, \psi)$ and PE_{diff} were calculated according to the stage 2 using the displacement data u_x in the area marked in blue (fig. 8, b), and as a result the $PE_{diff}(1/\psi)$ has been found (Fig. 8, c, d). These graphs are similar to those obtained by mathematical modelling (Fig. 6, 7) and they show presence of significant plastic deformations around the hole. FE simulation has shown that the slope k_ψ of the curve (Fig.8, d) fit values of the ratio $\sigma_{rs}/\sigma_{02} = 0.92$. Thus, in the point of drilling the residual stresses $\sigma_{xx} = 169MPa$.

Summary

A new approach for evaluation of the plasticity effect after drilling a hole in the objects being under high residual stresses ($> 0.6\sigma_{02}$) is developed. This is achieved by analysis of the displacements data in the points around drilled holes that are measured by the ESPI method. The procedure for calculation of RS with plasticity effects has been proposed. Effectiveness of the ESPI-HD method for measuring of RS at the object in the area of high stress state has been demonstrated.

a)

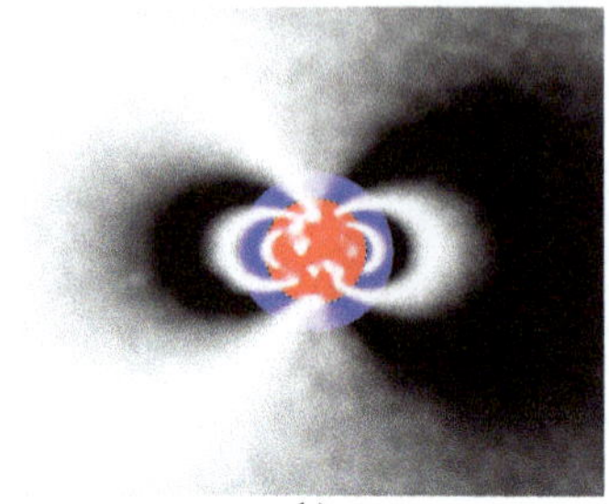

b)

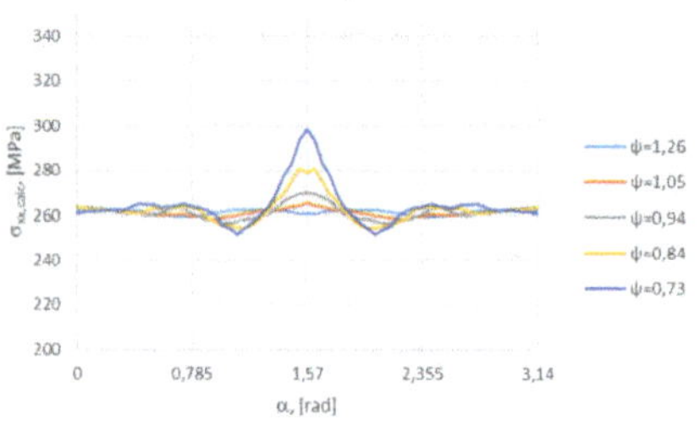

c)

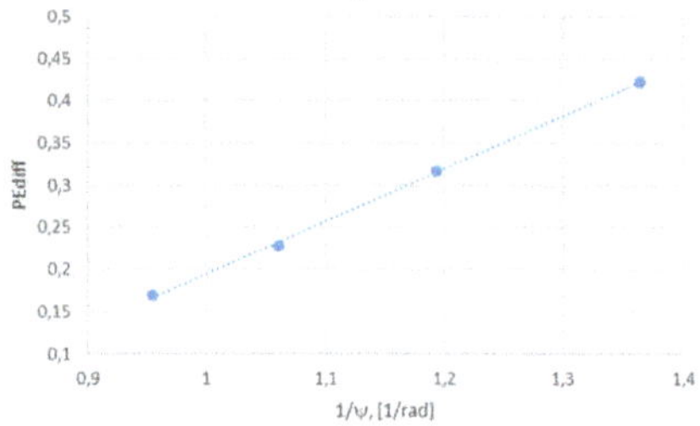

d)

Fig.8. Determination of RS by the ESPI-HD method taking into account the plasticity effect: a) – the ESPI-device for determination of RS installed on the welded specimen; b) – interference fringe pattern recorded by ESPI after drilling of the hole in the sample; c) – dependence $\sigma_{xx,calc}$ on α for different values of ψ; d) – dependence PE_{diff} on $1/\psi$.

References

[1] ASTM E837-13a, "Standard test method for determining residual stresses by the hole-drilling strain-gage method," in Annual Book of ASTM Standards, Am. Soc. Test. Mat. (2013).

[2] Lin YC, Chou CP. Error induced by local yielding around hole in hole drilling method for measuring residual stress of materials. Journal of Materials Science and Technology 1995;11:600–4. http://dx.doi.org/10.1179/mst.1995.11.6.600

[3] D. Vangi and S. Tellini Hole-Drilling Strain-Gauge Method: Residual Stress Measurement With Plasticity Effects J. Eng. Mater. Technol 132(1), 2009.

[4] L.M. Lobanov, V.A. Pivtorak, V.V. Savitsky, G.I. Tkachuk Procedure for determination of residual stresses in welded joints and structural elements using electron speckle-interferometry/ The Paton Welding Journal. 1/2006, pp. 24-29.

[5] M. R. Viotti, W. Kapp and A. Albertazzi Jr., "Achromatic digital speckle pattern interferometer with constant radial in-plane sensitivity by using a diffractive optical element," App. Opt. 48(12), 2275-2281 (2009). http://dx.doi.org/10.1364/AO.48.002275

[6] L. Lobanov, V. Pivtorak, V. Savitsky, G. Tkachuk Technology and Equipment for Determination of Residual Stresses in Welded Structures Based on the Application of Electron Speckle-Interferometry. Materials Science Forum, -2014, Vol. 768, pp. 166-173.

Residual Stresses 2016: ICRS-10 Materials Research Forum LLC
Materials Research Proceedings **2** (2016) 395-400 doi: http://dx.doi.org/10.21741/9781945291173-67

Numerical Modelling and Mitigation Technique of Welding Distortion for Fillet Welding of Aluminum Plate

M. Tsunori[1, a*], M. Mouri[2,b], S. Saso[2,c] and H. Kusumoto[3,d]

[1]IHI, Structure and Strength Department, Research Laboratory, Yokohama, Japan

[2]IHI, Welding Technology Department, Production Engineering Center, Yokohama, Japan

[3]Japan Marine United, Gas Project Department, Offshore and Engineering Division, Tokyo, Japan

[a]mitsuyoshi_tsunori@ihi.co.jp, [b]masashi_mouri@ihi.co.jp,

[c]shun_sasou@ihi.co.jp, [d]kusumoto-hiroki@jmuc.co.jp

Keywords: Weld Distortion, Residual Stress, In-Process Heating, Finite Element Method

Abstract. In marine and offshore industries, aluminium welding processes are often used especially for floating LNG (Liquefied Natural Gas) production, storage and LNG fuelled vessels. Welding distortions of aluminium plates are known to be larger than that of steel, so it is desired to mitigate the distortion. Also, numerical models of welding distortion prior to a manufacturing of components are expected in order to improve products. Finite element models of welding residual stresses and distortions on aluminium plates by using the commercial finite element code ABAQUS were developed in this study. A mixed material hardening model was employed in order to simulate aluminium material behaviours. In addition, a mitigation technique of distortions by in-process additional heating of plate bottom side was investigated by using the developed numerical models. The finite element models were validated by experiments which distortions and temperature histories of MIG fillet welding were measured. Experiments of in-process additional heating of plate bottom side were also conducted and results were compared with the simulation. The simulations and the experiments showed that the mitigation technique proposed in this study was able to achieve a reduction of distortions for aluminium fillet welding.

Introduction

A5083-O aluminium alloy is often used for marine and offshore industries especially for floating LNG (Liquefied Natural Gas) production, storage and LNG fuelled vessels since it has better strength to weight ratio and corrosion resistances compared with steel. Welding distortion of aluminium alloy is larger than that of steel and more difficult to remedy after the weld. Therefore, a prediction of distortions prior to manufacturing processes and mitigation techniques are desired.

A development of finite element models for welding distortion was originated from 1970's [1] and many researchers contributed the development [2,3]. Throughout the studies of the prediction of weld distortion, prediction accuracies were dependent on weld heat source models used in finite element models. Effect of material constitutive model was studied by the round robin project [4]. This project investigated experimental and numerical techniques for characterisation of residual stresses on a single bead on plate problem. This research concluded that predicted residual stresses and distortion are sensitive to material constitutive models.

Demands of mitigation techniques of welding distortion are increased. Several mitigation methods, which are mechanical constraint, pre-heating, in-process cooling and in-process heating [5,6,7,8], have been proposed, then it is important to choose an appropriate method for individual components. The mechanical constraint technique is the simplest method, but this technique requires

Materials Research Proceedings **2** (2016) 395-400 doi: http://dx.doi.org/10.21741/9781945291173-67

a huge mechanical load coping with a thermal load due to weld for large components. In case of aluminium components, it is difficult to apply thermal heating processes after the weld to correct the weld distortion because of a difficulty of controlling temperature distributions by the large thermal conductivity of aluminium alloy. Therefore, in-process technique is more desirable. In-process method of MIG welding of thin aluminium plates was studied and TIG heating ahead of MIG welding was effective to reduce weld distortions [6]. Therefore, the in-process heating technique is applicable but a heating condition should be optimised for different case because temperature field is sensitive with welding condition and plate thickness.

In this study, finite element models of the welding distortion and in-process additional heating of bottom side of aluminium plates were developed. Additional heating conditions to minimise weld distortions were optimsed by using the numerical models. Experiments of additional heating processes were conducted and a reduction of the weld distortion was achieved.

Experimental and numerical procedure

Experimental procedure. Dimensions of A5083-O aluminium plates used in the experiment are shown in Figure 1. Chemical compositions and mechanical properties of A5083-O aluminium are listed in Table 1 and Table 2. The web plate was fillet welded to the skin plate by MIG welding with A5183-WY filler metal. Both sides of the web plate were simultaneously fillet welded. During the welding process, the skin plate was rested on a jig and there was no clamp in order to avoid mechanical constraining effect. A heat source model of the TIG heating was investigated by using a thick plate, which is shown in Figure 1. Thick plate was used to reduce unexpected distortions by the TIG heating. The MIG welding and the TIG heating conditions are shown in Table 3.Weld distortion mitigation technique introducing in this study is schematically illustrated in Figure 2. The TIG torch was selected for the additional heat source. The heat input and velocity of the MIG welding and the TIG heating were kept constant during the welding. Weaving technique was used for the TIG torch in order to broaden heating area. A distance between the MIG welding torch and the TIG torch was selected as a parameter to be optimsed. In situ temperature measurements by thermocouples were made during the welding and heating experiments. The locations of thermocouples are shown in Figure 1. After the fillet welding experiments, plate distortion was measured by a photogrammetric measurement system. The photogrammetric measurement was done by taking digital photos of specimens with attached targets from various directions and the photos were processed to obtain three dimensional coordinates of the targets. The angular distortions at the plate edge were calculated from the results of the photogrammetric measurement.

Finite element model. The commercial finite element code ABAQUS [9] was selected to develop numerical models. A symmetry boundary condition was considered at the middle of the web plate so only half of the specimens were modelled. Dense and graded mesh distributions close to the fillet weld and additional heating area was made in the models. Effect of deformation field to thermal field was assumed to be small and neglectable therefore a sequentially coupled thermal-stress analysis was performed. The transient temperature field previously calculated by the heat transfer analysis was used as thermal load within the mechanical analysis.

A simple moving heat source model that assumes uniform distribution in the fillet weld metal area was utilized in the analysis. The ABAQUS user subroutine called DFLUX was implemented in order to model this moving heat source model. Efficiencies of the MIG welding and the TIG heating were determined by comparing with the result of temperature measurements of several cases in this research and listed in Table 3. It is necessary to take into account temperature dependent material properties [6]. They were measured up to 550 °C to improve an accuracy of the finite element models in this study. A heat transfer coefficient at the ambient temperature was determined as 10 $W/m^2/°C$ and an emissivity used in the simulation was 0.4, respectively.

A choice of material constitutive model is one of the most important aspects of mechanical analysis. Isotropic hardening rule is the simplest model and often used in weld simulations although it is well known that it gives overestimated residual stresses. A mixed isotropic and kinematic

Materials Research Proceedings **2** (2016) 395-400 doi: http://dx.doi.org/10.21741/9781945291173-67

hardening model is more accurate if the model is fitted to both a monotonic response and a cyclic response. In the present study, mixed hardening rule called Chaboche model implemented in ABAQUS was employed in order to overcome the shortcomings of isotropic hardening rule. The Chaboche model in ABAQUS is represented by the following equations.

$$f(\boldsymbol{\sigma}-\boldsymbol{\alpha})=\sqrt{\frac{3}{2}(\mathbf{S}-\boldsymbol{\alpha}^{dev}):(\mathbf{S}-\boldsymbol{\alpha}^{dev})}=\sigma^0 \tag{1}$$

$$\dot{\boldsymbol{\alpha}}=\sum_k\left[C_k\frac{1}{\sigma^0}(\boldsymbol{\sigma}-\boldsymbol{\alpha})\dot{\bar{\varepsilon}}^{pl}-\gamma_k\boldsymbol{\alpha}\dot{\bar{\varepsilon}}^{pl}\right] \tag{2}$$

$$\sigma^0=\sigma|_0+Q_\infty(1-e^{-b\bar{\varepsilon}^{pl}}) \tag{3}$$

where $\boldsymbol{\sigma}$ is the stress tensor, $\boldsymbol{\alpha}$ is the back stress tensor, $f(\boldsymbol{\sigma}-\boldsymbol{\alpha})$ is the equivalent von Mises stress, σ^0 is the size of the yield surface, $\mathbf{S}$ is the deviatoric stress tensor, $\boldsymbol{\alpha}^{dev}$ is the deviatoric part of the back stress tensor, $\bar{\varepsilon}^{pl}$ is the equivalent plastic strain, $\dot{\bar{\varepsilon}}^{pl}$ is the equivalent plastic strain rate, $\sigma|_0$ is the yield stress at zero plastic strain, and C_k, γ_k, Q_∞ and b are material parameters to be calibrated. Cyclic loading tests were carried out in addition to standard tensile tests in order to identify the parameters. One representative result, whose temperature was at room temperature, was shown in Figure 3. The result of the mixed hardening model shows better agreement with experimental results than that of the isotropic hardening model for the case of the cyclic test.

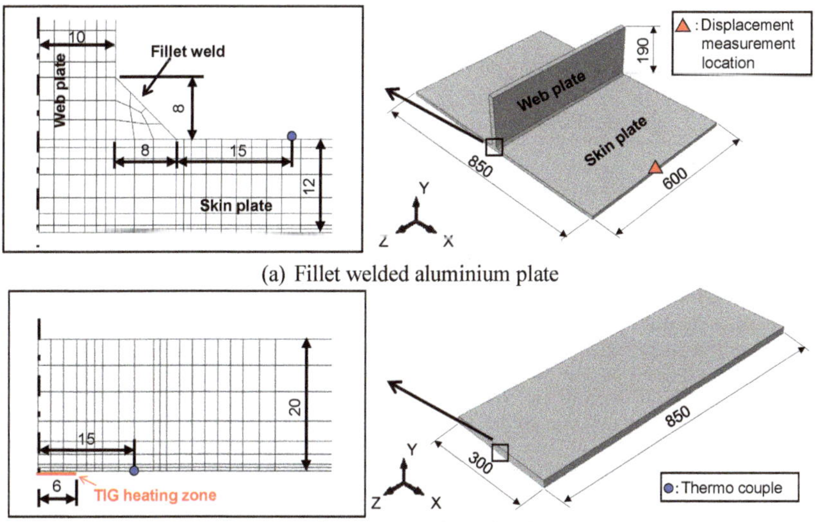

(a) Fillet welded aluminium plate

(b) Aluminium plate of TIG heating

Figure 1. *Schimatic and finite element models of the specimens with dimensions and thermocouple position*

Materials Research Proceedings **2** (2016) 395-400 doi: http://dx.doi.org/10.21741/9781945291173-67

Table 1. *Chemical composition of A5083-O plate*

Element	Si	Fe	Cu	Mn	Mg	Cr	Zn	Ti	Al
Weight(%)	0.09	0.24	0.04	0.6	4.5	0.08	0.01	0.01	Bal.

Table 2. *Mechanical properties of A5083-O plate*

Yield stress MPa	Tensile stress MPa	Elongation %
196	323	21.4

Table 3. *MIG welding and TIG heating conditions*

Type	Heat input (J/mm)	Efficiency
MIG	878	0.65
TIG	437	0.47

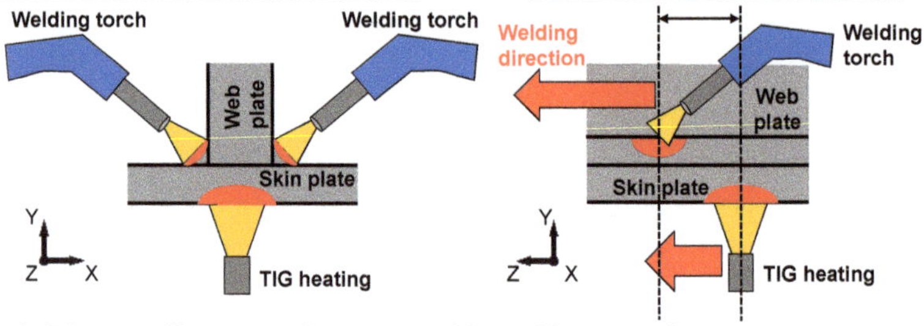

Figure 2. *Schematic illustration of in-process additional heating technique*

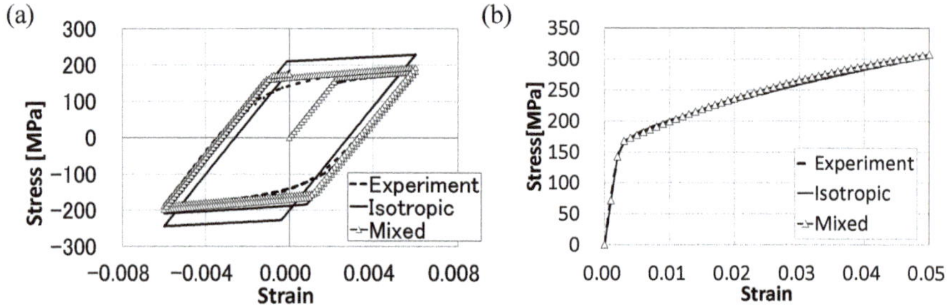

Figure 3. *Stress-strain relationship used in finite element model compared with experiments. (a) Stress-strain relationship by cyclic loading (b) Stress-strain relationship by tensile test*

Validation ofthe finite element models

Results of numerical models and thermocouple measurements of the MIG welding and the TIG heating are shown in Figure 4. As seen from the Figure 4 (a), a good agreement of temperature histories was obtained for the MIG welding. On the other hand, a discrepancy of the maximum temperature of the TIG heating was observed. One reason was that the numerical model did not take into account the weaving process of the TIG heating and it was affected to the results. Temperature histories and cross sections of the welded plate with the TIG additional heating were compared and results are shown in Figure 5. The TIG torch was placed -40 mm behind the MIG torch. A good agreement was observed for the temperature history. Larger melt pool was observed by the finite element analysis in comparison with the experiment. The difference was also considered by absence of the weaving effect of the TIG torch in the finite element model.

The out of plane displacement of the plate edge, which is indicated in Figure 1, obtained by the finite element model was listed in Table 4. There was a good agreement between the numerical model and the experiment for the MIG welding. The reduction rate of distortion by the additional

Materials Research Proceedings **2** (2016) 395-400 doi: http://dx.doi.org/10.21741/9781945291173-67

heating was approximately 62 % in the experiment. Predicted reduction rate was larger than the experiment. This was due to the predicted larger melting pool in the numerical model.

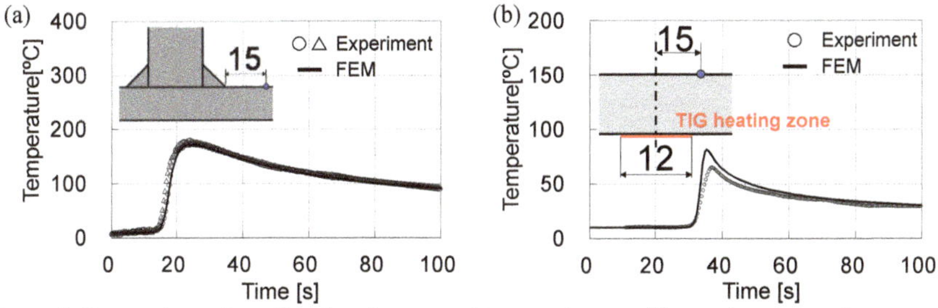

Figure 4. *Comparison of predicted and measured temperature profiles at thermocouple locations. (a) MIG welding (b) TIG heating*

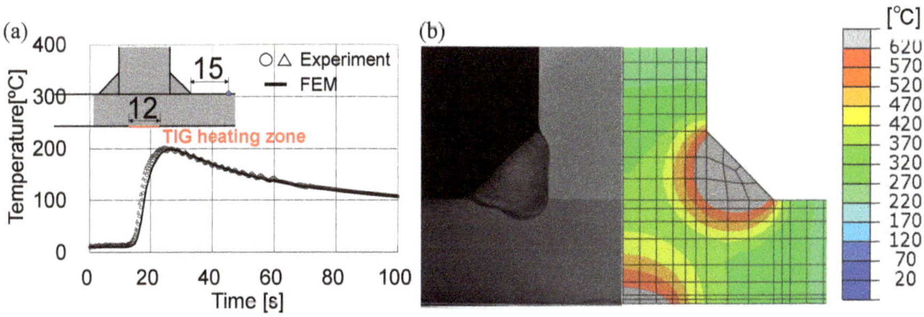

Figure 5. *Comparison of experiment and finite element model of additional TIG heating (-40 mm). Temperature history (b) Cross section of welded plate and melt pool by finite element model*

Table 4. *Results of out of plane displacement by experiment and finite element model*

Case	Out of plane displacement (mm)	
	Experiment	FEM
MIG	8.3	8.0
MIG+TIG (-40 mm)	3.1	1.7

Weld distortion mitigation

A distance between the MIG welding torch and the TIG torch and heat input of the TIG heating are important parameters. The heat input was determined by the preliminary performed numerical analysis so as to keep the melting pool size small. A relationship between distances of the two torches was investigated by numerical models and experiments. The finite element analyses were performed by placing the heat source distances from +60 mm to -80 mm. The positive value was defined the TIG torch travelling in front of the MIG welding torch. The results were shown in Figure 6. It was found that effect of the additional heating was sensitive to the distance of the additional heating position and -40 mm was the optimised distance. Figure 7 shows transverse residual stresses through the skin plate by the numerical analysis. The -40 mm additional heating made smaller transverse residual stress compared with +60 mm case. It is considered that heat input by preceding MIG welding make high temperature field at plate bottom and this temperature field reduces plate bending stress and angular distortions.

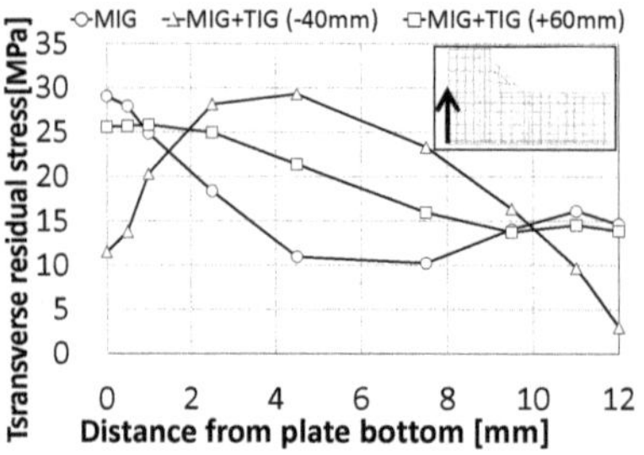

Figure 6. Relationship between TIG heating position and welding distortion

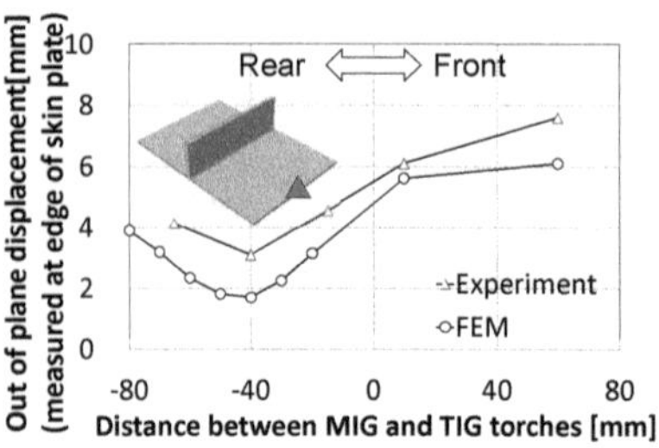

Figure 7. Comparison of transverse residual stress

Conclusions

In this study, welding distortion and its mitigation technique for fillet welding of aluminium plates were developed. The conclusions are summarized as follows.

(1) Finite element models of MIG welding, TIG heating and additional heating were developed and compared with experiments. Good agreement was obtained for MIG welding case. Larger melt pool was predicted by the finite element analysis. This difference was considered by absence of the weaving effect of TIG torch.

(2) Mitigation technique for fillet welding of aluminium plates was investigated by finite element models and experiments. The results showed that the optimised distance between the MIG welding and the TIG torch was -40 mm. An approximately 62 % reduction was achieved by the proposed method.

(3) Additional heating behind a welding torch is more effective than the method proposed by reference [6] in our welding condition. This result indicates that the optimised heating condition is sensitive to weld sequences, heat input and plate thicknesses.

References

[1] R. Kamichika, T. Yada, A. Okamoto, Internal stresses in thick plates weld-overlaid with austenitic stainless steel (Report 2), Transactions of the Japan Welding Society, 5 (1974) 295-302

[2] J. A. Goldak, M. Akhlaghi, Computational welding mechanics, Springer, 2005

[3] L. E. Lindgren, Computational welding mechanics, Woodhead publishing, 2007

[4] M.C. Smith, A.C. Smith, R. Wimpory, C. Ohms, A review of the NeT Task Group 1 residual stress measurement and analysis round robin on a single weld bead-on-plate specimen, International Journal of Pressure Vessels and Piping, 120-121 (2014) 93-140. http://dx.doi.org/10.1016/j.ijpvp.2014.05.002

[5] P. Michaleris, Minimization of welding distortion and buckling, Woodhead publishing, 2011. http://dx.doi.org/10.1533/9780857092908

[6] M. Mochizuki, M. Toyoda, Weld distortion control during welding process with reverse-side heating, Journal of Engineering Materials and Technology, 129 (2007) 265-270. http://dx.doi.org/10.1115/1.2400275

[7] S. Okano, M. Mochizuki, M. Toyoda, Thermal conductivity theoretical examination concerning characteristics of welding distortion produced by multiple heat source, Journal of the Japan welding society, 24 (2006) 324-330 (in Japnaese)

[8] S. Okano, M. Mochizuki, M. Toyoda, Study on angular distortion by temperature distribution control using back heating source, Journal of the Japan welding society, 25 (2007) 95-105 (in Japnaese)

[9] Simulia, Abaqus User's Manual, 2011

Residual Stress in Metal-Matrix Composite Cylinder Measured by Neutron Diffraction and Contour Method

Vladimir Luzin[1, a *], Kevin Thorogood[1, b], John R. Griffiths[2, c], Cameron J. Davidson[2, d] and Trevor R. Finlayson[3, e]

[1] Australian Nuclear Science and Technology Organisation, Locked Bag 2001, Kirrawee DC, NSW, 2232 Australia

[2]CSIRO Manufacturing, P O Box 883, Kenmore, QLD 4069, Australia

[3]School of Physics, University of Melbourne, VIC 3100, Australia

[a] vll@ansto.gov.au, [b] kjt@ansto.gov.au, [c] john.griffiths@csiro.au, [d] cameron.davidson@csiro.au, [e] trevorf@unimelb.edu.au

Keywords: Residual Stress, Neutron Diffraction, Contour Method, Composite

Abstract. A cylindrical sample, 14.8 mm in diameter, was machined from an Al-Mg-Si casting and then heat-treated. For the purposes of this research, the casting can be regarded as a two-phase composite of aluminium and 6 vol.% of near-spherical Si particles ~3 μm in diameter. Residual stresses in the cylinder are (i) long-range macrostresses resulting from the transient temperature gradients formed during heat-treatment, and (ii) short-range microstresses resulting from differences in the coefficients of thermal expansion between Al and Si. Neutron diffraction has been used to measure the stress tensors in each phase of the composite as a function of radial position with 2 mm spatial resolution and the microstress and macrostress components have been successfully separated. The contour method was applied to measure the axial component of the macrostress and the results are in good agreement with the neutron diffraction data.

Introduction

Al-Mg-Si castings are used in the aerospace and automotive industries due to their excellent castability, corrosion resistance, weldability and machinability, all combined with good mechanical properties. Typically, components are used in the heat-treated condition to achieve optimum mechanical properties.

Long-range residual stresses, or macrostresses, occur in the castings as a result of both the casting process and the subsequent heat-treatment ("long-range" means that the stresses vary over distances comparable to the size of the component). Heat-treatment comprises two steps, namely solution-treatment and ageing. The solution-treatment comprises an anneal at 540°C followed by quenching into water, the resulting transient temperature gradients being responsible for long-range residual stresses. The macrostress produced through quenching usually results in a tensile core and compressive periphery, though the actual stress distribution obviously depends on the sample geometry.

Short-range residual stresses, or microstresses, are the result of differences in the coefficients of thermal expansion of the microstructural constituents of the microstructure. The stresses vary over distances comparable to the scale of the microstructure.

In the current work, one such system that produces macro- and microstress is studied: a two-phase composite of near-spherical Si particles with 6 vol.% and ~3 μm in diameter in an aluminium matrix. In this paper only stresses caused by the heat-treatment are considered.

Several methods exist for the measurement of residual stress [1], each having its advantages and suitability for particular applications. Neutron diffraction is particularly suitable for measuring both the micro- and macrostresses in composite materials [2] although analysis of the data is notoriously

Materials Research Proceedings **2** (2016) 401-406 doi: http://dx.doi.org/10.21741/9781945291173-68

difficult, requiring the solution of the d_0 problem (i.e., a knowledge of the stress-free lattice parameter of each phase of a composite), the choice of elastic constants [3] and, finally, the separation of the micro- and macrostress fields [4]. With all these experimental difficulties, a reliable alternative method is highly desirable for validation.

Although being destructive, the contour method [5] provides a straightforward means of measuring macrostresses in materials and has many advantages over other residual stress measuring techniques, including insensitivity to texture and grain size effects, chemical variations and is able to measure extremely large samples with thick cross section and complex shapes [6]. The only real limitation is the electrical resistivity of the material, which determines if the sample can be cut using the electrical discharge machining (EDM), and the spatial resolution of the technique is too coarse to measure microstress fields. Although the contour method can be applied to a variety of material types including multi-phase alloys of similar nature, e.g γ/γ' Ni alloy, dual-phase steels or any other metal-metal composites, the applicability to composite materials with vastly different electrical, thermal and elastic properties, such as the Al-Si composite, can also be demonstrated.

In the current study, such an example for cross-validation between two techniques, neutron diffraction and contour method, using a cylindrical Al-Si composite sample is described.

Sample and material

The material was alloy A356, with Sr added to spheroidise the eutectic Si inclusions. The composition was (wt%): 6.6 Si, 0.4 Mg, 0.05 Fe, 0.18 Ti, 0.019 Sr with Cu, Mn, Zn all <0.01. Plates, measuring 140 × 160 × 25 mm^3, were made by sand-casting. Slices 25 mm wide were cut from the plates and were solution heat-treated at 540°C for 6 hours followed by a room-temperature water quench and ageing at 170°C for 6 hours. A 135 mm long cylindrical "dog-bone" sample was machined from one of the slices for tensile testing, with a 14.8 mm gauge diameter and 50 mm gauge length. A series of samples were produced for deformation tests, but one sample was studied in the as-heat-treated condition, with no deformation applied, therefore only thermal stresses caused by quenching and heat-treatment exist.

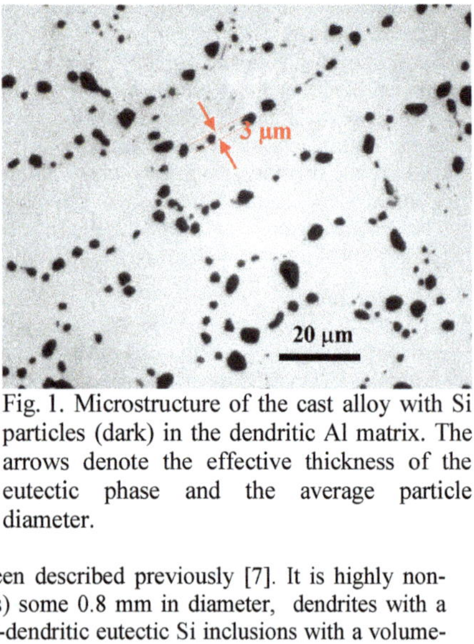

Fig. 1. Microstructure of the cast alloy with Si particles (dark) in the dendritic Al matrix. The arrows denote the effective thickness of the eutectic phase and the average particle diameter.

The microstructure is shown in Fig.1 and has been described previously [7]. It is highly non-uniform, comprising coarse dendrite colonies (grains) some 0.8 mm in diameter, dendrites with a secondary dendrite arm spacing of ~60 μm, and inter-dendritic eutectic Si inclusions with a volume-equivalent diameter ~3 μm and an aspect ratio of ~1.6 [8] together with age-hardening nano-precipitates of MgSi. The volume fraction of the Si inclusions is 0.062.

Experimental method I: neutron diffraction measurements and data analysis

Neutron residual stress scanning was carried out on the KOWARI neutron diffractometer at the OPAL research reactor (ANSTO) [9]. Both phases, Al matrix and Si particles, were measured at 90°-geometry using wavelengths appropriate for each phase: λ = 1.73 Å for the Al(311) reflection and λ = 1.58 Å to work with the Si(422). Three principal directions, radial, hoop and axial, were measured across the specimen diameter. A gauge volume of 2×2×2 mm^3 was used for lattice spacing measurements in the axial direction, while, in order to utilize the geometry of the long cylindrical sample, the gauge volume was enlarged to 2×2×10 mm^3 for radial and hoop measurements (the 10 mm dimension is parallel to the cylinder axis). Due to this difference, depending on component and phase, the measurement time was changed accordingly. A typical measurement time for hoop and radial directions was 20 minutes for Si and 3 minutes for Al. These times provided strain accuracy of ~70 μstrain for the Si(422) and ~30 μstrain for the Al(311).

A crucial capability of the instrument utilised for the measurements in the coarse-grained Al matrix was continuous sample rotation around the cylindrical axis to cover a physically larger sampling volume (i.e., number of grains) while keeping the same radial localization. This improves the grain statistics dramatically and yields statistically stable results. Additionally, and for the same reason, while measuring the axial component the cube gauge volume was moved in the axial direction by 10 mm, thus sweeping the same volume as for the hoop and radial components.

To overcome the critical d_0 problem associated with measuring the residual stress in a composite material, standard pure Si powder was used to measure the strain-free lattice parameter for the same reflection Si(422). This is a unique advantage offered by the alloy since the eutectic precipitates are pure Si. Unfortunately, the same does not apply to the Al phase: here the Al contains Si and other impurity elements in solution so that a stress-free d_0 is not available. Accordingly, the aluminium d_0 was assumed to be constant across the sample and was set in such a way that provided fulfilment of the macrostress balance conditions in the case of hoop and axial as well as the boundary condition for the radial component. This was achieved through performing several steps to separate micro- and macrostresses [4].

Firstly, components for the total stress of each phase, Si and Al, and each radial location r were calculated from the strain in the usual way:

$$\sigma_i^{ph}(r) = \frac{1}{S_2^{ph}}\left[\varepsilon_i^{ph}(r) - \left(\frac{S_1^{ph}}{3S_1^{ph} + S_2^{ph}}\right)\left\{\varepsilon_{hoop}^{ph}(r) + \varepsilon_{radial}^{ph}(r) + \varepsilon_{axial}^{ph}(r)\right\}\right] \tag{1}$$

where ph can be Si or Al, i can be "hoop", "radial" or "axial", and S_1^{ph}, S_2^{ph} are the (hkl)-dependent diffraction elastic constants that were evaluated from the single crystal constants in the isotropic approximation [3]. They are reported in Table 1 together with (hkl)-dependent Young modulus, E, and Poisson ratio, v, that can be compared with the bulk calculated counterparts.

Table 1. Elastic properties of the composite two phases, Al and Si

	$S_1(hkl)$, TPa^{-1}	$S_2(hkl)$, TPa^{-1}	$E(hkl)$, GPa	$v(hkl)$	$E(bulk)$, GPa	$v(bulk)$
Si(422)	-1.285	7.260	167.4	0.215	163.1	0.222
Al(311)	-4.992	19.366	69.4	0.358	70.1	0.356

Secondly, the macrostress was derived from the total phase stresses according to the volume fraction of the phases,

$$\sigma_i^{M}(r) = V_f^{Si} \cdot \sigma_i^{Si}(r) + (1 - V_f^{Si}) \cdot \sigma_i^{Al}(r) \tag{2}$$

where $V_f^{Si} = 0.06$ is the volume fraction of the Si particles in the composite.

Thirdly, with the calculated macrostress, the two force balance conditions, one for hoop and the other for axial components, and the boundary condition for the radial component are applied:

$$\int_0^R \sigma_{hoop}^M(r)\cdot dr = 0; \qquad \int_0^R \sigma_{axial}^M(r)\cdot r\cdot dr = 0; \qquad \sigma_{radial}^M(R) = 0 \tag{3}$$

It is not possible to fulfil all three conditions (3) exactly, using only a single d_0-value for aluminium, and so a best-estimate value was used to ensure that they comply in a statistical sense, minimising the overall error which involves all three conditions (3).

The final step was to derive the microstress for both phases:

$$\sigma_i^{ph,\mu}(r) = \sigma_i^{ph}(r) - \sigma_i^M(r), \tag{4}$$

thus achieving separation of macrostress and microstresses.

Experimental method II: contour method measurements and data analysis

After the neutron measurements were completed, the sample was sectioned transverse to the long axis to assess the axial component of the macrostress by the contour method. Although not conducted for the purpose of this research, macrostresses in other principal directions could also be measured using the multiple-cut method [10]. The cut was made with a Sodick AQ535L wire electrical discharge machine (WEDM) with 0.25mm diameter uncoated brass wire and skim cut settings. Both cut surface contours were then measured using a Brown and Sharpe Global Performance 7.10.7 co-ordinate measuring machine (CMM) at ANSTO. The machine was equipped with a Renishaw PH10M/TP200 touch probe system consisting of a 2 mm diameter ruby tipped stylus, with a measurement precision of ±0.7 µm. The cut surfaces were sampled using a rotary scan method on 0.25 mm point spacing, producing approximately 2700 data points for each surface. The residual stresses were calculated from raw contour data using a well-proven technique consisting of MATLAB scripts and ABAQUS Finite Element code. Details of data processing methods, smoothing, fitting, etc. are reported by Prime *et al.* [5].

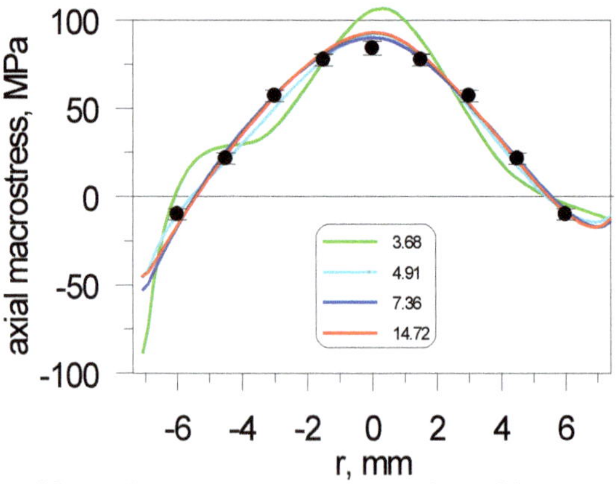

Fig. 2. Comparison of the axial component measurements obtained by neutron diffraction (symbols) and contour method (lines). Different contour method profiles correspond to the degree of smoothing (smoothing spline knot spacing parameter [mm] is given in the legend).

Results and discussion

The comparison of data obtained by the neutron diffraction and the contour methods is shown in Fig. 2. Several profiles representing results of the contour method are distinguished by the amount of regularization (smoothing) applied in the stress reconstruction algorithm. Although apparently different, they demonstrate a quick convergence to a stably smooth solution when using a smoothing spline knot spacing of ~5 mm, which is in good agreement with the neutron results, $\chi^2 \sim 2.0$. The only significant deviations from the general trend and symmetry in the contour method data are in the areas close to the perimeter and are apparently related to non-uniform EDM cutting.

While the contour method can provide a single axial macrostress component in one cut, all three principal stress components can be obtained with the neutron diffraction technique. The full set of results, three components of macro- and microstress, are shown in Fig. 3.

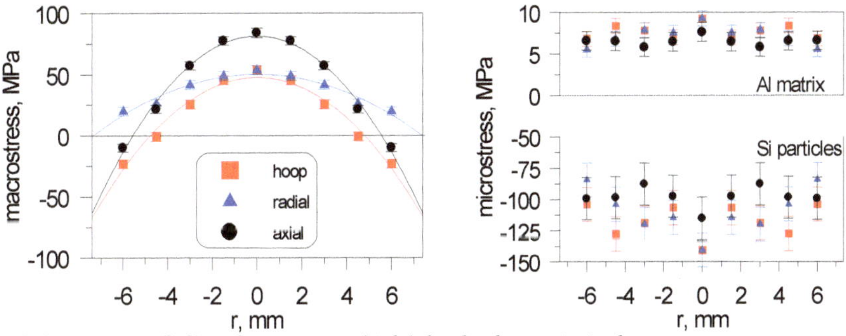

Fig. 3. Macrostress (left) and microstress (right) for the three principal stress components obtained by neutron diffraction. (The lines connecting symbols on the plot of macrostress are polynomials of the second order given for visual guidance only.)

The macrostress field shown in Fig. 3 is typical of stress fields produced by quenching although constructing a detailed anisotropic model [11] is beyond the scope of this paper. The microstress is, as expected for a composite, due to the difference in the coefficient of thermal expansion of Si particles and Al matrix. The experimentally observed hydrostatic microstress of ~-110 MPa in Si and ~7 MPa in Al would be generated by cooling through a temperature of 75°C (evaluated by an isotropic self-consistent model [12]), which is far less than the quench from the ageing temperature of 170°C to ambient. Therefore, the matrix surrounding the particles must yield locally and a detailed model for the hydrostatic stress requires a knowledge of the variation of the time-temperature history and of the dependence of yield stress on temperature.

A component of the hydrostatic microstress is also caused by the macrostress field. The maximum effect is expected at the centre of the specimen where the hydrostatic macrostress is ~50 MPa. Using the isotropic self-consistent model it is expected that the contribution to the microstress at the centre is +12 MPa for the Si particles and +0.8 MPa in the Al matrix. Thus, the macrostress-to-microstress coupling is small in comparison with the thermally generated microstress and the phase microstress values are close to the experimental error in the measurements of ±13 MPa and ±0.8 MPa, respectively for Si and Al. The deviatoric component of the macrostress also contributes to the microstress producing non-hydrostatic effects, that potentially can be detected, but its contribution is even smaller and certainly beyond accuracy of the current method.

Summary

Residual stress was measured in an Al-Si composite (A356 aluminium alloy) cylinder by neutron diffraction and the contour method. The cross-validation of the two methods using the axial macrostress component demonstrated a good agreement between the two methods. For the neutron diffraction technique, this ensures the proper approach to resolve the d_0 problem and correct data

Materials Research Proceedings **2** (2016) 401-406 doi: http://dx.doi.org/10.21741/9781945291173-68

treatment. For the contour method, this establishes applicability of the method to composite materials.

Using neutron diffraction data, a separation of microstress and macrostress was accomplished correctly, validated by an alternative technique. The quenching mechanism of macrostress formation was evidenced by the stress distribution and thermal mismatch between Si particles and the Al matrix and this is responsible for the microstress developed.

References

[1] M. Shokrieh, Residual Stresses in Composite Materials, Burlington: Elsevier Science, 2014.

[2] I.C. Noyan, J.B. Cohen, Residual stress: measurement by diffraction and interpretation, Springer, 1987. http://dx.doi.org/10.1007/978-1-4613-9570-6

[3] T. Gnaupel-Herold, P.C. Brand, H.J. Prask, J. Appl. Crystallogr., 31 (1998) 929-935. http://dx.doi.org/10.1107/S002188989800898X

[4] R.A. Winholtz, Separation of Microstresses and Macrostresses, in: M. Hutchings, A. Krawitz (Eds.) Measurement of Residual and Applied Stress Using Neutron Diffraction, Springer Netherlands, 1992, pp. 131-145. http://dx.doi.org/10.1007/978-94-011-2797-4_8

[5] M.B. Prime, R.J. Sebring, J.M. Edwards, D.J. Hughes, P.J. Webster, Exp. Mech., 44 (2004) 176-184. http://dx.doi.org/10.1007/BF02428177

[6] M.B. Prime, A.T. DeWald, The Contour Method, in: Practical Residual Stress Measurement Methods, John Wiley & Sons, Ltd, 2013, pp. 109-138. http://dx.doi.org/10.1002/9781118402832.ch5

[7] C.H. Caceres, C.J. Davidson, J.R. Griffiths, Q.G. Wang, Metallurgical and Materials Transactions A, 30 (1999) 2611-2618. http://dx.doi.org/10.1007/s11661-999-0301-8

[8] Q.G. Wang, C.H. Caceres, J.R. Griffiths, Metallurgical and Materials Transactions A, 34 (2003) 2901-2912. http://dx.doi.org/10.1007/s11661-003-0190-1

[9] O. Kirstein, V. Luzin, U. Garbe, Neutron News, 20 (2009) 34-36. http://dx.doi.org/10.1080/10448630903241175

[10] P. Pagliaro, M.B. Prime, H. Swenson, B. Zuccarello, Exp. Mech., 50 (2010) 187-194. http://dx.doi.org/10.1007/s11340-009-9280-3

[11] V. Luzin, Mater. Sci. Forum, 768-769 (2014) 193-200. http://dx.doi.org/10.4028/www.scientific.net/MSF.768-769.193

[12] Z. Hashin, Journal of Applied Mechanics, 50 (1983) 481-505. http://dx.doi.org/10.1115/1.3167081

Residual Stresses 2016: ICRS-10 Materials Research Forum LLC
Materials Research Proceedings **2** (2016) 407-412 doi: http://dx.doi.org/10.21741/9781945291173-69

Simulative Investigations of the Influence of Surface Indentations on Residual Stresses on Inner Raceways for Roller Element Bearings

J. Kehl*, R. Drafz, F. Pape, G. Poll

Institute for Machine Design and Tribology (IMKT), Leibniz University Hannover
Welfengarten 1A, 30167 Hannover, Germany

*Tel.: +49(0) 511 762 2267 E-mail: kehl@imkt.uni-hannover.de

Abstract. Resource-efficient machine elements are in the focus of current research. One of the most widely used machine elements are roller bearings. Thus, the optimization of bearings and their tribological properties promises to result in significant resource savings. Special focus is set on the bearing fatigue life, which may be significantly reduced by indentations on the raceways. The reduction in fatigue life can be caused by processes such as rolling over particles or by brinelling. These processes induce local stress peaks and lead to elastic-plastic deformations of the raceways. During the subsequent operation, the pile up of material around the indentations is flattened and hence the residual stresses change. Inside these so called shoulders stress peaks, residual stresses and hardening effects occur possibly resulting in crack initiation, crack growth under cyclic loading, and eventually spalling of material. For deeper and more sharp-edged indentations the bearing fatigue life is reduced more. To quantify the influence of an indentation on the bearing rating life a calculation model was developed based on the approach of IOANNIDES, BERGLING and GABELLI. For this, a 3D-FE model is used to calculate the three dimensional stress fields by superposition of residual and load stresses.

Introduction

When LUNDGREN and PALMGREN published their life theory in 1947 they focused on the fatigue life of ball bearings which is defined as the time until the first spalling appears [1]. These surface damages are due to cracks that are initiated in the subsurface area and propagate to the surface where they result in the typical fatigue damage. The cracks begin at material defects such as impurities and are initiated foremost by the shear stresses that occur due to the Hertzian pressure in the contact area. Their theory based on the previous works of WEIBULL who introduced the probabilistic approach to engineering [2]. Throughout the 20th century, however, the quality of steels has significantly improved and these material defects could be reduced. Classical subsurface fatigue life became less important [3].

ZARETSKY et. al. conducted a study with approximately 224,000 roller bearings. He estimated the percentage of fatigue failures to only 3% of all failures whereas particles are held responsible for over 20 % [4]. There are less conservative studies that estimate these failures to over 40 % [5]. GABELLI et al. demonstrated that indentations from hard materials such as corundum can reduce the bearing fatigue life by up to 95 % [6]. To account for such influences on the bearing fatigue life IOANNIDES and HARRIS introduced their advanced theory of bearing life calculation in 1985 [7], [8]. They introduced an analytical approach to consider the cleanliness of the lubricant and indirectly the contamination by particles. Eventually their work was incorporated as easy to handle equations in the ISO 281:2007 [9] and ISO/TS 16281:2008 [10] that set the standards for the advanced bearing life calculation. However, this approach does not explicitly consider the influence of particle contamination. It does not regard the increase in stress caused by the altered geometry of the surface.

To avoid particle damages the present countermeasures solely focus on the cleanliness of the lubricant. Primarily seals and filter systems are used to prevent particles from entering the bearing, but in some cases particles nevertheless enter the contact zone.

Materials Research Proceedings **2** (2016) 407-412 doi: http://dx.doi.org/10.21741/9781945291173-69

NEUBAUER investigated the fatigue life of roller bearings in which residual stresses were artificially induced by hard turning and deep rolling [11]. These residual stresses were designed to affect the stresses that occur from the Hertzian pressure and cause subsurface fatigue. He observed in this way that the fatigue life of roller element bearings can be increased by the factor of 2.3. In this study his approach is applied to dented surfaces and to examine the effect on the bearing fatigue life.

FE Model

Roller bearings of the type NU 206 are investigated in particular in this study. The inner ring of these bearings has the highest risk of failure and hence is in the focus of this study, see Figure 1. The calculation model can be divided into two main steps, see Figure 2. Firstly a FE analysis is conducted that simulates a sphere penetrating a flat surface. This simulation is comparable to a Brinell hardness test, since the said particle is modeled as a rigid undeformable body, because of that the results can be expected to be slightly exaggerated. In earlier simulations a sphere with a radius 150 μm was found to achieve the best results. The shape of the flat surface is a cube with an edge length of 300 μm. The ball penetrates the surface by 15 μm and leaves the surface afterwards at the same amount.

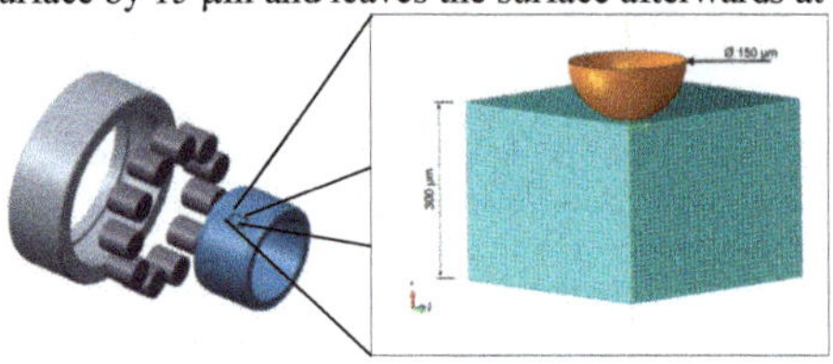

Fig. 1: Initial configuration of model 1

This quasi-static analysis was done twice. Once with residual stresses from the manufacturing process and the second time without. The distribution of the residual stresses is depicted in Figure 3. The commercial CAE software Abaqus 6.11 is used. To save calculation time the axis symmetry was used. Therefore symmetry boundary conditions were used on the y and z symmetry surfaces. On the bottom the model is encastre while on the outer surfaces the degrees of freedom in the x and y directions were locked. Standard C3DR elements were used. The model has 35 elements along each edge and thus contains 42875 elements in total. To save more calculation time only the highly loaded volume was investigated in this paper. This is justifiable since the volume elements that will fail and thusly contribute to the fatigue Life are expected to be in the highly loaded volume only.

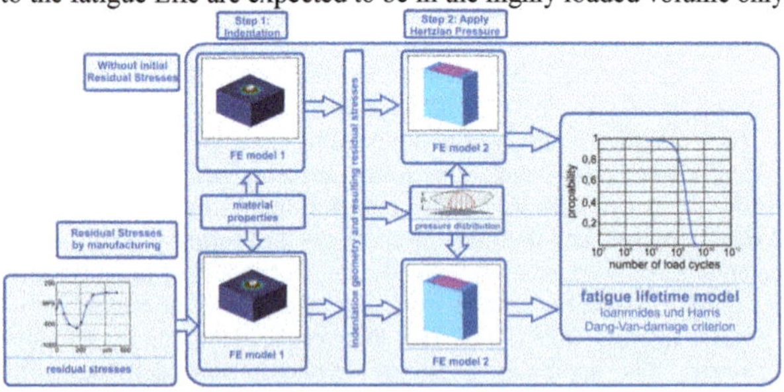

Fig. 2: Calculation scheme of the fatigue life prediction. The calculations are done with and without residual stresses as initial condition. Step 1: the indentation is simulated. Step 2: the resulting residual stresses from step 1 are taken as initial conditions. The indentation diameter and depth are taken to calculate the altered Hertzian contact pressure. The results are evaluated with the bearing fatigue life approach of Ioannides and Harris using the Dang-Van-criterion.

Materials Research Proceedings **2** (2016) 407-412 doi: http://dx.doi.org/10.21741/9781945291173-69

The material chosen is AISI 52100 (100Cr6) standard bearing steel. The elastic plastic material data were obtained by investigations by HACKE at the TU Clausthal in a common research project, see [12]. The yield strength R_p is 1830 N/mm^2 and the tensile strength R_m is 2620 N/mm^2. As investigated in [13–17] the hardening behavior of the material influences the development of residual stresses by indentations significantly.

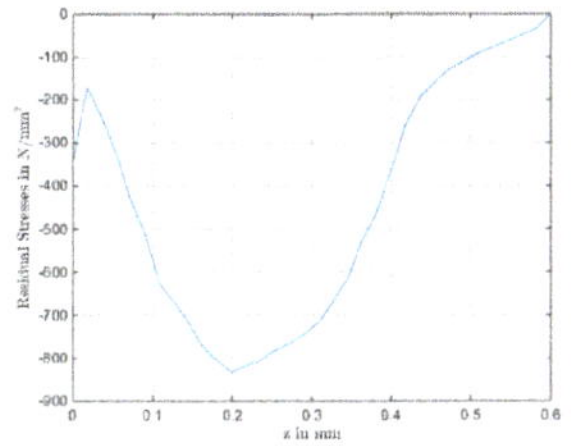

Fig. 3: Residual stress distribution used in the simulation [11].

In the aforementioned papers the authors investigated the plastic deformation depending on different hardening behaviors such as isotropic and kinematic hardening. The results of these studies prove, that the more distinctive the hardening behavior of a material is, the lower the remaining residual stresses and plastic strains are and the smaller the material pile up around an indentation. Therefore a simple isotropic hardening law is established to account for these effects. Contact is standard surface-to-surface-contact with a friction coefficient of 0.2 as described in [14]. From the results of this first step the indentation depth and diameter and the residual stresses that are produced by the particle indentation are obtained to feed the second step.

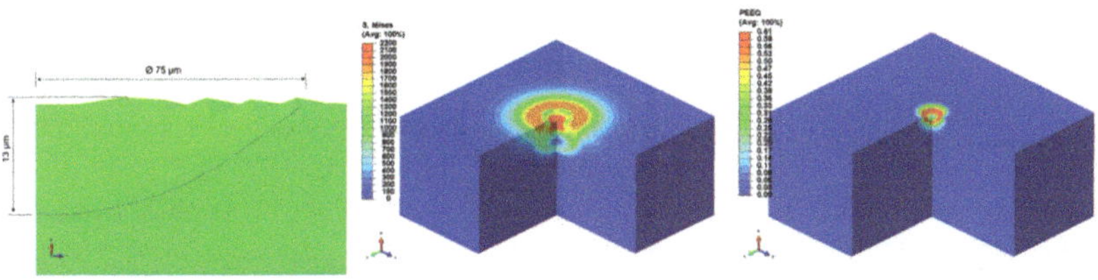

Fig. 4: Results of simulation step 1. The dimensions of the plastically deformed geometry of the surface (left) are used to feed equation one. From that the altered Hertzian contact pressure is calculated. The residual stresses are directly applied to the second model step 2 as initial condition (middle). In the area of contact a significant amount of plastic deformation is observed illustrated by the equivalent plastic strain (left).

The results of step 1 can be seen in Figure 4. Here only the results for the case without residual stresses are depicted. The von Mises stresses show a good agreement with the simulations in the literature mentioned before. There is significant plastic deformation in the contact area of the surface and the ball. The remaining plastic deformation left an indentation of 75 µm in diameter with a depth of 13 µm.

The second step of the calculation consists again of two quasi-static FE analyses. To represent the contact pressure of the roller element a Hertzian contact pressure is applied to the surface of a second model similar to the first one. The Hertzian pressure was adapted to the altered geometry of the surface. To do so, the analytical approach by COULON [18, 19] was used to avoid complex Elastohydrodynamic calculations of the lubricant film and contact calculations. The altered Hertzian pressure distribution can be seen in Figure 5. The plot shows an increase of the Hertzian pressure by approximately 20 %. The governing equation is

$$\Delta P = -\frac{\Delta P_{sh}}{\left(\frac{\Phi}{2}\right)^2}\cdot\left(\sqrt{X^2+Y^2}-\Phi\right)^2 + \Delta P_{sh} \text{ with} \Delta P_{sh} = 0.7\,\frac{\Phi}{D_{th}}. \quad (1)$$

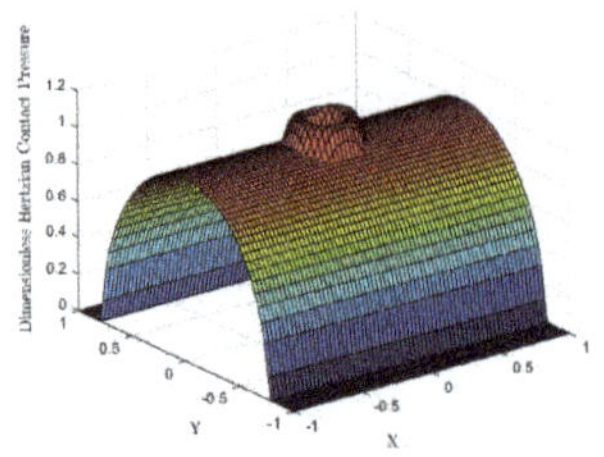

Fig. 5: Dimensionless Hertzian contact pressure at line contact on a dented surface with an indentation diameter of 75 µm and a depth of 13 µm.

With Φ as the dimensionless indentation diameter and D_{th} the dimensionless indentation depth.

The peak Hertzian pressure is 2516 $\frac{\mathrm{N}}{\mathrm{mm}^2}$. The material model is the same as in step one. The residual stresses that remain after the indentation are applied as initial conditions. As the first step has shown, the relevant volume for the second step is much smaller than the actual bearing model. The model for the second step is thus adapted to the size of the first step. The analysis is run both with and without residual stresses.

To evaluate the stresses considering the bearing fatigue life the analytical approach of IOANNIDES and HARRIS for advanced life service is used [8]. Here the classical fatigue life concept of PALMGREN and LUNDBERG is extended by implementing the actual stresses that exist below the raceway surface, see equation (1).

$$\ln\left(\frac{1}{S_i}\right) = AN^e\left(\int_{V_r} H(\sigma_i - \sigma_u)\frac{(\sigma_i-\sigma_u)}{z_i'^h}\,dV\right). \quad (2)$$

These stresses are evaluated by the failure criterion by DANG VAN [20] with which the hydrostatic stress state, hoop and residual stresses can be considered.

Results and discussion

After the second step the residual stresses follow as presented in Figure 6. The peak von Mises stresses concentrate in three big areas around the indentation. Two of them are directly at the surface to the left and the right of the indentation. The third area is located approximately 100 µm below the indentation. The areas are slightly shifted to the direction rolling direction. In the case of the analysis with residual stresses (left), the stresses drop almost to zero at 300 µm and rise again at the bottom at 400 µm. The peak value is 1885 N/mm^2. In the case without residual stresses (right), the stresses do not fade as much and stay constant from 100 µm to 400 µm of depth. The peak value is 1846 N/mm^2. Below that the hydrostatic stresses are presented. Here the overall stress level in the case with residual stresses is increased. The peak for the residual stress case is 3896 N/mm^2 whereas in the case without residual stresses the peak is 3685 N/mm^2.

These stresses are evaluated with the equations shown above and the resulting bearing fatigue life is calculated, see Figure 7. The L_{10} fatigue life of the analysis with residual stresses is $1.97\cdot10^7$ revolutions whereas the case without residual stresses reaches a L_{10} of $2.06\cdot10^7$ revolutions. This means even a decrease in fatigue life of around 4.6 %, which is not significant. This indicates that there is no benefit from the residual stresses induced by manufacturing with the parameters chosen in [11] which were optimized for the classical subsurface fatigue mechanism and not for surface initiated fatigue triggered by indentations. The fatigue life from [11] is illustrated for comparison. The L_{10} here is $3.61\cdot10^8$ revolutions.

Materials Research Proceedings **2** (2016) 407-412 doi: http://dx.doi.org/10.21741/9781945291173-69

The decrease is due to the complexity of the stress field that arises from superposition of the different stresses. Due to the particle indentation a residual stress state that varies vastly between high tensile and compressive hydrostatic stresses. The residual stresses induced during the manufacturing decrease the compressive stresses while the tensile stresses are enhanced.

Conclusion and future works

A two-step FE model was developed that accounts for plastic-elastic behavior and isotropic hardening of bearing steels. In the first step residual stresses emerging from particle indentation and the remaining indentation depth and diameter were calculated. In the second step these residual stresses were applied to the bearing model as initial conditions as well as the Hertzian contact pressure. The distribution of the pressure was altered using an analytical approach using the indentation diameter and depth of the indentation. Step two, then, was conducted with and without artificially induced residual stresses and the resulting stresses were evaluated with the fatigue life concept of IOANNIDES and HARRIS.

The study has shown that artificially induced residual stresses with the amount and depth as presented in [11] can decrease the fatigue life of roller element bearings in the case of particle indentations on the bearing raceway. However, residual stresses that are specifically customized for these stresses in the case of particle indentation by an appropriate choice of amplitude and depth of the stresses might have a positive effect. A combination of different sphere sizes during the deep rolling procedure could induce residual stresses in a wider range and closer to the surface.

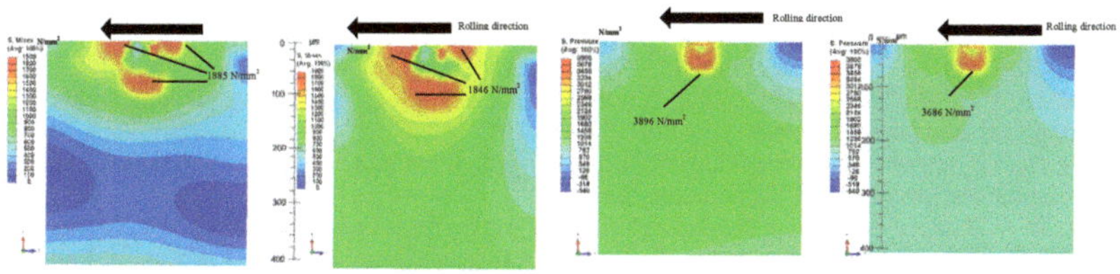

Fig. 6: Stresses of the second step. Analysis with residual stresses (left) and without residual stresses (right). Comparison of von Mises stresses (top) and hydrostatic stresses (bottom).

In [21] HAMER et. al. found similar results as presented in this study in which they investigated different geometries of indentations such as grooves. In [22] KO and IOANNIDES found opposite results to those presented in this study. They used larger indentations and did not consider the hardening behavior of the material. In return they examined a bigger variety of parameters.

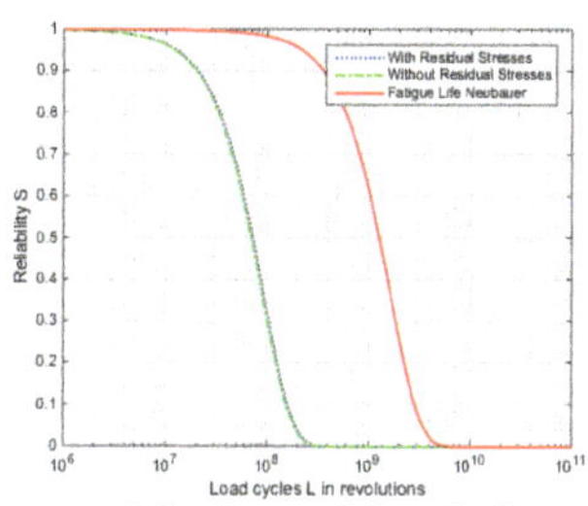

Fig. 7: Comparison of the evaluation of the fatigue life calculation.

Since the model focused on the most heavily loaded volume of the bearing and ignored the major part of the bearing volume it can only provide a tendency. To validate the results, experiments will be performed both on bearings with and without residual stresses that have indented surfaces and on reference bearings for comparison. Furthermore the FE model will be enhanced to allow for the calculation of the entire loaded volume of the bearing. Different distributions of manufacturing induced residual stresses will be studied.

Acknowledgement
Thanks go to Mr. NEUBAUER on whose works this study is based and whose data and FE Models were provided by the author. This work was sponsored in part by the DFG (German Research Center) within the SPP1551 in the context of the research program "Resource efficient Machine Elements".

References

[1] LUNDBERG, G.; PALMGREN, A.: Dynamic Capacity of Rolling Bearings. In: Generalstabens Litografiska Anstalts Förlag, 1947.

[2] WEIBULL, W.: Strength of Materials. In: Generalstabens Litografiska Anstalts Förlag, 1939.

[3] VOSKAMP, A.: Microtructural changes during rolling, 1996.

[4] ZARETSKY, Erwin V. ; BRANZAI, Emanuel V.: Rolling-Bearing Service Life Based on Probable Cause for Removal – A Tutorial. In: Tribology Transactions, 2016.

[5] SIRCAR, S. D.: *Rolling Contact Fatigue of Hypoeutectoid Steel.* Karlsruhe, Karlsruher Institut für Technologie. Dissertation. 2015

[6] GABELLI, A., IOANNIDES, E., MIGLIETTA, E.: Increased Life Performance of Rolling Element Bearings in Gearboxes and Transmissions. In: VDI-Berichte 1230, 1996, pp. 631–645.

[7] BERGLING, G. ; GABELLI, A. ; IOANNIDES, E.: An analytical formulation for the life of rolling bearings, Bd. 137. In: *Acta polytechnica Scandinavia: Mechanical engineering series.*

[8] IOANNIDES, E. ; HARRIS, Tedric A.: *A new Fatigue Life Model for Rolling Bearings.* In: *Journal of Tribology* 107, 1985, No. 1, pp. 367–378

[9] ISO 281:2007. 2007-02-15. Rolling bearings - Dynamic load ratings and rating life

[10] ISO/TS 16281:2008. 2008-06-15. Rolling bearings - Methods for calculating the modified reference rating life for universally loaded bearings

[11] NEUBAUER, T.: Betriebs- und Lebensdauerverhalten hartgedrehter und festgewalzter Zylinderrollenlager. Leibniz Universität Hannover, IMKT. Dissertation. 2015

[12] HACKE, B.: Wälzlagerlebensdauer-Windgetriebe : FVA 541. In: FVA-Forschungsheft, 2011, No. 967

[13] KRAL, E. R. ; KOMVOPOULOS ; BOGY, D. B.: Elastic-Plastic Finite Element Analysis of Repeated Indentation of a Half-Space by a Rigid Sphere. In: Journal of applied mechanics 60, 1993, pp. 829–841

[14] ANTALUCA, Eduard ; NÉLIAS, Daniel: Contact Fatigue Analysis of a Dented Surface in a Dry Elastic–Plastic Circular Point Contact. In: Tribology Letters 29, 2008, No. 2, pp. 139–153

[15] XU, Gang ; SADEGHI, Farshid ; HOEPRICH, Michael: *Residual Stresses Due to Debris Effects in EHL Contacts.* In: *Tribology Transactions* 40, 1997, No. 4, pp. 613–620

[16] WARHADPANDE, Anurag ; SADEGHI, Farshid ; EVANS, Ryan D. ; KOTZALAS, Michael N.: *Influence of Plasticity-Induced Residual* Stresses *on Rolling Contact Fatigue.* In: *Tribology Transactions* 55, 2012, No. 4, pp. 422–437

[17] KO, C. N. (Hrsg.); IOANNIDES, E. (Hrsg.): *Debris denting—the associated residual stresses and their effect on the fatigue life of rolling bearing: an FEM analysis.* The University of Leeds : Elsevier, 1989 (Tribology Series Proceedings of the 15th Leeds-Lyon Symposium on Tribology held at Bodington Hall)

[18] BIBOULET, N.: *Influence of indentations on rolling bearing life.* Lyon, Institut National des Sciences Appliquées. Dissertation. 2008

[19] COULON, S. ; VILLE, F. ; LUBRECHT, A. A.: *Effect of a Dent on the Pressure Distribution in Dry Point Contacts.* In: *Journal of Tribology* 124, 2002, No. 1, pp. 220–223 – Überprüfungsdatum 2016-06-07

[20] DANG VAN, K. (Hrsg.): *Multiaxial Fatigue Limit - A New Approach.* New Dehli, 1984 (Int. Conf. Fract. Mech., ICF 6)

[21] HAMER, J. C. ; LUBRECHT, A. A. ; LOANNIDES, E. ; SAYLES, R. S.: Surface damage on rolling elements and its subsequent effects on performance and life, Bd. 14. In: *Proceedings of the 15th Leeds-Lyon Symposium on Tribology held at Bodington Hall, The University of Leeds*, pp. 189–197

[22] KO, C. N. ; LOANNIDES, E.: Debris denting-The associated residual stresses and their effect on the fatigue life of rolling bearing: An FEM analysis, Bd. 14. In: *Proceedings of the 15th Leeds-Lyon, Tribology Series*, pp. 199–207

Residual Stresses 2016: ICRS-10 Materials Research Forum LLC
Materials Research Proceedings **2** (2016) 413-418 doi: http://dx.doi.org/10.21741/9781945291173-70

Benchmarking studies of the MPISI Material Science Diffractometer at SAFARI-1

Andrew M Venter[1, 2, a*], Deon Marais[1, b] and Vladimir Luzin[3, c]

[1]Research and Development Division, Necsa SOC Limited, Pretoria, South Africa

[2]DST-NRF Centre of Excellence in Strong Materials, South Africa

[3]ANSTO, Bragg Institute, Lucas Heights, NSW, Australia

[a]Andrew.Venter@necsa.co.za, [b]Deon.Marais@necsa.co.za, [c]Vladimir.Luzin@ansto.gov.au

Keywords: Neutron Strain Scanner, Benchmark, Laser Shock Peening

Abstract. The measurement performance of the new material science instrument MPISI at the SAFARI-1 research reactor in South Africa has been comprehensively benchmarked for strain scanning applications. In addition to the traditional VAMAS Ring and Plug specimen, the benchmarking was extended to a project that required sub-millimeter spatial resolution. For the latter the MPISI performance has been compared with experimental investigations on the same sample set at the KOWARI instrument at ANSTO. Overall, good data quality and agreement have been observed between the two instruments.

Introduction

A new material science neutron diffractometer named MPISI (Materials Probe for Internal Strain Investigations; Zulu name for spotted hyena) has recently been established at the SAFARI-1 research reactor in South Africa. As part of the commissioning program the well characterized VAMAS Ring and Plug specimen [1] was investigated to confirm the general instrumental performance in terms of attainable strains and positional resolution.

Various projects have since been performed on MPISI with residual stress investigations in standard geometries using fully embedded and large gauge volumes. Resolving steep stress gradients close to surfaces however require higher precision setups and measurement strategies to achieve the desired positional accuracies. This entails positioning the apertures as close as possible to the sample, in conjunction with very accurate directional alignment to the instrument center of rotation (CoR) to ensure precise definition of the measurement gauge volume. In addition the sample reference point should be determined to within 10% of the minimum dimension of the gauge volume since poorer accuracy may lead to significant spurious strains in data points treated for partially submerged gauge volumes.

By performing comparative studies on the same samples the correspondence between the data quality, stress resolving ability and overall instrument performance can be determined.

Experiment design

To assess our high-spatial resolution measurement strategy, the depth dependences of the in-plane stresses were determined in thin laser-shock peened (LSP) aluminium plates. LSP is a new laser surface treatment technique that can introduce compressive residual stresses extending beyond 1 mm in depth into metallic components to enhance fatigue life and stress corrosion performance [2]. Investigations were done on the MPISI instrument using the smallest practically achievable gauge volume that was selected as a trade-off between the strain positional resolution and measurement times. The investigations were subsequently extended to the KOWARI neutron strain scanner [3] at ANSTO which is a well-established high performing instrument that, with its higher neutron flux, enabled achieving even smaller through-depth spatial resolution.

Sample description. Four AA6056-T4 aluminium plate samples treated to different laser-shock peen parameters have been investigated on both the MPISI and KOWARI neutron strain scanners. The plate sizes were 200 mm (along the rolling direction) x 50 mm x 3.2 mm where the top surface of each sample was laser-shock peened over an area of 25 x 25 mm^2. The LSP treatment was performed at the CSIR National Laser Centre (South Africa) using a Quanta Ray Pro-270 laser system from Spectra Physics. This was accomplished without a sacrificial ablative coating [4] using a laser power intensity of 3 GW/cm^2 in conjunction with a laser spot size of 1.5 mm that was applied in a raster pattern with spot overlap as indicated in the schematics of Table 1. Differences in the LSP treatment of the samples pertain to the laser spot coverage densities as summarised in Table 1. In all cases the laser track lines were along the plate widths.

***Table 1**: Parametric variation of spot density in samples*

Sample name	G	H	I	A (reference)
Spot density	100 spots per cm^2	500 spots per cm^2	2000 spots per cm^2	Untreated
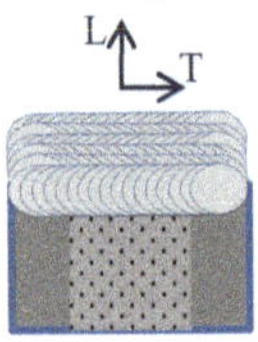				

Principal sample (and strain) directions are defined in terms of the plate geometry as longitudinal (L) parallel to the long dimension of the plate which coincides with the rolling direction, transverse (T) across the plate length, and normal (N), normal to the plate surface. The as-received parent material plate was used as stress reference that was investigated with the same setup and measurement protocol.

MPISI experimental procedure. MPISI is a constant wavelength diffraction instrument that delivers a 1.659 Å wavelength neutron beam onto the instrument CoR when diffracting from the (331) plane of a silicon multiwafer single crystal monochromator at a fixed take-off angle of 83.5°. This places the diffraction angle from the (311) planes of aluminium at approximately 85.5° as measured with the 2D neutron detector. A matchstick-shaped gauge volume was established with the primary beam aperture width set at 0.6 mm and 15 mm in height, with the secondary aperture width at 0.6 mm this rendered a nominal gauge volume of 5.4 mm^3. Both apertures were positioned 20 mm from the CoR to minimize beam divergence and precisely aligned by means of through-thickness intensity scans from an aluminium pin located at the CoR. Fig. 2a-d shows the experimental setup and measuring directions of the samples. Alignment of the sample with respect to the neutron gauge volume was achieved by performing through-wall-thickness intensity scans from the back surfaces (opposite to the peened faces) of all samples and determining the surface position using the procedures described in [5]. Alignment accuracies are estimated to be better than ±10 μm.

Strain measurements have been performed through the sample thickness in the central part of the LSP treatment patch as line scans by applying different measurement point densities. Close to the peened surfaces measurements were done in 0.2 mm steps, as overlapped gauge volumes, over the first millimeter to capture the near-surface peening effect, followed by step sizes of 0.5 mm to cover the remainder of the sample thickness. A total of 11 points (per strain direction) were thus measured through thickness. In order to compensate for spurious strains resulting from partially submerged gauge volumes at the sample surfaces, each sample was rotated 180° in the horizontal plane and the depth-dependence of the strain component re-measured at corresponding positions to the first measurements. The d-values at each corresponding position were averaged for the 0° and 180°

Materials Research Proceedings **2** (2016) 413-418 doi: http://dx.doi.org/10.21741/9781945291173-70

orientations as described in [6]. Data acquisition times for all data points were 3600 s that rendered *Δd/d* better than 10^{-4}.

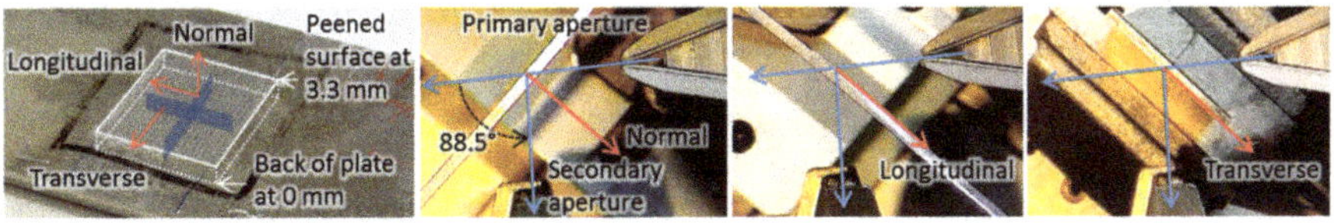

(a) (b) (c) (d)

***Fig.1**: Schematic depictions of the orientations and through-thickness measurement regions (a). Photographs showing the sample setups on MPISI for the measurement of the three principal strain directions normal (b), longitudinal (c) and transverse (d). The longitudinal and transverse components were measured with the sample in transmission geometry whereas the normal component was measured in reflection geometry.*

KOWARI experimental procedure. KOWARI is a constant wavelength diffraction instrument with a variable take-off angle. A take-off angle setting of 79° delivered 1.76 Å wavelength neutrons from the (400) refection of a silicon monochromator that put the (311) reflection of aluminium at ~90°. A nominal gauge volume of 0.2 x 15 x 0.2 mm^3 was established with the primary and secondary apertures positioned at 10 mm from the CoR to minimize the beam divergence.

The same instrument and sample setup procedures as used on MPISI were employed, though with faster measurement times. Through-thickness points were measured with regular 0.2 mm steps resulting in 16 overall points plus one additional point on the very surface. Data acquisition times of 1000 s for the T-component and 1500 s for the L-component were applied to compensate for texture related intensity variations in the plates, and 1200 s for the N-component. Acquisition times were increased by 50% for the first measurement position (closest to the peened surfaces) where data were taken with ½ submerged gauges. This gave diffraction peak intensities to provide *Δd/d* better than 10^{-4}. At each depth, measurements were taken by stepwise translation of the sample parallel to the surface at each depth (equivalent to "oscillating the sample") over a distance of 2 mm in 0.5 mm steps to improve the grain statistics. This protocol is depicted in the measurement grid shown in Fig. 2a-c. The 5 data points at each depth were averaged in the data analysis.

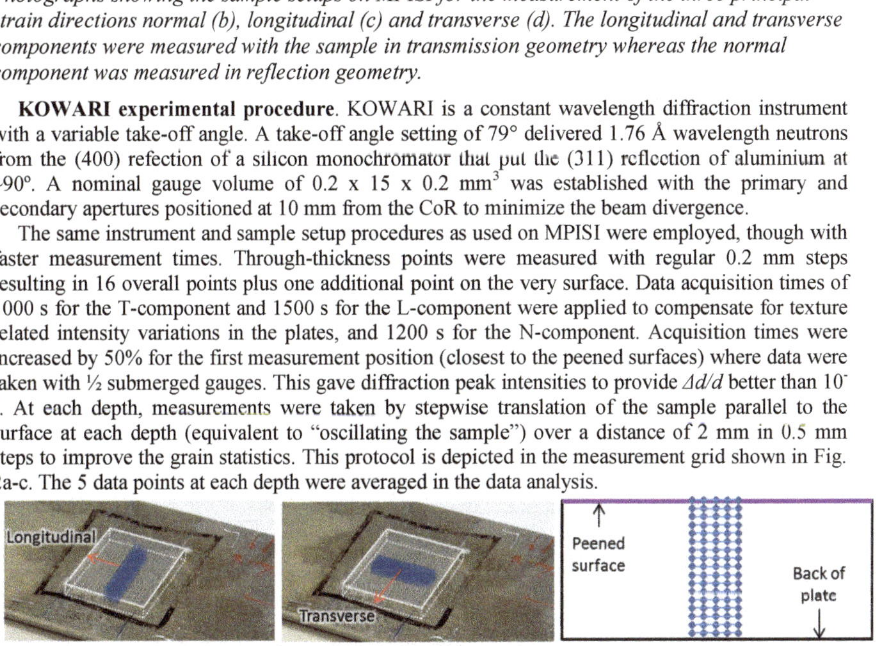

(a) (b) (c)

***Fig. 2**: Depiction of the measurement grid employed for the longitudinal (a) and transverse (b) component measurements on KOWARI. A cross sectional view of the in-plane and through-thickness measurement grid is given in (c).*

Results

A reasonable assumption for thin plates is that the stress normal to the surface $\sigma_{\perp}$ is zero, from which the in-plane stress ($\sigma_{//}$) component can be determined by enforcing a bi-axial stress condition

$$\sigma_{//} = \frac{1}{½S_2\,(hkl)}\left(\frac{d_{//} - d_{\perp}}{d_0}\right) \tag{1}$$

where *½S₂(hkl)* is the corresponding diffraction elastic constant for the material and reflection, $d_{//}$ (d_L or d_T) and $d_{\perp}$ (d_N) respectively the in-plane and normal components of the lattice plane spacing at the corresponding depth positions. The values for the diffraction elastic constants of aluminium used are $S_1(311)$ = -5.158 TPa^{-1} and $½S_2$ (311) = 19.574 TPa^{-1} as determined from the single crystal elasticity constants using the Kröner approximation. Under the plane-stress approximation, the stress-free lattice plane spacing per point was calculated from

$$d_0 = \left(\frac{½S_2(hkl) + 2S_1(hkl)}{½S_2(hkl)}\right)d_N - \frac{S_1(hkl)}{½S_2(hkl)}(d_L + d_T) \quad (2)$$

Stress results, with partial illumination corrections applied, are indicated in Fig. 3. In all curves the MPISI results are indicated as points only, in blue, whereas the data points for the KOWARI results are linked with lines (guides to the eye) in red. In the parent material significant differences are evident between the L and T stress components, especially at mid-plate thickness where distinctively large compressive stresses exist in the L-component (parallel to the rolling direction), that are balanced by slight tensile stresses at the surface regions. This is reminiscent of the deep (hot) rolling manufacture of the plate with consecutive cold rolling. The T-component does not reflect any large variation with thickness. A very good correlation is observed between the results determined with MPISI and KOWARI. These results on the reference sample were later used to obtain stresses purely associated with the peening action in the peened samples through point-by-point subtraction at each depth of the averaged L- and T-component stresses in the parent material from the stress values in the respective peened plates.

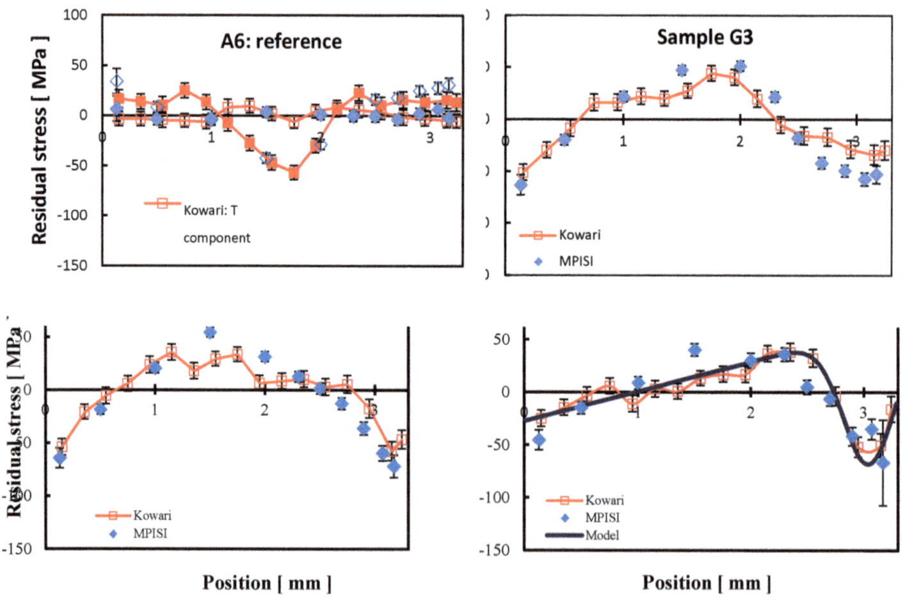

Fig. 3: Depth-resolved in-plane residual stresses determined for the four samples of this study. Sample A serves as reference that represents the parent material. Results are depicted as the in-plane stresses where the L-component refers to the stress along the plate rolling direction. Stresses depicted for the shock peened plates are the stresses purely due to the peening action. The peened surfaces are at 3.3 mm.

Materials Research Proceedings **2** (2016) 413-418 doi: http://dx.doi.org/10.21741/9781945291173-70

The calculated stress components L and T (not shown) in the peened plates have similar trends, but the magnitudes of the L-component are larger than the T-component, approximately 50% in sample G and 200% for samples H and I. This anisotropy is most likely related to the peening raster applied to cover the complete surface area that was done along lines over the plate widths and progressively translated along the plate length to accomplish the area coverage as shown in the schematic in Table 1. For the comparative characterization purposes, the stress fields are depicted by one profile, the in-plane stress. Hereafter, the individual longitudinal and transverse stress components were averaged and are reported in Fig. 3.

Sample I shows prominent reverse stress yielding at the peened surface and to a much lesser effect in sample H. Sample G shows a gradual stress gradient which becomes progressively more steep in samples H and I. In addition the results for sample I indicate the calculated peening stress distribution determined from the KOWARI data as input to an eigenstrain analysis [7]. Since the peen spot densities are progressively increased from sample G to H to I, it is expected that the residual stress results should reflect a similar interrelationship. This is though not seen. This may be ascribed to the complicated and distributed stresses that already exist in the parent material, or over-peening with the LSP treatments. Measurements were repeated at two different parent material samples using KOWARI, but no significant differences were observed.

Another parameter that has been considered in the benchmarking is the full-width-at-half-maximum of the Bragg peaks as a qualitative identifier of the plastic deformation regions. Results for the four samples are summarized in Fig. 4. In the parent material the peak widths are constant through thickness within the experimental errors. Sample G shows a slight increase in peak width as the peened surface is approached, whereas distinct peak broadening is observed in samples H and I. The peak broadening extends up to approximately 0.75 mm and 1.3 mm from the peened surface in samples H and I respectively. This indicates direct correlation between the peen spot density and the extent of the plastic yielding zone. Comparing the results between the two instruments reveals that the peak widths far from the peened regions are of similar resolution. In the peen-influenced region, the KOWARI results show better depth resolution between the different samples due to the smaller gauge volume employed.

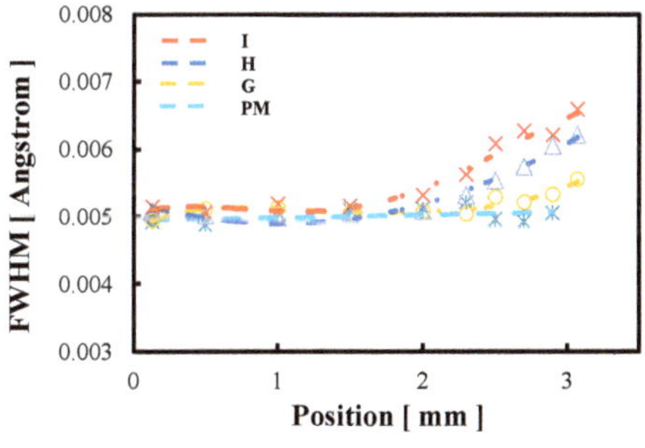

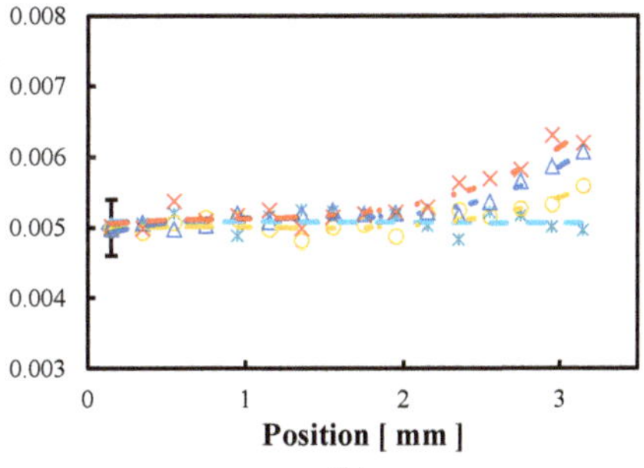

Fig. 4: Depth dependence of the FWHM for the four samples of this study. Curves have been determined from the diffraction data taken: (a) MPISI and (b) KOWARI. The solid lines indicate polynomial fits through the data points that serve as guides to the eye to indicate the trends.

Discussion

Notwithstanding the larger gauge volume used with the MPISI investigations and subsequent averaging of data over larger depth dimensions compared to KOWARI, essential features in the stress and FWHM profiles were captured. Due to the higher flux of KOWARI, nuances such as the reverse yielding profile in the near-to-surface region and discrimination of the plastically deformed regions could be resolved to greater detail.

Using KOWARI results as a reference, analysis of the peak intensities versus stress values, uncertainties and spatial resolution obtained, either the data acquisition times on MPISI or the size of the gauge volume could be reduced by 50%. For these high-resolution investigations the latter would be more advantageous and could be reduced to a gauge volume of 0.4 x 15 x 0.4 mm^3 instead of 0.6 x 15 x 0.6 mm^3. In addition the depth resolution could be further improved by increasing the vertical dimension of the beam to 20 mm with a comparative reduction in the depth dimension. A similar "oscillation" measurement strategy should also be enforced to improve the grain statistics.

Summary

With the MPISI neutron strain scanner having been recently commissioned, the investigation formed part of a benchmarking study of MPISI against the KOWARI neutron strain scanner in the high-spatial resolution regime. High-resolution depth-resolved measurements have been taken in an untreated and three LSP treated AA6056-T4 aluminium plates. Two in-plane components (longitudinal and transverse) as well as normal strain components were measured with a similar protocol and both attempting the best spatial resolution to provide cross-validation of the experimental results.

The latter investigation served to verify the experimental results obtained, as well as to yield some practical recommendations for optimizing high resolution measurement and to facilitate experimental unraveling of the residual stresses close to the peened surfaces. Overall, very good correlation is observed between the MPISI and KOWARI investigations.

Acknowledgments

AM Venter acknowledges financial support from Necsa SOC Limited and the use of the neutron strain scanner instruments MPISI and KOWARI under ANSTO proposal P4639. D Glaser is thanked for providing the LSP samples used in these studies.

References

[1] VAMAS report No 38 'Neutron diffraction measurements of residual stress in a shrink-fit ring and plug', G.A. Webster (Ed.), NPL, Teddington, UK, Jan 2000.

[2] K. Ding and Y Lin, Laser shock peening: Performance and process simulation, Woodhead Publishing, 2006.

[4] Y. Sano et al., Improvement in fatigue performance of friction stir welded A6061-T6 aluminium alloy by laser peening without coating, Materials & Design 36 (2012) 809-814.

[3] Kirstein, O., Luzin, V., and Garbe, U., 2009, The Strain-Scanning Diffractometer Kowari: Neutron News, v. 20, no. 4, p. 34-36.

[5] P.C. Brand, H.J. Prask, New methods for the alignment of instrumentation for residual-stress measurements by means of neutron diffraction, J. Appl. Cryst. 27 (1994) 164-176.

[6] M.T. Hutchings, P.J. Withers, T.M. Holden, T. Lorentzen, Introduction to the Characterization of Residual Stress by Neutron Diffraction, Taylor and Francis, London, 2005.

[7] A. M. Korsunsky, 2005, On the modelling of residual stresses due to surface peening using eigenstrain distributions: Journal of Strain Analysis for Engineering Design, v. 40, no. 8 (2005) p. 817-824.

Residual Stresses 2016: ICRS-10 Materials Research Forum LLC
Materials Research Proceedings **2** (2016) 419-424 doi: http://dx.doi.org/10.21741/9781945291173-71

Discussing of Deformation of Additive Manufacturing due to External Bending

Ya-Ju Lin[1,a], Jia-Chang Wang [2,b], Cherng-Yuh Su[3,c], Che-Hua Yang[4,d]

[1,2,3]Department of Mechanical Engineering, National Taipei University of Technology, Taipei, TAIWAN

[4]College of Mechanical and Electrical Engineering, National Taipei University of Technology, Taipei, TAIWAN

[a,*]parrbtaiwan@gmail.com, [b]jcw@mail.ntut.edu.tw, [c]cysu@mail.ntut.edu.tw, [d]chyang@ntut.edu.tw

Keywords: Additive-Manufacturing, 3D Printing, Ceramic, Multilayer, Residual Stress

Abstract. Default and deformations are some of issues for Additive manufacturing, as known as the 3D printing. When it processes multilayer thin films stacking, this process will introduce higher residual stress that causes deformation. This phenomenon makes the product faulty such as delamination, crack and blisters. The reason could be included ununiformed casting, temperature, cubical contraction etc. However, it is not easily to seem the difference form the films in the present. For ceramic materials, if deformation was existed at interface between layers, the crack would appear after the sintering was completed. For elastic material, the film will be bended, if the structure of product is unstable or the thickness is not thick enough to resist bending. In order to observe the deformation from material, residual stress and external bending must be considered. The purpose of the present study is to find the best approximation of formula for 3D printing and to predict the residual stress for the products. Later, the producing process could be adjusted until the most deformation is eliminated. The methods for present study are considered for two multilayer formulas. One is derived from Hsueh's closed-form, the anther is Timoshenko Theory. The results are combined simulation and experimental for two material and their accuracy is examined. Except considering the effect of the thermal stress, the acceptability of assuming for analyses in the 3D printing system is discussed.

Introduction

Additive manufacturing (AM), an exclusive mechanical technology, has introduced in past few decade. It is different from conventional manufacturing process that limited for creating a small and complex product. There are fewer restrictions for use of machining, molding, costing. Its potential can benefit designer to improve their artwork easily. The model was designed form Computer Aided Design (CAD) as 3D digital data and completed by AM. It simplified many complex preprocess steps from conventional manufacturing such as preprocess and post process. Therefore, the material waste and production time can be reduced. Based on different of material was used for AM, the produce of process can be widely different. For metal additive manufacturing, the processes contain power bed process, blown powder and wire feed processes [1].

Ceramic material is hardly to fabricate a small and complex product when using conventional manufacturing process. AM is an alternative way to product a high quality product such as creating dentures. Also, it does not need to prepare molds for model as preprocess. Therefore, to improve those advantage such as strengths and efficiency, there are many kinds of 3D-Printing devices were introduced recently such as Sterelolithography(SL) (Griffith *et al.*, 1996; Hinczewski *et al.*, 1998; Chartier *et al.*, 2002) ; Michosterelothograph (In-Baek Park *et al.*, 2009; Young-Myoung Ha *et al.*, 2010); Selective Laser Sintering (SLS) (Subramanian *et al.*, 1995; Liu *et al.*, 2007); Fused Deposition of Ceramics (FDC) (Agarwala *et al.*, 1996); Laminated Object Manufacturing (LOM) (Klosterman *et al.*, 1998), Slurry-Based Three Dimensional Printing (S3DP) (Grau *et al.*, 1997).

However, there are unexpected and irregular deformation appeared when fabricate the product layer by layer. The reason can be caused by incomplete adhesion between layers or uncompleted rigidified surface. Fig. 1 shows deformation in the alumina ceramic product. Bending due to shrinkage is also happened to both materials. Fig. 2 illustrated the bending stress generate in each layer due to shrinkage and illustrated layer additive process section where (a) casting slurry on the platform, (b) digital light processing (DLP) projector exposed onto object. The exposed area is consolidated and constrained form lower layer, (c) compressive stresses are then imposed on the individual layers to achieve displacement compatibility, (d) the asymmetric stresses system occurred bending product. Therefore, the process of AM can be described as a multilayer system. The multilayered systems are strongly influenced by residual stress. Most analyses are generally based on classical beam bending theory and strain continuity at interface between layers is required [2-3]. Hsueh presented an analysis for multilayer systems that the deformation was due to external bending [4].

Solvent-based slurry stereolithogtaphy (3S) [5] was the main AM for all experimental specimens in this paper. The purpose of the present study is to observe shrinkage of 3S product in multilayer systems. In order to detect inner stress and measure deformation, it would be necessary to observe eigenstrain. A semi-empirical procedure for observing eigenstrain from additive manufacturing was presented. Based on detected curvature form bending system, the eigenstrain can be simply defined. The predicted and measured result can be compared. The material was used photopolymer resin and alumina ceramic. Both materials were mixed photoinitiator that changes properties when exposed to the ultraviolet light. All products were made at room temperature. For alumina ceramic, sintering process was completed after printing process.

Method

The interfacial shearing stress can be predicts form Thimoshenko's joined beams bimetal thermostat model [6]. Sherman et al. extend theory to multilayer system as the eigenstrain-thickness function [7]. understand the bending stress in multilayer material for 3D printing, the theoretical analysis was considered. The assumption of eigenstrain-thickness function for multilayer film addressed as follows:

1. The material for each layer is homogeneous, isotropic and linearly elastic.
2. The thickness of each layer is uniform.
3. Perfectly bonded interface exists between each layer.
4. The peeling stress assumed as zero.

For all n^th of linear elastic multilayer films, each denoted by subscript *i*, which has their own thermal eigenstrain, $\varepsilon_i^{thermal}$, eigenstrain due to growth within the layers ε_i^{bend}, and the axial eigenstrain ε_i^{axial}. Thus the total eigenstrain, ε_i^{total}, as shown as below:

$$\varepsilon_i^{total} = \varepsilon_i^{thermal} + \varepsilon_i^{bend} + \varepsilon_i^{axial} \tag{1}$$

For each of thickness force and moment is defined as *P, M*. The resultant force is zero due to the uniform strain component. Moreover, the sum of bending moment is equilibrium. Therefore the equilibrium equation descripted as below:

$$\sum_{n=1}^{n} P_i = 0 \tag{2}$$

$$\sum_{i=1}^{n} M_i - \sum_{i=1}^{n-1} P_i \left[\frac{t_i - t_n}{2} + \sum_{k=i+1}^{n-1} t_k \right] = 0 \tag{3}$$

Materials Research Proceedings **2** (2016) 419-424 doi: http://dx.doi.org/10.21741/9781945291173-71

where t is layer thickness. The moment for each layer is related to the curvature and contained bending rigidity as shown as below:

$$\sum_{i=1}^{n} M_i = \sum_{i=1}^{n-1} M_i + M_n = \frac{1}{\rho\left(1-v_i^2\right)}\left(\sum_{i=1}^{n-1} E_i I_i + E_n I_n\right) \tag{4}$$

where ρ, v, E, I is radius of curvature, Poisson's ratio, Young's modulus and Moments of inertia respectively.

The unit elongation between two layers which along the $n-1$ interfaces

$$\varepsilon_{i-1}^{total} + \frac{P_{i-1}}{E_{i-1}t_{i-1}} + \frac{t_{i-1}}{2\rho} = \varepsilon_i^{total} + \frac{P_i}{E_i t_i} - \frac{t_i}{2\rho} \tag{5}$$

combined Eq.3, 4, 5, then the curvature can be defined as

$$\frac{1}{\rho} = \frac{\Delta\varepsilon^{total}}{\frac{P_{i-1}}{E_{i-1}t_{i-1}} - \frac{P_i}{E_i t_i} + \frac{t_i + t_{i-1}}{2}} \tag{6}$$

The discrete values of residual stress before unconstrained is calculated from

$$\sigma_i = \frac{P_i}{t_i} + \frac{t_i E}{2\rho} \tag{7}$$

Experimental measurement

There are two measurements to measure Young's modulus for materials. For specimen of photopolymer resin, Young's modulus can be measured by using tensile test. The other material couldn't be detected correctly form tensile test. For specimen of alumina ceramic, the ultrasound signal generator is an alternative method to measure Young's modulus, as shown in Fig. 4. The velocity of a transverse wave (C_T) and longitudinal wave (C_L) can be obtained by using ultrasound signal generator. The equation of transverse wave and longitudinal wave can be founded below [8].

$$C_T = \sqrt{\frac{E}{2D(1+v)}}, \quad C_L = \sqrt{\frac{\lambda + 2G}{D}} \tag{8}$$

where D is density of the specimen. G and λ is shear modulus, wavelength respectively. Therefor Young's modulus can be calculated from this equation shown as:

$$G = \frac{E}{2(1+v)} \tag{9}$$

According to bending system, the total curvature needs to obtain by experimental data. The 3D measurement system is a way to measure the total curvature of specimens for both materials.

Result

Table 1 lists 6 different total thickness of photopolymer resin. The specimen of photopolymer resin is designed as 30 mm (wide) x 50 mm (length). The thickness is 50 μm for each layer. After completed printing process, specimens exposed in the UV light again. This post-process is to make sure the boundary and surface are fully solidified binder and the specimens were not dissolved in methanol at room temperature. After irradiating process, the shrinkage of specimens is around in 90% to 95 %. The Young's modulus is 294 MPa, which was obtained by using tensile test. The measurements of curvature, which presented in black color in Fig 5., shows declined when total thickness of specimens was increased. Using curvature data, the eigenstrain can be obtained individually form eigenstrain-thickness function. The result shows in Fig. 6. The average eigenstrain calculated form each results can be used for predicted curvature of specimen. By linking resolution of each eigenstrain, the increasing trend can be created as a function that used for prediction in different thickness of specimens.

The specimen of alumina ceramic (Al_2O_3) is designed as 30 mm (wide) x 30 mm (length). The total thickness of Al_2O_3 is 4 mm and 8 mm respectively. The number of film layers is 200 layers and 400 layers. Each layer of the thickness was set up to 20 μm for 3S. The measurement of alumina ceramic can be separated in two parts from sintering. For sintering process, furnace temperature was raised to 1600 °C at a rate of 5 °C/min and was retained for 2 hours. Before sintering the product, the Young's modulus was around form 13.2 GPa to14.5 GPa, which are depended on the thickness of specimen. The average of Young's modulus is 238.27 GPa after sintering process. The total shrinkage of Al_2O_3 is 79% after post-process. The curvature for each specimen is 292.3 mm and 253.85 mm. The changed of curvature is not larger than that in photopolymer resin. By calculated from eigenstrain-thickness function, the average of eigenstrain can be obtained which is $3.91x10^{-4}$ and $4.51x10^{-4}$ respectively. The prediction of curvature using average eigenstrain is closed to the measurement. However, the result of eigenstrain was contained extrinsic and intrinsic phenomena that are shrinkage and phase transformation. This is not easily to detect those eigenstrain individually. Still, the eigenstrain can be simply described the bending system in two effects.

Summary

It is hardly to obtain eigenstrain and curvature at the same time. Therefore, in this paper, a semi-empirical procedure for 3D printing was introduced. This method relies on an experimental procedure to determine the eigenstrain. The eigenstrain-thickness is available to understand the bending system for 3D printing product. This result can be useful for predict curvature and know the thickness would be affected to the product bending. In this study, different of material has different Young's modulus that need to establish a reasonable experimental process. Shrinkage can be roughly described one of the reason that caused the product bending. More detail reasons can be found from preprocessing which is included the mixture of material, the slurry mixing, partial size of the material and the power of UV light. For the material of alumina ceramic, sintering process is an important process to discuss. Phase transformation is one of resolution that caused bending system. In present study, the eigenstrain of Al_2O_3 is described for total eigenstrain that thermal stress was included in the product. As the reason discussed above, more detail should be investigated in the future.

Acknowledgments

This research was sponsored by Ministry of Science and Technology R.O.C under contract MOST103-2622-E-027-029-CC2.

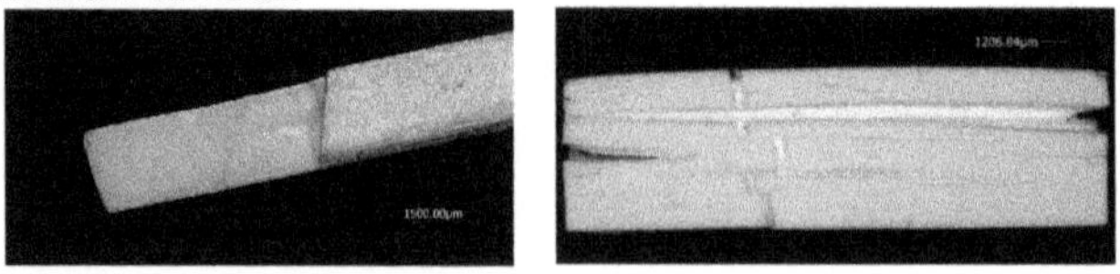

(a) (b)

Fig. 1 (a) The inner force caused the quality of product unstable and cracked form internal; (b) The inferior attachment between each layer of product appeared along the boundary of alumina ceramic after sintering.

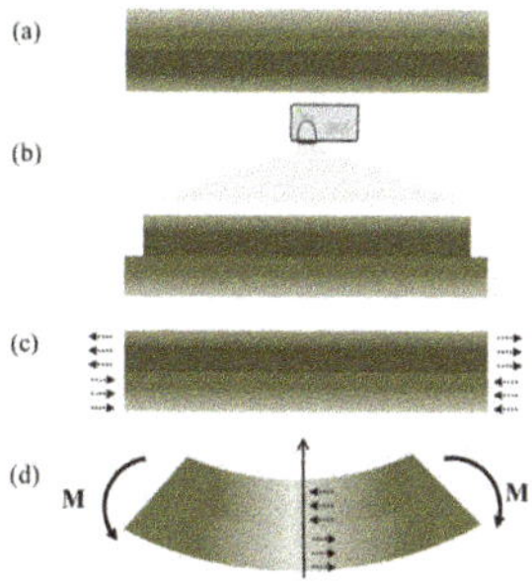

Fig. 2 After finish the printing, bending stress develop to balance bending moment. (a) Initial stress free condition (costing slurry); (b) constrained strain (mask exposing); (c) constrained in-plane strain; (d) banding induced by asymmetric stresses (complete rigid part)

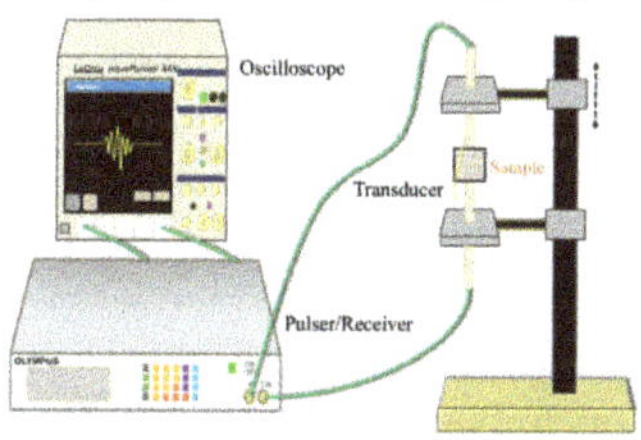

Fig. 3 The illustration for ultrasound signal generator.

Table 1. The initial experimental data of photopolymer resin.

No. of layers	Total thickness (mm)	Film thickness (µm)	Young's modulus (MPa)	Curvature (mm)
80	3.65	45.00	294.4	596.25
100	4.67	46.70	294.4	474.30
120	5.7	47.50	294.4	550.97
140	6.56	46.85	294.4	459.40
160	7.6	47.50	294.4	627.91
180	8.44	46.88	294.4	435.10

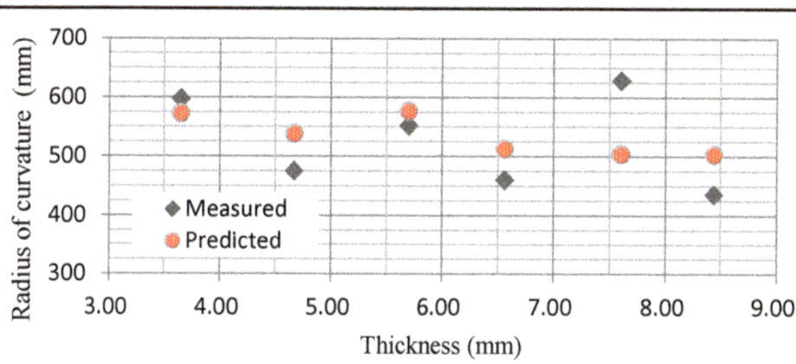

Fig. 5 Comparison of measured and predicted radius of curvature in 6 different thickness.

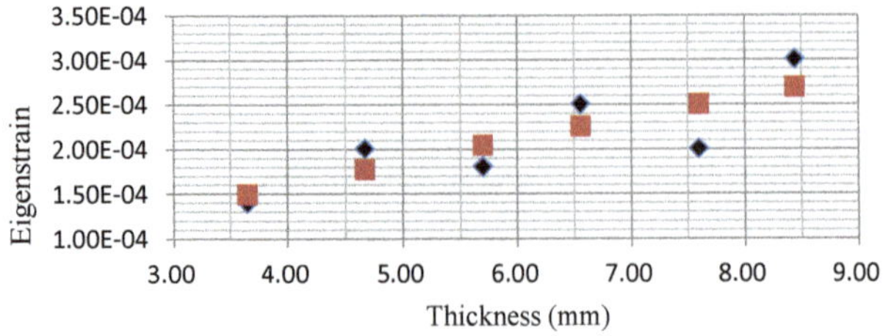

Fig. 6 Comparison of measured and predicted eigenstrain in 6 different thickness.

References

[1] Mustafa Megahed, Hans-Wilfried Mindt, Narcisse N'Dri, Hongzhi Duan, Olivier Desmaison, Metal additive-manufacturing process and residual stress modeling, Int. Materials and Manufacturing Innovation, (2016).

[2] R.H. Saul, Effect of GaAsx P1-x transition zone—GaAs substrates, J. Appl. 40, pp. 3272 (1969). https://doi.org/10.1063/1.1658174

[3] G.H. Olsen, M. Ettenberg, Calculate stresses in multilayered heteroepitaxial structures, J. Appl. 48, pp. 2543 (1977). https://doi.org/10.1063/1.323970

[4] C.H. Hsueh, Modeling of elastic deformation of multilayers due to residual stresses and external bending. J. Appl., pp. 9652 (2002).

[5] Jia-Chang Wang, A novel fabrication method of high strength alumina ceramic parts based on solvent-based slurry stereolithography and sintering, Int. J. Precis. Eng. Man., 14, (2013) 485 – 491. https://doi.org/10.1007/s12541-013-0065-3

[6] S. Timoshenko, Analysisof bi-metal thermostats, J. Opt. Soc. Amer., vol. 11, pp. 233-255 (1925). https://doi.org/10.1364/JOSA.11.000233

[7] A. Ni, D. Sherman, R. Ballarini, H. Kahn, B. Mi, S. M. Phillips, A. H. Heuer, Optimal design of multilayered polysilicon films for prescribed curvature, J. Materials Sci., vol. 38, pp. 4169 (2003). https://doi.org/10.1023/A:1026333707011

[8] Viktorov I.A., Rayleigh and lamb waves: physical theory and applications, Plenum Press, New York (1967). https://doi.org/10.1007/978-1-4899-5681-1

Residual Stresses 2016: ICRS-10 Materials Research Forum LLC
Materials Research Proceedings **2** (2016) 425-430 doi: http://dx.doi.org/10.21741/9781945291173-72

Residual Stresses in Dengeling-Treated Aluminum Alloy AA 7050

Linnéa Selegård[1,a], Ru Lin Peng[2,b*], Annethe Billenius[2,c], Gert Petersén[1,d], Markus ESS[3,e], Mattias Jonsson[1,f]

[1]Saab Aeronautics, S-58188 Linköping, Sweden

[2]Dept of Management and Engineering, Linköping University, SE-581 83 Linköping, Sweden

[3]Starrag AG, Seebleichestrasse 61, 9404 Rorschacherberg, Switzerland

[a]linnea.selegard@saabgroup.com, [b]ru.peng@liu.se, [c]annethe.billenius@liu.se

[d]gert.petersen@saabgroup.com, [e]markus.ess@starrag.com, fmattias.jonsson2@saabgroup.com

* corresponding author

Keywords: Dengeling, Mechanical Surface Treatment, Aluminum Alloy, Residual Stress, Roughness

Abstract. Dengeling is a new method for mechanical surface treatment of metallic parts to enhance fatigue properties. The treatment produces consecutive lines of indents by a spherical indenter, resulting in compressive residual stresses in a surface layer. This paper investigated residual stresses, surface roughness and near surface deformation, in specimens of AA 7050 T7451 after different dengeling treatments. The effect of indent overlap (0%, 25% and 50%) and indenter size (Ø3 and Ø8 mm) were studied. X-Ray diffraction measurements revealed that plastic deformation and compressive residual stresses were generated in a surface layer of about 1 mm by treatments using the Ø3 mm indenter and about 1.2 mm by treatment using the Ø8 mm indenter. Anisotropic residual stress fields were observed with higher compressive residual stresses (-360 MPa for 50% indent overlap, independent of indenter size) perpendicular to indent lines and lower (max -270 MPa for all the treatments) parallel to indent lines. Increasing the overlap between indents gave higher subsurface compressive residual stresses only in the transverse direction of indent lines. It also reduced the surface roughness. Best surface finish (Ra below 1 μm) was obtained when using the Ø8 mm indenter and 50% indent overlap.

Introduction

Dengeling is a new method for mechanical surface treatment of metallic parts to enhance fatigue properties. The treatment is carried out on a numerically controlled machine using a special indenter which moves and strikes the surface of the metal part in a given pattern and at a high frequency, up to 600 Hz. The transfer of kinetic energy from the indenter upon impact causes in-plane plastic deformation, producing consecutive lines of indents in the treated surface. Similar to the widely used shot peening process in which the surface is bombarded by hard shots, the dengeling treatment induces compressive residual stresses in a surface layer. As a considerable number of publications on shot peening have demonstrated [1-3], compressive stresses in near surface regions can greatly improve the fatigue resistance of a part. Dengeling has certain benefits over shot peening. The process is NC controlled, thus the operator can control exactly the location and magnitude of the generated residual stresses. Furthermore, the dengeling treatment is performed on the same machine that is used for machining the part; it can therefore be integrated into the machining process of metallic parts. That machining and surface mechanical treatment are carried out on the same machine means a great reduction of production cycle time and cost.

Materials Research Proceedings **2** (2016) 425-430 doi: http://dx.doi.org/10.21741/9781945291173-72

Process parameters, including the diameter of indenter, center-to-center distances between neighboring dimples, and stroke distance of the indenter, can be controlled in a dengeling treatment to vary the induced surface modification such as surface roughness, depth distributions of residual stresses, and near surface hardness. Such surface changes are considered to be important for the fatigue performance of the treated part, as researches with respect to shot peening have shown [4,5]. However, little knowledge exists, which relates the treatment results to the dengeling process parameters used. The purpose of the current work is to characterize depth distribution of residual stress, surface roughness, and subsurface deformation in dengeling treated specimens of aluminum alloy 7050 in T7451 condition with regard to the effect of overlap between dimples and indenter size.

Experimental details

Materials and Dengeling treatment. A 6 cm thick plate of aluminum alloy, AA 7050 in T7451 condition, was used. Before the dengeling treatment, the plate surfaces were levelled by milling, which induced a surface residual stress of 33±21.7 MPa in the cutting direction and -135.9±9.7 MPa in the transverse direction. The XRD measurements for residual stresses also revealed plastic deformation in a surface layer of about 50 µm.

The dengeling treatments were performed by producing consecutive lines of indents on the milled surfaces. The process parameters were selected to show the influence of coverage rate and indenter size while the indenting frequency, 200 Hz, and stroke distance, 0.5 mm, remained unchanged. Two indenter sizes, one 3 mm and the other 8 mm in diameter, were used. As can be seen in Table 1, the diameter of dimples produced by the smaller indenter was 0.75 mm and for the larger indenter 0.72 mm. The center-to-center distance between dimples in the moving direction of the indenter was used to calculate overlapping rate between dimples. They were 0%, 25% and 50% as given in the last column of Table 1. The line feed defining the distance between neighboring lines of dimples was approximately equal to the selected dimple distance.

Table 1 Dengeling process parameters

Specimen	Feed rate (mm/min)	Indenter diameter (mm)	Dimple diameter (mm)	Line feed (mm/line)	Dimple distance (mm)	Dimple overlap (%)
S1	9000	3	0.75	0.760	0.7500	0
S2	6750	3	0.75	0.559	0.5625	25
S3	4500	3	0.75	0.380	0.3750	50
S4	4320	8	0.72	0.369	0.3600	50

Characterization experiments. Residual stress measurements were performed using the standard $\sin^2\psi$ method [6] with 9 ψ-angles spreading between ±55°. Diffraction peaks from the {311} lattice planes of Al were measured using the Cr-Kα radiation and the X-Ray elastic constant for residual stress calculation was $\frac{1}{2}S_2 = 19.54\ MPa^{-1}$. Residual stresses parallel to the indent lines (LD) and the line feed direction (LFD) were determined. As they showed a significant difference, residual stresses in the 45° direction with respect to LD were also investigated. Layer removal by electrolytic polishing was employed to facilitate measurements of in-depth residual stresses and thus the desired depth profiles could be obtained. No correction was made for possible stress relaxation due to the electrolytic polishing.

A cross-section cut in 75° to LD was prepared for each specimen for examination in Optical Microscope (OM) and Scanning Electron Microscope (SEM) to investigate changes in surface topography and microstructure after the dengeling treatments.

Results and discussions

Surface roughness and deformation. Vertical lines of indents with a difference in overlap between the indents within the lines and between the lines are clearly identified in the optical micrographs in Fig. 1. At 0% dimple overlap, the S1 surface was not fully covered by dimples; a

Materials Research Proceedings **2** (2016) 425-430 doi: http://dx.doi.org/10.21741/9781945291173-72

calculation using the geometry values in Table 1 gave a theoretical area coverage of 78.5%. An almost full area coverage was observed for S2 which was treated with 25% dimple overlap. When the Ø8 mm indenter was used (S4), the produced indents seem to be much shallower; a comparison to the milled surface confirmed that marks from the milling operation were still visible on the surface.

Results from the profilometer measurements were listed in Table 2. A small difference between the measurements along LD ($Ra_{,LD}$) and LFD ($Ra_{,LFD}$) was related to the different arrangements of indents in the two directions. With a decreased dimple distance from 0% to 50% overlap, $Ra_{,LD}$ was reduced almost by half, from 8.6 μm to 4.45 μm, while $Ra_{,LFD}$ was decreased from 6.5 μm to 4.8 μm. The use of the large indenter (S4) with an overlap of 50% resulted in very fine surface roughness with a Ra below 1 μm. Micrographs of the 75° cross-sections were compared in Fig. 2, which showed the evolution of the surface towards a more homogenous profile when the dimple overlap was increased. Roughness parameters, Ra and Rz, estimated from 4 micrographs (a total width of about 8 mm) over the cross-section of each specimen were also shown in Table 2. Except for S1 which had too few indents for the OM measurements, the obtained Ra values were similar to those from the profilometer measurements.

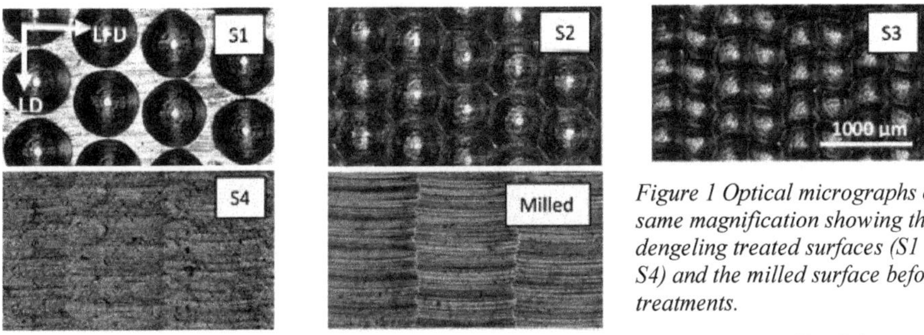

Figure 1 Optical micrographs of same magnification showing the dengeling treated surfaces (S1 to S4) and the milled surface before treatments.

Table 2 Surface roughness measured on surface ($Ra_{,LD}$, $Ra_{,LFD}$) and on cross-sections (Ra, Rz)

	S1	S2	S3	S4
$Ra_{,LD}$ (μm)	8.60	6.28	4.45	0.92
$Ra_{,LFD}$ (μm)	6.50	6.35	4.80	0.73
Ra (μm)	5.71±1.29	7.12±0.40	4.94±0.38	0.84±0.28
Rz (μm)	26.82±6.24	30.58±2.47	22.92±3.03	5.87±2.01

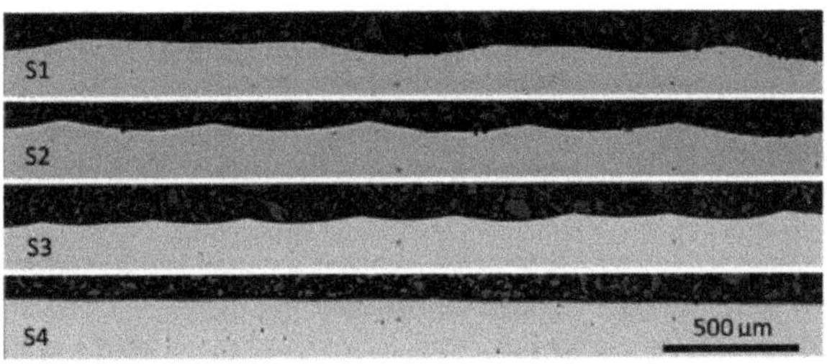

Figure 2 Surface roughness profile became more homogeneous with increased dimple overlapping: 0% (S1), 25% (S2) and 50% (S3). S4 with the Ø8 mm indenter and 50% overlap gave a very fine surface finishing.

Plastic deformation in the treated surface layer was studied by electron channeling contrast imaging in SEM and some images were shown in Fig. 3 as examples. A direct comparison between the specimens was not trivial; the small grains in the material requires examinations under high magnification while the surface profile varies in sub-millimeter scale and the deformation extends to a millimeter depth. Nevertheless, it was revealed that plastic deformation occurred not only close to the surface under dimples but also in the regions between dimples. It was also shown that the Ø8 mm indenter produced a more gentle effect on the surface; the near surface plastic deformation appeared to be lower (S4 in Fig. 3) than that in the treatment using the Ø3 mm indenter (S3 in Fig. 3), evident by the stronger strain contrast in S3.

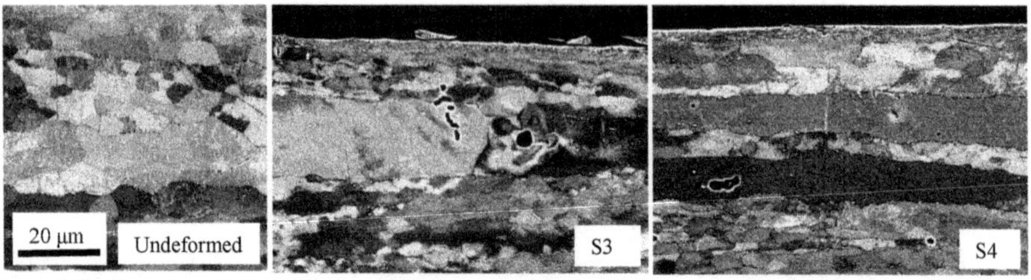

Figure 3 Backscatter electron images of same magnification showing stronger strain contrasts in S3 than in S4, indicating that the small indenter induced larger plastic deformation in the near surface region.

Residual stresses. The employed dengeling treatments were found to have induced compressive residual stresses in a significant depth, about 1 mm when using the Ø3 mm indenter and 1.2 mm for the Ø8 mm indenter. Typical depth-profiles of residual stresses can be seen in Fig. 4. For this particular specimen treated using the Ø3 mm indenter and 50% dimple overlap, the surface residual stresses were measured to be -198.6±5.8 and -301.8±14.3 for the LD and LFD directions, respectively. The residual stresses in the 45° direction were in between the LD and LFD stresses. About 100 to 400 μm below the surface, all the three stress components showed a maximum stress level with the highest compressive stress values, -273 MPa in the LD and -356 MPa in the LFD direction. At the 1 mm depth, the residual stress profiles changed to low tensile values. The thicknesses of the surface compression layers from the dengeling treatments studied here appear to be large, about two to four times of that commonly found for shot peened high strength aluminum alloys, see for example [3-5]. The large penetration depth of the dengeling treatment could partly be explained by a more efficient transfer of kinetic energy from the indenter to the part surface, in comparison with shot peening in which free bouncing of the shots are allowed. Another feature distinct from shot peening is a significant difference between the in-plane stress components; higher compressive residual stresses appeared in the LFD than in the LD direction. The largest difference, about 100 MPa, was in the surface, which sustained up to the 400 μm depth. The same trend existed for the other specimens, see Fig. 5 and Fig. 6, the difference between the LD and LFD stresses was however smaller for the S1 and S2 specimens with a lower overlapping rate of dimples. This stress anisotropy can be related to an anisotropic in-plane plastic deformation as the dengeling treatment was carried out by producing lines of indents.

Broadening of the diffraction peaks is related to the amount of plastic strain in the measured volume. As the peak width profile (PW) in Fig. 4 indicated, the plastic deformation penetrated to almost the same depth as the compressive stresses. It should be noted that the subsurface residual stress plateau corresponds to a small plastic deformation gradient in the region and the compressive residual stresses dropped as the plastic deformation decreased more rapidly at the larger depth.

Materials Research Proceedings **2** (2016) 425-430 doi: http://dx.doi.org/10.21741/9781945291173-72

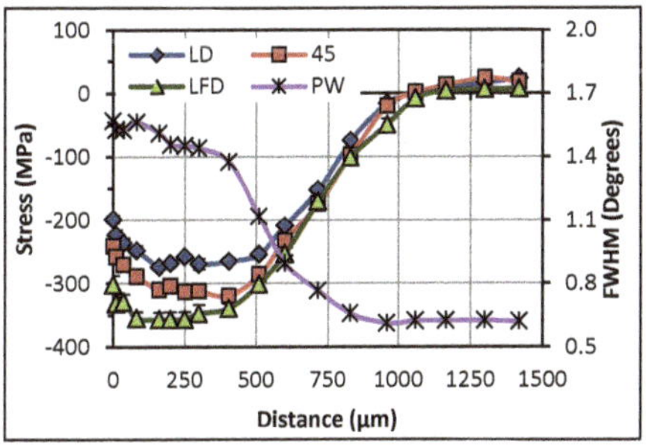

Figure 4 Depth profiles of residual stresses measured in S3 treated using a Ø3 mm indenter, 0.5 mm stroke distance and 50% overlapping. LD: line direction; LFD: line feed direction; 45: 45° to LD; PW: peak width as FWHM. The lines are guide to the eye.

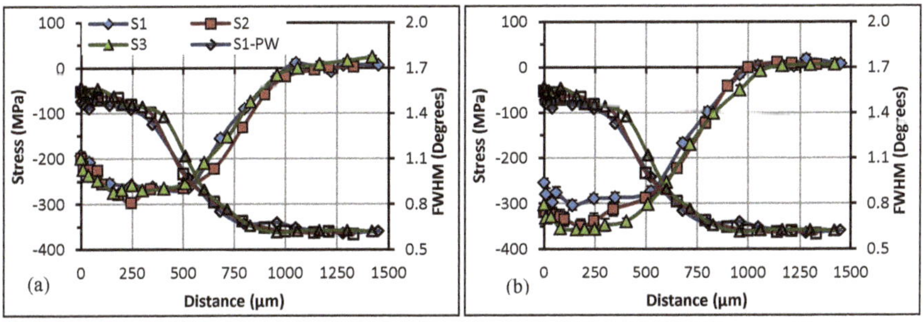

Figure 5 Influence of dimple overlapping with 0% (S1), 25% (S2) and 50% (S3) on residual stresses. (a) Residual stresses parallel to the LD direction and (b) residual stresses in the LFD direction. The lines are guide to the eye.

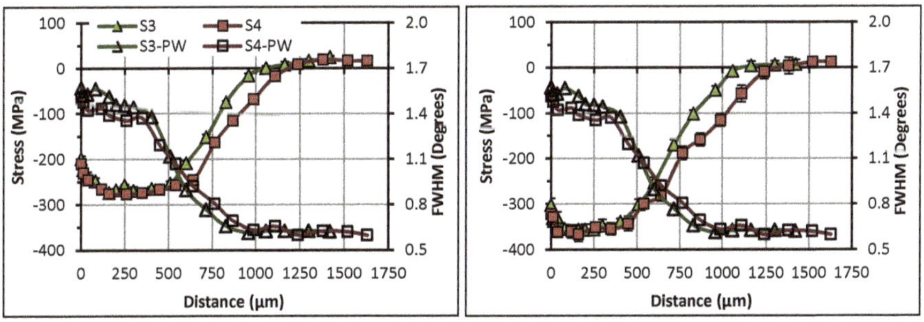

Figure 6 Influence of indenter size, Ø3 mm (S3) and Ø8 mm (S4), on residual stresses. (a) Residual stresses parallel to indent lines, (b) perpendicular to indent lines. The lines are guide to the eye.

The influence of overlap between dimples was illustrated in Fig. 5 where residual stress profiles were compared for specimens treated using different dimple distances while the indenter diameter (Ø3 mm) and stroke distance (0.5 mm) were the same. As Fig. 5(a) showed, the residual stresses parallel to the LD direction was essentially the same independent of dimple overlap. In the LFD direction, however, an increase of dimple overlap from 0% (S1) to 25% (S2) resulted in higher surface and subsurface compressive residual stresses although a further increase from 25% to 50% (S3) overlap produced a smaller effect. As a result of increased LFD stress, the stress anisotropy increased with increased dimple overlap as aforementioned. The surface and subsurface plastic deformations were also increased slightly for a higher overlapping rate, which might be the cause of higher compressive residual stresses in the LFD direction. However, the dimple overlap had little effect on the depth of plastic deformation and compressive residual stresses; it was approximately 1 mm for all the three specimens shown in Fig. 5.

In terms of the influence of indenter size, Fig. 6 revealed that a larger depth of compressive residual stresses was induced by treatment using the Ø8 mm indenter (S4) while the surface and subsurface compressive residual stresses did not seem to vary much with the indenter size. The effect on the plastic deformation was obvious: the Ø8 mm indenter gave a larger deformation depth but somewhat less surface and subsurface deformation which agrees with the lower subsurface deformation observed in SEM (Fig. 3).

Summary

Dengeling treatments with different dimple overlap and indenter size were performed on an aluminium alloy, AA 7050 T7451. The treatments were observed to induce compressive residual stresses to a significant depth, 1 mm or larger, which is two to four time of that for conventional shot peening of high strength aluminum alloys. This large thickness of compression could give a great increase in fatigue resistance. Other main results with respect to the employed process parameters are given below.

The affected depth was increased with the indenter size increasing from Ø3 to Ø8 mm but was almost independent of the overlap between dimples which varied from 0% to 50%.

The induced residual stress fields were anisotropic, characterized by a larger compressive residual stress transverse to indent lines (maximum -360 MPa) and a lower compressive stresses parallel to indent lines (max -270 MPa). This could result in a variation of fatigue properties with sample direction. The stress anisotropy was hardly affected by the indenter size but increased with increased dimple overlap.

Reducing the distance between dimples improved the surface roughness. The best surface finish, Ra < 1 μm, was found with the treatment using the Ø8 mm indenter with 50% dimple overlap.

References

[1] K.A. Soady, Life assessment methodologies incoroporating shot peening process effects: Mechanistic consideration of residual stresses and strain hardening: Part 1 - Effect of shot peening on fatigue resistance, Mater. Sci. Technol. 29 (2013) 637-651. http://dx.doi.org/10.1179/1743284713Y.0000000222

[2] A.T. Bozdana, On the mechanical surface enhancement techniques in aerospace industry - A review of technology, Aircr Eng Aerosp Technol. 77 (2005) 279-292. http://dx.doi.org/10.1108/00022660510606349

[3] M. Benedetti, V. Fontanari, P. Scardi, C.L.A. Ricardo, M. Bandini, Reverse bending fatigue of shot peened 7075-T651 aluminium alloy: The role of residual stress relaxation, Int. J. Fatigue. 31 (2009) 1225-1236. http://dx.doi.org/10.1016/j.ijfatigue.2008.11.017

[4] W. Zinn, B. Scholtes, Mechanical Surface Treatments of Lightweight Materials - Effects on Fatigue Strength and Near-Surface Microstructures, J Mater Eng Perform. 8 (1999) 145-151. http://dx.doi.org/10.1361/105994999770346972

[5] L. Wagner, Mechanical surface treatments on titanium, aluminum and magnesium alloys, Mater. Sci. Eng. A. 263 (1999) 210-216. http://dx.doi.org/10.1016/S0921-5093(98)01168-X

[6] I.C. Noyan, J.B. Cohen, Residual Stress Measurement by Diffraction and Interpretation, Springer-Verlag, (1987).

Residual Stresses 2016: ICRS-10 Materials Research Forum LLC
Materials Research Proceedings **2** (2016) 431-436 doi: http://dx.doi.org/10.21741/9781945291173-73

In situ X-Ray Diffraction Investigation of Surface Modifications in a Deep Rolling Process under Static Condition

Heiner Meyer[1, a*], Jérémy Epp[1, b] and Hans-Werner Zoch[1, 2, c]

[1]Stiftung Institut fuer Werkstofftechnik, Bremen, Germany

[2]MAPEX Center for Materials and Processes, University of Bremen, Bremen, Germany

[a]hmeyer@iwt-bremen.de, [b]epp@iwt-bremen.de, [c]zoch@iwt-bremen.de

Keywords: 2D Strain Mapping, Process Signature, in situ X-Ray Diffraction, Deep Rolling, Internal Material Load, Material Modifications

Abstract. The deep rolling process is widely used as a finishing step, improving the surface properties through cold working and creation of residual strains. In the present investigations, in situ X-ray diffraction experiments were performed with a self-built deep rolling device at the European Synchrotron Radiation Facility (ESRF) in Grenoble, France. The measurements were performed with a cylindrical roller-tool on steel samples at beamline ID11 with a monochromatic beam of 50 × 50 µm in transmission. Several properties could be investigated based on the diffraction data. In particular, 2D-strain maps were determined in the range of several millimeters around the deep rolling tool. Based on the data collected during loading and after unloading, knowledge about the transient state leading to the resulting remaining property modifications like residual strains and plastic deformation were generated.

Introduction

The new Collaborative Research Centre "Process Signature" of the University of Aachen and Bremen concentrates on the study of process-independent surface modification mechanisms in order to achieve predictive manufacturing processes. After the process, a given volume near the surface shows property modifications including microstructural changes, increased hardness and generation of compressive residual strains. The internal strain field created during the process can be predicted by FEM simulations as well as contact mechanics theory depending on the applied load and contact geometry. However, an experimental verification of the real strain fields during the process does not exist up to now. The characterization of material modifications in mechanical processes is often constrained to ex situ analysis of the end state and a theoretical evaluation of the external factors that lead to the observed modifications. This approach often shows deviations in simulations and theoretical concepts applied to the process, because of unknown parameters. From external factors, however, a specific internal material load is found in the surface near region and in the region below, which directly influences the materials response to the applied mechanical load. Measurements of the internal material load and of the material modifications during the process show several problems depending on the utilized method. Optical methods with digital speckle photography allow measurements of surface strain fields but no direct analysis of the volume state as can be seen in the works of Johnson [1]. Another method is the integration of sensors inside the material, as done by Tausendfreund et al. [2]. The bondings between the resistive thin film and the material as well as the whole integration however introduce errors in the propagation of the strain field, shown by Dumstorff and Lang [3]. One further method for in situ process analysis is based on the use of high energy synchrotron radiation, which can penetrate steel samples up to a depth of several millimeters or even centimeters and give diffraction patterns from the illuminated crystalline structure along the beam path with sufficient intensity, like already used by Uhlmann et al. [4] for the analysis of an

Materials Research Proceedings **2** (2016) 431-436 doi: http://dx.doi.org/10.21741/9781945291173-73

orthogonal cutting process. The method makes the measurement of a strain state during mechanical loading possible and gives additional information about texture, grain size, phase and specific elastic strain depending on the analysis as shown by Liss et al. [5]. In the present study monochromatic synchrotron radiation at the European Synchrotron Radiation Facility (ESRF) has been used to analyze the strain field and further material modifications of samples in transmission geometry during and after static loading with a deep rolling tool cylinder, to determine internal material load contribution and resulting residual strain state in situ respectively.

Experimental methods

Materials. The experiments were performed on quenched and tempered state of the 42CrMo4 (AISI 4140H) material. The chemical composition is given in Table 1.

Table 1 Chemical composition in wt.% of steel AISI 4140H (EN 42CrMo4) in the samples

Steel	AISI	Heat treatment	C [%]	Cr [%]	Mo [%]	Mn [%]	Si [%]	S [%]	Hardness [HRC]
42CrMo4 (Q+T)	4140H	850°C/2h Quenched to 60°C (Oil)/ Tempered at 400°C/4h	0.43	1.09	0.25	0.74	0.26	<0.001	47 ± 2

Experimental deep rolling device. For the experiments performed at the ESRF on beamline ID11EH1 a frame was constructed from aluminium profiles, which made the installation of an industrial deep rolling system with the full pressure range of up to 400 bar, provided through an external hydraulic fluid system, possible. Application of pressure was done in a top-down way and the sample positioning and movement was provided through a linear stage. Since the tool head is fixed in respect to the sample stage and the material state is homogenous isotropic, gauge volumes at different positions in respect to the contact point can be scanned during processing. The basic setup, a close up view of the sample with the deep rolling tool and the schematic layout are shown in Fig. 1.

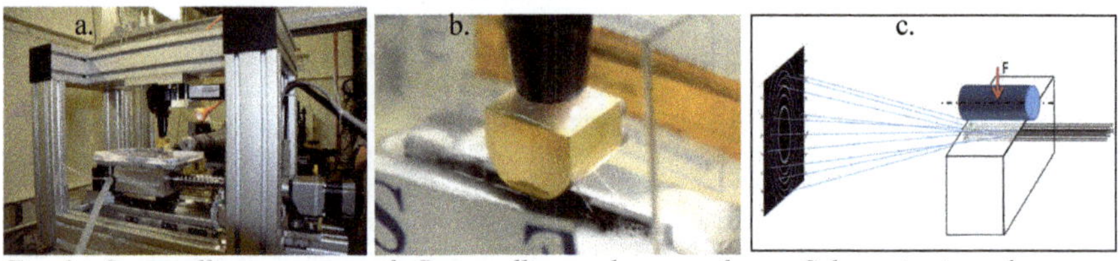

Fig. 1a. Deep rolling setup frame mounted in ID11 *b. Deep rolling tool on sample during process* *c. Schematic view of geometry*

Experimental and measurement parameters. The experiment was carried out at beamline ID11 on experimental hutch 1 at the ESRF with high energy monochromatic radiation. The parameters for the measurement with the detector and the slit system settings are shown in Table 2.

Table 2 Experimental parameters for the measurements at ID11 EH1

Photon Energy	Beam size [µm]	Detector	Measurement time [s/ meas.]	2Theta range
100 [keV] 0.0123984 [nm]	50 x50	FReLoN 2D CCD (2048x2048 pixel)	0.2s (+1.0s for axis movement)	Complete diffraction rings 0 - 10°

The detector was positioned at a distance of 273 mm from the sample behind the frame and the diffraction rings were measured in transmission geometry. A photon energy of 100 keV was used to allow steel samples of sufficient width to be measured in transmission so that mechanical stability during the processing in the setup is ensured, while small beam dimensions are best suited for the spatial resolution of the strain field, which has a high gradients below the contact point in this geometry. Sample geometry was approximately cuboid with a 20 mm by 70 mm height and length, where the initial thickness of 3 mm from the wire-cut electrical discharge machining was further reduced by 100 µm in the experimentally scanned area by electrolytic surface removal, giving a final sample thickness of 2.8 mm width .

The deep rolling tool for the static loading had a cylindric geometry with diameter of 13 mm and a length of 15 mm, housed inside a brass tool head, and was machined from tungsten carbide (WC) material. It was applied at a pressure of 300 bar, corresponding to a force of 3100±100 Newton along the width of the sample.

Methodical approach. With these parameters measurements of the 2D internal material load field around the deep rolling contact point were possible. Scans of the strains during static loading and the residual strain state after unloading were performed. As an example, the standard field in a static test consisted of a matrix of points with up to 50µm spatial resolution in the region near the contact point, with maximum grouping of points at the center along the y- and z-axis, which is shown in Fig. 2.

A full scanning field consists of 946 points in these measurements, where a range of 7.8 mm was scanned in horizontal direction and a zone of 4.15 mm depth from the surface of the sample was analyzed in vertical direction.

The movement of the setup frame on the diffractometer table for the different positions increases the total time for each point with 0.2 s measuring time + 1.0 s moving time to a mean value of 1.2 s/ point, giving an effective measurement frequency of 0.78 Hz.

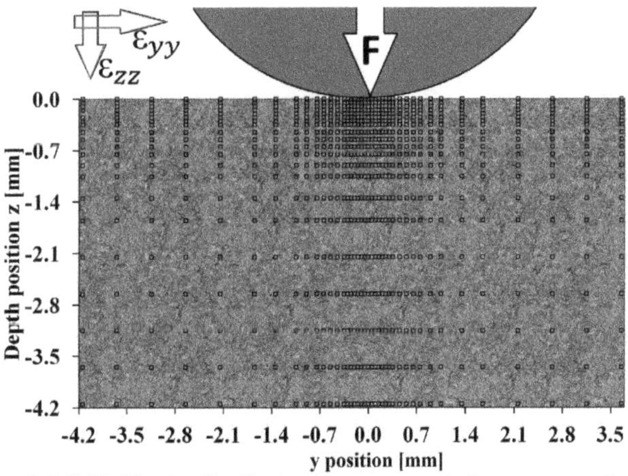

Fig. 2 Full-Field point distribution in scan range below contact point

Data analysis. Information from the gauge volume of 50 µm x 50 µm x 3 mm = 0.0075 mm³ at each measurement point contains the information about the strain state, phase content and other parameters of the microstructural condition of the diffracting crystallites. Because of geometric and material parameters, uniform loading and deformation of the material by the rolling tool is achieved in the region of the beam path, which is along the x-direction.

$$\varepsilon_{yy,zz}^{hkl} = \left(d_{yy,zz}^{hkl} - d_0^{hkl}\right)/d_0^{hkl} . \quad (1)$$

Strain evaluation along orthogonal directions relies on caking, meaning azimuthal integration of a part of the diffraction ring that corresponds to the strain directions, as is shown in Fig.3. Using Eq. 1 the strain in the direction corresponding to the azimuthal section can be determined by using a d_0-value of the unstrained material state, which was determined from a scan of the region before contact and compared with values from the edge points during and after loading. The strain evaluation is obtained from the α{211} reflex of the tempered martensite structure, as seen in the azimuthal integration in Fig. 4. This reflex is used because of its position around diffraction angle 6.05 deg with an intensity compared to the α{110} of $I_{\{211\}}/I_{\{110\}}$= 0.25, achieving a mean error of $\Delta\varepsilon$= 35 μstrains and $\Delta FWHM$= 0.0007 deg. Further analysis based on determination of lattice spacing from all available reflexes were not done yet but are planned in the next steps.

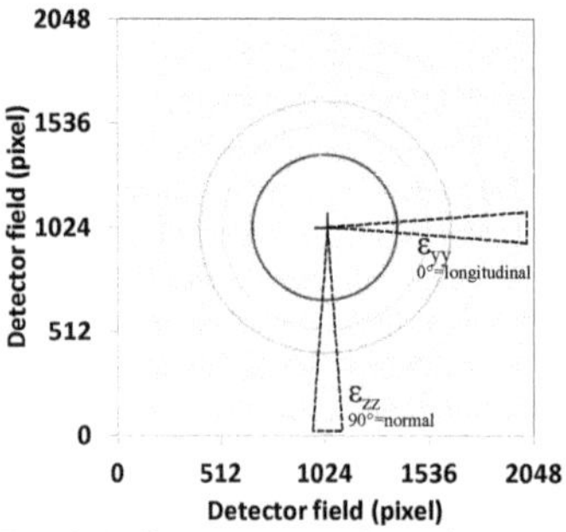

Fig. 3 Diffraction rings from detector readout with caking area in directions of $\varepsilon_{yy,zz}$

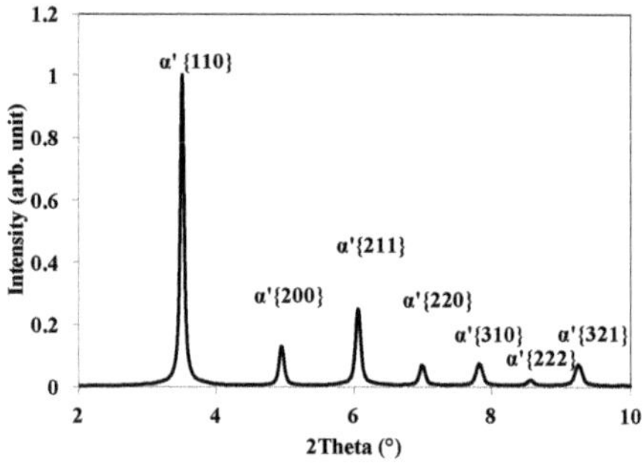

Fig. 4 Integrated diffraction pattern

Results

2D strain maps. In order to observe the local variations of the strain state under the contact point, the obtained strain maps are only plotted for a region of -1 mm < y < +1 mm and z < -1 mm. It can be seen in Fig. 7a and 7b that high load strains are measurable in a region around the contact point. Main strain field components lie in the direction of the applied force. The reconstruction of the strain maps is done point by point using Eq. 1 and yield during loading the following strain 2D fields on the Q+T material. Loading strains of the ε_{zz} component show a maximum of internal material load of -7500 μstrains at contact point, decreasing in depth. The strain field expands laterally in the material at lower strain state, which can be observed along contour lines. In the ε_{yy} direction compressive strains of up to -2500 μstrains are localized in the surface region to the side of the tool and tensile strains of up to 1500 μstrains are found in depth of -0.5 mm directly under the contact point. The fields show symmetric properties, were minor deviations from symmetry can be explained through experimental errors and slight variations in the local microstructure at the different measurement positions.

The resulting residual strains after unloading as seen in Fig. 8 are concentrated in the same region, while compressive residual strains up to -580 μstrains were measured in the ε_{yy} direction to both sides of the contact point and tensile strains of up to 600 μstrains under the contact point in the ε_{zz} component. Material modifications are not only found in depth direction under the roller but also sideways displaced material by plastic flow exhibits a compressive residual strain state.

Determination of the residual stress state in this case is impeded by the possibility of a triaxial stress state with non-zero components along the sample thickness. A modified plain stress state approach will be tested, while further ex situ measurements of the processed samples and simulations are currently ongoing and will give better understanding of the stress state.

Materials Research Proceedings **2** (2016) 431-436 doi: http://dx.doi.org/10.21741/9781945291173-73

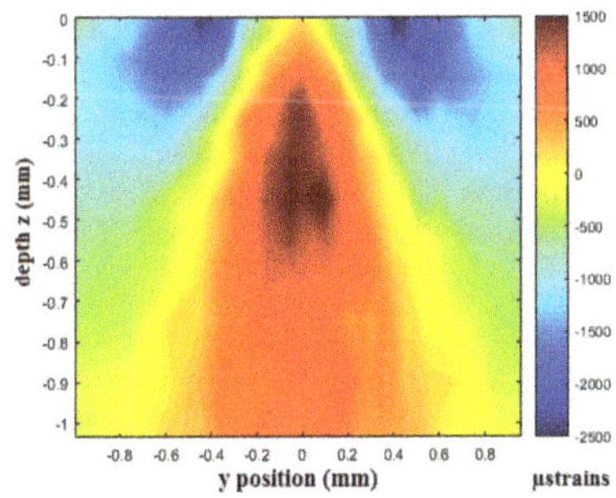

Fig. 7a 2D strain map of ε_{yy} during loading (300 bar)

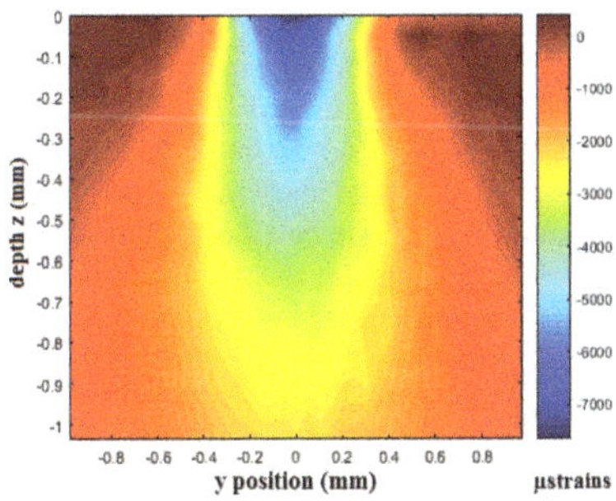

Fig. 7b 2D strain map of ε_{zz} during loading

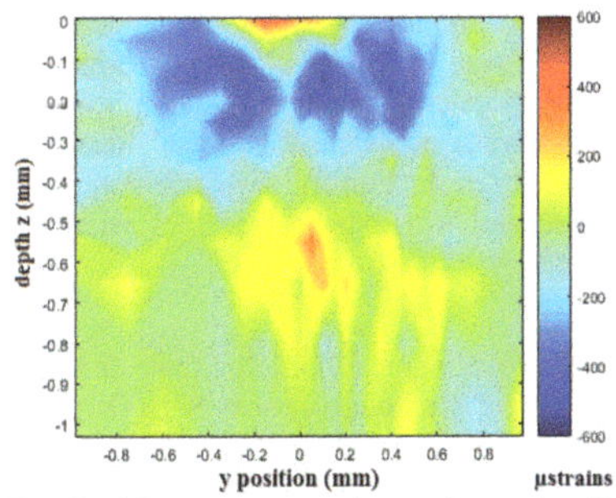

Fig. 8a 2D strain map of ε_{yy} after unloading

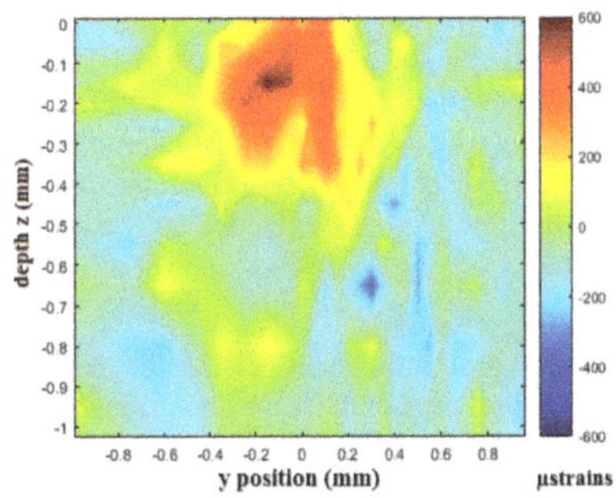

Fig. 8b 2D strain map of ε_{zz} after unloading

FWHM analysis. The integrated {211} peaks in 90° (ε_{zz}) direction were fitted with a pseudo-voigt function (PV) at every point of the 2D map and the variation of Full-Width at Half-Maximum (FWHM) of the peaks is displayed in the same way as the strains.

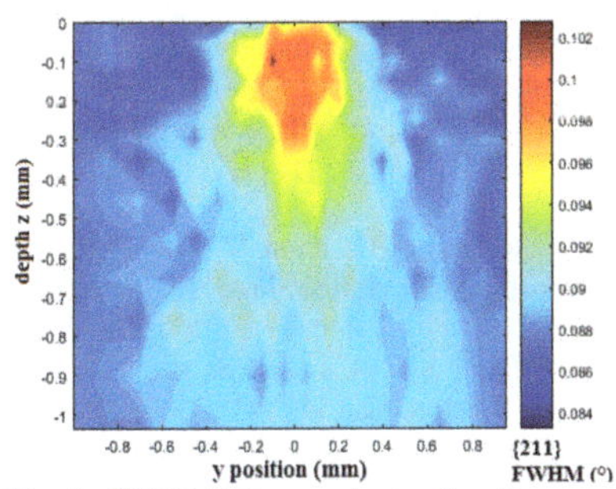

Fig. 9a FWHM 2D map during loading

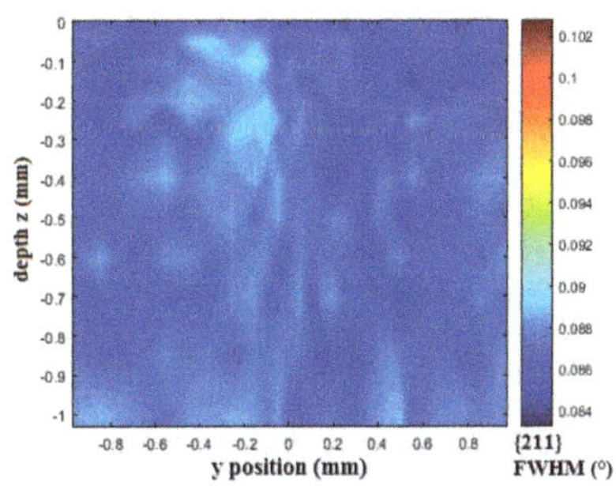

Fig. 9b FWHM in the same region after unloading

As can be seen in Fig. 9 a strong difference in FWHM values during loading and after unloading is present. If the FWHM increase would only have been resulting from plastic deformation, no change should have take place after unloading. Therefore, the strong increase of FWHM below the contact point during loading can be attributed to inhomogeneous elastic strains, either of 1st kind (strain inhomogeneity in the illuminated zone, e. g. over thickness or within the 50 µm beam size) or/and of 2nd kind (intergranular strains between crystallites during load, as seen by Morooka et al.

Materials Research Proceedings **2** (2016) 431-436 doi: http://dx.doi.org/10.21741/9781945291173-73

[6]).Further evaluations are needed and currently ongoing for deeper analysis and interpretation of these results. On the other hand, no significant increase of FWHM is observed after unloading. In this case, a similar behavior as reported by Zinn et al. is expected, where no pronounced peak width variation could be observed for quenched and tempered steel with an initial hardness close to 500 HV (49 HRC) [7]. Due to the mechanical energy input, already present dislocations can decrease due to annihilation effects, while new dislocations are created, which results in an almost constant FWHM for this material state.

Conclusion

It was shown that using monochromatic high energy synchrotron radiation, spatially-resolved measurements of internal material load during the deep rolling process can be performed with high spatial resolution. First results showing the 2D strain fields from spatially-resolved gauge volumes distributed around the point of contact inside the material were presented. The symmetric field with characteristics predicted by theory and simulation was found during the contact of the cylindrical deep rolling tool under static condition. The available data gives the prospect of detailed comparison of the internal material load state with the resulting residual strains as well as further material modifications. In further analyses of the data, deeper investigation of the FWHM distribution and resulting dislocation density will be performed. As well, stress calculations will be done by using different assumptions, based on FEM simulation and ex-situ investigations of the residual stress state of the processed samples. These are currently ongoing for deeper evaluation and interpretation of the experiments and will be extended to varying tool pressure and to dynamic processes with tool movement.

Acknowledgements

The authors gratefully acknowledge the German Research Foundation (DFG) for their financial support of the project C01 in the Transregional Collaborative Research Center SFB/TRR 136 “Process Signatures”. The in situ experiments were performed on beamline ID11 at the ESRF, Grenoble, France. We are grateful to J.P. Wright at the ESRF for providing assistance in using beamline ID11.

References

[1] P. Johnson, Strain field measurements with dual-beam digital speckle photography, Optics and lasers in engineering 30(3) (1998) 315-326. http://dx.doi.org/10.1016/S0143-8166(98)00021-9

[2] A. Tausendfreund, D. Stöbener, G. Dumstorff, M. Sarma, C. Heinzel, W. Lang and G. Goch, Systems for locally resolved measurements of physical loads in manufacturing processes, CIRP Annals - Manufacturing Technology 64(1) (2015) 495-498.

[3] G. Dumstorff, W. Lang, Investigations on the Impact of Material-Integrated Sensors with the Help of FEM-Based Modeling, *Sensors 15* (2015) 2336-2353. http://dx.doi.org/10.3390/s150202336

[4] E. Uhlmann, R. Gerstenberger, S. Herter, T. Hoghé, W. Reimers, B. Camin, T. Fischer, In situ strain measurement in the chip formation zone during orthogonal cutting, Prod. Engineer. 5(1) (2011) 1-8. http://dx.doi.org/10.1007/s11740-010-0266-x

[5] K.D. Liss, A. Bartels, A., Schreyer, H. Clemens, High-energy X-rays: a tool for advanced bulk investigations in materials science and physics, Textures Microstruct. 35(3-4) (2003) 219-252. http://dx.doi.org/10.1080/07303300310001634952

[6] S. Morooka, Y. Tomota, T. Kamiyama, Heterogeneous deformation behavior studied by in situ neutron diffraction during tensile deformation for ferrite, martensite and pearlite steels, ISIJ international 48(4) (2008) 525-530. http://dx.doi.org/10.2355/isijinternational.48.525

[7] M. Lebsanft, M. Tiffe, A. Zabel, W. Zinn, D. Biermann, B. Scholtes, Residual Stresses in Different Heat Treated Workpieces after Turning, Advanced Materials Research 996 (2014) 652-657. http://dx.doi.org/10.4028/www.scientific.net/AMR.996.652

Residual Stresses 2016: ICRS-10 Materials Research Forum LLC
Materials Research Proceedings **2** (2016) 437-442 doi: http://dx.doi.org/10.21741/9781945291173-74

Analysis of the Residual Stress in ARMOX 500T Armour Steel and Numerical Study of the Resultant Ballistic Performance

Michael Saleh[1,5,a*], Vladimir Luzin[2,b], Muhammad M. Kariem[3,c], Dong Ruan[4,5,d]

[1] Nuclear Fuel Cycle, ANSTO. Locked Bag 2001, Kirrawee DC, NSW, 2232. Australia

[2] Australian Centre for Nuclear Scattering, ANSTO, Locked Bag 2001, Kirrawee DC, NSW, 2232. Australia

[3] Faculty of Mechanical and Aerospace Engineering, Bandung Institute of Technology, Ganesha 10 Street, Bandung, West Java, 40132, Indonesia

[4] Faculty of Engineering and Industrial Sciences, Swinburne University of Technology, John Street, Hawthorn, VIC 3122, Australia

[5] DMTC, Level 2, 24 Wakefield St. Hawthorn. VIC 3122. Australia

[a] mis@ansto.gov.au, [b] vll@ansto.gov.au, [c] kariem.itb@gmail.com, [d] druan@swin.edu.au

Keywords: Residual Stress, ARMOX, Ballistic, FEA, Johnson-Cook

Abstract. Armour steels and their response to ballistic and blast threats have been dutifully studied in light of increased conflict and advances in protection levels. The strength of these quenched and tempered martensitic steels is a result of micro alloying, Ni Cr, and Mn, a combination of hot and cold rolling of the steel and proprietary heat treatments. The resultant hardness and toughness are pronounced and are reflected in the ballistic performance of the steel, although the role of the residual stress (RS) has not been unambiguously confirmed. To elucidate the effects of the RS on ballistic properties a two-step study was performed. Firstly, stress measurements were carried out on ARMOX 500T on the RS diffractometer KOWARI at ANSTO on a 8.3 mm thick plate. Stress components in rolling and transverse directions were calculated. Using the experimentally measured stress profiles as an input, numerical analysis was carried out on the ballistic response of the plate to the 7.62 mm APM2 round. Since numerical modelling inherently require the evaluation of material properties at elevated strain rates, to gauge the impact driven stress-strain response, the material's responses were derived using experiments utilising quasi-static testing and instrumented high strain rate experiments using the Split Hopkinson Pressure Bar (SHPB) at Swinburne University, Australia. These experimentally determined high strain rate data were incorporated into the Johnson-Cook (J-C) computational models for the flow stress along with literature sourced parameters for the failure model of the plate. Analysis of the two starting conditions, with and without residual stress, allows the authors to draw some conclusion about the role of the residual stress on the ballistic performance of ARMOX 500T armour steel.

Introduction

The evaluation of ballistic impact performance is often done through experimental testing. Analogously FEA (Finite Element Analysis) can be used as a predictive tool for the optimisation and validation of armour designs and to investigate the likely failure mechanism. RS are known to play a significant role in many material failure processes (e.g. fatigue, fracture, and stress corrosion cracking) [2], which are not characterised by high strain rate deformation of dynamic failure as in the case of ballistic impact. Based on a literature review of hot rolled high strength steels, the role of RS on their dynamic failures is yet to be fully elucidated, studies of HSLA-100 [3] under impact exhibited a pronounced stress field, for the as-received material, in the through plate thickness, analogous to the current study although with a dissimilar tensile stress magnitude. Another example can be found for SS 316 weldments [4] under impact. In a previous study the design of multi-layered armour systems has been found to be beneficial [5] in reducing the aerial density with a commensurate increase in performance when compared to homogenous rolled armour alone, RS was not a part of that study. Since all armour materials (plates) have RS of varying magnitude as a result of the manufacturing routes, a study of the role of the RS is desirable. To better clarify the ballistic behaviour and performance of armour steels the current study looks to measure the RS in 8.3 mm thick ARMOX 500T plate, quantify high strain rate material behaviour through SHPB testing and combine these into numerical FEA modelling to evaluate ballistic performance in several regimes.

Apart from the ballistic performance, the knowledge of the RS in armour plates is important for a number of other reasons. Commonly the compressive residual stresses at the surface can improve the resistance to crack initiation when a tensile load is applied. This may also improves fatigue strength and resistance to stress-corrosion cracking for vehicle mounted amour steels. Conversely, tensile RS at the surface are undesirable as they increase the overall stress level. It is widely accepted that these can lead to stress-corrosion cracking and fatigue failure by reducing the fatigue life and strength of the part.

Microstructure and material properties

Fig 1. SEM image of microstructure of as received ARMOX 500T (2000×).

ARMOX 500T belongs to a class of quenched and tempered high strength steel which are predominately used in amour applications [6-9]. The material is characterised by a high chemical purity, as shown in Table 1, with high Mn and Ni, to stabilise the steel in quenching, reduce distortion of the plate and to increase the hardenability and hardness depth. Typically the production of these armour steels involves hot rolling at a homologous temperature of $T^*_= 0.8$. The cut plates are then evenly quenched from a temperature of $T^*_= 0.65$ to form a largely martensitic microstructure. Low carbon steels such armour steel have small amounts of retained austenite but these steels are often tempered at $T^*_= 0.1$-0.3 to increase ductility. Secondary processes such as welding or cutting of ARMOX 500T are done below the tempering temperature of 200°, $T^*=0.1$ as the steels strength is negatively impacted. These low tempering temperature allow for the decomposition of the small amount of retained austenite to martensite albeit with a smaller reduction in the internal stresses when compared with higher tempering temperatures [10]. The resultant fine grained lathe martensite structure can be seen in Fig 1.

C %	Si %	Mn %	P %	S %	Cr %	Ni %	Mo %	B %
0.32	0.4	1.2	0.015	0.010	1.0	1.81	0.7	0.005

Table 1 Typical composition of ARMOX 500T.

High-strain rate testing

One method to quantify the high strain rate performance, which is readily used is armour testing, is the SHPB [11, 12]. The test involves a striker bar being fired at the incident bar with a pre-specified velocity. This induces stress waves, which propagate through the incident bar. Upon reaching the interface with the specimen the stress pulse is partially reflected back through the incident bar, whilst a portion of the pulse is transmitted through the specimen and into the transmitter bar. By placing strain gauges on both the incident bar and the transmitter bar the strain histories can be evaluated. The stress and strain can be evaluated through the Kolsky relation [13].

Materials Research Proceedings **2** (2016) 437-442 doi: http://dx.doi.org/10.21741/9781945291173-74

Neutron diffraction experiment

Neutron RS measurements in the 8.3 mm thick ARMOX 500T plate were performed on the KOWARI neutron diffractometer at the OPAL research reactor at ANSTO [14]. The Fe(211) reflection was used at 90°-geometry employing a neutron wavelength of $\lambda = 1.67$ Å. Three principal directions, rolling (RD), transverse (TD) and normal (ND), were measured scanning through thickness in 0.5 mm steps with a nominal gauge volume of size of $0.8\times0.8\times20$ mm^3 and resulting in overall 17 point strain profiles. The high flux of KOWARI, optimised for this wavelength, yielded average strain accuracy better than 50 μstrain (or ~10 MPa in stress scale after calculations and error propagation) in ~7 minutes of beam time. Of the three strain measurements two stress components were recalculated for RD and TD Fig3. (a), as well as d_0 values Fig.3 (c), assuming the zero plane stress condition. In the authors' opinion this is the most valid approach for stress calculation in the current case. The applicability of the plane-stress condition in thin plates can be based on theoretical consideration using the equation of equilibrium for the normal component, $\partial\sigma_{33}/\partial x_3 = -(\partial\sigma_{13}/\partial x_1 + \partial\sigma_{23}/\partial x_2)$, where x_3 is the through-thickness dimension and x_1 *and* x_2 are the in-plane coordinates. In case of large uniform plate, with no gradient in x_1 *and* x_2 dimensions, there is also no gradient of the σ_{33} in the through-thickness direction and since $\sigma_{33} = 0$ exactly on the both surfaces, this holds true through the whole thickness.

Constitutive modelling

The experimentally derived stress strain curves are fit to the three parameter J-C constitutive model [15] and is used to compute the flow stress of materials under high strain:

$$\sigma_{eq} = [A + B\varepsilon^n][1 + C ln\dot{\varepsilon}^*][1 - (T^*)], \quad (1)$$

where σ_{eq} is the flow stress, ε is the effective plastic strain, $\dot{\varepsilon}^* = \dot{\varepsilon}/\dot{\varepsilon}_0$ is the dimensionless plastic strain rate, $T^* = \dfrac{T - T_0}{T_{melt} - T_0}$ is the homologous temperature and A, B, n, C and m are material constants. This phenomenological model accounts for (i) the increase in yield stress due to strain hardening through dislocation pile up (ii) the increase in yield and tensile strength with increased strain rate and (iii) the thermal softening of the material due to thermal-plastic instability. The temperature increase is assumed adiabatic. Johnson and Cook [16] also developed a fracture model for ductile materials focusing on the stress tri-axiality sensitivity. The failure strain is computed from a multiplicative relation that expresses the stress tri-axiality, $\sigma^* = \frac{P}{\sigma_{eq}}$, where P is the hydrostatic stress, strain rate dependence and the homologous temperature, where D_1, D_2, D_3, D_4, D_5 are the J-C Fracture strain model constants:

$$\varepsilon_f = (D_1 + D_2 \exp(D_3\sigma^*))(1 + D_4 \ln(\dot{\varepsilon}^*))(1 + D_5 T^*) \cdot \quad (2)$$

The 7.62 mm projectile, Fig. 2, is similarly modelled using a modified version of the J-C model as outlined in [8]. A breakdown of the derivation and the applicability of the newly derived high strain data will be analysed in a future publication. Preliminary strength model constants for ARMOX 500T are listed in Table 2. Damage model contstants are reproduced from [17].

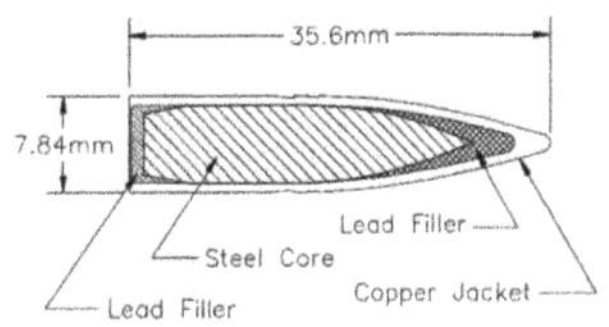

Fig. 2 Schematic of 7.62 mm APM2 [1].

A [GPa]	B[GPa]	n	C	m
0.385	**1.405**	**0.0263**	**0.0087**	**1.044**
D_1	D_2	D_3	D_4	D_5
0.068	**5.328**	**-2.554**	**0**	**0**

Table 2 J-C flow stress and failure model constants for ARMOX 500T.

Materials Research Proceedings **2** (2016) 437-442 doi: http://dx.doi.org/10.21741/9781945291173-74

Results of residual stress

The macro-stress profile in the through thickness of the steel plate is shown in Fig. 3 (a), and exhibits large oscillatory behaviour as a result of the production route. It characterised by large compressive stresses, -200 MPa, present at the surface, tensile stress zones ~1 mm below the surfaces, ~120 MPa, and a moderately compressive zone in the center, -100 MPa. Despite a more pronounced stress in the rolling direction (RD), the similarities of the two profiles, rolling direction and transverse direction (TD), indicate cross rolling was a most likely manufacturing route.

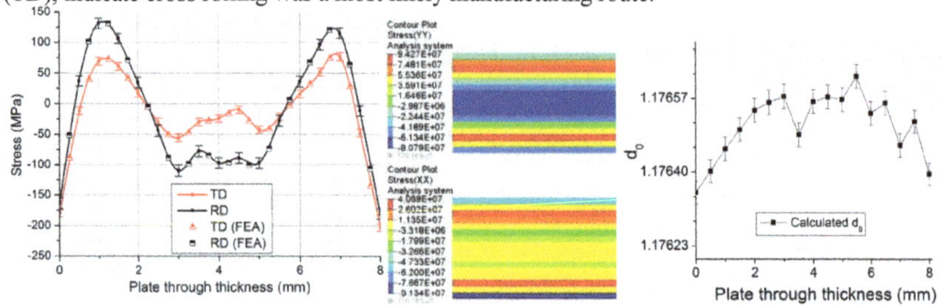

Fig. 3 (a) Residual stress measured using neutron and mapped to FE model (b) contours of the stresses after dynamic relaxation step (c) d_0 value profile evaluated for the zero stress condition in the through thickness approach.

Results for modelling simulation

Utilising a numerical routine the resultant stresses shown in Fig. 3 (a) were mapped in the plate's through thickness for the RD stress, σ_{yy} and the TD stress, σ_{xx} using MATLAB, the stresses were initialised at the integration points using a dynamic relaxation step. The FEA model is built using 3D half symmetry with 500,000 single-point integrated solid elements in LS-DYNA, with a plate through thickness element size of 0.5 mm. Shown in Fig. 3 (b) are the results of the precursor dynamic relaxation simulation in which sample equilibrium is achieved through nodal displacement based on the user defined initial stress. Note that some discrepancies are apparent between these dynamic relaxation stresses and those of the experimental measurements. Greater discretisation of the through thickness mesh would alleviate this issue and will be tackled in future studies.

The numerical LS-DYNA simulation of bullet collision with stressed and stress-free armour plate was performed for four bullet velocities of 450, 600, 750 and 900 m/s and with 0° of obliquity. These present velocities above and below the steel's V_{50} limits [18].

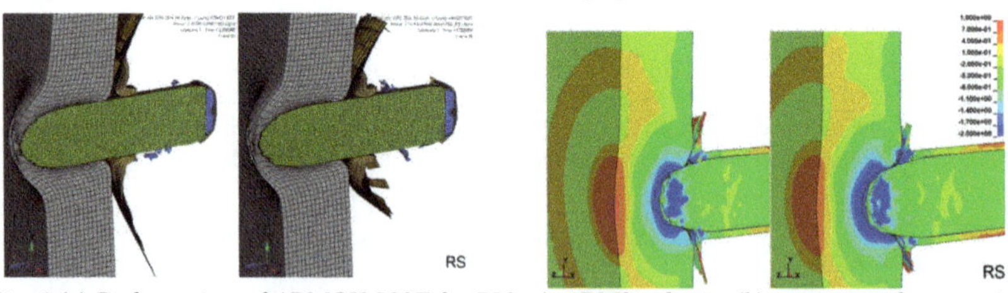

Fig. 4 (a) Deformation of ARMOX 500T for 750m/s APM2 velocity (b) contours of stress tri-axiality.

The model results demonstrates the classical ductile hole growth failure mode, whereby the projectile penetrates into the plate by pushing the target material aside, Fig.4(a). There is a small amount of localised bulging with a largely straight penetration channel and no bullet tumbling. The penetration process at t=32 μs shows the brass jacket stripped from bullet for the higher velocities.

Materials Research Proceedings **2** (2016) 437-442 doi: http://dx.doi.org/10.21741/9781945291173-74

The lead filler atop the steel core is largely eroded due to adiabatic heating and to a lesser extent the plastic work, the lead filler to the rear of the bullet is damaged by the collapsing brass jacket. This damage is induced by the complex contact which creates disparities in the bullet component's velocities. Some bullet tip erosion is apparent but no pronounced shape variations are shown between the steel with and without RS. The von Mises stress, Fig.5, in the bullets impacting at higher velocities is more apparent and more distributed for the RS plates. We observed a reduction of the residual velocity of the bullet after perforation, at an incident velocity of 900m/s, from 453m/s to 444 m/s for the RS plate.

For all the scenarios analysed the authors observed a reduction in the back face stress for the RS plates with a correspondingly small reduction in the back face deformation at velocities of 450m/s and 650m/s, Fig.6. At higher velocities, 750m/s there is greater propensity for the plate without RS to fail, in penetration, on the back face as seen in Fig 4(a). This stems from the increase in stress tri-axiality in the plates without RS, as the large compressive surface stresses in the RS plates improve the ballistic performance by reducing the magnitude of the material with positive tri-axiality, Fig.4(b) while the internal compressive stresses increase the volume of material with negative tri-axiality, also a mitigating mechanism to failure.

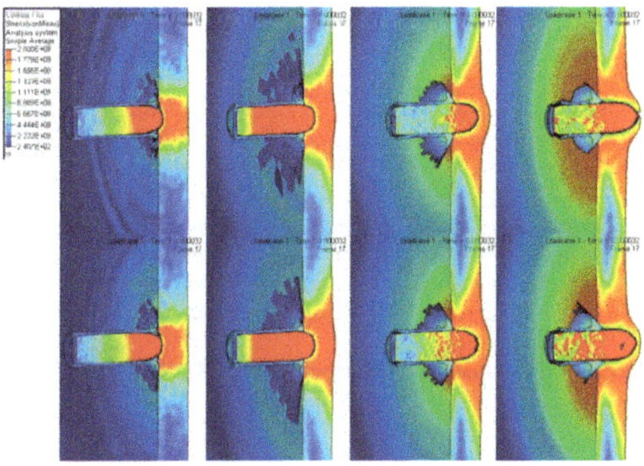

Fig5 von Mises stress for L-R 450,600,750,900 m/s at t=32μs. The bottom 4 results are the models initialised with RS.

material ahead of the bullet with negative tri-axiality. There is a small reduction in the residual velocity of the penetrating projectile for RS plates at higher velocities, above V_{50}. The reduction could become more pronounced where multiple sandwiched plated are used and were the cumulative RS effect may not be a simple additive relationship.

Conclusion

Residual stress for hot rolled ARMOX 500T has been characterised using neutron diffraction techniques with pronounced compressive and tensile stresses inherited from the production route. Stress mapping of the RS was used to model the ballistic performance of the plate against 7.62mm APM coupled with newly parametrised data for J-C flow stress model. RS has a small beneficial effect in reducing back face deformation in lower velocity projectile impact, below the ballistic limit. This is through a reduction in the positive tri-axiality experienced by the plate's back face and an increase in the volume of

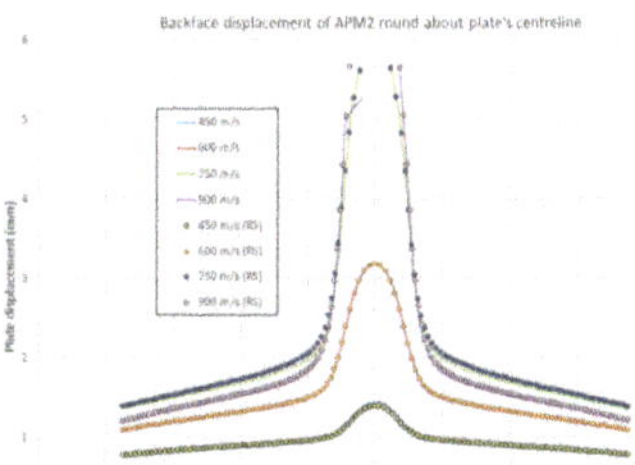

Fig. 6 Back face displacement for the ARMOX 500T.

Acknowledgement

D. Shanmugam and R. Smith from THALES for the supply of the ARMOX 500T plate.

References

1.Showalte, D.D., et al., *Ballistic Testing of SSAB Ultra-High-Hardness Steel for Armor Applications*, 2008, Army Research Laboratory.

2.Noyan, I.C. and J.B. Cohen, *Residual Stress Measurement by Diffraction and Interpretation*. 1987, New York: Springer Science+Business Media.

Materials Research Proceedings **2** (2016) 437-442 doi: http://dx.doi.org/10.21741/9781945291173-74

3.Prime, M.B. and R.L. Martineau, *Mapping Residual Stresses after Foreign Object Damage Using the Contour Method.* Materials Science Forum, 2002. **404-407**: p. 521-526.
4.Flores-Johnson, E.A., et al., *Numerical analysis of the effect of weld-induced residual stress and plastic damage on the ballistic performance of welded steel plate.* Computational Materials Science, 2012. **58**: p. 131-139.
5.Flores-Johnson, E.A., M. Saleh, and L. Edwards, *Ballistic performance of multi-layered metallic plates impacted by a 7.62-mm APM2 projectile.* International Journal of Impact Engineering, 2011. **38**(12): p. 1022-1032.
6.Jena, P.K., P. Senthil P, and S.K. K, *Effect of tempering time on the ballistic performance of a high strength armour steel.* Journal of Applied Research and Technology, 2016. **14**(1): p. 47-53.
7.Ponguru Senthil, P., et al., *Effect of heat treatment on ballistic performance of an armour steel against long rod projectile.* International Journal of Impact Engineering, 2015. **80**: p. 13-23.
8.Børvik, T., S. Dey, and A.H. Clausen, *Perforation resistance of five different high-strength steel plates subjected to small-arms projectiles.* International Journal of Impact Engineering, 2009. **36**(7): p. 948-964.
9.Maweja, K. and W. Stumpf, *The design of advanced performance high strength low-carbon martensitic armour steels: Microstructural considerations.* Materials Science and Engineering: A, 2008. **480**(1–2): p. 160-166.
10.Bawa, H.S., *Manufacturing Processes - Ii.* 2004: McGraw-Hill Education (India) Pvt Limited.
11.Lee, W.-S. and G.-W. Yeh, *The plastic deformation behaviour of AISI 4340 alloy steel subjected to high temperature and high strain rate loading conditions.* Journal of Materials Processing Technology, 1997. **71**(2): p. 224-234.
12.Børvik, T., et al., *Strength and ductility of Weldox 460 E steel at high strain rates, elevated temperatures and various stress triaxialities.* Engineering Fracture Mechanics, 2005. **72**(7): p. 1071-1087.
13.Kaiser, M.A., *Advancements in the Split Hopkinson Bar Test*, 1998, Virginia Polytechnic Institute and State University.
14.Kirstein, O., V. Luzin, and U. Garbe, *The Strain-Scanning Diffractometer Kowari.* Neutron News, 2009. **20**(4): p. 34-36.
15.Johnson, G.R. and W.H. Cook. *A constitutive model and data for metals subjected to large strains, high strain rates and high temperatures.* in *Proc. 7th Int. Symp. on Ballistics.* 1983. The Netherlands.
16.Johnson, G.R. and W.H. Cook, *Fracture characteristics of three metals subjected to various strains, strain rates, temperatures and pressures.* Engineering Fracture Mechanics, 1985. **21**(1): p. 31-48.
17.WIŚNIEWSKI, A. and P. ŻOCHOWSKI, *Building and Validation of the Johnson–Cook Constitutive Model of Nano-Composite NANOS-BA® Steel for Armour Applications* in *9th International Armament Conference on Scientific Aspects of Armament and Safety Technology*2012: Poland.
18.DoD, *Department of Defence Test method standard: V50 ballistic test for armor*, 1997: USA.

Structural Characterization of Ancient Japanese Swords from MAAS Using Neutron Strain Scanning Measurements

Filomena Salvemini[1, a*], Vladimir Luzin[1, b], Francesco Grazzi[2, c], Sue Gatenby[3, d], Min-Jung Kim[3, e]

[1] ANSTO, Bragg Institute, Lucas Heights, NSW, Australia

[2] CNR, Istituto dei Sistemi Complessi, Sesto Fiorentino, Italy

[3] Museum of Applied Arts and Sciences, Sydney, Australia

[a*]filomena.salvemini@ansto.gov.au, [b]vladimir.luzin@ansto.gov.au, [c]francesco.grazzi@fi.isc.cnr.it, [d]sue.gatenby@maas.museum, [d]min-jung.kim@maas.museum

Keywords: Neutron Residual Stress, Non-Invasive Analysis, Archaeometallurgy, Japanese Sword

Abstract. The current paper presents a forensic study that aims to characterize non-invasively the laminated structure of a set of Samurai's swords, part of the East Asian Collection of the Museum of Applied Arts and Sciences (MAAS) in Sydney, via strain scanning measurements. Neutron residual stress analyses were undertaken on samples of well-known origin, time period and authorship to create a reference database on the main lamination methods developed by Japanese swordsmiths. The benchmark data were cross-matched with results obtained from a *mumei* (no-signature) blade in order to attribute its manufacturing tradition based on qualitative and quantitative data rather than stylistic criteria. Maps of two stress components and d_0-values in the transverse cross-section of each sword were determined as a result of the neutron diffraction experiment with sub-millimeter spatial resolution. Since these two material characteristics are induced by the manufacturing process, analysis and comparison of the results allows drawing conclusions about variability or similarity of the actual production techniques of the Japanese swords.

Introduction

Japanese swords are famed among all others all over the world as the most effective in terms of hardness, resilience and, last but not least, aesthetics [1]. Their forging technique was almost unique; steel lumps, obtained from the furnace, were strongly pre-treated to obtain a homogenous and purified multilayered sheet. Distinctive carbon steels, characterized by different hardness, were shaped and specifically used for different parts of the blade components to optimize their mechanical feature (Fig. 1) [1].

Since the Koto age, 10th-17th century AD, five different traditions developed distinctive construction techniques that evolved during the following historical periods [2]. However, the actual techniques that were used by the early sword-smiths were never documented and the necessary information was orally transmitted from the master to his most skilled pupils. In spite of the large amount of studies published on the subject, different manufacturing techniques are still not fully understood.

Until recently, only expendable samples were investigated by standard analytical techniques, which mainly require sample cutting or are based on surface analysis. Nowadays, neutron diffraction [4] and neutron imaging methods [5] have been demonstrated to be the most suitable tools to qualitatively and quantitatively characterize composition and micro-structural properties of metal artifacts in a non-destructive way, mandatory for well conserved museum collections.

In the current study we attempted systematic research of the Japanese blades in order to investigate whether the laminated structure of a set of Samurai's swords can be determined in non-destructive manner by means of neutron tomography, neutron diffraction stress analysis or both. The research is based on part of the East Asian Collection of the Museum of Applied Arts and Sciences (MAAS) in Sydney. The Japanese sword collection contains both samples of well-known origin, time period and authorship (group 1) as well as a group of *mumei* (no-signature) blades (group 2). Group 1 allowed us to build up a reference database on the main lamination methods adopted by Japanese swordsmiths and therefore to be used as benchmark data that can be cross-matched with the results obtained from group 2 in the attempt of attributing the corresponding manufacturing tradition to the unknown blades.

Although a number of different blades were analyzed, in the current work we report only the neutron diffraction stress analysis on blades which were classified as *katana*, while the neutron tomographic data and results on *wakizashi* will be published separately somewhere else.

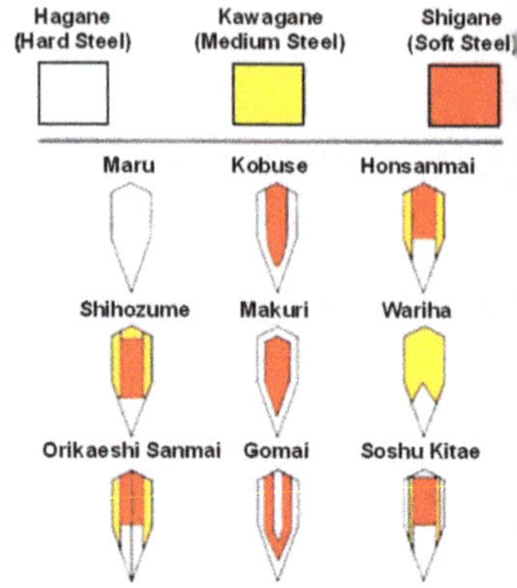

Fig. 1. The most common lamination structures of the Japanese swords.

Samples

The attributes of the four *katana* from of the East Asian Collection of MAAS that were used in the forensic investigation are reported in Table. 1. Origin, time period and authorship are known only for three of them by transliteration of the signature engraved on their hilt. According to stylistic study, only the time period can be assigned for blade H6856 while the manufacturing tradition it belongs to is still uncertain.

The physical description of the swords is given in Table 2, where the blades' total length and thickness and width, measured in the mid length (where neutron diffraction measurement were made) are reported.

Experimental: neutron diffraction stress analysis

The neutron diffraction stress analysis in objects like Japanese swords is difficult because of two reasons. First, a technical aspect, a high spatial resolution is required, ~0.5 mm, since individual multiple layers can be of sub-millimeter size while a typical thickness of a blade is 5-7 mm. At the same time many points are required to be measured since the exact number and location of layers are not quite known. Therefore a compromise between resolution, number of measurement points and experimental time per point needs to be found.

Second, a theoretical aspect, since the experimentally measured peak shift combines two effects, variation of (macrostress-free) d0 and elastic macrostress, and both of them are expected to be present due to manufacturing process, identifying and separation of these two contributions is essential for the aim of the experiment. However, there is no chance for proper resolution of d0 since only non-destructive analysis is allowed for the museum items. Thus,

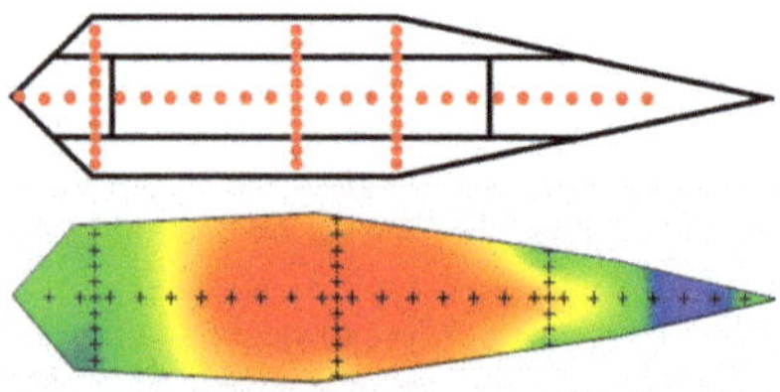

Fig. 2. A generic mesh in the transverse cross-section of the sword for neutron diffraction mapping and a real example of the longitudinal stress map for sword H4839 (min: -446 MPa, max: 291MPa).

Materials Research Proceedings **2** (2016) 443-448 doi: http://dx.doi.org/10.21741/9781945291173-75

the approach described in the following was used as the best possibility to resolve d0 problem.

The first problem was solved by use of ANSTO's neutron residual stress diffractometer KOWARI [6], which can provide necessary sub-millimeter (0.5 mm) spatial resolution and feasibility of that was proved in the past experiments [7]. Although a full 2D mapping would be very desirable, due to time limitations only a number of one-dimensional scans were performed. However, the measurement points are selected in such a manner and such numbers, >50, that would allow distinguishing unambiguously between possible structures as shown in Fig. 2 and at the same time some interpolation can be applied to produce 2D maps. At each mesh point spaced by 0.5 mm through thickness and by 1.0 mm along the central line, strain measurement was carried out in three orthogonal directions (longitudinal, transverse, normal) as in a traditional stress scanning experiment. A nominal gauge volume as small as 0.5×0.5×0.5 mm3 was used for measurements of the longitudinal strain component (symbols in Fig. 2 are scaled to this gauge volume size), while it was enlarged to 0.5×0.5×20 mm3 for measurement of the normal and transverse, since this was allowed by blade geometry, with extension of the gauge volume along the longitudinal direction. Strain measurement has been done using wavelength of 1.67 Å that provides Fe(211) reflection at $2\theta = 90°$ and typical accuracy of ±30 µstrain that was achieved for the normal and transverse components after an acquisition time of 4 minutes. For the longitudinal strain component, however, a typical accuracy was ±50 µstrain with a measurement time of 20 minutes. Overall, approximately 1.5 days of beam-time was used for each blade.

Table 2: Swords characterisation. Thickness and width are taken at the point of maximum curvature of the blade. The length is measured as a straight line from the point to the notch where the back of the katana meets the mounting of the hilt.

ID	Dimensions, mm	Full view
H1360	Thickness = 5.7 Width = 24.5 Length = 69	
H5378	Thickness = 7.4 Width = 28.5 Length = 76.6	
H4839	Thickness = 5.4 Width = 25.2 Length = 72.5	
H6856	Thickness = 5.7 Width = 26.8 Length = 63.5	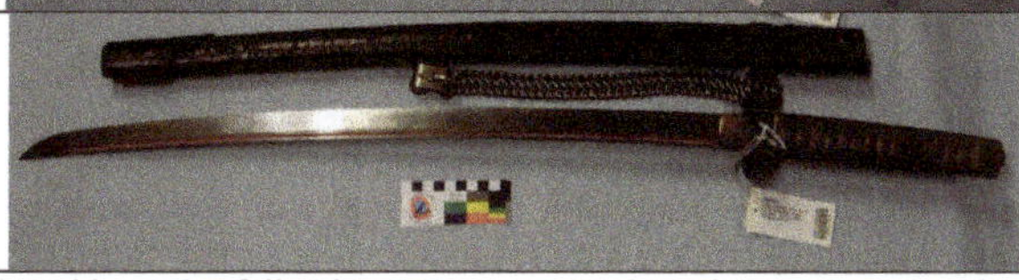

In order to resolve the second problem, the following strategy was adopted. In contrast to the traditional stress experiment, where d_0 must be provided and three stress components are calculated, in the current experiment, the zero-through-thickness-stress condition was applied enabling calculation of two stress components and one d_0-value out of the d-spacings for three measured directions. This plane-stress condition should be fulfilled with good accuracy (within our experimental errors) because of sample planar geometry, but the major concern would be the anisotropic behaviour of d_0. Until slicing of a blade is allowed, this is the only practical assumption. However, some checks can be applied to verify the assumption. If the abovementioned assumption is accurate enough and stress and d_0 are resolved correctly then the transverse stress component should satisfy with boundary conditions on top and bottom edges of the blades and the longitudinal integral stress balance condition can be (at least approximately) evaluated.

Results and Discussion

The neutron diffraction experiment resulted in maps of two stress components and d_0-values in the transverse cross-section of each sword which were determined with sub-millimeter spatial resolution. These two material characteristics are the imprint of the production process and their study can help to identify the actual manufacturing techniques based on qualitative and quantitative data rather than stylistic criteria.

Although 2D maps give a good idea about spatial distribution, in order to facilitate the interpretation of data only detailed 1D views of the central line are provided in Fig. 3. To make results geometrically comparable all curves are plotted in the scale normalized to 1. The 2D maps omitted in this paper will be reported and more in-depth discussed in a separate publication.

Somewhat contrary to expectation no blade clearly showed a layered structure even with experimental spatial resolution of 0.5 mm. Through thickness profiles of stress, d_0 and peak width are all smooth functions (see Fig. 2b as an example). Although features with smaller dimensions are not detectable due to the gauge volume averaging, the main laminae (differing in C content and microstructure) were expected to be detectable since their size, 1-2 mm, is larger than the gauge volume used in the measurement (Fig. 2). The most evident features for all blades are associated with the cutting edge area. The peak width and d_0-values peak at this area which is obviously attributed to the special thermo-mechanical treatment the edge receives in the production process with typical water quenching of the edge and formation of martensite. Since in the angle-dispersive neutron diffraction experiment with a single diffraction peak, Fe (211) in our case, there is no possibility to distinguish between ferrite and martensite, the presence of martensite is manifested as both peak broadening and rapid variation of d_0. There are some more subtle variations of d_0 and peak width (for example very small changes through thickness in blade H4839), that can be attributed to different degree of cementation and formation of various pearlitic/bainitic or even martensitic microstructures depending on carbon content. Since the process of carbon diffusion from different original pieces is complex, formation of gradients and variations of microstructures in a blade steel is not surprising in quenched areas. In principle, this should be reflected in the patterns of d_0 and peak width through mechanism of microstress due to thermal, elastic and plastic mismatch between ferrite and cementite composing pearlite/bainite. However, in comparison with the effects of martensite, those effects are much smaller, e.g. d_0 shift of -0.0003 Å due to typical microstress in ferrite of -120 MPa and almost no peak broadening in ferrite [8]. Therefore, the dominant features are due to the presence of martensite.

Yet another commonality for all blades is a very weak transverse stress component which has almost no changes along the central line, though some variation in through thickness is clearly present, most likely due to forging action leaving surface layers in slight relative compression in respect to the inner layers (the corresponding figure is omitted here to streamline the paper). The longitudinal stress component is dominant and stress state can be considered as almost one-dimensional. Thereafter, only longitudinal component is considered, often referred to as simply "residual stress".

The martensitic pattern at the cutting edge can be considered as a distinctive feature of one manufacturing school and, in some cases, its unique profile was purposely created as a visual instantaneously recognizable signature of the author. Therefore the martensitic area can be more or less extended from blade to blade according to manufacturing tradition or blacksmith's intention.

Materials Research Proceedings **2** (2016) 443-448 doi: http://dx.doi.org/10.21741/9781945291173-75

Therefore, the pattern of martensitic microstructure can be used as a distinct signature of the sword making school/author. Visual inspection of the patterns plotted in Fig. 3 suggests a similarity of the blades H5378 and H4839 (blue and green). They share almost identical pattern for the stress, d_0 and peak width parameter. It cannot be said the same for the blades H1360 and H6856 (black and red in Fig. 3) whose d_0 don't match in any of the characteristics. However, the following considerations can indicate otherwise. It is known that some of old blades with long history of usage were subjected to refurbishment and renewal of the cutting edge and blade surfaces, especially if damaged. If we assume that blade H1360 over its lifetime lost some 3 mm of material due to repetitious polishing and/or restoration upon possible damage of the cutting edge, after dimensional re-normalization (so that the modern cutting edge actually corresponds to 3 mm into depth location of the original blade), the patterns demonstrate remarkable similarity (Fig. 4). In fact, the match of the stress patterns might have an even better match if corrections for force and moment balance are introduced, since material under compressive stress was probably removed by re-polishing therefore distorting the original stress distribution to some extent. Although this consideration cannot be considered as the final proof, the experimental observation cannot be disregarded as a mere speculation.

Comparing the two groups, the following differences are the most obvious and most significant. For the H5378/H4839 the effect of martensite presence is more pronounced resulting in a wider diffraction peak and much larger d_0. Quenching, accompanied with the martensitic transformation, also results in formation of the compressive stress zone in the area of cutting edge, some first 5 mm starting from the cutting edge (Fig. 3). These compressive stresses (and stress momentum) are compensated by tensile stresses in the core of the blade. In the pair H1360/H6856 this happens in a way similar to a simple bending with a typical almost linear slope on the opposite side (Fig. 4). However, for the pair H5378/H4839, another compressive zone is created on the back side of the sword, which is more deeply extended into the interior, up to 10 mm from the back (Fig. 3).

Although this can be associated with quenching, (without martensite transformation, since there is no apparent change either in d_0 or peak width) but it is most probably related to the use of soft steel at the back of the blades (e.g. Honsanmai style, Fig. 1) that acquired a compressive stress during deformation in combination with harder steel of the adjacent areas.

Based on these comparative data, the *mumei* blade H6856 might date back to the oldest of the historical periods (Koto age, 987–1596 A.D.), to which blade H1360 is certainly attributed. The period the blade is stylistically attributed (c. 1600 A.D.) is known as a time of transition when the Shogun extended his control also to the manufacture of weapons in the frame of its political strategy of unification (Shinto age, 1596–1781 A.D.). As a consequence, the differentiation among various traditions that were developed during the previous age slowly dissolved, only remaining as a sort of merely stylistic differentiation with no real connection to the working process [2].

Summary

For the first time a detailed strain mapping of the Japanese *katana* blades was achieved with 0.5 mm resolution using non-destructive neutron diffraction technique. Three quantities were analyzed, stress d_0-value and peak width, and interpolated 2D maps were acquired making possible mutual comparison of the blades based on the signature patterns of these measured quantities.

The analysis of four blades revealed two groups of blades, each with a characteristic pattern of stress, d_0 and peak width. While one blade H6856 (c. 1600) was of unknown origin, through its characteristic pattern association it was attributed to the tradition common with the oldest blade in the collection H1360 (Sadatsugu, 1346-1370).

This study lays down a foundation for a future analysis of the Japanese blades and proves feasibility of the approach with 0.5 mm spatial resolution mapping.

These results will provide the MAAS with new valuable information on these swords which will be available to researchers, museum collection managers, the general public and future generations.

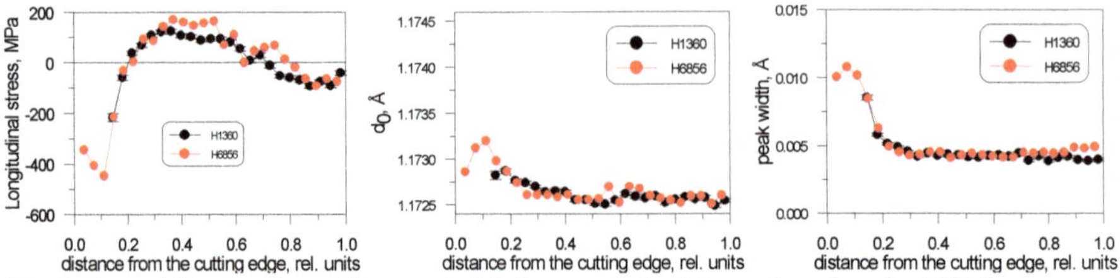

Fig. 4. Central line profiles of three material characteristics (stress, d_0 and peak width) for two blades, H1360 and H6856, with re-normalisation of the width of blade H1360.

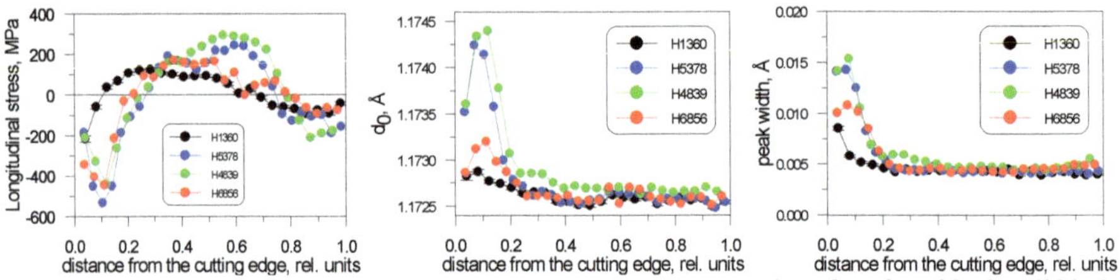

Fig. 3. Central line profiles of three material characteristics (stress, d_0 and peak width) for all blades.

References

[1] J.M. Yumoto, The Samurai Sword. A handbook. Tuttle, Tokyo (1958)

[2] K. Nagayama, The Connoisseurs Book of Japanese Swords, Kodansha International (1997)

[3] W.M. Hawley, Laminating Techniques in Japanese Swords (1974)

[4] F. Grazzi et al., From Koto age to modern times: Quantitative characterization of Japanese swords with Time of Flight Neutron Diffraction, Anal. At. Spectrom. 26 (5) (2011) 1030-1039. http://dx.doi.org/10.1039/c0ja00238k

[5] F. Salvemini et al., Neutron computed laminography on ancient metal artefacts, Anal. Methods 7 (2015) 271-278. http://dx.doi.org/10.1039/C4AY02014F

[6] O. Kirstein, V. Luzin, U. Garbe, The Strain-Scanning Diffractometer Kowari, Neutron News 20 (4) (2009) 34-36. http://dx.doi.org/10.1080/10448630903241175

[7] Ruiz-Hervias et al., Residual Stress Distributions in Bi-Metal (Ferritic to Austenitic Steel) Joints Made by Laser Welding, Mater. Sci. Forum. 772 (2013) 181-185 http://dx.doi.org/10.4028/www.scientific.net/MSF.772.181

[8] P. Van Houtte, K. Van Acker, J. Root, 1999, Residual Stress Determination in the Cementite and Ferrite Phases of High Carbon Steel, Textures and Microstructures 33 (1-4) (1999) 187-206. http://dx.doi.org/10.1155/TSM.33.187

Residual Stresses 2016: ICRS-10 Materials Research Forum LLC
Materials Research Proceedings **2** (2016) 449-454 doi: http://dx.doi.org/10.21741/9781945291173-76

A Portable Optical DSPI System for Residual Stresses Measurement by Hole Drilling Using the Integral Method in Terms of Displacement

Armando Albertazzi G. Jr.[1, a*], Matias Viotti[1, b], Celso Veiga[1,c]

[1] Mechanical Engineering Dept. - Universidade Federal de Santa Catarina, Campus da UFSC, Trindade, Florianópolis, SC, Brazil, CEP 88.050-005, Tel.: +55 (48) 3239 2030

[a]a.albertazzi@ufsc.br, [b]mov@labmetro.ufsc.br, [c]clv@labmetro.ufsc.br

Keywords: Residual Stress Measurement, Incremental Hole Drilling, ESPI, DSPI, Integral Method

Abstract. A portable optical device for residual stresses measurement is presented here. It combines an achromatic digital speckle pattern interferometer (DSPI) with radial sensitivity with a high speed pneumatic drilling unit. The system automatically acquires a 1280 x 960 pixel image for each step of incremental hole drilling. The number of drilling steps, as well the drilling increments, is fully programmable. The data is processed and fitted to a set of displacement curves obtained by the Finite Element Method according to the integral formulation. This paper describes the device as well as the data reduction procedure. The system has been applied to measure residual stress profiles in four to eight points of cross-sections of pipes and, from that, to compute the amount of bending moment induced in pipelines used for oil transportation. The paper presents results of a calibration test in a 12 m long test bench with a 200 mm steel diameter pipe.

Introduction

Residual stresses measurement using the incremental hole drilling approach with strain gauges is a well-established technique.[1] Despite it is normalized by the ASTM E837 standard [2], the evaluation of a residual stress profile in a single point is a quite time-consuming task. Surface preparation, strain gauge alignment and installation, wiring, drilling alignment, incremental drilling are some required steps that make a complete measurement typically to take an hour or more.

Optical methods are very attractive alternatives for residual stress measurements by the hole drilling method since almost no surface preparation or tedious installations are required. However, most optical systems already available for residual stress measurement are designed to be used inside labs.[3] Optical techniques usually require that the part to be measured must be transported to a laboratory.

The authors have developed a portable optical device for residual stress measurement using the incremental hole drilling approach. [4-6] A highly stable and achromatic digital speckle pattern interferometer (DSPI), capable to measure in radial polar coordinates, is integrated with a high speed drilling unit in a fully computer controlled device for residual stress measurement by incremental hole drilling. A robust construction is designed in order to make it appropriate for in-field measurement. The total measuring time is about 15 to 20 min, which is four times faster than a strain gauge measurement. Similar uncertainty level is achieved. The optical principle, constructive aspects, data reduction approach and one practical application are presented in this paper.

The Radial in-Plane Interferometer

A 30 mm diameter binary diffractive optical element (DOE) is the key element of this digital speckle pattern interferometer (DSPI). It consists of a circular binary diffraction grid with a pitch of about 1.30 μm. Collimated light if diffracted, as it passes through this grid, and bent about 30°, forming a circular double illuminated area of about 10 mm diameter as it is shown in the left part of Fig. 1.

Materials Research Proceedings **2** (2016) 449-454 doi: http://dx.doi.org/10.21741/9781945291173-76

The right part of this figure describes the working principle of the DSPI interferometer. The light coming from the laser (L) is expanded by a lens (E). It passes through the elliptical hole of the mirror M1, reaches mirrors M2 and M3 and is reflected back to the mirror M1. Mirror M1 directs the expanding light to the collimating lens (CL) in order to obtain an annular collimated beam. Finally, the light is diffracted by the DOE mainly in the first diffraction order towards the specimen surface. Mirrors M2 and M3 are two special concentric circular mirrors. The former is joined to a piezoelectric actuator (PZT) and the later has a circular hole with a diameter slightly larger than diameter of M2. Furthermore, mirror M3 is fixed while M2 is mobile. The PZT actuator moves the mirror M2 along its axial direction generating a relative phase difference between the beam reflected by M2 (central beam) and the one reflected by M3 (external annular beam). Every point over the illuminated area receives only one light ray coming from M2 and only other one from M3. Thus, PZT enables the introduction of a relative phase shift for efficient optical phase calculation by means of phase shifting algorithms. The central hole placed at M1 has several functions: (a) to allow that the light coming from the laser source reaches mirrors M2 and M3, (b) to prevent that the laser light reaches directly the specimen surface having triple illumination and (c) to provide a viewing window for the camera (CCD).

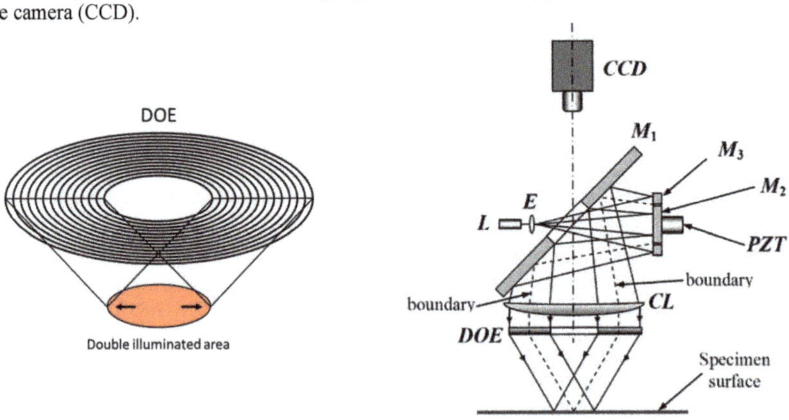

Fig. 1 - Double illumination speckle interferometer with radial in-plane sensitivity.

Since each point on the circular illuminated area is double illuminated, the sensitivity direction is parallel to the surface and aligned with the radial direction. Therefore, only the radial displacement component is measured. This configuration results in an achromatic interferometer, i.e., its sensitivity is not wavelength-dependent. [4] Therefore, cheap and compact diode lasers can be used as light source without stability loss.

The Integrated Device

Fig. 2 shows in-field applications of the radial in-plane DSPI interferometer integrated with a pneumatic high-speed drilling unit in a portable residual stresses measurement device. [4-6] The out-of-lab application is possible since the relative motions between the device and the part to be measured are minimized by a highly stiff clamping. Three legs, with sharp conical tips of hardened steel, are provided for leveling and stablishing a well-defined and firm binding interface to the specimen surface. A set of four rare earth magnets develops the attraction forces to firmly press the three legs against the specimen surface. A motorized rotating stage is used for automatically exchanging the measuring unit and the drilling unit. An integrated measurement software automatically controls all steps of an incremental hole-drilling measurement and processes all necessary images to compute the combined stresses profile as a function of the surface depth.

Materials Research Proceedings **2** (2016) 449-454 doi: http://dx.doi.org/10.21741/9781945291173-76

Fig 2. In-field applications of the integrated DSPI/hole drilling device.

The surface to be measured needs to be clean, without any thick coating and reasonably smooth. If the surface is very dark, or shining, a very thin layer of light spray painting is recommended.

Image Processing and Data Reduction

The left part of the Fig. 3 shows a typical DSPI phase difference pattern. It was obtained after drilling a hole with 1.6 mm in diameter and 1.0 mm in depth into a specimen with uniaxial residual stresses along the horizontal axis. The corresponding radial displacement signal was extracted along the constant radius line (light blue) drawn in the left part this figure. The elliptical line corresponds to the measured signal expressed in polar coordinates in a much-enlarged scale. The right part of the figure shows the same radial signal now expressed in Cartesian coordinates. The horizontal axis corresponds to the angle in degrees and the vertical axis is the radial displacement component. Note that, due to image symmetry, the radial displacement signal is sinusoidal with two periods along 360°.

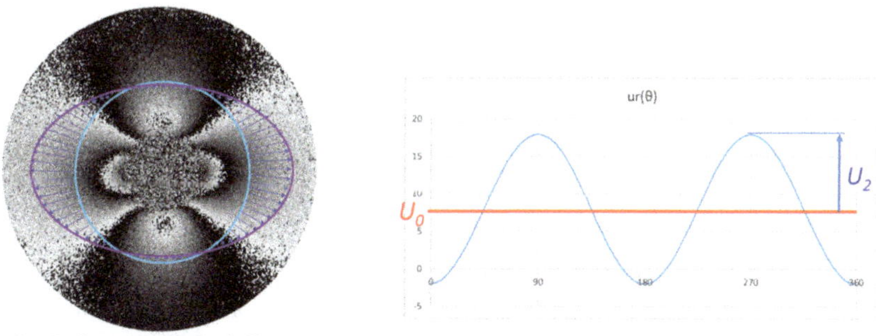

Fig 3 – Left: typical hole drilling image. Right: Signal extracted along a constant radius line.

The general equation for the radial displacement component (u_r) is like Eq. 1. For a given radius, this equation corresponds to the right part of Fig. 3. There are two signal components: U_0, the mean value and a sinusoidal component with two waves and amplitude U_2. The signal component U_0 is further referred as the zero harmonic and is related to the residual stresses sum $\sigma_1 + \sigma_2$. The signal U_2, is the second harmonic and is related to the residual stresses difference $\sigma_1 - \sigma_2$. Both quantities vary for different values of the radius r. Therefore, they are functions of r. The angle β corresponds to the principal stress direction.

$$u_r(r,\theta) = U_0(r) + U_2(r)\cos(2\theta - 2\beta) \quad (1)$$

Materials Research Proceedings **2** (2016) 449-454 doi: http://dx.doi.org/10.21741/9781945291173-76

Eq. 2 has a more convenient relationship, where U_0 is replaced by $a\cdot(\sigma_1+\sigma_2)$ and U_2 by $b\cdot(\sigma_1-\sigma_2)$. Both *a(r)* and *b(r)* are the normalized radial displacement responses for the unitary stress sum and difference, respectively. They depend on the hole diameter, the hole depth and the material properties.

$$u_r(r,\theta)=a(r)(\sigma_1+\sigma_2)+b(r)(\sigma_1-\sigma_2)\cos(2\theta-2\beta) \quad (2)$$

Residual Stresses Calculation

The functions *a(r)* and *b(r)* can be determined numerically to a set of discrete radius values (r_k). To quantify the amount of residual stresses from the measured quantities $U_0(r_k)$ and $U_2(r_k)$ a linear behavior is assumed. Therefore, the stress values are proportional to the amount of radial displacement. As an example, let's suppose that $U_0(r_k)$ is *P* times bigger than $a(r_k)$. In this case, the sum of residual stresses will be *P* times bigger than 1.0 MPa. Since random errors are expected in the experimental values $U_0(r_k)$ as well as an unexpected bias *C*, the relationship between $U_0(r_k)$ and $a(r_k)$ is in the form of Eq.3, where $\varepsilon(r_k)$ is a random error component. The *k* sub index was omitted here for simplicity. The values of *P* and *C* can be determined by a least square fitting. [7] A similar approach is used to relate $U_2(r_k)$ and $b(r_k)$, but it will not be described here.

$$U_0(r)=P\cdot a(r)+C+\varepsilon(r) \quad (3)$$

Indeed, a set of discrete numerical normalized signals are required to apply the integral method for the residual stresses determination by the incremental hole drilling. Each numerical signal is related to the hole depth (*i*) and the stress layer (j). Let $a_{ij}(r_k)$ denotes the numerical signal computed for the *i*-depth and *j*-stress layer [8]. For the first hole depth, the relationship between the experimental data $U_0(r_k)$ and numerical signal $a_{11}(r_k)$ is given by Eq. 4. Also here, the *k* sub index was omitted.

$$U_{01}(r)=P_1\cdot a_{11}(r)+C_1+\varepsilon_1(r) \quad (4)$$

For the second incremental hole depth, the integral method expects to relate the experimental signal to the numerical one according to Eq. 5. Since Eq. 4 alone can determine the value of P_1, Eq. 5 involves two additional unknowns: P_2 the stress value for depth 2 and the additive constant C_2.

$$U_{02}(r)=P_1\cdot a_{21}(r)+P_2\cdot a_{22}(r)+C_2+\varepsilon_2(r) \quad (5)$$

It is straightforward extending Eq. 5 for an integer number of *S* hole increments. The least squares solution for the P_i values can be derived and results in a *S* x *S* system with the *S* unknowns: P_1 to P_S. The resulting matrix system can be written as Eq. 6 where the elements of the matrix *[X]* and vectors *{P}* and *{Y}* are given in Eq. 7.

$$[X]\{P\}=\{Y\} \quad (6)$$

$$\{P\}=\begin{Bmatrix}P_1\\ \vdots \\ P_S\end{Bmatrix} \qquad X_{m,n}=\sum_{i=Q}^{S}\left[\sum_{all\ r}(a_{im}\cdot a_{in})-\frac{1}{N}\sum_{all\ r}a_{im}\cdot\sum_{all\ r}a_{in}\right]$$

Materials Research Proceedings **2** (2016) 449-454 doi: http://dx.doi.org/10.21741/9781945291173-76

$$Y_m = \sum_{i=Q}^{S}\left[\sum_{all\,r}(a_{im}\cdot U_{0i}) - \frac{1}{N}\sum_{all\,r} a_{im}\cdot\sum_{all\,r} U_{0i}\right] \tag{7}$$

The integer numbers m and n are the row and column indexes of each matrix element. The quantity Q is defined as the minimum integer between m or n: $Q = min(m, n)$. N is the number of discrete radial values related to the index k (omitted in Eq. 6 and 7).

Bending Moment Calculation

The combined hole drilling DSPI system was used to determine the amount of bending stress developed in a 12 m long and 200 mm (8") diameter steel pipe [7, 9]. Left part of Fig. 4 shows a sketch of the test bench with two parallel pipes. They were clamped at both ends and a hydraulic jack was used to apply a transverse force measured by a load cell as (37 621 ± 58) N at the central point. Bending moments were measured by the hole drilling method in eight different cross-sections of the right pipe in Fig. 4 and compared with the reference values computed from the central load value. The sample pipe is the right one in the figure.

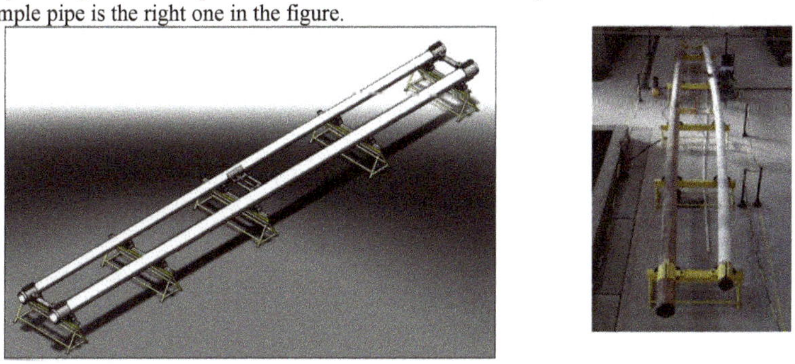

Fig 4 – 12 m long steel pipe loaded by a known bending moment.

To determine the bending moments, a set of eight hole drilling measurements was carried out in each cross-section at different angles separated by 45°. The holes were drilled in 20 steps, from 0.00 to 1.00 mm with depth increments of 0.05 mm. The phase images obtained in each final drilling step are shown in Fig. 5.

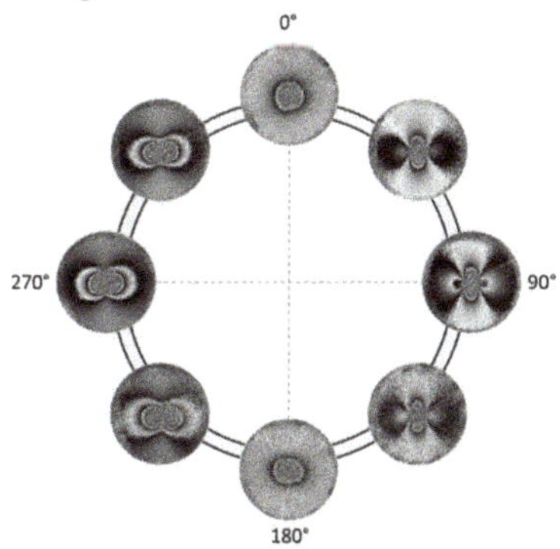

Fig 5 – Different residual stress measurements at eight different angles of a same pipe cross-section.

Materials Research Proceedings **2** (2016) 449-454 doi: http://dx.doi.org/10.21741/9781945291173-76

A compressive residual stresses peak value of -200 MPa was typically obtained at the depth of 0.25 mm due to the pipe manufacturing process. The different residual stresses profile offsets in the pipe axis direction were used to determine the amount of bending stresses. The bending stress in each cross-section varied from 70 MPa to 340 MPa, depending on the angular location of the measurement. The agreement between the central values determined by the hole drilling method and the reference ones were in the range of ± 20 MPa.

Conclusions

The authors developed a portable optical system with a radial in-plane achromatic interferometer for residual stresses measurement by the incremental hole drilling technique using the integral method. The total setup and measurement time ranged from 15 to 20 min for a complete 20 steps incremental hole drilling measurement. No special skills are required from the user. The paper presented results from bending stresses measurement in a 12 m long and 200 mm diameter pipe subjected to a known amount of bending moment. The central values for the bending stresses obtained for eight different cross-sections were compared with the reference values computed from the load cell and bending bench geometry. The deviations were all inside the range of ± 20 MPa.

References

[1] Schajer, G. S. (Editor), "Practical Residual Stress Measurement Methods", Wiley 2013, 7, ISBN 9781118342374.

[2] ASTM E837-13. "Standard test method for determining residual stresses by the hole-drilling strain-gage method", Annual Book of ASTM Standards. American Society for Testing and Materials (2013).

[3] Schajer, G.S. and Rickert, T.J., "Incremental Computation Technique for Residual Stress Calculations Using the Integral Method", Exp. Mech., 51(7), 1217-1222 (2011). http://dx.doi.org/10.1007/s11340-010-9408-5

[4] Viotti, M. R., Kapp W. and Albertazzi Jr., A. "Achromatic digital speckle pattern interferometer with constant radial in-plane sensitivity by using a diffractive optical element," App. Opt. 48(12), 2275-2281 (2009). http://dx.doi.org/10.1364/AO.48.002275

[5] Viotti, M. R., Albertazzi Jr., A. and Kapp, W. A., "Experimental comparison between a portable DSPI device with diffractive optical element and a hole drilling strain gage combined system," Opt. Lasers Eng. 46 (11), 835-841 (2008). http://dx.doi.org/10.1016/j.optlaseng.2008.05.020

[6] Viotti, M. R. and Albertazzi Jr., A. "Compact sensor combining digital speckle pattern interferometry and the hole-drilling technique to measure non-uniform residual stress fields," Opt. Eng. 52 (10), 101905.1-101905.8 (2013).

[7] Albertazzi Jr, A.; Zanini, F. ; Viotti, M. R. ; Veiga, C. L. N. "Residual stresses measurement by the hole-drilling technique and DSPI using the integral method with displacement coefficients". In: 5th International Symposium on Experimental Mechanics e 9th Symposium on Optics in Industry, 2015, v. 1. p. 1-7.

[8] Schajer, G. S., "Measurement of Non-Uniform Residual Stresses Using the Hole-Drilling Method. Part I - Stress Calculation Procedures", J. Eng. Mater. Technol.. 1988; 110(4):338-343. doi: 10.1115/1.3226059. http://dx.doi.org/10.1115/1.3226059

[9] Pacheco, A.; Fontana, F.; Viotti, M. R. ; Veiga, C. L. N. ; Lothhammer, L. R. ; Albertazzi Jr, A. "Bending moment evaluation of a long specimen using a radial speckle pattern interferometer in combination with relaxation methods". In: SPECKLE 2015: International Conference on Speckle Metrology, 2015, Guanajuato. SPECKLE 2015. Guanajuato: CIO, 2015. v. 1.

Residual Stresses 2016: ICRS-10 Materials Research Forum LLC
Materials Research Proceedings **2** (2016) 455-460 doi: http://dx.doi.org/10.21741/9781945291173-77

Residual Stress Characterization and Control in the Additive Manufacture of Large Scale Metal Structures

Jan Roman Hönnige[1,a*], Stewart Williams[1,b], Matthew J. Roy[2,c], Paul Colegrove[1,d], Supriyo Ganguly[1,e]

[1]Welding Engineering and Laser Processing Centre, Cranfield University, MK43 0AL Cranfield, UK

[2]School of Mechanical, Aerospace and Civil Engineering, University of Manchester, M14 9PL Manchester, UK

[a]j.honnige@cranfield.ac.uk, s.williams@cranfield.ac.uk, [c]matthew.roy@manchester.ac.uk, [d]p.colegrove@cranfield.ac.uk, [e]s.ganguly@cranfield.ac.uk

Keywords: Additive Manufacturing; Neutron Diffraction; Contour Method; Cold Rolling; Pinch Rolling

Abstract. Additive Manufacture of metals is an area of great interest to many industrial sectors. All metal additive manufacturing processes suffer with problems of residual stresses and subsequent distortion or performance issues. Wire + Arc Additive Manufacture (WAAM) is a metal additive manufacture process that is suitable for the production of large scale engineering structures. Paramount to the successful industrial application of WAAM is the understanding and control of residual stress development and their subsequent effects. Vertical inter-pass rolling can be used to reduce these residual stresses, but its potential is limited due to the absence of lateral restraint of the wall. So it deforms the wall in its transverse direction rather than reducing longitudinal tensile residual stresses, which is the main source of the distortion. The potential of a new *pinch-roller* concept is currently being investigated at Cranfield University with very promising preliminary results: It was possible to entirely eliminate the distortion of a Ti-6Al-4V WAAM wall.

Introduction

Wire + arc additive manufacturing (WAAM) is a robotic and welding equipment based high-deposition-rate additive manufacturing (AM) process, which can be used for the manufacture of large-scale aerospace parts. Fig. 1 shows a Ti-6Al-4V landing gear rib near-net-shape demonstrator part manufactured with WAAM. Compared to conventional subtractive machining, the main benefits of WAAM are the significant time, material, and cost savings.

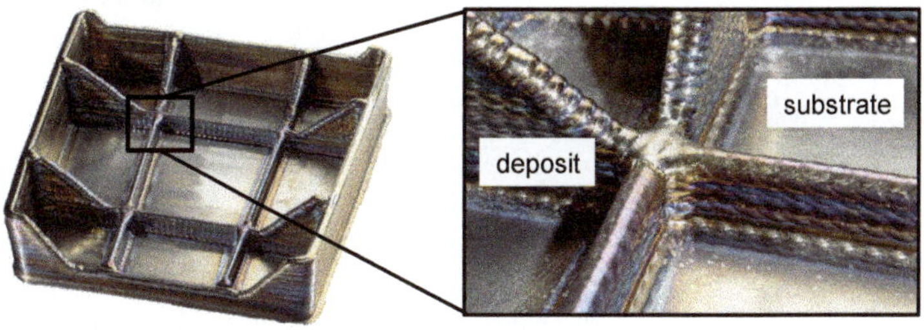

Figure 1: WAAM Demonstrator: Bombardier Landing Gear Rib; Ti-6Al-4V; 0.7 m; 20 kg

The process can be used to deposit a range of materials including stainless steel and Al-alloys. However Ti-6Al-4V is the material to benefit most from the advantages of WAAM, due to the commonly high material and process costs. Suitable equipment to deposit Ti-6Al-4V wire is TIG and Plasma heat sources. A high deposition rate is generally achieved at the expense of the minimum feature size. The genuine targets of a high-deposition rate process are material-, lead time- and cost savings of large parts with medium complexity rather than design or feature freedom [1].

The deposition of liquid metal on a solid substrate and its' constrained shrinkage during cooling causes elastic and plastic deformation. In the longitudinal direction, the tensile residual stresses appear to be highest. The temperature-dependent flow strength and thermal properties of the metal determine the magnitude of these residual stresses [2]. In steel the residual stresses can be as high as the materials yield stress at room temperature. After unclamping, the section shows a significant out-of-plane distortion and the longitudinal residual stresses redistribute from being constant along the height into dropping linearly [3]. The redistribution and resulting distortion due to unclamping is illustrated in Fig. 2 a) to c).

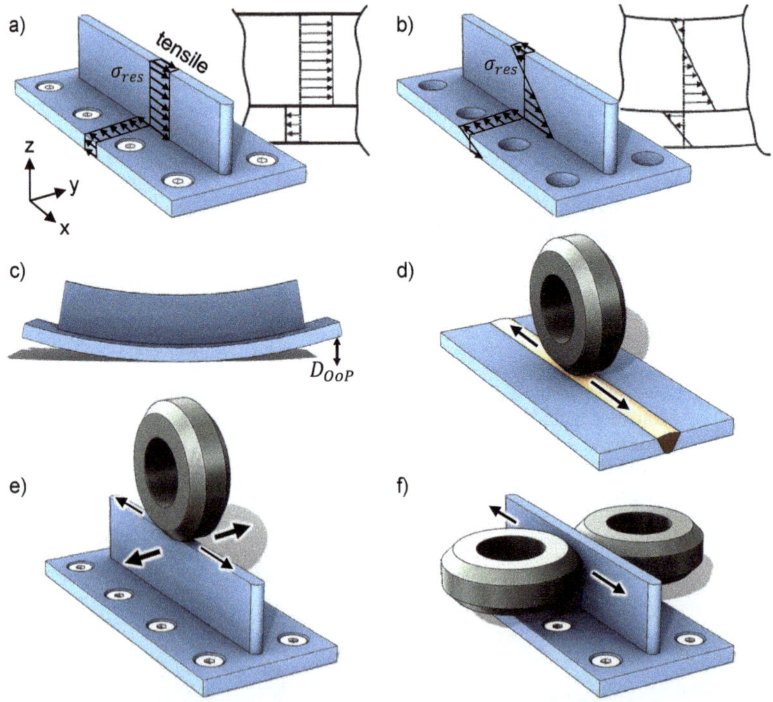

Figure 2: Residual Stress and Distortion in WAAM: Stress Field in a clamped a) and un-clamped Section [3]; c) Out-of-Plane Distortion D of the un-clamped Section; d) inducing Longitudinal Strain into Butt-Welds by vertical Rolling; e) inducing predominantly undesired lateral Strain in WAAM wall by vertical Rolling; f) Concept of Pinch-Rolling

The relatively young and destructive Contour Method (CM) is the only concurrent residual stress measurement technique to provide a full 2-D stress map over the cross section with the required

Materials Research Proceedings **2** (2016) 455-460 doi: http://dx.doi.org/10.21741/9781945291173-77

spatial resolution. The more established Neutron Diffraction (ND) technique can profile full residual stress tensors along characteristic lines with limited resolution, driven by the material. Both, CM and ND are powerful tools to investigate the development of the residual stresses in WAAM, which is vital to evaluate potential applications to control them [4].

A promising mechanical technique to manipulate the residual stresses and distortion is cold rolling, which initially has been used on butt welds (Fig. 2 d), because it can eliminate residual stresses and distortion efficiently and entirely by inducing compressive strain in the longitudinal welding direction [5]. In additive manufacturing the deformation is mainly induced in the transverse direction instead, because of the absence of lateral restraint (Fig. 2 e). Hence vertical rolling can only reduce residual stress and distortion in Ti-6Al-4V WAAM, but not eliminate them entirely [4]. In Colegrove et al. [3] this lateral deformation was successfully attempted to restrain for the first time by using "slotted roller". It shows a greater improvement on the distortion compared to other "flat" or "profiled" roller geometries. However, the use of the slotted roller is very cumbersome, time consuming and limited.

A new concept is pinch-rolling (Fig. 2 f), which would roll the wall from both sides. This would prevent the lateral deformation and keep the rolling forces in the tool instead of the entire frame. Before developing this tool, the existing vertical rolling equipment will be used to roll on one side of the wall as a preliminary study. In the present work we show the effect of side rolling a Ti-6Al-4V single-path WAAM part on the residual stress and distortion.

Methodology

Nine Ti-6Al-4V walls were built using a plasma power supply. After the final layer, the wall remained clamped on the aluminium backing bar and cooled to room temperature for cold side-rolling. The specimen was turned by 90°, so that one side of the deposit evenly rested on a tool steel block, as shown in Fig. 3. A hydraulic cylinder was used to roll both sides of four specimens with 100 kN, 130 kN, 160 kN and 190 kN respectively using a large roller. The same was done on four other samples with a smaller roller diameter. The ninth sample was the control sample. After unclamping, a height gauge was used to measure the out-of-plane distortion D of the unclamped sample.

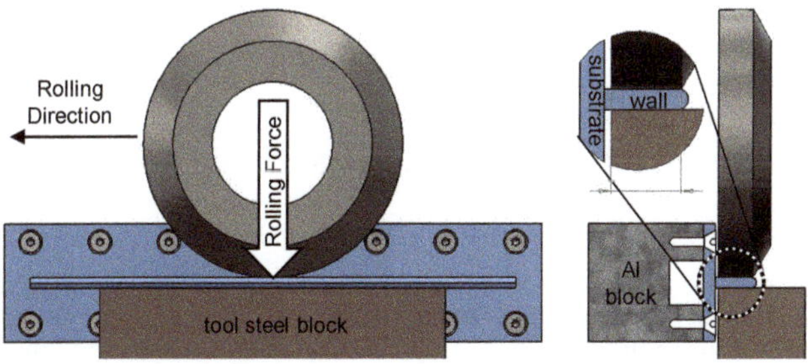

Figure 3: Schematic of Side-Rolling

Neutron Diffraction strain analysis was carried out at the ENGIN_X instrument at ISIS Rutherford Appleton Laboratory in Didcot/UK. Ten gauge volumes along the vertical centreline of 2x2x2 mm each were exposed to the monochromatic beam for 40 μA counting time. The single peaks of the $\{103\}_\alpha$ and $\{101\}_\alpha$ hkl planes were fitted individually to determine their spacing. With these, residual stresses were calculated using Hooks law with the following plane specific values: $d_{0,103} = 1.3253$ Å, $d_{0,101} = 2.2238$ Å, $E_{103} = 108.3$ MPa, $E_{101} = 126.96$ MPa, $\nu_{103} = 0.285$ and $\nu_{101} = 0.39$. The longitudinal residual stresses calculated from both planes agreed well and were averaged.

Afterwards, the CM was employed on the same specimens at the University of Manchester to generate a stress map on the longitudinal yz-plane in the centre of the sample by cutting specimens

Materials Research Proceedings **2** (2016) 455-460 doi: http://dx.doi.org/10.21741/9781945291173-77

via Electro-Discharge Machining (EDM) and numerically analysing the resulting relaxation. With Neutron diffraction, three profiles were generated: The first two profiles were taken from the unrolled control specimen in two different conditions. The first condition was the still-clamped condition and the second the unclamped and distorted condition. One wall that has been side rolled was also measured in the unclamped condition. With CM, the same samples were measured, aside from the clamped control. To extract the most detail from CM, an analysis route was selected with a very fine numerical description of the cross-section, approaching twice that of the wire diameter (250 µm) employed for cutting. While very fine spatial resolution was obtained, this was at the expense of potentially incorporating small variations in the cut surface in the resolved stress.

Results and Discussion

Distortion. The Out-of-Plane distortion of the control sample is 6.8 mm, which is comparable to what has been found before [4]. Fig. 4 shows the distortion of the eight rolled samples vs. the rolling load. Post-weld side rolling shows a greater reduction in distortion with increasing rolling loads and decreasing roller diameter as known from previous work [4], [6]. Comparing the roller diameters, the smaller roller always showed a greater reduction in distortion for the same load by approximately 1 mm. For both rollers, the distortion closest to zero has been achieved using 130 kN load. The result suggests that ~85%...90% of the rolling load applied by the bigger roller achieves comparable results when using the smaller roller. Forces higher than 130 kN even resulted in negative distortion. Given the fact, that conventional inter-pass rolling has to be applied after each layer and still does not achieve comparable reduction in residual distortion, it can be said that side rolling met the expectations and is a much more effective tool than vertical rolling in respect to the distortion. The reason is the successfully restrained lateral deformation by actively applying the load on both sides of the wall.

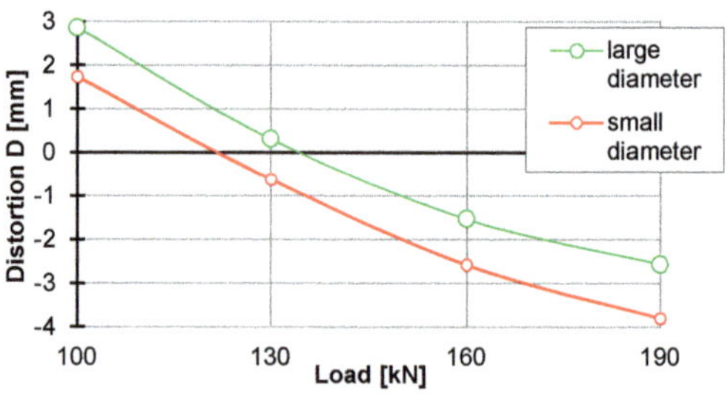

Figure 4: Out-of-Plane Distortion after rolling with a small and a big roller with different loads

Residual Stress. Fig. 5 a) shows all longitudinal residual stress profiles vs. the distance above the substrate. While still clamped, the residual stress of the unrolled control sample along the centre line is fairly constant at around 550 MPa, which is significantly lower than the yield value of Ti-6Al-4V deposit (810-870 MPa). This constant tensile stress has only been modelled before in mild steel [7], but never measured. After unclamping, the profile shows a constant drop of residual stress with increasing distance to the substrate, with being compressive in the top layers. This trend is also supported by this model and also reported elsewhere [3], [4], [8]. Side Rolling took place from 3 mm above the substrate to the top of the wall. In this region, tensile residual stresses were reduced significantly, due to imposed plastic deformation. The stress profile dropped from being 550 MPa and varies between -200 to +200 MPa. This ultimately was the reason for the successful reduction of distortion. Prior to rolling, the contact surface of the wall was uneven, varying by ± 0.05 mm over a nominal width. Therefore, the roller imposed non-uniform strain across the wall, particularly near the surface; where the wall was wider, significantly higher compressive stress can be observed.

Materials Research Proceedings **2** (2016) 455-460 doi: http://dx.doi.org/10.21741/9781945291173-77

Fig. 5 b) shows the contour map of the distorted control and a side-rolled sample. The wall in the control agrees well with the ND profile, with stress varying nearly linearly from tensile to a compressive state towards the top of the wall. The compressive upper and tensile lower region in the substrate shows the same trend. The stress field outside of the HAZ in the substrate is a result of the out-of-plane distortion of the section. Also, rolled plot agrees well with the ND result in this area. The residual stresses within the area of both the wall and the substrate are more uniform. Unlike ND, the resolution of the CM is high enough to show that side rolling did not induce unintentional gradients in the transverse direction, as it is very uniform along the y-axis.

However, employing a contour analysis with very little smoothing is also sensitive to artefacts induced by the EDM-cut. This type of artefact is very likely the cause of the high gradient in both plots that appears at the interface between the wall and the substrate. The source is attributed to a change in cutting length, as the cutting power density per unit length decreases, causing the cut to proceed slowly through this part of the cross-section. It is surmised that this is also complicated by plasticity effects, when the EDM wire cuts through a region with high tensile stress. Sacrificial material could obviate these artefacts during the cut or a second stress-free control cut near the first contour can be performed to identify the cutting-length-caused power-density artefacts afterwards. As a result, the current CM analysis has an unknown uncertainty near the interface of wall and substrate in the HAZ. In the same region, the ND did not show sufficient peak intensity in the spectrum for diffraction peak fitting due to texture and path length issues. However, the ND results in the region affected by processing are without concern and compare well with those obtained by CM.

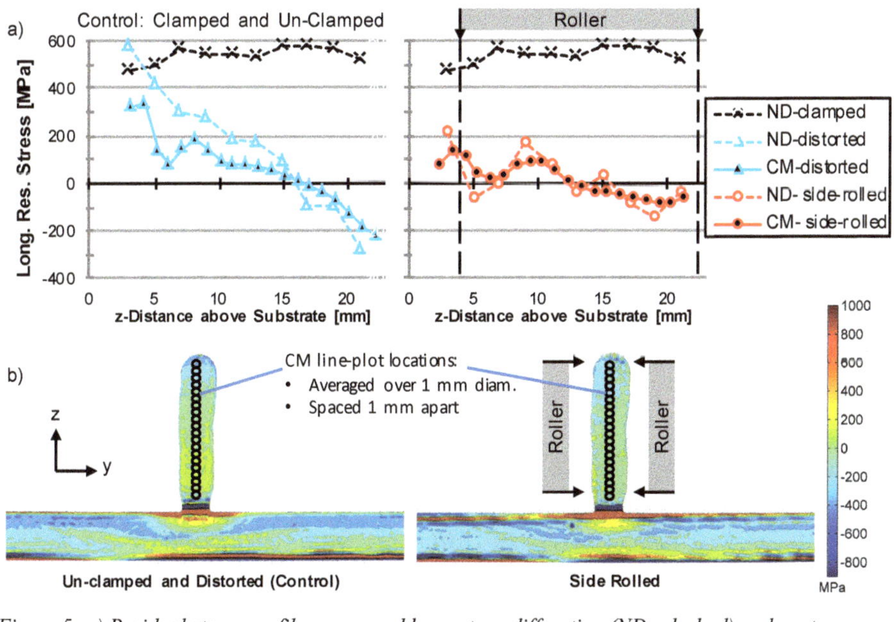

Figure 5: a) Residual stress profiles measured by neutron diffraction (ND - dashed) and contour method (CM - solid): clamped control (black); unclamped and distorted control (blue) and side-rolled (red); b) contour maps of unclamped control and a side-rolled sample

Conclusion

In this study we have shown that:

1. Side Rolling has the potential to eliminate longitudinal residual stresses and distortion in straight AM parts entirely. The effect on both is remarkably higher than vertical rolling, which also is being applied after each layer.
2. A smaller roller diameter has a greater effect on both using a constant force. The effect can also be estimated analytically.
3. The compressive strain induced by a flat roller from the side on a WAAM wall is very sensitive the unevenness of the rolled surfaces' profile. A wall profile variation of only ±0.05 mm can cause residual stress variations around approximately ±200 MPa.
4. Before unclamping (which is the condition where cold work is identified to be applied), the longitudinal tensile residual stress has been proven to be constant in each layer. The peak stress is significantly lower than the material's yield value in a single path Ti-6Al-4V WAAM wall.
5. The contour method analysis of the T-shaped single WAAM wall is likely to be sensitive to EDM-cut artefacts caused by steep cutting length changes and plasticity effects. Both must be investigated further.

Acknowledgement
The authors wish to acknowledge the help and practical support from Joe Kelleher and Clement Bühr at the ENGIN-X Rutherford Appleton Laboratory during the neutron diffraction measurement.

References
[1] S. W. Williams, F. Martina, a. C. Addison, J. Ding, G. Pardal, and P. Colegrove, "Wire+Arc Additive Manufacturing," *Mater. Sci. Technol.*, vol. 00, no. 0, p. 1743284715Y.000, 2015.
[2] J. Ding, P. Colegrove, J. Mehnen, S. Williams, F. Wang, and P. S. Almeida, "A computationally efficient finite element model of wire and arc additive manufacture," *Int. J. Adv. Manuf. Technol.*, vol. 70, pp. 227–236, 2014. http://dx.doi.org/10.1007/s00170-013-5261-x
[3] P. a. Colegrove, H. E. Coules, J. Fairman, F. Martina, T. Kashoob, H. Mamash, and L. D. Cozzolino, "Microstructure and residual stress improvement in wire and arc additively manufactured parts through high-pressure rolling," *J. Mater. Process. Technol.*, vol. 213, pp. 1782–1791, 2013. http://dx.doi.org/10.1016/j.jmatprotec.2013.04.012
[4] F. Martina, M. J. Roy, B. A. Szost, S. Terzi, P. A. Colegrove, S. W. Williams, P. J. Withers, J. Meyer, and M. Hofmann, "Residual stress of as-deposited and rolled wire+arc additive manufacturing Ti–6Al–4V components," *Mater. Sci. Technol.*, vol. 0836, no. April, 2016. http://dx.doi.org/10.1080/02670836.2016.1142704
[5] H. E. Coules, "Contemporary approaches to reducing weld-induced residual stress," *Mater. Sci. Technol.*, vol. 29, no. 1, 2012.
[6] P. Colegrove, J. Ding, M. Benke, and H. Coules, "Application of High-Pressure Rolling to a Friction Stir Welded Aerospace Panel," in *Mathematical Modelling of Weld Phenomena 10*, C. Sommitsch and N. Enzinger, Eds. Graz, 2013, pp. 691 – 702.
[7] J. Ding, P. Colegrove, J. Mehnen, S. Ganguly, P. M. S. Almeida, F. Wang, and S. Williams, "Thermo-mechanical analysis of Wire and Arc Additive Layer Manufacturing process on large multi-layer parts," *Comput. Mater. Sci.*, vol. 50, no. 12, pp. 3315–3322, 2011. http://dx.doi.org/10.1016/j.commatsci.2011.06.023
[8] B. A. Szost, S. Terzi, F. Martina, D. Boisselier, A. Prytuliak, T. Pirling, M. Hofmann, and D. J. Jarvis, "A comparative study of additive manufacturing techniques: Residual stress and microstructural analysis of CLAD and WAAM printed Ti-6Al-4V components," *Mater. Des.*, vol. 89, pp. 559–567, 2016. http://dx.doi.org/10.1016/j.matdes.2015.09.115

Materials Research Proceedings **2** (2016) 461-466 doi: http://dx.doi.org/10.21741/9781945291173-78

Tensile Residual Stress Mitigation Using Low Temperature Phase Transformation Filler Wire in Welded Armor Plates[1]

YU Xinghua [1, a], TZELEPIS Demetrios[2,b], BUNN Jeff [3,c], PAYZANT E. Andrew[3,d] and FENG Zhili [1,e]

[1] Materials Science and Technology Division, Oak Ridge National Laboratory, Oak Ridge, TN, 37831, USA

[2] U.S. Army TARDEC RDTA-SIE-PLE-MEC, Warren, MI 48397, USA

[3] Chemical and Engineering Materials Division, Oak Ridge National Laboratory, Oak Ridge, TN, 3783, USA

[a]yux1@ornl.gov, [b]demetrios.a.tzelepis.civ@mail.mil, [c]bunnjr@ornl.gov, [d] payzanta@ornl.gov, [e] fengz@ornl.gov,

Keywords: Hydrogen Induced Cracking, Armor Steel, Residual Stress, Low Temperature Transformation

Abstract. Hydrogen induced cracking (HIC) has been a persistent issue in welding of high-strength steels. Mitigating residual stresses is one of the most efficient ways to control HIC. The current study develops a proactive in-process weld residual stress mitigation technique, which manipulates the thermal expansion and contraction sequence in the weldments during welding process. When the steel weld is cooled after welding, martensitic transformation will occur at a temperature below 400 °C. Volume expansion in the weld due to the martensitic transformation will reduce tensile stresses in the weld and heat affected zone and in some cases produce compressive residual stresses in the weld. Based on this concept, a customized filler wire which undergoes a martensitic phase transformation during cooling was developed. The new filler wire shows significant improvement in terms of reducing the tendency of HIC in high strength steels. Bulk residual stress mapping using neutron diffraction revealed reduced tensile and compressive residual stresses in the welds made by the new filler wire.

Introduction

Hydrogen induced cracking (HIC), also called cold cracking or delayed cracking has been a persistent issue in welding of high-strength steels. The main feature of this type of crack is that it occurs in the welds of high-strength martensitic steels, and generally occurs a short time after welding, usually within 48 hours. Fabricating HIC-free structures of advanced high-strength steels (AHSS), particularly the ultra-high-strength grade such as fully hardened martensitic steels used in armored vehicles, can be difficult in field. The existence of HIC poses serious threat to the structural integrity and safe operation of the armor plates.

Four factors contribute to the HIC: susceptible microstructure (usually martensitic microstructure), residual stress, hydrogen content and near ambient temperature. Mitigating residual stress is one of the most efficient ways to control HIC. Weld residual stress is a result of non-

[1] *This manuscript has been authored by UT-Battelle, LLC under Contract No. DE-AC05-00OR22725 with the U.S. Department of Energy. The United States Government retains and the publisher, by accepting the article for publication, acknowledges that the United States Government retains a non-exclusive, paid-up, irrevocable, world-wide license to publish or reproduce the published form of this manuscript, or allow others to do so, for United States Government purposes. The Department of Energy will provide public access to these results of federally sponsored research in accordance with the DOE Public Access Plan (http://energy.gov/downloads/doe-public-access-plan).*

Materials Research Proceedings **2** (2016) 461-466 doi: http://dx.doi.org/10.21741/9781945291173-78

uniform expansion and contraction in a welded structure. The non-uniform deformations can have a detrimental effect on the integrity of welded structure. Tensile residual stress is very common and may exceed the yield strength of the material in the weld [1,2]. Many studies have shown that tensile residual stress contributes to the high susceptibility of pre-mature fatigue failure [3,4] and hydrogen induced cracking (HIC) in steel welds [5-8]. Post weld heat treatment (PWHT) are the most common way to mitigate tensile residual stress [9]. These techniques include grinding, air hammer peening, shot peening, needle peening and TIG re-melting. Weld toe stress concentration can be reduced by grinding and TIG dressing, however, the tensile residual stress still cannot be entirely eliminated. Furthermore these process have little if any effect on the internal tensile residual stress. On the other hand, the peening processes can both improve weld profile and mitigate tensile residual stress. However, peening requires an extra post weld processing which increases the overall cost of the welded structure. In addition, all of these methods are restrained by the geometry of the weld structure.

In-process residual stress control, which manipulates the thermal expansion and contraction sequence in the weldments during welding process, has recently been developed to mitigate residual stress in steel welds. When the steel weld is cooled after welding, the martensitic transformation will occur at a temperature below 400 °C. Volume expansion in the weld due to transformation will reduce tensile residual stress and in some cases even produce compressive residual stress in the weld. The first LTT filler wire with 10 wt.% Cr and 10 wt.% Ni was developed by Ohta et. al in 2000 [10]. The LTT welds have shown a significant improvement of fatigue lives compared to that of conventional joints [11-13]. However, research on HIC mitigation using LTT filler wire is very limited. Current study applied LTT filler wire principles to mitigate HIC in high-strength armor steels. A special filler wire was designed with martensitic transformation to eliminate HIC in the weld of a high-strength armor steel. Y-groove tests (aka Tekken test) was applied to evaluate the HIC tendency of designed welding filler wire. Neutron diffraction was applied to study residual stress distribution.

Experimental work

Materials and Welding. The chemical composition of high-strength steel base plate MIL-DTL-12560 is shown in Table 1. After initial thermomechanical processing, the steels were normalized at a temperature of 904°C for 21 minutes followed by water quenching. Then the plates were tempered in a furnace at a temperature of 482°C for 39 minutes, followed by cooling in air to room temperature.

Y-groove tests for HIC cracking were performed in accordance with international standard ISO 17642-2. Testing plates are 12.7 mm in thickness. Gas metal arc welding (GMAW) with 95% argon and 5% carbon dioxide cover gas was performed on the plates with one single pass. Welding current, voltage and speed were 230 A, 25.7 V and 3.4 mm/s respectively. LTT experimental filler wires HV1766 with 1.2 mm in diameter were developed and evaluated in current study (see composition range in Table 1). For comparison, commercial ER100 filler wire with similar ultimate tensile strength, SuperArc LA-100 from Lincoln Electric was also used. The wire feed rate was 140 mm/s. Since the formation of cracks in LA-100 Y-groove sample will release tensile residual stress in the plate, the residual stress measured in Y-groove plate with a crack would not accurately represent the residual stress state of the weld made by LA-100 filler wire. As a result, for residual stress analysis, pre-heating at a temperature of 150°C was performed for the plate welded by LA-100 filler wire.

Residual Stress Analysis. The neutron diffraction measurements for the residual stress analysis on Y-groove samples were carried out at the Neutron Residual Stress Mapping Facility (NRSF2) beam line [14] of the High Flux Isotope Reactor, Oak Ridge National Laboratory. The diffraction of neutrons with a wavelength of $\lambda \approx 1.72$ Å on the {211} lattice planes of ferrite/martensite was analyzed. These planes were chosen as the {211} reflection best represents the macro-mechanical behavior of steels [1]. In order to compute the multiaxial residual stress state, the strain of each point was determined in transverse, normal and longitudinal direction, with the assumption that the principal stresses lie along these directions. The gauge volume was 2×2×10 mm^3 for transverse and normal directions with exposure time of 200 seconds and 2×2×2 mm^3 for longitudinal direction with exposure time of 600 seconds, taking advantage of sample symmetry. Area of 27×5 mm^2 was mapped for each sample with a grid spacing of 1 mm.

Table 1. Base metal and weld metal composition (in wt.%) and mechanical properties

	C	Mn	Cu	Ni	Cr	Mo	Si	V	Yield Strength [MPa]	Ultimate Tensile Strength [MPa]
MIL-DTL-12560	0.23	1.2	0.17	0.12	0.12	0.45	0.25	0.003	920-980	1000-1040
SuperArc LA-100	0.1	1.8	0.35	0.8	0.2	0.15	0.35	<0.01	730-750	780-790
HV1766	0.05-0.3	0.1-2	0.05-0.5	6-16	6-16	0.1-1	0.1-1	0-1	950-1020	1040-1090

Results

Y-groove Testing Results. The Y-groove test is used to determine HIC sensitivity of welding filler wires. MIL-DTL-12560 Y-grooved plates were both welded by conventional ER100 (LA-100) filler wire and LTT filler wire. The metallographic images of corresponding weld cross section are shown in Fig. 1. A crack throughout the weld in conventional ER100 filler wire indicates ER100 is very susceptible to HIC for MIL-DTL-12560. On the other hand, no crack is seen using HV1766 filler wire. Though the LTT filler wire in the current study shows improved HIC resistance, whether the improved resistance is due to the mitigation of tensile residual stress in the weld is still unclear. As a result, it is important to understand residual stress distribution in the Y-groove welds.

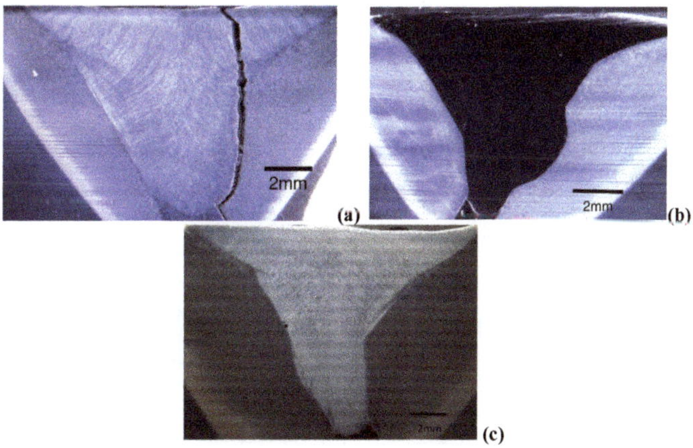

Figure 1. Y-groove test result of (a) SuperArc LA-100 filler wire (b) LTT wire showing cracks in ER100 weld but not in LTPT weld.(c) SuperArc LA-100 filler wire with 150°C preheat

Residual Stress Distribution. Since the weld metal and base metal have very different compositions, a significant spatial variation in stress free lattice spacing, d_0, is expected in the weld, largely due to the change of weld metal dilution from location to location. As a result, accurate

Materials Research Proceedings **2** (2016) 461-466 doi: http://dx.doi.org/10.21741/9781945291173-78

determination of d_0 is very challenging. Zhang et al. [15] found the normal stress in 15.2 mm plate with a dissimilar weld is negligible except near fusion boundaries As a result, plane stress can be used as an assumption for the welded plate with thickness of 15 mm or thinner without accurate determination of d_0.

The present analysis procedure is based on the plane stress assumption that the normal stress component σ_Z is equal to zero. This makes it possible to determine the two in-plane stress components σx and σy and stress free lattice spacing using the equations described in [15]. With the plane stress assumption, the residual stress distribution in transverse direction and longitudinal direction within the weld can be calculated. The residual stress map of the area indicated in Fig. 1 (a) is shown in Fig. 2. In addition, the distribution of d0 and full-width at half maximum (FWHM) is also shown in Fig. 2. Due to grain size differences between the weld metal (WM), heat-affected zone (HAZ) and base metal (BM), the FWHM can be used to distinguish these three regions. In Fig. 2 (d), FWHM clearly shows the shape of the weld and these three distinct regions: WM, HAZ and BM. Fig. 2 (a) shows transverse residual stress up to 1000 MPa. Since HIC was observed mainly in the HAZ at a location about 0.2 mm from the fusion boundary (Fig. 1), the transverse residual stress in this location is very important. According to Fig. 2 (a), the transverse residual stress in the HAZ around the fusion boundary is very uniform with a value of 700 MPa in tension.

Maximum longitudinal residual stress in LA-100 is about 1400 MPa in tension, which is much greater than the tensile strength of MIL-DTL-12560 (~1000 MPa). After the welding process, newly formed martensite with high yield strength is expected in the HAZ of MIL-DTL-12560. As a result, the high residual stress of 1400 MPa in HAZ is possible. The WM shows tensile longitudinal residual stresses which range from 200 to 600 MPa. Fig. 2 (c) shows fairly uniform distribution of d_0 in the WM and the BM, which indicates the plane stress assumption is reasonable for the current study in both regions. In the HAZ near the fusion boundaries, two regions show relatively low d0 values. Normal stress may not be zero in these two regions.

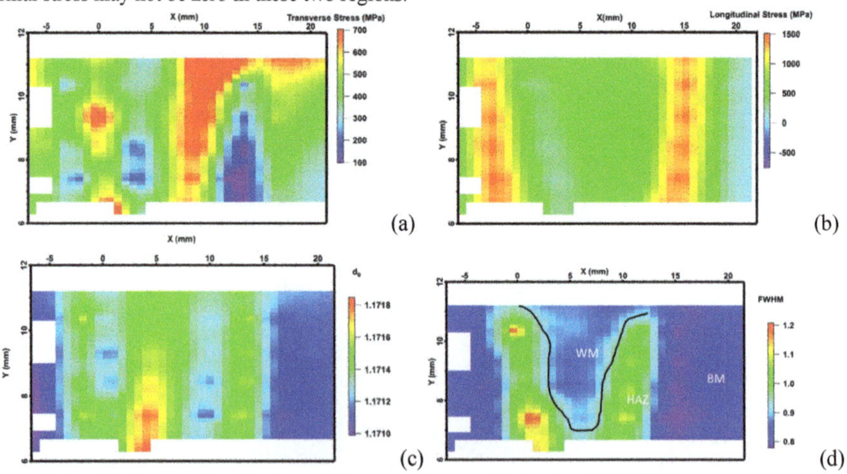

Figure 2. Residual stress distribution of Y-groove plate welded using LA-100 (a) transverse direction (b) longitudinal direction (c) distribution of stress free lattice spacing d_0 (d) distribution of FWHM

Fig. 3 shows the residual stress calculation result of the Y-groove plate welded using LTT filler wire HV1766. Though Fig. 3d exhibits a significant spatial variation of FWHM, the WM cannot be identified in the map since the FWHM of the WM and HAZ is very similar. The transverse residual stress map shows a uniform distribution of stress of 400 MPa except a slight compressive transverse residual stress the plate surface and high tensile at the bottom. For the regions which are susceptible to HIC, the transverse residual stress is below 300 MPa, which is much lower than that in the plate welded by LA-100. Fig. 3 (b) shows high compressive longitudinal residual stress in the center of the weld. Compressive residual stresses in both the transverse and longitudinal directions should be a result of the low temperature martensitic transformation of the HV1766 filler wire during cooling. For the plate welded by HV1766, the maximum longitudinal stress has a similar value and was observed at similar locations as the plate welded by LA-100. It is noted even though HV1766 could eliminate HIC when it is used to weld MIL-DTL-12560 high strength steel, high longitudinal residual stress is still observed in HAZ close to weld root. The reason of high longitudinal residual stress is not clear at this moment. The high tensile residual stress may be detrimental to the mechanical integrity of the weld, which need to be investigated in the future.

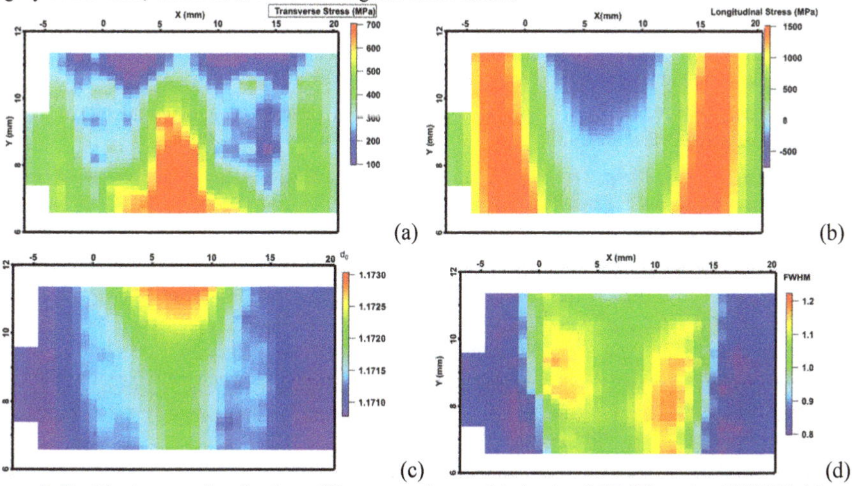

Figure 3. Residual stress distribution of Y-groove plate welded using LTT filler wire HV1766 (a) transverse direction (b) longitudinal direction (c) distribution of stress free lattice spacing d_0 (d) distribution of FWHM

Summary and Conclusions

Low temperature transformation (LTT) weld filler wires have been developed and show a reduced tensile residual stress and an improved HIC resistance. The current study used the Y-groove test to determine HIC sensitivity during welding. MIL-DTL-12560 plates were both welded by conventional ER100 (SuperArc LA-100) filler wire and LTT filler wire HV1766. A crack throughout the weld in conventional ER100 filler wire indicates ER100 is very susceptible to HIC. No crack is seen using LTT filler wire HV1766. Preliminary neutron diffraction results have shown the LTT filler wire HV1766 produced a compressive residual stress in the weld in longitudinal direction. In transverse direction, the HV1766 plates shows lower tensile residual stress compared that in ER100 plate. This lower tensile residual stress contributes to HIC resistance of LTT filler wire.

Acknowledgement

The authors acknowledge the research funding from *U.S. Army TARDEC RDTA-SIE-PLE-MEC*. A portion of this research used resources at the High Flux Isotope Reactor, a DOE Office of Science

User Facility operated by the Oak Ridge National Laboratory. The assistance of P.A. Cornwell with the strain mapping experiments is gratefully acknowledged.

References

[1] P. J. Withers, H. K. D. H. Bhadeshia, Residual Stress Part 1 - Measurement Techniques, Mater Sci Tech 17 (2001) 355-365. http://dx.doi.org/10.1179/026708301101509980

[2] P. J. Withers, H. K. D. H. Bhadeshia, Residual Stress Part 2 - Nature and Origins, Mater. Sci. Tech. 17 (2001) 366-375. http://dx.doi.org/10.1179/026708301101510087

[3] G. Bussu, P. E. Irving, The role of residual stress and heat affected zone properties on fatigue crack propagation in friction stir welded 2024-T351 aluminium joints, Int. J. Fatigue. 25 (2003) 77-88. http://dx.doi.org/10.1016/S0142-1123(02)00038-5

[4] G. A. Webster, A.N.Ezeilo, Residual stress distributions and their influence on fatigue lifetimes, Int J Fatigue 23 (2001) 375-383. http://dx.doi.org/10.1016/S0142-1123(01)00133-5

[5] N. Okuda, Y. Ogata, Y. Nishikawa, T. Aoki, A. Goto, T. Abe, Hydrogen-induced cracking susceptibility in high-strength weld metal, Weld. J. 66 (1987): 141s-146s.

[6] M.D. Rowe, T. W. Nelson, J. C. Lippold, Hydrogen-induced cracking along the fusion boundary of dissimilar metal welds, Weld. J. 78 (1999): 31s-37s.

[7] H.W. Lee, S. W. Kang, The relationship between residual stresses and transverse weld cracks in thick steel plate, Weld. J. 82 (2003) 225s-230s.

[8] J. L. Otegui, A. Rivas, C. Manfredi, C. Martins. Weld failures in sleeve reinforcements of pipelines. Eng. Fail. Anal. 8 (2001) 57-73. http://dx.doi.org/10.1016/S1350-6307(99)00049-7

[9] X. Cheng, J. W. Fisher, H. J. Prask, T. Gnäupel-Herold, B. T. Yen, S. Roy, Residual stress modification by post-weld treatment and its beneficial effect on fatigue strength of welded structures, Int. J. Fatigue, 25 (2003), 1259-1269. http://dx.doi.org/10.1016/j.ijfatigue.2003.08.020

[10] A. Ohta, O. Watanabe, K. Matsuoka, T. Shiga, S. Nishijima, Y. Maeda, N. Suzuki and T. Kubo, Fatigue Strength Improvement by Using Newly Developed Low Transformation Temperature Welding Material, Weld. World 43 (1999), 38–42.

[11] J. Eckerlid, T. Nilsson, L. Karlsson. Fatigue properties of longitudinal attachments welded using low transformation temperature filler. Sci. Technol. Weld. Joi. 8 (2003) 353-359. http://dx.doi.org/10.1179/136217103225005525

[12] Z. Barsoum, M. Gustafson. Spectrum fatigue of high strength steel joints welded with low temperature transformation consumables. In:2nd International Conference on Fatigue Design, Senlis, France, 2007.

[13] P. P. Darcis, H. Katsumoto, M. C. Payares-Asprino, S. Liu, T. A. Siewert. Cruciform fillet welded joint fatigue strength improvements by weld metal phase transformations. Fatigue. Fract. Eng. M. 31 (2008) 125-36. http://dx.doi.org/10.1111/j.1460-2695.2007.01205.x

[14] W. Woo, V. Em, C.R. Hubbard, H.J. Lee, K.S. Park, Residual stress determination in a dissimilar weld overlay pipe by neutron diffraction, Mat. Sci. Eng. A 528 (2011) 8021-8027 http://dx.doi.org/10.1016/j.msea.2011.07.059

[15] W. Zhang, Z. Feng, P. Crooker. Improved procedure for computing residual stresses from neutron diffraction data and its application to multipass dissimilar welds. Sci. Technol. Weld. Joi. 16 (2011) 254-260. http://dx.doi.org/10.1179/1362171810Y.0000000023

Residual Stresses 2016: ICRS-10 Materials Research Forum LLC
Materials Research Proceedings **2** (2016) 467-472 doi: http://dx.doi.org/10.21741/9781945291173-79

In-situ Monitoring of Laser Surface Line Hardening by Means of Synchrotron X-Ray Diffraction

Kiefer D.[1,a*], Gibmeier J.[1,b], Beckmann F.[2,c] and Wilde F.[2,d]

[1] Karlsruhe Institute of Technology (KIT), Institute for Applied Materials (IAM-WK), Engelbert-Arnold-Str. 4, 76131 Karlsruhe, Germany

[2] Helmholtz-Zentrum Geesthacht (HZG), Institute of Materials Research, Max-Planck-Str. 1, 21502 Geesthacht, Germany

[a]Dominik.Kiefer@kit.edu, [b]Jens.Gibmeier@kit.edu, [c]Felix.Beckmann@hzg.de, [d]Fabian.Wilde@hzg.de

Keywords: in-situ X-Ray Stress Analysis, Synchrotron X-Ray Diffraction, Laser Surface Hardening, Real Time Monitoring

Abstract. An in-situ X-ray diffraction investigation of a power controlled laser line hardening experiment has been carried out at the synchrotron beamline P05 at the German electron synchrotron (DESY) in Hamburg. During the process the local strain and stress evolution is monitored using synchrotron radiation with a time resolution of 10 Hz. Samples made of steel grade AISI 4140 were line hardened by means of a high-power diode laser (HPDL) unit at constant laser feed. A specially designed process chamber was used, allowing the control of the inert gas atmosphere to avoid oxide scale formation. Through the symmetric application of 4 fast micro-strip line detectors various {hkl} - diffraction lines were recorded during the short-time heat treatment by means of the single exposure approach. Thermal and elastic strains were separated and time resolved stress analysis was carried out according to the $\sin^2\psi$-method. Stress and strain evaluation during the complete laser hardening process lead to unprecedented experimental insights.

Introduction

Structural steel components (e. g. cutting matrices, forging dies and affiliated closing edges) experience high mechanical and thermal loads. Due to these loads component failure occurs through fatigue crack initiation and oxidation at the surface. Special attention is given to the improvement of surface layers through increase of wear resistance and fatigue strength by generation of favorable compressive residual stress states. In this regard, thermal and thermo-chemical processes possess large potential. Over the last decade, with the development of high power diode laser (HPDL) systems, laser surface hardening has gained increased interest. Present-day optimization of the laser hardening process is mainly based on case studies of the final state of hardened samples [1, 2, 3]. Real time insights into strain, stress and microstructural (e. g. phase transformations) evolution were only possible by means of numerical simulations [4, 5]. Process prediction of the laser surface hardening is mainly based on case studies for a certain material and hence collecting real time data could be used to improve process prediction and to validate numerical simulations. Synchrotron X-ray diffraction is a viable option to gain real time insights into complex thermal and thermo-chemical processes. An experimental set-up for in-situ monitoring of phase transitions and strain evolutions is presented in [6] and its reliability proven in [7]. In the present work, a similar set-up upgraded with two additional fast micro-strip line detectors and a motor actuator to move the laser beam over the sample surface at constant feed was used. Using the upgraded instrumentation laser surface line hardening experiments were carried out and in-situ X-ray synchrotron diffraction data was monitored at a time resolution of 10 Hz. The thermal and elastic strains were separated for each exposure and the stresses were determined according to the $\sin^2\psi$-method [8] using the approach presented in [7]. The results of the phase-, strain- and stress-evolution are carefully discussed and compared to the results of a point hardening experiment [7].

Materials Research Proceedings **2** (2016) 467-472 doi: http://dx.doi.org/10.21741/9781945291173-79

Experimental

The in-situ X-ray diffraction experiments were carried out on steel type AISI 4140 in a quenched and tempered state. The sample geometry is a cuboid (80 × 50 × 10 mm^3). The sample faces were ground mechanically to provide a clean reproducible surface. Subsequently a heat treatment at 510 °C for 90 min was carried out to provide a stress-free sample state. The experimental set-up is established at the HZG undulator beamline P05@PETRAIII at the German electron synchrotron (DESY) in Hamburg, Germany. Monochromatic synchrotron radiation was provided by a double crystal monochromator (DCM) at a wavelength of $\lambda = 0.11384$ Å (10.899 keV). The X-ray spot is adjusted by means of a cross-slit system to a nominal spot size of 1 × 1 mm^2. For laser hardening a fiber coupled 10 kW high power diode laser unit of type LDF 1000-60 from Laserline GmbH, Mühlheim-Kärlich (Germany) was used in combination with a homogenizing optic with a spot size of 8 × 8 mm^2. The laser feed is set to 0.2 m/min through a motorized linear actuator by well-defined tilting of the homogenizing optic. The power was selected to 2.5 kW. The temperature in the process zone was measured using a one-color (monochromatic) pyrometer. The upgraded experimental set-up is schematically shown in Fig. 1 and the corresponding photographs in Fig. 2. Valves and an affiliated controller allow the adjustment of the atmosphere inside the process chamber. The experiment was carried out under a low He-gas excess pressure to avoid oxidation of the sample surface.

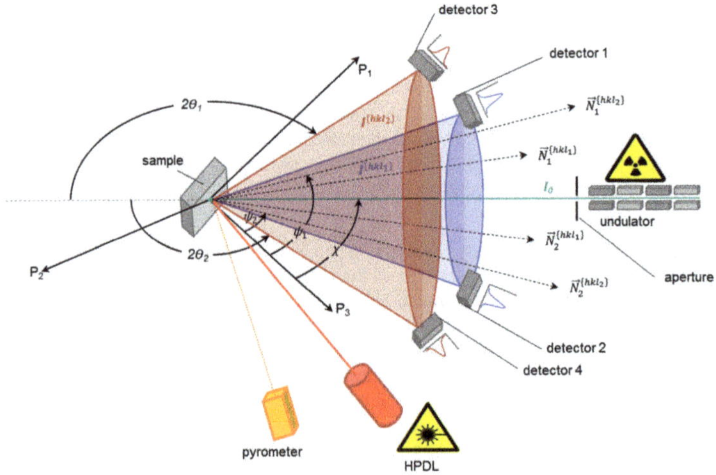

Figure 1. *Schematic view of the single exposure set-up for fast in-situ X-ray diffraction analysis of the laser surface hardening process. The indicated angles ψ_1 and ψ_2 apply to detector 3 and 4.*

The sample is tilted by an angle $\chi = -35°$ with respect to the primary beam axis. As key components 4 fast micro-strip detectors of type Mythen 1K from Dectris ltd., Switzerland, are arranged radially around the process chamber. By this means the diffracted beam can be detected through a polyimide-window attached to the process chamber. Detectors 1 and 2 were placed on the detector wing at a distance of 200 mm from the process zone covering a 2θ-range from approx. 143°-161°. Detectors 3 and 4 were installed at a distance of 250 mm covering approx. 128°-139° in 2θ. The detectors were arranged in a way corresponding to the single exposure technique described in [8]. The set-up in Fig. 1 allows the contemporaneous measurement of the reflex {hkl} from the identical diffraction cone under two different ψ angles ψ_1 and ψ_2. By definition the tilt angle ψ is the angle between the sample surface normal P_3 and the diffraction vector $N_i^{\{hkl\}}$. Powder samples of

Materials Research Proceedings **2** (2016) 467-472 doi: http://dx.doi.org/10.21741/9781945291173-79

LaB_6 and α-Fe were used for detector calibration. A flat field scattering background correction derived from diffraction measurements of long-term flat illumination of an amorphous glass sample was carried out individually for each detector module. The flat field correction was followed by an absorption correction and a linear background subtraction. Subsequently, the diffraction peaks are fitted using a Pseudo-Voigt function.

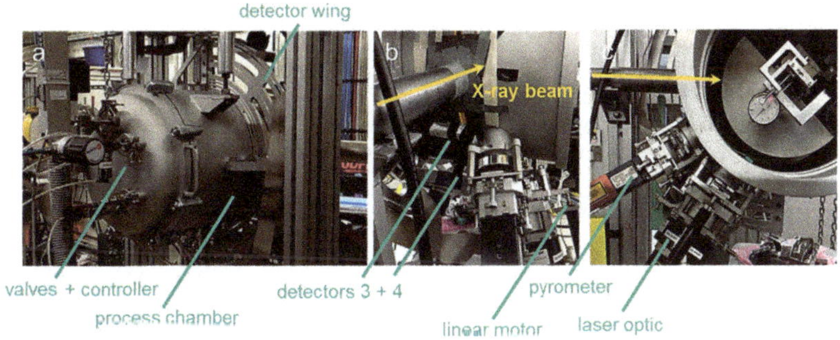

Figure 2. *Upgraded process chamber designed for in-situ X-ray diffraction experiments during laser surface hardening. (a) Overview of the process chamber. (b) In-line view on key components. (c) Vertical view into opened process chamber.*

At the chosen synchrotron radiation wavelength detectors 1 and 2 measured diffraction lines of {422}-α-Fe ($2\theta_0 = 153.206°$) and {620} γ-Fe ($2\theta_0 = 161.263°$) respectively- whilst detectors 3 and 4 measured the {332}-α-Fe ($2\theta_0 = 137.300°$), {531}-γ-Fe ($2\theta_0 = 134.719°$) and {600}-γ-Fe ($2\theta_0 = 138.790°$) peaks. In a 2θ-$\sin^2\psi$ line plot thermal strains result in a vertical shift due to the fact that they affect the hydrostatic part of the stress tensor, while elastic strains result in a change in slope. Hence, elastic and thermal strains can be separated and stress evolution can be evaluated as described in [7]. The stress state is assumed to be biaxial ($\sigma_{33} = 0$) and rotationally symmetric, consequently the strain free direction $\sin^2\psi^*$ can be calculated from the X-ray elastic constants according to Eq. 1.

$$sin^2\psi^* = \frac{-2s_1}{\frac{1}{2}s_2}. \qquad (1)$$

The X-ray elastic constants used in the determination of $\sin^2\psi^*$ and σ^{dev} were calculated from the temperature dependent macroscopic Young's moduli E and Poisson ratios ν listed in [9]. E and ν are interpolated from the measured temperature. The X-ray elastic constants are calculated according to Eq. 2 and Eq. 3.

$$s_1 = \frac{-\nu(T)}{E(T)}. \qquad (2)$$

$$\frac{1}{2}s_2 = \frac{\nu(T)+1}{E(T)} \qquad (3)$$

The determined in-situ X-ray stresses were compared with ex-situ X-ray residual stress measurements at the identical measuring point using the classic $\sin^2\psi$-method and conventionally generated X-rays. Those analyses were carried out on a ψ-diffractometer using CrK_α-radiation at the Institute for Applied Materials (IAM-WK) at KIT. Here, the {211} diffraction line ($2\theta_0 = 156.394°$)

of the α/α′-phase was determined at 15 equidistant ψ angles in the range -60°< ψ <60°. For stress evaluation X-ray elastic constants, $s_1^{\{211\}} = -1.27\cdot10^{-6}$ MPa and $\frac{1}{2}s_2^{\{211\}} = 5.82\cdot10^{-6}$ MPa, are used.

Results and Discussion

Fig. 3a shows the laser line hardened sample as well as the measuring point. A detailed cross sectional view into the hardening lens is given in Fig. 3b and the microstructure detail in Fig. 3c. The laser track has a length of 34 mm. The spot like end of the laser track follows from a not perfect synchronization of laser feed and power. At the measuring location (indicated by a square in Fig. 3a) the hardening lens has a width of 6.9 mm and a maximum depth of approx. 818 µm. The microstructure is fine-grained with rounded martensitic needles, which is due to a slight annealing effect due to a global sample warming caused by the progressing laser beam. In contrast, cross sectional images from point hardening experiments [4, 7] show a sharper martensitic microstructure because there is no annealing effect from the laser.

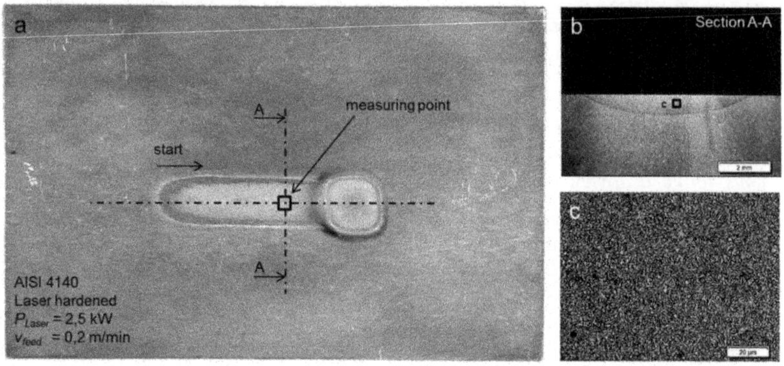

Figure 3. Laser line hardened sample. (a) Overview of the 34 mm long laser line. (b) cross section view at the measuring point (view A-A). (c) microstructure inside the hardening lens.

In Fig. 4 the measured diffraction data during the laser surface line hardening experiment are plotted versus the processing time in 2D-contour plots. The process temperature measured by the monochromatic pyrometer is also plotted. The intensities are normalized for every single detector module and flat field corrected. At the beginning of the process detector 1 and 2 (Fig. 4 left) show a shift of the {422}-α-Fe interference line to lower 2θ values with increasing temperature. Simulaneously, the peak intensity decreases. This shift is due to the temperature rise. At a processing time of 6.1 s and a temperature of about 890°C no {422}-α-Fe signal could be further detected. The point indicates the ferrite-austenite phase transformation. This statement is supported by the appearance of the {620}-γ-Fe peak at the upper 2θ range of the detectors 1 and 2. The temporal progression of its diffraction signal is also following the temperature. At temperature of 410°C and a processing time of about 9.7 s a broad {422}-α'-Fe peak appears marking the austenite-martensite transformation.

Fig. 4 (right) shows similar intensity plots for detectors 3 and 4. At approx. 6.1 s the ferrite-austenite transformation is marked by the appearance of the {600}-γ-Fe reflection and a slight jump shift of interference in 2θ from the {332}-α-Fe signal to {531}-γ-Fe. As a consequence of cooling down by self-quenching the {531}-γ-Fe interference line shifts to larger 2θ values accompanied by increasing intensity and peak broadening. The peak broadening could be explained by adaption deformation preparing the martensite-formation. Subsequently, the austenite peak vanishes and a broad {332}-α'-Fe reflection remains. The determined thermal and elastic strain and stresses are depicted in Fig. 5 top, mid and bottom. The thermal strain evolution in the ferrite region corresponds with the 2D-contour plots in Fig. 4. An interference shift to lower 2θ values with temperature

Materials Research Proceedings **2** (2016) 467-472 doi: http://dx.doi.org/10.21741/9781945291173-79

increase results in higher thermal strains. The transformation to austenite is marked by an abrupt drop of the thermal strains to nearly 0.

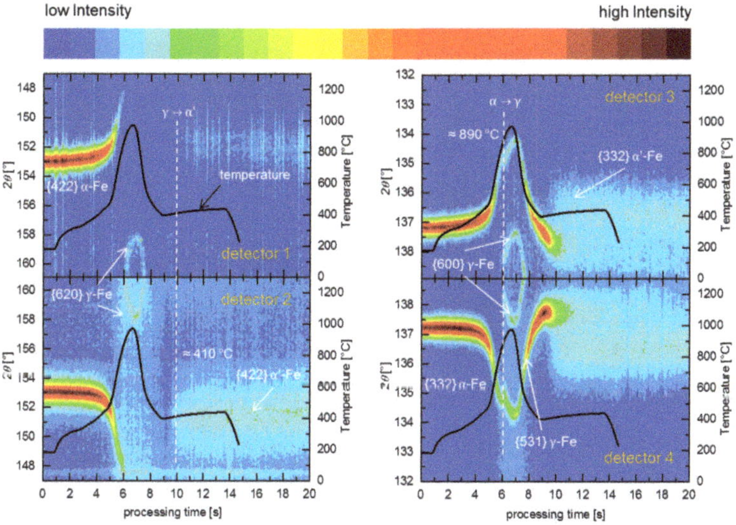

Figure 4. *Normalized 2D-contour-plots of the interference profiles recorded with a time resolution of 10 Hz for the detector pairs 1-2 (left) and 3-4 (right) and plot of the measured temperature vs. process time for laser surface line hardening of steel AISI 4140 at constant feed of 0,2 m/min and laser power of 2,5 kW.*

Subsequently, the thermal strains follow the temperature. At approximately 600 °C thermal strains change their sign to compression for the {531}-γ-Fe reflection. With the beginning of martensite formation an abrupt rise to positive thermal strains of the {332} and {422}-α' diffraction lines occurs. Furthermore, a slight decrease is observed. In a time region from about 9.8 s until 17 s both austenite and martensite peaks allow evaluation.

The time developement of the elastic strains and corresponding deviatoric stresses is qualitatively in good agreement to simulations in the literature [5]. The thermal expansion of the material leads at first to compressive stresses vertical to the sample axis due to the constraint of the surrounding cold material. When these stresses exceed the warm yield strength of the material, plastic deformation starts, decreasing the compressive stresses despite increasing temperature. Furthermore, material yields towards the free sample surface. The volume decrease due to the formation of austenite amplifies this effect and forces the stresses towards the tensile region. The result is a first maximum in compressive stress. After reaching the maximum temperature cooling starts and the tensile stresses increase, due to plastic deformation, in which the displaced material volume is no longer available for volume contraction. Additionally, the steep temperature gradient at the surface results in higher shrinkage and the larger coefficient of thermal expansion of austenite over ferrite, both increase the tensile stresses. Upon reaching a temperature of about 410 °C the martensite transformation occurs. Here, the stresses change their sign to compressive stresses, due to the volume expansion [5]. Further fluctuations in the stresses are based on statistical effects disappearing with higher X-ray exposure times (processing times > 50 s). The additionally determined ex-situ residual stress value is also presented in Fig. 5 (bottom) showing good agreement with the in-situ X-ray diffraction data after cooling down to ambient temperature. Hereby, a good validation of experimental robustness is demonstrated. In comparison with a similar point hardening experiment using only two line detectors

[7], the upgraded set-up allows information to be gained about elastic strains and stresses in the austenite region, so a full process analysis can be conducted.

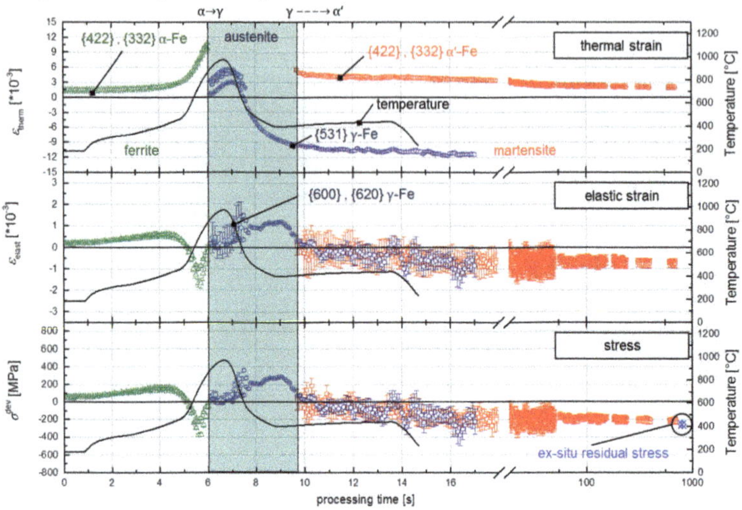

***Figure 5.** .Thermal strain (top), elastic strain (middle), and deviatoric stress component calculated from elastic strain (bottom) vs. processing time for laser surface line hardening of steel AISI 4140 at constant feed of 0.2 m/min and laser power 2.5 kW. Error bars calculated on the basis of 95% confidence interval of the peak fit.*

Summary

Systematic studies of laser surface line hardening experiments were carried out by means of synchrotron X-ray diffraction. The upgrade of the measuring set-up for high-speed X-ray strain and stress analysis (tilt of optics holder in combination with motorized linear actuator) was successfully tested for laser surface line hardening. The use of 4 fast micro-strip detectors allowed coverage of a much larger 2θ range. Hence, it was possible to gain more information about the high temperature region especially in the austenite regime. The α-γ phase transformation point could be exactly determined by means of the synchrotron X-ray data. However, the martensite formation is not instantaneous inside the measurement volume.

Acknowledgements

Financial support by the German Research Foundation (DFG) through the project GI376/10-1 | BE5341/1-1 is gratefully acknowledged.

References

[1] T. Miokovic, V. Schulze, O. Vöhringer, D. Löhe, Härterei-Techn. Mitt.,60 (2005) 142-149
[2] K. Obergfell, V. Schulze, O. Vöhringer, Surface Engineering, 19 (2003) 359-363
[3] K. Müller, H.W. Bergmann, Z. Metallkunde, 90 (1999) 881-887
[4] T. Miokovic, Doctoral Thesis, University of Karlsruhe, (2005)
[5] K. Müller, C. Körner, H. W. Bergmann, Härterei-Techn. Mitt.,51 (1996) 19-28
[6] V. Kostov, J. Gibmeier, S. Doyle, A. Wanner, Mater. Sci. For., 638-642 (2010) 2423-2428
[7] V. Kostov, J. Gibmeier, F. Wilde, P. Staron, R. Rössler, A. Wanner, Rev. Sci. Instrum.,83 (2012) 115101 1-11
[8] E. Macherauch, P. Müller, Z. angew. Phys., 13 (1961) 305-312
[9] P. Graja, Doctoral Thesis, University of Karlsruhe, (1987)

Residual Stresses 2016: ICRS-10 Materials Research Forum LLC
Materials Research Proceedings **2** (2016) 473-478 doi: http://dx.doi.org/10.21741/9781945291173-80

Consideration of Tool Chamfer for Realistic Application of the Incremental Hole-Drilling Method

Nicola Simon[1, a *], Jens Gibmeier[1, b]

[1]Karlsruhe Institute of Technology (KIT), Institute for Applied Materials (IAM), Kaiserstr. 12, D-76131 Karlsruhe, Germany

[a]nicola.simon@kit.edu, [b]jens.gibmeier@kit.edu

Keywords: Residual Stress Measurement, Incremental Hole-Drilling Method, Hole Geometry, FEM

Abstract. The incremental hole-drilling method is well-established in residual stress analysis. A small blind hole is drilled incrementally and the strain relief is measured on the sample's surface. In order to calculate residual stresses from the measured strain calibration data is necessary. Typically, the calibration data is determined numerically and is based on the assumption of an ideal cylindrical blind hole. However, widely used six-blade milling bits have rather large chamfers at the cutting edges, which result in hole geometries that clearly differ from the ideal cylindrical blind hole. Especially in the first drilling increments a significant impact on the calibration data by the real hole geometry is expected. In this work, a numerical model is set up that allows for consideration of tool chamfers. A systematic finite element study is carried out to analyze the effect on relieved strains. Calibration data is computed for the ideal blind hole as well as for the realistic hole geometry. Finally, numerical results are compared with experimental results gained by defined uniaxial loading experiments. The results clearly indicate a significant impact of the tool chamfer geometry for strain relief and stress data close to the surface. Hence, based on the results it is highly recommended to consider the real tool geometry to provide accurate stress evaluation by means of the incremental hole-drilling method in particular for the first drilling increments.

Introduction

The incremental hole-drilling method is a widely used mechanical method for residual stress depth profiles analysis due to its simple instrumentation and fast execution. Since J. Mathar [1] first proposed the test method in 1933 it has been under constant development. State of the art is incremental drilling of blind holes of small diameters in order to determine local stress gradients. The method is based on redistribution of residual stresses due to local material removal. Entailed strain relaxations can be measured on the surface area around the hole e.g. by means of strain gages. Due to the fact that strains are only partly released when introducing a blind hole, calibration data is needed to evaluate the residual stress profile from measured strains. Usually, finite element (FE) simulations are used to calculate these calibration data. In commercially available evaluation routines generally an ideal cylindrical blind hole is considered. Flaman [2] showed that the experimental setup of high speed drilling in combination with inverted cone end mills lead to the best approximation of the ideal cylindrical blind holes, while inducing a negligible amount of machining stresses. The widely used six-blade tungsten carbide milling bits with nominal diameters of 0.8 or 1.6 mm, that are used in commercially available pneumatic high speed drilling devices, have large chamfers at the cutting edges, which result in hole geometries that differ from the ideal cylindrical blind hole.
Two shortcomings arise out of this geometrical deviation (see also Fig. 1):

- a remaining bottom fillet for each single drilling increment
- an increasing hole diameter for the first drilling increments

Materials Research Proceedings **2** (2016) 473-478 doi: http://dx.doi.org/10.21741/9781945291173-80

In Scafidi, et al. [3] the first impact, i.e. the bottom fillet, was investigated by considering a bottom fillet radius in numerical simulations. It was figured out, that deviations of strain relaxations can cause erroneous results in case of large bottom fillet radii. In Nau, et al. [4] different kinds of end mills and their influence on the occurring hole geometries for a nominal hole diameter of 1.6 mm were studied. A numeric model was set up considering the effect of increasing hole diameters in the first drilling steps by using the mean hole diameter over the actual removed depth increment. The conclusion of both studies was that the consideration of more realistic tool geometries for calibration results in more meaningful stress data. However, in [3] and [4] only the general effects were discussed without taken into account the real chamfered cutting edge of the most often applied end mills. Furthermore, up to now only chamfered tools with a nominal diameter of 1.6 mm were considered.

In our project we have the inherent necessity to apply small hole diameters using a nominal tool diameter of 0.8 mm for residual stress depth distributions showing a relative steep gradient close to the surface. Hence, in this work a FE model is set up for a more accurate simulation of the hole geometry by considering the real chamfered edges using a conventional tapered end mill with 0.8 mm diameter. Thus, both shortcomings of the non-ideal cylindrical hole geometry are taken into account (remaining bottom fillet and unsteady hole diameter). The FE model can be used to calculate chamfer-considering calibration data for reliable stress evaluations using the differential method [5]. Finally, the numerical results are validated by experimental findings of a defined 4-point bending test.

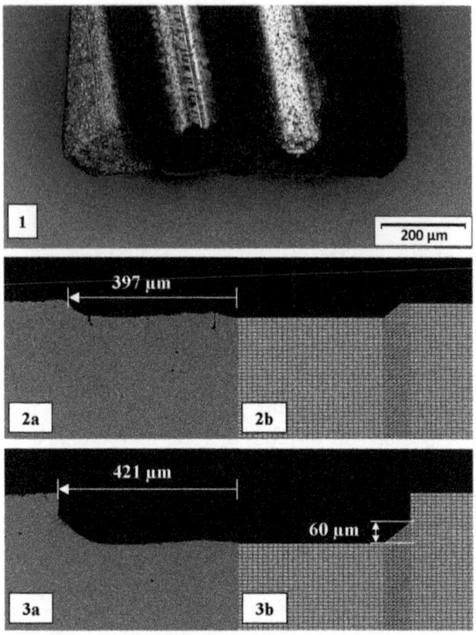

Fig. 1: *Six-blade TiN coated tungsten carbide end mill with nominal diameter of 0.8 mm (1), drilled holes of 40 µm depth (2) and 140 µm depth (3), cross section of drilled sample (a), side view of FE model (b).*

Finite element simulation

Finite element (FE) model. The 3D FE model was defined in ABAQUS and consists of 750,000 elements of type C3D8R and C3D6. This hybrid element model is needed to take the chamfer geometry into account. An inverted cone tungsten carbide six-blade end mill of 0.8 mm diameter with a chamfer height of 0.06 mm at an angle of 45° was assumed for the numeric model (Komet, Gebr. Brasseler GmbH & Co. KG). Fig.1 shows the cutting edges of the end mill and the cross sections of a steel sample with two drilled blind holes of different depths. On the right hand side a side view of the FE model is presented. It can be seen that it is in good agreement with the actual blind hole geometry, even for a small drilling depth of 40 µm. Due to the symmetry of the calculated problem only a quarter model was used. A full view of the model is shown in Fig. 2. The drilling process is simulated by stepwise removing the elements in the region of the hole (drilling increments are highlighted in Fig. 2). For each drilling step the released strains on the surface area around the hole were calculated and averaged at element surface nodes. In a post processing step, the strains were integrated and averaged over three virtual strain gage areas at the positions 0°, 45° and 90° leading to three single strain values $\varepsilon_{0°}$, $\varepsilon_{45°}$, $\varepsilon_{90°}$, comparable to those, gained in experimental studies.

Strain gage rosette dimensions considered in the model are similar to type A residual stress strain gage rosette according to the ASTM E837 [6].
For the hole-drilling simulation a biaxial nominal stress state of $\sigma_{nom} = \sigma_x = \sigma_y = 300$ MPa was applied on the outside surfaces of the quarter model (see Fig. 2). Linear elastic material behavior of isotropic steel with a Young's modulus of 210 GPa and a Poisson's ratio of 0.3 was applied. The chosen hole diameter of 0.8 mm is the minimum achievable hole diameter in practical application by the use of an end mill of the assumed size. Two identically meshed numeric models have been used to simulate the resulting strain data $\varepsilon_{chamfered}$, for the chamfered drilled hole and $\varepsilon_{cylindrical}$, for the ideal cylindrical hole, respectively.

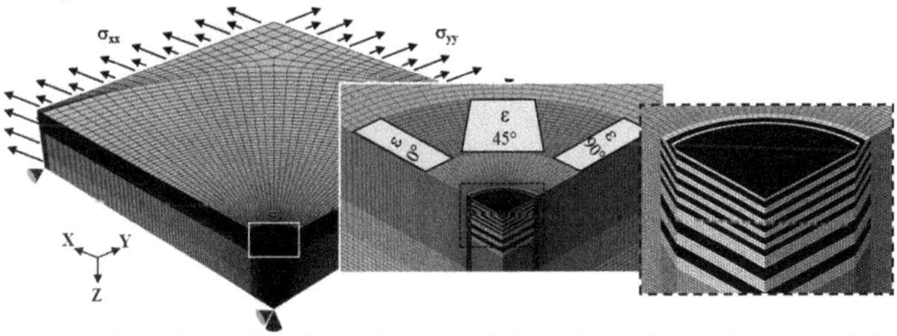

Fig. 2: *ABAQUS 3D FE-model with detailed views of the position of the strain gages and the drilling increments.*

Stress calculation was carried out for the strain data $\varepsilon_{chamfered}$ of the hole-drilling simulation considering chamfered holes, i.e. the better approximation of the real hole geometry. The stress was evaluated using calibration data $C_{cylindrical}$, which are typically provided by commercially available hole-drilling evaluation software, assuming ideal cylindrical holes. In comparison, a stress evaluation based on the consideration of the chamfer using the calibration data $C_{chamfered}$ was carried out. Since we are most likely interested in the near surface data, the results are only presented up to a drilling depth of 0.4 mm.

Results and discussion. In order to show the impact on measured strain relaxations, absolute and relative strain deviations (referred to the ideal hole geometry) between the two drilled hole geometries are shown in Fig. 3a. Due to the axi-symmetric stress state, only one strain gage component is plotted. Drilling steps with increasing hole diameters (up to 0.06 mm depth) are shaded in a grey zone. In general, for the loading state considered strain relaxation is higher for the assumption of an ideal blind hole. However, a significant decrease of the relative strain deviation can be seen. From 30 % strain deviation in the first increment of 0.01 mm, values are decreasing to less than 5 % for the last drilling depth of 0.4 mm. The absolute strain deviation is showing a maximum of 10 µm/m at about 0.24 mm drilling depth. This behavior can be explained by the interaction of two effects that are accompanied by the geometrical deviation of the drilled holes. The first effect is the remaining hole-bottom fillet, which is leading to a significant deviation of the removed material volume (see Fig. 3b). Especially for the first drilling increments, where the final hole diameter is not yet reached, the removed volume for the ideal hole geometry is 1.2 the size of the chamfered hole volume. This difference of volume clearly has an impact on the large relative strain deviation in the first drilling increments. Furthermore, the hole geometry influences the notch factor of the drilled hole (second effect that depends on the stress state considered), leading to a higher maximum stress in case of a sharp edged ideal cylindrical hole. From a drilling depth of 0.24 mm this effect is declining, leading to a decrease of the absolute strain deviation. It must be mentioned that in practical application, typically larger hole diameters occur due to slight asymmetric end mills or slightly eccentrically installed turbine shafts. Numeric simulations regarding larger hole diameters up to about 0.96 mm showed that larger hole diameters lead to smaller relative strain deviations for the same strain gage positions.

Thus the simulation with a drilled hole of 0.8 mm in diameter gives a limit of the maximal relative strain error that can occur by disregarding the chamfered hole geometry.

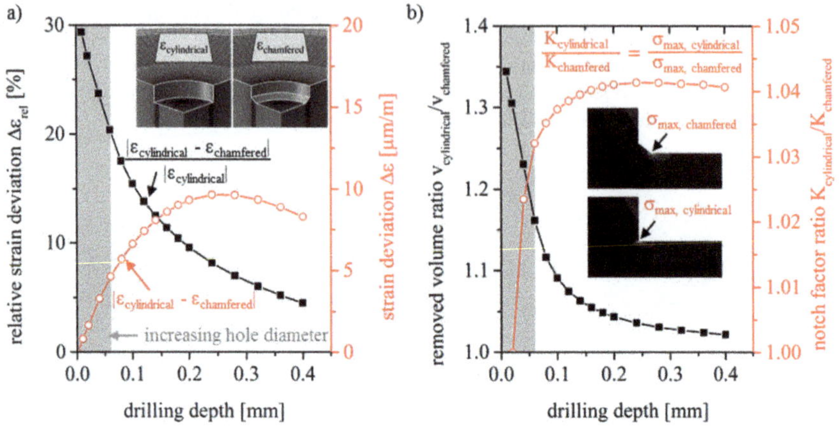

Fig. 3: strain deviation of $\varepsilon_{cylindrical}$ and $\varepsilon_{chamfered}$ (a), removed volume ratio and notch factor ratio (b).

The stress evaluation according to the differential approach of the chamfered hole-drilling simulation with calibration data considering a cylindrical hole is showing high stress deviations from the nominal stress (see Fig. 4). The large strain deviations in the first increments lead to a significant stress error of about 27 % in maximum. In contrast, stress evaluation with calibration data considering the chamfer geometry leads to small stress deviations of less than 2 %, due to numerical errors. Following the numerical results, the use of chamfer-considering calibration data is highly recommended for an accurate stress evaluation.

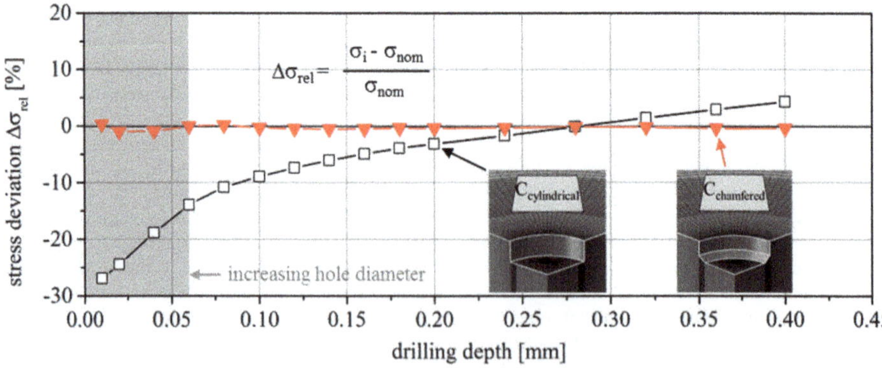

Fig 4: Stress evaluation of the chamfer-considering hole-drilling simulation with calibration data $C_{chamfered}$ and $C_{cylindrical}$.

Materials Research Proceedings **2** (2016) 473-478 doi: http://dx.doi.org/10.21741/9781945291173-80

Experimental application

Experimental setup. Experimental tests were carried out on a fine grained construction steel S690QL bar with length l = 160 mm, width w = 40 mm and thickness t = 6 mm. A stress relief annealing (1h, 600°C) was conducted prior to the hole-drilling experiments. XRD residual stress analyses according to the $\sin^2\psi$-method were carried out on the bottom and top surface after the stress relief treatment. Residual stress values were less than 13 MPa, hence the sample can be assumed to be stress′free′. The sample has been loaded by means of a four point bending device, with a distance of the inner supports of 50 mm. A uniaxial bending stress gradient from -400 MPa on the top surface to 400 MPa on the bottom surface over the bending height of 6 mm was applied. The load does not exceed more than about 58 % of the materials minimum yield strength of 690 MPa. For the elastic material properties a Young's modulus E = 210 GPa and a Poisson's ratio ν = 0.3 have been assumed. Three hole-drilling experiments have been carried out with measuring points in a distance of 10 mm to each other. After the three experiments each final hole diameter has been measured and was implemented exactly in the numeric model to calculate chamfer-considering calibration data. Finally the experimental strain data has been smoothed with cubic spline functions and the stress gradient has been evaluated by means of conventional calibration data $C_{cylindrical}$ and by means of chamfer-considering calibration data $C_{chamfered}$, respectively. For presentation of the results the mean values of the 3 single experiments are used.

Results and discussion. Fig. 5 shows the evaluated stress gradients as a function of the drilling depth. The mean stress value of the three drilling experiments for each single depth increment is shown. Standard deviation of the mean stress value is given by surrounding error lines. Whereas the conventional approach (calibration data for cylindrical hole geometry) leads to a stress deviation of approx. 70 MPa in the first increments, the calibration for considering the chamfer geometry leads to a stress value that is in good agreement with the nominal stress with deviation of less than 26 MPa. From a drilling depth of 0.08 mm on, the stress deviation of both evaluation methods is tolerable. Due to a slightly larger experimental hole diameter, the stress errors in the first increments do not exceed the large relative stress deviations of 30 % that appear in the simulation for a hole diameter of 0.8 mm (Fig. 4).

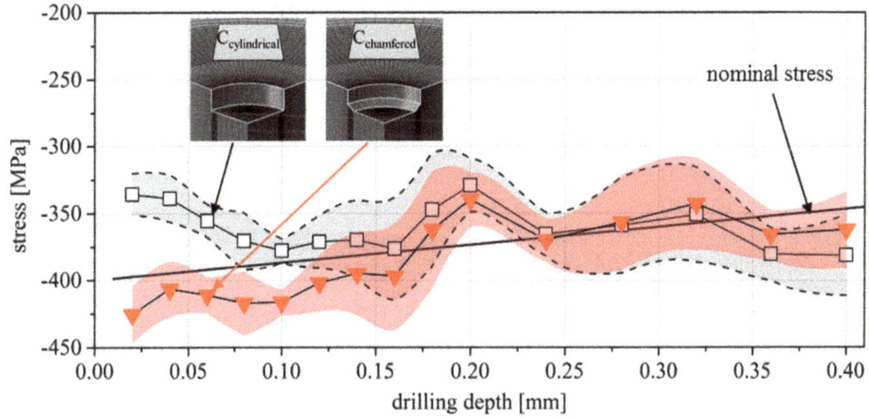

Fig. 5: *Stress evaluation of the experimental loading test with calibration data $C_{chamfered}$ and $C_{cylindrical}$, mean stress values with standard deviations for the 3 drilling experiments.*

It has to be taken into account that in experimental application strain measurement errors always appear due to scattering in the strain signal. However, for high strain relaxation in the first drilling

Materials Research Proceedings **2** (2016) 473-478 doi: http://dx.doi.org/10.21741/9781945291173-80

increments, the absolute strain measurement errors are negligible. The data clearly show that the discrepancies indicated through the numerical simulations are also obvious for the near surface residual stress data determined by the defined uniaxial loading experiment. Due to the significant stress deviations in the surface near area the use of chamfer-considering calibration data is recommended, if this experimental setup is used and accurate stress values in the first increments are needed.

Summary

In this study, a FE model was set up that considers realistic tool chamfers for evaluation of residual stresses by means of the incremental hole-drilling method. The two effects of increasing hole diameter and a reduced material removal are taken into account. Strain data of the simulated hole-drilling experiment for a hole diameter of 0.8 mm showed significant deviations to the conventional approach, based on ideal cylindrical hole geometry. Especially in the first increments, the increasing hole diameter due to chamfered end mill geometry is leading to large relative strain deviations. By use of conventional calibration data, erroneous stress data arise. Experimental hole-drilling tests have been carried out for a defined 4-point-bending loading test that confirmed the numerical findings. Chamfer-considering calibration data obtain a smaller stress deviation from nominal stress for the sensitive depth increments in the surface near area. Thus it can be concluded in the presence of large residual stresses in surface near areas, the chamfer-considering calibration is highly recommended to achieve accurate stress evaluation in particular for the first drilling increments.

Acknowledgement

The Graduate School 1483 “Process Chains in production: modeling, interactions and assessments of process zones” and the German research foundation (DFG) are gratefully acknowledged for funding.

References

[1] J. Matar, Determination of initial stresses by measuring the deformation around drilled holes, Transactions ASME Vol.56 No. 4 (1934) 249-254.

[2] M. Flaman, Brief investigation of induced drilling stresses in the center-hole method of residual-stress measurement, Experimental Mechanics 22(1) (1982) 26-30. http://dx.doi.org/10.1007/BF02325700

[3] A. Nau, B. Scholtes, Evaluation of the high-speed drilling technique for the incremental hole-drilling method, Experimental Mechanics 53 (2013) 531-542. http://dx.doi.org/10.1007/s11340-012-9641-1

[4] M. Scafidi, E. Valentini, B. Zuccarello, Effect of the hole-bottom fillet radius on the residual stress analysis by the hole drilling method, ICRS-8 The 8th International Conference on Residual Stress – Denver (2008) 263-270.

[5] T. Schwarz, H. Kockelmann, Die Bohrlochmethode – ein für viele Anwendungsbereiche optimales Verfahren zur experimentellen Ermittlung von Eigenspannungen, Messtechnische Briefe 29 (1993) 324-333.

[6] ASTM, Determining residual stresses by the hole-drilling strain-gage method, ASTM Standard Test Method E837-13a (2013)

Materials Research Proceedings **2** (2016) 479-484 doi: http://dx.doi.org/10.21741/9781945291173-81

Full Stress Tensor Determination during Phase Transformation of a Metal Matrix Composite by in situ High Energy X-Ray Diffraction and Micromechanical Simulations

Guillaume Geandier[a,1,*], Lilian Vautrot[b,1], Benoît Denand[c,1], Moukrane Dehmas[d,1], Elisabeth Aeby-Gautiera[e,1], Julien Teixeira[f,1]. Sabine Denis[g,2]

[1]Institut Jean Lamour, CNRS - Université de Lorraine, Parc de Saurupt - CS 50840, 54011 Nancy Cedex – France

[2]LabEx DAMAS - Université de Lorraine, Parc de Saurupt - CS 50840, 54011 Nancy Cedex – France

[a]guillaume.geandier@univ-lorraine.fr, [b]lilian.vautrot@univ-lorraine.fr, [c]benoit.denand@univ-lorraine.fr, [d]moukrane.dehmas@univ-lorraine.fr, [e]Elisabeth.Gautier@univ-lorraine.fr, [f]julien.teixeira@univ-lorraine.fr, [g]sabine.denis@univ-lorraine.fr

Keywords: Metal Matrix Composite, in situ X-Ray Diffraction, Stresses, Phase Transformation, Microstructure

Abstract. In situ high energy X-ray diffraction using a synchrotron source performed on a steel metal matrix composite reinforced by TiC allows following the evolutions during cooling of internal stresses thanks to the development of a new original experimental device (a transportable radiation furnace with a controlled rotation of the specimen). Results from experiments as well as from finite element micromechanical simulations allow to understand the stress genesis in the composite.

Introduction

Mass reduction (for example in transportation applications where reduction of fuel consumption and pollution are aimed) can be obtained, by using new lighter materials with at least the same mechanical properties as the former ones. Metal Matrix Composites (MMC) reinforced by ceramic particles allow reaching this goal. In our study, steel matrix composite reinforced by TiC particles obtained by powder metallurgy (mixture of 75 % steel powder and 25 % TiC powder) allows reducing the density (7 g/cm^3 for this composite instead of 7.8 g/cm^3 for steel alone i.e. a decrease of 11.4 % in mass). Final properties of MMC depend on the chemical composition, on the nature of the interfaces, on the microstructure of the matrix and on the stresses in the reinforcements and in the matrix. These stresses are generated during the heat treatment and result from the differences in the coefficient of thermal expansion between matrix and reinforcements [1-3] and also from the phase transformations of the matrix that occur during cooling and induce volume changes. The residual stress levels and distributions are a key factor for the final properties of the MMC [4]. In a previous study [5], the evolution of the matrix and the reinforcements were analysed during the heat treatment using in-situ high energy X ray diffraction focusing on the structural aspects. In this paper, we focus on the internal stress analysis.

Sample

A metal matrix composite has been elaborated using powder metallurgy by Mecachrome. 75 % (vol.) steel powder and 25 % TiC powder were milled and consolidated by hot isostatic pressing (HIP). After hot isostatic pressing, the microstructure of the steel matrix composite (see fig. 1) is non-homogeneous and presents two different areas: a steel area (pearlitic) without TiC particles called unreinforced area (lighter area) and a darker area being a mixture of TiC particles and steel. Details on the microstructures and chemical compositions can be found in [5].

Materials Research Proceedings **2** (2016) 479-484 doi: http://dx.doi.org/10.21741/9781945291173-81

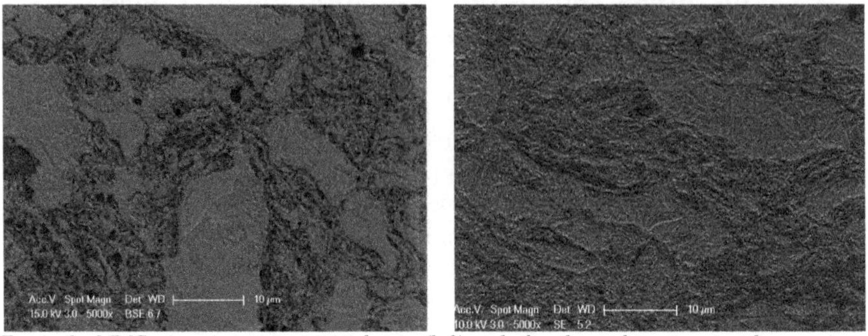

Figure 1 : MMC microstructure at initial state (left) and after thermal treatment with a martensitic microstructure (right) in the matrix

High-energy X-ray diffraction.

Set-up : The high energy X-ray diffraction (XRD) experiments were performed at the European Synchrotron Radiation Facility (ESRF, Grenoble, France) on the ID15B beam line. The in situ measurements were conducted with a monochromatic beam of 87 keV. The high energy beam allowed to analyse a large volume of the sample (due to the low absorption of the sample) thus being representative of the bulk and lessen the surface effect. The transmitted signal is collected by a large area 2D detector that records the whole Debye-Scherrer rings. The important flux gives a high quality diffraction signal quickly, thus XRD frames can be recorded at high rate (up to 10 fps). By this method, we can follow Debye-Scherrer rings evolution during a thermal treatment. More details on the setup can be found in [6].

For our experiments, the sample was a cylinder of 3mm diameter and 30mm length and the beam size was 0.4 x 0.4 mm^2. The sample was heated up to 900°C thanks to a radiation furnace specially developed to perform~~ed~~ a thermal treatment with a controlled rotation of the sample. The temperature was held during 5 minutes. Then, the sample was air cooled (at about 5 °C/s). XRD frames were recorded every 0,1 s during cooling. Using the sample rotation during in situ cooling and a fast acquisition rate allows collecting the data to determine the full stress tensor as we obtain all necessary orientations of the sample by combining the rotation in plane of the sample and 2D diffraction in the perpendicular plane.

Phase analysis : As the sample presents small grains and no texture during the whole thermal treatment, a Rietveld analysis has been conducted to extract phase fractions and mean cell parameters from the XRD frames using Fullprof software [7-8]. Data have been corrected and reduced to (2θ, intensity) patterns using fit2D software [9].

Stresses analysis : For stress analysis, 2D images caking has been made with sectors of 1° along the rings using fit2D software. Corrections for taking into account the variations of frame center during the experiments (due to a lack of stability of the synchrotron beam) [10] as well as the variations of the sample position during rotation have been done. Thus, we obtain 360 diffractograms (2θ, intensity) from each image. In order to extract the stresses in each phase from these data, we have determined the 2θ position of one peak in each phase: steel-α'(200), steel-γ(200) and TiC(220), as they do not overlap with other peaks (fig. 2). Diffraction peaks are approximated by a Pearson VII function. In order to apply the "$\sin^2\Psi$" method [11-12], we have converted our configuration to classical (Φ, Ψ) configuration [6].

Materials Research Proceedings **2** (2016) 479-484 doi: http://dx.doi.org/10.21741/9781945291173-81

By selecting the rotation position corresponding to a specific orientation of the sample, we can extract the σ_{11}-σ_{33} and σ_{22}-σ_{33} differences from positions (0°,180°) and (90°,270°) versus temperature. An automatic procedure (about 10 million peaks to be analysed per phase, 150GB raw data) has been developed.

To obtain the full strain and stress tensor, we need the stress free parameters of the phases a_0 and their evolutions with temperature. For steel-γ and TiC, we have used the value of the cell parameter at high temperature assuming that stresses are fully relaxed at 900°C and for martensite values from literature [13]. Evolution of stress free cell parameter versus temperature has been calculated using the thermal expansion coefficient of each phase, extracted from experimental studies [5] or literature for steel [14] and TiC [15].

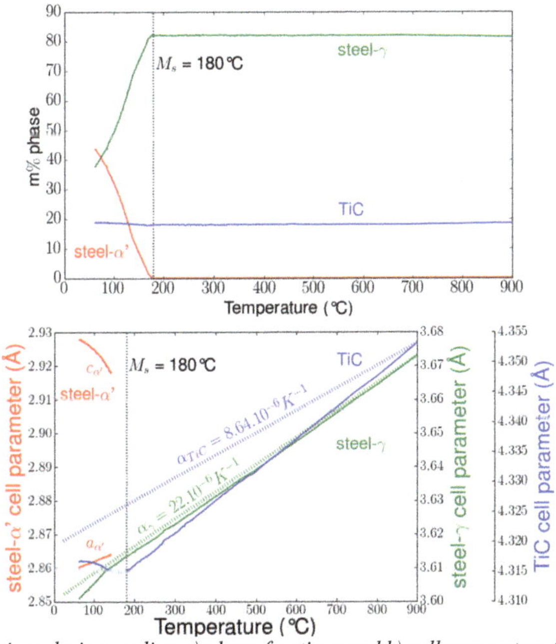

Figure 2 : Evolutions during cooling a) phase fractions and b) cell parameters (thermal expansion coefficients of TiC and austenite are shown too)

Results

Kinetics of phase transformation: on figure 2, we present the evolutions of the phase fractions and the mean cell parameters during the cooling from 900 °C to 50 °C, leading to the formation of martensite in the matrix (see fig. 1). At the beginning of the cooling, only austenite (γ) and TiC are present with 81m% γ and 19m% TiC. The martensitic transformation (α' tetragonal structure) begins at 180°C. During the cooling, the amount of TiC remains constant and the amount of austenite decreases quickly during the martensitic transformation.

Mean cell parameters evolutions: from the beginning of cooling and until the beginning of martensitic transformation, the cell parameters of the different phases decrease continuously without changes of slope (fig. 2b). From previous analysis [16] of the apparent CTE (coefficient of thermal expansion), it was concluded that during cooling, thermal stresses are generated in the phases. TiC is under a mean compression state since the CTE of austenite is about 3 times larger than the one of TiC. The mean stress state in austenite seems very small. As the martensitic transformation occurs, TiC cell parameter increases indicating a decrease of the mean compression in TiC. As the martensite content reaches about 15%, austenite cell parameter decreases indicating an increase in compression. We can observe an increase of the tetragonality (c/a ratio) of the martensite (increase of the c parameter whereas a parameter decreases linearly).

Stress evolutions: using $\sin^2\psi$ method, we are able to determine all the components of the stress tensors in the different phases during cooling. First, it comes out that the mean stress states remain hydrostatic (no shear stresses) in all the phases during cooling even during the martensitic transformation. Fig. 3 presents the stress evolutions during cooling from 900°C to 50°C.

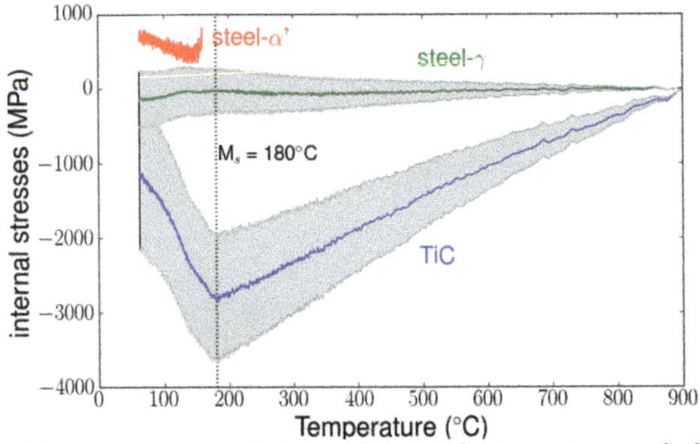

Figure 3 : internal stresses evolutions during cooling. (As the stress states are hydrostatic, only one component of the stress tensor is presented)

We can remark the high level of compressive stress in TiC increasing from the high temperature until the appearance of martensite. Stress level in austenite is quite low. The martensitic transformation induces a large decrease of the compressive stress in TiC and small compressive stresses in austenite. For martensite, firstly the tensile stress decreases and then increases as temperature decreases.

We can notice that the calculated macroscopic stress for the composite, is not zero even before the phase transformation. It should be pointed out that stress values are highly dependent on the input parameters of each phase mainly a_0 and CTE. On figure 3, the grey envelope presents the variations on the stress levels for austenite and TiC including experimental uncertainties on the thermal expansion coefficient of both phases. (For martensite, the stress variations are not shown as they will overlap the stress scattering). It comes out that variations on the stress levels are quite large and a more equilibrated macroscopic stress state could be reached by adjusting the stress free cell parameters and their evolutions with temperature.

Micromechanical model

In order to better understand the internal stress evolution during cooling of the MMC, a 3D micromechanical finite element model has been used (software ZeBuLoN [17]). We have introduced a thermo-elasto-visco-plastic behaviour law for the steel matrix including the strains associated with the phase transformation (volumic variation and transformation plasticity). All the thermomechanical data for the steel are taken from previous studies. For the reinforcement, a thermo-elastic behaviour law is taken into account based on literature data [14]. Martensite phase fraction is calculated by Koistinen-Marburger relation [18] with parameters determined from experimental data.

Calculations have been made on a 3D simplified microstructure: a periodic distribution of spherical TiC particles (representing 25vol%) embedded in a steel matrix.

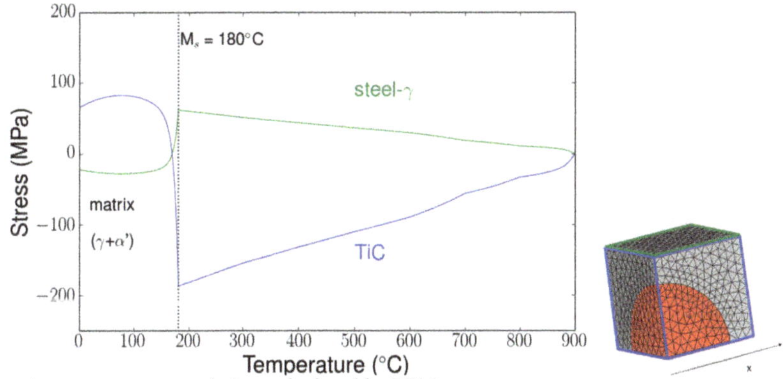

Figure 4 : mean stresses evolution calculated by FEM.

The calculated results allow to analyse the stress and strain gradients in the reinforcement and in the matrix during cooling as well as the permanent deformations in the matrix. Here, we show only the mean stresses in the particles and in the matrix (figure 4) that are hydrostatic. Their evolutions are in accordance with the experimental results; in particular, the martensitic transformation of the matrix (with its volumic expansion) leads to a large stress relaxation in the TiC particles and to a lower one in the matrix. But the calculated levels differ largely from the experimental values in particular for TiC. The thermo-elasto-visco-plastic behaviour of the matrix seems to overestimate the stress relaxation by plastic deformation. As a matter of fact, if we do not consider plastic deformation of the matrix, considering an elastic behaviour, we obtain calculated mean stress values for TiC closer to experimental ones. Moreover, from the calculations, the cumulated equivalent plastic deformation reaches a value of 3% in the matrix whereas experimentally we due not observe a change on the full width at half maximum (FWHM) of the austenite peaks during cooling.

Thus, the question arises concerning the strain accommodation between the matrix and particles in the MMC during cooling. One important aspect is that the real microstructure is much more complex that the simulated one. Indeed, as mentioned before, the real microstructure is very heterogeneous (see fig. 1) with a non reinforced area (no particles) and a reinforced area where the amount of TiC can reach 50vol%. Calculations are on the way to take the real microstructure into account.

Concluding remarks

We have developed an original experimental device and a methodology for robust and fast stress analysis starting from 2D images (i.e. Debye-Scherrer rings) obtained by in situ synchrotron X ray diffraction experiments during heat treatment. From the present experiments performed during heat treatment of a steel matrix composite reinforced by TiC particles, we have determined the evolutions of the full stress tensor in the different phases austenite, martensite and TiC. The role of the martensitic transformation of the matrix on the stress states in the particles has been clearly evidenced. In addition, a 3D micromechanical approach has allowed to better interpret the experimental evolutions of the mean internal stress states in the different phases during cooling. Further investigations are necessary to understand the discrepancies between experimental and calculated stress levels.

Materials Research Proceedings **2** (2016) 479-484 doi: http://dx.doi.org/10.21741/9781945291173-81

Aknowledgments
The authors gratefully acknowledge Mecachrome for supplying the materials, the Direction Générale des Entreprises (DGE) for financial support in the AMETIS program and the European Synchrotron Radiation Facility (ESRF) for provision of beamtime at beamline ID15B. This work was supported by the French State through the program "Investment in the future" operated by the National Research Agency (ANR) and referenced by ANR-11-LABX-0008-01 (LabEx DAMAS).

References
[1] Meixner M., Fitzpatrick M. and Reimers W.: Composite Science and Technology, Vol. 71 (2011), p. 167-176. http://dx.doi.org/10.1016/j.compscitech.2010.11.003

[2] Lee R., Chen G. and Hang B.: Composites, Vol. 26 (1995), p. 425-429. http://dx.doi.org/10.1016/0010-4361(95)90915-M

[3] Pagounis E., Haimi E., Pietikaïnen J., Talvitie M., Vahvaselkä S. and Lindroos V.K.: Scripta Materialia, Vol. 34 (1996), p. 407-413. http://dx.doi.org/10.1016/S0956-716X(95)00529-5

[4] Wu Y., Wang M., Chen Z., Ma N. and Wang H.: Materials Science, Vol. 48 (2013) p. 2928-2933. http://dx.doi.org/10.1007/s10853-012-6933-x

[5] Mourot M., Courleux A., Dehmas M., Gautier E., Geandier G., Dezellus O., Viala J.C., Martin O., Karnatak N. and Danoix F.: Solid State Phenomena, Vol. 172-174 (2013), p. 747-752. http://dx.doi.org/10.4028/www.scientific.net/SSP.172-174.747

[6] Vautrot, L.; Geandier, G.; Mourot, M.; Dehmas, M.; Aeby-Gautier, E.; Denand, B. & Denis, S. - François, M.; Montay, G.; Panicaud, B.; Retraint, D. & Rouhard, E. (Eds.), Trans Tech Publications, Switzerland, 2014, 996, 944-950

[7] Rietveld H.M.A. : Journal of Applied Crystallography, Vol. 2 (1969), p. 65-71. http://dx.doi.org/10.1107/S0021889869006558

[8] Rodriguez-Carvajal J.: Physica B, Vol. 192 (1993), p. 55-59. http://dx.doi.org/10.1016/0921-4526(93)90108-I

[9] Hammersley A.P., Svensson S.O., Thompson A., Graafsma H., Kvick A. and Moy J.P.: Review of Scientific Instrument, Vol. 66 (1995), p. 2729-2733. http://dx.doi.org/10.1063/1.1145618

[10] Le Bourlot, C., Caractérisation de l'hétérogénéité du champ des déformations élastiques dans les matériaux polycristallins par diffraction des rayons X et des neutrons - application à un acier duplex - PhD thesis (2012) *Université Paris 13, France*

[11] Noyan I.C. and Cohen J.B.: Measurement by diffraction and interpretation in residual stress, Springer-Verlag, (1987)

[12] Hauk V., Structural and reisudual stress analysis by nondestructive methods, ELSEVIER, (2006)

[13] Roberts C.S.: Journal of Metals, 5 (1953) 203-204

[14] Vidal G., Lescop P. and Raymondin L.: Revue de métallurgie 48, Vol. 11 (1951), p. 864 874

[15] Wall J., Choo H., Tiegs T. N. and Liaw P. K.: Materials Science and Engineering A, Vol. 421 (2006), p. 40-45. http://dx.doi.org/10.1016/j.msea.2005.10.002

[16] Geandier G., Dehmas M., Mourot M., Aeby Gautier E., Denis S., Martin O. and Karnatak N.: Materials Science Forum, Vols. 768-769 (2014), p. 313-320. http://dx.doi.org/10.4028/www.scientific.net/MSF.768-769.313

[17] Besson J., Leriche R., Foerch R. And Cailletaud G., Revue Européenne des Éléments, 7:5 (1998) 567-588

[18] Koistinen D.P., Marburger R.E., Acta Metallurgica, 7 (1959) 59-60. http://dx.doi.org/10.1016/0001-6160(59)90170-1

Residual Stresses 2016: ICRS-10 Materials Research Forum LLC
Materials Research Proceedings **2** (2016) 485-490 doi: http://dx.doi.org/10.21741/9781945291173-82

In-vitro Investigation of Air Plasma-Sprayed Hydroxyapatite Coatings by Diffraction Techniques

Ntsoane P. Tshepo[1, 2, a*], Theron Chris[1, b], Venter Andrew[2, c], Topic Mira[3, d], Härting Margit[4, e], Heimann Robert[5, f]

[1]Department of Physics, University of Pretoria, Pretoria, South Africa

[2]Research and Development Division, NECSA SOC Limited, Pretoria, South Africa

[3]Materials Research Department, iThemba LABS/National Research Foundation, Cape Town, South Africa

[4]Department of Physics, University of Cape Town, Cape Town, South Africa

[5]Am Stadtpark 2A, D-02826 Görlitz, Germany

[a]tshepo.ntsoane@necsa.co.za, [b]chris.theron@up.ac.za, [c]andrew.venter@necsa.co.za, [d]mtopic@tlabs.ac.za, [e]margit.harting@uct.ac.za, [f]robert.heimann@ocean-gate.de

Keywords: Plasma-Sprayed Hydroxyapatite Coatings, High-Energy Diffraction, Quantitative Phase Analysis, Depth-Resolved Residual Stress

Abstract. The influence exposure to simulated body fluid (SBF) has on plasma-sprayed hydroxyapatite (HAp) coatings of medical-grade Ti6Al4V rods was investigated by quantifying the depth dependence of the phase composition and residual stress through the coating thickness using diffraction techniques. Chemical phase identification showed HAp existing together with its thermal decomposition products, tetra-calcium phosphate (TTCP), tri-calcium phosphate (TCP) and calcium oxide. With depth, the HAp content decreases with a corresponding increase in TTCP. The near surface stress condition comprised ~50 ± 10 MPa hoop stress with the radial stress being close to zero. With depth the hoop stress decreases linearly to ~-50 ± 25 MPa at the substrate interface, whilst the radial stress increases with depth. Upon exposure to SBF, the coating composition reveals an increase in HAp from ~80.0 ± 0.5 to ~86.0 ± 0.5 wt%, accompanied by a decrease of TTCP from ~10 ± 2 to ~6 ± 2 wt% wt%. A change in stress state occurred within the first day of incubation; where after, with further exposure time the stress state converted back to values similar to that of the as-sprayed condition.

Introduction

The second-generation biomaterial hydroxyapatite (HAp), owing to its similarity to the inorganic component of bone, has been extensively studied as a candidate material in biomedical applications. Applications range from filling bone cavities [1] and artificial eye components [2] to coatings for hip endoprostheses and dental root implants for improved biological fixation [3]. However, the poor mechanical properties of the material limit its utilisation in fully load-bearing applications. In such cases, the material is applied as a coating on metallic substrates such as Ti, Ti alloys and CoCrMo, where the excellent mechanical properties of the metal are combined with the osseoconductive ability of the coating [4]. With the plethora of coating techniques available for deposition [5], thermal spray remains the method of choice on industrial scale [6]. It is known that the interaction of the high temperature droplets formed in the plasma, and subsequent rapid cooling on interaction with the cold substrates, leads to thermal decomposition. This causes the introduction of undesirable thermal decomposition products [7], such as TCP, TTCP, and sometimes calcium oxide. In addition it leads to a reduction in crystallinity [8]. These phases are known to be susceptible to dissolution in SBF [9]. This, together with stresses generated by the differential thermal mismatch (CTE) and quenching of

Materials Research Proceedings **2** (2016) 485-490 doi: http://dx.doi.org/10.21741/9781945291173-82

the droplets, may compromise the mechanical stability and integrity of the coating. Notwithstanding extensive investigations of the effect of incubation of HAp coatings in SBF [10], the bulk of the work focused on the near-surface region that has contact with living tissue. The present study deals with depth-resolved investigations of the effect that exposure to SBF has on HAp coatings.

Materials and Methods

Sample preparation. Hydroxyapatite powder (CAPTAL 90, batch P215, Plasma Biotal Limited, Tideswell, Derbyshire, UK), 120 ± 20 μm, was plasma-sprayed onto 7 mm diameter medical grade Ti6Al4V alloy rods supplied by Biomaterials Limited, North Yorkshire, UK. Details on the samples, HAp deposition procedure and spray parameters can be found in [11]. Coating thicknesses ranged from 150 to 180 μm. Subsequent to spraying, the samples were incubated for 1 and 7 days in a revised simulated body fluid (rSBF) based on Kokubo's formulation [12] to mimic the physiological environment with the solution temperature and pH kept at 36 °C and 7.4 respectively during the incubation period. The rSBF solution had an ionic concentration similar to that of human blood plasma, but free of non-collagenous proteins and enzymes. After incubation, disks approximately 5 mm in thickness were cut from the central part of the coated rod for investigation. The samples are shown in Fig. 1. Factors considered in determining the optimum disk thicknesses included minimization of the synchrotron beam attenuation, which, based on the energy of the beam and the mass absorption of the coating material was minimally influenced by the disk thickness, adequate diffracted intensities for good counting statistics, as well as limiting stress relaxation due to the cutting. Notwithstanding that the cutting would lead to relaxation of the axial stresses, due to the circular geometry, the hoop and radial stress components should not be influenced.

Through-thickness characterisation. High–energy synchrotron radiation, 70-130 keV, generated at the 3rd generation Advanced Photon Source's X-ray Operation and Research 6-ID-D beamline at Argonne National Laboratory (ANL), USA was used for the depth-resolved investigations of the phase composition and residual stresses. The experimental details and measurement procedure are reported in [11]. Measurements were done in transmission mode through the cut disks using an incident beam size of 0.035(V) x 0.400 (H) mm^2. The diffracted beams were recorded with an area detector located ~ 1m downstream. Supplementary near-surface investigations were carried out in reflection geometry, using a Bruker D8 Advance and a Discover instrument equipped with a Cu-anode that gave 8 keV energy X-rays that rendered ~0.015 mm penetration into HAp.

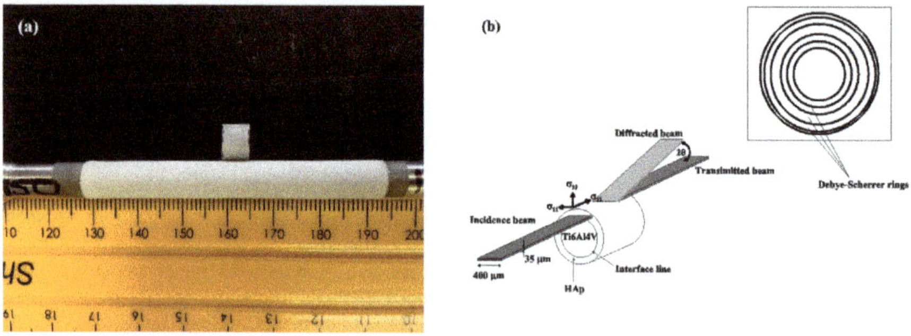

Figure 1: ***a)*** *Geometry and size of the HAp-coated (white) Ti6Al4V rod. Also shown is the cut disk; b) Schematic diagram illustrating the measurement geometry of the synchrotron beam in relation to the sample and the principal stress directions. This geometry in conjunction with an area detector enabled simultaneous measurement of the hoop (= σ_{11}) and radial (= σ_{33}) stress components at each depth.*

Materials Research Proceedings **2** (2016) 485-490 doi: http://dx.doi.org/10.21741/9781945291173-82

Results and discussion

Chemical phase analysis. Figure 2 shows diffraction patterns, displaced vertically for clarity, taken at different depths through the coating thickness from the as-sprayed and incubated samples. The phase determination shows the presence of the thermal decomposition products of HAp caused by the plasma temperature, being present throughout the coating thickness. The two top patterns, taken at the coating-substrate interface region, primarily show Ti6Al4V, indicate that the measurements extended up to the substrate. The CaO phase dissolved within the first week of incubation.

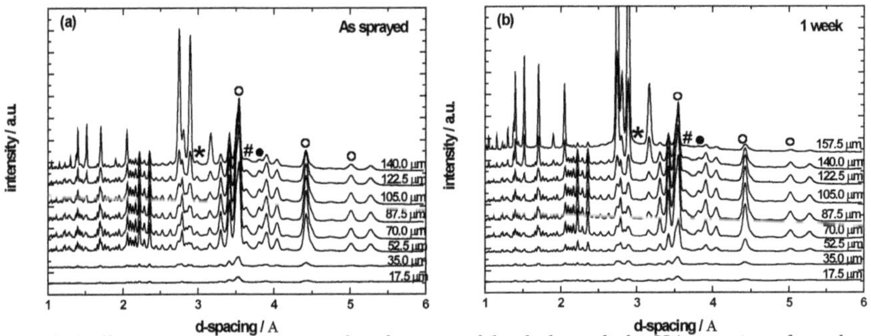

***Figure 2**: Diffraction patterns measured as function of depth through the HAp coatings from the coating surface (bottom pattern) to the substrate (top pattern) for: (a) As-sprayed and (b) Incubated for 1 week. The patterns show the presence of HAp (o), tetracalcium phosphate (#), tricalcium phosphate (•) and CaO (*). The diffraction patterns are displaced vertically for comparison purposes and respectively represent depth steps of 17.5 µm*

Figure 3 shows the depth dependence of the primary phase, HAp, and the main thermal decomposition product, TTCP, for the as-sprayed and incubated samples. Also shown are their dependences as function of incubation times. Due to weak intensities from the coating, associated to the partial illumination of the coating, the first two diffraction patterns in Fig. 2 have been excluded in all analyses.

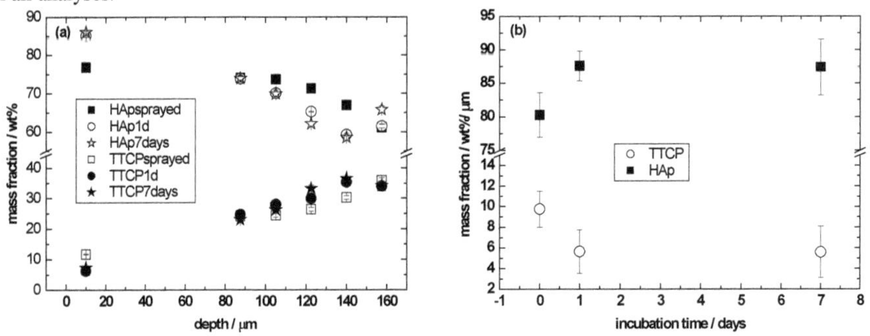

***Figure 3:** Variation of the chemical phase composition of the major phases (a) with depth and (b) with incubation time are shown for the as-sprayed and incubated samples, respectively.*

Materials Research Proceedings **2** (2016) 485-490 doi: http://dx.doi.org/10.21741/9781945291173-82

The HAp content of the as-sprayed coating (solid squares) systematically decrease with depth from 76 wt% to ~60 wt%. A similar trend is observed for the incubated samples where the HAp content decrease from ~86 to 62 wt% (open circles, open stars). The observed relative increase in HAp content with incubation time (Fig. 3b) can be attributed to the dissolution of thermal decomposition products, leaving the more stable HAp. Correspondingly, in the as-sprayed coating the TTCP phase (open squares) increases with depth from ~7 wt% to 35 wt%, whereas the incubated ones (solid dots, solid stars) increased from ~11 wt% to ~35 wt%. A cross-over point of the as-sprayed and incubated slopes occurs around mid-coating thickness. Although the preferential dissolution of TTCP in simulated body fluid has been widely reported [13], the results of this study indicate that this does not happen uniformly across the coating thickness. This implies that the solution only penetrates the surface region of the coating. The variation of the surface mass fraction with incubation time shown in Fig. 3, (right) reveals that the two phases show changes within the first day of incubation, with HAp increasing sharply from 80 wt% levelling off at ~86 wt% during further incubation; the opposite trend is observed for TTCP, decreasing from ~10 to ~6 wt% with time.

Residual stress profile

The HAp (213) diffraction peak was used to evaluate the residual strain. Assuming the Kroner-Eshelby grain interaction model, the X-ray elastic constants (XECs) $S_1(213) = -2.48 \cdot 10^{-6}$ MPa^{-1} and $\frac{1}{2}S_2(213) = 11.5 \cdot 10^{-6}$ MPa^{-1} [14] were used for the stress determination. Since only one azimuth orientation was measured, the stress components σ_{11}(= hoop), and σ_{33} (= radial) could be determined from the respectively measured d-values; the axial component, σ_{22} couldn't be determined with the measurement geometry. The stress-free reference value for strain determination was obtained from measurement of the starting powder. The d-spacing in the starting powder and ground coating flakes were within ± 0.00005 Å that correlates to ± 5.00 MPa stress uncertainty. The quantity (σ_{11}-σ_{33}), called the stress amplitude, is a useful parameter to consider since it is independent of possible sample misalignment contributions. The observed stress distribution comprises the accumulated effects of droplet quenching (also known as intrinsic stress), differential thermal mismatch between the coating and substrate, and stresses arising from possible volume changes due to phase transformations. Figure 4 summarises the depth distribution of the residual stress components and associated stress amplitude.

The results show that the hoop stress in the as-sprayed condition does not exceed 50 ± 10 MPa on the surface, whereas the radial stress is close to zero. The hoop stress steadily decreases with depth crossing the zero stress level around the coating mid-point to become ~ -50 ± 25 MPa compressive at the substrate interface. The radial stress steadily becomes tensile with depth reaching a maximum ~ 40 MPa at 157.5μm. Based on the intensities of diffraction patterns in Fig2, the measurements at 17.5, and 35 microns were carried out with only partially submerged beam hence the stress values at these depths are considered artefacts and therefore excluded; the same effect is expected at 175 and possibly at 157.5 microns. This effect in addition to possible relaxation and or overlapping substrate peak can be attributed to the observed large scatter at the interface. The observed radial stress distribution is consistent with previous synchrotron results and FEM analyses [15] which showed in general small tensile radial stresses near the surface region that increased with depth. The distribution can be attributed to various factors, including thermal gradient effects across the coating thickness during deposition [16], quenching stresses and effects associated with different inter-laminar bonding, intra-laminar micro-cracking, and pores.

Upon incubation, a change in stress state is observed in the hoop component. It becomes compressive on the surface and decreases with depth, i.e. an opposite trend to that of the as-coated condition. With further incubation, it reverts back to a trend similar to the as-sprayed condition. The radial component remains unaffected except at the interface region. The observed change in stress state during incubation can be attributed to the volume change based on the dissolution of thermal

Materials Research Proceedings **2** (2016) 485-490 doi: http://dx.doi.org/10.21741/9781945291173-82

decomposition products and onset of formation of an apatite-like octocalcium phosphate phase, with the former resulting from the existence of an ionic concentration gradient and the latter due to precipitation that forms due to supersaturation of the simulated body fluid with Ca2+ and HPO42- ions, respectively. Although Nimkerdphol et al. [17] observed an opposite stress state, a similar behaviour upon incubation was observed.

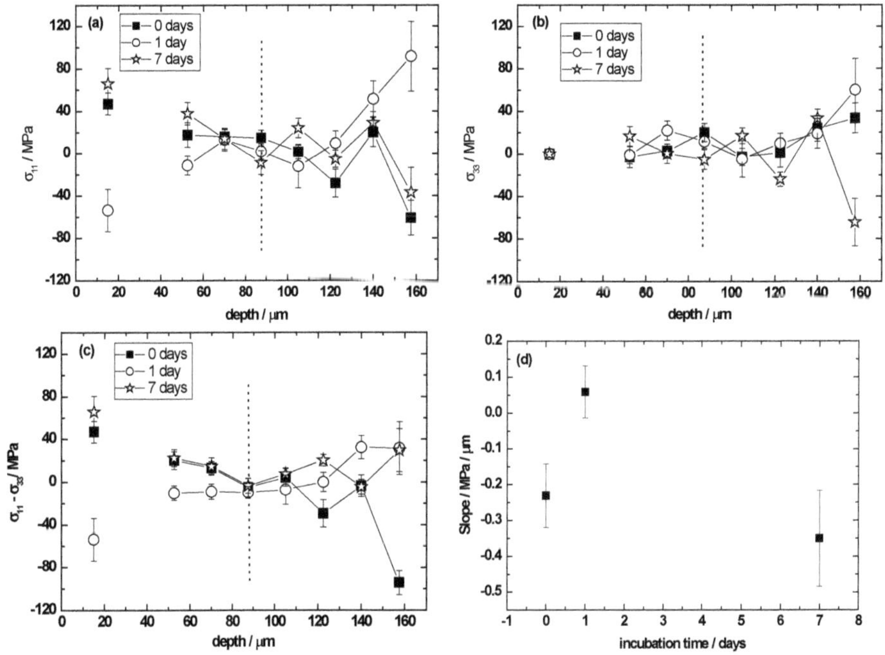

Figure 4: *Behaviour of the principal residual stress components with depth and incubation time, respectively for: (a) hoop, σ_{11}; (b) radial, σ_{33}; (c) σ_{11}-σ_{33}; and (d) average through-thickness stress gradient with incubation time.*

Conclusions

Depth-resolved in-vitro studies employing conventional X-ray and synchrotron radiation were carried out on air plasma-sprayed HAp coatings that revealed the following:

Phase identification results showed thermal decomposition products to be present throughout the as-sprayed condition with CaO dissolving completely within 7 days of incubation. Quantitative Rietveld analyses showed a near linear decrease of HAp with depth from ~76 ± 1 to ~60 ± 1 wt% with a corresponding linear increase of TTCP from ~7 ± 2 to ~35 ± 2 wt% for as-sprayed coatings. A similar trend was observed for samples incubated between 1 and 7 days in that HAp decreased with depth from ~86 ± 1 to ~62 ± 1 wt% and TTCP increased from ~11 ± 2 to ~35± 2 wt%. The coating composition as a function of immersion time reveals a steep increase of HAp from ~80 ± 0.5 to a constant ~86 ± 0.5 wt%, accompanied by a decrease of TTCP from ~10 ± 2 to ~6 ± 2 wt%.

The as-sprayed condition revealed that the hoop stress varied linearly from a tensile value at the surface to compressive at the substrate interface. The radial component is low at the surface and steadily increases to a maximum tensile value of ~ 30 MPa at 157 μm. The observed change in stress state, of the hoop component, upon immersion into simulated body fluid can be attributed to both dissolution and onset of bone-like apatite precipitate, octacalcium phosphate.

Acknowledgements
Use of the 6-ID-D beamline at Argonne National Laboratory (ANL) was supported by the U.S. DOE Contract No. DE-AC02-06CH11357. The authors are indebted to Mrs Margitta Hengst, Department of Mineralogy, Technische Universität Bergakademie Freiberg, Germany for sample preparation, Drs Douglas Robinson and Jonathan Almer, ANL/APS for assisting with the experimental set-up, data acquisition and analysis. The research was sponsored by the German Federal Ministry of Education and Research and the South African National Research Foundation within the research project "Characterisation and determination of residual stress in bioactive coatings and layered structure" (Project code 39.6.G0B.6.A).

References
[1] T. Yamamoto, T. Onga, T. Marui, K. Mizuno, J. Bone Joint Surg. (Br.), 82B (2000) 1117. http://dx.doi.org/10.1302/0301-620X.82B8.11194

[2] T. Colen, Ophthalmology, 107 (2000) 1889. http://dx.doi.org/10.1016/S0161-6420(00)00348-1

[3] R.G.T. Geesink, K. de Groot, C.P.A.T. Klein, Clin. Orthop. Relat. Res., 225 (1987) 147.

[4] K. de Groot, R. Geesink, C.P.A.T. Klein, P. Serekian, J. Biomed. Mater. Res., 21 (1987) 1375. http://dx.doi.org/10.1002/jbm.820211203

[5] R.B. Heimann, H.D. Lehmann, Bioceramic Coatings for Medical Implants. Trends and Techniques. Wiley-VCH, Weinheim, Germany. 2015. pp 467.

[6] M. Topić, T. Ntsoane, T. Hüttel, R.B. Heimann, Surf. Coat. Technol.,201(6) (2006) 3633. http://dx.doi.org/10.1016/j.surfcoat.2006.08.139

[7] Y.C. Yang, B.C. Wang, E. Chang, J.D. Wu, J. Mat. Sci.: Mater. Med., 6 (1995) 249. http://dx.doi.org/10.1007/BF00120267

[8] K.A. Gross, C.C. Berndt, H. Herman, J. Biomed. Mater. Res., 39 (1998) 407. http://dx.doi.org/10.1002/(SICI)1097-4636(19980305)39:3<407::AID-JBM9>3.0.CO;2-N

[9] P. Ducheyne, S. Radin, L. King, J. Biomed. Mater. Res., 27 (1993) 25. http://dx.doi.org/10.1002/jbm.820270105

[10] S. R. Radin, P. Ducheyne, J. Materials Science: Materials in Medicine, 3, 1992, 33. http://dx.doi.org/10.1007/BF00702942

[11] T.P. Ntsoane, M. Topic, M. Härting, R.B. Heimann, C. Theron, Surf. Coating Technol, 294 (2016) 153. http://dx.doi.org/10.1016/j.surfcoat.2016.03.045

[12] T. Kokubo, H. Kushitani, S. Sakka, T. Kitsugi, T. Yamamuro, J. Biomed. Mater. Res., 24 (1990) 721. http://dx.doi.org/10.1002/jbm.820240607

[13] L. Sun, C.C. Berndt, K.A. Khor, H.N. Cheang, K.A. Gross, J. Biomed. Mater. Res., 62(2) (2002) 228. http://dx.doi.org/10.1002/jbm.10315

[14] T.N. Gardner, J.C. Elliot, Z. Sklar, G.A.D. Briggs, J. Biomech., 25 (1992) 1265. http://dx.doi.org/10.1016/0021-9290(92)90282-6

[15] A. Manescu, G. Albertini, P. Fogarassy, A. Lodino, N. Markocsan, J. Neutron Research, 12 (2004) 147. http://dx.doi.org/10.1080/10238160410001734603

[16] L. Pawlowski, M. Vardelle, P. Fauchais, Thin Solid Films, 94 (1982) 307. http://dx.doi.org/10.1016/0040-6090(82)90492-8

[17] A.R. Nimkerdphol, Y. Otsuka, Y. Mutoh, J. Mech. Behaviour Biomed. Mater., 36 (2014) 98. http://dx.doi.org/10.1016/j.jmbbm.2014.04.007

Residual Stresses 2016: ICRS-10 Materials Research Forum LLC
Materials Research Proceedings **2** (2016) 491-496 doi: http://dx.doi.org/10.21741/9781945291173-83

Residual Stress Behavior in Hardened Shot Peened 42CrMo4 Specimens during Fatigue Load

David Cseh[1, a*], Valeria Mertinger [1,b] and Janos Lukacs [2,c]

[1]Institute of Physical Metallurgy, Metalforming and Nanotechnology, University of Miskolc, H 3515 Miskolc-Egyetemvaros, Hungary

[2]Institute of Materials Science and Technology, University of Miskolc, H 3515 Miskolc-Egyetemvaros, Hungary

[a]cseh88david@gmail.com, [b]femvali@uni-miskolc.hu, [c]janos.lukacs@uni-miskolc.hu

[*]Tel: +36-46-450-111/10-97

Keywords: Fatigue, Residual Stress Relaxation, X-ray Diffraction

Abstract. The advantages of applied compressive residual stress on fatigue properties of materials is a well-described topic, but not in all respects. Compressive macro residual stresses in the surface region with medium and high hardness increase the fatigue life and the fatigue limit compared to materials that are free from designed compressive residual stresses because of their increased resistance against crack initiation and propagation. For this aim various surface compressing methods such as burnishing, shot peening, rolling have been developed. The monitoring of residual stress variation during fatigue is important. All properties, which exert lifetime, should be analyzed. The residual stress state of machine elements can change during application, therefore it is necessary to describe how these changes are related to the operational parameters. The surface residual stress state evolution of hardened (quenched and tempered)-and shot peened-42CrMo4 steel during fatigue tests was investigated nondestructively by X-ray diffraction. Four fatigue stress levels were applied. The residual stress state was recorded in shot peened state and monitored during the fatigue tests. The fatigue test was stopped after certain cycles until the specimens fractured. The stress state was measured after each fatigue test stops and the stress relaxation is given in the percentage of the initial stress state in function of cycle number.

Introduction

Shot peening is a widely used surface compressing method to create compressive residual stress in machined elements [1, 2] The compressive residual stress state has a beneficial effect on the fatigue durability of metals, especially on fatigue strength [3, 4, 5]. For the assessment of the influence of residual stresses on fatigue behavior, the stability of the residual stress state is very important. During fatigue, the macro and micro residual stress states interact with the cyclic loading stresses and the work hardening and softening processes. For example the surface compressive residual stress improves the fatigue resistance properties of all materials.

The stability of the surface compressive residual stress during low and high cycle fatigue in annealed micro-alloyed, medium carbon, perlitic and austenitic steels has been already investigated [6,-9]. The authors have shown that the residual stress relaxation is not monotonic in these alloys. However, in general, machine elements, which are tipically exposed to high cycle fatigue loads; (gears, shafts) are mainly made of high strength hardened steels, therefore it is worth to investigate the stability of residual stress in such material. The stability of surcafe compressive residual stress on hardened, shot peened CM45MV specimens during fatigue was investigated in one of our eralier project [10]. Residual stress relaxation was not observable in that steel, because of the effect of microstructure defects and specimen geometry. In this paper,

Materials Research Proceedings **2** (2016) 491-496 doi: http://dx.doi.org/10.21741/9781945291173-83

our aim was to analyze the variation of compressive residual stress state in shot peened hardened *42CrMo4 steel specimens* during fatigue load. This is a widely used steel in the vehicle industry especially in the case of surface compressed machined elements. These measurements are based on our earlier project. Since then, the specimen geometry was improved and the material selection has changed.

Materials

The specimens were made of EN:42CrMo4 type steel. The chemical composition is given in Table 1.

Table 1. Chemical composition of the examined steel

Composition wt.%					
C	Si	Mn	Cr	Mo	Ni, V, W, Co others
0.41	0.3	0.7	1.1	0.2	-

Fatigue specimens were cut from a hot rolled bar with 20 mm diameter (Fig.1). The yield strength (1200 MPa) and tensile strength (1250 MPa) were measured by three parallel tensile tests on hardened test specimens made of the same piece of bar.

The faults of the heat treatments could strongly influence the residual stress state [11, 12, 13,]. Our specimen geometry (long, thin) is especially sensitive on distortions, so the heat treatment was carried out with care. The specimens were immersed in oil by hand, used a pliers.

The specimens were induction hardened and oil quenched. To measure the temperature of the specimens and to control the inducer, Ni-NiCr thermocouple was soldered to the surface. First, the specimens were rapidly heated to *1000°C* during a few seconds then the temperature was held for 2 seconds When the sample reached 850°C, the thermocouple was cut from the specimens and the specimens were immersed in *quenching oil media*, During the immersion the longitudinal axis was perpendicular to the surface of the media. The tempering was carried out at *600°C* in an inductive furnace for half an hour, and specimens were subsequently cooled in air. The Fig. 2 shows the homogeneous martensitic microstructure of the cross section of the quenched and tempered sample. The HV10 (subscript) hardness data on the cross section (332, 337, 339, 361, 341, 333) confirms that the quenching was successful.

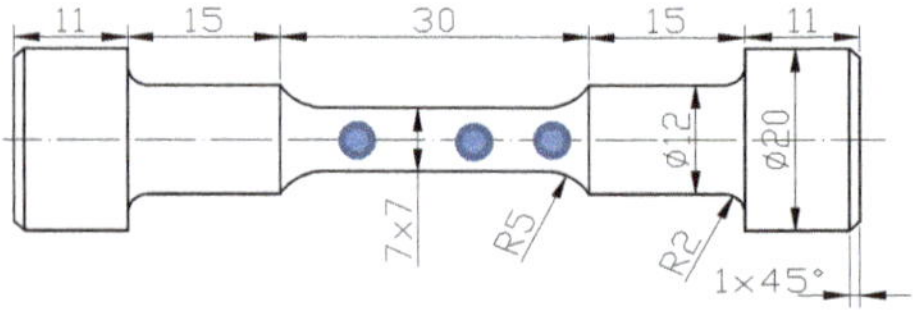

Figure1. Size of the shot peened specimens
The monitored areas of the surface is indicated with blue spheres

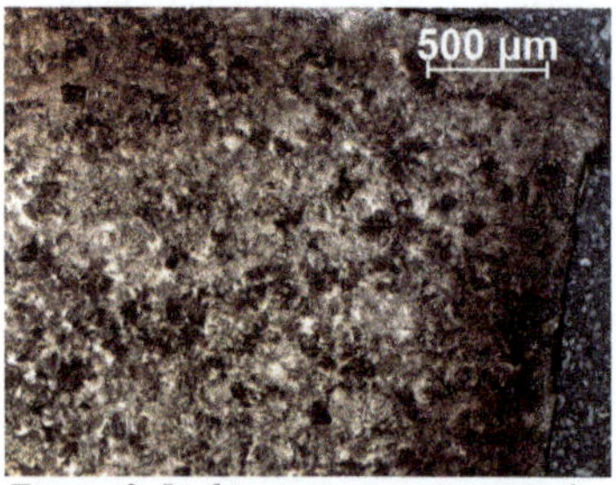

Figure 2. Light microscopy image of the hardened sample, cross section

The next step was the shot peening. It was performed in the Rába Automotive Holding Plc. The specimens were treated together with Rába manufactured shafts in the same time. The parameters of the shot peening were given by the requirements of the shaft. This means that the parameters of the applied peening method were industrial parameters. The shot peening was carried out in a WMKD 3

Materials Research Proceedings **2** (2016) 491-496 doi: http://dx.doi.org/10.21741/9781945291173-83

type shot peening equipment (Fig.3). The peening time was 9 minutes. Two Almen C strips were fixed on the sample holders to control the peening. The values of the Almen C probes were 0.26 and 0.24. The used peening balls were made of 1.8 mm diameter steel wire (430 HV), cut in 1.8 mm length pieces and conditioned for 4 hours. The minimal peening coverage was 150%. The average speed of the balls, which left the peening wheels was 55 m/s. As an average 117 J/mm^2 energy was imparted with the surface. The performance of the peening wheels was 252 kg/min.

Figure 3. Right: the specimens in the sockets before shot peening, Left: specimens after shot peening, with the Almen C holder. The heads of the specimens were shadowed during the peening.

Experiments

The residual stress was measured non-destructively by a Stresstech Xstress 3000 G3R X-ray diffractometer, resident at the Institute of Physical Metallurgy, Metalforming and Nanotechnology at the University of Miskolc. Our surface results regard to residual stress values from approximatelly *10 μm* depth of the surface layer. All of our residual stress values refer to this layer. The stress was calculated from the shift of the {211} reflection of ferrite according to the $\sin^2\psi$ method using Young's modulus (E) of 210000 MPa and Poisson's ratio of 0.29. Reflections were obtained in ψ geometry from 3 tilting positions in the -45°/+45° range according to Fig. 4. The spot size was 3 mm in diameter. The stress was measured with less error (scatter) than ±10 MPa in every point [10]. Fig 4 shows an example of the measured results in one point of sample No.1. The shear stress was under the 7 MPa value in every cases.

All four sides of the specimens were monitored. The axial residual stress state was recorded at 3 spots with *3 mm* spot size areas, on each side of each specimen in the initial shot peened state and during the fatigue loads. The same areas were investigated in each step of the monitoring. The fatigue schedule is given in Table 2. The represented stress value is always an average of twelve data after each step. First, the residual stress field of the shot peened state was recorded, (Table 2- in.RS) then, as the specimens were fatigued and stopped (as it is given is Table 2.) the residual stress state was measured again and again until the fracture of the specimen was reached. The first six specimens were fatigued until fracture without stop. In the possession of the results of the preliminary fatigued six specimens, the expected lifetime was determined. The cycle number at the first stop was defined based on the result of the first six specimens.

The fatigue load was tensile-compressive, the stress ratio (R_σ) was *-1* ($R_\sigma = \sigma_{min}/\sigma_{max}$; σ_{max} = *350; 400; 450; 500 MPa*); so the absolute values of the tensile and compressive stresses were equal. The cyclic load was implemented by an MTS type universal electro-hydraulic testing machine of the Institute of Materials Science and Technology at the University of Miskolc.

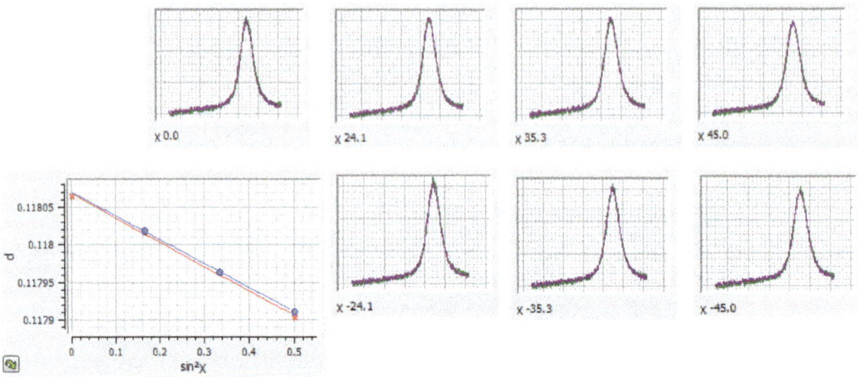

Figure 4.The result of one measured data of sample No. 1, χ is the tilting angle [°] , d lattice distance [nm]

Table 2. The schedule of the fatigue tests

No.	size [mm]	in. R.S [MPa]	σ [MPa]	cycle numbers in the first step	additional cycle numbers	No.	size [mm]	in. R.S [MPa]	σ [MPa]	cycle numbers in the first step	additional cycle numbers
1	7x7	-450	510	Load to fracture without stop, (preliminary fatigued specimens for life time estimations)		10	5x5	-407	500	20000	12500
2	7x7	-391	450			11	5x5	-360	450	30000	20000
3	7x7	-279	450			12	5x5	-348	400	45000	30000
4	7x7	-373	400			13	5x5	-332	350	70000	45000
5	7x7	-264	400			16	7x7	-225	500	45000	12500
6	7x7	-356	400			17	7x7	-254	450	70000	20000
7	7x7	-244	500	20000	12500	18	7x7	-351	400	11000	30000
8	7x7	-392	450	30000	20000	19	7x7	-244	450	50000	20000
9	7x7	-282	400	45000	30000	20	7x7	-259	350	70000	45000
H	7x7	12	350	70000	45000						

Results

The initial average residual stress values of the specimens showed a great deviation (Table 2.) The average values of the different samples were located between 225-450 MPa interval, this scattering is caused by the rough shot peening. The surface compressive residual stress relaxation of the shot peened specimens was measured at four applied stress amplitudes (500, 450, 400, 250 MPa). One specimen was fatigued in as-hardened state, so it was given no shot peening treatment. This specimen is indicated with a "H" letter and was fatigued at the lowest stress amplitude. It survived around 80000 cycles.

Fig. 5. shows the residual stress relaxation at different stress levels during fatigue loads. (Specimens 14 and 15 were notched then, investigated in another way and not presented here.) Each illustrated point is an average value of twelve performed measurements on the four sides of one specimen. Lines in group 450 MPa and 400 MPa show large scattering. The larger residual stress relaxation is not related to the larger absolute value of initial average residual stress. This scattering is acceptable in such kind of measurements.

Materials Research Proceedings **2** (2016) 491-496 doi: http://dx.doi.org/10.21741/9781945291173-83

In order to monitor the stress variation, the residual stress values measured during the fatigue tests, are presented as percentage values of the initial stress states. This unintentional scattering of the average surface residual stress measured in our specimens let us to investigate the role of it during the inspected phenomenon.

The significance of the fatigue stress amplitude is multiple: the higher stress level results shorter lifetime, and higher extent of residual stress relaxation.

To test the sample size effect on the residual stress relaxation during the fatigue load, the thickness of six specimens (No. 10-15) was decreased (5x5 mm) and shot peened again with the same peening parameters. No systematic effect is connected to the sample size. The stress relaxation rate shows an increased level by increasing the fatigue stress amplitude. It is evident that the rate of the stress relaxation is very fast in the first thousands of the cycles.

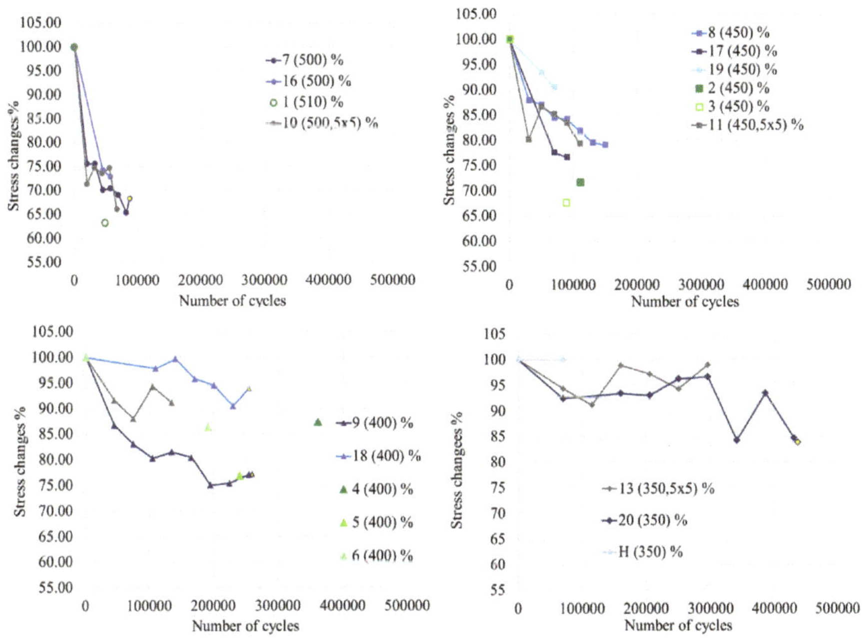

Figure 5. The residual stress relaxation at single stress levels during fatigue loads

Summary

Stress relaxation of quenched and tempered 42CrMo4 steel was investigated during fatigue test. Fatigue specimens were cut and shot peened with industrial parameters. The surface compressing residual stress stability was investigated using different fatigue stress levels. Non-destructive residual stress measurements were carried by X-ray diffraction method. The fatigue load was stopped several times (3-10 stops depend on specimens) and the residual stress state is monitored in each steps. The stress relaxation is given in the percentage of the initial stress state in function of cycle number.

No sample size effect in the rate of relaxation was observable. The beneficial impact of shot peening appears in small number of cycles until the fracture of the as-hardened sample. The rate

Materials Research Proceedings **2** (2016) 491-496 doi: http://dx.doi.org/10.21741/9781945291173-83

of the stress relaxation is faster in the first thousands of the cycles. The higher fatigue stress amplitude indicates larger extent of residual stress relaxation. The absolute value of the initial residual stress state does not have systematic effect during the relaxation. The higher relaxation has no correlation with the value of the initial residual stress.

Our concept regards to high strength steel specimens. None of the applied fatigue stress levels reached the half of the yield strength. Our results are related to others (Dalaei and Kim) with differences in the material selection and in the quotient of the applied fatigue stress levels and the yield strength. Our results confirm the permission of the expansibility of the declared observations in case of steel, independently by material quality and heat treatment state.

Acknowledgement

This work has been carried out as part of the TÁMOP-4.2.1.B-10/2/KONV-2010-0001 project within the framework of the New Hungarian Development Plan. The realization of this project is supported by the European Union, co-financed by the European Social Fund.

References

[1] J. Davis, Surface hardening of steels, USA: ASM International, 2002.

[2] S. Timoshenko, History of strength of materials, Dover: Dover Publications, 1983.

[3] ASM International, Handbook of Residual Stress and Deformation of Steel, USA: 2002.

[4] R. I. Stephens, A. Fatemi, R. R. Stephens és H. O. Fuchs, Metal fatigue in Engineering 2nd Edition, USA: John Wiley & Sons, 2001.

[5] E. Macherauch és V. Hauk, Residual stress in Scinence Technology Vol 2, Németország: I.R. Publications, 1987.

[6] K. Dalaeia, B. Karlssona és L.-E. Svenssona, „Stability of shot peening induced residual stresses and their influence on fatigue lifetime," *Mat. Science and Engineering A,* pp. 1008-1015, 2010.

[7] J.-C. Kim, S.-K. Cheong és H. Nog, „Residual stress relaxation and low- and high-cycle fatigue behavior of shot-peened medium-carbon steel," *International Journal of Fatigue,* pp. 114-122, 2013. http://dx.doi.org/10.1016/j.ijfatigue.2013.07.001

[8] J.-C. Kim, S.-K. Cheong és H. Nog, „Proposed fatigue damage measurement parameter for shot-peened carbon steel..." *International Journal of Fatigue,* pp. 97-106, 2014.

[9] I. Nikitin és M. Besel, „Correlation between residual stress and plastic strain amplitude during low cycle fatigue of mechanically surface treated austenitic stainless steel AISI 304 and ferritic–pearlitic steel SAE 1045," *Materials Science and Engineering A,* pp. 297-303, 2008.

[10] D. Cseh, V. Mertinger és J. Lukács, „Residual Stress Evolution During Fatigue Test of a Shoot Peened Steel Sample," *Materials Science Forum ,* pp. Vol. 752 95-104, 2013.

[11] ASM International, Cast Irons ASM Specialty Handbbok, USA: ASM International, 1996.

[12] ASM International, Heat treating ASM Handbook Vol. 4,, USA: ASM International, 1991.

[13] R. Haimbaugh, Induction Heat Treating, USA: ASM International, 2001

Residual Stresses 2016: ICRS-10 Materials Research Forum LLC
Materials Research Proceedings **2** (2016) 497-502 doi: http://dx.doi.org/10.21741/9781945291173-84

Stress in Thin Wall Structures Made by Layer Additive Manufacturing

Vladimir Luzin [1, a *], Nicholas Hoye [2, b]

[1] Australian Nuclear Science and Technology Organisation, Locked Bag 2001, Kirrawee DC, NSW, 2232 Australia

[2] Faculty of Engineering and Information Sciences, University of Wollongong, NSW 2522 Australia

[a] vll@ansto.gov.au, [b] nicholas.hoye@gmail.com

Keywords: Residual Stress, Additive Manufacturing, Titanium

Abstract. Manufacturing of thin wall structures is one of the main applications of additive manufacturing, where it has significant advantages over traditional milling and machining techniques or welded analogues. Such thin walled structures are common in structural aerospace components, and are also frequently made from titanium alloys. For such large-scale components, layer deposition strategy is more advantageous rather than a pixel-wise deposition approach due to the demand for high productivity and size requirements. Several techniques can be used to produce layer-wise build-ups, including laser-powered Direct Metal Deposition (DMD) process or gas tungsten arc welding (GTAW). Although, in the general case of arbitrary thin wall structures the stress distribution is complex, for some simple geometries, the stress state is simple and can be well characterized within a model by a single parameter representing a layer deposition stress in the steady-state regime. The model calculations were verified by experimental results on a thin-walled sample component that was manufactured from Ti-6Al-4V by GTAW with the residual stresses measured using KOWARI neutron strain scanner at the OPAL research reactor (ANSTO).

Introduction

Titanium based alloys are very widely used in aerospace industry due to their high specific strength, fatigue properties and excellent corrosion/oxidation resistance [1], and frequently need to be shaped into thin-wall structures, such as wing ribs and spars, various structural elements, gear boxes, etc. Traditional metal forming (e.g. rolling, extrusion) and machining methods are difficult, labour/cost intensive [2] and frequently extremely wasteful, especially when thin-wall structures are to be fabricated, resulting in unacceptably high buy-to-fly ratios. In comparison, additive manufacturing (AM) of titanium components seems to be the most attractive manufacturing technology, dramatically improving manufacturing costs and reducing waste to minimum with almost no limitation on the component shape. While some components for aerospace applications are small scale (<300 mm) and can be readily manufactured by powder-bed techniques such as Selective Laser Melting (SLM) or Electron Beam Melting (EBM), large-scale major airplane components present challenges due to size limitations of the build chamber. In this case, other techniques such as blown-powder based Direct Metal Deposition (DMD) or wire-fed based Wire-Arc Additive Manufacturing (WAAM) are used. The latter is also often called Wire-Arc Additive Layer Manufacturing (WAALM) since fabrication route involves multiple-pass deposition to build wall-like structures in a layer-by-layer manner to produce the 'near net shape' profile. Development of these techniques into industrial scale for high quality production of engineering components from titanium alloys in an economically efficient way is a general challenge for AM technology at present.

Although many technological process parameters such as deposition energy and speed, feedstock deposition rate, built-up trajectory, clamping system, resulting microstructure and defect structure are

Materials Research Proceedings **2** (2016) 497-502 doi: http://dx.doi.org/10.21741/9781945291173-84

to be considered when optimising product quality and production costs, the residual stress is one of the most serious issues, being the core reason for significant bowing, bending and deflection, often resulting in compromised dimensional tolerances. For experimental studies, a single wall structure (T-shape sample) of a constant thickness and height, built on a rectangular base plate appears to be a standard choice [3-6]. Depending on the material, exact dimensions of the build-up and the process, the deflection of the back side of the base plate can vary greatly, but a typical reported deflection of the base plate in case of Ti-6Al-4V is 14 mm per 1000 mm longitudinal base length [6]. In addition to causing distortion, residual stresses may have detrimental effects on mechanical properties, especially fatigue behaviour, thereby degrading the performance of the component in service.

Considering AM process of a T-shape sample, two distinct steps or episodes in the overall stress formation can be isolated. When first layer is deposited a stress distribution in the base plate is created which is essentially very similar in nature and stress distribution to a single-bead welding path [5]. Although second and third passes can modify the stress distribution created by the first pass, all consecutive passes outstanding several mm from the base plate do not affect the base stress distribution. Instead they build-up stress in a different manner dictated by the geometry of the thin-wall structure. If the wall thickness is much smaller than the wall height and length, the zero plane stress condition is applicable with transverse stress (through thickness) being equal to zero.

The aim of the current study is to investigate residual stress build-up in a T-shaped sample made from Ti-6Al-4V by WAALM in order to quantify residual stress process within an empirical model. One of the multiple possibilities is to consider different scenarios for clamping and its influence on the resultant residual stress. This quantification will allow certain conclusions to be drawn about the mechanism(s) of residual stress formation, hence allowing the prediction of residual stresses in samples of different dimensions and, possibly, to evaluate residual stress mitigation strategies.

Sample production, process and materials

WAALM, as an arc-based deposition process can be realised in several ways, it can utilize either the gas metal arc welding (GMAW) or the gas tungsten arc welding (GTAW). The latter version has been developed in the University of Wollongong as a practical AM method with high deposition rate enabling production of large components [7]. In this process, build-up of a three-dimensional near-net shape freeform is achieved through deposition of a single row of successive weld beads onto a substrate to produce a component in a layer-wise manner. The process utilizes robotic automation that can be programmed to suit the design of the future component [8].

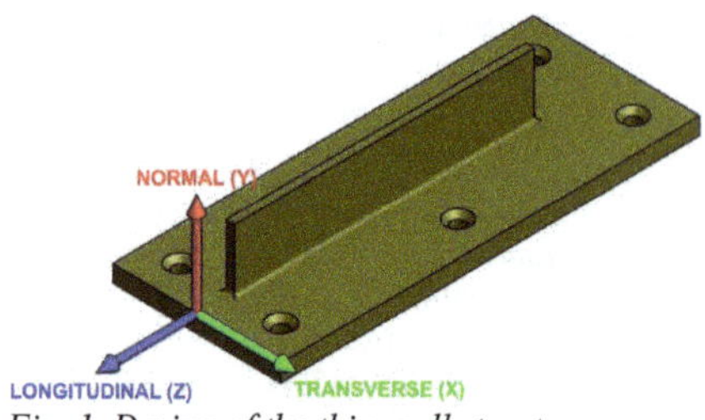

Fig. 1. Design of the thin-wall structure produced by WAALM with indication of the principal directions.

Using the reported WAALM, a thin-wall of Ti-6Al-4V was built to the full length along the centre line of a Ti-6Al-4V base plate (L250xW100xT12, mm). The resulting build-up thin wall was approximately 8 mm in thnessi and 40 mm in height. The as-deposited was further machined via conventional milling to a wall of high precision dimensions with 5 mm thickness, 36 mm height and 178 mm length. The final shape of the component is shown in Fig. 1.

Parameters of deposition GTA welding process have been already reported [9]. The base plate, attached to a linear actuator, was moved with a travel speed of 150 mm/min, while the welding torch and wire feed were held stationary. Deposition was conducted using a current-controlled power source operating at a steady state current of 110 A and giving an average arc energy of 485 J/mm. Filler wire of 1.0 mm diameter was fed at a rate of 2000 mm/min to provide a specific energy input of approximately 10 kJ per gram of deposited material. Another sample was produced at conditions

Materials Research Proceedings **2** (2016) 497-502 doi: http://dx.doi.org/10.21741/9781945291173-84

of 20 kJ per gram of deposited material by reducing the travel speed by a factor of two Both, the base plate and the filler wire materials were commercially sourced and produced to ASTM B265 and B863 standards respectively.

Neutron diffraction experiment and data analysis

Neutron residual stress measurements of the wall-structure were performed on the KOWARI neutron diffractometer at OPAL research reactor at ANSTO [10]. For the measurement the Ti(103) reflection was used at 90°-geometry employing a neutron wavelength of approximately λ = 1.7 Å. Three principal directions were measured, the normal, transverse and longitudinal across the wall height (from base plate to the top of the wall) in the middle portion of the 250 mm long sample with variable density of the measurement point (1 mm close to the base plate and 2 mm away from the base plate). More than 20 experimental points along the wall height were obtained by scanning.

To use efficiently sample geometry, a gauge volume with size of 2×2×20 mm^3 was used for measurements of the normal and transverse strain components, while this gauge volume was reduced to 2×2×2 mm^3 when the longitudinal component was measured. To acquire better grain statistics, the gauge volume was moved during measurements of the longitudinal component to cover equivalent volume of 2×2×20 mm^3. For the given experimental conditions, an average accuracy of ~100 µstrain was achieved providing stress accuracy of ~20 MPa in terms of calculated stresses (errors only due to the neutron counting statistics).

With respect to determining the sample d_0, the more standard approach of cutting small coupons that can be assumed stress free proved unreliable due to poor statistics associated with the comparatively large prior β-Ti grain size. As only the stress distribution in the thin wall section was of interest, an alternative approach was employed where the through thickness stress was assumed to be zero in the thin wall section. This assumption of a plane stress condition is considered valid due to the 5 mm wall thickness being of similar size to the ~3 mm gauge volume (or spatial resolution), and was seen to give results with great accuracy. Based on this condition, three d-spacings for three directions could be resolved into two stress components (longitudinal and normal) and d_0.

Modelling of stress profile in a wall

A simple model can be considered based on the established experimental facts that the transverse and normal components are insignificant in comparison to the dominant longitudinal component. The wall build-up process can be considered as a progressive deposition of the infinitesimal layers of material characterised by a single parameter, the deposition stress σ_d, just as the progressive deposition of layers can be applied to describe stress formation in coatings [11]. Although a generalized model can deal with dissimilar materials of the wall and the base plate, for our particular purpose, we consider only similar materials for both (same Young's modulus and Poisson ratio).

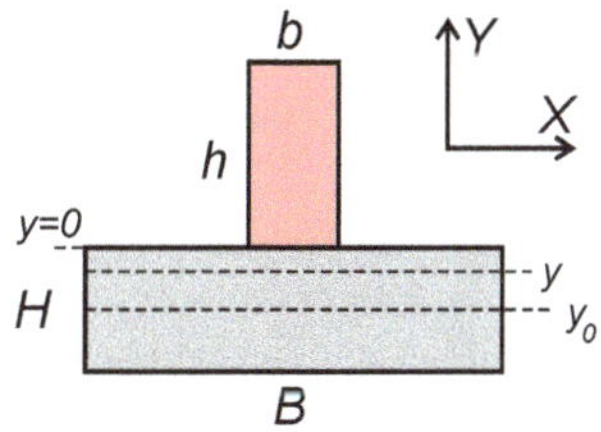

Fig. 2. Geometry and dimensions of the wall-on-base sample.

While a newly deposited layer cools down, shrinks and acquire some intrinsic quenching stress or deposition stress in general, the T-beam bending theory can be applied to find the resultant stress distribution required to satisfy the force balance and the moment balance [12]. The resultant stress distribution can be computed in few steps. The deposition stress σ_d, in a wall element, of width b and height h, built on the base plate, of width B and thickness H, causes the appearance of a pair of equal and opposite forces, F, (one in the wall, one in the base)

$$F = \sigma_d \frac{bh \cdot BH}{bh + BH} \quad \text{and a moment, } M, \quad M = F\frac{(h+H)}{2}.$$

The induced stress distribution in the wall is according to

$$\sigma(y) = \frac{F}{bh} - \frac{M}{I}(y - \bar{y}),$$

Materials Research Proceedings **2** (2016) 497-502 doi: http://dx.doi.org/10.21741/9781945291173-84

where position of the neutral axis $\bar{y}$ and area moment of inertia I can be calculated for the (composite) beam as follows

$$\bar{y} = \frac{bh\frac{h}{2} - BH\frac{H}{2}}{bh + BH},$$

$$I = \frac{BH^3}{12} + BH\left(\frac{H}{2} + \bar{y}\right)^2 + \frac{bh^3}{12} + bh\left(\frac{h}{2} - \bar{y}\right)^2.$$

The same forces also produce curvature, κ,

$$\kappa = \frac{M}{IE} = \frac{\sigma_d}{E} \cdot (geometrical\ factor),$$

which is proportional to the deposition stress.

Using the above formulae, the two extreme scenarios can be considered that can be called, "fully constrained" and "fully free" base plate. In the former case, the full wall height is first build up and then the mechanical equilibrium is achieved, as would be expected when the thin-walled section is released from the constrains of the base plate. In the latter case, the equilibrium is found after each episode of the infinitesimal layer deposition, simulating continuous sample curvature change due to the stress build-up.

Results

The stress distributions of two stress components and two analysed samples are shown in Fig. 3. Although no assumption was made for the normal component, as it can be seen, it is very close to zero for both samples. Therefore, it can be concluded that not only the transverse component is zero, but also normal, what makes stress state very simple with only longitudinal component.

Although the specific energy input of two samples is different, there is no clear indication about correlation between stress and heat input. The only distinct feature of the high specific energy input sample, 20 kJ/g, is larger statistical oscillations most evidently related to the coarser-grained microstructure.

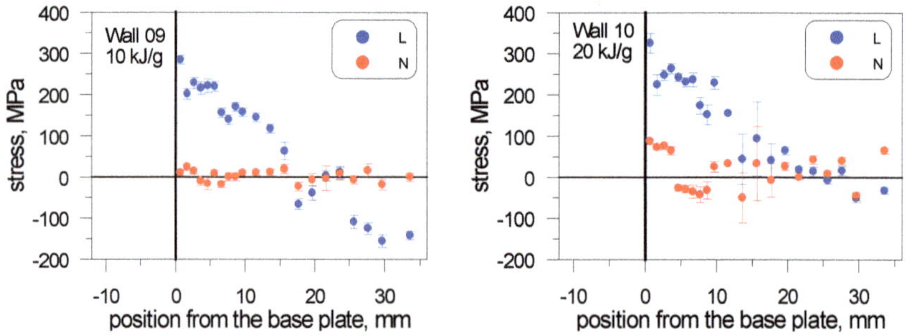

Fig. 3. Residual stress analysis for the samples of two heat input, 10kJ/g and 20 kJ/g.

The neutron diffraction stress distributions in the walls were analysed with the modelling approach discussed above to produce a fit to the experimental dataset. Two scenarios, fully constrained sample and fully free, were considered and their comparison is shown in Fig. 4 together with the experimental results from the both samples. Clearly, the "fully constrained" sample assumption works the best and provides the linear trend of stress dependence exhibited in the experimental data, while "no constrain" sample obviously does not follow the trend and has a distinct non-linearity. In both cases, the deposition stress parameter was 400 MPa.

Materials Research Proceedings **2** (2016) 497-502 doi: http://dx.doi.org/10.21741/9781945291173-84

The same approach was applied to other datasets taken from the published literature. In one case the Ti-6Al-4V alloy 6 mm thin wall was additively manufactured by a similar technique and measured using neutron diffraction [6]. In the other case, a stress analysis of a steel 5 mm thin wall on the steel substrate was carried out also with neutron diffraction [5]. The experimental data from these studies also agrees with "full constrain" profiles, as shown in Fig. 5. The corresponding deposition stress is 400 MPa for the Ti-6Al-4V samples and 500 MPa for the steel sample.

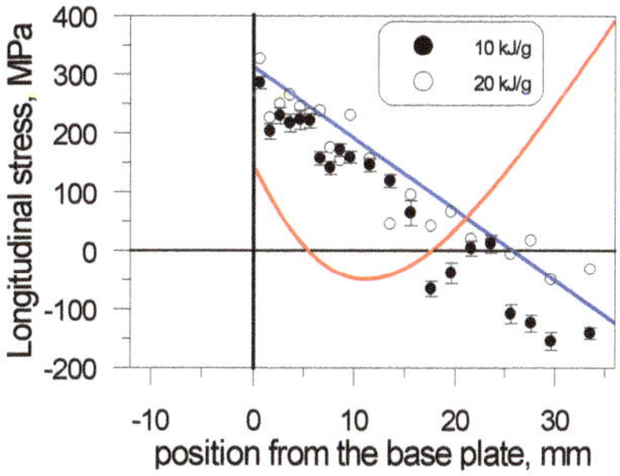

Fig. 4. Residual stress distribution in the wall for the samples (10 kJ/g and 20 kJ/g specific energy input). Experimental points are overlapped with predicted stress profiles. Blue line correspond to "full constrain" scenario while red line is associated with "no constrain" model.

Discussion

Three Ti-6Al-4V samples, considered in terms of deposition model, demonstrate remarkably close outcome, 400 MPa deposition stress. This result suggests that material properties (i.e. yield stress) are the most important factor for the stress formation. Although details of the deposition process such as deposition rate, travel speed, energy input, etc., might vary, the thermo-mechanical history of the deposited material seems to remain within a narrow envelop of physical parameters providing conditions for a stable and efficient build-up of the wall. Considering this, it is evident that the deposition stress is a result of quenching with accumulation first of very high stress, up to the yield point, followed by some partial annealing by subsequent passes.

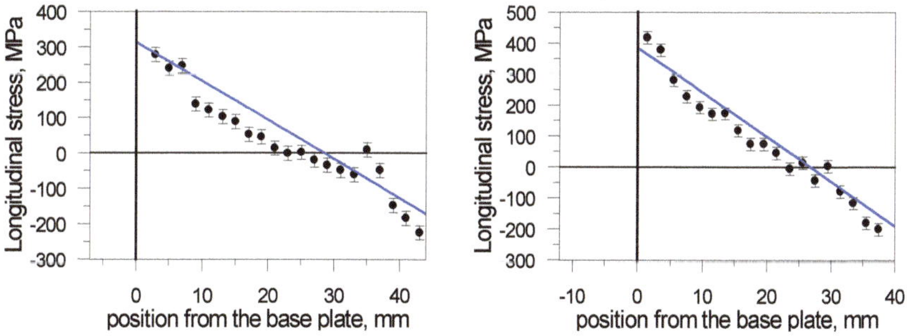

Fig. 5. Experimental residual stress distributions in walls two samples, a Ti-6Al-4V thin wall (left, adopted from [6]) and a steel thin wall (right, adopted from [5]). Experimental points are overlapped with simulated stress profiles (blue lines) calculated using "full constrain"

The stress build-up can be easily and successfully modelled within a simple model based on a beam bending theory and an assumption (and practical realisation) of the full constraint of the base plate. The model assumes very simple stress state with only the longitudinal component being non-zero and it is fully characterised by a single parameter, the deposition stress σ_d or, equivalently, the curvature κ. The fact that the normal and transverse stresses can be accepted as zero for all practical purposes has a solid experimental corroboration in this study as well as in other studies [4, 6]. While not every detail can be derived with the simple analytical model, it is clear that the main trend and magnitude of stress distribution in the wall can be reproduced. It is worth noting that this simplified

Materials Research Proceedings 2 (2016) 497-502 doi: http://dx.doi.org/10.21741/9781945291173-84

modelling approach is most likely unable to accurately characterise the stress field at the joint point with the base plate, as well as at the very top layer of deposited material, which would have a different history to all other deposited layers.

The tensile deposition stress due to quenching, as high as 500 MPa, is potentially detrimental to mechanical properties and most certainly leads to significant curvature of the base plate. Some stress mitigation approaches have been attempted to reduce stresses [4], though the definition of the most practical and most efficient solution remains one of the challenges for additive manufacturing technologies.

Summary

In this study, an experimental and modelling approach were attempted to characterize residual stress distributions in a thin wall deposited by additive manufacturing (WAALM). It was demonstrated that the stress state may be characterised only by the longitudinal component and that the model can provide a general steady-state linear behaviour of the longitudinal stress distribution in the wall observed experimentally. Typically very high tensile deposition stresses are produced by WAALM, 400 MPa for Ti-6Al-4V and 500 MPa, as in the reported examples, highlighting the requirement for stress mitigating strategies if component distortions during additive manufacturing are to be avoided.

References

[1] S. Seong, O. Younossi, B.W. Goldsmith, T. Lang, M. Neumann, Titanium: Industrial Base, Price Trends, and Technology Initiatives, RAND Project Air Force, 2009.

[2] J.P. Davim, Machining of Titanium Alloys, Springer Berlin Heidelberg, 2014. http://dx.doi.org/10.1007/978-3-662-43902-9

[3] B. Baufeld, O.V.d. Biest, R. Gault, Additive manufacturing of Ti–6Al–4V components by shaped metal deposition: Microstructure and mechanical properties, Mat. Des., 31, Supplement 1 (2010) S106-S111. http://dx.doi.org/10.1016/j.matdes.2009.11.032

[4] P.A. Colegrove, H.E. Coules, J. Fairman, F. Martina, T. Kashoob, H. Mamash, L.D. Cozzolino, Microstructure and residual stress improvement in wire and arc additively manufactured parts through high-pressure rolling, J. Mater. Process. Technol., 213 (2013) 1782-1791. http://dx.doi.org/10.1016/j.jmatprotec.2013.04.012

[5] J. Ding, P. Colegrove, J. Mehnen, S. Ganguly, P.M. Sequeira Almeida, F. Wang, S. Williams, Thermo-mechanical analysis of Wire and Arc Additive Layer Manufacturing process on large multi-layer parts, Computational Materials Science, 50 (2011) 3315-3322. http://dx.doi.org/10.1016/j.commatsci.2011.06.023

[6] B.A. Szost, S. Terzi, F. Martina, D. Boisselier, A. Prytuliak, T. Pirling, M. Hofmann, D.J. Jarvis, A comparative study of additive manufacturing techniques: Residual stress and microstructural analysis of CLAD and WAAM printed Ti–6Al–4V components, Mat. Des., 89 (2016) 559-567. http://dx.doi.org/10.1016/j.matdes.2015.09.115

[7] D. Ding, Z. Pan, D. Cuiuri, H. Li, A multi-bead overlapping model for robotic wire and arc additive manufacturing (WAAM), Robotics and Computer-Integrated Manufacturing, 31 (2015) 101-110. http://dx.doi.org/10.1016/j.rcim.2014.08.008

[8] D. Ding, C. Shen, Z. Pan, D. Cuiuri, H. Li, N. Larkin, S. van Duin, Towards an automated robotic arc-welding-based additive manufacturing system from CAD to finished part, Computer-Aided Design, 73 (2016) 66-75. http://dx.doi.org/10.1016/j.cad.2015.12.003

[9] N. Hoye, H. Li, D. Cuiuri, A. Paradowska, Measurement of residual stresses in titanium aerospace components formed via additive manufacturing, Mater. Sci. Forum, 777 (2014) 124-129.

[10] O. Kirstein, V. Luzin, U. Garbe, The Strain-Scanning Diffractometer Kowari, Neutron News, 20 (2009) 34-36.

[11] Y.C. Tsui, T.W. Clyne, An analytical model for predicting residual stresses in progressively deposited coatings .1. Planar geometry, Thin Solid Films, 306 (1997) 23-33. http://dx.doi.org/10.1016/S0040-6090(97)00199-5

[12] S. Timoshenko, J.M. Gere, Mechanics of materials, Van Nostrand Reinhold Co., 1972.

Materials Research Proceedings **2** (2016) 503-508 doi: http://dx.doi.org/10.21741/9781945291173-85

Effect of Element on Porosity and Residual Stress Distribution of A7N01S-T5 Aluminum Alloy Welded Joints in High-Speed Trains

Yali Liu[1,a] , Jia Chen[1,b] ,Guoqing Guo[1,c *],Jiangli An[1,d],Hui Chen[1,e],

[1]School of Materials Science and Engineering, Southwest Jiaotong University, Chengdu 610031, China

[2]Chengdu Industry and Trade college,611731, China.

[a]15184362059@163.com,[b]15574423@qq.com,[c*]gouguoqing1001@163.com, [d]1559499460@qq.com, m,[e] xnrpt@163.com

Keywords: A7N01S-T5 Aluminum Alloys, Chemical Element, Porosity Distribution, High Speed Trains

Abstract. In this paper, four types of A7N01S-T5 aluminum alloys with different chemical elements were investigated. Welded joints of the alloys were fabricated under 70%environmental humidity conditions at 10°C with single pulse GMAW welding technology. The alloys and their joints were tested and examined with impact toughness, porosities distribution and Synchrotron Radiation. The element results showed that the elements of Zn and Mg were the main factors that affect impact properties of the alloys. The #2 alloy which has the element of Zn-4.29 Mg-1.56 Mn-0.22 Cr-0.14 Zr-0.01 Ti-0.027 had the best impact properties which were 20.76J.The results indicated that the #2 had the lest stomatas only 22 and had the lowest porosity rate of 0.009% . Weld zone compared with the base material, the content of Mn and Ti had reduced.the#2 residual stress evenly distributed,the maximum tensile stress was 99Mpa in weld and the maximum compressive stress -66Mpa in base metal.forming .

Introduction

A7N01S-T5 aluminum alloy belongs to the Al–Zn–Mg alloy series. It has been widely used in high-speed train bodies and welding structures, such as corbels, beams and under frames due to its high strength, low density and good welding properties [1-2]. Porosity is one of the main defects produced in gas metal arc welding (GMAW) of aluminum–alloy structures during fabrication of high speed trains. Hydrogen (H) is the main contributor to porosity during welding . The solubility of H ions decreases as the temperature of the weld pool decreases during cooling. Thus, the H ions will escape from the weld pool during cooling by forming bubbles and floating to the surface. Bubbles that do not escape the weld pool become porosity. Research results conclude that porosity results a reduction in strength.

In addition to Al and the main alloying elements Zn and Mg, A7N01S-T5 alloy contains minority elements and impurity elements, such as Mn, Cr, Zr, Ti, Fe and Si. It has generally been recognized that alloying element and treatment determine the grain type, grain size, and distribution of precipitated phase, which affect the strength and fracture toughness of the alloy.Although extensive studies on one or two elements have been made to affect various mechanical properties and microstructural changes in Al–Zn–Mg alloys, published works on multiple elements of the same is rather limited. In this paper, the elements which have a similar effect on the properties would be taken into consideration as the same one factor. Therefore, it is important to understand the effect of elements on porosity distribution, impact toughness ,residual stress distribution of A7N01S-T5 aluminum alloy welded joints in high-speed trains.

Materials Research Proceedings **2** (2016) 503-508 doi: http://dx.doi.org/10.21741/9781945291173-85

Materials and experimental work

Four types of 8mm thick aged Al–Zn–Mg alloys (A7N01) with T5 aging were provided andnamed #1, #2, #3 and #4. The detailed composition were listed in Table 1. Welded joint samples were made by Metal Inert-Gas (MIG) welding method with PHOENIX 421 EXPERT welding machine with ER5356 welding wire with a diameter of 1.6mm. In order to remove the oxides and decrease the porosity of the joints, the surface of the base metal was chemical cleaned before the welding process. During welding experiments, 99.999% pure argon was used .The room temperature was 10 °C and the room humidity is 70%.The elements of the A7N01S-T5 base material and ER5356 are listed in Table 2.The welding technology was listed in Table 3.

Table 1 Base metal chemical element (wt%)

Sample **No.**	Si	Fe	Cu	Mn	Mg	Cr	Zn	Ti	Zr	V
#1	0.10	0.13	0.12	0.35	1.41	0.27	4.14	0.02	0.01	0.17
#2	0.08	0.14	0.11	0.28	1.56	0.14	4.29	0.02	0.01	0.17
#3	0.09	0.16	0.06	0.36	1.55	0.26	4.59	0.06	0.01	0.13
#4	0.11	0.26	0.28	0.35	1.56	0.26	4.60	0.07	0.02	0.14

Table2 Chemical element of ER5356 filler wire (wt%)

element	Si	Fe	Cu	Mn	Mg	Cr	Zn	other	Al
element	≤0.25	≤0.10	≤0.10	0.05-0.20	4.5-5.5	0.05-0.20	≤0.10	≤0.15	margin

Table 3 Welding process parameters

Material Type	thickness (mm)	Passes	Welding current(A)	Welding voltage(V)	Welding speed(mm/min)	Soldering Temperature (°C)	Welding humidity (%)
#1 #2 #3 #4	8	1	198-220	23.3-25.2	600-656	10℃	70%
		2	205-225	24.1-25.6	636-656		
		3	208-232	24.5-25.9	500-525		

The residual stresses evaluation of welded joints were evaluated using ultrasonic residual stress measurement. The measuring equipment is HT1000 residual stress evaluation machine, as shown in Fig. 1.

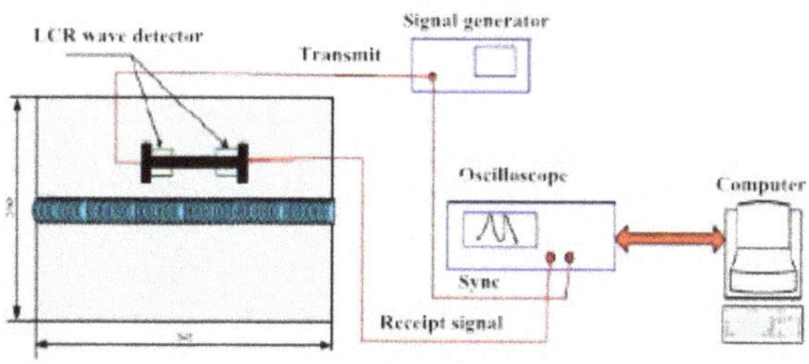

Fig.1 Ultrasonicresidual stress measurement system

The samples that were used for optical microscopy observations were prepared by Keller reagent (1% HF + 1.5% HCl + 2.5%HNO3 + 95% H2O). The microstructures of the samples welded joints were examined using a Zeiss AX10 optical microscope. The cross sections of the welded joints first were examined using a VHX digital microscope and then the porosity quantity was analyzed with Image-Pro Plus software based on a color-difference analysis. method.

The residual stresses evaluation of welded joints were evaluated using ultrasonic residual stress measurement. The measuring equipment is HT1000 residual stress evaluation machine, as shown in Fig. 1.

Results and discussion

The impact toughness of the base metal and the weld zone are shown in Fig.2. The difference was tiny and the H#2 alloy had the highest Impact with 20.67J. The H#1 alloy had the lowest Impact ,lower 7J.

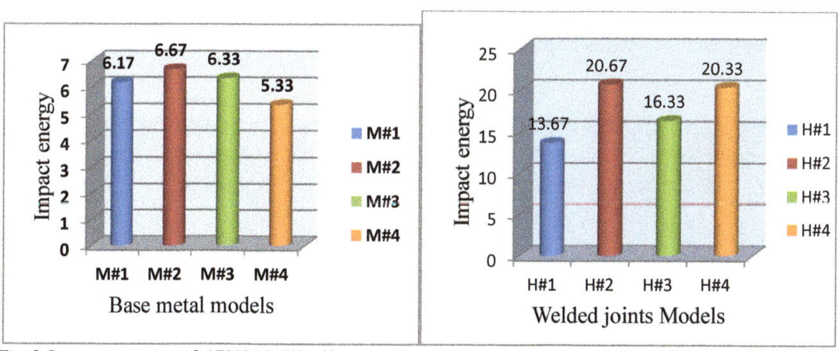

Fig.2 Impact energy of A7N01S-T5 alloy at various composition

By comparing the cross-sectional morphologies at various element.in Fig.3.The porosity quantity and area ratio of #2 were the lowest and the values of them were the highest in #3. From Fig.4, the concentration of Mn was apparently different between the weld and HAZ, and the concentration in the weld is lower than that of base metal. The reduction of Mn could lead to generate some harmful phases, e.g. b-AlFeSi phase with gray-needle shape, and Mg_2Si phase with black-striation shape, which usually can be found at the grain boundaries. Therefore, the mechanical properties can be seriously affected by the vaporization of Mn. Ti can help to form $TiAl_3$ dispersion particles and refine the weld structure.

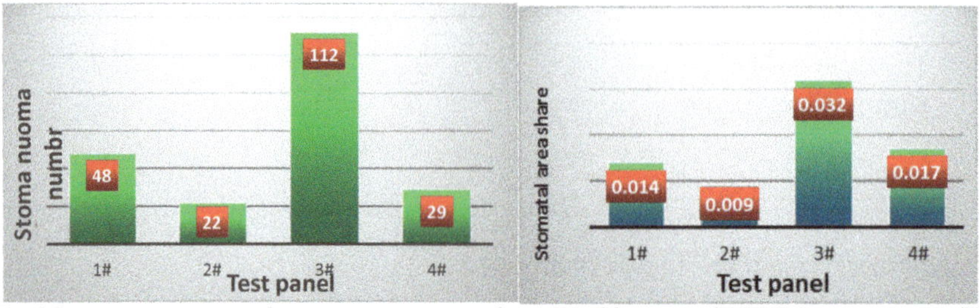

Fig.3 Porosity quantity and possession rate as a function of four samples(a) Porosity quantity(b) possession rate.

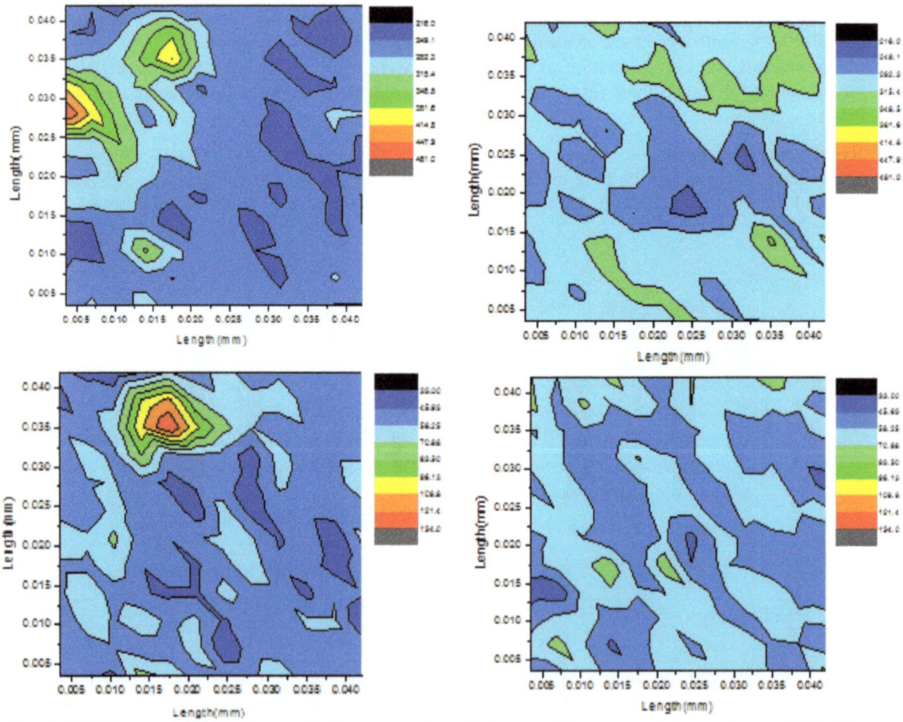

Fig.4 LXRF maps of elements inside base metal and joints, (a) Base metal element Mn, (b)Weld element Mn, (c)Base metal element Ti, and (d) Weld element Ti

Materials Research Proceedings **2** (2016) 503-508 doi: http://dx.doi.org/10.21741/9781945291173-85

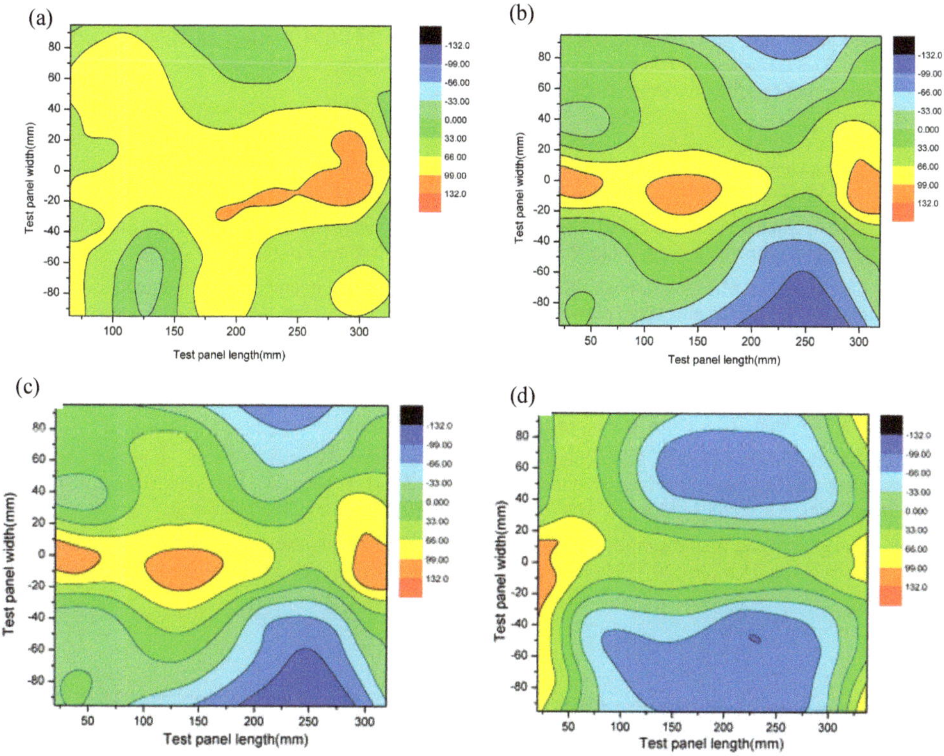

Fig.5 X axis the residual stress distribution of total area:(a)#1,(b)#2,(c)#3,(d) #4.

For # 1 and # 2, the area occupied by tensile stress is significantly greater than that by compressive stress. For #3 and # 4, the area of tensile and compressive stresses is quite different. the # 1 alloy maximum tensile stress was 132Mpa in weld, the maximum compressive stress was -33Mpa in base metal. The # 2 plate maximum tensile stress was 99Mpa in weld and maximum compressive stress -66Mpa in base metal. The 3# alloy maximum tensile stress was 120 Mpa in weld, the maximum compressive stress was -132Mpa in base metal. The # 4 alloy the maximum tensile stress is 66Mpa in weld, the maximum compressive stress -99Mpa in base metal.

Summary

(1)A7N01 different element of the alloy has a great influence on mechanical properties.Combination with welding defects and content of alloying elements can be found in the impact properties #2 sample is preferable.

(2) With the help of SR-IXRF, the distribution of these strengthening alloy elements in the base metal d and welded joint was obtained. From this distribution, the vaporization of Mn and Ti should also take responsibility for the different of the impact performance of the base metal and the welded joint.

(3)Under the same condition of welding surroundings and constraints.Different components of welded plates have different weld residual stress distribution.2# (Zn-4.29 Mg-1.56 Mn-0.22 Cr-0.14 Zr-0.01 Ti-0.027)stress distribution was evenly,the maximum tensile stress was 99Mpa in weld and the maximum compressive stress -66Mpa in base metal.

References:

[1] Qin C, Gou G Q, Che X L, et al. Effect of element on tensile properties and fracture toughness of Al–Zn–Mg alloy (A7N01S-T5) used in high speed trains [J]. Materials & Design, 2015, 91:278-285.

[2] J. Li,M. Zhang, Y. Zhao, Z. Lin, Control of porositis in aluminum alloys welds and research of welding joint property, J. East China Shipbuilding Inst. (Nat. Sci. Ed.) 18 (5)(2004) 78–81.

[3] Gou G, Zhang M, Chen H, et al. Effect of humidity on porosity, microstructure, and fatigue st Rength of A7N01S-T5 aluminum alloy welded joints in high-speed trains [J]. Materials & Design, 2015, 85: 309-317.

[4] Fang H C, Chen K H, Chen X, et al. Effect of Zr, Cr and Pr additions on microstructures and properties of ultra-high strength Al–Zn–Mg–Cu alloys[J]. Materials Science & Engineering A, 2011, 528(25–26):7606-7615. http://dx.doi.org/10.1016/j.msea.2011.06.018

[5] J-Y. Buffie`re, S. Savelli, P.H. Jouneau , E. Maire , R. Fouge. Experimental study of porosity and its relation to fatigue mechanisms of model Al–Si7–Mg0.3 cast Al alloys[J]. `Materials Science and Engineering A316 (2001) 115–126. http://dx.doi.org/10.1016/S0921-5093(01)01225-4

[6] Bray DE, Tang W. Subsurface stress evaluation in steel plates and bars using the LCR ultrasonic wave. Nucl Eng Des 2001;207:231–40. http://dx.doi.org/10.1016/S0029-5493(01)00334-X

Residual Stresses 2016: ICRS-10 Materials Research Forum LLC
Materials Research Proceedings **2** (2016) 509-514 doi: http://dx.doi.org/10.21741/9781945291173-86

Evaluation of Residual Stress Effects in Aft Pressure Ring Frame of a Royal Australian Air Force P-3C Maritime Patrol Aircraft

Kevin Walker[1, a *], Michael Ryan[1,b] and James Ayling[2,c]

[1]DST Group, 506 Lorimer St Fishermans Bend, Victoria, Australia

[2]Airbus Group Australia Pacific, NSW, Australia

[a]kevin.walker@dsto.defence.gov.au, [b]michael.ryan@dsto.defence.gov.au, [c]james.ayling@airbus.com

Keywords: X-Ray Diffraction, Finite Element Analysis, Thermal Control, P-3C Orion, Fuselage Pressure Frame

Abstract. Aft pressure bulkhead ring frames on the RAAF P-3C Orion aircraft are being replaced due to in-service fatigue cracking. Installation of the new 7075-T6511 extrusion L-section circular frame requires local adjustment of the included angle at various positions around the circumference. The adjustment should be carried out under a controlled temperature in the range of 135 – 160° C. However in some recent cases the temperature was either not well controlled and/or was below that range, raising concerns about the potential for adverse residual stresses in a region already known to be susceptible to fatigue. This paper details an investigation including 3-D non-linear finite element (FE) modelling and X-Ray diffraction based residual stress measurements on representative frame segments. The FE modelling was performed with NASTRAN using 3-D solid elements. Plastic behaviour was modelled using a Ramberg-Osgood approach. The X-Ray diffraction was performed using a Proto iXRD Combo instrument with a Cobalt X-ray source and a rectangular aperture (1.0, 1.5 or 2.0 x 5.0 mm depending on the location) and also electro-polishing for depth data. Permanent resultant angles of the frame segments were also measured using a "MarSurf CD 120" contact profiler. Typically low residual stresses were found for both the within and the below specification temperature cases. The FE and measurements were in reasonable agreement giving confidence in the results. The results from this work supported a decision to clear affected aircraft for a return to service.

Introduction

The aft pressure bulkhead structure at FS 1117 on the P-3 Orion aircraft, Figure 1, includes an approximately "L" shaped forward frame segment, Figure 2. The frame, manufactured from 7075-T651 alloy extrusion has a nominal included angle of 100°. Due to cracking which has occurred in service at the location of interest shown in Figure 2, the frames are being replaced on some RAAF aircraft. During installation of the new frames, the included angle needs to be adjusted in some areas. In the upper region of the fuselage the angle is tightened to about 94°, and in the lower end the angle is increased to about 102°. The angle adjustments are performed mechanically as shown in Figure 3 with locally applied heating [1]. The temperature range for the heating should be carefully controlled and thermocouple measurements are to be taken to ensure that the temperature range is controlled to the range 135 – 160° C. In some cases however, thermocouple measurements were not taken and it was thought that the hand forming in those cases may have occurred at a temperature below the specified range.

Materials Research Proceedings **2** (2016) 509-514 doi: http://dx.doi.org/10.21741/9781945291173-86

Figure 1 : Aft pressure bulkhead location at FS 111, RAAF AP-3C aircraft

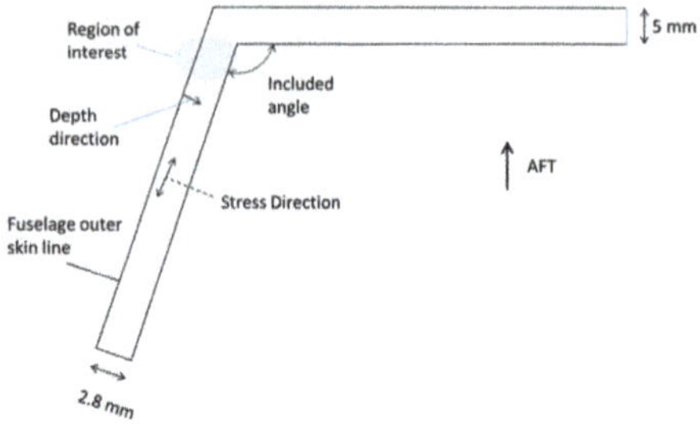

Figure 2 : Cross section of forward ring frame of FS 1117 aft pressure bulkhead

Figure 3 : Hand forming process for FS 117 forward ring frame included angle adjustment

Even if the forming process is carried out correctly, residual stresses are expected to be present in the frame. But there was a concern that if the process was carried out incorrectly in terms of temperature control in particular, then the final residual stress state might be such that fatigue crack growth and fatigue life would be significantly degraded.

This paper provides a summary of experimental and analytical activities to investigate this issue. The objective was to determine if performing the ring frame adjustments without correct temperature control would result in a significantly detrimental outcome in terms of residual stress. Samples of ring frame segments were obtained including the following scenarios; upper and lower locations, parts removed from a service aircraft, adjusted parts with and without proper temperature control, and nominal as manufactured parts. This resulted in a total of eight test specimens. The main area of concern is the upper frame where the nominal angle is reduced. That scenario is expected to result in tensile residual stress at the inner surface which is the location where fatigue cracking has been experienced in service.

X-Ray residual stress measurements were conducted using a Proto iXRD Combo with a cobalt x-ray source and a rectangular aperture (1.0, 1.5 or 2.0 x 5.0 mm). Measurements were taken wherever possible on the inner and outer surfaces at the location of interest, and also into the depth using electro-polishing. The analytical work included three-dimensional elastic-plastic finite element

Materials Research Proceedings **2** (2016) 509-514 doi: http://dx.doi.org/10.21741/9781945291173-86

modelling using MSC Nastran. These results were compared and used to determine if performing the frame adjustments without controlling the temperature would result in a detrimental outcome in terms of residual stress.

Finite Element Modelling

Finite element modelling was performed using NASTRAN with six-sided, eight-noded solid non-linear elements. Plastic behaviour was modelled using a Ramberg-Osgood approach. Further details are provided in [1]. The most significant result for comparison later with the measurements relates to the Specimen 1 scenario of an upper location with a nominal angle reduction of 6°. The FE predicted stress in that case is 385 MPa with room temperature forming, and 300 MPa with 160° C forming The FE analyses also determined that if the stress was tensile at the inner surface, then the stress is compressive at about the same magnitude at the outer surface. This finding was very important for the assessment of the X-Ray based residual stress measurements because many of them were conducted from the outer surface due to texturing and access difficulties with inner surface measurement. These were the key findings from the FE modelling to be compared with the X-Ray rsidual stress and final angle measurements which are discussed next.

Measurement details

Specimen descriptions. A total of eight test specimens were available. These covered both the upper and lower locations, combined with the three scenarios of forming with and without proper temperature control, and no forming at all (as manufactured). Another two were the upper and lower sections removed from a service aircraft, Tail Number A09-662. It was also noted whether the section was from the left or right hand side (LH, RH), but that was not expected to be a significant variable here and was not explicitly investigated.

The specimens provided have an arc length of about 500 mm. A typical example is shown in Figure 4.

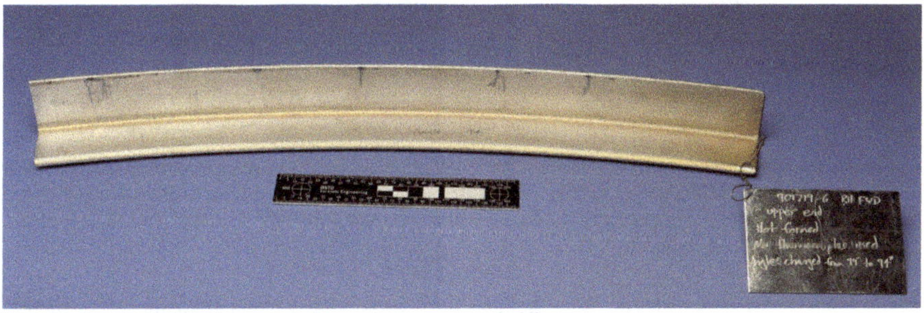

Figure 4 Typical FS1117 ring frame specimen provided for investigation

X-Ray diffraction details. X-Ray diffraction measurements were performed with a Proto iXRD Combo with a Cobalt X-Ray source and a rectangular aperture (1.0, 1.5 or 2.0 x 5.0 mm). Profile measurements into the depth were made wherever possible. Depth measurements were performed by taking progressive electro-polishing cuts using the Proto system. Most measurements were taken from the outer surface, but some readings were also undertaken where possible on the inner surface. The typical set-up for an outer surface measurement is shown in Figure 5.

Materials Research Proceedings **2** (2016) 509-514 doi: http://dx.doi.org/10.21741/9781945291173-86

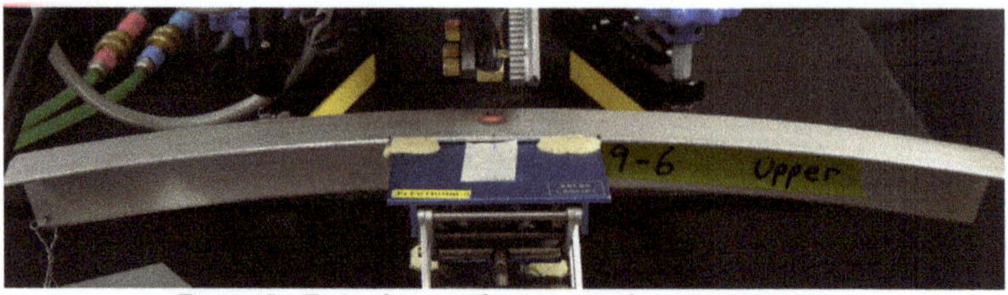

Figure 5 : Typical set-up for outer surface measurement

When measurements were taken on the inner surface it was necessary to cut away part of the thick flange to ensure that the X-Rays were not shielded from the region of interest. A sample with the flange cut away is shown in Figure 6

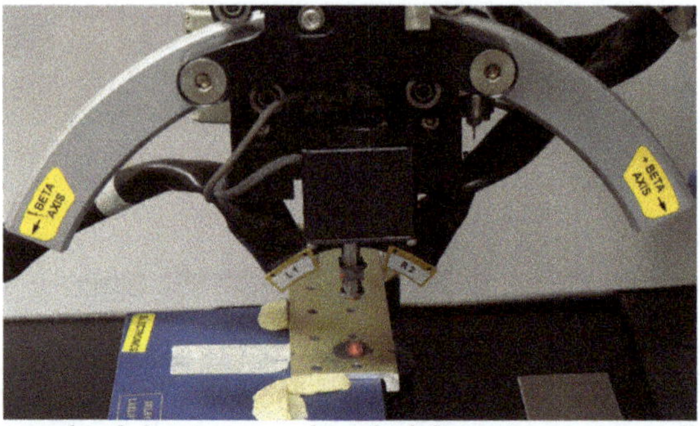

Figure 6 : XRD set up detail showing case where thick flange is cut away to allow inner surface measurement

Included angle measurement details. The included angle measurements were performed using a Mahr "MarSurf CD 120" contact profiler. Measurements were taken at each end of each sample, and also at the mid-region. The precision of the instrument is such that angle measurement to at least the nearest 0.1° or better is expected.

Results

The residual stress results into the depth from the outer surface are detailed in Figure 7. In addition, the following surface only results were obtained for the residual stress measurements:

Specimen 8 (lower frame, as manufactured) Outer surface Stress -50 +/- 10 MPa	Specimen 3 (lower frame, removed from service aircraft) Inner Surface Stress -13 +/- 10 MPa

Materials Research Proceedings **2** (2016) 509-514 doi: http://dx.doi.org/10.21741/9781945291173-86

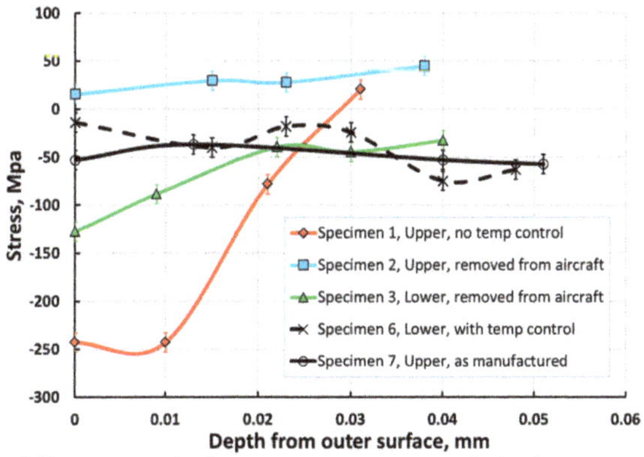

Figure 7 : X-Ray diffraction residual stress measurements with depth

The included angle measurements are detailed in Table 1.

Table 1 : Included angle measurement results

Specimen #	Description	Formed, Y/N	Thermocouples used, Y/N	Desired angle change, degrees	Measured Angle, degrees		
					Tag end	Middle	Other end
1	RH Upper	Y	N	100 to 94	92.7	91.0	93.0
2	RH Upper, removed from A09-662	N/A	N/A	N/A	95.1	94.4	94.1
3	RH Lower, removed from A09-662	N/A	N/A	N/A	102.2	101.8	103.1
4	RH Lower	Y	N	100 to 102	102.8	104.3	103.2
5	LH Upper	Y	Y	100 to 94	94.8	94.0	95.0
6	LH Lower	Y	Y	100 to 102	102.6	103.1	104.7
7	RH Upper	N	N/A	None	98.6	98.7	98.0
8	RH Lower	N	N/A	None	98.4	98.2	97.2

Discussion

The objective of this work was to measure the state of residual stress and final included angle for a range of representative scenarios in P-3 aft pressure bulkhead forward ring frames. It was not possible to obtain residual stress measurements in all the samples. This was attributed to texture effects in the elongated "pancake" type grain structure which is present in the 7075-T651 extruded ring frame segments. The texturing makes it more challenging for the X-Ray approach to obtain acceptable diffraction signal to determine the stress condition.

The peak residual stress in the tested ring frames were found to be mostly low, i.e. less than 50 MPa in magnitude. The most significant residual stress measured was about -250 MPa (compression) at the outer surface on sample 1 which was at the RH upper location (no temperature control) with a nominal angle change from 100° to 94°. In comparison, the FE result for a nominal

Materials Research Proceedings **2** (2016) 509-514 doi: http://dx.doi.org/10.21741/9781945291173-86

6° angle change is -385 MPa (compression) at room temperature, and -300 MPa at 160° C. Given that some heat was applied to Specimen 1, the comparison between the FE and the measurement is considered reasonable, with the FE being more conservative. In this case, the "spring back" effect explains why the stress state would be significant compression on the outer surface. The measured final angle was in the range 91-93° (see Table 1), which is slightly lower than the "target" angle of 94°. So the residual stress is expected to be larger in magnitude than if the final angle was higher (less change). These results can be compared with Specimen 2 which was removed from a service aircraft. The upper location nominally requires a reduction in the angle from 100 to 94°. For Specimen 2, the measured angle was about 94-95° which implies that less change was applied compared to Specimen 1. The peak residual stress at the outer surface for Specimen 2 was very low, 15 MPa, which is consistent with the observations just described.

Specimen 3 was removed from an aircraft and was at the lower location, where the angle is nominally increased from 100 to 102°. The "spring back" effect mentioned earlier suggests that a state of residual tension would be expected at the outer surface, but in fact the measured stress was about -13.0 MPa (compression), which is very low. The final angle was measured at about 102-103°, which is very close to the target value. But the angle change in this case was likely very small.

The highest absolute stress measured in any sample was a tensile stress close to +50 MPa at a depth of about 0.04 mm in Specimen 2, which was removed from the upper location in a service aircraft A09-662. The final angle was measured to be 94-95°. Being the upper location, like Specimen 1, the nominal condition would be decreasing the angle from 100 to 94°. In this case it appears likely that the angle change was therefore in the range 5-6°, which is less than for Specimen 1 where the nominal change was 7-9°. It is therefore consistent that Specimen 1 would exhibit larger magnitude residual stress than Specimen 2.

It is important to note that the measured residual stress even in new ring frame segments with no angle change were around -50 MPa (compression) at the outer surface (Specimens 7 and 8). This is a measure of the typical residual stress in the as-manufactured condition, so this gives an indication of the possible starting point for residual stress in these frames.

Conclusions

The residual stress state and final angle for a range of representative FS1117 forward ring frame samples were measured and reported. The residual stresses were generally low. The results suggest that the residual stress state is more likely to be related to a combination of the initial state and the angle change itself rather than whether the temperature was controlled or not. The FE analysis showed that although the temperature condition had some effect, the main parameter which influenced the resiudal stress condition was the permanent angle change. The X-Ray measurements showed that the FE analysis was conservative. It is therefore concluded that the ring frame segments where poor temperature control was applied would be expected to be not significantly worse in terms of residual stress than the cases where the specified temperature range was applied. This supported a decision to allow the affected aircraft to be returned to service.

Acknowledgements

The authors appreciate the assistance of Mr Bruce Grigson, DST Group, who conducted the included angle measurements using the Mahr system.

References

1. Scholz, L., *FS1117 residual stress investigation and analysis*, A09-ER-003002, 2015, Airbus Group Asia Pacific.

Residual Stresses 2016: ICRS-10 | Materials Research Forum LLC
Materials Research Proceedings **2** (2016) 515-520 | doi: http://dx.doi.org/10.21741/9781945291173-87

A Non-Destructive Investigation of two Cypriot Bronze Age Knife Blades using Neutron Diffraction Residual Stress Analysis

Christopher Davey[1, a*], David Saunders[1, b], Vladimir Luzin[2, c], Joseph Bevitt[2, d], Jennifer Webb[3, e], James Donlon[3, f] and Mihail Ionescu[2, g]

[1] Australian Institute of Archaeology, La Trobe University, VIC 3086, Australia

[2] Australian Nuclear Science and Technology Organisation, Locked Bag 2001, Kirrawee DC, NSW, 2232 Australia

[3] Department of Archaeology and History, La Trobe University, VIC 3086, Australia

[a]director@aiarch.org.au, [b]spedtwo@gmail.com, [c]vll@ansto.gov.au, [d]jbv@ansto.gov.au, [e]jenny.webb@latrobe.edu.au, [f]jamesrobertdonlon@gmail.com, [g]mio@ansto.gov.au

Keywords: Ancient Bronzes, Neutron Stress Measurements

Abstract. This paper presents the results of a residual stress analysis that is part of a wider study of Cypriot Bronze Age knife and other weapon blades from a corpus of artefacts held by a number of institutions in Australia. The current focus is on knives from Early/Middle Bronze Age burial sites at Bellapais *Vounous*, Cyprus; a significant number of the blades were found on excavation to be bent. The aim of the study was to provide, by means of non-destructive neutron residual stress analysis, likely insights into fabrication methodologies of the knives and determine the stage in the life of each knife blade at which bending occurred. Two *Vounous* knives from the Australian Institute of Archaeology collection, one measurably bent and the other severely bent and broken, were studied using neutron diffractometer KOWARI to establish the residual stress profiles through the thickness of the knives at several locations. Since the knives were 1 - 2 mm thick at their thinnest sections, a very high through-thickness spatial resolution of 0.1 mm was used to resolve the residual stress profiles. The experimental data from the knives suggested forging/hammering as a possible method of fabrication of functional (hard edge) knife blade. Most significantly, however, the post fabrication bending of both knives at ambient temperature was established. The residual stress data for the two knives were considered in the context of reported metallurgical studies and the archaeological information from Cypriot Bronze Age sites.

Introduction

The work reported in this paper brings together some aspects of the physical metallurgy of a specific corpus of artefacts from the well-documented burial site Bellapais *Vounous*, Fig. 1, an Early/Middle Bronze Age site in Cyprus (c 2450-1700 BCE). The site is a cemetery of over 160 tombs that were excavated by a number of archaeologists; notably Dikaios [3] and Stewart and Stewart [4]. Post excavation, the metallic artefacts, (including spearhead, knives and razors) and others of pottery were distributed to a small number of museums in England and Australia. Many of the *Vounous* artefacts available for study in Australia are provenanced to specific tombs at the site; and thus can be used to

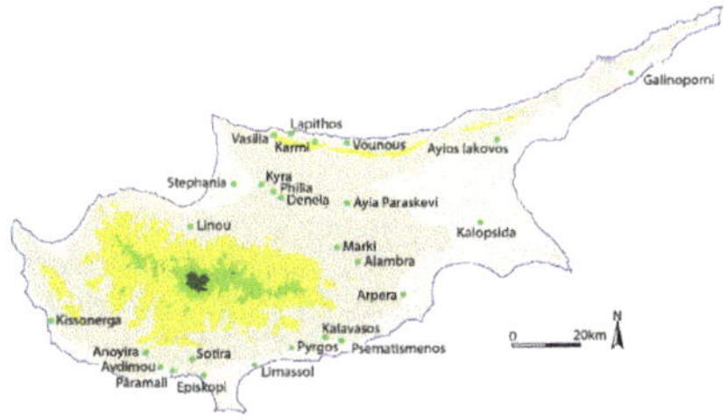

Fig. 1. Map of Cyprus showing the location of Bellapais Vounous and other Early/Middle Bronze Age sites (after [1, 2]).

provide insights into origins of raw material, use of the material, fabrication methods and treatment of the artefacts when used as burial items. These objects belong to early phase metallurgy in Cyprus and therefore offer insight into the nature and origins of its ancient copper industry. One particular feature of the Early/Middle Cypriot Bronze Age burials is that a significant number of knife and spear blades from this, and other, Bronze Age sites were found, on excavation, to be bent. Many scholars have noted the deliberate destruction or disabling of metal weapons in burial contexts (e.g. [2, 5-7]). Often referred to as 'ritual killing', the act of disabling is believed to have taken place at the burial site and assumed to have been intended to enable the weapons to accompany the dead and perhaps ensure that they could not be used as weapons after burial. Any insights into the fabrication route of the knives and determining the stage in the life of each knife blade at which bending occurred are highly desirable.

There have been limited metallurgical studies of *Vounous* metallic artefacts and much of the work to date has provided elemental analyses [1, 8], Pb isotope analysis [1] and metallography [9]. Through the metallographic analysis [9], there is a notion that the blades/weapons had been cold worked from original states, most likely cast blanks, annealed and then additionally cold worked to their final form. Since residual stresses are intrinsically linked to each these processes, there is a potential to provide new and corroborating information about the fabrication and treatment or handling of the Early/Middle Bonze Age bronze artefacts from fabrication and post-fabrication perspectives and possibly to support the view of ritual bending practices at some ancient Cypriot burials. Although application of neutron diffraction to study ancient bronzes has been reported [10], there are currently limited data and, to the authors' knowledge, no studies of residual stresses or textures within the corpus of *Vounous* artefacts. The current paper reports the results of the application of non-destructive neutron diffraction methods to two Bronze Age knife blades from *Vounous*.

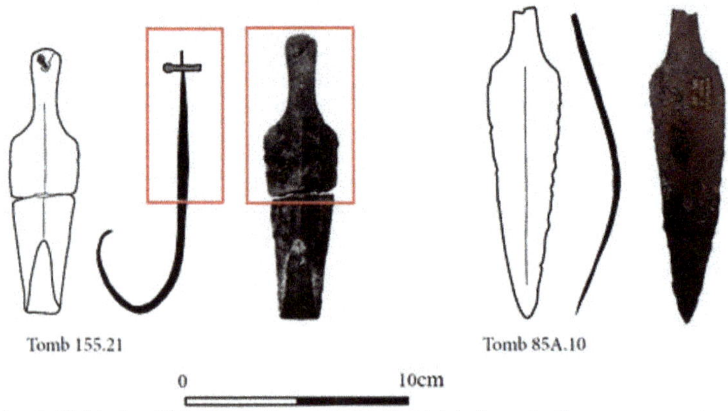

Fig. 2. Two knife blades, IA 2.525 (left) and IA 2.268 (right), from the Stewart excavations at Bellapais Vounous [4]. For IA 2.525, only the section illustrated within the red boundary was available for study while the severely bent section of the blade was lost some time after photographic recording for reference 4.

Samples

Two knives from the Australian Institute of Archaeology, AIA, *Vounous* artefacts IA 2.525 and 2.268 (Fig. 2), were submitted for neutron diffraction residual stress study and chemical analysis at the Australian Nuclear Science and Technology Organisation, ANSTO. Both knives were found bent in situ. Artefact IA 2.525 was bent through 180° and was found fractured. Artefact IA 2.268, was bent to approximately 40° at the mid-section of the blade.

Materials Research Proceedings **2** (2016) 515-520 doi: http://dx.doi.org/10.21741/9781945291173-87

Elemental characterization of IA 2.525 and IA 2.268

The chemical compositions of the two *Vounous* knives were determined by combined Proton Induced X-Ray Emission spectroscopy (PIXE) and Proton Induced Gamma-ray Spectroscopy (PIGE) using 2.5 MeV proton beam at the ANSTO Centre for Accelerator Science. The chemical analyses of other *Vounous* artefacts studied in work of Webb et al. [1] and Craddock [8] demonstrated that the majority of blades analyzed were essentially Cu-As alloys with, usually, 2.5-5 wt.% As. The two *Vounous* blades studied in this work were significantly different alloys, with As < 0.5 wt%.; IA 2.525 contained 1.30 wt% Zn and only a trace of Sn and IA 2.268 contained 0.56 wt % Zn and 0.66 wt% Sn.

The absence As suggests that the knifes were struck from a low As indigenous copper ore. With the arsenic content of IA 2.525 and IA 2.268 at < 0.5 wt%, much lower than threshold of 2.0 wt%, which some archaeologists believe is an indicator of deliberate alloying with this element, the ability to develop a hardened cutting edge would have been limited. As modern chemical analyses have shown, for copper blades from Cypriot Early/Middle Bronze Age sites, including *Vounous*, As was most likely added as an ore in the smelting process the result of which aided casting and increased propensity for work hardening [11]. The presence of Zn in IA 2.525 confirms an earlier analysis reported in Stewart and Stewart [4]. Balthazar [9] and other researchers have noted the presence of Zn in some Early/Middle Bronze Age artefacts but no conclusions about its presence have been drawn, other than questioning some earlier analytical methods. Further chemical analyses of Cypriot artefacts in Australian institutions are proposed and these may provide additional insights regarding Zn and other elements as alloying elements.

Neutron residual stress measurements

The residual stresses within the two *Vounous* knives were determined using the ANSTO KOWARI neutron stress diffractometer. [12]. Neutron diffraction stress measurements were undertaken at three gauge locations (L1, L2 and L3) on IA 2.268 (Fig. 3) and at one location (L1) on IA 2.525 (Fig. 4), to determine the stress profiles through each of the blades. With thickness of the blades varying from 1 to 2 mm, through-thickness stresses were measured with a relatively high spatial resolution of 0.1 mm in through-thickness dimension. The strongest Cu(311) reflection was used for the residual stress measurements at wavelength of 1.55 Å. A gauge volume of $0.1\times0.1\times10$ mm^3 was used for the measurements in the three principal directions (longitudinal, normal and transverse) at the central line of the blades, at L1 and L2, however for measurements in the thinnest section of IA 2.268 the gauge volume, at L3, was reduced to $0.1\times0.1\times7$ mm^3.

To fully resolve the stress state, a zero normal stress condition was used to calculate stresses with high accuracy at the same time making possible non-destructive determination of each lattice parameter d_0 of the alloy. A large number of measurement points through thickness, 9-13, with 0.1 mm steps, was considered necessary to determine several possible contributions to the total residual profile (hot forging, cold hammering, bending, etc.) which were resolved and separated. With measurement time of only 10-20 minutes per through-thickness position, a strain accuracy of 50 μstrains was routinely achieved.

Fig. 3. Approximate locations of the three positions used for residual stress measurements of IA 2.268.

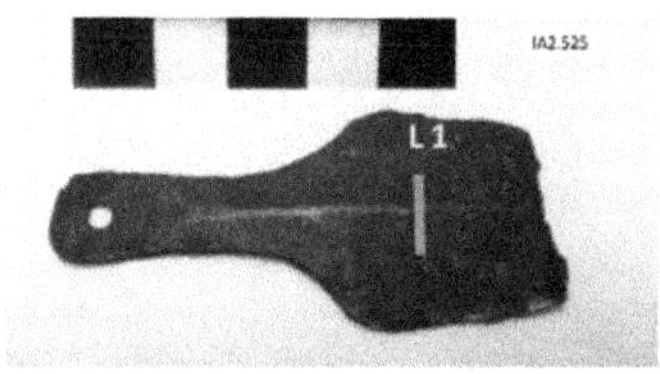

Fig. 4. Approximate locations of the position used for residual stress measurements of IA 2.525

Materials Research Proceedings **2** (2016) 515-520 doi: http://dx.doi.org/10.21741/9781945291173-87

Results and discussion

The results of the residual stress measurements at the gauge locations L1, L2 and L3 for IA 2.268 are shown in Fig. 5 a,b,c and for gauge location L1 for IA 2.525 in Fig. 6.

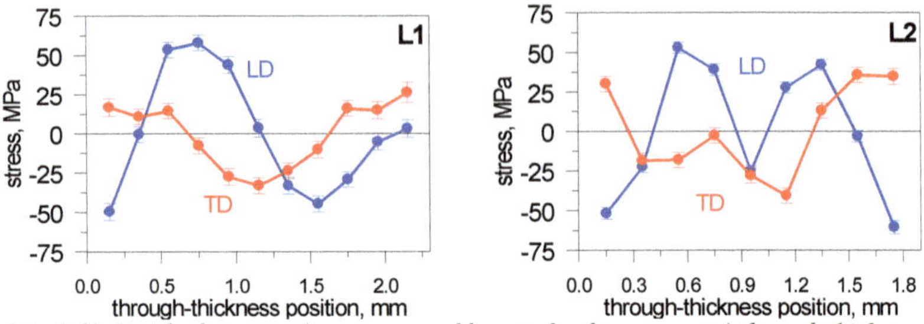

Fig. 5 (a & b). Residual stresses (transverse and longitudinal components) through thickness of knife IA 2.268 at positions L1 and L2.

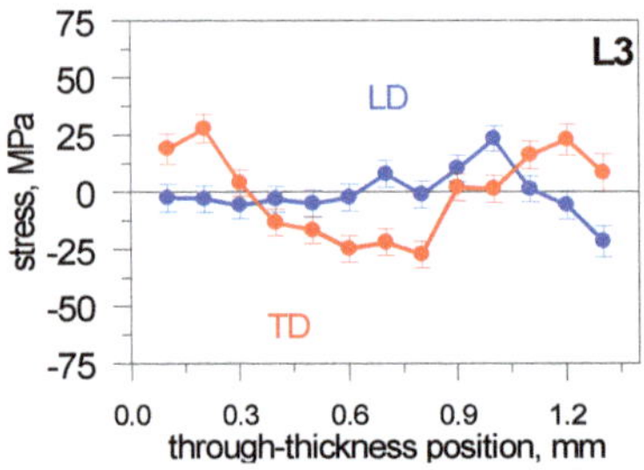

Fig. 5c. Residual stresses (transverse and longitudinal components) through thickness of knife IA2.268 at position L3.

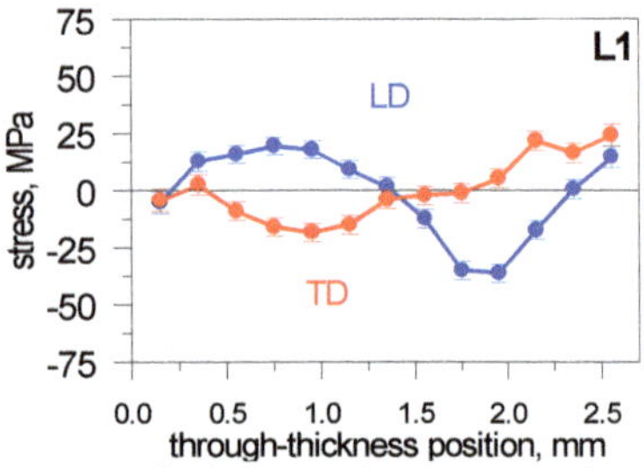

Fig. 6. Residual stresses (transverse and longitudinal) through thickness of knife IA 2.525 at position L1.

The through-thickness residual stress profiles of the transverse direction are similar for both blades and all locations. However, the transverse component of the thicker knife IA 2.525 at position L1 seems to incorporate also a bending stress component (slope) that is superimposed by the surrounding parts of the blade other than the location of measurements. Assuming that the plastic bending in the blades does not significantly affect the other stress components, most likely it is representative of the pre-bending treatments that can be associated with combination of hot forging

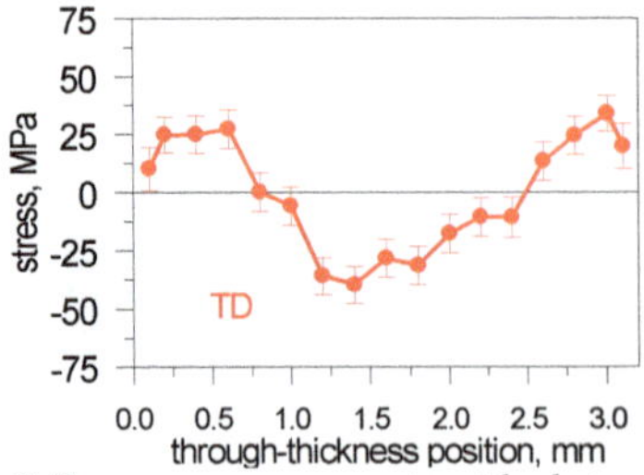

Fig. 7. Transverse component residual stress profile of a hot rolled Cu strip.

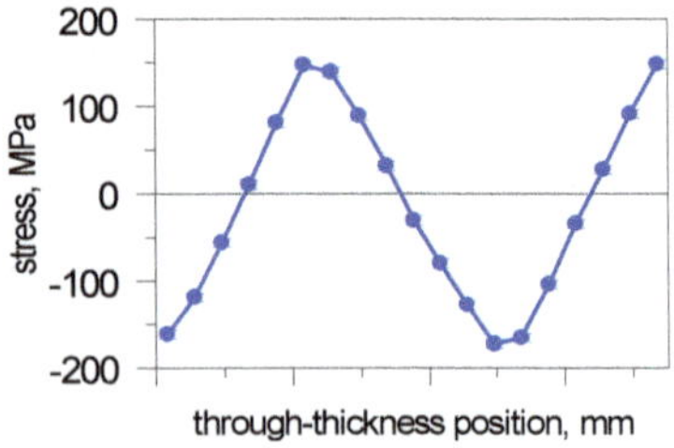

Fig. 8. Longitudinal component residual stress profile of a cold bent Al plate.

or hammering (producing typical U- or V-shape stress profiles) and surface cold hammering (producing flattening of the stress profiles close to the surfaces due to compressive peening contributions). A contemporary analogue of a similar production route is given in Fig. 7. Therefore, the results show the expected route of production of similar bronze blades with possible heating and hammering a cast metal blank into shape as discussed below.

The through-thickness profiles of the longitudinal stress components are clearly a sequence of bending applied to both samples. A typical for bending "z-shaped" profile of the longitudinal component (Fig. 8) appears in the thickest part (L1) of the both blades. Thus, the through-thickness residual stress profiles of the longitudinal direction strongly indicate that the blade had been bent in the cold state. The longitudinal stress profile in the location of maximum curvature (L2) is most probably indicative of double bending, although the experimental data might be affected by microstructural variations and casting defects. This could be resolved at a later time through metallographic studies although currently this is not planned. Neutron imaging methodologies may also prove effective in identifying highly deformed material and the presence of inclusions, Cartechini et al [13].

Finally, the thinnest part of the bent knife IA 2.268, L3 evidently does not show the same bending pattern as for the thick part but rather bending in the opposite direction suggesting that the bending support point was between the point of maximum bending and the location of measurements. Alternatively, the profile could be interpreted rather as an evidence that one side of the blade (right side in Fig. 5a) was worked by forging/hammering at ambient temperature as the last episode of the treatment producing compressive zone in the surface region balanced by tension zone but most likely this treatment would produce the same feature in the transverse component that is not corroborated by the experimental data at L2.

Overall, the residual stress signatures for IA 2.525 at L1 and IA 2.268 at L1 were similar and consistent (although sample IA 2.268 is represented with just one profile) and suggested that both blades were subjected to bending at ambient temperature after fabrication.

Historical context

Webb and Frankel [2] p 132 note, (the) "Intentional breakage or distortion of copper-based spearheads and knives clearly occurred across Cyprus during the Early and Middle Bronze Age. While the practice was not universal, the number of examples is by no means insignificant." Their analysis showed that, of the knives excavated from Early/Middle Bronze Age sites, some 22.3% of 301 known knives from Early/Middle Bronze Age burials were bent to greater or lesser degrees, and of interest is that 33.8% of the knives recovered at *Vounous*, were damaged in some way. It is, of course, possible that some of the damage was accidental and residual stress measurements will not differentiate this aspect of damage, particularly where there are only small amounts of bending, possibly for bends of up to 20 degrees.

The residual stress results of IA 2.268 and IA 2.525 does not contradict the suggestion by Balthazar [9] based on metallographic evidence that the knife blades had been forged or hammered from what was most likely a "blank" (as cast) to achieve the final shape with evidence of annealing after hammering. This is also consistent with the results of excavations at other Early/Middle Bonze Age Cypriot sites, such as Sotira *Kaminoudhia* [14] where both a dagger and cast blank for a dagger were found. Casting moulds for ingots have been found at Alambra *Mouttes* and Marki *Alonia* [15].

The residual stress results suggest that the objects were not knives specifically made as burial objects, but rather normal functional tools with work hardened blades. The knives IA 2.268 and IA 2.525 may have been used at or around the site of *Vounous*, but certainly, as functional tools, they must have had sufficient significance to be included in burials that warranted "removal" of the knives from utilitarian circulation. Knife IA 2.268, for example, was most probably of significance to either the person buried in Tomb 85A or/and as part of a burial ritual. Additionally, the deliberate damage to the knife blade by bending suggests that the functionality of the blade was destroyed or disabled prior to its deposition in the tomb. As deliberate damage is observed across a number of Early/Middle Bonze Age burial sites this suggests a consistent burial practice or ritual, see [1, 7] for discussions on this matter.

Summary

This preliminary study of residual stresses in two Bronze Age knives from the Australian Institute of Archaeology collection was undertaken using the KOWARI neutron diffractometer at ANSTO. The results provided evidence for the hammered "working" of one blade edge of one of the knives and suggested that both knives were subjected to bending after fabrication.

This information, when used with data from *Vounous* and other similar sites and other compositional and metallurgical analyses, provide a basis for placing these artefacts within Early/Middle Bronze Age material and social context, particularly their use in burials as part of mortuary practices.

This work has demonstrated the value of using neutron diffraction residual stress measurements for the non-destructive study of ancient metallic artefacts, which is particularly important from a cultural heritage and artefact preservation perspective.

References

[1] J.M. Webb, D. Frankel, Z.A. Stos, N. Gale, Early Bronze Age metal Trade in the Eastern Mediterranean. New Compositional and Lead Isotope Evidence from Cyprus. Oxford Journal of Archaeology, 25 (3) (2006), pp. 261 – 288. http://dx.doi.org/10.1111/j.1468-0092.2006.00261.x

[2] J.M. Webb, D. Frankel, Coincident Biographies, Bent and Broken Blades in Bronze Age Cyprus. in K. Harrell, J. Driessen, (Eds.) Contextualising Intentional Destruction of Objects in the Bronze Age Aegean and Cyprus, Publ. Louvain, 2015, pp. 117 – 142.

[3] P. Dikaios, The Excavations at Vounous Bellapais in Cyprus, 1931-2, Archaeologia 88 (1940), pp. 1-174. http://dx.doi.org/10.1017/S0261340900014557

[4] E. Stewart, J. R. Stewart, Vounous 1937 -38. Field Report on the Excavations Sponsored by the British School of Archaeology at Athens. Publ. Lund, 1950.

[5] P. Åström, Intentional destruction of grave goods, in: R. Laffineur (Ed.), Thanatos. Les coutumes funéraires en Egée à l'Age du Bronze (Aegaeum I), Liège, 1987, pp. 213–217.

[6] G. Georgiou, J.M. Webb, D. Frankel Psematismenos-Trelloukkas. An Early Bronze Age Cemetery in Cyprus, Publ. Department of Antiquities, Nicosia, 2011.

[7] P. Keswani Mortuary Ritual and Society in Bronze Age Cyprus, Monographs in Mediterranean Archaeology, No. 4, Equinox Publishing, London and Oakville, 2004.

[8] P.T. Craddock, Report on the composition of metal tools and weapons from Ayia Paraskevi, Vounous and Evreti, Cyprus in: A Catalogue of Cypriot Antiquities in the Birmingham Museum and Art Gallery, Birmingham, 1981.

[9] J.W. Balthazar, Copper and Bronze Working in Early through Middle Bronze Age Cyprus Studies, Mediterranean Archaeology and Literature, Pocket Book 84, Paul Åström's Förlag, Göteborg, 1990.

[10] S. Siano, L. Bartoli, J.R. Santisteban, W. Kockelmann, M.R. Daymond, M. Miccio, M., G. De Marinis, Non-Destructive Investigation Of Bronze Artefacts From The Marches National Museum Of Archaeology Using Neutron Diffraction, Archaeometry, 48 (1) (2006), pp. 77-96. http://dx.doi.org/10.1111/j.1475-4754.2006.00244.x

[11] H. Lechtman H. Arsenic Copper: Dirty Copper or Chosen Alloy? A View from the Americas, Journal of Field Archaeology, 23 (4) (1996), pp 477 – 514.

[12] O. Kirstein, V. Luzin, U. Garbe, The Strain-Scanning Diffractometer Kowari, Neutron News, 20 (4) (2009), pp. 34-36.

[13] L. Cartechini, R. Arletti, R. Rinaldi, W. Kockelmann, S. Giovannini, A. Cardarelli Neutron scattering material analysis of bronze Age metal artefacts, Journal of Physics of Condensed Matter, 20 (2008), pp. 1-8. http://dx.doi.org/10.1088/0953-8984/20/10/104253

[14] S. Swiny, G. Rapp. E. Hersher, E. (Eds.), Sotira Kaminoudhia: An Early Bronze Age Site in Cyprus, Cyprus American Archaeological Research Institute Monograph Series, Vol. 4, American Schools of Oriental Research, Boston, 2003.

[15] A.B. Knapp, Metallurgical production and Trade on bronze Age Cyprus: Views and Variations in: V. Kassianidou, G. Papasavvas (Eds.) Eastern Mediterranean Metallurgy and Metalwork in the 2nd Millennium B.C., Publ. Oxbow, 2009, pp. 14 – 125.

Materials Research Proceedings **2** (2016) 521-526 doi: http://dx.doi.org/10.21741/9781945291173-88

Study of Stress Partitioning in a 0.68 wt%C Pearlitic Steel Using High Energy X-Ray Synchrotron Radiation

BRAHAM Chedly[1,a*], BACZMAŃSKI Andrzej[2,b], GONZALEZ Gonzalo[3,c], SIDHOM Habib[4,d], GADALIŃSKA Elżbieta[5,e], WROŃSKI Sebastian[2,f], BUSLAPS Thomas[6,g], WAWSZCZAK Roman[2,h]

[1]Laboratoire Procédés et Ingénierie en Mécanique et Matériaux (PIMM, CNRS UMR 8006) Arts et Métiers-ParisTech, 151 Bd de l'Hôpital, 75013 Paris, France

[2]AGH-University of Science and Technology, Faculty of Physics and Applied Computer Science, Al. Mickiewicza 30, 30-059 Kraków, Poland.

[3]Instituto de Investigaciones en Materiales, Universidad Nacional Autónoma de México Circuito exterior S/N, Cd. Universitaria, A.P. 70-360, Coyoacán, C.P. 04510, Mexico

[4]Mechanical, Materials and Processes Laboratory (LR99ES05), ENSIT 5, Avenue Taha Hussein 1008, University of Tunis, Tunisia

[5]Institute of Aviation, Materials and Structures Research Center, al. Krakowska 110/114, 02-256 Warszawa, Poland

[6]ESRF, 6, rue J. Horowitz, 38500 Grenoble Cedex, France

[a] chedly.braham@ensam.eu, [b]Andrzej.Baczmanski@fis.agh.edu.pl, [c] joseggr@unam.mx, [d] habib.sidhom@gmail.com, [e] Ela.gadalinska@gmail.com, [f] Sebastian.Wronski@fis.agh.edu.pl, [g]buslaps@esrf.fr, [h]Roman.Wawszczak@fis.agh.edu.pl

Keywords: Pearlitic Steel, Residual Stress, X-Ray Diffraction, Synchrotron, Elasto-Plasticity, Self-Consistent Model

Abstract. In the present work, the evolution of the phase stresses in a 0.68 wt%C pearlitic steel is analyzed by synchrotron diffraction during uniaxial tensile loading, at room temperature. The diffraction measurements were done at ESRF beamline ID15B (Grenoble, France). The microstructure of the studied material, obtained after an austenitizing at 1050°C for 7 minutes followed by cooling under blowing air, corresponds to fully pearlitic steel with a cementite volume fraction of about 12.5%. As expected, the results show a clear effect of elastic and plastic anisotropy in the both phases. For the interpretation of the diffraction data, different models are compared. In elastic range and for small plastic deformation, the self-consistent model presents the best agreement with the experimental data. For large plastic deformation, this model does not predict correctly the stress partitioning between the phases as well as the macro behavior of the studied steel. Therefore a mixture model "(1-x)*self-consistent model + x*Taylor" was used to take into account the interaction between the phases.

Introduction

Offering an excellent combination of ductility, strength and cost, the fully pearlitic steels are the most used plain carbon steels in manufacturing to produce wires for reinforcing tires, cables for suspension bridges, engineering springs for automotive and railroads. The role of the microstructure and specially the interlamellar space in the mechanical behavior and fatigue resistance of pearlitic steels has been studied in previous works [1, 2]. In-situ diffraction technique during mechanical loading is a powerful method to investigate the mechanical behavior of the phases and the stress partitioning between the cementite and the ferrite [2-5]. Only few results concerning the role of the

Materials Research Proceedings **2** (2016) 521-526 doi: http://dx.doi.org/10.21741/9781945291173-88

lamellar cementite in the hardening of the fully pearlitic steel have been reported [1, 2]. We propose in this study to apply to fully pearlitic steel an approach based on the analysis of the mechanical properties of the polycrystalline material at the grain and phase scale using synchrotron X-ray diffraction technique and elasto-plastic models.

Material

The pearlitic steel EN C70 (SAE 1070) investigated in this study was produced by ASCOMETAL France Company in the form of bars of 80 mm in diameter. Its chemical composition is given in Table 1. The bars were submitted to an austenitizing at 1050°C for 7 minutes followed by cooling under air blown. The resulting microstructure consists of entirely pearlite colonies having an average size of 7- 8 μm with cementite plates lamellae spaced by 170 nm (Fig. 1). The cementite fraction was estimated at 12.5 vol % using Rietveld phase quantification from X-ray diffraction data. The mechanical properties of the obtained microstructure are reported in Table 2.

Table 1- Chemical composition of EN C70 pearlitic steel (wt%)

C	Si	Mn	S	P	Ni	Cr	Mo	Cu	Al	Fe
0.68	0.192	0.846	0.010	0.010	0.114	0.160	0.027	0.205	0.042	balance

Table 2- Mechanical properties of the annealed EN C70 pearlitic steel

Yield Stress [MPa]	Ultimate Stress [MPa]	Elongation [%]	Hardness [HV_{50}]
498	997	15	270 ± 15

Experimental setup

X-ray diffraction measurements in a transmission mode were carried out during in situ tensile test on the studied steel at the European Synchrotron radiation Facility (ESRF, ID15B beamline). Monochromatic high energy X-ray synchrotron radiation (~90 keV, λ=0.14256 Å) with a beam size of 100 μm x 100 μm was used for measurements of lattice strain evolution in both phases (ferrite and cementite) independently during increasing of macroscopic applied tensile stress.

Measurements were performed for numerous ferrite and cementite reflections in order to study the anisotropic effects on the loading transfer. Two dimensional diffraction patterns, measured in the range 2θ = 1.9°-15.5°, were recorded by the Thales PIXIUM 4700 detector and integrated using the FIT2D software [6] and converted in d-spacing. Then, a pseudo-Voigt function was fitted to the measured diffraction profiles and the peak positions were determined using the MULTIFIT software [7].

Results and discussion

Lattice strain evolution under tensile loading. The elastic lattice strains in the axial loading direction (ε^e_{11}) and also in the transverse loading direction (ε^e_{22}) were determined in both phases, for various reflections (Fig. 2). In the range of elastic deformation, the plots of phase lattice strains vs macroscopic applied stress (Σ_{11}) show similar values of lattice strains for both phases of the studied steel ($\Sigma_{11} \leq 500$ MPa). The slopes were quite similar but the responses of phases, under loading, depend on the (hkl) planes (Fig. 2). Under larger imposed strain corresponding to macroscopic stress level higher than 500 MPa, the lattice strains of ferrite are approximately constant, while those of cementite increase significantly with the increasing applied tensile stress (Figure 2). This indicates

Materials Research Proceedings **2** (2016) 521-526 doi: http://dx.doi.org/10.21741/9781945291173-88

clearly the occurring of load transfer from ferrite to cementite in plastic range which continues until fracture. A low hardening of ferritic phase during plastic deformation leads to slight strengthening of this phase. The different responses of lattice reflections indicate that anisotropy is also present in plastic range for both phases.

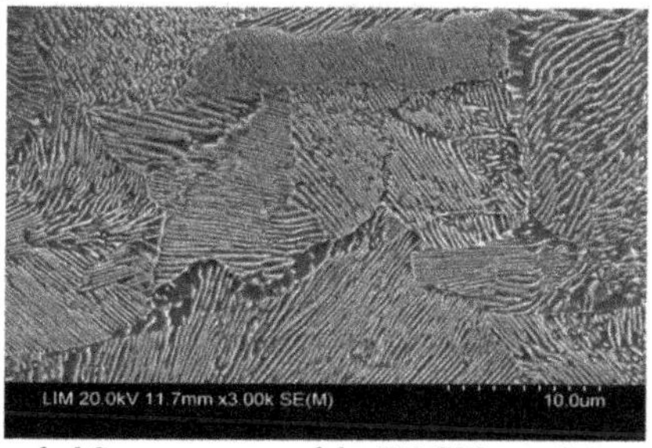

Figure 1- Microstructure of the EN C70 pearlitic steel

Elastoplastic behavior modeling. To interpret the experimental results related to the individual behavior of phases and macroscopic tensile curves, different elasto-plastic models were used and compared: Reuss, Voigt, and self-consistent model. For the self-consistent model, calculations were carried out for 10,000 inclusions representing the grains of ferrite and cementite. It was assumed that ferritic crystallites are deformed plastically due to slips on the <111> {211} and <111> {110} systems, while the cementite grains are deformed only elastically. The formulation of the self-consistent elastoplastic model proposed by Berveiller and Zaoui [8] and developed by Lipinski and Berveiller [9] was used in our calculations. In this model the interaction of ellipsoidal inclusion with the homogenous medium is described by the Eshelby-type tensor.

For the ferrite phase in the elastic range of deformation, the comparison between calculations with the three chosen models and our experimental results shows a good agreement only in the case of the self-consistent model. The Voigt model assumes homogeneous deformations does not consider the elastic anisotropy and do not agree with the experimental results. In the case of the Reuss model (assuming homogeneous stresses) the effect of anisotropy was overestimated. For the cementite, also the self-consistent model gives the best range agreement but significant differences are observed in the slopes of the curves for different hkl reflections. The calculations for ferritic phase were performed using elastic constants given in Table 4. However in the case of cementite the best agreement between experimental and theoretical values was obtained when 80% of *ab initio* values shown in Table 4 were used in modeling. Two different assumptions concerning macroscopic polycrystalline aggregate were considered: one phase and two phase materials. The comparison between these assumptions did not show any difference due to the similar elastic behavior of the ferrite and the cementite.

Until a small range of plasticity, up to 700 MPa, the classical self-consistent model is still valid and a good agreement between experimental and calculated results are obtained using 80% of *ab initio* elastic constants for the cementite [10] (Fig. 3).

For large plastic deformation, the self-consistent model does not predict correctly the stress partitioning between the phases as well as the macro behavior of the studied steel (Fig. 4a). Therefore a mixture model is proposed in this study to better take into account the interaction between the phases (Fig. 4b). In this model the strain rate concentration tensor A_{ijkl} relating macrostrain rate $\dot{E}_{kl}$ with strain rate at the polycrystalline grain $\dot{\varepsilon}_{ij}$:

$$\dot{\varepsilon}_{ij} = A_{ijkl}\dot{E}_{kl} \qquad \text{and} \qquad A_{ijkl} = (1-x)A^{sc}_{ijkl} + x\, I_{ijkl} \qquad (1)$$

Materials Research Proceedings **2** (2016) 521-526 doi: http://dx.doi.org/10.21741/9781945291173-88

where A^{sc}_{ijkl} is the concentration tensor calculated using self consistent method, I_{ijkl} is the unit forth range tensor representing Taylor model in which the strain is homogenous, while x represent fraction of the Taylor model used in the calculations.The variable Taylor fraction x% in this model is adjusted to account for the different phenomena occurring during the tensile test: the strong phase interaction (i.e. x=0.2) appeared from the beginning of the ferrite plastic deformation followed by a softening stage (decreasing x-value) for very high range of plastic deformation. The strong interaction can be attributed to the localization of dislocations around the cementite particles and dislocation pile up phenomenon. The decrease of this interaction in the followed stage may be attributed to the rearrangement of dislocations and the fragmentation of cementite when the cracking is initiated. This mixture model was tested successfully for a large plastic range as shown in Fig. 4b and 5. A good agreement was found between experimental results and the mixed model calculations as well for the macroscopic behavior of the pearlitic steel (Fig. 5b) as for the evolution of lattice strains vs applied macro stress Σ_{11}, for the different reflections of the ferrite (Fig.5a) and cementite (Fig.5b).

Table 3- Single crystal elastic constants of ferrite used in modeling

	C11	**C44**	**C12**
[GPa]	231	116	134

Table 4- Cementite ab initio elastic constants [10]

	C11	**C22**	**C33**	**C44**	**C55**	**C66**	**C12**	**C13**	**C23**
[GPa]	385	341	316	13	131	131	157	162	167

Conclusion

Synchrotron X-ray diffraction measurements were applied to study the evolution of lattice strains in both phases (ferrite and cementite) of EN C70 pearlitic steel containing 12.5 (vol%) of cementite lamellae under tensile loading. Diffraction was performed for numerous reflections, in order to evaluate the effect of microstructural anisotropy, as well as the load partitioning between ferrite and cementite during tensile test. It has been demonstrated that in elastic range cementite and ferrite are loaded similarly and there is no load transfer due to their nearly equivalent Young modulus. However at the plastic range, load is transferred gradually from ferrite to cementite. The effect of elastic/plastic anisotropic occurs in different deformation ranges.

To reproduce the experimental results related to the individual behavior of phases and macroscopic tensile curves, different elasto-plastic models were used and compared. The classical self-consistent model presented a good agreement for elasticity and small plasticity. A mixture model "(1-x)*self-consistent model + x*Taylor" was tested successfully for a larger plastic range.

Acknowledgement

The work was supported partly by NCN - Polish National Center for Science, grants: No. UMO - 2011/03/N/ST8/04058 and DEC - 2013/11/B/ST3/03787; and partly by the MNiSW - Polish Ministry of Science and Higher Education. We also acknowledge the ESRF for the use of the synchrotron radiation facilities.

Materials Research Proceedings **2** (2016) 521-526 doi: http://dx.doi.org/10.21741/9781945291173-88

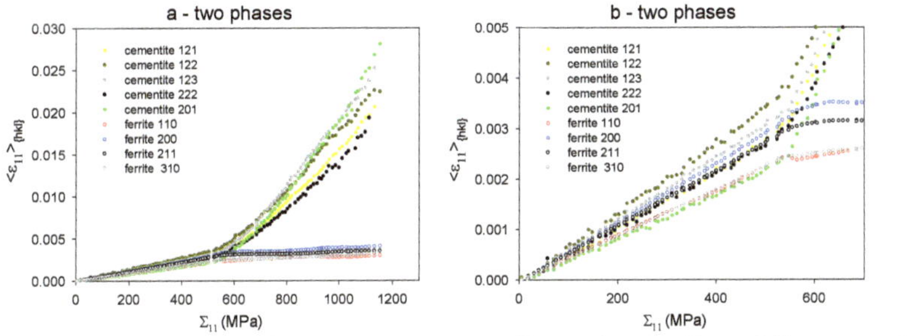

Figure 2- Variation of the lattice strains measured by synchrotron X-ray diffraction vs the applied macro stress Σ_{11}, shown for large (a) and small (b) stresses applied to the sample.

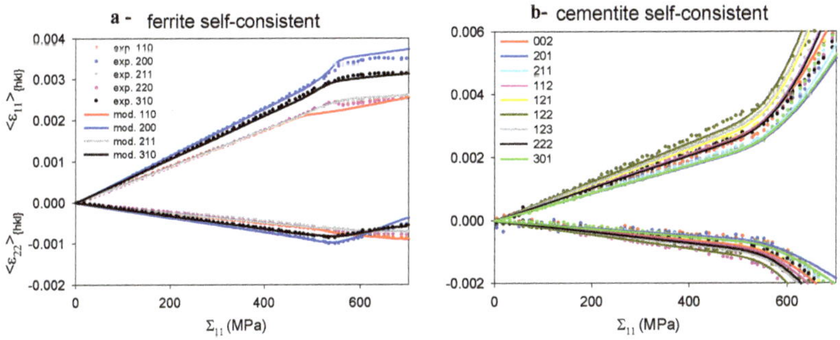

Figure 3- Longitudinal $<\varepsilon_{11}>$ and transversal $<\varepsilon_{22}>$ lattice strains vs applied macro stress Σ_{11}, comparison between experimental results and self-consistent model calculations for the ferrite (a) and the cementite (b). In the case of cementite we used 80% of ab initio elastic constants.

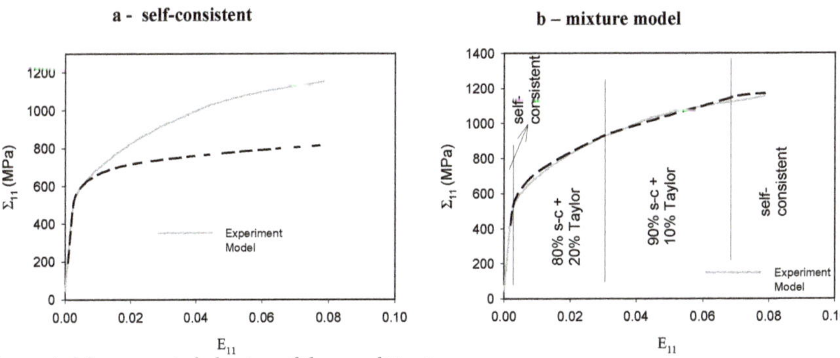

*Figure 4- Macroscopic behavior of the pearlitic steel - comparison of the experimental tensile curve with the curves calculated by (a) the self-consistent model and (b) the mixed model " (1-x)*self-consistent + x*Taylor).*

Materials Research Proceedings **2** (2016) 521-526 doi: http://dx.doi.org/10.21741/9781945291173-88

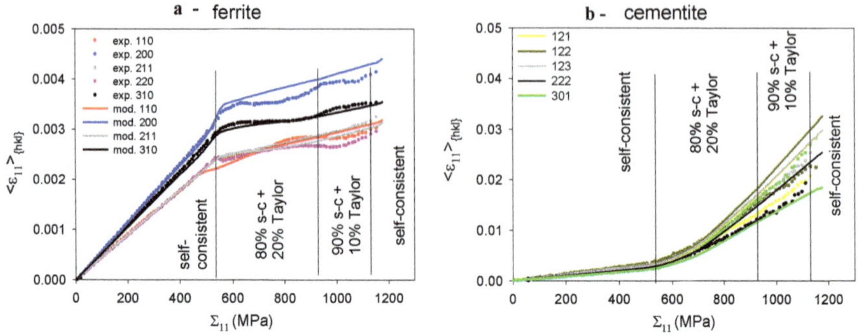

Figure 5- Longitudinal $<\varepsilon_{11}>$ *lattice strains vs macro applied stress* Σ_{11} *- comparison between experimental results and mixed model calculations for the ferrite (a) and the cementite (b).*

References

[1] H. Sidhom, H. Yahyaoui, C. Braham, G. Gonzalez, J.Mater. Eng. Perform 2015, 24: 2586-2596. http://dx.doi.org/10.1007/s11665-015-1537-7

[2] H. Yahyaoui , H. Sidhom , C. Braham , A. Baczmanski , Materials & Design 2014, 55 : 888-897. http://dx.doi.org/10.1016/j.matdes.2013.10.062

[3] M.R. Daymond, H.G. Priesmeyer, Acta Mater 2002, 50:1613–26. http://dx.doi.org/10.1016/S1359-6454(02)00026-5

[4] V. Taupin, R. Pesci, S. Berbenni, S. Berveiller, R. Ouahab, O. Bouaziz, Mater Sci Eng 2013, A561:67–77. http://dx.doi.org/10.1016/j.msea.2012.10.086

[5] M. Nikolussi, S.L. Shang, T. Gressmann, A. Leineweber, E.J. Mittemeijer, Y.Wangb and Z.K. Liub, Scripta Materialia 2008, 59: 814–817. http://dx.doi.org/10.1016/j.scriptamat.2008.06.015

[6] A.P. Hammersley, S.O. Svensson, M. Hanfland, A.N. Fitch and D. Häusermann, High Pressure Research, 14 (1996), 235-248, http://dx.doi.org/10.1080/08957959608201408

[7] S. Merkel, Multifit/ Polydefix Polycrystal Deformation using X-rays, https://code.google.com/p/multifit-polydefix/

[8] P. Lipinski and M. Berveiller, Int. J. of Plasticity, 5, 149 (1989). http://dx.doi.org/10.1016/0749-6419(89)90027-2

[9] M. Berveiller, A. Zaoui, J. Mech. Phys. Solids 26, 325–344 (1978). http://dx.doi.org/10.1016/0022-5096(78)90003-0

[10] M. Nikolussi et all. Scripta Materialia 59 (2008) 814-817

Residual Stresses 2016: ICRS-10 Materials Research Forum LLC
Materials Research Proceedings **2** (2016) 527-532 doi: http://dx.doi.org/10.21741/9781945291173-89

Effect of Thermal and Mechanical Loadings on the Residual Stress Field in a Nickel Based Superalloy using X-Ray Laue Microdiffraction

Gader Altinkurt [1,2,a *], Mathieu Fèvre [1,b], Guillaume Geandier [2,c], Odile Robach [3,4,d], Soufiane Guernaoui [2,e] and Moukrane Dehmas [2,f]

[1] LEM, UMR 104 CNRS-Onera, Châtillon, France

[2] IJL, UMR 7189 CNRS-Université de Lorraine, Nancy, France

[3] CRG-IF BM32 at ESRF Grenoble, France

[4] CEA, INAC-SP2M, Grenoble, France

[a] gader.altinkurt@onera.fr, [b] mathieu.fevre@onera.fr, [c] guillaume.geandier@univ-lorraine.fr, [d] odile.robach@cea.fr, [e] soufiane.xr@gmail.com, [f] moukrane.dehmas@univ-lorraine.fr

Keywords: Laue Microdiffraction, Residual Deviatoric Strain, Shot-Peening, Fatigue, Coarse-Grained Crystal, Nickel Based Superalloy

Abstract. The shot-peening operation is used to improve the fatigue lifetime of mechanical components through the introduction of compressive residual stresses and plastic deformation in a surface layer. In this study, the Laue microdiffraction technique is used to investigate deviatoric strain fields caused by the shot-peening operation and their redistribution after fatigue testing in a nickel-based polycrystalline superalloy with a 40 μm average grain size.

Introduction

Nickel-based polycrystalline superalloys elaborated by powder metallurgy are used in high pressure turbine discs of aircraft engines. Specific compositions, microstructures and mechanical treatments are developed to enhance crack growth, creep and fatigue resistance during service. The shot-peening process is used extensively to delay crack initiation in the surface layer of components by introducing compressive residual stresses and plastic deformation. The determination of such stresses is generally performed with diffraction based techniques using the $\sin^2\Psi$ method and the effect of temperature and fatigue conditions on stresses relaxation has been deeply investigated in fine-grained microstructures. Most outcomes are associated with macroscopic properties of the average alloy. At the grain scale, electron backscattered diffraction and transmission electron microscopy were employed to characterize plastic deformation and dislocation densities after the shot-peening operation [1]. Even if compressive residual stresses have been relaxed, high dislocation densities persist in the microstructure providing hardening and resistance. A complete understanding of the relationship between the microstructure (precipitate and grain sizes) and residual stresses is still missing. With the Laue microdiffraction technique, stress states and geometrically necessary dislocation densities can be investigated at the grain and phase scales. The technique has already been used to understand the dislocation mechanism in grains under a tensile loading [2], near to grain boundary during a fatigue loading [3] or in zone heavily deformed by friction stir welding [4]. Local deviatoric strain or stress tensors have been also determined in areas close to scratches [5] and cracks [6] or under a tensile loading [7]. The Laue microdiffraction technique has not been employed to characterize strain fields in shot-peened nickel-based superalloys. This method is in principle a suitable tool for macroscopic and microscopic stress determination in coarse grained microstructures since the $\sin^2\Psi$ method is not well adapted to investigations in such microstructures.

In this paper, deviatoric strain fields generated by the shot-peening operation and then redistributed by a low-cycles fatigue test at 450 °C are characterized in a nickel-based superalloy with a 40 µm average grain size.

Experimental conditions

Sample features. Cylindrical specimens were removed using Electro Discharge Machining in a sector of a turbine disk of the N18 superalloy [8] produced at Snecma (SAFRAN group). The microstructure consists in γ' hardening precipitates with a simple cubic structure embedded in a γ matrix phase with a face-centered cubic structure. A coarse grained microstructure with one population of γ' precipitates is obtained after a supersolvus solution treatment (4 hours at 1205°C and air quench) followed by an ageing treatment (1h at 900°C and air quench). At the millimetre scale, the analysis of the Electron Back Scattered Diffraction (EBSD) images acquired in the Scanning Electron Microscope (SEM) shows an average γ grain size of 40 µm with a wide distribution ranging from 5 µm to 200 µm (Fig. 1b). At the micrometer scale, SEM micrographs show cubic shape γ' precipitates with an average size close to 200 nm homogeneously distributed in the γ matrix (Fig. 1c). In order to simplify the x-ray data analysis and to dissociate physical mechanisms in such materials, this microstructure sensibility differs from those employed in aircraft engines (5-20 µm grain sizes and three populations of γ' precipitates [8]). Three 6.22 mm in diameter fatigue test specimen were then machined. One was used as a reference. One was subjected to shot-peening, one to shot-peening and interrupted fatigue testing. The peening operation was done using an ultrasonic technique with 3 mm-diameter 100Cr6 steel shots during 1080s. The fatigue test was performed at imposed deformation ($\Delta\varepsilon/2 = 0.5$ %, $R_\varepsilon = 0$) at 450°C during 300 cycles. 10 mm height samples were finally cut perpendicular to the shot-peened surface in the gauge length. X-ray measurements were performed in a circular cross section which has been mechanically and chemically polished to reduce stresses introduced by the cutting process.

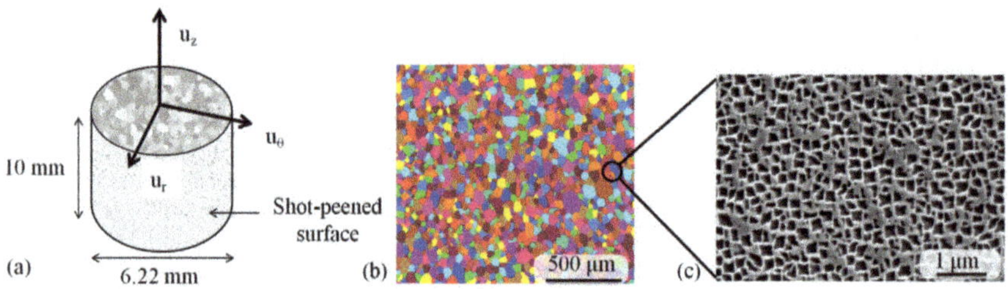

Fig. 1: (a) sample geometry with the local cylindrical coordinate system ($\vec{u_r}$, $\vec{u_\theta}$, $\vec{u_z}$). (b) EBSD map revealing grains γ (random colours). (c) SEM micrograph showing γ' precipitates (dark grey) embedded in the γ matrix (light grey).

µ-Laue technique. Laue microdiffraction was performed at the BM32 beamline of the European Synchrotron Radiation Facility at Grenoble (France). A polychromatic beam with a 5-17 keV energy spectrum is focused to a cross section of 0.35 x 0.6 µm² at the sample position. Emerging X-rays are collected on a CCD 2D detector oriented in 90° with respected to the incident beam in reflection geometry (see Ref. [9] for a more detailed description of the setup). A xyz motorized sample stage enables to record Laue patterns arising from different locations at the sample surface. With this setup, the characteristic penetration depth of the beam in the material ranges from 4 µm to 15 µm and up to one frame per 5 second can be acquired. Due to the grain size and the cross section of the beam, our samples behave like single crystals and the Laue pattern consists in spots. The Miller indices of each spot can be retrieved from the crystal symmetry and peak positions in pattern. Based on the formulation proposed by Hung and Ice [10], the indexation of four diffraction peaks permits to determine the orientation of the crystal lattice regarding to the coordinate system of the experiment and five of the six unit cell parameters: two lengths ratio and three angles usually denoted by α, β, γ and b/a, c/a respectively. The knowledge of these parameters for a non-deformed structure (a_0, b_0, c_0, α_0, β_0, γ_0) enables to determine the deviatoric elastic strain tensor ε^d within the small transformation approximation. Its coefficients are given by:

$$\varepsilon^d_{ij} = \varepsilon_{ij} - Tr(\varepsilon).\delta_{ij}/3 \quad i,j=1,2,3 \tag{1}$$

where ε is the total strain tensor and δ_{ij} the Kronecker delta function. ε^d is associated with the distortion of the crystal unit cell. When the stiffness tensor of the single crystal is known, the deviatoric stress tensor can be deduced by applying the generalized Hooke's law.

Data processing

Due the highly coherent interfaces (lattice mismatch $< 10^{-3}$), fundamental reflections (*h*, *k*, *l* with the same parity) correspond to the convolution of the intensities belong to the γ and γ' phase whereas superstructure reflections (*h*, *k*, *l* with different parities) correspond only to the contribution of the γ' phase intensities. Laue patterns provide thus quantitative information about the average alloy (γ + γ') and the γ' phase. Data analysis has been performed using the XMAS [11] and LaueTool [12] softwares. The main output of the calculations is the coordinates of the reciprocal lattice vectors in a given system of coordinates (sample, laboratory) for each crystallite founded in the Laue pattern. This is then used to determine the crystal orientation, its unit cell parameters and finally deviatoric strain and stress components. The strain-free lattice-parameters of the γ and γ' phases have been determined for the fine grain N18 alloy [8]. A Vegard's law is used to determine the strain free lattice-parameter associated with the average alloy (a_0=0.35917 nm). With the Laue technique, the strain sensitivity can be 5.10^{-5} when optimal conditions are reached [13] but generally, it is closer to $1\text{-}2.10^{-4}$ [5,10]. In this study, the assessment of uncertainties in a sample which has not been subjected to shot-peening results in a 2.10^{-4} uncertainty on strain values.

Results and discussions

In the following, results associated with the analysis of the fundamental reflections (γ + γ') are presented. The crystal orientation is expressed with respect to the fixed laboratory coordinate system within the axis–angle representation. Based on the sample geometry and the applied external loadings, the deviatoric strain components ε^d_{ij} are given with respect to the local cylindrical coordinate system ($\overrightarrow{u_r}$, $\overrightarrow{u_\theta}$, $\overrightarrow{u_z}$). $\overrightarrow{u_r}$, $\overrightarrow{u_\theta}$ correspond to radial and circumferential axis and $\overrightarrow{u_z}$ is the axis aligned with the free surface normal (Fig.1a). The single-crystal elastic constants of the N18 are unknown. To avoid bad physical interpretations, only strain tensor components are discussed below. Since the deviatoric strain components associated with the $\overrightarrow{u_z}$ direction could be affected by the presence of the free surface, only ε^d_{rr}, $\varepsilon^d_{\theta\theta}$ and $\varepsilon^d_{r\theta}$ are discussed below. Orientation and strain xy maps are obtained from meshes with steps corresponding to the beam size.

<u>Reference sample.</u> Measurements in the sample which has not been subjected to shot-peening have been realized on a 1x3 mm² area. No strain heterogeneities between identified grains were observed. Strain components smoothly evolve with values close to the sensitivity limit of the technique ($\pm 2.10^{-4}$) excepting for the ε^d_{zz} component which is between -3.10^{-4} and -6.10^{-4}. This is probably due to the surface polishing realized during the sample preparation. Such values of deviatoric strains have also been obtained by Chao et al [6] with the same technique in a nickel-based Alloy 600 sample before tensile deformation. The analysis also shows that strains in the order of $\pm 1.5\text{x}10^{-3}$ caused by the grinding machine used develop over a layer with a thickness lower than 100 µm.

<u>Effect of shot-peening.</u> Measurements in the sample which has been subjected to shot-peening have been realized on a 1.02x0.225 mm² area close to its edge. The xy-orientation map and the maps associated with the components of the deviatoric strain tensor are represented in figure 2 (a-d). Lines in figure 2a correspond to radius contours and the edge of the sample is located at the right side of the figure (last contour line). In the first 300 µm wide area close to sample edge (white color), the

Materials Research Proceedings **2** (2016) 527-532 doi: http://dx.doi.org/10.21741/9781945291173-89

material hardening introduced by the shot-peening has caused large intensity spread in the Laue pattern. It was not possible to accurately quantify strain tensor components. Figures 2b to 2d show different components of the strain field. Along the average peening direction $\overrightarrow{u_r}$ (Fig. 2b), the deviatoric strain component strongly depends on the distance from the sample centre (negative for 2.0 mm < x < 2.5 mm and positive for 2.5 mm < x < 2.7 mm) and has a less pronounced variations along contours. Strain amplitudes range is from -4.10^{-3} to 4.5×10^{-3}. Along the circumferential direction $\overrightarrow{u_\theta}$ (Fig. 2c), the deviatoric strain component has also this dependency with a negative sign. Some variations can be observed close to several grain boundaries. Strain amplitudes range is from -4.5×10^{-3} to 1.10^{-3}. In figure 2d, the $\varepsilon^d_{r\theta}$ component has a heterogeneous behaviour with strain levels much more dependent on the microstructure and a range from -2.10^{-3} to 2.10^{-3}. For all components, we clearly observe that after shot peening, high shear strains take place deeply in the material. Indeed, at a distance of 1mm, all components differ significantly from zero. In order to compare Laue microdiffraction results to those which could be obtained in the laboratory with a beam section close to square millimetres, strain values can be averaged along the radius contour lines of figure 2a. Results are represented as a function of the distance from the shot-peened edge in figure 3a. Because of the dependency of the deviatoric strain at this scale, this representation is justified. The ε^d_{rr} and $\varepsilon^d_{\theta\theta}$ components follow trends described previously. A large variation of strains is visible in a layer located between 0.3 mm and 0.6 mm depths and they evolve smoothly further. The $\varepsilon^d_{r\theta}$ component is close to zero for all depths and the presence of strong local heterogeneities is not discernable with this average.

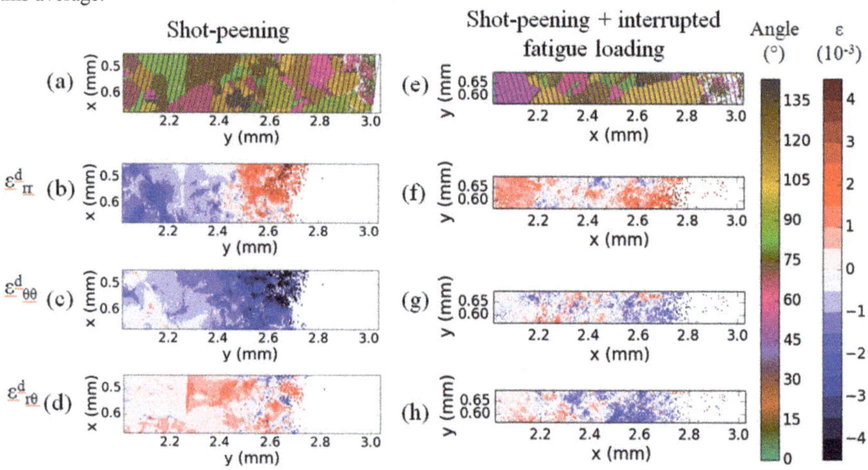

Figure 2: Local orientation and deviatoric elastic strain component maps for a sample which has been subjected to shot-peening on the left and to shot-peening and an interrupted fatigue test at 450°C on the right. Orientation maps are determined within the axis-angle representation.

Effect of fatigue. Figure 2e-h represents the orientation and the deviatoric strain maps obtained from measurements in the sample which has been shot-peened and then subjected to low-cycle fatigue at 450°C during 300 cycles. The investigated area is 1.01x0.125 mm² close to a sample edge. The deviatoric strain fields observed just after the shot-peened in figures 2b to 2d have been strongly modified by the thermal and mechanical loading. All components exhibit positive and negative heterogeneous strains in the whole area with amplitudes comparable to what has been determined previously ($-3.10^{-3} < \varepsilon^d < 4.10^{-3}$). By contrast with the shot-peened state, the axial and the circumferential deviatoric strains are independent on the distance from the sample edge and more

Materials Research Proceedings **2** (2016) 527-532 doi: http://dx.doi.org/10.21741/9781945291173-89

influenced by each grain features. Strain values averaged along the radius contour lines are represented as a function of the distance from the shot-peened edge in figure 3b. On the few considered grains, the $\varepsilon^{d}_{r\theta}$ component behaves similarly for these two states. However, some relaxation is visible for the two other components after fatigue. To have a better statistic and probe a large number of grains, strain profiles from a 2x2mm² map (not shown) with a smaller spatial resolution are plotted in figure 3c. At this scale, all components differ significantly from zero up to a distance of 0.4 mm from the sample edge. The highest strain amplitudes are close to $1\text{x}10^{-3}$ and take place close to the shot-peened surface. Most of the fluctuations visible in figure 3b are removed by the averaging procedure and the $\varepsilon^{d}_{r\theta}$ component behaves differently close to the sample edge in figure 3b and 3c. These results highlight the fact that large local strains occur in the microstructure even though no strain is determined with a method involving averages on many grains as with the $\sin^2 \Psi$ method. Consequently, a good understanding of the role of the microstructure on the fatigue properties of materials requires investigations at the grain scale.

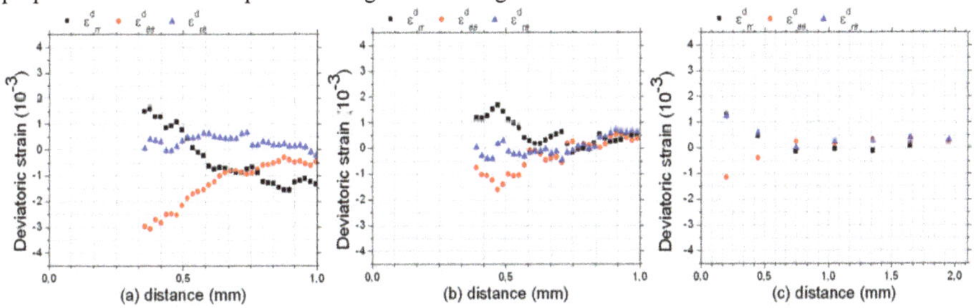

Figure 3: Deviatoric strain components obtained after averaging along radius contour lines (a) after the shot-peening operation, (b) and (c) after the interrupted fatigue test at 450°C. Distances are given from the shot-peened sample edge.

Conclusions

The Laue microdiffraction technique has been used to investigate deviatoric strain fields in nickel-based superalloy samples with a coarse grained microstructure and subjected to ultrasonic shot-peening and fatigue testing at 450°C. This method is suitable for investigations in coarse grained microstructures as an alternative to conventional techniques and provides additional information at the grain scale. The method should able to study the plastic deformation induced by shot-peening with the study of peaks shape. It should able to provide new data on the influence of the γ' precipitates on internal stresses in superalloys. However, information from the highly strained 300 µm thick region close to the surface could not be obtained as Laue patterns could not be indexed with a good accuracy. Moreover, the mapping of the full strain tensor is still a challenging task.

Measurements at the grain scale have shown that residual elastic strains with amplitudes ranging from -4 to 4.10^{-3} take place at depths larger than 1 mm. The redistribution of strain fields associated with the shot-peening operation was also clearly observed after 300 cycles of fatigue at 450°C. In particular, the dependence of axial and circumferential strains with the distance from the shot-peened surface observed just after the shot-peening operation has completely disappeared after fatigue leaving to heterogeneous positive and negative fields with a strong dependence on the underlying microstructure.

Acknowledgments

This work benefited from the support of the project REMEDDIES ANR-13-BS09-016 of the French National Research Agency (ANR). The experiments were performed on beamline BM32 at the European Synchrotron Radiation Facility (ESRF), Grenoble, France. We are grateful to Local Contact at the ESRF for providing assistance in using beamline BM32.

Reference

[1] O.M.D.M. Messe, S. Stekovic, M.C. Hardy, C.M.F. Rae, Characterization of plastic deformation induced by shot-Peening in a Ni-Base superalloy, JOM 66 (2014) 2502-251. https://doi.org/10.1007/s11837-014-1184-8

[2] R. I. Barabash, E.W Huang, J.J. Wall, J.H. Wilkerson, Y. Ren, W. Liu, S.C. Vogel, G.E. Ice, L.M. Pike, P.K. Liaw, Texture crossover: Trace from multiple grains to a subgrain, Mater. Sci. Eng. A528 (2010) 3-10. https://doi.org/10.1016/j.msea.2010.07.035

[3] E.-W. Huang, R.I. Barabash, G.E. Ice, W.J. Liu, Y.-L. Liu, J.-J. Kai, P.K. Liaw, Cyclic-loading-induced accumulation of geometrically necessary dislocations near grain boundaries in an Ni-based superalloy, JOM 61-12 (2009) 53-58. https://doi.org/10.1007/s11837-009-0181-9

[4] O.M. Barabash, R.I. Barabash, G.E. Ice, Z.Feng, D. Gandy, X-ray microdiffraction and EBSD study of FSP induced structural/phase transitions in a Ni-based superalloy, Mater. Sci. Eng. A524 (2009) 10-19. https://doi.org/10.1016/j.msea.2009.03.086

[5] M.L. Suominen Fuller, R.J. Klassen, N.S. McIntyre, A.R. Gerson, S. Ramamurthy, P.J. King, W. Liu, Texture, residual strain, and plastic deformation around scratches in alloy 600 using synchrotron X-ray Laue micro-diffraction, J. Nucl. Mater. 374 (2008) 482-487. https://doi.org/10.1016/j.jnucmat.2007.10.015

[6] J. Chao, A. Mark, M.L. Suominen Fuller, N.S. McIntyre, R.A. Holt, R.J. Klassen, W.J. Liu, Study of residual elastic- and plastic-deformation in uniaxial tensile strained nickel-based Alloy 600 samples by polychromatic X-ray microdiffraction (PXM) and neutron diffraction methods, Mater. Sci. Eng. A524 (2009) 20-27. https://doi.org/10.1016/j.msea.2009.05.033

[7] J. Chao, M.L. Suominen Fuller, N. Sherry, J. Qin, N. Stewart McIntyre, J. Ulaganathan, A.G. Carcea, R.C. Newman, M. Kunz, N. Tamura, Plastic and elastic strains in short and long cracks in Alloy 600 studied by polychromatic X-ray microdiffraction and electron backscatter diffraction, Acta Mater. 60 (2012) 5508-5515. https://doi.org/10.1016/j.actamat.2012.06.060

[8] S.T. Wlodek, M. Kelly, D. Alden, The Structure of N18, Proceedings of Superalloys 1992 (1992) 467-476. https://doi.org/10.7449/1992/superalloys_1992_467_476

[9] O. Ulrich, X. Biquard, P. Bleuet, O. Geaymond, P. Gergaud, J.S. Micha, O. Robach, F. Rieutord, A new white beam x-ray microdiffraction setup on the BM32 beamline at the Europea Synchrotron Radiation Facility, Rev. Sci. Instr. 82 (2011) 033909-1-6. https://doi.org/10.1063/1.3555068

[10] J. Chung, G. Ice, Automated indexing for texture and strain measurement with broad-bandpass X-ray microbeams, J. Appl. Phys. 86 (1999) 5249-5255. https://doi.org/10.1063/1.371507

[11] N. Tamura, A. A. MacDowell, R. Spolenak, B. C. Valek, J. C. Bravman, W. L. Brown, R. S. Celestre, H. A. Padmore, B. W. Batterman, J. R.J. Patel, Scanning X-ray microdiffraction with submicrometer white beam for strain/stress and orientation mapping in thin films, Synchrotron. Radiat. 10 (2003) 137-143. https://doi.org/10.1107/S0909049502021362

[12] Information on https://sourceforge.net/projects/lauetools/

[13] F. Hofmann, S. Eve, J. Belnoue, J.S. Micha, A. M. Korsunsky, Analysis of strain error sources in micro-beam Laue diffraction, Nuclear Instrum. Methods Phys. Res. A660 (2011) 130-137. https://doi.org/10.1016/j.nima.2011.09.009

Residual Stresses 2016: ICRS-10 Materials Research Forum LLC
Materials Research Proceedings **2** (2016) 533-538 doi: http://dx.doi.org/10.21741/9781945291173-90

Comparison of Submerged Arc Welding Process Modification Influence on Thermal Strain by in-situ Neutron Diffraction

Rahul Sharma[1, a*], Uwe Reisgen [2,b] and Michael Hofmann [3,c]

[1,2]RWTH Aachen University, Welding and Joining Institute, Pontstr. 49, 52062 Aachen, Germany

[3] Heinz-Maier-Leibniz-Zentrum, FRMII, Technical University Munich, Lichtenbergstr. 1, 85747 Garching, Germany

[a*]sharma@isf.rwth-aachen.de, [b]reisgen@isf.rwth-aachen.de, [c]michael.hofmann@frm2.tum.de

Keywords: Neutron Diffraction, Strain Analysis, in-situ Experiment, Submerged Arc Welding, Cold Wire

Abstract. In this paper the application of neutron diffraction for measuring the thermal strain field in the vicinity of the weld pool during submerged arc welding is described. The aim of the research was to determine the influence of a welding process modification on the thermal strain within a sample rod. The welding experiment was carried out on the instrument STRESS-SPEC at the MLZ FRMII facility in Garching, Germany. Submerged arc welding equipment with additional cold wire feeding was adapted to the diffractometer and thereon single layer bead-on-plate welds were carried out. Sample rods made of the nickel base alloy 625 were used. The measured strain values are presented and discussed with respect to the resulting weld bead geometry and the thermal profile within the sample.

Introduction

The application of arc welding processes on metallic materials implies high gradients of temperature, strain and stress due to the high power density of the moving heat source. Besides the deterioration of mechanical properties, the buildup of residual stresses follows up the thermal manufacturing processes (Figure 1 left). For the estimation of residual stresses within welded components, it is therefore inevitable to know the transient temperature- and strain fields in the weld zone. Furthermore thermomechanical models for the description of the materials behavior are needed to understand the development of stress and distortion.

The description of thermomechanical behavior of materials subject to welding relies on continuum mechanics principles which are well established. Nevertheless it is still difficult to estimate the resulting distortion and residual stress in welded structures due to the complex superposition of influencing factors. Experimental investigations to analyze the stress and strain during welding have been conducted in the past to achieve a deeper knowledge of the physical processes leading to buildup of residual stresses. These experiments relied on various surface strain measurement principles as strain gauges on the base metal surface, e.g. [1] or Digital Image Correlation [2]. Another approach is the application of neutron diffraction for the determination of the in-depth-strain state. A first experiment using TIG remelting of a 316L sample has been described in [3] and [4], which showed the feasibility of using a quasi-stationary welding setup.

A central issue in welding of high alloyed steels and nickel based alloys is the formation of hot cracks. These welding defects occur if a critical combination of chemical composition, mechanical load and welding parameters are used. Therefore it is difficult in high performance welding processes to hit the suitable parameter window. The submerged arc welding process (SAW) is such a high performance welding process and it is often applied for manufacturing of thick walled components because of its high deposition rate, deep penetration and process stability. In some nickel base alloys it is difficult to set up even a SAW process with sufficiently low heat input to prevent hot crack formation. Furthermore the loss of alloying elements through burn-off to the slag limits the feasibility of SAW for creep resistant nickel based alloys.

Materials Research Proceedings **2** (2016) 533-538 doi: http://dx.doi.org/10.21741/9781945291173-90

To overcome these restrictions the submerged arc welding with cold wire addition (SAW-CW) process has been investigated in the past [5-7]. Slight evidence for lower hot crack formation when using SAW-CW led to the conduction of a in-situ-diffraction experiment to compare the process induced elastic strains in the hot surrounding zone of the weld pool.

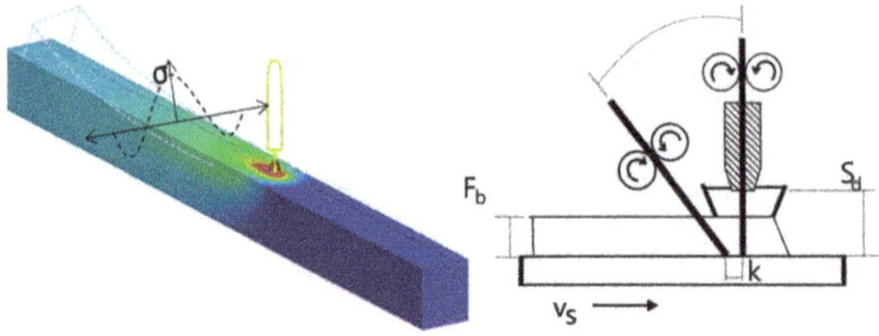

Figure 1: colour mapped temperature distribution in a welded sample rod (left), basic principle of SAW-CW process with parameters Fb (flux height), Sd (contact tip distance), k (cold wire distance) (right)

Experimental Approach

The common SAW process is carried out using a continuously fed wire electrode. By applying the welding voltage between the contact piece and the base metal, an electric arc burns under the welding flux which is melted to form the slag pool. Metal vapour and gasified slag components form the cavity wherein the electric arc burns and the metal droplet transfer takes place.

The SAW-CW process complements an additional welding wire, which is fed into the weld pool behind the welding electrode. Since there is no voltage applied between additional wire and base metal, the melting is exclusively induced by the excessive heat of the weld pool. In this arrangement the loss of alloying elements is reduced as the droplet does not pass the arc zone. The electrical process characteristics are not influenced by the cold wire feeding, so that no change in heat input occurs.

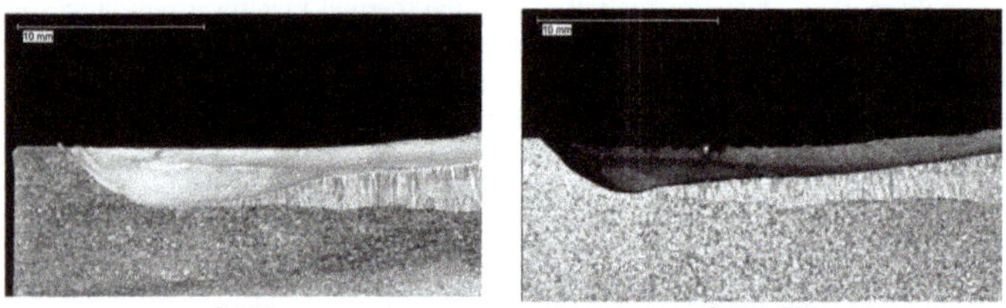

SAW SAW-CW

Figure 2: longitudinal centerline section of a SAW weld bead (left) and a SAW-CW weld bead with a cold wire speed of 50 % oft he electrode feed speed

A change of weld pool geometry and thermal field is the result of the additional wire feeding. The weld bead buildup increases with additional wire feeding whereas the penetration depth is reduced [5].

The SAW-CW leads to reduced peak temperatures near to the fusion line, but the overall cooling time of the weld metal is mostly uninfluenced [6,8]. This could be caused by the reduced heat dissipation area between the weld pool and the base metal, which is a result of the weld pool migration out of the base metal.

In this study a high temperature nickel base alloy 625 has been used for the experiments. It consists of a fully austenitic microstructure with solid solution hardened matrix. The composition is given in Table 1.

Table 1: chemical composition of Alloy 625 base metal rods (measured by spark OES)

[%]	C	Si	Mn	P	S	Cr	Mo	Fe	Al	Nb	Ti	Co	Ta	Ni
Base metal	0,027	0,12	0,17	0,010	0,002	22,3	9,1	4,0	0,14	3,4	0,15	0,2	0,01	bal.

The welding setup was installed on the sample table of the materials diffractometer STRESS-SPEC at the FRMII neutron source in Garching, Germany (Figure 3). Due to the high flux of thermal neutrons

Materials Research Proceedings **2** (2016) 533-538 doi: http://dx.doi.org/10.21741/9781945291173-90

and the resolution of the area detector, this instrument is intended for such experiments. The basic parameters and performance characteristics have been described elsewhere [9].

Figure 3: installed welding setup at STRESS-SPEC instrument (left), alignment of beam guidance, detector, welding head and sample table (right)

For the in-situ-diffraction experiment steady state boundary conditions were needed as the attentuation of the neutron beam while passing through the sample volume led to a minimum time of 30 seconds to achieve an evaluable diffraction peak. This was done by installing the sample rod of the dimensions 20 x 20 x 500 mm³ on a linear sample table. The position of the welding head was chosen so that the position of diffraction gauge volume with regard to the electrode tip stayed constant. After completion of a run-on stage the gauge volume stayed along the same isothermal line for the whole data acquisition time. This arrangement implies that for every measuring point and spatial orientation, a whole sample rod had to be welded.

Table 2: welding parameters used within the experiment

	Heat input [kJ/cm]	Voltage [V]	Current [A]	welding speed [cm/min]	welding wire speed [m/min]	cold wire speed [m/min]
SAW	10,5	34,6	325	60	6	0
SAW-CW	10,5	34,6	323	60	6	1,5

To compare the influence of the welding process on the strain field two welding parameter sets were defined (Table 2). The only difference was the cold wire addition in the second set, while all other welding parameters stayed constant.

For both the electrode wire and the cold wire, a commercial welding wire S NI 6625 (ISO 18274) with a diameter of 1.6 mm was used. The welding flux was SA-AB 2 (ISO 14174).

For every measuring point the strain in longitudinal direction (parallel to welding direction) and transverse direction were measured. Since the high thermal gradient within the gauge volume of 3 x 3 x 2 mm³ was suspected to cause extreme peak broadening, every measurement was repeated three times, which sums up to six samples per measuring point.

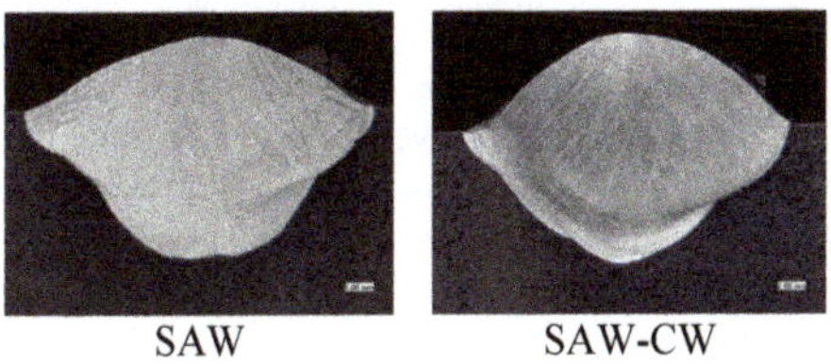

Figure 4: cross section of SAW (left) and SAW-CW (right) weld beads on Alloy 625 sample rods, welding parameters acc. to Table 2

For the determination of the measuring point location, weld width and penetration depth were determined in preliminary welding experiments using the same welding parameters and materials. It was found that only a slight reduction in penetration depth was achieved when cold wire was fed into the weld pool with 25 % of the electrode speed (Figure 4). Since the SAW process comprises a flux covered welding zone, the length of the weld pool can only be determined by removing the weld pool.

Materials Research Proceedings **2** (2016) 533-538 doi: http://dx.doi.org/10.21741/9781945291173-90

	weld pool length [mm]	std deviation (n=5)
SAW	69,0	0,3
SAW-CW	67,2	0,8

Table 3: weld pool length of SAW and SAW-CW on used Alloy 625 sample rods, welding parameters acc. to Table 2

One possible method consists of blowing out the liquid metal with a high pressure gas stream, while the welding process is shut down [10]. The results are given in Table 3. Four gauge volume locations (MP1-MP4) were selected taking into account the position of the fusion line (Figure 5).

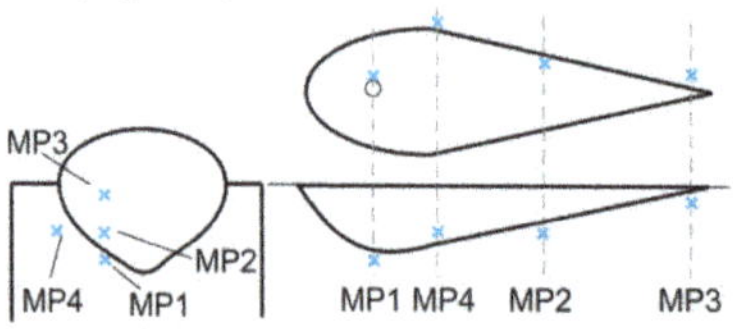

Figure 5: schematic overview of the gauge volume locations

Results

For both welding parameter sets, the chemical analysis of the weld metal is given in Table 4.

A slight increase of chromium and molybdenum content in the weld metal could be observed due to the lower burn-off in SAW-CW. The lower iron content however is a result of the lower dilution compared to the SAW process. Due to the relatively low cold wire addition, there is no substantial difference in chemical composition between both variants.

Table 4: chemical composition of the weld metal (measured by spark OES)

	C	Si	Mn	P	S	Cr	Mo	Fe	Al	Nb	Ti	Co	Ta	Ni
SAW	0,011	0,20	0,10	<0,0002	<0,0003	20,5	8,3	3,1	0,14	3,2	0,18	0,05	<0,005	bal.
SAW-CW	0,012	0,19	0,09	<0,0002	<0,0003	20,8	8,5	2,4	0,13	3,2	0,18	0,04	<0,005	bal.

The temperature measurement was carried out using type K thermocouple wires, which were welded on the ground of cylindrical holes inside the sample rods. The hole depth corresponded to the position of the gauge volume. Since the geometrical extent of the gauge volume led to thermal gradients of several 100 K, the measured values could only be processed as calibration data for a numerical model of the temperature distribution within the sample. Therefore the software simufact.welding has been used. However the temperature to consider is the spatial integration over the gauge volume, since the low thermal conductivity of Alloy 625 leads to steep thermal gradients (Figure 6).

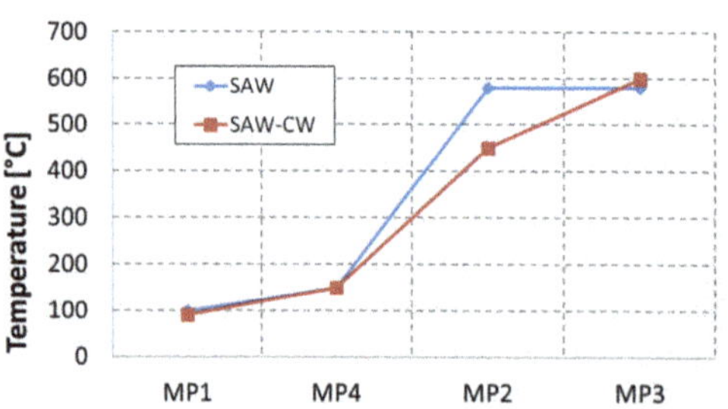

Figure 6: simulated steady state temperature integrated over the gauge volume

The neutron experiment was carried out using a constant wavelength of 1.55 Å which was achieved through the application of a Si(400) monochromator. For the measurement of strain by neutron diffraction at steady state sources, the 311-peak seems to be the most eligible, since it is least affected to integranular stresses [11]. The average value of lattice plane distance for the 311 peak is shown in Figure 7. However these results show the combined influence of thermal expansion and load or restraint induced elastic strain. A methodology to separate thermal strains and load strains was applied by performing a furnace experiment on the same diffractometer. Therefore a cylindrical sample (d = 6 mm) was mounted within the diffraction furnace and heated stepwise up to 1200 °C. During every temperature step of 25 °C a diffraction pattern was acquired. The strains were calculated from the diffraction data, using Eq 1, where d refers to the actual lattice plane distance, d0 to the load-free and homogenous tempered lattice plane distance for the respective temperature.

$$\varepsilon = \frac{d-d_0}{d_0} = \frac{\sin\theta_0}{\sin\theta}-1 \quad (1)$$

The calculated strain values are given in Figure 8. For the SAW strain in longitudinal direction it can be seen that there is a zone of tensile strain behind the welding electrode (MP4). With increasing distance to the electrode, the tensile strain relaxates to nearly zero. For the SAW-CW process, the tensile strain in longitudinal direction is significantly lower. Furthermore compressive strain is the leading influence afterwards. In transverse direction the strain distribution appears to be different. The SAW-CW process induces high tensile strain behind the solified weld pool in here. This could be caused by a geometrical effect which is due to the small width of the sample. Since the weld penetration area and the peak temperatures in SAW are higher compared to SAW-CW, a larger proportion of the total width of the sample is subject to softening. The lower dilution and higher weld buildup in SAW-CW will result in a higher restraint intensity. Furthermore the zone of deep penetration of the weld pool is much smaller in SAW-CW. For moving heat sources a compressive plastic zone is found near to the weld pool before the peak temperature is reached. After the peak temperature has been reached the strain state changes to plastic tensile strain [12]. When considering Figure 8 this is reflected by both welding processes. Even though the integral temperature of the gauge volume is lower at position MP4, the formation of hot cracks will originate from this position on near to the fusion line. Since the longitudinal strain is compressive for SAW-CW a lower risk of hot crack formation is suspected. For wider samples, the transverse tensile strain for SAW-CW is expected to become compressive, as stated before. This has to be proved within further investigations.

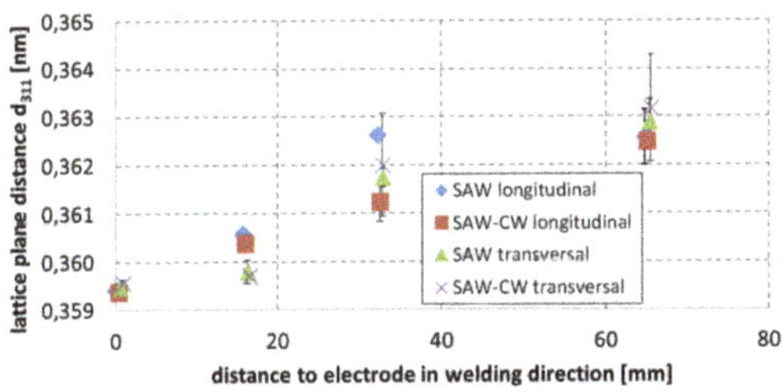

Figure 7: average lattice plane distance in welded samples, gauge volume position axis parallel to welding direction

Conclusion

The measured strain field during welding with both processes shows a direct relationship to geometrical and thermal influencing factors. It complies with fundamental descriptions of stress and strain in the zone around the weld pool [12]. Based on the experimental results a slightly lower portion of tensile strain in the deep penetration area of the weld pool could be observed but it is still unclear if there is a direct relationship to the formation of a lower hot cracking risk.

However this reflects the limitation of the method stemming from the necessary use of relatively large gauge volume. To overcome this restriction in future experiments, the use of synchrotron radiation for in-situ strain measurement will be taken into account. A transient measurement of strain at the fusion line could give clearer results.

Summary

Within this work, neutron diffraction measurements during welding have been carried out to compare the influence of SAW and SAW-CW on the resulting strain field in the base metal near to the fusion line. For the considered nickel base alloy 625 it was shown that the strain field in longitudinal direction has a higher tensile proportion for SAW. The higher tensile strain in transverse direction was put down to the restraint condition and sample geometry in SAW-CW. For future work a smaller gauge volume is needed to enhance the current results.

Acknowledgement

All presented investigations were conducted in the context of the Collaborative Research Centre SFB1120 "Precision Melt Engineering" at RWTH Aachen University and funded by the German Research Foundation (DFG). For the sponsorship we wish to express our sincere gratitude. This work is based upon experiments performed at STRESS-SPEC instrument operated by HZG, FRMII and TU Clausthal at the Heinz Maier-Leibnitz Zentrum (MLZ), Garching, Germany.

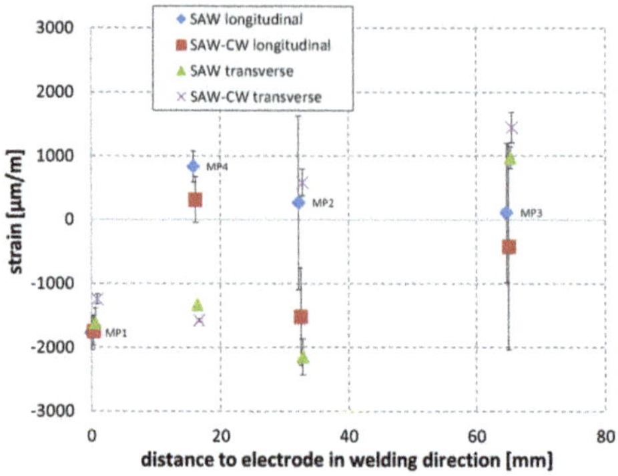

Figure 8: average strain in welded samples, gauge volume position axis parallel to welding direction

References

[1] W. Dahl, U. Dilthey in D. Aurich, K.H. Kloos (Eds.):"Eigenspannungen und Verzug durch Wärmewirkung", S. 175-201, Wiley-VCH Weinheim, 1999

[2] M. Shibahara, K. Ikushima and S. Itoh STWJ 2012 Vol 17 No 6 pp. 511-517 doi 10.1179/1362171812Y.0000000027

[3] U. Reisgen, C. Geffers, R. Sharma, J. v. d. Heydt, MSF Vols. 768-769 (2014) pp 644-651, doi:10.4028/www.scientific.net/MSF.768-769.644.
http://dx.doi.org/10.4028/www.scientific.net/MSF.768-769.644

[4] U. Reisgen, R. Sharma, J. v. d. Heydt, AMR Vol. 996 (2014) pp 424-430, doi:10.4028/www.scientific.net/AMR.996.424.
http://dx.doi.org/10.4028/www.scientific.net/AMR.996.424

[5] U. Reisgen, U. Dilthey, I. Aretov, . in T. Böllinghaus, H. Herold, C. E. Cross, J. C. Lippold (Eds.): Hot Cracking Phenomena in Welds, Pt. 2. Berlin, Heidelberg, Springer , 2008 pp. 215- 240

[6] U. Reisgen, K. Willms, S. Jochindke, IIW-DOC: XII-2207-15, 2015

[7] R. Sharma, IIW-Doc II-1847-13 (II-C-458-13), 2013

[8] L. Karlsson, H. Arcini, P Dyberg, S. Rigdal, M. Thuvander, Proc. Stainless Steel World Conference Maastricht, NL, 2003, pp. 283–294

[9] Heinz Maier-Leibnitz Zentrum. STRESS-SPEC: Materials science diffractometer. Journal of large-scale research facilities, 1, A6., 2015, http://dx.doi.org/10.17815/jlsrf-1-25.

[10] U. Reisgen, A. Schiebahn, O. Mokrov, O. Lisnyi, R. Sharma, CWA Journal 11, 2015, 6, pp. 90 - 94

[11] M.R. Daymond, B. Clausen, Journal of Applied Physics 82, No. 4, pp. 1554 - 1562, 1997 http://dx.doi.org/10.1063/1.365956.

[12] Radaj, D.: Heat effects of welding, Springer Berlin, 1992. http://dx.doi.org/10.1007/978-3-642-48640-1

Residual Stresses 2016: ICRS-10 Materials Research Forum LLC
Materials Research Proceedings **2** (2016) 539-544 doi: http://dx.doi.org/10.21741/9781945291173-91

Residual Stresses on Electro-Deposited NiCo-Al and NiCo-Zr Composite Coatings

Fei Cai[1,a], Chuanhai Jiang[1,b*] and Bevin Pan[2,c]

[1]School of Materials Science and Engineering, Shanghai Jiaotong University, 800 Dongchuan Road, Minhang District, Shanghai, China 200240

[2]Shanghai High School International Division, 400 Shangzhong Road, Shanghai, China200231

[a,b]chjiang@sjtu.edu.cn, [c]bevinpan@yandex.com

Keywords: Composite Coating, Electro-Deposition, Texture, Residual Stress

Abstract: This experiment examined the residual stresses on electrodeposited NiCo-Al and NiCo-Zr composite coatings as well as the impact of varying Co^{2+}concentrations on other properties of the coating, namely grain size and texture. To prepare for the experiments, this project employed modified Watt Baths containing $NiSO_4\cdot 6H_2O$ (240 g/L), $NiCl_2\cdot 6H_2O$ (40 g/L), H_3BO_3 (30 g/L) and $C_{12}H_{25}NaSO_4$ (0.2 g/L) in order to electrically deposit them on to Aluminum and Zirconium and to produce the two types of aforementioned specimen: NiCo-Al and NiCO-Zr. The amount of Co in the end product is regulated via adjusting the concentration of $CoSO4\cdot 7H_2O$, providing a range of concentrations from 5 to 40 g/L. Furthermore, structural analyses of the coatings were performed via X-ray diffraction at 40kV and 30mA in standard 2θ–θ mode. The Voigt method was used to calculate grain size and micro-strain of coatings using the integral breadth of the (200) peak. Residual stresses on the as-deposited composite coatings were measured by using stress analyzer and classical $\sin 2\psi$ method. Peaks of (331)α of Ni were selected to calculate the residual stresses. Results show that, for both NiCo-Al and NiCo-Zr composite coatings, grain size decreases and micro-strain increases with increasing Co contents. Pole figure experiments shown typical fiber texture for composite coatings. Texture coefficient results show that the (111) texture coefficient increases while (111) texture coefficient decreases with increasing Co content. In addition, results show that the residual stress of NiCo-Al and NiCo-Zr composite coatings increases as Co content increases, which could be attributed to decreasing grain size and texture evolution from (200) texture to (111) texture.

Introduction

In recent years, there has been increasing interest focused on the electro-deposited NiCo and NiCo based metal matrix composite (MMCs) coatings due to their combined superior properties, such as higher hardness [1-5], improved anti-wear [2, 4], better corrosion resistance [1, 3, 6] and oxidation resistance [5, 7] as compared with the pure Ni and Ni based composite coatings. The properties of the NiCo or NiCo based composite coatings were mainly dependent on the microstructures of the NiCo matrix and the incorporated particles.

The properties of pure Ni coatings could also be enhanced by alloying with Co element [1, 6]. Meenu Srivastava et al [1] found that the microstructures and properties of the NiCo alloy were dependent on the Co content and the coating attained the maximum hardness at Co content of 50 wt %. Jianqiang Kang et al. [6] pointed out that the presence of Co element increased the thickness of the space charge layer, which enhanced greatly the stability of the passive film than that of the pure Ni coating. Similar to the pure Ni coatings, the particle -reinforced Ni based composite coatings could also be enhanced by alloying with Co [5, 7, 8]. Meenu Srivastava et al. [5] found that the presence of 25 wt % cobalt in the Ni–CeO_2 matrix increased the hardness and improved its wear resistance. Also in Meenu Srivastava's study [7], the NiCo-Al composite coatings were fabricated and addition of 30 wt % Co in the NiCo-Al composite coating exhibited high temperature oxidation and corrosion resistance. In reference [8], addition of 28 wt % Co content in the Ni-SiC matrix showed a higher hardness than other coatings.

In our early works, the fabrication and characterization of Ni-Al and Ni-Zr were reported [9, 10]. However, the effects of Co on the microstructures and properties, especially the texture, residual stress and corrosion resistance of the NiCo-Al composite coatings need further investigation. In this study, the NiCo-Al and NiCo-Zr composite coatings with different Co contents were prepared. Then, the effects of the Co content on the microstructure, texture, grain size, micro-strain and residual stress of the NiCo-Al and NiCo-Zr composite coatings were investigated in detail.

Experiments

Coating processes

NiCo-Al and NiCo-Zr composite coatings prepared by the electrodeposition method from the modified Watt baths containing $NiSO_4 \cdot 6H_2O$ (240 g/L), $NiCl_2 \cdot 6H_2O$ (40 g/L), H_3BO_3 (30 g/L) and $C_{12}H_{25}NaSO_4$ (0.2 g/L). Co was added as the CoSO4·$7H_2O$ and the additions ranged from 5 to 40 g/L of Co concentrations. Al and Zr particles with a mean diameter of 1 μm were added into the solutions at concentration of 20 g/L and 10 g/L, respectively. The solution temperature and pH value were maintained at 50 °C and 4.2, respectively. Pure nickel plate and stainless steel plate with an area of 1 × 1 cm^2 were used as the anode and cathode, respectively. The applied current density was maintained at a constant of 4 A/dm^2 for 1 hr.

Coating characterization

The chemical composition was examined by using the Energy Dispersive X-ray Spectroscopy (EDX) method attached to the FESEM(FESEM, JSM-7600F). Structure analysis of the coatings was performed by a Rigaku Ultima IV X-ray diffractometer (XRD, Cu Kα radiation, λ = 1.54056 Å) in standard 2θ–θ mode and the voltage and current were 40 kV and 30 mA, respectively. The Voigt method [11] was used to calculate the grain size and microstrain of the coatings using the integral breadth of (200) peak. Residual stress data on the as-deposited Ni-Zr composite coatings were collected on a Proto LXRD Residual Stress Analyser and analyzed using the classical $\sin^2\psi$ method [12]. The voltage and current were 30 kV and 25 mA, respectively. The peaks of (331)α of Ni were selected to calculate the residual stress.

Results and Discussion

Fig. 1 show the chemical contents of the NiCo-Al and NiCo-Zr alloy coating as a function of the $CoSO_4 \cdot 7H_2O$ concentrations in the bath. It was obvious that the Co contents linearly increased while Ni contents linearly decreased with the increasing $CoSO_4 \cdot 7H_2O$ concentrations in the bath.

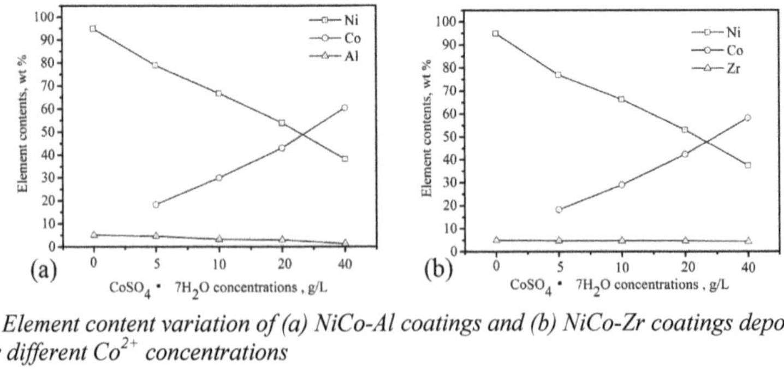

Fig.1 Element content variation of (a) NiCo-Al coatings and (b) NiCo-Zr coatings deposited under different Co^{2+} concentrations

Materials Research Proceedings **2** (2016) 539-544 doi: http://dx.doi.org/10.21741/9781945291173-91

Fig. 2 show the XRD patterns of the NiCo-Al and NiCo-Zr composite coatings with different Co contents. As shown in Fig. 2, the NiCo matrix peaks corresponding to the (111), (200), (220), (311) and (222) planes were detected. It could also be found from the XRD patterns that, as the Co contents reached 60.5 wt%, the hcp Co phase appeared. The similar results were also reported in reference [13].

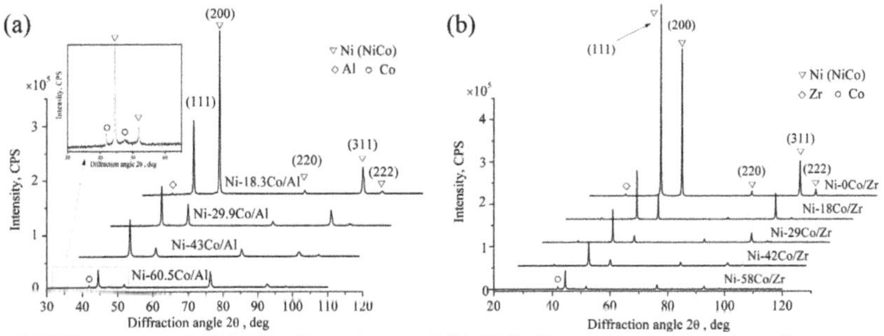

Fig.2 XRD patterns of (a) NiCo-Al coatings and (b) NiCo-Zr coatings with various Co contents

Fig. 3 show the texture coefficients of the NiCo-Al and NiCo-Zr composite coatings. It could be found that, as the Co contents increased, the (200) texture coefficients decreased and (111) texture coefficients increased.

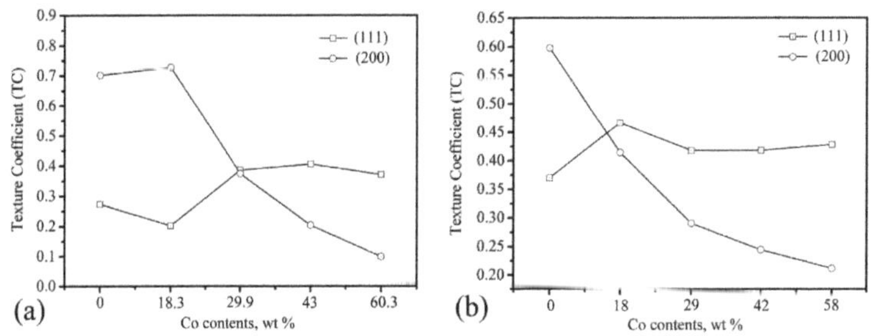

Fig.3 Texture coefficient of (a) NiCo-Al coatings and (b) NiCo-Zr coatings with various Co contents

Fig. 4 demonstrate the calculated grain size and microstrain of the NiCo-Al and NiCo-Zr composite coatings with different Co contents. As seen in Fig. 4, increasing the Co contents led to the decrease in grain size of the composite coatings. From Fig. 4, it could also be found that the microstrain of the composite coating increased with increasing Co contents.

Materials Research Proceedings **2** (2016) 539-544 doi: http://dx.doi.org/10.21741/9781945291173-91

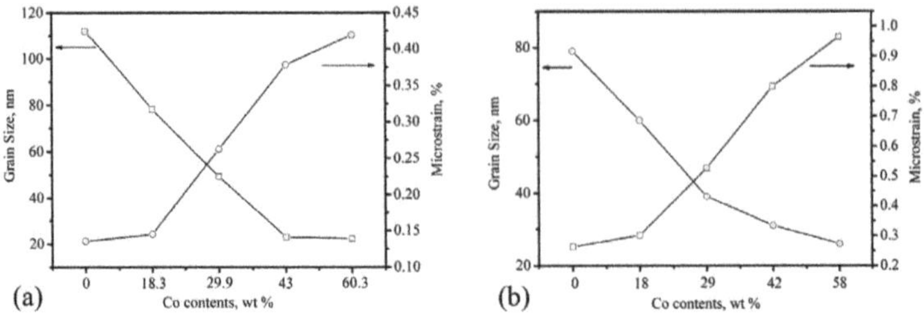

Fig.4 Grain size and microstrain of (a) NiCo-Al coatings and (b) NiCo-Zr coatings with various Co contents by using single peak method

Fig.5 shows the FWHM of (331) diffraction peak as a function of NiCo-Al and NiCo-Zr coatings with various Co contents and it could be found that the FWHM increased with increasing Co content for the NiCo-Al and NiCo-Zr coatings. Fig. 6 present the residual stresses of composite coatings with different Co contents. It could be found that all the coatings exhibited the tensile residual stress. From Fig. 6, it could be found that the residual stresses of the composite coatings increased from with increasing Co contents.

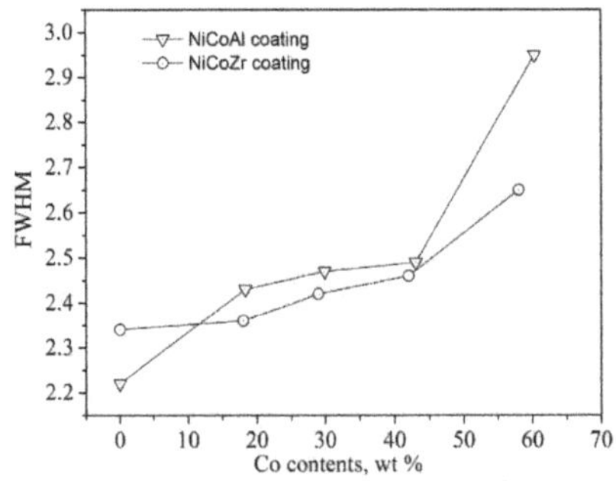

Fig.5 FWHM as a function of NiCo-Al and NiCo-Zr coatings with various Co contents

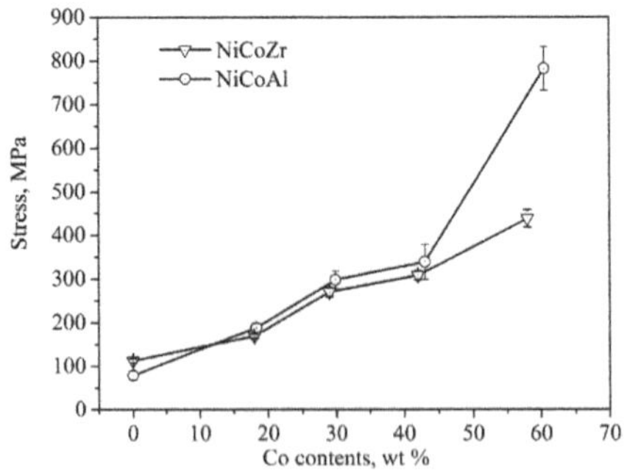

Fig.6 Stresses of composite coatings with various Co contents

The increase in residual stresses was closely related to the formation of solid solution NiCo and the texture evolution from the (200) preferred orientation to a random orientation or slight (111) orientation. As mention above, the substitution of Ni atoms by Co atoms resulted in the distortion of the composite coatings, which could increase the residual stress of the coatings. Therefore, the residual stresses increased with increasing microstrain by increasing the Co contents in the composite coatings. Moreover, the Young's moduli in different directions are different and the average Young's moduli are 303 GPa and 137 GPa for (111) and (200), respectively [14]. The higher strain energy density in (111) texture film likely led to the higher tensile internal stress in the (111) texture coatings compared with that in (200) texture coating, as reported in references [15]. However, the higher stresses in NiCo-Al and NiCoZr composite coatings might be due to a contribution from magnetostriction since NiCo is definitely ferromagnetic.

Conclusions

NiCo-Al and NiCo-Zr composite coatings with different Co contents were prepared by the conventional direct current electrodeposition from modified Watt baths containing different Co^{2+} concentrations. The Co content in the deposits increased with increasing the Co^{2+} concentration. The crystal structures of the composite coatings were mainly dependant on the Co contents. At low Co contents, the composite coatings exhibited solid solution NiCo crystal structure. However, hcp Co was detected with further increasing the Co content. Increasing Co contents in the NiCo-Al composite coatings resulted in the texture evolution from the (200) preferred orientation to the random orientations or the slight (111) preferred orientation. The grain size decreased and microstrain increased with increasing Co content. The residual stress also increased with increasing Co content.

References

[1] Meenu Srivastava, V. Ezhil Selvi, V.K. William Grips, K.S. Rajam, Corrosionresistance and microstructure of electrodeposited nickel-cobalt alloy coatings, Surf. Coat. Technol. 201 (2006) 3051-3060. http://dx.doi.org/10.1016/j.surfcoat.2006.06.017

[2] L. Shi, C.F. Sun, P. Gao , F. Zhou, W.M. Liu, Electrodeposition and characterization of Ni-Co-carbon nanotubes composite coatings, Surf. Coat.Technol. 200 (2006) 4870-4875. http://dx.doi.org/10.1016/j.surfcoat.2005.04.037

[3] B. Ranjith, G. Paruthimal Kalaignan, Ni-Co-TiO_2 nanocomposite coating prepared by pulse and pulse reversal methods using acetate bath, Appl. Surf. Sci. 257 (2010) 42-47. http://dx.doi.org/10.1016/j.apsusc.2010.06.029

[4] Lei Shi, Chufeng Sun, Ping Gao, Feng Zhou, Weimin Liu, Mechanical properties and wear and corrosion resistance of electrodeposited Ni-Co/SiC nanocomposite coating, Appl. Surf. Sci. 252 (2006) 3591-3599. http://dx.doi.org/10.1016/j.apsusc.2005.05.035

[5] Meenu Srivastava, V.K. William Grips, K.S. Rajam, Electrodeposition of Ni-Co composites containing nano-CeO_2 and their structure, properties, Appl. Surf. Sci. (2010) 717-722. http://dx.doi.org/10.1016/j.apsusc.2010.07.046

[6] Jianqiang Kang, Yifu Yang, Xi Jiang, Huixia Shao, Semiconducting properties of passive films formed on electroplated Ni and Ni-Co alloys. Corros. Sci. 50 (2008) 3576-3580. http://dx.doi.org/10.1016/j.corsci.2008.09.005

[7] Meenu Srivastava, J.N. Balaraju, B. Ravisankar, C. Anandan, V.K. William Grips, High temperature oxidation and corrosion behaviour of Ni/Ni-Co-Al composite coatings, Appl. Surf. Sci. 263 (2012) 597-607. http://dx.doi.org/10.1016/j.apsusc.2012.09.115

[8] Meenu Srivastava, V.K. William Grips, K.S. Rajam, Electrochemical deposition and tribological behaviour of Ni and Ni-Co metal matrix composites with SiC nano-particles, Appl. Surf. Sci. 253 (2007) 3814-3824. http://dx.doi.org/10.1016/j.apsusc.2006.08.022

[9] Fei Cai, Chuanhai Jiang, Influences of Al particles on the microstructure and property ofelectrodeposited Ni–Al composite coatings, Appl. Surf. Sci. 292 (2014) 620-625. http://dx.doi.org/10.1016/j.apsusc.2013.12.021

[10] Fei Cai, Chuanhai Jiang, Xueyan Wu, X-ray diffraction characterization of electrodeposited Ni–Al composite coatings prepared at different current densities, J. Alloys Compd. 604 (2014) 292-297. http://dx.doi.org/10.1016/j.jallcom.2014.03.063

11] Th.H. De Keijser, J.I. Langford, E.J. Mittemeijer, A.B.P. Vogels, Use of the Voigt function in a single-line method for the analysis of X-ray diffraction line broadening. J. Appl. Cryst. 15 (1982) 308. http://dx.doi.org/10.1107/S0021889882012035

[12] P.J. Withers, H. Bhadeshia, Part 1 – Measurement techniques. Mater. Sci. Tech. 17 (2001) 355. http://dx.doi.org/10.1179/026708301101509980

[13] C. Lupi, A. Dell'Era, M. Pasquali, P. Imperatori, Composition, morphology, structural aspects and electrochemical properties of Ni-Co alloy coatings, Surf. Coat. Technol. 205 (2011) 5394-5399. http://dx.doi.org/10.1016/j.surfcoat.2011.06.002

[14] M. A. Meyers and K. K. Chawla, Mechanical Behavior of Materials P. 275, Prentice Hall, New Jersey, US, 1999.

[15] C. V. Thompson, STRUCTURE EVOLUTION DURING PROCESSING OF POLYCRYSTALLINE FILMS, Annu. Rev. Mater. Sci. 30 (2000) 159. http://dx.doi.org/10.1146/annurev.matsci.30.1.159

Neutron Diffraction Measurements of Residual Stress and Mechanical Testing of Pressure Sintered Metal-Ceramic Composite Systems

Karl Toppler[1,a*], Vladimir Luzin[1,b], Michael Saleh[1,c], Andrew Ruys[2,d], Kaveh Kabir[3,e] and Dorji Chavara[1,f]

[1] Australian Nuclear Science and Technology Organisation. Locked Bag 2001, Kirrawee DC, NSW, 2232. Australia

[2] School of Aerospace, Mechanical and Mechatronic Engineering, University of Sydney. NSW 2006. Australia

[3] Dept. of Engineering, Macquarie University, NSW.2109. Australia

[a]kto@ansto.gov.au, [b]vll@ansto.gov.au, [c]mis@ansto.gov.au, [d] andrew.ruys@sydney.edu.au, [e] kaveh.kabir@mq.edu.au, [f]drc@ansto.gov.au

Keywords: Functionally Graded Materials, FGM, Residual Stress, Metal-Ceramic Composite

Abstract. Functionally graded materials (FGMs) are composite materials which vary in phase composition, microstructure and properties over one or more dimensions. They are a good potential choice for nuclear reactor components as they can be engineered to effectively resist corrosion and radiation damage. In the case of a metal-ceramic FGM, they can mate the strength and ductility of a metal with the hardness and toughness of a ceramic. A series of composite samples of variable metal-ceramic ratios was manufactured by hot uniaxial pressing in cylinders. Bulk uniform samples of a certain composition were manufactured as a more efficient way of studying FGMs without the extreme gradient required in practical applications. Thermally and mechanically generated stresses, inherent in composites, frequently create conditions for micro-cracking development, depending on the material's micro-structural characteristics and the thermo-mechanical processing route. Bulk stress measurements in the prepared samples were carried out on the Kowari diffractometer on the OPAL reactor at ANSTO. Both phases - metal matrix and ceramic inclusions - were measured in both axial and lateral directions for full characterisation of the composite stress state. When compared against analytical evaluation, experimental results, for some samples, demonstrated significant stress relaxation with micro-cracking being the main suspect.

Introduction

Functionally graded materials (FGMs) were first proposed as an advanced engineering material in 1972, and have been actively pursued since then. They are proposed in areas such as biomaterials, aerospace, chemical plants, mining and building material commodities[1, 2]. Within the nuclear industry FGMs can be engineered to effectively resist corrosion, radiation and be used as potential nuclear reactor components; first wall for fusion reactors and fuel pellets. They have been proposed as potential plasma facing components (PFCs) in which the ideal PFC would gradually vary from a refractory material (e.g. tungsten, plasma face) to a heat sink material (e.g. copper). In the case of a metal-ceramic FGM, the composite can be engineered to mate the strength and ductility of a metal with the hardness and toughness of a ceramic [3-5]. FGMs have been successfully fabricated using various methods to produced thin film/interfacial FGMs or bulk FGMs (layered/continuous) [3].

Materials Research Proceedings **2** (2016) 545-550 doi: http://dx.doi.org/10.21741/9781945291173-92

The FGM manufacturing routes can be broadly classified as layered or continuous based on their gradient profile. As the name suggests, layered FGMs have distinct step changes of material composition/properties across the layered axis, in contrast to continuous gradient FGMs, which do not [3].

Being a composite material, FGMs have all the attributes of composites; with inhomogeneous microstructure, material properties, etc., while characterisation of their mechanical state requires description in terms of micromechanics. Experimental studies of FGMs are difficult due to the gradients of properties. A more economical way of doing this is through the preparation and characterisation of bulk samples of uniform and pre-determined composition and properties, suitable for mechanical testing as well as for neutron diffraction studies of the micro-stress state. This study examines the micro-mechanical behaviour of a series of composite samples of variable composition, to predict the likely properties of a FGM.

Fig. 1. Experimental setup and samples on the KOWARI diffractometer.

Sample synthesis and characterisation

The system analysed in the current study is a Molybdenum-Yttrium Oxide system. The pertinent material properties are shown in Table 1 and the as-received powder morphology was imaged using a scanning electron microscope. The samples were manufactured using a traditional powder processing route. To produce the required volume based composite analogues, the required mass of powder for each portion (i.e. ceramic or metallic) was calculated using theoretical densities (Table 1). Both portions were mixed with a tubular mixer for 10 min. With the aid of a uniaxial press the mixed powder was pressed with a tool-steel die to 50 MPa. These pellets were hot pressed at 1600°C at a pressure of 40 MPa for 1 h.

Sample densities were measured as per AS1774.5:2014 [6] and compared to theoretical densities as a function of sample composition.

Light microscopy images were taken of these analogues to understand the fabricated macrostructure (Figure 3b). The elastic moduli were measured through the impulse exciation technique according to standards [7-9] using a GrindoSonic MK4 appartus. The Young's modulus for a rectangular bar sample is determined from the natural frequency corresponding to the first flexural mode of vibration.

Table 1. Material properties of the Mo-Y2O3 powders [3, 10]

	Molybdenum (Mo)	Yttrium Oxide (Y_2O_3)
Thermal Expansion (CTE), [$\times 10^{-6}$ K^{-1}]	4.8-5.1	8.1
Density, [g/cc]	10.2	5.03
Melting Point, [°C]	2610	2410
Crystal Structure	body centered cubic	cubic

To assess residual stress, both phases of the mixed composition samples were measured. Pure material samples were used as the d_0 reference, as all samples were synthesised through the same thermo-mechanical treatment (i.e., pressure and temperature characteristics of the sintering process). A wavelength of λ =1.67Å was used to measure the Mo(211) and Y_2O_3(440) reflections at scattering angles of 81° and 53°, respectively. The samples were stacked in series and measured while spinning continuously to achieve better statistics, as shown in Figure 1.

Materials Research Proceedings **2** (2016) 545-550 doi: http://dx.doi.org/10.21741/9781945291173-92

For the cylindrical samples of diameter 25 mm and height 8–10 mm, a relatively large gauge volume of 5x5x5 mm^3 was used; it was assumed there was no macro-stress (samples were uniform and isotropic). However, the samples were measured in two principal directions (axial and lateral). Strains were measured with an accuracy of 5E10-5 in order to get an accuracy of ~20 MPa in the stress scale for metals, which was somewhat larger for the ceramics. In order to achieve this, the measurement time was adjusted according to the volume fraction and material, varying from 1 to 30 min.

Results and Discussion

The metallic powder used to fabricate the FGMs had a significantly larger particle size than its ceramic counterpart (Figure 2). Ceramic particles were flaky and had a fractured appearance in contrast to the more cleaved spherical morphology of the metallic powders.

Fig. 2. Powder morphology for (a) Mo and (b) Y_2O_3 powders.

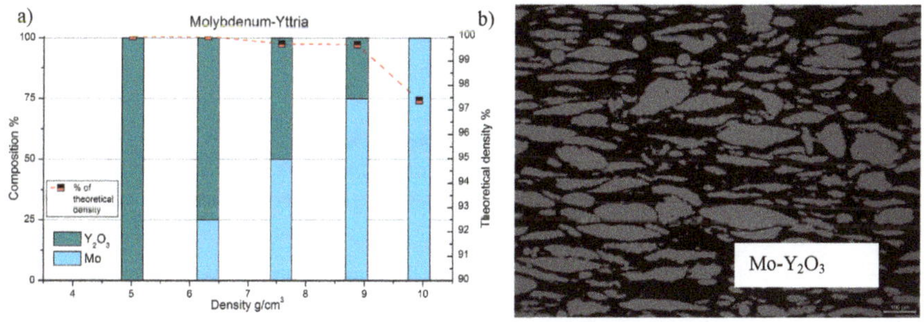

Fig. 3. (a) Density measurements of samples as function of the composition (b) Light microscopy micrograph with 50%Mo (light grey phase) and 50%Y_2O_3 (dark phase).

The results of neutron residual stress analysis on the composite samples are shown in Figure 5. The overall trend of compressive stress in the Mo, with Y_2O_3 being in tension, suggests the stresses stem from the thermal mismatch [12]. However, the stress state is anisotropic with distinctly different axial and lateral components. Although the lateral component Mo compressive stress seems to be over evaluated (especially for the 25% sample), the stress balance, $f \cdot \sigma_i^{Ytt} + (1-f) \cdot \sigma_i^{Mo}$, shows it is within experimental uncertainties for all components.

Interphase micro-stresses explain the results, as no macro-stresses were detected in the composite samples, which were uniform, but not isotropic. Considering the manufacturing route (high-temperature uniaxial pressing), material properties of the different phases (coefficient of thermal expansion (CTE) and Young's Modulus (E) mismatch) and the microstructure details (anisotropic elongation with micro-cracking), a qualitative interpretation of the stress state in the composite was possible, as follows:

(i) The thermal stresses due to the mismatch in CTE were largely isotropic/hydrostatic, however, the oriented micro-cracking suggested a deviatoric stress component. The samples exhibited compressive stresses in Mo and tensile stresses in Y_2O_3 attributable to the discrepancy in the CTE (~5.0 for Mo vs. 8.1 for Y_2O_3). The hydrostatic stress can be evaluated for the quasi-isotropic approximation using the Hashin-Shtrikman relation [11] shown in Figure. 6. When compared with the experimental data, stress relaxation is observable for the samples with $f(Y_2O_3)$ = 0.5 and 0.75 which indicates the presence of micro-cracking. This conclusion is backed up by the reduced Young's modulus measurements (Figure. 4).

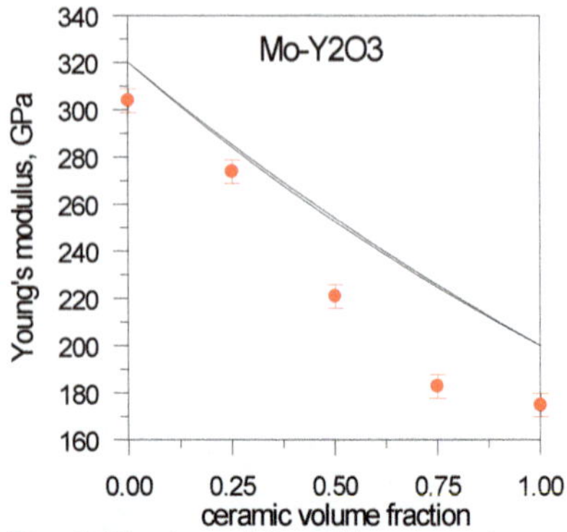

Fig. 4. Elastic modulus of the composite as a function of the ceramic-metal composition. Experimental values (symbols) are compared with theoretically calculated Hashin-Shtrikman bounds.

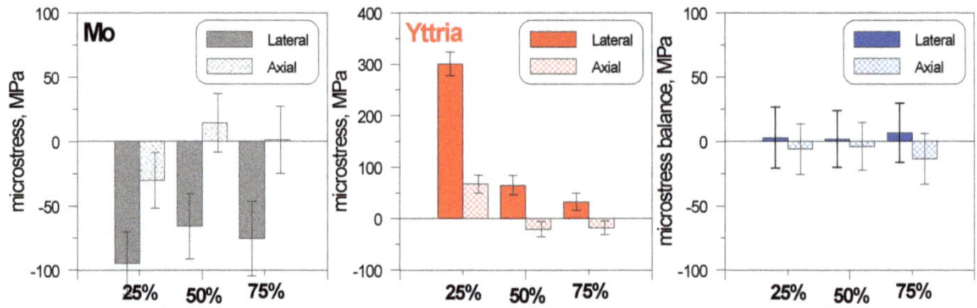

Fig. 5. a) Micro-stress distribution in the lateral and axial directions as functions of ceramic composition for the metal and ceramic phases. b) micro-stress balance across each of the samples.

(ii) Anisotropy in the stress state could not reasonably be explained through the thermal mechanism only, as it generates mostly hydrostatic stress. The difference in lateral and axial stress can be attributed to the plastic deformation mechanism. Although quantitative estimations to accurately evaluate this effect are complex, simplified qualitative considerations support the concept of a deformation-induced stress factor. In general, for metal-ceramic composites, two phases are defined by the elastic-plastic metal and the semi-rigid ceramic particles. Uniaxial compression leads to an anisotropic stress state with the ceramic particle phase in compression in the axial direction and in tension for the lateral direction. Thus, combined with the thermally generated tensile stresses, the axial stress for the ceramic phase was expected to have more tensile stresses in the lateral direction and a reduced tensile stress profile for the axial direction. The same trend was observed experimentally at higher volume fractions $f(Y_2O_3)$ = 0.5 and 0.75, though the exact effect was determined through a combination of these competing mechanisms.

(iii) Stress relaxation through micro-cracking was another mechanism at play. This was supported by the elastic modulus measurements which indicated stress relaxation in the ceramic-rich samples. However, to result in anisotropic stress relaxation, the micro-cracking system is assumed to also be anisotropic with a preferred orientation of the crack plane normal to the axial direction, so the axial stress component is largely relaxed while there is less effect in the lateral direction. This assumption of the anisotropic micro-crack system is also consistent with the anisotropic microstructure observed in Figure. 3(b). The determination of the exact amount and type of micro-cracking is a matter of further study and will be investigated in a future publication.

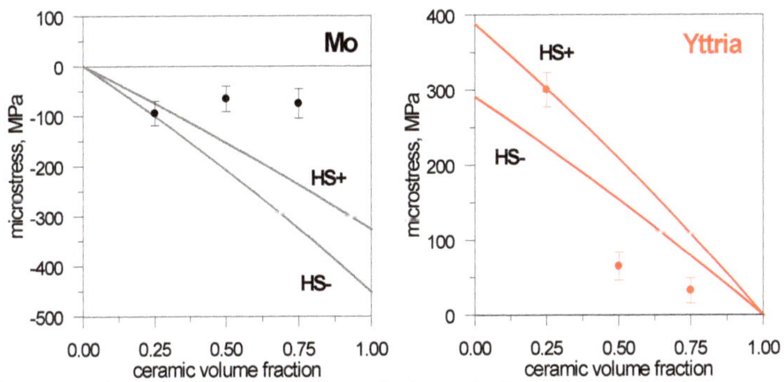

Fig. 6. Micro-stress in each phase due to thermal mismatch: theoretical evaluation vs. experimental data (the axial component was taken as the experimental reference). The Hashin-Shtrikman bounds correspond to ΔT = -450°C.

Conclusion

This study assessed the residual stresses in a series of metal-ceramic Mo-Y_2O_3 FGM analogue samples by means of neutron diffraction. The micro-stresses that formed were predominantly a result of the thermal coefficient mismatch; the micro-stresses were largely isotropic/hydrostatic. The presence of anisotropic stresses highlighted by the difference in the lateral and axial stresses were attributed to plastic deformation, induced in part by the production process. Stress relaxation through the development of micro-cracking was also detected by two independent experimental techniques; these accounted for the reduction in Young's modulus. Overall, the final stress state of the composite samples of various phase compositions was determined experimentally to be a combination of these stress-inducing mechanisms, which are intrinsically linked to the nature of the materials and the technological manufacturing process. Future work will be centred on altering the processing route with the use of cold and hot isostatic pressing, in an attempt to re-evaluate the resultant residual stress by eliminating the directional plastic deformation.

References

1. Kieback, B., A. Neubrand, and H. Riedel, *Processing techniques for functionally graded materials.* Materials Science and Engineering: A, 2003. **362**(1–2): p. 81-106. http://dx.doi.org/10.1016/S0921-5093(03)00578-1

2. Miyamoto, Y., *Functionally Graded Materials: Design, Processing and Applications.* 1999: Springer US. http://dx.doi.org/10.1007/978-1-4615-5301-4

3. Chavara, D.T., M. University of Sydney. School of Aerospace, and M. Engineering, *Synthesis and Testing of Metal-ceramic Functionally Graded Materials.* 2009: University of Sydney.

Materials Research Proceedings **2** (2016) 545-550 doi: http://dx.doi.org/10.21741/9781945291173-92

4. Ling, Y.-H., et al., *Fabrication and evaluation of SiC/Cu functionally graded material used for plasma facing components in a fusion reactor.* Journal of Nuclear Materials, 2002. **303**(2–3): p. 188-195. http://dx.doi.org/10.1016/S0022-3115(02)00801-2

5. Nawkaa, S., et al., *Synthesis, Characterization and FEM-simulation of W/CuCrZr-Composites for Extreme Thermal Applications.*

6. Australian-Standard, *AS 1774.5:2014 Refractories and refractory materials-physical test methods. Method 5: determination of bulk density, apparent porosity and true porosity (ISO 5017:2013, MOD)*, 2014, Australian Standard.

7. Latella, B.A. and T. Liu, *High-Temperature Young's Modulus of Alumina During Sintering.* Journal of the American Ceramic Society, 2005. **88**(3): p. 773-776. http://dx.doi.org/10.1111/j.1551-2916.2005.00082.x

8. ASTM-International, *ASTM E1875-13 Standard test method for dynamic young's modulus, and poisson's ration by sonic resonance*, 2013, ASTM International.

9. ASTM-International, *ASTM C1259-15 Standard test method for dynamic young's modulus, shear modulus, and poisson's ratio for advanced ceramics by impulse excitation of vibration*, 2015, ASTM International.

10. *www.matweb.com.* [cited 2016].

11. Hashin, Z., *The Elastic Moduli of Heterogeneous Materials.* ournal of Applied Mechanics, 1962. **29**(1): p. 143-150. http://dx.doi.org/10.1115/1.3636446

12. Clyne, T.W. and P.J. Withers, *An Introduction to Metal Matrix Composites.* 1993: Cambridge University Press. http://dx.doi.org/10.1017/CBO9780511623080

Residual Stresses 2016: ICRS-10 Materials Research Forum LLC
Materials Research Proceedings **2** (2016) 551-556 doi: http://dx.doi.org/10.21741/9781945291173-93

The Role of Intergranular Stresses in Plastic Deformation Studied Using a Diffraction and Self-Consistent Model

E. Gadalińska[1,a], A. Baczmański[2,b], M. Wróbel[2,c], S. Wroński[2,d], M. Wroński[2,e], R. Wawszczak[2,f], C. Braham[3,g], Y. Zhao[4,h], L. Le Joncour[4,i], T. Buslaps[5,j] and Ch. Scheffzük[6,7,k]

[1]Institute of Aviation, al. Krakowska 110/114, 02-256 Warszawa, Poland

[2]AGH-University of Science and Technology, WFiIS and WIMiIP al. Mickiewicza 30, 30-059 Kraków, Poland

[3]Arts et Métiers ParisTech (ENSAM), PIMM, 151, Bd de l'Hôpital 75013, Paris, France

[4]Université de Technologie de Troyes (UTT), 12 rue Marie Curie, 10010 Troyes, France

[5]ESRF, 6, rue J. Horowitz, 38500 Grenoble Cedex, France

[6]Frank Laboratory of Neutron Physics, Joint Institute for Nuclear Research, 141980 Dubna, Russia

[7]Karlsruhe Institute of Technology, Adenauerring 20b, 76131 Karlsruhe, Germany

[a]elzbieta.gadalinska@ilot.edu.pl, [b]Andrzej.Baczmanski@fis.agh.edu.pl, [c]Miroslaw.Wrobel@fis.agh.edu.pl, [d]Sebastian.Wronski@fis.agh.edu.pl, [e]Marcin.Wronski@fis.agh.edu.pl, [f]Roman.Wawszczak@fis.agh.edu.pl, [g]chedly.braham@ensam.eu, [h]yuchen.zhao@utt.fr, [i]lea.le_joncour@utt.fr, [j]buslaps@esrf.fr, [k]christian.scheffzuek@kit.edu

Keywords: Diffraction Methods, Intergranular Stresses, Self-Consistent Model, in-situ Tensile Test

Abstract. Diffraction methods are commonly used for the determination of the elastic lattice deformation from the displacement and broadening of the diffraction peak. The measurements are performed selectively, only for crystallites contributing to the measured diffraction peak. When several phases are present in the sample, measurements of separate diffraction peaks allow the behaviour of each phase to be investigated independently [e.g. 1-4]. Comparison of experimental data with a multi-scale model allows us to understand the physical phenomena which occur during sample deformation at the level of polycrystalline grains. In the present work the methodology combining diffraction experiment and self-consistent calculation was used to study the mechanical behaviour of groups of grains within stainless duplex steel and Al/SiC composite. Special attention has been paid to the role of second order stresses on the yield stresses of the phases, as well as on the evolution of these stresses during the deformation process. The inter-granular stresses were determined from lattice strains measured "in situ" during tensile tests. The diffraction measurements were done using synchrotron (ID15B, ESRF, Grenoble, France) and neutron (EPSILON, FLNP, JINR, Dubna, Russia) radiations.

Experimental methodology

To study the elastoplastic behavior of two phase materials the diffraction measurements were performed "in situ" during a tensile test. Two materials, i.e., an aged duplex steel UR45N (50% of ferrite and 50% of austenite; microstructure, texture and composition are given in [3,4]) and a particle reinforced Al/SiC$_p$ composite (Al2124 matrix with 17% of SiC particles with size of 0.7 μm subjected to the T1 treatment) were studied using diffraction methods of two types. The Al/SiC$_p$

Materials Research Proceedings **2** (2016) 551-556 doi: http://dx.doi.org/10.21741/9781945291173-93

composite was obtained by powder metallurgy processing followed by the T1 thermal treatment (air cooled from elevated temperature forming process). For the steel specimen monochromatic synchrotron radiation with an energy of about 90keV ($\lambda = 0.14256$Å) and a beam size of $100 \times 100\ \mu m^2$ was applied. The measurements were carried out at the European Synchrotron Radiation Facility in Grenoble on the ID15B beamline. The diffraction pattern in the range of $2\theta = 1.7 - 7.5°$ was collected on two-dimensional detector PIXIUM4700 in the form of concentric rings corresponding to different *hkl* reflections for both phases: ferrite and austenite. The necessary conversion of two-dimensional images into 2θ diffractograms was performed with FIT2D software [5]. The peak positions were determined using the MULTIFIT software [6] and after that Braggs' law was employed to determine the interplanar spacings d_{hkl}.

The second experiment was performed for the particle reinforced Al/SiC$_p$ composite specimen using the time-of-flight (TOF) neutron diffraction method enabling simultaneous measurement of different *hkl* reflections for both phases in the studied composite. For data acquisition two of nine detector banks covering a 2θ-range of $82° \leq 2\theta \leq 98°$ were employed on the EPSILON-MDS diffractometer at the Joint Institute For Nuclear Research, Dubna, Russia [7]. A geometry of this kind allowed to determine stress tensor if the lattice strains are measured for two sample orientations with respect to the experimental setup. The measurement was performed for the initial material and after deformation of the sample during a tensile test. The incident beam of 10 mm width was pointed at the sample of 4.4 mm x 4.4 mm cross-section.

Evolution of deviatoric stresses in duplex steel during plastic deformation

Initial state of the specimen

The analysis of the initial stresses and of the evolution of the lattice parameter for both phases was performed for the duplex steel measured using synchrotron radiation. Firstly the stress tensor was determined for the initial non-deformed sample. The principal stresses were decomposed into two parts: hydrostatic (p) and deviatoric (q, r, s), according to Eq. 1. The deviatoric stresses were determined directly from the measured lattice strains, while the additional assumption $p = 0$ (zero values of hydrostatic stresses) was introduced in order to calculate the initial values of the lattice parameter a_0 for both phases. The results are presented in Table 1.

$$\begin{bmatrix} \sigma_{11} & 0 & 0 \\ 0 & \sigma_{22} & 0 \\ 0 & 0 & \sigma_{33} \end{bmatrix} = \begin{bmatrix} p & 0 & 0 \\ 0 & p & 0 \\ 0 & 0 & p \end{bmatrix} + \begin{bmatrix} q & 0 & 0 \\ 0 & r & 0 \\ 0 & 0 & s \end{bmatrix} \quad (1)$$

Table 1. Initial stresses and lattice parameters determined for both phases of the studied steel assuming zero value of hydrostatic stresses. The parameters used in the self-consistent model are also presented.

		Austenite	*Ferrite*
Initially measured values	*(q)* $\sigma_{RD}[MPa]$	134 ± 15	-155 ± 19
	(r) $\sigma_{TD}[MPa]$	84 ± 15	-44 ± 19
	(s) $\sigma_{ND}[MPa]$	-218 ± 15	199 ± 18
	a_0 [Å]	3.6102 ± 0.0001	2.8791 ± 0.0001
Single crystal elastic. constants	$c_{11}, c_{12}, c_{44}[GPa]$	198 , 125 , 122	231, 134, 116
Slip systems		⟨111⟩{110} and ⟨111⟩{211}	⟨110⟩{111}
Parameters of Voce law	τ_0 $[MPa]$	170	370
	θ_0 $[MPa]$	380	190

As illustrated in Fig. 1 and presented in Table 1, the initial deviatoric stresses acting in ND direction on austenite grains are compressive, while the stresses acting on the ferrite grains are tensile. Meanwhile the initial stresses in direction RD are tensile for austenite grains and compressive for the ferrite ones. The absolute values in both cases (for ND and RD directions) are approximately equal but they have opposite signs for each phase, therefore the specimen containing the same fractions of ferrite and austenite is in a state of equilibrium. The initial stresses were generated during cooling of the material, after aging process (details of the thermal treatment are given in [3,4]). The deviatoric type of stress state would be explained probably due to various interaction of the grains in different directions caused by the ellipsoidal shape of inclusions (see Fig.1) and the difference in thermal expansion coefficients of austenite and ferrite.

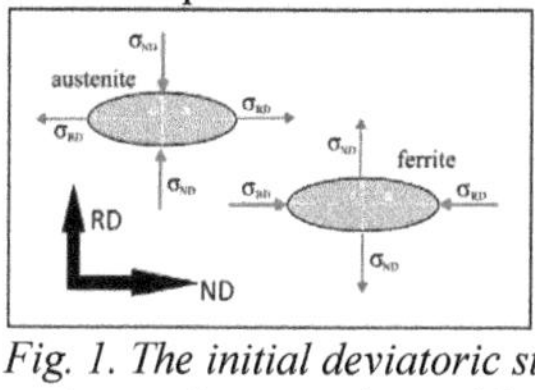

Fig. 1. The initial deviatoric stresses acting on the austenite and ferrite grains in the initial sample.

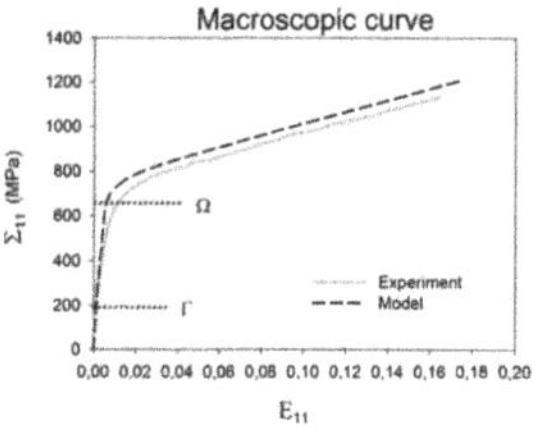

Fig. 2. The dependence of macrostress $\mathbf{\Sigma_{11}}$ vs. macrostrain $\mathbf{E_{11}}$ for experimental data and model predictions.

The initial deviatoric stresses (shown in Table 1) and the previously measured crystallographic textures [3,4] were used as the input data for elasto-plastic self-consistent model [8]. The calculation were performed for the single crystal elastic constants and slip systems given in Table 1. Additionally, the ellipsoidal shape of inclusions corresponding to microstructure presented in [3,4]: $a/b = 5$, $a/c = 10$ (the a and c axes are defined in Fig.1, while b axis is parallel to TD direction) was assumed for both phases. The aspect ratios of inclusions were determined from the EBSD orientation maps shown in [3,4].

Lattice parameter evolution

Due to the small range of deformation (about 17 %) and the almost linear character of hardening in the plastic range of deformation (Fig. 2) only two parameters of Voce law (Eq. 2) were adjusted for each phase of the studied steel.

$$\tau_c^{gr} = \tau_0^{ph} + \left(\tau_1^{ph} + \theta_1^{ph}\xi^{gr}\right)\left[1 - exp\left(-\frac{\theta_0^{ph}}{\tau_1^{ph}}\xi^{gr}\right)\right] \tag{2}$$

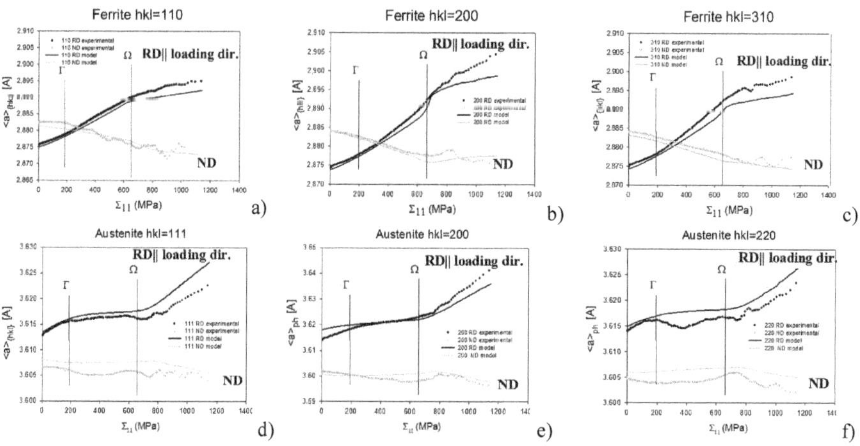

Fig. 3. The evolution of lattice parameter during the in-situ tensile test of duplex steel for different hkl reflections for a-c) ferrite d-f) austenite.

Materials Research Proceedings **2** (2016) 551-556 doi: http://dx.doi.org/10.21741/9781945291173-93

The model results were fitted simultaneously to the macroscopic stress-strain curve (Fig. 2), as well as to the dependence of lattice parameters ($\langle a \rangle_{hkl}$) measured vs. applied stress (Σ_{11}) for different *hkl* reflections (Fig. 3). The values of τ_0 and θ_0 obtained by trial and error optimization are shown in Table 1 for both phases of the studied steel. The information about the value of the lattice parameter was obtained for the experimental data directly using Eq. 3 and for the model data according to Eq. 4.

$$\langle a \rangle_{hkl} = \langle d \rangle_{hkl} \sqrt{h^2 + k^2 + l^2} \quad (3)$$

$$\langle a \rangle_{hkl} = a_0 \left(\langle \varepsilon \rangle_{hkl}^{model} + 1 \right) \quad (4)$$

where a_0 is the initial lattice parameter calculated from the initial sample (Table 1), $\langle d \rangle_{hkl}$ is the lattice strain measured using diffraction and $\langle \varepsilon \rangle_{hkl}^{model}$ is the model predicted lattice strain.

Figs. 2 and 3 show a very good agreement of model and experimental data both for the macroscopic and microscopic/grain states. We can distinguish three different stages, the elasticity of both phases (before Γ point), the elasticity of ferrite and the plasticity of austenite (between Γ and Ω points) and the plasticity of both phases (after Ω point).

The lattice parameter for initial and deformed sample

The dependence of the lattice parameter measured for different *hkl* direction vs. orientation factor $3\Gamma = 3\frac{h^2k^2 + h^2l^2 + k^2l^2}{(h^2+k^2+l^2)^2}$ is shown in Fig. 4 for the initial (a and c) and the deformed (a and c) samples (where the deformed/residual state is after unloading of the applied stress).

Ferrite

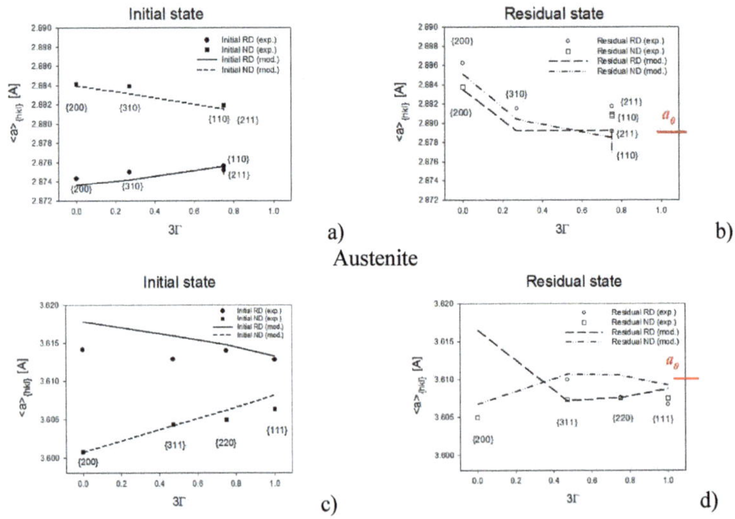

Fig. 4. The measured lattice parameter as a function of the orientation factor 3Γ *for a) ferrite in the initial state, b) ferrite in the residual state, c) austenite in the initial state and d) austenite in the residual state.*

Analysing the obtained results we can see a good agreement between experimental and model results. We can conclude that important phase deviatoric stresses exist in the initial sample (observed large difference between the lattice parameter in RD and ND directions), and that the initial stresses do not

Materials Research Proceedings 2 (2016) 551-556 doi: http://dx.doi.org/10.21741/9781945291173-93

depend significantly on grain orientations (linear dependence of $\langle a \rangle_{hkl}$ vs. 3Γ). On the other hand, for the majority of grain orientations, the residual stresses decreased in the deformed and unloaded sample, i.e. for many *hkl* reflections in each phase the $\langle a \rangle_{hkl}$ values are similar in RD and ND directions, and they approach the value of the initial a_0 parameter. It means that in general the deviatoric stresses decrease and an insignificant evolution of hydrostatic stresses was observed. However, the residual stresses after plastic deformation depend strongly on grain orientations and the $\langle a \rangle_{hkl}$ vs. 3Γ dependence is no longer linear (significant deviations were observed for the 200 reflection in both phases).

Thermal stresses in Al/SiC$_p$ composite during plastic deformation

In the previous section the evolution of deviatoric stress was studied, but insignificant changes of hydrostatic stress were measured and predicted using a self-consistent model. The second experiment described in this section will show a significant evolution of hydrostatic type of stresses. The experiment was performed using TOF diffraction and the measurement was performed for the Al/SiC$_p$ composite subjected to the T1 treatment. It is well known that a significant difference in the thermal coefficient of aluminium and silicon carbide (Table 2) causes the appearance of initial thermal stresses between those phases in the composite. In stress analysis the method allowing determination of stress tensor as well as *c/a* parameter was applied [9], assuming single crystal elastic constants for the 6H polytype of SiC [10] (Rietveld analysis of X-ray diffraction results shown about 80% of this polytype in the SiC powder).

Table 2. Material data of Al/SiC$_p$ composite.

	Al	SiC (6H polytype)
Single crystal elastic constants (GPa)	C_{11}= 106.8, C_{12}= 60.4, C_{44}= 28.3	C_{11}= 501 , C_{12}= 111 , C_{13}= 52, C_{33}= 553 , C_{44}= 163
Thermal expansion coefficient (K^{-1})	$\boldsymbol{\alpha_a = 23.6 \cdot 10^{-6}}$	$\boldsymbol{\alpha_a = 4.3 \cdot 10^{-6}}$, $\boldsymbol{\alpha_c = 4.7 \cdot 10^{-6}}$

Table 3 contains information about the thermal hydrostatic stresses in SiC (denoted by p in Eq. 1) and about the average response of the matrix Al for both stages: for the initial and the deformed specimen (tensile test). In the case of the Al matrix we have calculated the hydrostatic component of the average stress tensor seen by the diffraction experiment due to averaging over many grains, although local stresses in the matrix are not of purely hydrostatic nature. The main problem in evaluating the absolute p values is that the measurement performed on the reference stress free powder specimen is not reliable. That is why we show also a relative value Δp which is the difference between the value for the deformed specimen and for its initial state. We can conclude that an important increase of hydrostatic stress occurred in the SiC reinforcement corresponding to the relaxation of initially large compressive thermal stress. Also, the decrease of the hydrostatic part of average stress in the Al matrix has been observed as a consequence of stress relaxation of the stress around SiC inclusions.

As seen in Table 3 the values of Δp are not correctly predicted by the self-consistent model in which the hydrostatic stresses do not influence plastic behavior and do not change during plastic deformation. Therefore to explain the relaxation of the hydrostatic stresses occurring in the Al/SiC$_p$ composite the heterogeneity of the stresses in the matrix around reinforcement particles must be taken into account. In Table 3 the deviatoric stresses are presented as well. In this case we can observe the evolution which is qualitatively (but not quantitatively) predicted by the self-consistent model.

Table 3. Stress values of phases in Al/SiC composite in initial state and after deformation. Experimental (exp) and model predicted values (mod) are shown.

	Specimen	p $[MPa]$	Δp $[MPa]$	q $[MPa]$	r $[MPa]$	s $[MPa]$
Al	Initial	55±20 (exp)	-67±20 (exp)	25±7 (exp)	-8±8 (exp)	-18±8 (exp)

	Deformed	-12±80 (exp)	-3 (mod)	-75±9 (exp) -81 (mod)	42±10 (exp) 27 (mod)	34±10 (exp) 54 (mod)
SiC	Initial	-573±20(exp)	338±80 (exp) 13 (mod)	66±62 (exp)	-112±76 (exp)	46±75 (exp)
	Deformed	-235±80 (exp)		193±64 (exp) 374 (mod)	-72±77 (exp) -125 (mod)	-121±79 (exp) -249 (mod)

Summary

In the preset work the evolution of deviatoric and hydrostatic stresses during plastic deformation was studied using diffraction and a self-consistent model. At first the deviatoric stress in duplex steel was measured and compared with model calculations. It was found that the evolution of this type of stress is correctly predicted by the elatoplastic model for both phases of the studied steel. Deviatoric stresses cause a plastic process at the macroscopic level as well as at polycrystalline grains leading to the activation of slip systems. This is why the self-consistent model correctly accounts for the statistical role of such stresses. On the contrary, the evolution of hydrostatic stresses during plastic deformation, studied for the Al/SiC$_p$ composite, were not well predicted by the self-consistent model. These stresses cannot activate slips on crystallographic planes and in fact they are not seen in the crystallographic elastoplastic models based on statistical calculations in which the stress heterogeneity is not taken into account. It was found that in the self-consistent model hydrostatic stresses do not change, while in the real composite the relaxation of those stresses was determined using diffraction. The latter phenomena can be explained by the evolution of heterogeneous stresses in the Al matrix or/and by the damage processes initiated between the reinforcement and the metal matrix. This process could be predicted using for example a FEM model.

Acknowledgements

The work was supported partly by NCN - Polish National Center for Science, grants: No. UMO - 2011/03/N/ST8/04058 and DEC - 2013/11/B/ST3/03787; and partly by the MNiSW - Polish Ministry of Science and Higher Education. We also acknowledge the ESRF for the use of the synchrotron radiation facilities.

References

[1] Jia, N., Cong, Z.H., Sun, X., Cheng, S., Nie, Z.H., Ren. Y., Liaw, P.K., Wang, Y.D. Acta Mater., 57 (2009) 3965–3977. http://dx.doi.org/10.1016/j.actamat.2009.05.002

[2] Wroński, S., Baczmański, A., Dakhlaoui, R., Braham, C., Wierzbanowski, K., Oliver, E.C. Acta Mater., 55 (2007) 6219–6233. http://dx.doi.org/10.1016/j.actamat.2007.07.044

[3] Baczmanski, A., Le Joncour, L., Panicaud, B., Francois, M., Braham, C., Paradowska, A. M., Wroński, S., Amara, S. and Chirone, R., J. Appl. Cryst., 44 (2011) 966-982. http://dx.doi.org/10.1107/S0021889811025957

[4] Baczmański, A., Zhao, Y., Gadalińska, E., Le Joncour, L., Wroński, S., Braham, C., Panicaud, B., François, M., Buslaps, T., and Soloducha, K.. Int. J. Plast., 81 (2016) 102–122. http://dx.doi.org/10.1016/j.ijplas.2016.01.018

[5] Hammersley, A.P., Svensson, S.O., Hanfland, M., Fitch, S.O. and Häusermann, D. High Pressure Research, 14 (1996) 235-248. http://dx.doi.org/10.1080/08957959608201408

[6] Merkel. S (2011) Multifit/ Polydefix Polycrystal Deformation using X-rays, https://code.google.com/p/multifit-polydefix.

[7] Frischbutter, A., Neov, D., Scheffzuk, Ch., Vrana, M., Walther, K. J. Struct. Geology, 22 (2000) 1587-1600. http://dx.doi.org/10.1016/S0191-8141(00)00110-3

[8] Lipinski, P., Berveiller, M., Reubrez, E., Morreale, J. J. Arch. Appl. Mech., 65 (1995) 291–311. http://dx.doi.org/10.1007/BF00789222

[9] Marciszko, M., Baczmański, A., Braham, C. , Wróbel, M., Seiler, W., Wroński, S. and Berent, K., J. Appl. Cryst., 49 (2016) 85-102. http://dx.doi.org/10.1107/S1600576715021810

[10] Kamitani, K., Grimsditch, M., Nipko, J.C., Loong, C.-K.,J. Appl. Phys. 82 (1997) 3152-3154. http://dx.doi.org/10.1063/1.366100

Residual Stresses 2016: ICRS-10 Materials Research Forum LLC
Materials Research Proceedings **2** (2016) 557-562 doi: http://dx.doi.org/10.21741/9781945291173-94

Experimental Investigation of Welding Stresses in MWIC Weldability Test

Houman Alipooramirabad [1,a*], Anna Paradowska [2,b], Reza Ghomashchi[1,c], Nicholas Hoye[3,d] and M. Reid[2,e]

[1]School of Mechanical Engineering, the University of Adelaide, SA 5005, Australia

[2] Bragg Institute, Australian Nuclear Science and Technology Organisation (ANSTO), Lucas Heights, NSW 2234, Australia

[3] University of Wollongong, Wollongong, Australia

[a]houman.alipooramirabad@adelaide.edu.au, [b]anp@ansto.gov.au, [c]reza.ghomashchi@adelaide.edu.au, [d]nicholas.hoye@gmail.com, [e]markr@ansto.gov.au

Keywords: Residual Stress, Neutron Diffraction, HACC, MWIC Weldability, SMAW and MSAW Process

Abstract. The use of high-strength steels in the manufacture of energy pipelines, coupled with the transition to larger pipe diameters and greater wall thicknesses, has led to an increased potential for cracking including hydrogen assisted cracking of energy pipelines due to higher constraint induced stresses. In the present study, a modified version of the Welding Institute of Canada (MWIC) restraint test was used to simulate the constraint conditions of full-scale girth welds on energy pipelines, allowing the influence of welding process parameters on crack formation to be assessed. MWIC test samples of X70 grade high-strength low alloy pipeline steel were manually welded using two different welding processes, namely shielded metal arc welding (SMAW) and modified short-arc welding (MSAW). Residual strains, and hence stresses, in these samples were analysed quantitatively using neutron diffraction technique. Overall, results indicate that the modified WIC restraint test produces significant residual stresses and so is effective in constraining the root run and in consequence studying the hydrogen assisted cracking of high-strength pipeline steels.

Introduction

The construction of Australian oil and gas pipeline networks is carried out using high strength low alloy (HSLA) steel line pipe and employing Shielded Metal Arc welding (SMAW) in conjunction with hydrogen rich cellulosic consumables. The application of cellulosic electrodes at ambient temperature ensures good weld penetration and shorter construction lead time with huge cost saving [1]. The drawback however is the risk of hydrogen cracking emanating from high hydrogen content of cellulosic electrodes and the high levels of restraint as a result of clamping and lifting stresses the pipeline is subjected to during construction. The Welding Institute of Canada (WIC) test was developed more than 30 years ago to provide a simple and economical way for the assessment of weldability and risk of hydrogen assisted cold cracking. Over the past three decades it was extensively utilised by the industry to qualify pipeline welding procedures [2-5]. However, the geometry of WIC test has some shortcomings associated with the difficulty for instrumentation of the WIC for the physical measurements of strain and temperature distribution during welding. This is further exacerbated with a low success rate of achieving industry acceptable welds in the WIC test environment. The original WIC design has been modified to better represent stovepipe welding of pipelines, as shown in Fig.1. The modifications enabled better control of welding mechanics and thus achieving reasonable success rate in depositing an industry acceptable weld [6]. The new designed geometry enables easier access for instrumentation and more reproducible production of test specimens. To ensure that the modifications did not modify significantly the thermal and restraint

Materials Research Proceedings **2** (2016) 557-562 doi: http://dx.doi.org/10.21741/9781945291173-94

conditions of the test and the results are comparable with those achieved in other studies using the WIC test, physical and Finite Element studies for the original and modified test geometries have been undertaken [6, 7]. The results have shown that the modifications to the design have little or no influence on the thermal and mechanical properties of the test, but improve the ease with which consistently high quality samples are produced. The results have shown the modifications to the design have little or no influence on the thermal and mechanical properties of the test, but improve the ease with which consistently high quality samples are produced. The modified test procedure is now accepted for a wide range of investigations.

Due to the novelty of this weldability test, there is no report on the prediction and measurement of residual stresses, which is one of the influential factors in weld metal cracking including hydrogen assisted cold cracking (HACC), in the MWIC test. Moreover, no direct link has been established so far between the welding processes of the MWIC test and the welding stresses. To address the above issues, the current paper is focused on an investigation of the effects of welding process on residual stresses in the MWIC test. Residual stress measurements in these samples were analysed quantitatively using neutron diffraction techniques on the KOWARI strain scanner at the OPAL research facility operated by the Australian Nuclear Science and Technology Organisation (ANSTO).

Experimental procedure

Details of weld deposition procedure

The weld consumable was specified to be a E6010 electrode for the shielded metal arc welding (SMAW) with a diameter of 3.2 mm while for the MSAW the root pass was completed with a ER70s-6 electrode. The chemical composition of both electrodes is shown in Table.1. Four samples were fabricated, two samples was used to measure the lattice spacing ($d_{0,hkl}$) in a stress free mode for both welding processes and the other two samples were used to evaluate the residual stresses for SMAW and MSAW process. The yield strength of the parent metal is 490-520 MPa.

Table 1: Chemical composition of welding consumables and parent metal

Chemical Compositio	%C	%Mn	%S	%Si	%P	%Cu	%Cr	%Ni	%Mo	%V	%Ti	%NB
ER 70S-6	0.09	<1.6	0.007	0.9	0.007	0.20	0.05	0.05	0.05	0.05	-	-
E6010	0.16	0.62	0.009	0.1	0.009	-	0.02	0.2	0.01	<0.1	-	-
Parent metal	0.05	1.55	0.001	0.2	0.009	0.15	0.26	0.19	0.18	0.02	0.01	0.04

The macrographs of both welded specimens are shown in Fig. 1. The size of the root pass along with the HAZ are clearly visible on the macrographs. The MSAW welded joint appears to be more uniform with higher weld deposition volume in contrast to the SMAW with less uniform and lower weld deposition volume.

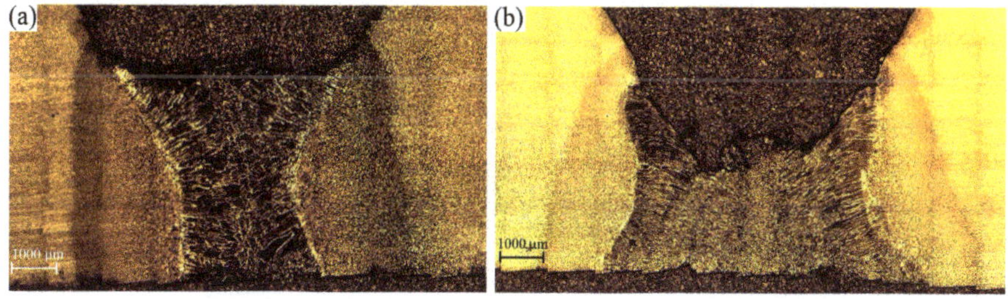

Fig.1: Overview of MWIC root weld bead produced with (a) MSAW and (b) SMAW process

Neutron diffraction

Neutron diffraction utilises a beam of neutrons with a momentum p, and associated wavelength λ:

$$\lambda = \frac{h}{p} \tag{1}$$

where h is Planck's constant. When the neutron beam penetrates crystalline materials, a diffraction pattern with sharp maxima is produced. The diffraction pattern can be described in terms of Bragg's law:

$$2d_{hkl} \cdot \sin\theta_{hkl} = n\lambda \ , \tag{2}$$

where d_{hkl} are the lattice spacing, n is an integer number representing the order of the reflection plane and θ is the angle between the incident ray and the scattering planes. A small change in the lattice spacing (Δd_{hkl}) will result in a change in the angular position of the diffraction peak ($\Delta\theta_{hkl}$) given by the following equation:

$$\Delta\theta_{hkl} = -\tan\theta_{hkl} \cdot \frac{\Delta d_{hkl}}{d_{hkl}} \tag{3}$$

The lattice normal strain ε is given by:

$$\varepsilon = \frac{d_{hkl} - d_{0,hkl}}{d_{0,hkl}} = -\Delta\theta_{hkl} \cdot \cot\theta_{0,hkl} \ , \tag{4}$$

where $d_{0,hkl}$ is the strain-free lattice spacing for the hkl planes, and $\theta_{0,hkl}$ is the diffraction angle of the unrestrained lattice. The strains ($\varepsilon_{xx}, \varepsilon_{yy}, \varepsilon_{zz}$) convert to the three-dimensional stress ($\sigma_{xx}, \sigma_{yy}, \sigma_{zz}$) state by using the generalised Hooke's law [5]. Measurement of residual stresses were performed using the KOWARI instrument with neutron diffraction for the α-Fe (2 1 1) reflection using a nominal gauge volume of 2x2x2 mm^3. SScanSS virtual laboratory software was used in experimental design to determine the optimum sample orientations to minimise path length and hence count times during measurements. It must be noted that due to the large thickness in the support plate we have implemented a necessary machining step to achieve the desired path length of less than 80 mm for measurements in the transverse direction, a small 10 mm blind hole at 45 degree (minimal to retain the fully constrained condition) was machined in the support material as shown in Fig. 2a. Measurement points were located in along a single line running parallel to the transverse (Y) direction at Z=1.6 mm in the middle of the specimen in X direction. A total of 14 measurement points were used at 2 mm intervals to cover a total range in the transverse direction of ±10 mm from the weld centreline. Unstrained (d-zero) measurements were performed using reference cubes (6x6x6 mm^3) from the weld and parent metal. Further details of the neutron diffraction analysis to evaluate the residual stresses during welding are given in the published works of [8, 9].

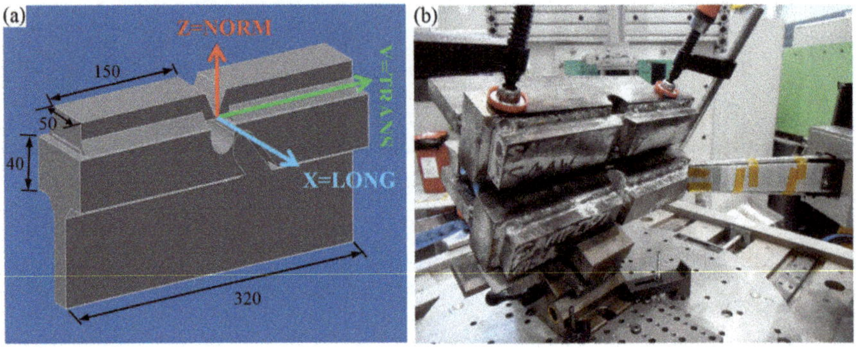

Fig.2: Experimental set up (a) Schematic and definition of coordinate system and (b) Neutron diffraction measurements of the MWIC using Kowari strain scanner at ANSTO

Results and discussion

Fig.3 (a) and (b) show the longitudinal, transverse and normal residual stresses for the MSAW and SMAW processes in the MWIC specimen. Results showed that for both processes, stresses in the transverse direction were tensile and dominant, with longitudinal stresses also predominantly tensile. In contrast, normal stresses were found to be oscillatory and comparatively low in magnitude. Measured peak stresses in the longitudinal direction were approximately equal and lower than transverse in both samples. This was due to the short weld length of 50 mm and lower constrained condition in that direction. These observed differences in residual stress profiles may be attributed to process effects such as energy density and deposition rate, but may also be due to observed differences in weld bead profile between the MSAW and SMAW samples causing a redistribution of stresses. In particular the apparent weld bead size is greater in the MSAW sample which can be clearly seen in Fig.1. Clearly such changes in weld size significantly change constraint conditions and hence residual stresses. As such, the measured values of stress are not directly comparable between the MSAW and SMAW samples despite them being produced with equivalent weld heat inputs.

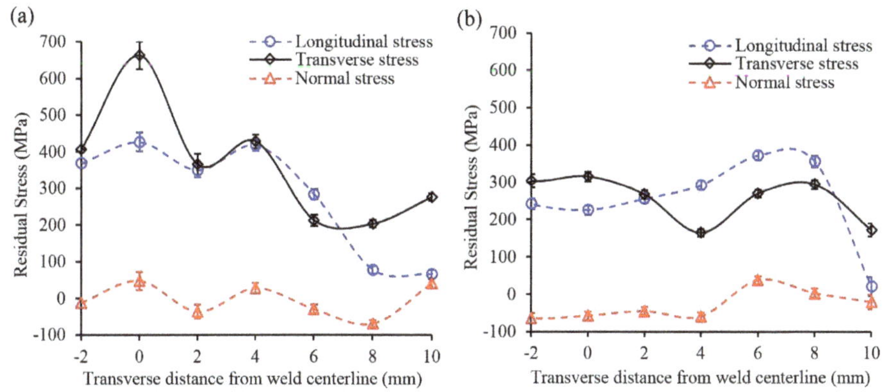

Fig.3: Residual stress measurements in samples produced with: (a) MSAW (b) SMAW processes

Conclusions

The key findings of this experimental study were:

- Highest tensile residual stresses were found in the transverse direction, while the magnitudes of the longitudinal stresses were slightly lower due to short length of the weld.
- The transverse residual stresses were of greatest magnitude in the MSAW sample. While this may be unexpected, it could be related to the significant differences in weld profile between the MSAW and SMAW samples. The larger weld bead size in the MSAW sample could significantly change the constraint conditions and hence residual stresses.

Acknowledgement

This work was conducted with the assistance of an Australian Nuclear Science and Technology Organisation (ANSTO) facilities access award (DB3748). The welding was carried out at AWS and mechanical Engineering workshop at Adelaide University. We would like to acknowledge with gratitude the support of Mr. Neville Cornish and Ashly Blanchard, welding technician from AWS, Rahim Kurji and Pascal Symons from Adelaide University. We are also grateful to EP-CRC and APIA-RSC for providing financial assistance during the course of this study (HA).

References

[1] Kotousov A, Borkowski K, Fletcher L, Ghomashchi R. A model of hydrogen assisted cold cracking in weld metal. 2012 9th International Pipeline Conference: American Society of Mechanical Engineers; 2012. p. 329-334.

[2] Alipooramirabad H, Kotousov A, Ghomashchi R. Numerical analysis of welding stresses in WIC weldability test. 8th Australasian Congress on Applied Mechanics: ACAM 8. Engineers Australia, 2014.

[3] Yurioka N, Suzuki H. Hydrogen assisted cracking in C-Mn and low alloy steel weldments. International Materials Reviews. 1990;35:217-249. http://dx.doi.org/10.1179/imr.1990.35.1.217

[4] Yurioka N, Yamasaki S, Morimoto H. Hydrogen effusion from high strength weld metal. Science and Technology of Welding & Joining. 2005;10:497-502. http://dx.doi.org/10.1179/174329305X46682

[5] Alipooramirabad H, Paradowska A, Ghomashchi R, kotooussov A, Hoye N. Prediction of welding stresses in WIC test and its application in pipelines. Materials Science and Technology. 2016. p. 1-9.

[6] Kurji R, Coniglio N. Towards the establishment of weldability test standards for hydrogen-assisted cold cracking. The International Journal of Advanced Manufacturing Technology. 2015;77:1581-1597. http://dx.doi.org/10.1007/s00170-014-6555-3

[7] Kurji R, Griggs J, Linton V, Barbaro F, Kotooussov A, Gamboa E, Ghomashchi R, Coniglio N. An improved Welding Institute of Canada test for evaluation of high-strength pipeline steel weldability. International Pipeline Technology Conference (06 Oct 2013-09 Oct 2013: Ostend) 2013.

[8] Alipooramirabad H, Ghomashchi R, Paradowska A, Reid M. Residual stress-Microstructure-mechanical property interrelationships in multipass HSLA steel welds. Journal of Materials Processing Technology. 2016; 231:456-467. http://dx.doi.org/10.1016/j.jmatprotec.2016.01.020

[9] Alipooramirabad H, Paradowska A, Ghomashchi R, Kotousov A, Reid M. Quantification of residual stresses in multi-pass welds using neutron diffraction. Journal of Materials Processing Technology. 2015;226:40-49. http://dx.doi.org/10.1016/j.jmatprotec.2015.07.002

Residual Stresses 2016: ICRS-10 Materials Research Forum LLC
Materials Research Proceedings **2** (2016) 563-568 doi: http://dx.doi.org/10.21741/9781945291173-95

Investigating the Effects of Mitigation Techniques on Residual Stress and Microstructure of HSLA Welds

Houman Alipooramirabad [1,a *], Reza Ghomashchi[1,b], Anna Paradowska [2,c] and Mark Reid[2,d]

[1]School of Mechanical Engineering, the University of Adelaide, SA 5005, Australia

[2] Bragg Institute, Australian Nuclear Science and Technology Organisation (ANSTO), Lucas Heights, NSW 2234, Australia

[a]houman.alipooramirabad@adelaide.edu.au, [b]reza.ghomashchi@adelaide.edu.au, [c]anp@ansto.gov.au, [d]markr@ansto.gov.au

Keywords: Microstructural Characterization, HSLA Welds, Neutron Diffraction, PWHT, Residual Stress

Abstract. Post-weld heat treatment is often required for pressure vessel and piping components for relaxing residual stresses and increasing the resistance to brittle fracture. The present study employed neutron diffraction to examine the effects of conventional post-weld heat treatment (PWHT) on the residual stresses in multi-pass, high-strength, low-alloy-steel, weld joints made by combined Modified Short Arc Welding (MSAW) and Flux Cored Arc welding (FCAW) processes. Residual stresses in excess of yield strength were developed in the Heat Affected Zone (HAZ) and the weld metal of the as-welded specimen (particularly the upper layers of the weld) which were reduced significantly as a result of applying PWHT. Also PWHT lead to substantial changes in the microstructural characteristics of high-strength, low-alloy-steel welds.

Introduction

Post-weld heat treatment (PWHT) is a stress relieving process where a uniform heating is applied at subcritical temperatures for a specified period of time until the desired stress relief is attained. The selection of the subcritical temperature is dependent on the alloy chemistry while the soaking time is thickness related [1]. In order to confirm the success of PWHT, an analysis of the microstructure and remaining residual stresses along with some basic mechanical testing is mandatory. An inappropriate stress relief heat treatment could lead to increasing residual stresses, distortion and degradation of the microstructure due to precipitation of carbides and resultant embrittlement [2]. Furthermore, from an economical view point, it is important to know if the annealing time is either too long (higher energy consumption) or unnecessarily short.

So far several experimental and numerical investigations have been carried out on the comparison of residual stress levels before and after PWHT. Paddea, et al. [3] used the neutron diffraction method to measure the residual stress distributions in a P91 steel-pipe girth weld before and after PWHT. They reported peak tensile residual stresses dropping from 600 MPa to 120 MPa after reheating to 750°C during PWHT. Smith and Garwood [4], using the more classic method of hole drilling, have reported a similar finding in a submerged-arc weldment in a 50 mm thick ferritic steel. They found significant reduction from 740 MPa to 140 MPa following PWHT. Cho, et al. [5] developed a 2D, thermal-elastic-plastic, finite-element model to evaluate the residual stresses for the as-welded condition and after PWHT. It was found that maximum residual stress for K- and V-type weld joints of thick plates were 316 and 256 MPa, respectively, which were reduced to 39.3 and 3.7 MPa. Mitra, et al. [6] developed experimentally validated, finite-element models to investigate the effects of temperature and time on the relieving of residual stresses after PWHT for 800 mm thick

Materials Research Proceedings **2** (2016) 563-568 doi: http://dx.doi.org/10.21741/9781945291173-95

steel weldment. Rapid decrease of residual stresses during the initial stages of holding time at the PWHT temperature was found. Moreover, it was found holding temperature had a significant role on the relief of residual stresses, particularly when the PWHT is carried out at relatively lower temperatures. Yu, et al. [7] investigated the effects of PWHT on microstructural and mechanical properties in laser welds of GH3535 superalloy. An increase of up to 90 MPa in the tensile strength of the joints was found after the PWHT at 871 °C for 6h, which was exceeding the strength of the original state of the base metal. The increase in the tensile strength of post-weld heat treated joints was found to be related with Mo-Si-rich, M6C-type carbides which were observed at solidification grain boundaries and solidification subgrain boundaries.

As briefly mentioned here, there is no report to examine the effects of PWHT on residual stress and microstructure of high-strength, low-alloy steel (HSLA) welds. Also the underlying mechanism of relaxation of residual stresses during PWHT process has not been identified. The current study is therefore carried out to characterize the PWHT process with respect to microstructure and residual stresses of the welded joint for multi-pass welds. In this paper we present comparative results of microstructural and residual stress measurements for the specimens in the as-welded condition and after successful PWHT.

Experimental procedure

Details of weld deposition procedure

The test specimens comprised two 20 mm thick steel plates (API 5L grade X70) with the dimensions of 250x200 mm^2. Two samples were fabricated, one sample was used to measure the lattice spacing ($d_{0,hkl}$) in the stress free mode and the other sample was used to evaluate the residual stresses for the as-welded condition and after the PWHT process. The yield strength of the parent metal is 490-520 MPa while the yield strength of the weld metal is about 545 MPa.

Microstructural analysis

The welded specimens were sectioned transversely and ground and polished down to 1 μm diamond paste for metallographic analysis of the weld metal (WM), the heat-affected zone (HAZ) and parent metal (PM). For microstructural analysis, the samples were etched in 5% Nital (5 % nitric acid in ethanol) or prepared by a double etching procedure using 2% Picral (2% picric acid in ethanol) and 2% Nital (2 % nitric acid in ethanol) solutions. An Axio Zeiss optical microscope and an SEM QUANTA 450 were used to examine the microstructure of the WM, HAZ and the PM. Also transmission electron microscopy (TEM) analysis was carried out to fully characterize the morphology of the microstructure, precipitates and dislocations in different regions of the weld. TEM sample preparation was carried out using focused Ion beam (FEI Helios Nanolab 600-SEM) at the Adelaide microscopy centre.

Neutron diffraction

Neutron diffraction measurements were performed at the Australian Nuclear Science and Technology Organization (ANSTO) using the KOWARI strain scanner. Details of the procedure can be found elsewhere [8, 9]. To perform strain measurements for the as-welded material and after PWHT, a $3 \times 3 \times 3\ mm^3$ gauge volume was used. The Si (400) type double focusing monochromators generated a neutron beam with the wavelength of 1.67Å. A detector angle, 2θ, was set at 90° corresponding to the αFe (211) diffraction peak.

Results and discussion

Residual stress

Fig.1 (a) and (b) show the longitudinal, transverse and normal residual stresses for the as-welded specimen. Tensile residual stress in excess of the yield strength of both the weld and parent metal was found at the depth of 3 mm from the weld surface and in the weld centreline for the as-welded specimen. The high tensile residual stress (longitudinal) at this region may be due to the fact that this is the last and final weld pass deposited and therefore the weld material could not have benefited from the tempering effect that previous passes had as the following pass was deposited. Therefore, thermal distribution is less uniform within this region as on one side the last pass is in contact with the already deposited passes and on the other side the heat is directly dissipated into the air. The formation of columnar grains is an indication of directional cooling of the last pass.

However as the depth of measurement increases (10 mm below the top surface) a decrease in the magnitude of residual stresses can be found due to tempering effects. The distribution of residual stresses across the weld after PWHT is shown in Fig. 2 (a) and (b). These data clearly reveal a substantial reduction in the magnitude of residual stresses after applying PWHT with a maximum of 142 MPa which is approx. 26% of the yield strength of the weld metal.

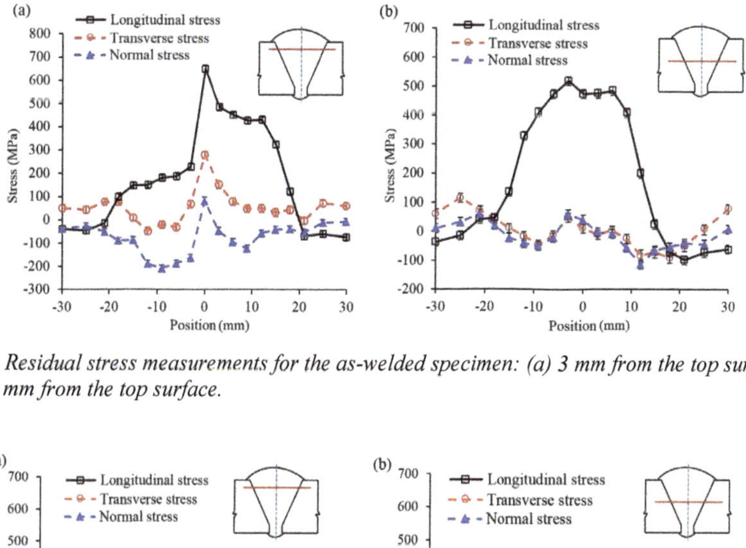

Fig.1: Residual stress measurements for the as-welded specimen: (a) 3 mm from the top surface; (b) 10 mm from the top surface.

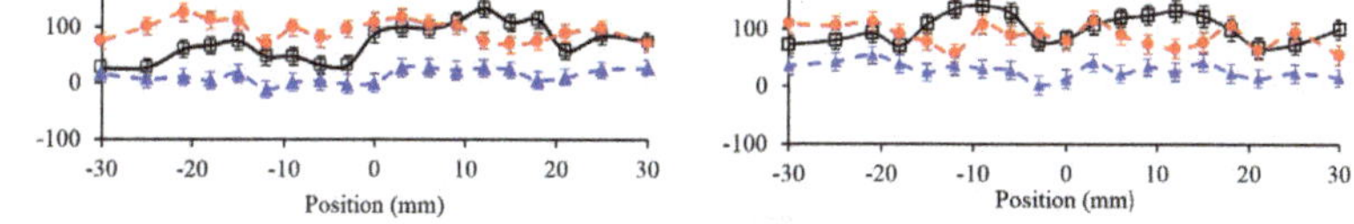

Fig.2: Residual stress measurements after PWHT: (a) 3 mm from the top surface; (b) 10 mm from the top surface.

Materials Research Proceedings **2** (2016) 563-568 doi: http://dx.doi.org/10.21741/9781945291173-95

Microstructural analysis

Fig.3 provides representative microstructures of the Coarse-Grain Heat-Affected Zone (CGHAZ, close to the fusion zone), for as-welded and post-weld heat treated specimens, examined in this work. The as-welded specimen however, contains Widmanstätten ferrite, bainite and polygonal ferrite but still is a mainly acicular ferritic weld [8]. While due to tempering effects in PWHT resulted in grain growth and coarsening of the microstructure, mainly equiaxed polygonal ferrite was found. It can also be seen that the microstructure in the CGHAZ is more homogeneous in terms of grain size after the PWHT process. In order to fully characterize the microstructure of the as-welded and post-weld heat treated specimens, TEM analysis was carried out. The TEM micrographs in Fig.4, which were taken from the weld centreline in the as-welded sample and after PWHT, clearly demonstrate the effect of thermal treatment on the rearrangement of dislocations and the formation of sub-grain boundaries. It is widely accepted that dislocation climb, a thermally activated process, is a prerequisite for sub-grain formation during creep of metal and alloys. Therefore existing sub-grains for the post-weld heat treated specimens, as shown in Fig. 4, indicate the dislocation climb and as a result, the development of creep strain during PWHT as a driving mechanism behind residual stress relaxation, which is in line with previous reports [10, 11].

(a) (b)

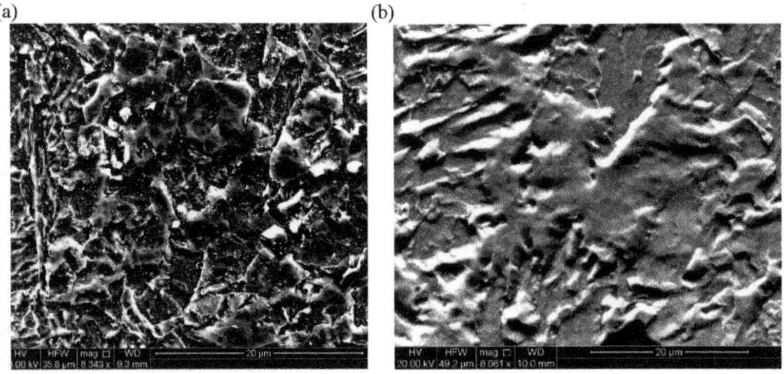

Fig.3: SEM micrographs showing the CGHAZ in the weld centreline in both (a) as-welded and (b) post-weld heat treated specimens.

Conclusions

The key findings of this experimental study were:

- The magnitude of the residual stresses decreased substantially after PWHT (about 26% of yield strength of the weld metal in the longitudinal direction and 20% in the transverse direction).
- High temperature leads to microstructural changes and an increase in the dislocation mobility, assisting dislocation climb and diffusion. This supports the creep-strain-induced, stress-relaxation mechanism during PWHT and explains the decrease of residual stress in the PWHT specimen.

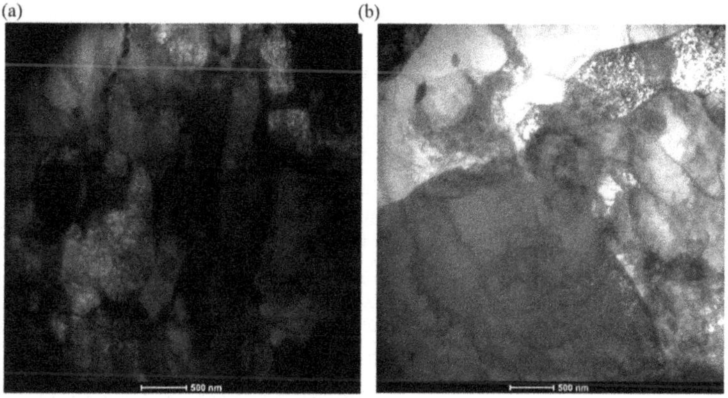

Fig.4: TEM micrographs showing the weld metal in (a) as-welded and (b) post-weld heat treated specimen.

Acknowledgments:
This work was conducted with the assistance of an Australian Nuclear Science and Technology Organisation (ANSTO) facilities access award (Award No. 4591). The welding was carried out at AWS. We would like to acknowledge with gratitude the support of Mr. Neville Cornish and Ashley Blanchard from AWS and Mr Pascal Symons at Adelaide University.

References

[1] Olabi A-G. Residual stresses and heat treatments for metallic welded components: Dublin City University; 1994.

[2] Aloraier A, Ibrahim R, Ghojel J. Eliminating post-weld heat treatment in repair welding by temper bead technique: role bead sequence in metallurgical changes. Journal of Materials Processing Technology. 2004;153:392-400. http://dx.doi.org/10.1016/j.jmatprotec.2004.04.383

[3] Paddea S, Francis JA, Paradowska AM, Bouchard PJ, Shibli IA. Residual stress distributions in a P91 steel-pipe girth weld before and after post weld heat treatment. Materials Science and Engineering: A. 2012;534:663-672. http://dx.doi.org/10.1016/j.msea.2011.12.024

[4] Smith DJ, Garwood SJ. Influence of postweld heat treatment on the variation of residual stresses in 50 mm thick welded ferritic steel plates. International Journal of Pressure Vessels and Piping. 1992;51:241-256. http://dx.doi.org/10.1016/0308-0161(92)90083-R

[5] Cho JR, Lee BY, Moon YH, Van Tyne CJ. Investigation of residual stress and post weld heat treatment of multi-pass welds by finite element method and experiments. Journal of Materials Processing Technology. 2004;155–156:1690-1695. http://dx.doi.org/10.1016/j.jmatprotec.2004.04.325

[6] Mitra A, Siva Prasad N, Janaki Ram GD. Influence of Temperature and Time of Post-weld Heat Treatment on Stress Relief in an 800-mm-Thick Steel Weldment. Journal of Materials Engineering and Performance. 2016;25:1384-1393. http://dx.doi.org/10.1007/s11665-016-1995-6

Materials Research Proceedings **2** (2016) 563-568 doi: http://dx.doi.org/10.21741/9781945291173-95

[7] Yu K, Jiang Z, Leng B, Li C, Chen S, Tao W, Zhou X, Li Z. Effects of post-weld heat treatment on microstructure and mechanical properties of laser welds in GH3535 superalloy. Optics & Laser Technology. 2016;81:18-25. http://dx.doi.org/10.1016/j.optlastec.2016.01.029

[8] Alipooramirabad H, Ghomashchi R, Paradowska A, Reid M. Residual stress- microstructure-mechanical property interrelationships in multipass HSLA steel welds. Journal of Materials Processing Technology. 2016;231:456-467. http://dx.doi.org/10.1016/j.jmatprotec.2016.01.020

[9] Alipooramirabad H, Paradowska A, Ghomashchi R, Kotousov A, Reid M. Quantification of residual stresses in multi-pass welds using neutron diffraction. Journal of Materials Processing Technology. 2015;226:40-49. http://dx.doi.org/10.1016/j.jmatprotec.2015.07.002

[10] Dong P, Song S, Zhang J. Analysis of residual stress relief mechanisms in post-weld heat treatment. International Journal of Pressure Vessels and Piping. 2014;122:6-14. http://dx.doi.org/10.1016/j.ijpvp.2014.06.002

[11] Zhang J, Dong P, Song S. Stress Relaxation Behavior in PWHT of Welded Components. ASME 2011 Pressure Vessels and Piping Conference: American Society of Mechanical Engineers; 2011. p. 673-679. http://dx.doi.org/10.1115/pvp2011-57826

Residual Stresses 2016: ICRS-10 Materials Research Forum LLC
Materials Research Proceedings **2** (2016) 569-574 doi: http://dx.doi.org/10.21741/9781945291173-96

Residual Stress Measurements in Vintage LPG Pressure Vessel Welds, via Neutron Diffraction

Kaan Sozen[1], Anna Paradowska[2,1 a], Mark Reid[2], Roger Griffins[3], John Daniels[1]

[1]UNSW, Sydney, NSW 2052, Australia

[2]ANSTO, Australian Centre for Neutron Scattering, Lucas Heights, NSW 2234, Australia

[3]WTIA, Unit 50, 8 Avenue of the Americans, Newington NSW 2127, Australia

[a]Tel: +61 (3)9719 3372 E-mail: anp@ansto.gov.au

Keywords: Welding, Residual Stress, Neutron Diffraction, Pressure Vessel, Submerged Arc, Liquid Petroleum Gas (LPG)

Abstract. Systems in power, petrochemical and refinery plants are subject to innumerable degradation mechanisms. Welds are the critical regions in such components. The focus of this project is on Liquid Petroleum Gas (LPG) storage vessels manufactured for Australian refineries in the 1960s. Residual stresses were measured in seam and circumferential welds extracted from the vessels. The aim of this project is to measure the residual stress storage vessel. This data will be used to engineer a procedure to repair the vintage steel plates of the pressure vessels with modern consumables and welding techniques.

Introduction

Residual stresses influence damaging mechanisms including; crack initiation and propagation, fatigue, creep, corrosion, stress corrosion cracking (SCC) and time to final fracture/failure of the component [1]. The residual stresses generated during welding process can be detrimental to the service life of the component, leading to premature and unexpected failure [2-6].

Until recently, the measurement of residual stresses in thick welded components was not practically feasible using non-destructive techniques. Conventionally, the residual stress in welded components was assumed to be equal to the yield of the material [8]. As many pressure vessels and similar components manufactured in the early to mid-twentieth century are still in service, it is important to measure these stresses, and in turn see how their actual values are similar or diverge from this assumption. Furthermore, it is important to understand how modern welding techniques can be employed, and how these repairs influence the remaining service life of the component.

There is a unique opportunity to have access to heritage steel from pressure vessels that were manufactured by the Australian pressure vessel industry in the 1960s. Furthermore, these vessels have been subject both thermal and pressure cycling over the past 50 years. Residual stress measurements in circumferential welded sections and seam welded sections were analysed, using neutron diffraction at the Australian Nuclear Science and Technology Organisation (ANSTO), using the KOWARI strain scanner. Due to the widespread use of aging pressure vessels around Australia and around the world, developing a clear understanding of the impacts of residual stress distributions is an important problem, and should contribute greatly to projected lifetimes of the pressure vessels in question.

Experimental Procedure

The GS9 pressure vessel was constructed in the 1960s with 29mm steel plate of B58A material. 11 plates were used for the body and 4 for the semi-elliptical heads. Plates were formed using a brake press, and welded using submerged arc welding. The seam and circumferential welds were not stress relieved by post weld heat treatment, however hydrostatic testing was employed in their manufacture.

The pressure vessels were designed for a pressure of 250 Psi, and hydrostatically tested to 425 Psi (being 40% and 67% of yield).

Circumferential and seam welds were sectioned from the (Table **1**: Test certificate material properties of parent metal) to final plate dimensions were 500x400x29 mm^3. This geometry was chosen to preserve the residual stress field in the weld at the centre of the plate. A stress free sample ($d_{0,hkl}$), was EDM cut from the cross section of the circumferential weld. The stress free reference sample was used to calculate the lattice distortion in the seam and circumferential samples, and in turn calculate the residual stress in these regions.

Table 1 - Test certificate composition and material properties of B58A (1965)

	Yield [MPa]	UTS [MPa]	El%	C %	P %	Mn %	S %
A1596	227	474	25	0.25	0.042	0.62	0.047

Neutron Diffraction

Neutron diffraction was used to measure the residual stresses in the welded samples. Experiments were conducted at Australian Nuclear Science and Technology Organisation (ANSTO), using the KOWARI strain scanner [10]. The nominal gauge volume was 5x5x5 mm^3, and 62 points were mapped individually for the circumferential and seam welds, in the longitudinal, transverse and normal directions. An exposure time of 300 seconds per measurement point was used. Neutrons with wavelength 1.67 Å were focussed using the Si (400) double focussing monochromator. Detector angle (2θ), was set to 90°, corresponding to α-Fe (211) diffraction peak.

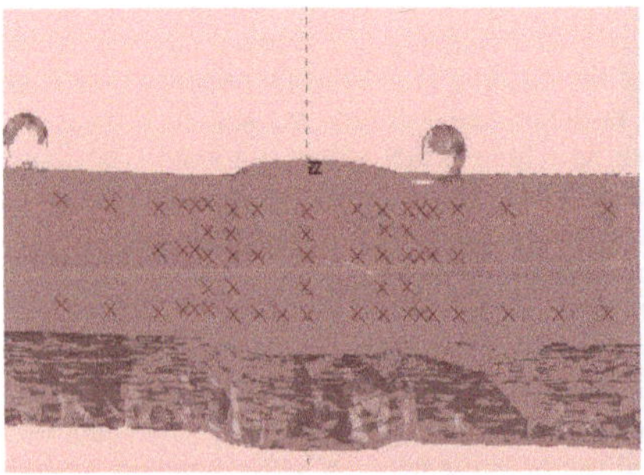

Figure 1 - SScans longitudinal simulation scan of seam weld specimen (62 points)

Figure 2 - Neutron Diffraction measurements of seam weld section, using KOWARI strain scanner at ANSTO

Materials Research Proceedings **2** (2016) 569-574 doi: http://dx.doi.org/10.21741/9781945291173-96

Results and Discussion

Figure 3 and Figure 4 show an etched macroscopic cross section of the circumferential and seam welds. The images showed that the vessels were welded with a total of 4 passes for each weld type. The first pass was from the inside. Through the measurement of the size of the weld cap and respective heat affected zones, the inside weld had a greater heat input with respect to the outside.

Figure 2 - Seam weld cross section microstructure (Inside of vessel is bottom of image)

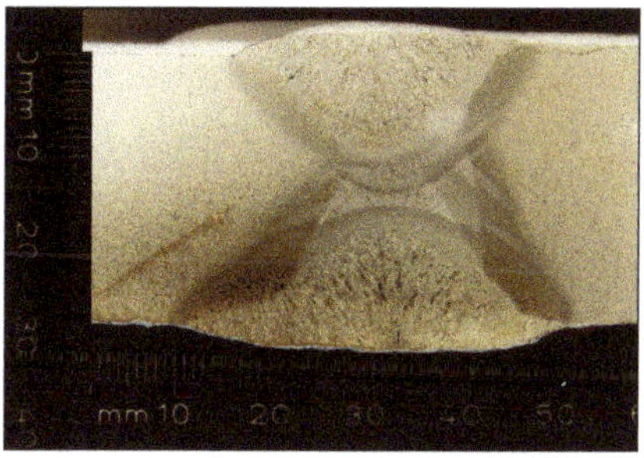

Figure 1 - Circumferential weld cross section microstructure (Inside of vessel is bottom of image)

Figures 5-8, below, show the longitudinal, transverse and normal residual stresses, at the center of the welds and 20 mm either side, for the circumferential and seam welds. Results showed that for both the welded samples, the maximum stresses were lower than the value used to calculate the critical crack size of 4mm by Moss in 1993 [7]. The seam weld has a normalized stress of up to 50% of the yield stress of the material, and the circumferential weld was larger reaching up to 90%.

For both welds, stresses in the longitudinal direction were tensile and dominant, with transverse stresses also predominantly tensile. The stresses in the radial direction were found to be compressive and low in magnitude. The magnitude of tensile stresses, are greater in the circumferential weld than the seam weld. Historic reports indicate that plates were formed into circular sections and seam welded first, and after the 11 sections of the body have been fabricated, they were circumferentially welded together [7]. This could be a reason to why circumferential weld residual stresses are greater than those seen in the seam welds.

Materials Research Proceedings **2** (2016) 569-574 doi: http://dx.doi.org/10.21741/9781945291173-96

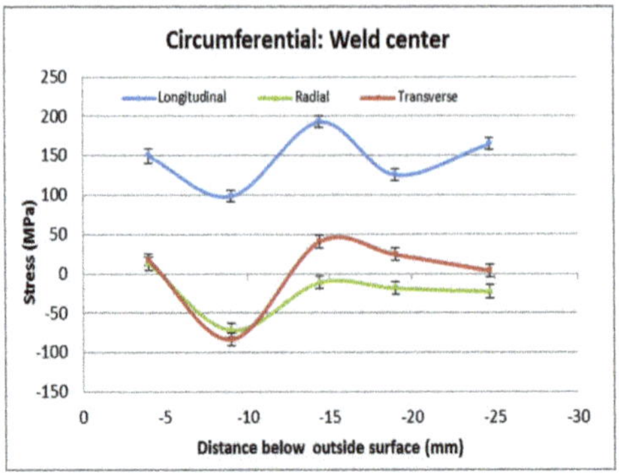

Figure 5 - Residual Stress at circumferential weld centre

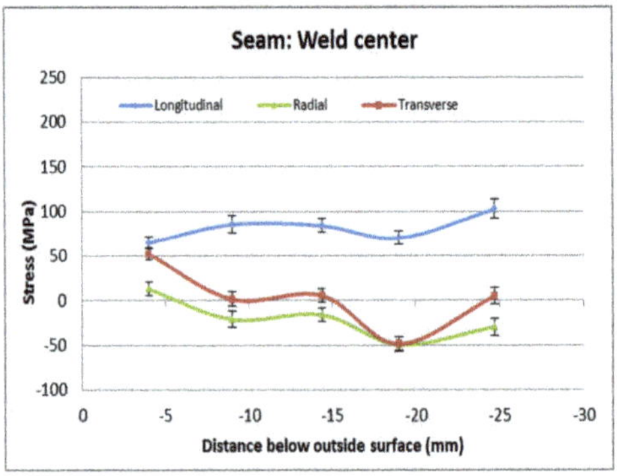

Figure 3- Residual Stress at seam weld centre

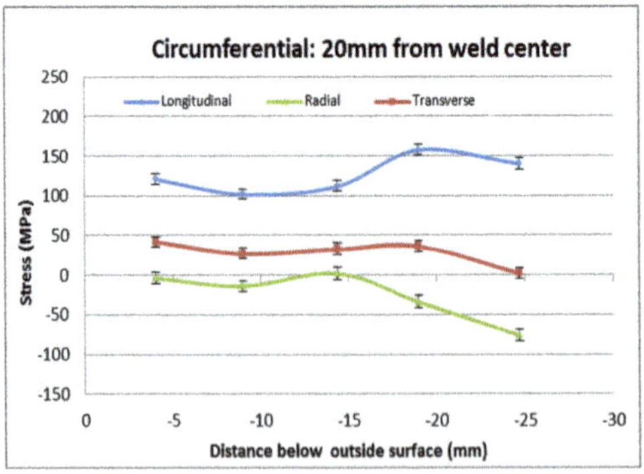

Figure x4- Residual Stress at the toe, 20mm from circumferential weld centre

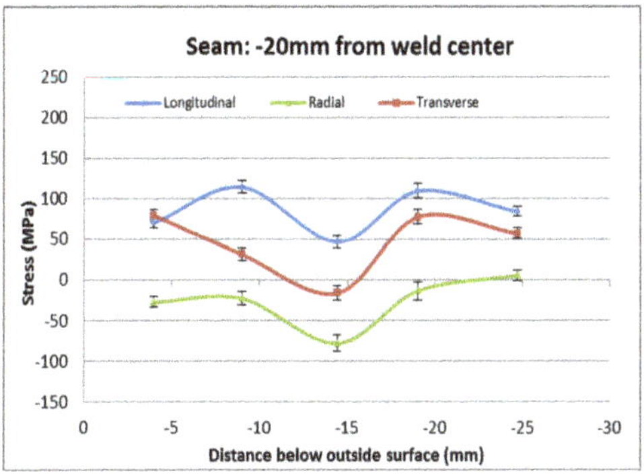

Figure 5- Residual Stress at the toe, -20mm from seam weld centre

Future work:

The above results will aid in the process to engineer a new repair procedure for the LPG pressure vessels, and other vessels manufactured using similar manufacturing practices. An incorrect welding procedure can induce severe undesirable stresses into components. A procedure must be designed to induce desirable stresses into the pressure vessels, to in turn not affect the remaining life time, and allow them to operate in a reliable and safe manner. A repair will be conducted on the inside of both circumferential and seam welded sections, and residual stresses generated through the repair procedure will be compared to the above data. A finite element model will be created for both cases to verify the results obtained through neutron diffraction.

Conclusions

Neutron diffraction measurements were successfully performed on the thick welded section extracted from the vintage LPG pressure vessel. The important results of this experimental study were:

- Residual stress distributions in the vessels are below the yield stress of the steel material.
- The magnitude of the tensile residual stresses were greater in the circumferential welds, particularly in the longitudinal direction.
- The stresses in the transverse direction were mainly tensile for both specimens, however stresses in the radial direction were found to be compressive and low in magnitude.
- The longitudinal stress in the center of the seam weld was around 50 % of yield.
- The longitudinal stress in the center of the circumferential weld was close to 90 % of yield.

Collected information will provide valuable information for the fitness-for-service assessment, in particular the critical crack size and potential repair procedure in the near future.

Acknowledgement

This work was conducted with the assistance of an Australian Nuclear Science and Technology Organisation (ANSTO) facilities access award. We are also grateful to the WTIA for providing access to the vintage steel components.

Materials Research Proceedings **2** (2016) 569-574 doi: http://dx.doi.org/10.21741/9781945291173-96

References

[1] M. T. Hutchings, Withers.P.J, T. M. Holden, and T. Lorentzen, "Introduction to characterization of residual stress by neutron diffraction," Mater. Today, 2005, 8: 57. http://dx.doi.org/10.1016/S1369-7021(05)00849-7

[2] P. J. Withers and H. K. D. H. Bhadeshia, "Residual stress Part 1 – Measurement techniques," 2001, Materials Science and Technology.17:355–365. http://dx.doi.org/10.1179/026708301101509980

[3] Alipooramirabad H, Paradowska A, Ghomashchi R, Kotousov A, Reid M. Quantification of residual stresses in multi-pass welds using neutron diffraction. Journal of materials processing technology. 2015;226:40-9. http://dx.doi.org/10.1016/j.jmatprotec.2015.07.002

[4] Balasubramanian V, Guha B. Effect of welding processes on toe cracking behaviour of pressure vessel grade steel. Engineering Failure Analysis. 2004;11:575-87. http://dx.doi.org/10.1016/j.engfailanal.2003.09.005

[5] V. Pearce and V. M. Linton, "Neutron diffraction measurement of residual stress in high strength, highly restrained, thick section steel welds," Phys. B Condens. Matter. 2006, 385–386: 590–593.

[6] A.M. Paradowska, J.W.H. Price, T.R. Finlayson, R.B. Rogge, R.L. Donaberger, R. Blevins, R. Ibrahim, Comparison of Neutron Diffraction Residual Stress Measurements of Steel Butt Welds with Current Fitness-For-Purpose Assessments, Transaction of the ASME - Journal of Pressure Vessel Technology, 2010, 132: 051503: 1-7

[7] C. J. Moss, "Investigation of the Structural Integrity of Eight LPG Bullets", ANSTO Report, 1993.

[8] T. Temperatures and W. R. Stresses, "PVP2007-26544 in Ferritic Steels,", 2007,949–956.

[9] J.A. James, L. Edwards, Application of robot kinematics methods to the simulation and control of neutron beam line positioning systems. Nuclear Instruments and Methods in Physics Research A., 2007, 571:709-718. http://dx.doi.org/10.1016/j.nima.2006.11.033

[10] A.M. Paradowska, C Shen, N Larkin, H Li, Z Pan, M Law, Residual stress measurements of welds made with a Tandem Gas Metal Arc Welding (T-GMAW) weld technique, Powder Diffraction, 2014, 29 (S1) 2014 24-27

Residual Stresses 2016: ICRS-10 Materials Research Forum LLC
Materials Research Proceedings **2** (2016) 575-580 doi: http://dx.doi.org/10.21741/9781945291173-97

Challenges of Measuring Residual Stresses in Large Girth Welded Pipe Spools by Neutron Diffraction

Yao Ren[1, a*], Anna Paradowska[2, b], Elvin Eren[3, c], Bin Wang[1, d]

[1]Brunel University London, Kingston Lane, Uxbridge, Middlesex UB8 3PH, UK

[2]ANSTO, Bragg Institute, Lucas Heights, NSW 2234, Australia

[3]TWI Ltd, Granta Park, Great Abington, Cambridge, CB21 6AL, United Kingdom

[a]yao.ren@brunel.ac.uk,[b]anp@ansto.gov.au, [c]elvin.eren@twi.co.uk,[d]bin.wang@brunel.ac.uk

Keywords: Residual Stress, Neutron Diffraction, Girth Welded Pipe

Abstract. In this paper, welding induced residual stresses in a welded API 5L X65 girth pipe spools are discussed in as-welded and in local post weld heat treated conditions. Stress measurements were carried out non-destructively using the neutron diffraction technique. For such large-scale components residual stress measurements require significant preparation and planning. First of all, a choice of stress free lattice spacing value, discussed extensively, is of great importance for the evaluation of residual strains and stresses correctly. Besides, the use of a virtual instrument (SSCANSS software) can optimize measurements for distorted or undistorted large components with or without complex details. Moreover, the well-planned "window" cut through the thickness greatly reduced measurement time. A number of points were measured across the weld, HAZ and the parent material. Measurement results showed that residual stresses in the as-welded condition was lower than the yield strength of the material, and significant relaxation was also observed in the post weld heat treated samples.

Introduction

Welding residual stresses are generated due to the thermal cycle induced incompatible internal strain that remained in the weld structure after cooling. The nature of the residual stresses and the factors affecting the stress conditions are discussed in Leggatt's paper [1]. With regard to the integrity of welded structures, high tensile residual stresses existing in pipeline weldments reduce the load capacity, increasing proximity to brittle fracture together with the applied external loads.

There are various measurement methods available, either destructive or non-destructive for evaluating residual stress states of componment. The neutron diffraction technique has been widely used in numerous research work [2-7] for the measurement of welding residual stresses due to its advantages. Basically, all three components of the residual stress can be measured non-destructively. Besides, the intensity of the beam enables penetration into the bulk material and through thickness stress profiles can be obtained [7-10]. The principle of measuring residual stress using neutron diffraction is to determine residual strains directly by quantifying changes in lattice spacing from their stress-free condition [11-13].

In this study, relatively large girth welded pipe spools were investigated. The welding procedure (representative of a procedure used in the industry) was described in the authors' previous work [14]. In this paper, the focus is on the experimental challenges in neutron diffraction procedure used for residual stress measurements. The emphasis in the rest of this paper will be on the choice of a representative strain-free reference for neutron diffraction to measure residual stresses before and after post weld heat treatment (PWHT).

Residual stress measurements by neutron diffraction

The principles of determining the residual stress magnitude are based on Bragg's Law and Hooke's Law. Equipped with Opal reactor neutron source at Australia Nuclear Science and technology Organization (ANSTO), Kowari strain scanning diffractometer utilizes the intense continuous monochromatic neutron beam to conduct strain scanning on samples.

The components investigated were made of API 5L X65 pipeline material. The dimensions of the girth welded pipe spool that used for stress evaluation by neutron diffraction were in 355.6mm (14 inch) outside diameter, 19.1mm wall thickness and 1000mm length. A six-pass narrow-gap girth weld was located in the middle of the pipe spool. The welding and PWHT procedures were described in a previous publication [14]. At this point, it is worth noting that both furnace and local post weld heat treatment methods were employed in this study but the results concerning local PWHT are discussed in this paper.

For the evaluation of the welding residual stresses, three line-scans parallel to the axis of the pipe spool were carried out. They were located 3mm below the pipe outer surface, 3mm above the pipe inner surface and at the mid-thickness. Through thickness measurements were conducted at weld center and weld toes. A set of measurement points across the weld in parent metal, heat affected zone and weld material are show in Fig. 2a. A window cut was introduced for measuring the axial strain component (Fig. 2b) before embarking on the measurements in as-welded condition. This window greatly reduced path length travelled by the neutrons. For the measurement of the pipe spool in post heat treated condition, another window was introduced for hoop strain measurements (Fig. 2c and 2d). Following PWHT, residual strain components were measured at the same locations as in Fig. 2a.

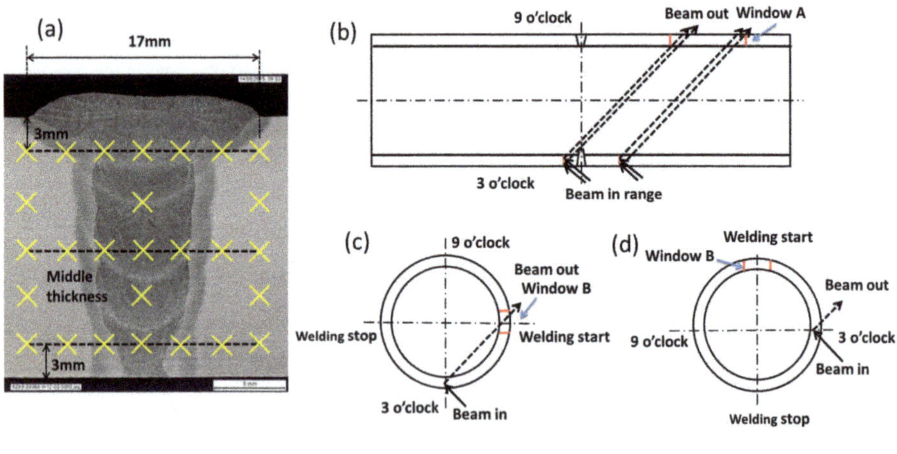

Fig. 2 Measurement locations and reduced travel path of neutrons after cutting a window on the pipe spool

Residual stress values were calculated using hkl specific material properties for the measured planes based on the Kröner model using the software DECcal [14]. The average reported errors range was ± 15 MPa. These errors are the peak fitting errors and do not take into account other possible sources of errors.

Stress-free comb sample (d_0 sample) was extracted from another pipe spool which was fabricated using the identical welding procedure. This sample was cut from the 3 o'clock position to make sure that the target micro-structure and geometry is replicated. The d_0 comb sample was 6mm thick and 95 mm long, containing the whole weld cross section and part of the parent material (Fig. 3). The strain component directions were defined the same as the corresponding orthogonal

Materials Research Proceedings **2** (2016) 575-580 doi: http://dx.doi.org/10.21741/9781945291173-97

(axial, hoop and radial) directions in the pipe. The slits in through pipe thickness direction were cut by Electrical Discharge Machining (EDM) and repeated every 6mm to create the 'teeth' on the sample. Additionally, two line-cuttings were introduced perpendicular to the "teeth" to manufacture 6x6x6 mm^3 grids on the comb sample. These procedures enabled the welding induced residual stresses in the comb sample to relax in hoop, axial as well as radial directions, and are supposed to provide the sample in a strain-free condition.

There were 24 points measured on the stress-free sample. The numbers shown in the Fig.3 demonstrate the scanning points sequence on the grids. With the given gauge volume 3x3x4mm^3, the measurements in three directions for each point were carried out focusing the beam at the center of each grid. At each scanning point, lattice spacing parameter d_{01} (supposed stress free lattice spacing in the as-welded condition) was measured in three orthogonal directions. The comb sample was subject to the same PWHT procedure as the pipe spool so that the same PWHT procedure is replicated on the comb. Lattice spacing parameters in it were measured again to obtain new lattice spacing d_{02} values (stress free spacing after PWHT) for the evaluation of the residual stress state.

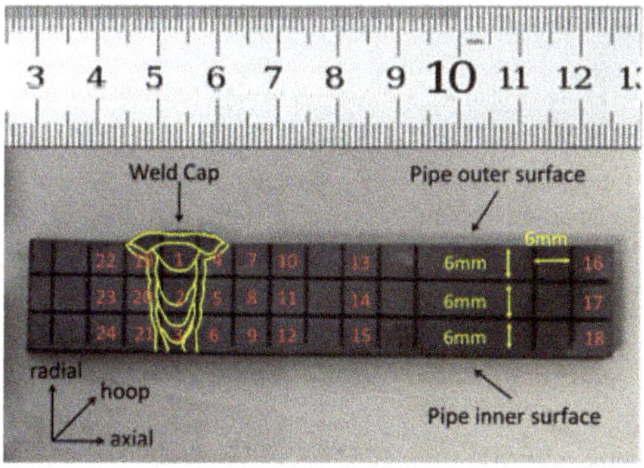

Fig. 3 The d_0 grid sample for calculating the strain-free inter-plane spacing

Assuming that hoop stress in the grid sample was completely relaxed during EDM cutting, 24 convoluted d-spacing values were calculated by using Hook's Law for each point. Averaging the values obtained from parent metal (from point 7 to point 18 and point 22 to 24), d_{01p} was calculated and selected as the strain-free d-spacing to calculate the relative strains in hoop, axial, normal directions as well as the convoluted strain variation in the parent material. These values provided the information regarding the strain variation hence the stress magnitude variation. After PWHT, all the data were analyzed following a similar procedure. An averaged strain-free d-spacing value was obtained from all the measurement locations in the parent material which was defined as d_{02p}. Strain variation in the weld metal and parent metal was compared as well. In this work, the lattice spacing in HAZ was not able to be accurately measured. This is because the HAZ was only 1mm wide in the narrow gap weld which was much smaller than the individual dimensions of the gauge volume.

Due to the large scale and complexity of the components to be measured, careful planning of the experiment was of great importance. KOWARI virtual instrument within Strain Scanning Simulation Software (SScanSS) [16] was adopted to plan and locate the measurement points accurately and maximize the efficiency of the experiment. The software was used in the planning and execution of measurement of the residual stress on both the pipe and the d_0 sample. Prior to neutron diffraction scanning (Fig. 4 left), 3D models of the pipe and the d_0 sample were created by laser scanning the surfaces. Measurement points were positioned on the sample models and the scanning sequence was displayed as the point number. Each scan was simulated to predict the measurement time and to check the orientation of the sample.

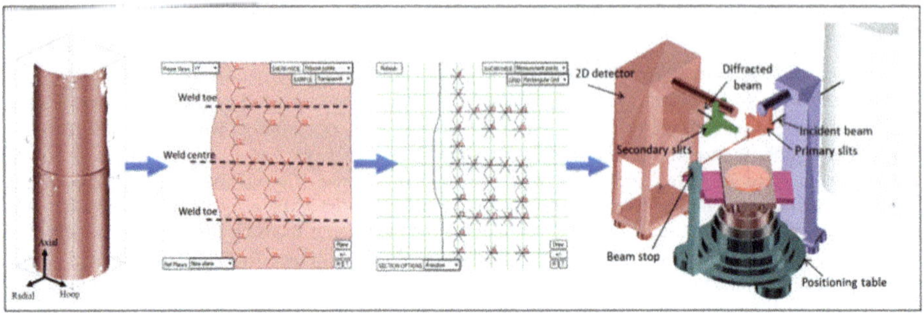

Fig. 4 The adoption of SScanSS software to pre-design the measurement points on the laser scanned 3D pipe model

Results and discussion

Reference (comb) sample

Strain values in the reference sample in as-welded condition were calculated using the d_{01p} values shown in Fig. 5a. The values following PWHT (Fig. 5b) were calculated using d_{02p}. These strain distributions illustrate the strain variation in the d_0 sample. In Fig. 5a and 5b, it also emphasizes the importance of selecting position dependent (weld metal, heat affected zone and parent metal) d_0 values for calculating the residual stresses.

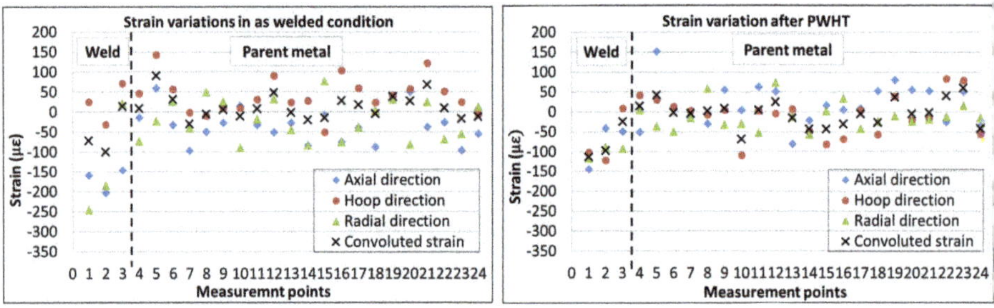

Figure 5. Strain variation based on selected reference d-scpacing values

In the as-welded condition, the averaged d_0 value, designated as d_{01p}, was selected to calculate strains in the parent material. There was approximately less than 100με variation and this leads to approximately 20MPa stress deviation in any single direction. However, in the weld metal, higher strain variations exceeding 200με occurred in axial and radial components resulting in up to 50MPa stress error. Therefore, position dependent d_0 values obtained from the weld of the stress free sample should be used to evaluate the stresses in the weld metal of the pipe in the as-welded condition.

In the PWHT condition, the use of averaged d_0 value, designated as d_{02p}, exhibited smaller strain variation in all directions as well as the convoluted strain. This d-spacing valued was used for the calculation of the stresses in the parent material after PWHT. The variation in the strain in the weld metal was less than 100με. However, it is also worth noting that the stresses remaining in the component following PWHT were significantly smaller. Therefore, for the evaluation of the stress state in the weld region, an averaged convoluted d_0 value obtained from weld metal measurements (measurement points 1, 2 and 3) was used, see Fig. 6.

Materials Research Proceedings **2** (2016) 575-580 doi: http://dx.doi.org/10.21741/9781945291173-97

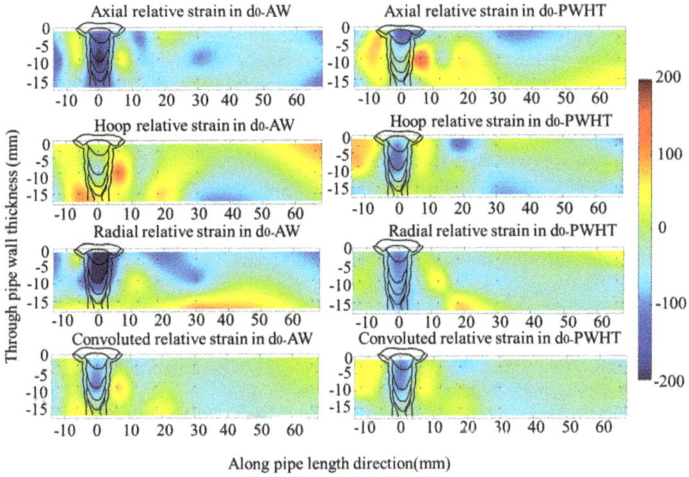

Fig. 6 Normalized strain variation maps of the comb sample before (left) and after (after) PWHT

Residual stress states before and after PWHT

Residual stress profiles in through thickness direction in the as-welded and post weld heat treated conditions are shown in Fig.7. Hoop stresses were normalized with respect to the greater (σ_y^+) of the weld and parent material, and axial stresses were normalized with respect to the lower value (σ_y^*) of the two [15] using the 0.2% proof strength of the parent material and weld metal. The weld was overmatching. Yield strength of weld metal (σ_y^+) was 657MPa and this value was used for normalizing the hoop stresses. For normalizing the axial stresses, parent material yield stress (σ_y^*) 510MPa of was used.

Through thickness residual stress profiles at weld center and toe were presented in Fig.7. Compared to the uniform, yield magnitude profile advised in BS7910 for the as-welded condition, much lower residual stress values were obtained from the measurements in the as-welded condition. Following PWHT, it was seen that axial stresses reduced to less than the 20% of yield strength of the parent material at room temperature. Hoop stresses were significantly lower than 30% of the yield strength of the weld metal at room temperature.

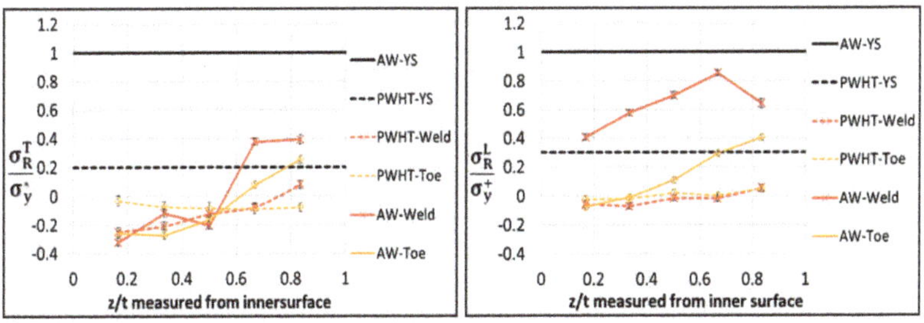

(a) Axial stresses (b) Hoop stresses

Figure 7. Normolised througth thickness residual stresses at weld center and toe before and after PWHT

Conclusions

1. Utilizing the virtual model to determine the neutron diffraction measurement points in such a large pipe spool improved the measurement location accuracy and enabled simulation of the scanning process as well as prediction of time required for each scan in advance. Cutting windows on the pipe greatly reduced measurement time.

2. The evaluation of the strain-free lattice spacing was essential to the subsequent residual stress calculation. Selecting appropriate d-spacing values for corresponding weld metal, heat affected zone and parent metal eliminated the strain variation resulting from chemical composition.

3. Residual stresses after local PWHT reduced in both axial and hoop directions in girth welded pipe spool. The results proved that the residual stress profile advised in BS7910 for furnace PWHT condition was conservative for the case presented in this paper.

References

[1] R. H. Leggatt, "Residual stresses in welded structures," *Int. J. Press. Vessel. Pip.*, vol. 85, no. 3, pp. 144–151, Mar. 2008. http://dx.doi.org/10.1016/j.ijpvp.2007.10.004

[2] T. Lorentzen, "Nondestructive evaluation of residual stresses by neutron diffraction," *NDT Int.*, vol. 21, no. 6, pp. 385–388, Dec. 1988. http://dx.doi.org/10.1016/0963-8695(88)90160-0

[3] G. Albertini, G. Bruno, P. Calbucci, F. Fiori, M. Rogante, and F. Rustichelli, "Non-destructive determination of residual stresses in welded components using neutron diffraction," *Weld. Int.*, vol. 12, no. 9, pp. 698–703, Jan. 1998. http://dx.doi.org/10.1080/09507119809452037

[4] M. J. Park, H. N. Yang, D. Y. Jang, J. S. Kim, and T. E. Jin, "Residual stress measurement on welded specimen by neutron diffraction," *J. Mater. Process. Technol.*, vol. 155–156, pp. 1171–1177, Nov. 2004. http://dx.doi.org/10.1016/j.jmatprotec.2004.04.393

[5] A. Paradowska, J. W. H. Price, R. Ibrahim, T. R. Finlayson, R. Blevins, and M. Ripley, "Residual stress measurements by neutron diffraction in multi-bead welding," *Phys. B Condens. Matter*, vol. 385–386, pp. 890–893, Nov. 2006.

[6] Y. V Taran, a M. Balagurov, J. Schreiber, and U. Stuhr, "Residual stresses in a shape welded steel tube by neutron diffraction," *J. Phys. Condens. Matter*, vol. 20, no. 10, p. 104258, Mar. 2008. http://dx.doi.org/10.1088/0953-8984/20/10/104258

[7] T. Neeraj, T. Gnäupel-Herold, H. J. Prask, and R. Ayer, "Residual stresses in girth welds of carbon steel pipes: neutron diffraction analysis," *Sci. Technol. Weld. Join.*, vol. 16, no. 3, pp. 249–253, Apr. 2011. http://dx.doi.org/10.1179/1362171810Y.0000000028

[8] S. V. Pearce and V. M. Linton, "Neutron diffraction measurement of residual stress in high strength, highly restrained, thick section steel welds," *Phys. B Condens. Matter*, vol. 385–386, pp. 590–593, Nov. 2006.

[9] W. Woo, V. Em, P. Mikula, G.-B. An, and B.-S. Seong, "Neutron diffraction measurements of residual stresses in a 50mm thick weld," *Mater. Sci. Eng. A*, vol. 528, no. 12, pp. 4120–4124, May 2011. http://dx.doi.org/10.1016/j.msea.2011.02.009

[10] W. Woo, V. T. Em, B. S. Seong, P. Mikula, and G. B. An, "Residual stress determination in a thick ferritic steel weld plate using neutron diffraction," *J. Mater. Sci.*, vol. 47, no. 14, pp. 5617–5623, Apr. 2012. http://dx.doi.org/10.1007/s10853-012-6456-5

[11] G. E. Bacon, *Neutron diffraction*, 3rd ed. United Kingdom: Clarendon Press., 1975.

[12] P. J. Withers, M. Preuss, A. Steuwer, and J. W. L. Pang, "Research papers methods for obtaining the strain-free lattice parameter when using diffraction to determine residual stress research papers," pp. 891–904, 2008.

[13] Paradowska, A., Finlayson, T., Price, J.W.H., Steuwer A., Ibrahim, R., Ripley, M., Investigation of Reference Samples for Residual Strain Measurements in a Welded Specimen by Neutron and Synchrotron X-ray Diffraction, Physica B: Condensed Matter, 2006, 385-389:890-893. http://dx.doi.org/10.1016/j.physb.2006.05.241

[14] Y. Ren, A. Paradowska, B. Wang, and E. Eren, "PVP2016-63378," in *Proceedings of the ASME 2016 Pressure Vessels and Piping Conference*, 2016, pp. 1–9.

[15] J.A. James, L. Edwards, Application of robot kinematics methods to the simulation and control of neutron beam line positioning systems. Nuclear Instruments and Methods in Physics Research A., 2007, 571: 709-718.T. Manns; B. Scholtes, "DECcalc - A Program for the Calculation of Diffraction Elastic Constants from Single Crystal Coefficients," *Mater. Sci. Forum*, vol. 681, no. 2, pp. 417–419. http://dx.doi.org/10.1016/j.nima.2006.11.033

[16] *BSI, "BS7910:2013 Guide to methods for assessing the acceptability of flaws in metallic structures," 2013.*

Residual Stresses 2016: ICRS-10 | Materials Research Forum LLC
Materials Research Proceedings **2** (2016) 581-586 | doi: http://dx.doi.org/10.21741/9781945291173-98

Design and Manufacture of Industrially Representative Weld Mock-ups for the Quantification of Residual Stresses in a Nuclear Pressure Vessel Steel

J. A. Francis[1, a *], M. C. Smith[1, b], B. Jeyaganesh[1, 2, c], A. N. Vasileiou[1, d] D. W. Rathod[1,e], M. J. Roy[1, f] and N. M. Irvine[3, g]

[1]School of Mechanical, Aerospace and Civil Engineering, The University of Manchester, Manchester M13 9PL, United Kingdom

[2]Nuclear AMRC, University of Sheffield, Rotherham S60 5WG, United Kingdom

[3]Dalton Nuclear Institute, The University of Manchester, Manchester M13 9PL, United Kingdom

[a]john.francis@manchester.ac.uk, [b]mike.c.smith@manchester.ac.uk, [c]jey.balakrishnan@namrc.co.uk, [d]anastasia.vasileiou@manchester.ac.uk, [e]dinesh.rathod@manchester.ac.uk, [f]matthew.roy@manchester.ac.uk, [g]neil.irvine@manchester.ac.uk

Keywords: Boundary Conditions, Distortion, Instrumentation, Laser Scanning, Narrow Groove Weld, Weld Design, Weld Restraint

Abstract. This paper describes work carried out under the NNUMAN research programme. This work focuses on the measurement and modelling of residual stresses in weld test pieces that have a thickness that is representative of primary components in a pressurised water reactor, such as the steam generators and the pressuriser. Weld test pieces at thicknesses of 30 mm and 130 mm have been and are being manufactured in SA508 Grade 3 Class 1 steel. Attention has been given to welding processes that are currently applied in nuclear manufacturing, such as narrow-groove arc-based welding processes, as well as to candidate processes for future build programmes, such as electron beam welding. The manufacture, characterisation and modelling of large test pieces each present challenges over and above those that arise when dealing with the smaller test pieces that are more typically manufactured in research laboratories. Some of those challenges, and the approaches that have been used to overcome them, are described. Plans for future work are briefly mentioned.

Introduction

Residual stresses can play a significant role in affecting the long-term structural performance of safety-critical components in nuclear power plants. Residual stresses can contribute to the driving force for crack growth [1] but, in nuclear environments, they can also activate degradation mechanisms such as creep [2] and stress-corrosion cracking [3] even in the absence of operating stresses. This is significant because many safety-critical components in a nuclear plant undergo welding during manufacture, and welding is known introduce substantial levels of residual stress [4].

The primary options for quantifying residual stresses in welds are to measure them, or to make predictions based on numerical models. However, both approaches are not straightforward since, on the one hand, many measurement techniques are either destructive or not suited to applications on large components and, on the other hand, numerical models must undergo rigorous validation before they can be used with confidence, through comparisons with measurements made with multiple and dissimilar techniques on test cases for which the manufacturing history has been documented in detail. This highlights the need for carefully designed weld mock-ups that can serve as weld modelling benchmarks. Such benchmarks enable weld models to be validated on well-defined test cases before they are applied to make predictions for real components.

As part of the NNUMAN research programme, we have designed and are manufacturing weld mock-ups that are larger than those that are typically manufactured in a laboratory environment. In doing so, our aim is to gain a better appreciation of the manufacturing challenges that are encountered when large nuclear components are manufactured in practice. Our hope is that, by manufacturing weld mock-ups that are more representative of welds that are made on an industrial scale, we will use welding procedures that are closer to those applied in practice, we will capture a greater range of relevant phenomena, and we will thereby reduce the uncertainties associated with inferring trends for industrial scale welded joints based on extrapolation from laboratory scale test pieces. While the motivation is straightforward, the manufacture of these weld mock-ups has presented different challenges to those that we encounter when making smaller weld test pieces. Thus the purpose of this paper is to highlight some of these challenges and the approaches we are employing to meet them.

Research Objectives

Welding research under the NNUMAN programme has the overarching aim of developing an understanding of how the choice of manufacturing process impacts on the performance of reactor components over the design life of the component, which for new build applications is typically 60+ years. This aim is quite distinct from the basic manufacturing requirement to produce a weld that will meet start-of-life weld quality requirements and pass inspection so that it will go into service. In this work, we have focused on welds made in SA508 Grade 3 Class 1 steel, which is a low-alloy steel that is typically used in the manufacture of primary components in a pressurised water reactor (PWR) such as the reactor pressure vessel or the steam generators. For this steel, and for such components, the parameters that will have the greatest influence of long-term structural integrity include the fracture toughness of the weld region and the weld residual stress distributions. Accordingly, the NNUMAN welding research programme has the following objectives:

- To manufacture industrially-representative weld mock-ups with welding processes that are currently applied to nuclear components, as well as with welding processes that may be used in future build programmes. The welding processes that have been chosen for investigation are narrow-groove gas-tungsten-arc welding (NG-GTAW), narrow-groove submerged arc welding (NG-SAW), electron beam (EB) welding and laser welding;
- To carefully record and document all aspects of the manufacturing processes and steps that are employed in each case so that the weld mock-ups can be used as the basis for validating numerical models for welding;
- To characterise the weld residual stress distributions in each of the weld mock-ups using multiple measurement techniques;
- To develop a methodology for the modelling of each weld and welding process so that residual stresses can be predicted and the interaction between the welding process and material can be understood.

Design Requirements for Weld Mock-Ups

The design requirements included (but were not limited to) the following:

- The thickness of the weld mock-ups should be representative of the thicknesses that are applicable to primary components in a PWR;
- The length and width of the weld mock-ups should be sufficient to enable steady-state welding conditions to be achieved along the length of the weld and for the residual stress distributions to be substantially unaffected by the width of the test pieces;
- The boundary conditions, particularly with respect to weld restraint, should be either clearly defined or well characterised.

In terms of the first requirement, a thickness of 130 mm was identified as being representative. The last requirement that is listed above is of great importance since welds usually need to be restrained in some way during manufacture in order to prevent excessive distortion. When this restraint is released, there will inevitably be some spring back and associated relaxation of residual stress. While

Materials Research Proceedings **2** (2016) 581-586 doi: http://dx.doi.org/10.21741/9781945291173-98

this can (in principle) be accounted for, unfortunately, common methods of restraint (e.g. clamping and tack welding of test pieces to backing plates) tend to provide levels of restraint that are not easily quantified. This means that such approaches can be difficult to represent accurately in numerical models. In addition to manufacturing welds at a thickness of 130 mm, it was decided that test pieces should also be manufactured at a thickness of 30 mm, as such an exercise would serve as a stepping stone in terms of the development of welding procedures and associated weld models, and it would also enable residual stresses to be characterised with neutron diffraction (not feasible at a thickness of 130 mm).

Basic Geometry

The basic geometries of the 30 mm and 130 mm thick specimens are shown in Figure 1. The 30 mm thick arc- and laser-welds had a geometry that was similar to that shown in Figure 1 (left). It can be seen that these 30 mm thick welds are not full-length groove welds, and that a ligament of parent material remained intact at either end. These ligaments served to provide a degree of self-restraint, so that the specimens did not need to be clamped or restrained with any significant force. Instead, light clamping could be employed simply to prevent the specimens from moving during welding. Furthermore, the specimen geometry could be captured in finite-element models so that the extent of self-restraint could be represented accurately. For the electron beam weld, it was possible to leave the weld free of restraint, due to the process requiring only a single weld pass. For the 130 mm thick specimens, a self-restraining approach was not feasible, and another method of restraint was developed.

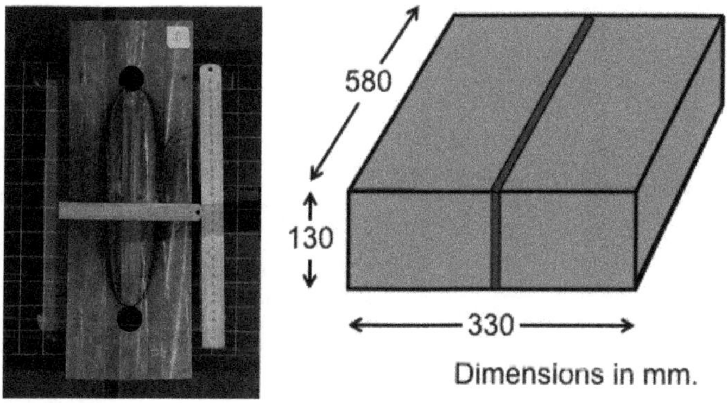

Fig. 1: Photograph of a 30 mm thick submerged arc welded specimen (left) showing ligaments of parent material that remain intact at each end of the specimen, and schematic representation of basic specimen geometry for the 130 mm thick weld mock-ups (right). The weld seam runs along the centre of the specimen in both cases.

Challenges Associated with Distortion

One of the major challenges in developing welding procedures for the 130 mm thick specimens was controlling the distortion of the welds, and designing weld grooves that would enable welds to be completed successfully in spite of this distortion. There are two components to this distortion, namely butterfly distortion, and transverse contraction, and both are represented schematically in Figure 2. Multipass welds made from one side experience both components of distortion. For these welds, butterfly distortion in particular can create problems with respect to the weld groove progressively closing as successive weld passes are deposited. If a narrow groove weld torch is used, then butterfly distortion can raise the possibility of the torch becoming trapped between the two

Materials Research Proceedings **2** (2016) 581-586 doi: http://dx.doi.org/10.21741/9781945291173-98

plates being joined. In this work, a restraint rig (Fig, 3) was designed, built and commissioned to mitigate butterfly distortion in multipass welds. Butterfly distortion is minimal in single-pass electron beam welds, so no restraint was applied with this welding process.

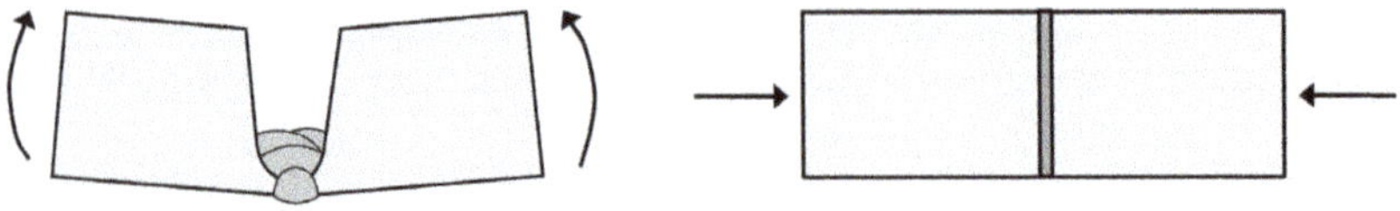

Fig. 2: Schematic representation of butterfly distortion (left), which is typically associated with one-sided multipass welds, and transverse contraction (right) as viewed on a cross-section through a single-pass electron beam weld. Transverse contraction occurs in all fusion welds to some degree. In both cases the welding direction runs into the page.

Fig. 3: Photograph of restraint rig that was designed, built and commissioned for the purpose of manufacturing the 130 mm thick weld mock-ups in the NNUMAN research programme. The weld assembly sits on a support plate which is pushed upwards so that the king plates apply restraint.

Although the restraint conditions differ for the multipass welds and the single pass welds, as well as for the 30 mm thick and 130 mm thick welds, this in itself does not create problems from the standpoint of directly comparing different welding processes. The most important requirement in the NNUMAN research programme is that the restraint conditions are understood and any restraint, if applied, can be quantified and represented in a finite-element model. In this respect, both the self-restraining specimen configuration (30 mm) and the employment of a restraint rig (130 mm) meet these criteria. In cases where the restraint rig was employed, the loads that were applied have been logged by recording the pressure of the hydraulic oil being supplied to the actuators (these raise the support plate so that it pushes the specimen assembly against the king plates in order to apply the restraint). Strain gauges have also been attached to the main structure of the rig to record the strains that develop as welding progresses. The distortions that develop during welding have also been

Materials Research Proceedings **2** (2016) 581-586 doi: http://dx.doi.org/10.21741/9781945291173-98

recorded before and after every weld pass in order to provide further information that may be of use in developing and assessing the performance of weld models. The distortions have been recorded with a 3-dimensional laser scanner, as well as by adding pockmarks to the specimens and manually measuring the movement of these marks between weld passes. Examples of the type of information that has been recorded are shown in Fig. 4.

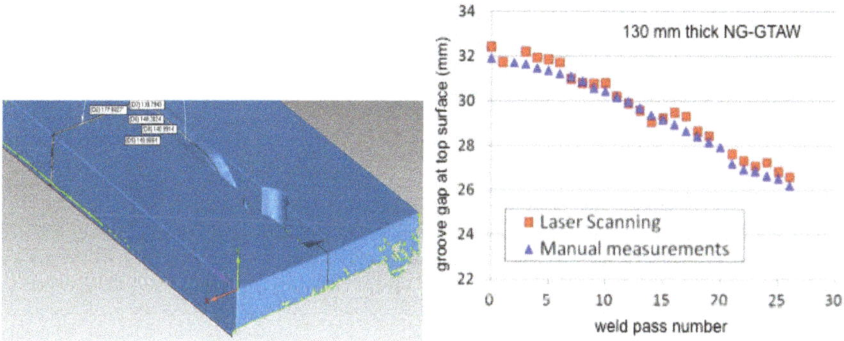

Fig. 4: An example of the three-dimensional information that was obtained from the hand-held laser scanner for a 30 mm thick weld test piece (left), and an illustration of the extent to which the groove gap at the top surface of a 130 mm thick GTAW specimen reduces as successive weld passes are deposited. Note that, with the welding parameters that were used for the GTAW specimen, approximately 80 weld passes are required to complete a 130 mm thick joint.

Fig. 5: Photograph of a 130 mm thick submerged arc weld specimen (left) placed up-side-down prior to the commencement of welding to reveal a backing channel, which allows an inert gas to shield the root of the weld bead from the atmosphere during welding, and which also provides access for thermocouples and preheating blankets at the base of the specimen. The thermocouples attached to the top surface of the weld are shown (right) prior to the commencement of welding.

Other Challenges

Other challenges associated with the 130 mm thick specimens have included the need to attach thermocouples and strain gauges to the specimens (for the purpose of validating models), whilst providing access for shielding gas to both sides of the plate (to prevent weld contamination), as well as preheating blankets (to prevent hydrogen cracking), all while not interfering with the ability of the restraint rig to apply the restraining loads. Figure 5 shows the way in which access to the underside of the 130 mm thick specimens was provided for these concurrent requirements. Finally, it should be mentioned that the need to apply multiple residual stress measurement techniques to each weld mock-up will result in the 30 mm thick specimens being characterised with neutron diffraction and the contour method, while the 130 mm thick specimens will be characterised using the incremental deep hole drilling technique and the contour method.

Summary

Weld mock-ups of a thickness that is representative of primary nuclear components (i.e. 130 mm) have been designed and are currently being manufactured in SA508 Grade 3, Class 1 steel. The purpose of these test pieces is to enable residual stresses to be evaluated and to enable a cross-process comparison to be made, which will involve both welding processes that are currently being applied on nuclear plants and candidate processes for future nuclear build programmes. A matching matrix of welds has also been manufactured at a thickness of 30 mm, since this thickness is amenable to residual stress measurements using neutron diffraction. The thermal histories, restraint boundary conditions and the evolution of distortion during manufacture have been and are being recorded in detail. Residual stresses are being measured in each specimen using multiple techniques. These welds will provide what we believe to be the most comprehensive comparison of current and candidate welding processes for nuclear pressure vessels. The authors hope that these mock-ups will provide invaluable insights relating to the generation of weld residual stresses, and serve as modelling benchmarks for years to come.

Acknowledgements

The authors would like to acknowledge the Engineering and Physical Sciences Research Council (EPSRC) for supporting this work through the NNUMAN Research Programme (Grant No. EP/J021172/1), and Dr. Alan Thompson, formerly with Tata Steel Europe, for valuable discussions.

References

[1] P.J. Withers: Residual Stress and its Role in Failure. *Reports on Progress in Physics*, Vol. 70 (2007), pp. 2211-2264. https://doi.org/10.1088/0034-4885/70/12/R04

[2] M. Turski; P.J. Bouchard; A. Steuwer; P.J. Withers: Residual Stress Driven Creep Cracking in AISI Type 316 Stainless Steel. *Acta Materialia*, Vol. 56 (2008), pp. 3598-3612. https://doi.org/10.1016/j.actamat.2008.03.045

[3] D.J. Hornbach; P.S. Prevey: The Effect of Prior Cold Work on Tensile Residual Stress Development in Nuclear Weldments', *Journal of Pressure Vessel Technology*, Vol. 124 (2002), pp. 359-365. https://doi.org/10.1115/1.1481035

[4] J.A. Francis; H. K. D. H. Bhadeshia; P. J. Withers: Welding Residual Stresses in Ferritic Power Plant Steels. *Materials Science and Technology* , Vol. 23 (2007), No. 9, pp. 1007–1020

Effects of Numerical Methods on Residual Stress Evaluation by the Incremental Hole-Drilling Technique Using the Integral Method

B. Gore[1,a*] and J.P. Nobre[2,b]

School of Mechanical, Industrial and Aeronautical Engineering, University of the Witwatersrand, Johannesburg, Private Bag 3, WITS 2050, South Africa

[a]bronwyn.gore@students.wits.ac.za, [b]joao.nobre@wits.ac.za

Keywords: Residual Stress, Incremental Hole-Drilling, Integral Method, Numerical Methods

Abstract. The incremental hole-drilling technique with Integral Method is widely used to estimate residual stresses. However, the fundamental nature of the Inverse Problem leads to numerically unstable results, which makes stress calculations very sensitive to measurement errors. The extent of this sensitivity is investigated by comparing different combinations of numerical methods for interpolating the input data and solving the equations. The methods are tested on both an experimental and simulated hole-drilling procedure to isolate inaccuracies due to numerical ill-conditioning and experimental errors. In the experiment, residual stresses determined by X-ray diffraction on a shot-peened sample were taken as the reference for comparison purposes. It was found that interpolating experimental data provides more stable results. Tikhonov Regularisation was more stable than direct methods, but susceptible to numerical errors introduced by the interpolation of input data.

Introduction

With current advancements in computing technology it is now possible to calculate residual stress more accurately. The complex nature of residual stress makes its quantification particularly difficult. The incremental hole-drilling technique is a well-established and widely accepted measurement method, yet it is limited by its sensitivity to errors – this is a fundamental drawback of the hole-drilling technique itself and not the associated calculation methods [1,2].

The incremental hole-drilling technique involves the drilling of a small, shallow hole in a number of increments while a strain gauge rosette measures changes in strain at the surface. Calculation procedures can be used to relate the measured strain to relieved stress. Even after simplifying assumptions have been applied there is still no closed-form solution for the residual stress calculation [3]. The fundamental nature of the Inverse Problem always leads to numerically unstable results, which makes stress calculations sensitive to measurement errors. Since its numerical solution is ill-conditioned, small errors in the input data cause large errors in the results [4]. As with all relaxation methods, residual stress can never be measured exactly where it is being relieved so there is always, by nature, an error in measurements. As the hole depth increases, the amount of relieved stress which can be sensed at the surface tends to zero independently of the residual stress magnitude existing in the deepest layers [5]. Therefore, for a depth equal to the hole diameter, strain relaxation cannot be measured anymore. Regarding the Integral Method in particular, numerically speaking, the stress calculation becomes strongly ill-conditioned for depths greater than half the hole diameter.

The Integral Method is the most widely used calculation method and the only one considered by the ASTM E 837 standard. Despite its sensitivity to errors, it is theoretically the most correct method since it considers that the strain relief at the material's surface is the accumulated result of the residual stresses originally existing in the zone of each successive increment. Therefore, the Integral

Materials Research Proceedings **2** (2016) 587-592 doi: http://dx.doi.org/10.21741/9781945291173-99

Method is the only one able to give accurate results if strong stress gradients exist [2]. According to the approach proposed by Schajer [2], which was accepted by the ASTM standard [3], the strains and stresses can be divided into hydrostatic and shear components and the residual stress can be determined for isotropic, linear elastic materials, solving the system of linear equations, Eq. 1, Eq. 2 and Eq. 3 – below. This takes into account the fact that the change in strain readings measured at the surface are due to both relieved stress and change in hole geometry.

$$\bar{\mathbf{a}}\mathbf{P} = \frac{E}{1+\nu}\mathbf{p}. \quad (1)$$

$$\bar{\mathbf{b}}\mathbf{Q} = E\mathbf{q}. \quad (2)$$

$$\bar{\mathbf{b}}\mathbf{T} = E\mathbf{t}. \quad (3)$$

Where:

$\bar{\mathbf{a}}$ is the coefficient matrix of the strain relaxation per unit depth caused by unit normal stress at each depth increment;

$\bar{\mathbf{b}}$ is the coefficient matrix of the strain relaxation per unit depth caused by unit shear stress at each depth increment;

E is the Young's modulus;

ν is Poisson's ratio;

p, **q**, and **t** are the strain components arrays, corresponding to the strains measured by a standard three elements strain-gauge rosette (1, 2 and 3 directions) that, for each depth increment, are given by:

$$\begin{cases} p = \dfrac{\varepsilon_3 + \varepsilon_1}{2} \\ q = \dfrac{\varepsilon_3 - \varepsilon_1}{2} \\ t = \dfrac{\varepsilon_3 + \varepsilon_1 - 2\varepsilon_2}{2} \end{cases}, \quad (4)$$

P, **Q**, and **T** are the stress components arrays to be calculted that, for each depth increment, are given by (directions 1, 2 and 3 are illustrated in Fig. 1):

$$\begin{cases} P = \dfrac{\sigma_3 + \sigma_1}{2} \\ Q = \dfrac{\sigma_3 - \sigma_1}{2} \\ T = \tau_{13} \end{cases}. \quad (5)$$

Since the stress arrays, **P**, **Q** and **T** have been calculated, the principal residual stresses and their directions, for each depth increment, can be determined by:

$$\begin{cases} \sigma_{\max} = P + \sqrt{Q^2 + T^2} \\ \sigma_{\min} = P - \sqrt{Q^2 + T^2} \\ \beta = \dfrac{1}{2}\arctan\dfrac{-T}{-Q} \end{cases}. \quad (6)$$

Materials Research Proceedings **2** (2016) 587-592 doi: http://dx.doi.org/10.21741/9781945291173-99

In this research, different methods for solving the Integral Method are applied and compared. The methods are applied to a simulated and an experimental hole-drilling procedure. The experiment was performed on a shot-peened steel sample. Results obtained from the developed numerical methods were compared to results from X-ray diffraction as well as the commercially-available software H-DRILL (which also uses the Integral method), developed by G. Schajer from the University of British Columbia. Even though the X-ray diffraction method has its own inaccuracies, it was determined accurate enough in a roughly-isotropic shot-peened sample. [6]

Numerical and Experimental Procedure

To solve the equations set out by the Integral Method, all the matrices must correspond to the same set of hole depth increments. However, the hole-drilling procedure will result in a set of data corresponding to experimental hole depth increments and not the standard increments used in the coefficient matrix determination. One of these sets of data will have to be interpolated to correspond to the other. In this research, a comparison was made between the final residual stress results, depending on which set (experimental or standard data) was interpolated. Once all the input data is in the correct form, the system of matrix equations can be solved. In this preliminary work, this was done using one of two numerical methods: Gauss-Seidel method or Tikhonov Regularisation.

Interpolating Standard Data. Bivariate Interpolation is a method of interpolating functions of more than one variable. It is recommended for this particular application by Schajer [2]. The coefficient matrices can be seen as tabulations of selected values of functions of hole depth and stress depth. Hole depth is the total depth of the hole at that increment and stress depth is the depth of the relieved strain component under consideration. Physically, stress depth always has to be less than or equal to hole depth which creates a lower triangular coefficient matrix. This feature requires that a triangular scheme of the Bivariate Interpolation be adopted. Using Bivariate Interpolation, the standard coefficient matrices can be interpolated to correspond to the experimental hole depth increments.

Interpolating Experimental Data. Each of the three strain measurements ε_1, ε_2 and ε_3 are plotted with hole depth and a cubic spline is calculated for each set. Cubic spline interpolation is chosen over polynomial interpolation because polynomial interpolations tend to oscillate for many data points whereas cubic splines do not. Cubic splines, therefore, give much smoother interpolating functions, especially if there are abrupt changes in the data. Cubic splines have been shown to be the more useful and common in engineering practice compared to its linear and quadratic counterparts [7].

Gauss-Seidel Method. This is a direct method for solving matrix equations. Gauss-Seidel is an iterative method so round off errors will be mitigated [7].

Tikhonov Regularisation. This is the most common method for solving ill-conditioned equations and is prescribed by the ASTM standard [3,7].

Using Python 2.7, two scripts were written to process the data before solving the system of equations:

- SDP (Standard Data Processor): standard data is interpolated.
- EDP (Experimental Data Processor): experimental data is interpolated.

Two scripts were written for solving the Integral Method equations once the input data had been processed:

- GSS (Gauss-Seidel Solver): solves the equations using the Gauss-Seidel method.
- TRS (Tikhonov Regularisation): solves the equations using Tikhonov Regularisation

These are combined to create four separate methods: SDP-GSS, SDP-TRS, EDP-GSS and EDP-TRS.

Simulated Hole-Drilling Procedure. A finite element analysis of the incremental hole-drilling method was carried out using ANSYS APDL code. A 100 MPa uniform tensile stress was applied to a plate in the x-direction to simulate a constant uniform residual stress state. Changes in strain on the surface were recorded and integrated for the geometry of a standard ASTM B strain gauge rosette.

Materials Research Proceedings **2** (2016) 587-592 doi: http://dx.doi.org/10.21741/9781945291173-99

The recommended hole depth increments (0.05 mm increments) were then removed. Because there were no experimental errors in the simulation, the numerical conditioning of each method could be tested.

Experimental Hole-Drilling Procedure. The methods were then applied to real experimental results of the hole-drilling procedure on a shot-peened piece of steel. X-ray diffraction was also performed on the sample and the results from the hole-drilling procedure were run through H-DRILL and the scripts developed in this research so they could all be compared.

Results

Simulated Hole-Drilling Procedure. Figure 1 shows the calculated maximum principal residual stress for each numerical method. The root mean square deviations (RMSD) values were 3.73 MPa for SDP-GSS, 1.74 MPa for SDP-TRS, 3.15 MPa for EDP-GSS and 0.88 MPa for EDP-TRS.

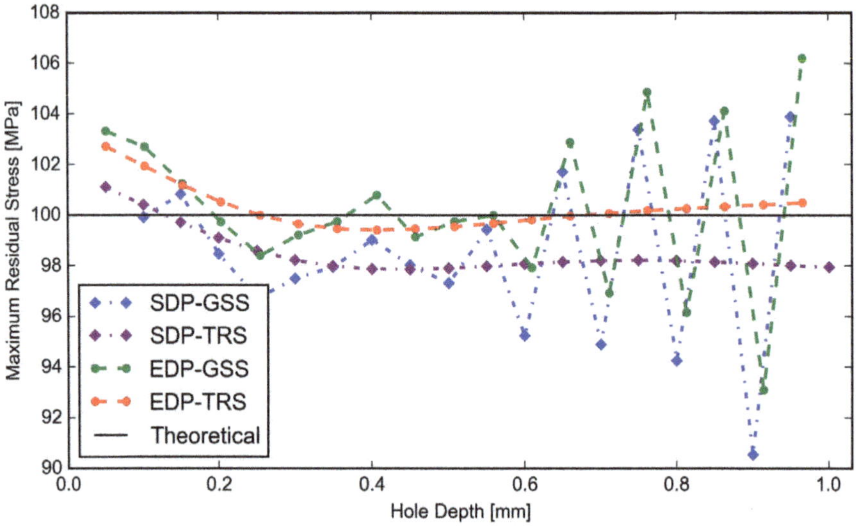

Figure 1 – Maximum Residual Stress for Simulated Hole-Drilling Procedure

Experimental Hole-Drilling Procedure. Figure 2 shows a comparison between the methods developed in this research and the X-ray diffraction results (left) and a comparison between the X-ray diffraction results, H-DRILL and the best performing numerical method EDP-TRS (right).

Discussion

Simulated Hole-Drilling Procedure. In the simulation, the Gauss-Seidel methods both oscillate about the true value with increased oscillation as hole depth increases. This is because as hole depth increases the problem becomes more numerically ill-conditioned. Because there are 20 hole depth increments, the matrices in the equations are large, which also contributes to ill-conditioning.

The experimental and standard data corresponded to the same hole depth increments so theoretically it shouldn't have mattered which one was interpolated. For the Gauss-Seidel methods it didn't make any difference but it is interesting to note that it made a big difference to the Tikhonov Regularisation method, with the bivariate interpolation introducing significant errors into this method. This suggests that Tikhonov Regularisation has a greater sensitivity to numerical errors in input data.

There is no accurate way to measure the actual error in these methods because even in the ANSYS simulation the actual residual stress is not known precisely.

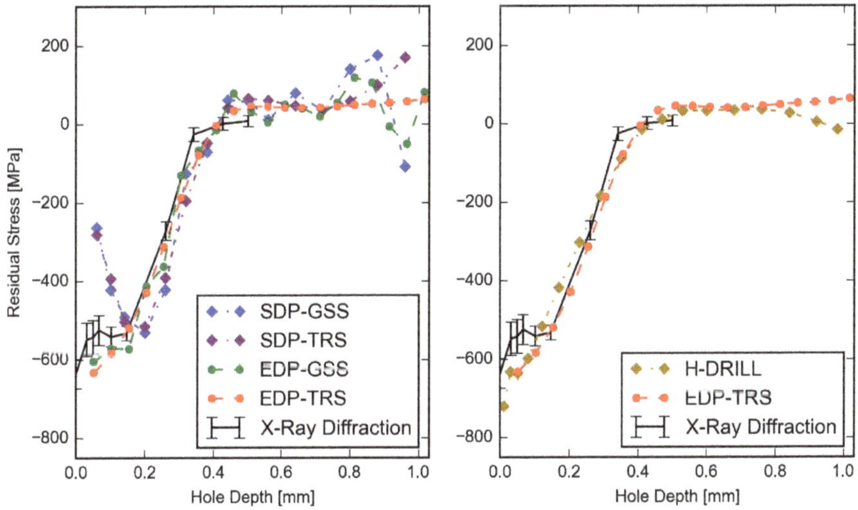

Figure 2 – Residual Stress for Experimental Hole-Drilling Procedure

Experimental Hole-Drilling Procedure. For the shot-peened sample, the ASTM E837 method recommends that if enough stress values have a magnitude greater than 60% of the yield stress of the material that the calculation is not accurate. In this case only one point falls into that category and since localised increases in yield stress due to the surface treatment has not been taken into account we can neglect it.

We took the X-ray diffraction values as the "true" values although this method also has its own inaccuracies – stresses in only two directions are measured so principal stress cannot be calculated. Although at each step the most compressive stress value was taken and the field is approximately isotropic, there will still be errors in this data.

The levelling off of stress values results in small change of strain measurements at the surface which makes it more sensitive to errors which accounts for the greater amplitude of oscillation with increasing hole depth. In the actual hole-drilling procedure the drill will never drill a perfectly cylindrical hole like in FEM simulation. There will inevitably be a chamfer around the bottom of the hole. This difference in geometry from the FEM used to calculate the coefficient matrices leads to errors in the computed stress. The stress relieved is actually smaller because it is acting on a smaller area (hole diameter is decreased at the chamfer). This causes significant errors at small hole depth increments where the non-cylindrical part of the hole makes up a large percentage of the total hole depth. This could account for the large errors seen at smaller hole depths.

In this procedure a large number of hole depth increments were used (21) which, again, leads to ill-conditioned matrix equations.

From Figure 2 we can see that the experimental data interpolation has consistently closer values. This could be a result of ill numerical condition from the bivariate interpolation. In these results, the experimental data interpolation has similar values for both Gauss-Seidel and Tikhonov Regularisation. The Gauss-Seidel results both oscillate more than the Tikhonov Regularisation

Summary
The combination of interpolating experimental data and using a Tikhonov Regularisation method for solving the residual stress profile has consistently produced the best results in terms of accuracy, smoothness and numerical stability. Tikhonov Regularisation makes a significant improvement to results from inputs which are numerically ill-conditioned due to small changes in strain. Interpolating the experimental data as opposed to the standardised data, significantly reduces the propagation of experimental errors and reduces numerical errors.

References

[1] A. Niku-Lari, J. Lu, J.F. Flavenot, Measurement of Residual-Stress Distribution by the Incremental Hole-Drilling Method, Experimental Mechanics, Vol. 25, No. 2, 1985, pp. 175-185. http://dx.doi.org/10.1007/BF02328809

[2] G.S. Schajer, Measurement of Non-Uniform Residual Stresses Using the Hole-Drilling Method. Part II – Practical Application of the Integral Method, Transactions of the ASME, Vol. 110, No. 4, 1988, Part I: pp. 338-343 and Part II: pp. 344-349.

[3] Determining Residual Stresses by the Hole-Drilling Strain-Gage Method, ASTM Standard E837, 2013.

[4] B. Zuccarello, Optimal Calculation Steps for the Evaluation of Residual Stress by the Incremental Hole-drilling Method, Experimental Mechanics, Vol. 39, No. 2, 1999, pp. 117-124. http://dx.doi.org/10.1007/BF02331114

[5] G.S. Schajer, E. Altus, Stress Calculation Error Analysis for Incremental Hole-Drilling Residual Stress Measurements, Journal of Engineering Materials and Technology, Vol. 118, No. 1, 1996, pp. 120-126. http://dx.doi.org/10.1115/1.2805924

[6] G.S. Schajer, Advances in Hole-Drilling Residual Stress Measurements, Experimental Mechanics, Vol. 50, No. 2, 2010, pp. 159-168. http://dx.doi.org/10.1007/s11340-009-9228-7

[7] S.C. Chapra, R.P. Canale, Numerical Methods for Engineers, fifth ed., McGraw-Hill, Singapore, 2006.

Residual Stresses 2016: ICRS-10 Materials Research Forum LLC
Materials Research Proceedings **2** (2016) 593-598 doi: http://dx.doi.org/10.21741/9781945291173-100

Residual Stress Redistribution due to Removal of Material Layers by Electrolytic Polishing

I. Surtee[1,a*] and J.P. Nobre[b]

School of Mechanical, Industrial and Aeronautical Engineering, University of the Witwatersrand, Johannesburg, Private Bag 3, WITS 2050, South Africa

[1]Graduate Mechanical Engineer, Degree and concerned research was carried out as a student at the University of the Witwatersrand

[a] isosurtee@gmail.com, [b] joao.nobre@wits.ac.za

Keywords: Electrolytic Polishing, Residual Stresses, X-Ray Diffraction

Abstract. The phenomenon of redistribution of residual stress due to the removal of material layers by electrolytic polishing can be attributed to the disturbance in the initial equilibrium state of the stress caused by the material removal, which must be considered when this technique is used together with X-ray diffraction for residual stress determination. This study investigates these redistributions and involves the development of a numerical simulation model on the ANSYS finite element platform, in which different material removal cases were simulated on a flat plate model. The cases simulated and compared includes the removal of complete layers in incremental depths, the removal of strip layers in incremental widths and depths as well as the removal of square sections in incremental sizes and depths. The main objective deals with the effect of removal depth for the full layer removal and the effect of plate geometry for the square and strip removal models. A mathematical correction technique that was shown to correct the redistributions within minimal limits for the square layer removal process is proposed. Moore and Evans correction as well as Pederson-Hanson correction techniques are currently being used, but both methods are not unique to square section removal corrections and were used for model validation. The correction proposed displayed promising results when applied to simulated data, but lacked enough data and experimental validation to be generalized. The findings show a lot of insight into the redistribution patterns of these stresses as well as the different parameters that affect their behavior.

Introduction

Residual stresses can be defined as the stresses locked in a material, which are in equilibrium in its interior without any external forces acting, due to any heterogeneity of deformation, at micro or macro scale, of different origins, such as mechanical, thermal and metallurgical. The knowledge of the residual stresses and their behavior in a material are of great importance as they affect the performance of the material and ultimately they are often responsible for the failure of mechanical components, or lifetime improvement when skillfully generated [1].

Residual stresses in crystalline materials can be analyzed non-destructively by the X-ray diffraction (XRD) method, measuring the induced lattice deformations using the lattice distance of specific {hkl} planes as gage length. In the presence of a given stress state, the lattice deformation is a function of the orientation of the diffracting {hkl}-planes relative to the sample's surface. The orientation-dependent variation of the lattice deformation can be related to the underlying stress state. Assuming a plane stress state (due to the small penetration depth of X-rays in metallic materials), the so-called $\sin^2\psi$ method [2] can be used to determine the corresponding residual stresses at material's surface, using specific X-ray elastic constants. One of the most common techniques for residual stress depth profiling is the use of the X-ray diffraction technique in conjunction with the electrolytic layer removal technique. The electrolytic polishing method of material removal allows successful depth profiling without introducing additional stresses [1].

Materials Research Proceedings **2** (2016) 593-598 doi: http://dx.doi.org/10.21741/9781945291173-100

The major concern of residual stress measurements of materials undergoing layer removal is that of redistribution. When material layers are removed, the equilibrium state residual stress profile inherent in the material is disturbed. This causes a distribution of stress as the stresses try to reach a new state of equilibrium. The concept of redistribution, whereby the removed layer disturbs the equilibrium stress state and the stresses redistribution, introduces the need for a correction of the measured stress to depict the stress state more correctly. The correction accounts for these redistributions. Current correction methods include the Moore and Evans theory [3], Pederson and Hansson theory [4] as well as the Savaria-Bridier-Bocher theory [5].

In this work, a model able to simulate the material layer removal process for residual stress profiles was developed. The effect of removal depth on the redistribution of residual stress for a full layer removal, strip layer removal and square layer removal model was investigated. In addition, the effect of specimen geometry on the redistribution of residual stress for the square and strip layer removal models was also investigated. The effect of removal geometry on residual stress redistribution was compared. Based on the observed trends, a correction methodology and attempt to generalize it has been studied and proposed.

Numerical Study

A numerical study using the finite element method (FEM) was carried out, using ANSYS code, to analyze and discuss the effect of removal depth on the redistribution of residual stress for a full layer removal, strip layer removal and square layer removal model. For the linear elastic and isotropic material, typical elastic constants for carbon steel, i.e., Young modulus of 210 GPa and a Poisson's ratio of 0.3, were considered. Since the residual stress model is known to be influenced by the plasticity of a material, a bilinear isotropic hardening feature of the material was included. This will ensure additional accuracy if the yield strength is exceeded. For this necessary numerical study, a 2D-dimensional axisymmetric finite element model using 4-node isoparametric elements was developed. To simulate the redistribution of stress due to the layer removal, a quarter model of a 5 mm thickness square plate was considered. The effect of a total 0.5 mm removal layer was analysed. The simulation of the layer removal was carried out in 0.1 mm thickness steps. A total of 5 simulation steps were thus considered for each case. The removal of each layer step was simulated using the so-called "birth and death" ANSYS [6] code features. The in-depth residual stress profiles, simulated by the finite element model, were generated imposing a corresponding temperature gradient to the restrained model, taking into account the elastic constants of the material and its thermal expansion coefficient. For each material layer, the elements at a given depth were subjected to the following in-depth temperature gradient:

$$\Delta T(z) = \frac{\sigma^{RS}(z)(1-\nu)}{\alpha E}, \qquad (1)$$

where, α is the thermal expansion coefficient, E the Young modulus, υ the Poisson coefficient, $\sigma^{RS}(z)$ the in-depth equi-biaxial residual stress profile to be generated and z represents the distance to the surface. The FEM model is full constrained and the simulated temperature, given by the equation above, is imposed to each element of the model. In this work a constant temperature gradient equal to 28 °C was imposed to obtain a uniform stress of 96 MPa. After each removal layer, the results were taken considering a path on the edge of the plate, through the plate's center, where the results are of interest, from the top surface to the bottom. The results were extracted along this path to simulate the residual stress profile along the components depth. In this way, the percentage distribution based on different material removal analyses can be compared and evaluated. It should be noted that even though a plastic deformation could have been simulated, the above mentioned method was used because the focus was not on simulating plastic deformations but rather simulating known common residual stress profiles in an attempt to study redistributions that occur. Modelling plastic deformation may also be done as an alternative for a better real life scenario model.

Results and Discussion

With reference to the results obtained for the full layer removal simulation shown in figure 1 the effect of removal depth on the residual stress redistribution can be observed. As the layer removal depth increased, the redistribution of the residual stress was more evident and pronounced.The findings regarding the effect of depth, is validated by the published data found in [7]. In the referenced study a shot peened profile was simulated using the ABAQUS software and the layer removal was simulated on the initial resulting profile [7]. A larger degree of redistribution is observed for the 15μm depth compared to the 5μm depth removal. This correlates well with the results obtained in the full layer removal simulation, shown in Figure 1.

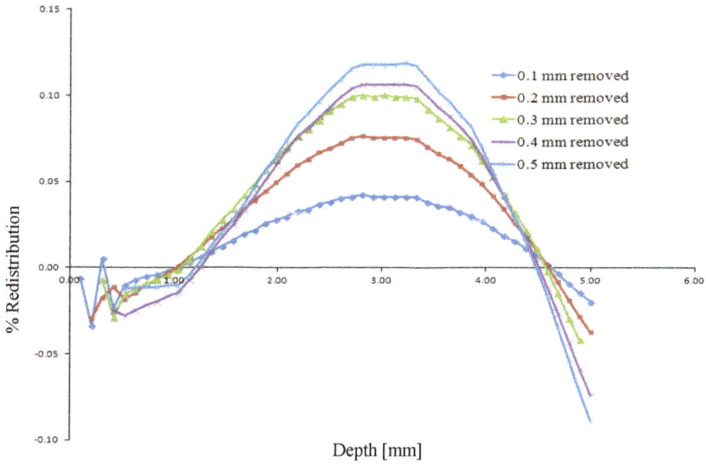

Figure 1- Percentage redistribution of residual stress due to complete layer removal

Considering the Moore and Evans full plate correction, which is a standard correction method for full layer removal, the following Equation 2 below can be written when expanding the integrands using Taylor series and considering that electrolytic polishing deals with small depths of material removal:

$$\sigma(z1) = \sigma_m(z1) + (-4\sigma_m(H)(\frac{\Delta Z1}{H}) \quad (2)$$

Where $\sigma(z1)$ is the stress at depth z1, $\sigma_m(z1)$ is the measure stress at z1 and H is the thivkness of the plate.This equation shows that as removal depth increases $\Delta Z1$ increases and, hence, the correction required increases, implying that the redistribution increases with the depth. This behaviour, once again, correlates with the results obtained in Figure 1.It was observed that for both the strip removal and square removal results, the redistribution was more pronounced at the surface of the specimen and seems to decrease with depth,shown in figure 2. The Pederson and Hansson study [5] observed a similar redistribution that was more pronounced at the surface for the strip removal simulated, explained it as a notch effect. The notch created by the edge of the removed geometry produces a non-linear distribution and together with the influence of the remaining material adjacent to the groove created, the effect of redistribution is larger at the surface of the material.

The removed geometry is no longer a complete layer but rather a local removal. This implies that a notch effect is created at the edge of the removal geometry creating increased stress gradients in that region, explains the redistribution trend observed. This can be seen in Figure 3.

Materials Research Proceedings **2** (2016) 593-598 doi: http://dx.doi.org/10.21741/9781945291173-100

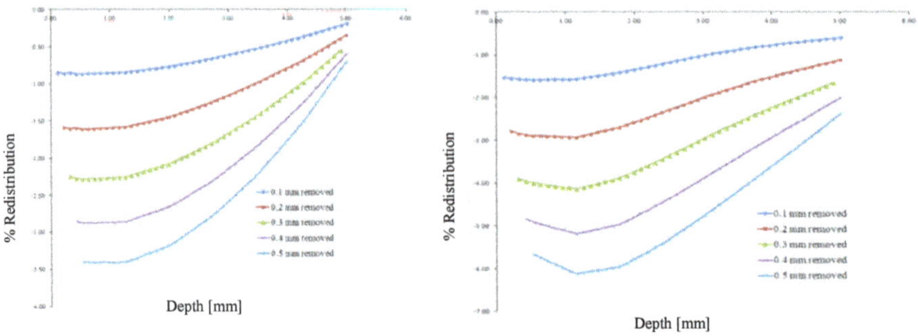

Figure 2-Effect of strip removal (left) and square removal (right).

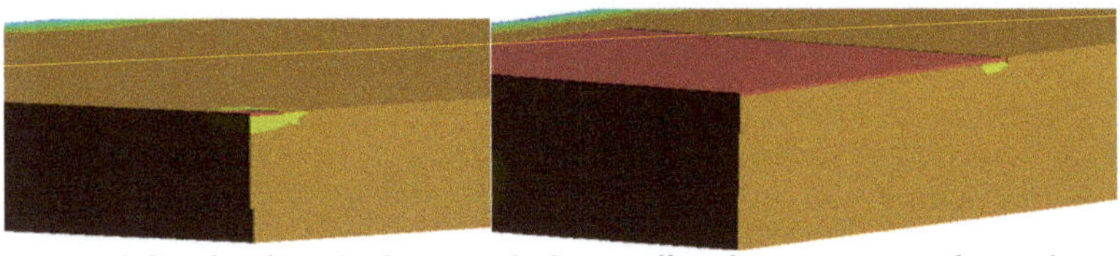

Figure 3-Simulated results depicting the larger effects being experienced at surface.

The explained reasoning regarding the notch effect and the fact that the local removals tend to be larger at the surface and decay with depth, could explain why the change in cross sectional geometry, with constant thickness, results showcased only slight redistribution differences, as shown in Figure 4. In addition, the Moore and Evans correction,although used for full layer removal, did not account for the cross sectional geometry of the specimen, motivating the conclusion that change in cross section for constant thickness does not affect the redistribution.This implies that the redistribution may be independent of the cross sectional geometry of the specimen.

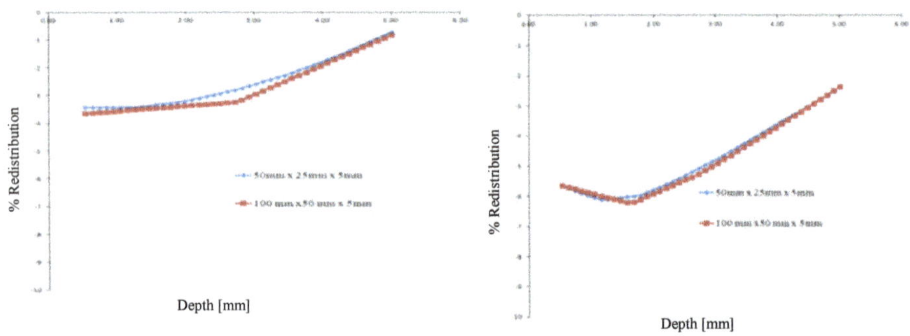

Figure 4- Minimal change in specimens of different cross sections for strip(left) and square(right) removals observed.

Secondly the effect of thickness change for the square and strip removal both depicted that the redistribution is dependent and affected by the thickness of the specimen, as shown in Figure 5. This implies that the redistribution is not independent of thickness of the specimen. The results correlate well with the findings that geometrical changes affect the redistribution of residual stress in a flat plate. In addition, the Moore and Evans correction although used for full layer removal did account

Materials Research Proceedings **2** (2016) 593-598 doi: http://dx.doi.org/10.21741/9781945291173-100

for original thickness H, motivating the conclusion that change in thickness does affect the redistribution of residual stress during material removal.

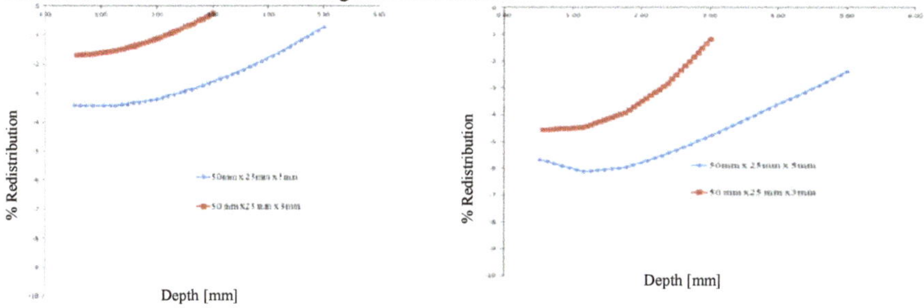

Figure 5- Effect of change in thickness in specimens of same cross section for strip (left) and square (right) removals.

A Correction method was develped for the square removal process. The difference between the full plate value and simulated XRD value was calculated and plotted as a function of depth. A linear regression was carried out for each plotted line and the gradient and intercept values of the lines were noted. The resulting average gradients and intercepts for each model thickness were plotted against corresponding square removal area and resulting trend lines were computed as shown in Figure 6 for a 5mm thick plate. The graphs will be used to extract a gradient and intercept, for example for a 1mm x 1mm removal the removal area is 1mm^2 hence the extracted gradient is 50.761 and intercept is 6.1536 from the graphs. The correction is then carried out by taking the XRD value + {extracted gradient (depth) + extracted intercept} the implemented correction is shown in Figure 7 for a 1mm x1mm square removal with the extraction equation 50.761(d) + 6.1536.

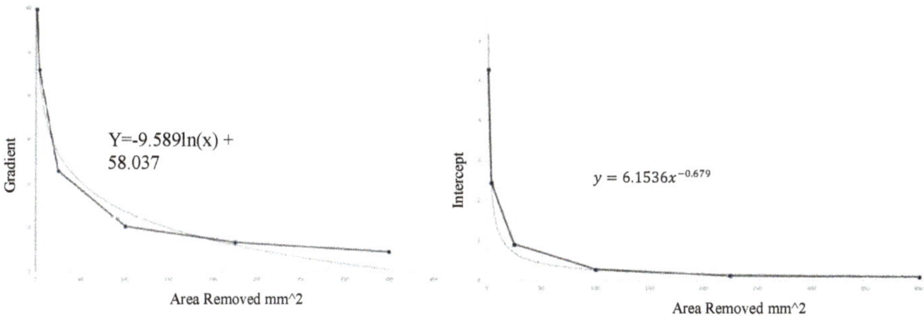

Figure 6 -Correction gradient and intercept extraction graphs proposed for 5mm plate with square removal.

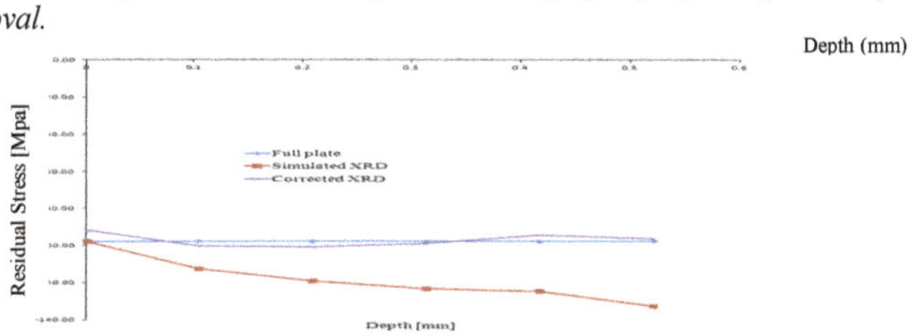

Figure 7-Performance of the correction method for a 1mm x 1mm square

Materials Research Proceedings **2** (2016) 593-598 doi: http://dx.doi.org/10.21741/9781945291173-100

It should however be noted that the simulation and correction was based on the simulated results of two cross sections each and additional work is required to confidently conclude its generality in correction of the data however the correction will serve usefull as the basis for future generalisation attempts. Experimental validation is lacking and experimental procedures should try and evaluate and validate the proposed correction. The correction and all attained results was derived for a constant initial residual stress state and its performance on different stress states such as shot peened samples or samples with bending stress states, is inconclusive.

Conclusions

1. The percentage redistribution increases as the removal depth increases regardless of removal geometry.
2. The removal geometry of full layer removal, square layer removal and strip layer removal do not cause identical redistribution. The redistribution is dependent on the removal geometry.
3. The geometry of the specimen affects the redistribution observed for square and strip removals; however specimens with constant thickness and different cross sections depicted minimal difference in redistribution. There was not sufficient data to comment on an independence of specimens cross sectional area.
4. The percentage redistribution for the square and strip removal geometries displayed larger distributions at the surface which decreased with depth of the specimen. It was proposed that this was due to high stress gradients caused by the notch and impact of the extra material adjacent to the groove, but this requires experimental validation.
5. The percentage redistribution for the square and strip removal geometries displayed larger distributions for smaller square geometries and smaller widths. The percentage redistribution decreased as the square size or strip width increased. It was proposed that this was due to high stress gradients caused by the notch and impact of the extra material adjacent to the groove being closer to the measurement point at the centroid for smaller squares and strip widths and further away for increased square size and strip widths. This requires experimental validation.
6. A correction methodology was proposed which produced promising results on the tested geometries. It requires experimental validation and possible refinement based on future findings. The correction derivation method itself may prove useful as well.
7. The model itself that was created successfully allows the simulation of the material layer removal process. It also offers a platform for simulating any initial stress state by manipulating the thermal stress feature in the setup.
8. More simulations and experiments are required to validate overall findings.

References

[1] Handbook of Measurement of Residual Stresses, Society for Experimental Mechanics (SEM), Edited by J. Lu, 1996, pp. 5-34/71-131.

[2] E. Macherauch und P. Müller, Das $\sin^2\psi$-Verfahren der Röntgenographischen Spannungsmessung, Zeitschrift für Angewandte Physik 13 (1961) 305-312.

[3] M. Moore and W. Evans, Mathematical Correction for Stress in Removed Layers in X-Ray Diffraction Residual Stress Analysis, SAE Technical Paper 580035 (1958).

[4] I.L. Hansson and T.F. Pedersen, Finite element calculations for correction of residual stress profiles of coated and uncoated materials measured by X ray diffraction, NDT 22 (1989) 347-352.

[5] V. Savaria, F. Bridier and P. Bocher, Computational quantification and correction of the errors induced by layer removal for subsurface residual stress measurements, Int. J. of Mech. Sci. 64 (2012) 184-195. http://dx.doi.org/10.1016/j.ijmecsci.2012.07.003

[6] ANSYS, 2010, ANSYS Advanced Analysis Techniques, SAS IP, Inc., Houston.

[7] R. F. Laubscher, Gary Styger, Residual stress depth profiling by layer removal, 9th South African Conference on Computational and Applied Mechanics, 2014.

Residual Stresses 2016: ICRS-10 Materials Research Forum LLC
Materials Research Proceedings **2** (2016) 599-602 doi: http://dx.doi.org/10.21741/9781945291173-101

Research on Corrosion Fatigue Crack Propagation Behavior of Welded Joints of A7N01P-T4 Aluminum Alloys

Jia Chen[1,2,a],Jian Xu[1,b],Maojuan Zhao[3,c*],Guoqing Gou[2,d]

[1]Chengdu Industry and Trade College, Chengdu 611731, China

[2] School of Materials Science and Engineering, Southwest Jiaotong University, Chengdu 610031, China

[3]Chengdu University, College of information science and engineering, Chengdu 610100, China

[a]15574423@qq.com;[b]jianxu1021@126.com;[c*] 1160698685@qq.com
[d]gouguoqing1001@163.com

Keywords: High Speed Train; Corrosion Fatigue; A7N01P-T4 Aluminum Alloy

Abstract: Corrosion fatigue crack propagation behavior of welded joints a of A7N01P-T4 aluminum alloys were investigated. The corrosion fatigue crack propagation rate of base metal is higher than that of the heat-affected zone and the welding seam has a better corrosion fatigue resistance.

Introduction

Corrosion fatigue failure is one of failure forms of structures which under the service of cyclic load with corrosive environment[1].This failure form often occurs in 2xxx, 7xxx series aluminum alloys of aircraft, vehicles and other important transportation structures and more than half of failures were in correlation with corrosion fatigue [3~6].The fatigue life of structures in corrosion environment is much shorter than that of in air, especially when the applied stress is much closer to the fatigue limit of the structures.

This paper reports the results from our investigation on the corrosion fatigue properties especially on the crack propagation behavior.

Material and experimental techniques

The experiment materials are A7N01P-T4aluminum alloy plates (P stands for plate and T4 stands for a steady state after solution treatment and natural cooling according to ISO XXXX[35]).Welding was performed by the Metal Inert-Gas (MIG) technique with a PHOENIX 421 EXPERT welding machine. The welding wires are ER5356 of1.6 mm diameter. The chemical composition of the A7N01P Al alloy base metal, A7N01P, and the welding wire are listed in Table 1. The welding parameters are listed in Table 2. To remove the oxides and reduce the porosity of the joints, the surface of the alloy was chemically cleaned before welding.

Table 1 *Chemical composition of base metal and welding wire*

Material	Zn	Mg	Cu	Mn	Ti	Si	Fe	Cr	Al
A7N01P-T4	4.0~5.0	1.0~1.8	0.10	0.20~0.70	0.01~0.06	0.35	0.40	0.06~0.20	Bal.
ER5356	0.10	4.5~5.5	0.1	0.05~0.20	0.2	0.25	0.10	--	Bal.

Note: 1. The chemical composition of A7N01P-T4 refers to GB/T 16475-2008: Temper designation system for wrought aluminum and aluminum alloy. 2. The chemical composition of ER5356, which is similar to filler metal, was listed for future analysis.

Materials Research Proceedings **2** (2016) 599-602 doi: http://dx.doi.org/10.21741/9781945291173-101

Corrosion fatigue crack propagation experiment was performed according to ISO 12108: Metallic materials—Fatigue experimenting—Fatigue crack propagation method. The specimen in this paper was improved from single edge notched tension specimen(SENT) in ISO 12108.The specimens of base metal and welding joints was machined along Y-X (X stands for the direction paralleling to the welding line and Y stands for the direction perpendicular to the welding line). BM, HAZ and WM respectively stand for the specimen in base metal, the heated-affected zone and the welded joint. Sampling method and dimension of corrosion fatigue specimen are shown in fig.1.

(a)

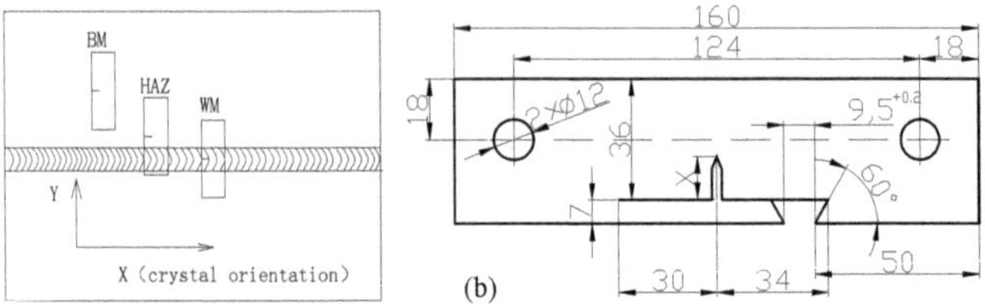

(b)

Fig.1 Sampling method and dimension of corrosion fatigue specimen (a)Sampling method (b)dim ension of corrosion fatigue specimen

Specimens in the direction of Y-X (see Fig.1) for corrosion fatigue cracking experimenting were cut and machined from the BM, HAZ and WM of the welded joints according to ISO 7539-6:2003 and ISO 7539-8:2000. A notch was cut at the edge of each specimens with the same way. The pre-setting notch was1~ 2 mm long and the pre-crack length was 1~2 mm long.

To assure the accuracy of the experimenting data, only the region of interest of each sample was immersed into the solution during experimenting, and the rest of the sample surface was masked with the BONLE type cyanoacrylate adhesive glue. The sharp incision and the extensometers were also isolated using glue and then covered with epoxy. A paper cup was used as the corrosion solution container. The entire set up for the corrosion fatigue cracking experimenting is shown in Fig. 2. The specimens were clamped through two fixtures.

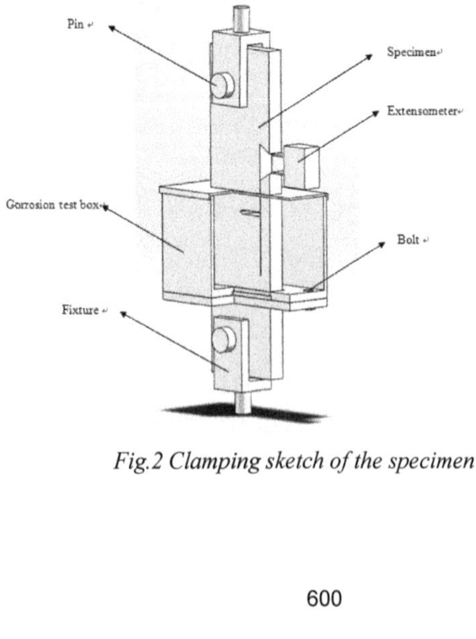

Fig.2 Clamping sketch of the specimen

Materials Research Proceedings **2** (2016) 599-602 doi: http://dx.doi.org/10.21741/9781945291173-101

Results and discussion

The relationship of corrosion fatigue crack propagation rate da/dN vs. stress intensity rangeΔK were shown in Fig.3 and Table 2. In order to get the statistics disciplinarian and trend disciplinarian, the seven points incremental polynomial method was used. The Paris equations (1) was used to process the measured data.

Where da/dN is the fatigue crack propagation rate;ΔK is the stress intensity range; C,m is the material constants.

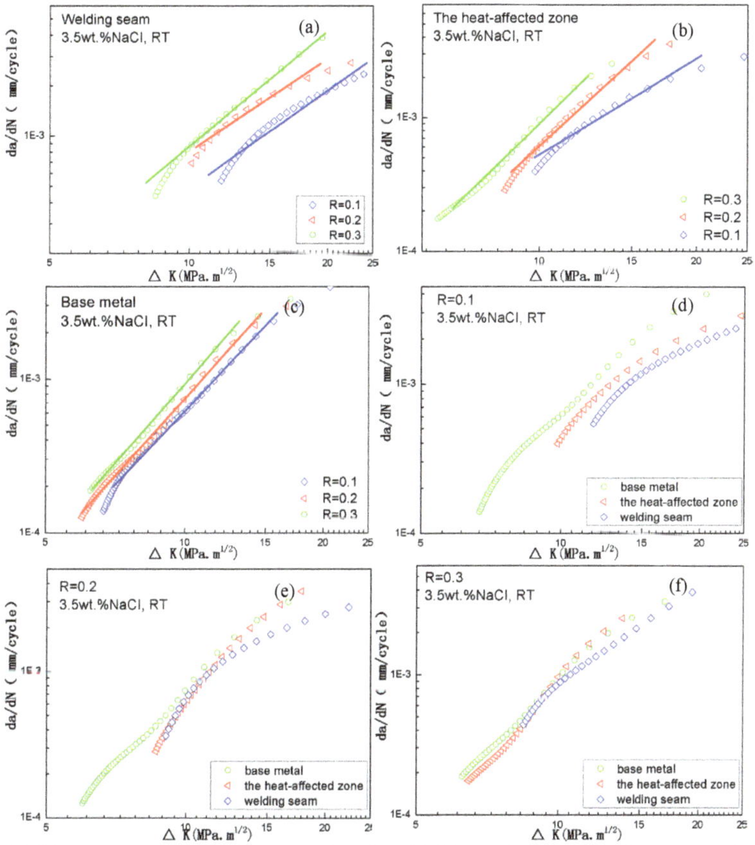

Fig.3 da/dN versus ΔK curves of base metal and welded joint: (a) corrosion fatigue crack propagation rate(CFCGR) curves of welding seam under different stress ratio, (b) CFCGR curves of the heat-affected zone under different stress ratio, (c) CFCGR curves of base metal under different stress ratio, (d) CFCGR curves of welding seam, the heat-affected zone and base metal under stress ratio R=0.1, (e) CFCGR curves of welding seam, the heat-affected zone and base metal under stress ratio R=0.2,(f) CFCGR curves of welding seam, the heat-affected zone and base metal under stress ratio R=0.3.

Table2 Fitting parameters of Paris formula of base metal and welded joint

Condition	C	m
WM, R=0.1	6.08×10^{-7}	3.0001
Welded seam, R=0.2	2.04×10^{-5}	1.5363
Welded seam, R=0.3	1.18×10^{-6}	2.6419
HAZ, R=0.1	2.74×10^{-6}	2.26123
The heat-affected zone, R=0.2	1.30×10^{-7}	3.65022
The heat-affected zone, R=0.3	1.39×10^{-7}	3.79517
Base metal, R=0.1	4.48×10^{-6}	2.02025
Base metal, R=0.2	4.24×10^{-7}	3.2281
Base metal, R=0.3	6.008×10^{-7}	3.14431

From the results, corrosion fatigue crack propagation rate of BM, HAZ and WM increased with the increase of stress ratio R. The propagation rate of BM was the highest, the propagation rate of HAZ was the medium and the WM was the lowest.

Conclusion

The behavior of corrosion fatigue crack propagation of base metal and welded joint of A7N01P-T4 aluminum alloy in 3.5%wt.NaCl under the different stress ratio was studied in this paper. On the basis of the experiment results, the conclusions are drawn as follows:

（1）The welding seam of A7N01P-T4 aluminum alloy has a better corrosion fatigue resistance than that of the heat-affected zone in 3.5%wt.NaCl and base metal has the worst corrosion fatigue-resistant performance.

References

[1] Lv Shengli, ZHANG youhong, LV guozhi.·research and evaluation on corrosion damage of aluminium alloy structure.Xian, Northwestern Polytechnic University Press, 2009.169.

[2] H. Seifert, S. Ritter, Corrosion Science 2016.

[3] GOU Guoqing, HUANG Nan, CHEN Hui, LI Da, MENG Lichun. Analysis on corrosion behavior of welded joint of A7N01S-T5 aluminum alloy for high-speed train. Transaction of the China Welding Institution. 2011,32（10）: 17-20.

[4] Frederic Menan, Gilbert Henaff. Influence of frequency and exposure to a saline solution on the corrosion fatigue crack propagation behavior of the aluminum alloy 2024[J]. Intenational Journal of Fatigue, 2009, 31: 1684-1695. http://dx.doi.org/10.1016/j.ijfatigue.2009.02.033

[5] U.Zupanca, J.Grumb. Effect of pitting corrosion on fatigue performance of shot-peened aluminium alloy 7075-T651[J]. Journal of Materials Processing Technology, 2010, 210: 1197-1202. http://dx.doi.org/10.1016/j.jmatprotec.2010.03.004

[6] ZHANG Zhenggui,ZHOU Zhaoyuan, LIU Changyong. Corrosion fatigue failure analysis of high-strength aluminum alloy［J］.Journal of Chinese Society for Corrosion and Protection, 2008, 28(1): 48-51.

Residual Stresses 2016: ICRS-10 Materials Research Forum LLC
Materials Research Proceedings **2** (2016) 603-608 doi: http://dx.doi.org/10.21741/9781945291173-102

Residual Stress and Critical Crack Size before and after Post-Weld Heat-Treatment

Law M[1], Paradowska[2]* A, Hoye N[3], Grace P[4]

[1, 2]ANSTO, Locked Bag 2001, Kirrawee DC NSW 2232, Australia

[3]University of Wollongong, Northfields Ave, Wollongong NSW 2522, Australia

[4]Jemena, 99 Walker St, North Sydney NSW 2060, Australia

[1]mlx@ansto.gov.au, [2]anp@ansto.gov.au, [3]nicholas.hoye@gmail.com, [4]Paul.Grace@jemena.com.au

Keywords: Welding, Neutrons, Residual Stress, PWHT

Abstract. Post-weld heat-treatment (PWHT) is performed to reduce residual stress, but is not always possible to perform. The residual stresses on a thick section weld on a gas pipeline were determined before and after PWHT to assess residual stress and critical defect sizes.

Introduction

Many essential pipeline modifications and repairs require welding. Codes mandate the use of PWHT above some thickness which varies depending on which code is followed [1]. Unfortunately, product flow within the pipeline carries away the heat necessary for PWHT and many pipelines cannot be shut down or depressurized, thus making PWHT impossible. The major reason for the use of PWHT is to avoid HACC (also known as cold cracking, or delayed cracking) which principally occurs in ferritic steels, most often during or a short time after welding. HACC is a risk only for a short period (approximately 1 day): once this period is past and the weld has been subject to delayed NDT, HACC is no longer a threat and any defect which is detected is subject to a failure assessment by a conventional defect assessment method.

Material and sample characterisation

The X70 grade spherical tee (1143mm OD x 41mm thickness) halves were placed around the run pipe, the halves welded together, then welded to a X60 grade run pipe (508mm OD x 8.5 mm thickness). The welding was carried out with water flowing through the run pipe to simulate the heat sink of the flowing product. After the residual stresses were measured, the sample was subject to a furnace PWHT and the measurements were repeated.

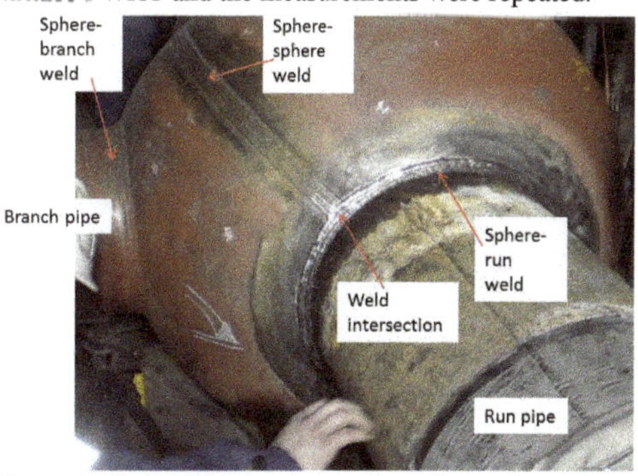

Figure 1 Weld identification in the spherical tee hot-tap junction.

Materials Research Proceedings **2** (2016) 603-608 doi: http://dx.doi.org/10.21741/9781945291173-102

As the entire tee could not be measured by neutron diffraction, a representative section containing the critical start and end of the welds was cut from the tee hot-tap junction. Finite element modelling was carried out determine a sample size which would not affect the residual stress values. A model was made of the main butt weld between the half spheres (the branch pipe was ignored) and pseudo stresses were introduced by Chill Modeling [2] (figure 2). The spherical tee was then reduced in size to assess the effect on the residual stresses. These pseudo-stresses do not represent the actual residual stresses but can assess any effects due to change of restraint from specimen cutting.

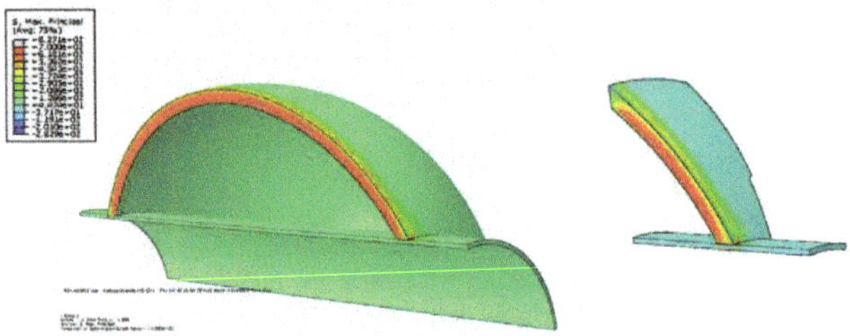

Figure 2. Maximum stresses in full model (only one quarter is shown for clarity) and cut sample

When stresses are measured along the weld, the area near the cut end (left end of figure 3) shows a large reduction in stresses. Measurements more than 68 mm (~2 thicknesses) away from the cut end will be within 5% of the un-cut values. Measurements closer to the runpipe will be affected less.

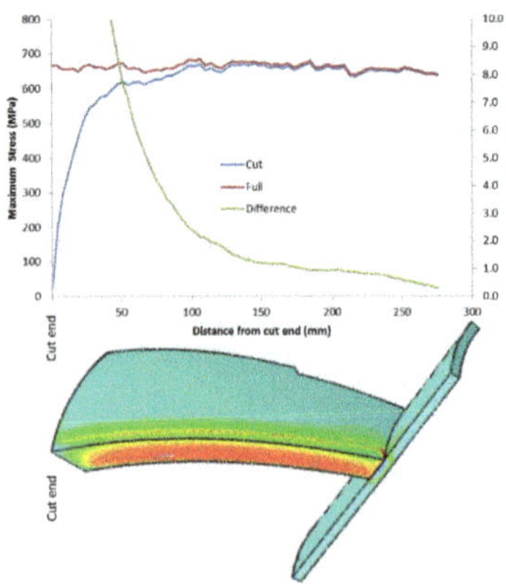

Figure 3. Stress results superimposed on cut model, with % difference between the uncut and cut stresses also shown.

Residual stress measurements

The residual stress measurements were performed on the Kowari strain scanning instrument [3] at ANSTO. Since the basic principles of this technique are well known [4,5,6] only details specific to this measurement will be reported. A nominal gauge volume of 3x3x4 mm^3 was used. Stress-free reference measurements were performed on slices 8mm thick slices. The errors in residual stress measurements were estimated to be ± 20 MPa.

Materials Research Proceedings **2** (2016) 603-608 doi: http://dx.doi.org/10.21741/9781945291173-102

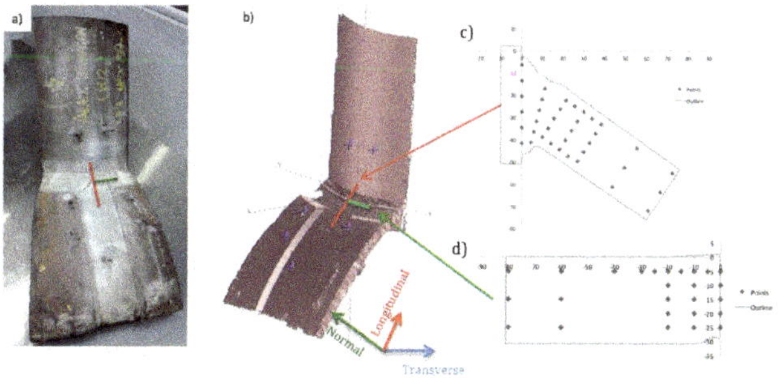

Figure 4. (a) Cut sample. (b) SCANNS model generated by laser profile scanning. The red line represents the sphere-sphere weld (Plane A) The green line represents the sphere-pipe weld (Plane B). Layout and measurement points in plane A (c) and plane B (d).

Results

The residual stresses were measured in the three directions shown in figure 4b to characterise the sphere-sphere and the sphere-pipe welds.

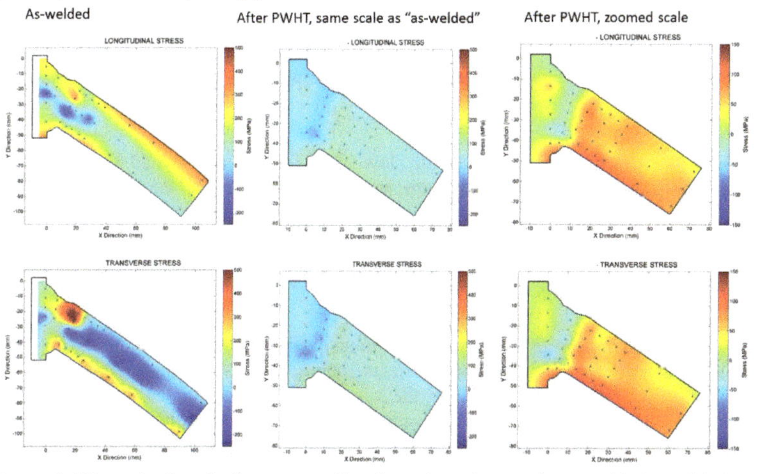

Figure 5. Plane A - Residual stress profile along the sphere-sphere weld. As-welded and PWHT stress maps are shown on the same scale for comparison, then right column shows rescaled PWHT results for more detail.

Materials Research Proceedings **2** (2016) 603-608 doi: http://dx.doi.org/10.21741/9781945291173-102

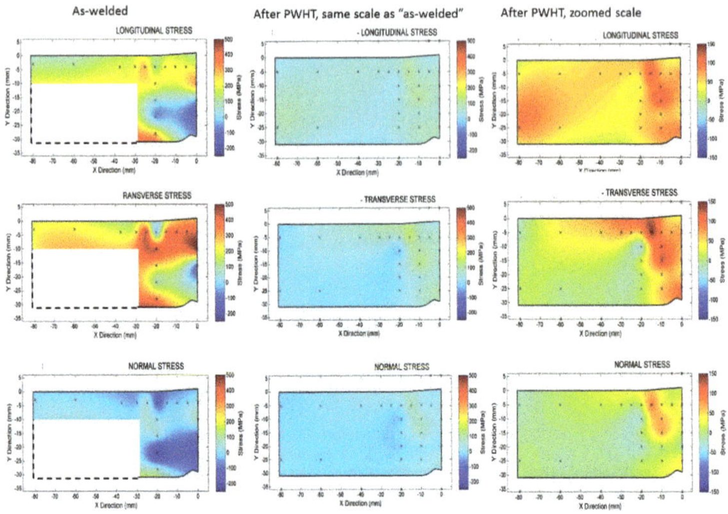

Figure 6. Plane B- Residual stress distribution along the sphere-pipe weld before and after PWHT (x=0 weld centre) the white area was not measured. As-welded and PWHT stress maps are shown on the same scale for comparison, then right column shows rescaled PWHT results for more detail.

The as-welded residual stresses were generally low, but approached the yield strength at one position. Through-thickness residual stress maps across the butt weld at the weld intersection (Figures 5, 6) shows relatively high residual, the largest residual stress was 10 mm below the surface. These as-welded residual stresses were lower than code-based estimates, except at the weld intersection, where the stresses approached code-based estimates. After PWHT the residual stresses were reduced to less than 30% of the as-welded values.

Structural integrity assessment of welds based on measured residual stresses

The critical defect depth for assumed inner wall defects 100 mm long were calculated using the R6 British Nuclear code [7]. To allow comparisons, the same stresses were assumed for both welds of 147 MPa membrane and 281 MPa bending. The weld was assumed to equal the sphere X70 tensile properties (485 MPa SMYS and 565 MPa SMTS) with a Charpy toughness of 40J.

The residual stress values are either code-based estimates or the measurements in this paper. Fitness for purpose assessments codes assume residual stresses either at yield, or some other distribution through-wall (Figure 7). In this work the BS 7910 Annex Q estimate was used for the as-welded, and the constant 110 MPa line was used in PWHT cases. Two sets of measured values were used, the low values in most positions in the weld ('Typical', Figure 7a) and the localised high values at the intersection of the 2 welds ('Intersection', Figure 7b). The 'typical' transverse residual stresses were lower than BS7910 estimate for high heat input welds. However, the residual stresses at the intersection of the two welds were higher than this code estimate at one point at the inner wall (Figure 7b).

Materials Research Proceedings **2** (2016) 603-608 doi: http://dx.doi.org/10.21741/9781945291173-102

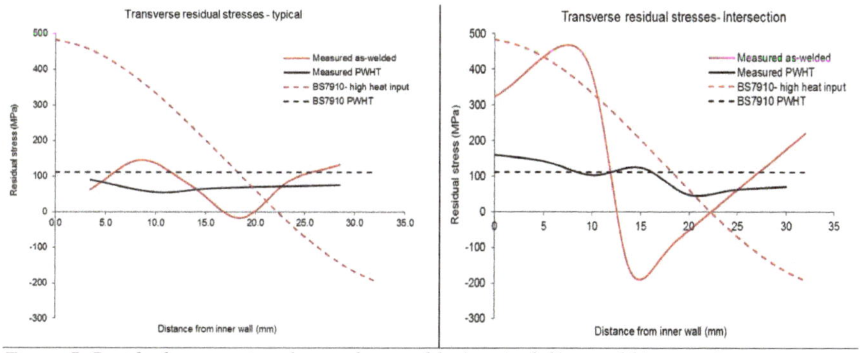

Figure 7. Residual stresses in sphere-sphere weld. a) typical. b) at weld intersection.

The critical crack depths for 100mm long defects were calculated (figure 8) PWHT increased the allowable depth by 60% for 'typical' locations (with low residual stress), but only 20% for the results at the weld intersection which had higher residual stresses. The use of measured residual stresses increased the critical crack depth by 15 to 30% over code-based residual stress estimates.

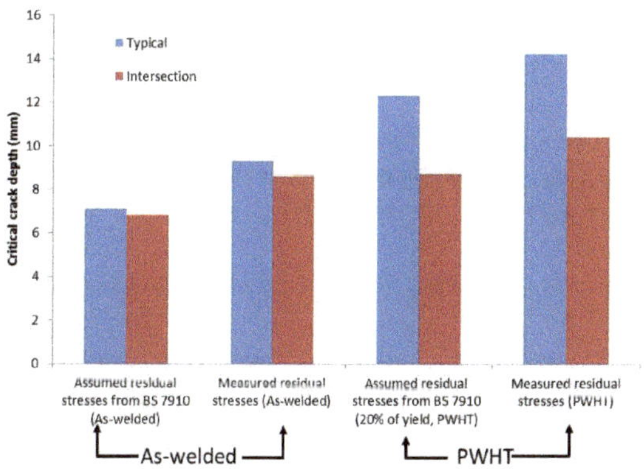

Figure 8. Critical crack depth for 100 mm long defect in typical weld and at weld intersection.

Materials Research Proceedings **2** (2016) 603-608 doi: http://dx.doi.org/10.21741/9781945291173-102

Conclusions

The main findings are:

- Residual stresses were highest at the weld toes and the weld root, and at the weld-weld intersection.
- All bar one residual stresses measurement was lower than code-based estimates.
- PWHT reduced the residual stress, which increased the critical defect size.
- In all cases the use of measured residual stresses increased the critical crack depth.

References

[1] Abson DJ et.al. “A review of postweld heat treatment code exemption - Part 1” Welding Journal, v.85, n.3, March 2006, pp.63-69

[2] Law M, Kirstein O, Luzin V ”An assessment of the effect of cutting welded samples on residual stress measurements by Chill Modelling” Journal of Strain Analysis for Engineering Design, V 45, Number 8 / 2010, pp 567-573

[3] O. Kirstein,V. Luzin, Materials Australia, Vol 41, 2008, 52-54

[4] Allen A, Hutchings MT, Windsor CG, Andreani C, "Neutron diffraction methods for the study of residual stress fields," Adv. Phys. 34, 445-473, 1985. https://doi.org/10.1080/00018738500101791

[5] Measurement of Residual and Applied Stress Using Neutron Diffraction, eds. M. T. Hutchings and A. D. Krawitz; NATO ASI Series 216E, Kluwer Academic Publishers, Dordrecht / Boston / London, 1992.

[6] Prask H.J, Brand PC "Residual Stress Determination by Means of Neutron Diffraction,", in Neutrons in Research and Industry, Proc. of Int. Conf. on Neutrons in Research and Industry, ed: G. Vourvopoulos, SPIE Proceedings Series, Vol. 2867, pp. 106-115 (1997). https://doi.org/10.1117/12.267879

[7] R6: Assessment of the Integrity of Structures Containing Defects, British Energy Generation Report R/H/R6, Revision 4, 2001

Keyword Index

Author Index

About the Editors

Between them the editors have approaching 100 years of experience relevant to the measurement, calculation, utilisation and, above all, understanding of both the beneficial and detrimental effects of Residual Stress. However, the more one studies residual stresses and its associated phenomena the more one realises how much there is still to learn in the field!

The ICRS series of conferences has been instrumental in developing the residual community that has greatly increased our understanding over the past thirty years and changing the understanding and utilisation of Residual Stresses from an *Art* to a *Science*.

ICRS-10 exemplified how far we have come and as it attracted a bright, able and engaging community who will continue to make progress in the field. It was a pleasure to edit these proceedings and the editors hope that they adequately represent the success that was ICRS-10.

Dr Tom M Holden
Dr Ondrej Muransky
Prof Lyndon Edwards

www.ingramcontent.com/pod-product-compliance
Ingram Content Group UK Ltd.
Pitfield, Milton Keynes, MK11 3LW, UK
UKHW020659041026
15590UKWH00001B/370

9 781945 291166